Biology and Evolution of

CROCODYLIANS

THIS BOOK IS DEDICATED TO HARRY MESSEL AC CBE
HIS IMAGINATION AND DRIVE MADE
SO MANY THINGS POSSIBLE

Biology and Evolution of
CROCODYLIANS

Text by GORDON GRIGG

Illustrations by DAVID KIRSHNER

Foreword by RICK SHINE

CSIRO
PUBLISHING

Comstock Publishing Associates
a division of
Cornell University Press
Ithaca and London

First published in the United States of America in 2015 by Cornell University Press with the ISBN 978-0-8014-5410-3
Simultaneously published in Australia by CSIRO Publishing

U.S. Librarians: Library of Congress Cataloging-in-Publication data is available.

National Library of Australia Cataloguing-in-Publication entry

Grigg, Gordon C. (Gordon Clifford) author.
Biology and evolution of crocodylians / Gordon Grigg and David Kirshner.

9781486300662 (hardback)
9781486300679 (epdf)
9781486300686 (epub)

Crocodilians.
Crocodilians – Evolution.
Crocodiles.
Crocodiles – Evolution.

Kirshner, David, illustrator.

597.98

CSIRO Publishing
Locked Bag 10
Clayton South VIC 3169
Australia

Telephone: +61 3 9545 8400
Email: publishing.sales@csiro.au
Website: www.publish.csiro.au

Cornell University Press
Sage House
512 East State Street
Ithaca, New York 14850
United States of America
Website: www.cornellpress.cornell.edu

Front cover: An estuarine or saltwater crocodile, *Crocodylus porosus*. Adelaide River, Northern Territory, Australia. (Photo DSK)
Title page: Broad-snouted caiman, *Caiman latirostris*. (Photo GCG)
Back cover: (clockwise from top left) Photos by GCG, Robinson Botero-Arias, DSK, DSK, Agata Staniewicz, Mitchell Eaton, DSK, DSK. See Fig. 3.20 on page 96 for species names.

Set in 10/14 Palatino and Trajan
Edited by Peter Storer Editorial Services
Cover design by Jenny Grigg, www.jennygrigg.com
Typeset by Thomson Digital
Printed in China by 1010 Printing International Ltd

CSIRO Publishing publishes and distributes scientific, technical and health science books, magazines and journals from Australia to a worldwide audience and conducts these activities autonomously from the research activities of the Commonwealth Scientific and Industrial Research Organisation (CSIRO). The views expressed in this publication are those of the author(s) and do not necessarily represent those of, and should not be attributed to, the publisher or CSIRO. The copyright owner shall not be liable for technical or other errors or omissions contained herein. The reader/user accepts all risks and responsibility for losses, damages, costs and other consequences resulting directly or indirectly from using this information.

Original print edition:
The paper this book is printed on is in accordance with the rules of the Forest Stewardship Council®. The FSC® promotes environmentally responsible, socially beneficial and economically viable management of the world's forests.

MIX
Paper from responsible sources
FSC® C016973

FOREWORD

Few animals are as charismatic as crocodylians, and as poorly understood by the general public. Everybody knows what a crocodile looks like, that they grow to awe-inspiring dimensions, and that they sometimes eat people. But once the conversation around the table goes beyond the danger of becoming a croc's dinner, all you are likely to hear is a succession of myths. Crocodiles, you are told, are just a special type of giant lizard; they are living fossils, unchanged since the Age of Dinosaurs. Cold-blooded, dim-witted, lazy, lying in wait for unwary African explorers. Simple and primitive.

This magnificent book explains why those statements are wrong, and why the truth about crocodylians is very different, and far more interesting. External appearances to the contrary, a crocodile is far more closely related to a bird than it is to a lizard. Look inside the scaly body, and you find a sophisticated, flexible physiology, exquisitely attuned to the challenges of life in the water, and at the water's edge. Recent research has transformed our understanding of crocodylian biology, and enabled us to view these leviathans with new eyes. Molecular studies have revealed new species, and clarified the evolutionary relationships among them. The unique internal organs of these spectacular reptiles have long been a source of debate, but physiologists finally are unravelling the mysteries of their roles. Studies on structure and function are showing us what a croc is capable of – such as how hard it can bite, and how long it can stay underwater. Dedicated field studies, and observations of captive crocodylians, are revealing an unsuspected complexity in the private lives of crocs, in their mating systems and devoted parental care. And the relationship between humans and crocodiles has been changed by new, bold and remarkably successful approaches to crocodylian conservation.

The book captures the excitement of those new findings, within the broader context of what it means to be a crocodylian. Gordon Grigg has conducted pioneering research on crocodiles and alligators, and has been a major contributor to the explosion of new understanding about these enigmatic creatures. David Kirshner combines his expertise in crocodylian biology, with an artist's ability to capture the essence of the animal. The collaboration between these two men has produced an extraordinary volume that celebrates some of the most misunderstood species on our planet. Grigg's text captures the excitement of fieldwork, and of scientific discovery; of questions asked and answered, and of new questions that still loom before us. Kirshner's art captures the awe-inspiring blend of power and elegance that defines a live crocodile.

In a culture where many people believe that you have to 'dumb down' your science to make it accessible to the public, this book is a delight. It's a visual feast, yet jam-packed with authentic information. It tells us what we know about crocodylians, and what we don't know. And in the process, it showcases the vital importance of conserving these complicated, intelligent, ecologically flexible predators.

Rick Shine AM FAA
Laureate Fellow of the Australian Research Council,
School of Biological Sciences,
The University of Sydney

Contents

PREFACE

The origin of this book goes back to 1970, when Harry Messel, Professor of Physics at the University of Sydney – and already legendary because of his annual Summer Science Schools – phoned Gordon in the Zoology Building to ask if he knew how to anaesthetise crocodiles. Gordon didn't, but reckoned he could find out and was happy to accept the invitation to meet over lunch the next day. Harry wanted to catch, anaesthetise and attach radio-transmitters to the largest estuarine crocodiles he could find. Over a bowl of soup, he told of his plans to develop telemetry equipment that would work in the difficult environment of northern Australia, and of his intention to involve biologists and students in a substantial research programme on crocodiles. His vision led ultimately to the establishment of a research base at Maningrida on the Arnhem Land coast 300 km east of Darwin, to the construction of a 21 m purpose-built research ship for surveys of crocodiles in rivers from the Kimberley to Cape York, to the purchase of a Cessna 206 for transport of personnel, resupply and aerial surveys of crocodile nesting habitat from the Coburg Peninsular near Darwin right across to the west coast of Cape York Peninsular, and to studies of reproductive biology, population ecology and even mangroves, and much more. It was a most extraordinary vision, and it led to a cohort of young biologists who suddenly and unexpectedly had the encouragement and also the logistic wherewithal to study crocodiles. Some of them went on with crocodylian studies throughout their careers. Harry published copiously on his

survey work in northern Australia and also chaired the Crocodile Specialist Group and championed crocodylian conservation. It is in recognition of his diverse contributions to crocodylian studies, including logistic support that made possible many of the studies featured in this book, that we take pleasure in dedicating it to him.

Gordon became involved directly or indirectly in many of these projects but was driven mostly by an interest in animal physiology and, in particular, a curiosity about crocodylians' body temperature and its regulation (as surviving dinosaur relatives), how their hearts work, how they live in salt water and, indeed, any topic in what might be called 'physiological ecology'. Funding from the Australian Research Council supplemented Harry's logistic support.

Somehow, despite the near arctic cold of Winnipeg, a young Canadian biologist had developed a lifelong passion for crocodylians (his earliest memory is of drawing alligators as a toddler) and was making enquiries from the other side of the globe about coming to Australia to work on crocs. David arrived in early 1981 and it was soon determined that he would study crocodylian buoyancy, because, at that stage, the only aquatic reptiles on which this had been studied were chelonians and sea snakes. His illustrative and artistic skills provided an unexpected, but welcome, addition to the laboratory, eventually embellishing the publication from the 1984 Australasian Herpetological Conference and also his PhD thesis. Somewhere in the early 1980s grew the idea that we would compile a book on crocodylians with Gordon's text and David's artwork. David meanwhile made a successful career as a biological illustrator, combining his zoology and art, then later as a zoo/aquarium interpreter, adding photography and videography to the mix. Gordon moved north in 1988 to head up a large Department of Zoology at the University of Queensland. But we both still had a book firmly in mind and, 30 years later, here it is.

HOW TO USE THIS BOOK

Although it relies heavily on direct experience with the estuarine or Indo-Pacific crocodile, *Crocodylus porosus*, plus some with Australia's 'other' crocodile, the freshwater crocodile or 'freshie', *C. johnstoni* and some on American alligators and caimans, the book aims to be a useful review of the biology of all crocodylians. Its focus is on the skills and capabilities that allow crocodylians to live how and where they do. It is a book about how they work, how they live and how they've evolved.

We have tried to put it together in a way that makes it accessible to anyone having a curiosity about them, as well as to the scientists with more specialised interests. The book is heavily illustrated, and captions to the figures often replicate or summarise material from the text in the hope of improving accessibility. Illustrations have been chosen to convey not only the information but also, if possible, the thrill and excitement that accompanies crocodylian research work, with an emphasis on work in the field. Personal experiences and anecdotes have been included, along with more than 1100 references to primary literature to facilitate more detailed study.

The book has a logical thread running through it, so it can be read chapter by chapter, skipping lightly over some of the detail, according to the reader's interest. Alternatively, it can be dipped into as a reference book, guided by the Contents and/or Index and, to facilitate that, there is some duplication of material between chapters and extensive cross referencing between them.

In addition to 'crocodile people', we hope the book's breadth will make it useful to herpetologists in general, ecophysiologists, and vertebrate palaeontologists thinking about the lifestyles and biology of dinosaurs.

Gordon Grigg and David Kirshner
January 2014

Gordon Grigg

David Kirshner

ACKNOWLEDGEMENTS

Gordon wishes to thank Jan for her patience and support throughout the last 7 years, during which more 'normal' retirement behaviour has been postponed. Thank you Jan.

David wishes to thank his friends and family for putting up with all of the times he's had to say 'I'm sorry but I can't, as I'm working on the croc book'. That goes double for you, Juleena.

We would both like to acknowledge all those colleagues and friends whose contributions to research projects and related experiences over the last few decades have made the book possible, particularly (in random order) Harry Messel, Lyn Beard, Peter Harlow, Laurence Taplin, Jon Wright, Grahame Webb, Bill Magnusson, Dean Stevens, Craig Franklin, Frank Seebacher, Marcos Coutinho, Tony Tucker, Winston Kay, Nancy Fitzsimmons, Jacob Gratten, Craig Moritz, Janet Taylor, Bill Green and Dan Lunney.

We are enormously grateful to Faye Bolton who read every chapter and exercised her editorial mouse, and to the following experts who read and made constructive comments on chapters or parts of chapters: David Booth, Peter Brazaitis, Chris Brochu, Pat Butler, Hamish Campbell, Mitch Eaton, Colleen Farmer, Ian Hume, Steve Johnston, Garry Lindner, Dave Lindner, Dan Lunney, Bill Magnusson, Harry Messel, Shaun New, Perran Ross, Steve Salisbury, Frank Seebacher, Roger Seymour, Matt Shirley, Laurence Taplin, Mike Thompson, Neil Todd, Anton Tucker, Grahame Webb, Rufus Wells, Saxon White and Jochen Zeil. Nevertheless, a book like this will certainly have errors and misunderstandings, for which Gordon takes full responsibility.

Adam Britton has been a regular source of advice and suggestions, and we are delighted that he will be hosting our supplementary videos on his website (see http://crocodilian.com/books/grigg-kirshner). He was also a generous host during our visit to 'The Territory' to photograph crocodiles in 2009. Adam Britton, Bill Magnusson, Roger Seymour, Laurence Taplin and Kent Vliet have been regular correspondents and Kent, Vladimir Dinets, Bill Green, Lou Guillette, Winston Kay, David Krause, Garry Lindner, Buck Salau, Steve Salisbury, Paul Sereno, Brandon Sideleau, Ruchira Somaweera, Merlijn van Weerd and Paul Weldon have been particularly generous with photographs and illustrations. Matt Shirley shared pre-publication material and he and Mitch Eaton contributed substantially to the section in Chapter 1 dealing with taxonomic revisions to African species. Alan Woodward, Grahame Webb, Harry Messel and Alistair Graham provided the data used in Fig. 1.36 and Laurence Taplin processed it. Craig Franklin and Mark Read made available unpublished satellite tracking data for Fig. 4.34. Hamish Campbell was generous with data and discussion for the chapter on crocodylian diving. Staff at the University of Queensland Library's Document Delivery Service deserve special mention: they were not stumped once. Donald Henderson at the Royal Tyrrell Museum of Palaeontology, Alberta, collaborated in exploring (and disproving) an idea about the role of gastroliths, discussed in Chapter 4. David's sister, Rena, provided room and board and the loan of a car on his frequent visits to Florida, making possible many of the alligator photographs. Régis Martin of the Museum and Art Gallery of the Northern Territory took photographs of skeletal material specifically for figures in Chapter 3. Weldon Owen Publishing allowed us to re-use images of crocodylians they had commissioned David to illustrate several of their books. To all of these people we say a very warm thank you.

One of the pleasures of putting the book together has been correspondence with so many people, crocodylian aficionados and other experts. They have been generous in answering queries and/or providing information, photographs and permission to use artwork. Any list will be incomplete, but must include: Carol Abraczinskas, Romain Amiot, César Luis Barrio Amorós, Don Ashley, Michael Axelsson, John Baldwin, Lyn Beard, Pam Beesley, Hilary Bierman, Chuck Booher, Robin Botero-Arias, Ken Bowler, Adam Bowman of Wallaroo Ecotours, Peter Brazaitis, Rob Bredl, Adam and Erin Britton, Chris Brochu, John Brueggen, Cássia Camillo, Hamish Campbell, Zilca Campos, Robert Chabreck, Keith Christian, Kelly Clark, Rohan Clarke, Kevin Coate, Roger Coles, Peter Cooke, Crocosaurus Cove, Marcos Coutinho, Tom Dacey, Tony de Groot, Vladimir Dinets, Hans-Rainer Duncker, Mitchell Eaton, Ruth Elsey, Greg Ericksen, Malcolm Evans, Colleen Farmer, Mark Ferguson, Richard Ferguson, Marta Fernández, Dennis and Joan Fitzsimmons, Nancy FitzSimmons, Craig Franklin, Yusuke Fukuda, Rob Gandola, Murray Garde, Dan Gist, Travis Glenn, Alistair Graham, Bill Green, Lou Guillette, Dennis Higgs, Ian Hume, John Hutchinson, Kate Jackson, Mona Lisa Jamerlan, Barrie Jamieson, Tim Jessop, Steve Johnston, Thomas Joyce, Winston Kay, Wayne King, David Krause, Darrel Lauren, Duncan Leitch, Mike Letnic, Chun Li, Dave Lindner, Garry Lindner, John Long, Milagros Lopez, Geoff McClure, Shane Bernard and The McIlhenny Co., Simon Maddock, Bill Magnusson, Charlie Manolis, Marine Biological Laboratory Archives, Todd Marshall, Régis Martin, Frank Mazotti, Nigel Monaghan, Geoff Monteith, Manuel Muñiz, Calvin Murakami, Luci Betti Nash, Shana Nerenberg, Sterling Nesbitt, Andrew Ness, Max Nickerson, Patrick O'Connor, Michael Patrick O'Neill, Russell Palmer, Louise Pastro, Jack Pettigrew, William Quatman, Mark Read, Amanda Rice Waddle, Stephen Richards, Perran Ross, Lazaro Ruda, Buck Salau, Paul Sereno, Rick Shine, Matt Shirley, John Sibbick, Brandon Sideleau, Luis Sigler, Vicki Simlesa, Daphne Soares, Ruchira Somaweera, Agata Staniewicz, Shaktri Sritharan, Michael Stern, George Swann, Douglas Syme, Laurence Taplin, Christine Tarbett-Buckley, Andrew Taylor, Janet Taylor, Marisa Tellez, Ravi Thomas, Mike Thompson, Neil Todd, Tony Tucker, Merlijn van Weerd, Luciana Verdade, Kent Vliet, Myrna Watanabe, Grahame Webb, Paul Weldon, Rufus Wells, Rom Whitaker, Craig White, John White, Paul Willis, Michael Winklhofer and Xiao-chun Wu.

Finally, we are both very grateful to Jenny Grigg for working her customary magic with the cover and to copy editor Peter Storer and to staff of CSIRO Publishing, notably Briana Melideo and Tracey Millen, without whose professionalism and patience this book would not have come to fruition.

1

INTRODUCTION

Would you be surprised to learn that a large estuarine crocodile, Crocodylus porosus, *would travel 20 km a day purposefully along a coastline for 3 weeks, to go home? I was, when we tracked a large male 'going home' after he had been translocated 120 km across Cape York by helicopter in mid August 2004. He spent the next 3 months or so swimming around in his new surroundings and then, on 4 December, started moving north. He rounded Cape York in mid-December and headed down the west coast of the Cape until he reached the same river where he was captured, arriving home on Christmas Eve! Two aspects are spectacular: he behaved as if he really knew where he was going and he covered >400 km in just 20 days (Fig. 4.35). Ocean currents were favourable for at least some of the journey but there must have been a lot of active swimming too. I have known crocs from close up for a long time, and already had an enormous admiration for their capabilities, but this observation showed them in a new light, exposing excellent navigational skills and challenging beliefs about their supposedly limited scope for sustained activity. It really jolted me!*

It is a good time to be writing a book about crocodylians. There is a huge amount of new information to review and incorporate and, like that long and deliberate journey home, much of it helps us to understand them a lot better.

They are amazing animals! This book is about what crocodylians are, where they came from, what they do and how they work.

INTRODUCING CROCODYLIANS

Everybody knows about crocodiles and alligators, and children learn about them very early (Fig. 1.1). Crocodiles, alligators, caimans and gharials, the Crocodylia, known collectively as crocodylians (or 'crocodilians': see below) are the world's largest living reptiles. The largest of them, probably the estuarine or saltwater crocodile, *Crocodylus porosus*, can grow to at least 6 m and weigh more than a tonne. They are creatures of great contrast. They can remain patiently still for ages, yet can also move like lightning to snap up a meal. They are formidably strong, active predators, with jaws that can tear apart a calf or a kangaroo, yet a mother (or a father) can gently assist her hatchlings out of the eggs, and carry them to the water in her mouth. Because large crocodylians can (and do) eat people, they invite fear and loathing, but they also inspire curiosity and admiration. Few people feel neutral about crocodylians.

Crocodile-like reptiles, the Crurotarsi (Pseudosuchia), diverged from the basal archosaurs in the Triassic more than 200 Mya (see Fig. 2.3) and the clade to which all the modern species belong, the Crocodylia, has been around for ~100 million years. Only a few of them survived the big extinction event that came at the end of the Cretaceous. That event, the Cretaceous–Palaeogene extinction (the K-Pg extinction, known previously as the K-T

The estuarine or saltwater crocodile, Crocodylus porosus, *showing the typical crocodylian form. (Illustration DSK, courtesy Weldon Owen Publishing)*

Whiteboard drawing of crocodile by Juleena, age 3 years

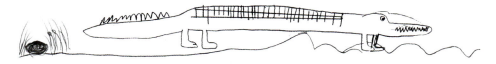

Drawing of crocodile and its nest by Gus, age 7 years

Fig. 1.1. *Elements of crocodylians captured by children. From a very young age, children are as familiar with crocodylians as they are with lions, zebras and kangaroos.*

extinction) spelt the end of the non-avian dinosaurs, leaving the surviving crocodylians as the world's dominant large reptiles (see Chapter 2). This book focusses on their many skills.

All crocodylians are amphibious and carnivorous, hunting particularly at the interface between water and land. They inhabit tropical and subtropical lakes, rivers and coasts, spending most of their time in or close to the water. Some alligatorids extend into temperate habitats. All are large by comparison with other reptiles: some are very large indeed but even these are small in comparison to some of their ancestors. The males are larger than females in all species. All come ashore from time to time, often to alter body temperature, sometimes for food and sometimes to avoid a strong current. Mating occurs in the water and females lay eggs in nests constructed or excavated on land and remain on land for long periods at the nest.

Crocodylians are excellent swimmers, with precise control over buoyancy, and they can also move fast on land; some can even gallop! Recent satellite telemetry shows that at least one species can make substantial coastal journeys. They usually return 'home' if translocated, demonstrating that they must have good navigational skills and sense of place. They can also remain submerged for long periods, probably assisted by the most elegant and complex heart of all the vertebrates. All of the crocodiles (as opposed to alligators and caimans) have salt glands on the tongue,

and at least two species (*Crocodylus porosus* and *C. acutus*) are at home in both the sea and in fresh water. Crocodylians are the only extant reptiles that show substantial maternal care.

Beyond all that, as archosaurs, the crocodylians are the sister group to dinosaurs, including birds. The Alligatoridae and Crocodylidae were separate groups by the Late Cretaceous, yet the anatomical, physiological and behavioural differences between them are surprisingly few. Many features of the crocodylians are more similar to birds than to other reptiles and this will be a common theme throughout the book. There is no doubt that the crocodylians provide the most reliable reference point for speculation about the probable behaviour and physiology of dinosaurs: a point emphasised by Hopson (1977), Senter (2008) and Brazaitis and Watanabe (2011).

Three groups of crocodylians have survived to modern times: the Crocodylidae, Alligatoridae and Gavialidae.

THE THREE 'FAMILIES': CROCODYLIDAE, ALLIGATORIDAE AND GAVIALIDAE

The surviving Crocodylia are remnants of three lineages that diversified from common ancestry some time in the Late Cretaceous: the gharial lineage (Gavialoidea); crocodile lineage (Crocodyloidea); and alligator and caiman lineage (Alligatoroidea) (Figs 1.2, 2.3). Because this classification reflects the

Fig. 1.2. Three families of Crocodylians: (top) estuarine, saltwater or Indo-Pacific crocodile, *Crocodylus porosus (Crocodylidae);* (centre) American alligator, *Alligator mississippiensis (Alligatoridae);* and (bottom) Indian gharial, *Gavialis gangeticus (Gavialidae). Living representatives of three separate clades of Crocodylia are strikingly similar in most of their anatomy, behaviour and physiology despite having had separate evolutionary paths since the late Mesozoic. (Photos Winston Kay, Louis Guillette, Vladimir Dinets)*

phylogeny, each can be recognised as a clade (see Box 2.1), and because the clades match the taxonomic category of 'superfamily', it is appropriate to refer to them as superfamilies, as their suffixes suggest. We are lucky to have survivors of each lineage. The survivors of these clades are members of the families Gavialidae, Crocodylidae and Alligatoridae, each being a sub-clade within the relevant superfamily, and each having extinct species as well.

DIFFERENCES BETWEEN CROCODYLIDS, ALLIGATORIDS AND GHARIALS

A common question is 'how do you tell the difference between an alligator and a crocodile?' Since salt glands were found in 1980, I have delighted in telling people that you just have to look at their tongues: the crocodiles are the ones with the salt glands (Chapter 11).

Apart from that, although snouts of alligatorids tend to be U-shaped and crocodylids V-shaped, this is a vast oversimplification as will be seen in Chapter 3. More reliably, the lower teeth fit into the top jaw differently (Fig. 1.3).

Another visible external difference is that, although all crocodylians have conspicuous integumentary sense organs (ISOs) on the head scales, densely on the upper and lower jaws, the crocodiles and gharials have them all over the body as well, usually just one per scale, while the alligators and caimans are 'cleanskins' behind the head (see Chapters 3 and 5). There is a lot more work to be done to resolve all the puzzles about this difference, and about the function of the ISOs. This is discussed in Chapter 5.

Considering that these three groups of crocodylians have had separate evolutionary histories for ~100 million years, these are actually quite small differences. As we shall see, their internal differences, and even differences in ecology and behaviour, are also surprisingly few. The successful body form and lifestyles evolved in the Cretaceous have remained.

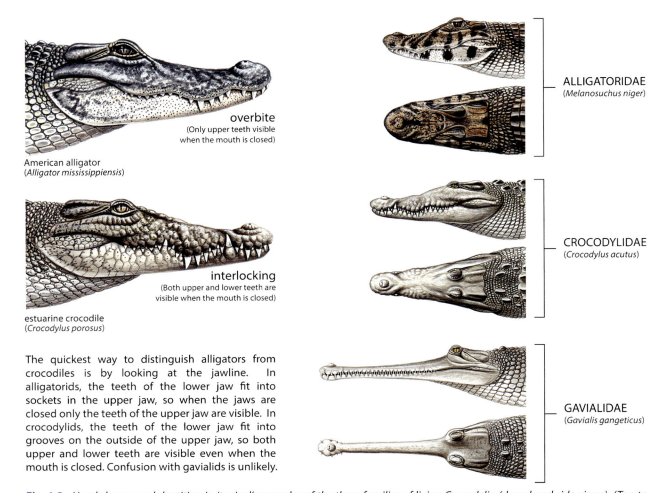

overbite
(Only upper teeth visible
when the mouth is closed)

American alligator
(*Alligator mississippiensis*)

interlocking
(Both upper and lower teeth are
visible when the mouth is closed)

estuarine crocodile
(*Crocodylus porosus*)

ALLIGATORIDAE
(*Melanosuchus niger*)

CROCODYLIDAE
(*Crocodylus acutus*)

GAVIALIDAE
(*Gavialis gangeticus*)

The quickest way to distinguish alligators from crocodiles is by looking at the jawline. In alligatorids, the teeth of the lower jaw fit into sockets in the upper jaw, so when the jaws are closed only the teeth of the upper jaw are visible. In crocodylids, the teeth of the lower jaw fit into grooves on the outside of the upper jaw, so both upper and lower teeth are visible even when the mouth is closed. Confusion with gavialids is unlikely.

Fig. 1.3. *Head shapes and dentition in 'typical' examples of the three families of living Crocodylia (dorsal and side views). (Top to bottom)* Melanosuchus niger *(Alligatoridae),* Crocodylus acutus *(Crocodylidae) and* Gavialis gangeticus *(Gavialidae). (Illustrations DSK, courtesy Weldon Owen Publishing)*

TERMINOLOGY

At the outset we need to clarify some terminology. Throughout the book, generalisations are made about groups of species, at various levels, and decisions had to be made about suitable terminology. The nomenclature used throughout is explained here.

Crocodylia, crocodylians

Rather than the term 'crocodilians', this book uses the less familiar 'crocodylians' as a collective term for all the surviving species; that is, the crocodiles, alligators and caimans, gavials and false gavials, as well as all the extinct species in the same clade, the Clade Crocodylia (Figs 2.1, 2.3 and see Box 2.1 for a discussion about clades and cladistics). Use of 'crocodylians' instead of 'crocodilians' is consistent with their membership of that clade, sometimes denoted as the Order Crocodylia. Another reason is to avoid ambiguity; 'crocodilians' is sometimes used to refer only to the extant species, sometimes to mean the Clade Crocodylia, sometimes to include all of the Crocodylomorpha (Fig. 2.3) and sometimes even to include all of the non-dinosaur archosaurs (Crurotarsi, or Pseudosuchia, Chapter 2), which are sometimes referred to as the 'crocodile-like reptiles'. The term 'crocodylians' is unambiguous. David Schwimmer, in his substantial 2002 book on *Deinosuchus*, 'King of the Crocodylians' made the same choice, as have some other recent authors.

[Note: The use of 'crocodylians' as a collective term follows the definition by Benton and Clark

(1988) of Crocodylia as a 'crown clade' comprising 'the last common ancestor of *Gavialis gangeticus* (Indian gharial), *Alligator mississippiensis* (American alligator) and *Crocodylus niloticus* (Nile crocodile) and all of its descendants' (Box 2.1). Martin and Benton (2008) argued later that 'crocodylia' should be broadened to include many more of the extinct 'crocodile-like' reptiles by equating it with the Crocodylomorpha minus Sphenosuchia; that is, with the Crocodyliformes (see Fig. 2.3). However, Brochu *et al.* (2009) responded with a strong case that the term should be restricted to the crown clade only, and this has become accepted.]

Use of the term 'crocodylian' in this way will not be without some controversy because the living species are commonly referred to as crocodilians, with an 'i', by both scientists and lay people alike, and many readers will wonder why this book departs from that practice. However, as well as 'crocodylians' being 'technically' more appropriate, and unambiguous, this book makes many physiological, ecological and behavioural generalisations about the surviving species and many of these generalisations are also likely to be applicable to the extinct members of the clade, but not to the extinct (mostly Mesozoic) non-crocodylian groups of the Crocodylomorpha (Fig. 2.3). Thus using the term 'crocodylians' matches the scope of the book well.

In summary, when the text is talking about crocodylians or Crocodylia, it will be saying something applicable to all the extant species and, perhaps, by implication, the extinct members of the clade as well.

Crocs, crocodiles, true crocodiles, alligators, caiman and caimans

Shorthand ways to refer to groups and sub-groups lead to a smoother text, but confusion must be avoided. Sometimes, and only when it is perfectly clear what is meant, the term 'crocs' is used to refer collectively to all crocodylians (Crocodylia).

Species in the genera *Crocodylus, Osteolaemus* and *Mecistops* (Crocodylidae) are referred to collectively or individually as 'crocodiles'. However, only members of *Crocodylus* are regarded as 'true crocodiles'. 'Caimans' is a collective term for the species of *Caiman, Melanosuchus* and *Paleosuchus*;

that is, Alligatoridae that are not either the American alligator, *Alligator mississippinesis*, or the Chinese alligator, *Alligator sinensis*. Note: the plural of 'caiman' is 'caiman' if they are of the same species, but 'caimans' if more than one species of caiman is being discussed (like 'fish' and 'fishes').

The word 'alligator' or 'alligators', when used without qualification, will refer to *Alligator mississippiensis* unless the text makes it clear that it refers to *Alligator sinensis*.

Another use of shorthand is the abbreviation of the generic name. Thus *Crocodylus porosus* can be written *C. porosus, Alligator mississippinesis* as *A. mississippinesis*. But we have two genera starting with 'C' so *C.* will always mean *Crocodylus* and *Caiman* will be written in full, as will *Melanosuchus* and *Mecistops*.

Crocodile-like reptiles

This term is sometimes used for the huge, diverse assemblage of Crurotarsi (Fig. 2.3), of which only a few of the Crocodylia survived the K-Pg extinction at the end of the Cretaceous.

Terminology for life stages

It is useful to have names for the various life-stage/size/age categories. These are somewhat loosely defined, but useful. Thus, hatchlings are hatchlings for the first year and yearlings for the next, after which they can be called juveniles, followed by sub-adults, usually after 5–6 years. Adulthood comes with sexual maturity, typically at ~8–10 years for females and 12–15 years for males. There are differences between species (smaller species grow a bit faster and mature a bit younger), between habitats (with different food supply) and between individual social status (which may influence access to food and warmth; see Chapter 10). These terms, despite their subjectivity, convey useful information about an individual's life stage.

THE LIVING SPECIES OF CROCODYLIANS

These are listed in Table 1.1, categorised taxonomically and geographically. Despite the three lineages having been separate for so long,

TABLE 1.1   LIVING SPECIES OF CROCODYLIAN

Species are grouped by family and, except Gavialidae, by geographic region. There is a mostly American collection of Alligatoridae (caimans and alligators), with collections of Crocodylidae in the Americas, in South-East Asia, in Australia and Papua New Guinea and in Africa. The recent splitting of *Osteolaemus*, *Mecistops* and *Crocodylus niloticus*, discussed in the text, is noted but not included in the table.

Species, size	Distribution	Habitat	Note of interest	Identifying characteristics
Family Gavialidae **(gharial and false gharial)**				
Gavialis gangeticus Gharial, gavial Males to 5–6 m	Northern India, Nepal. Ganges river drainage (but no longer in Bangladesh, Pakistan, Bhutan, Myanmar).	Slower pools of fast flowing rivers. Fish-eaters. Nest on sandbanks. Habitat shared with muggers (*Crocodylus palustris*) in places.	Aquatic. On land to bask and nest. Small limbs, only belly-slide on land but can do so rapidly. Mature males with odd protuberance on tip of snout. Focus of major captive breeding, conservation activities.	Fig. 1.2. Extremely long, narrow snout, parallel sided for most of its length (Figs 1.3, 2.2, 3.19, 3.20) and dramatically demarcated from base of the head/skull (Fig. 3.21). Uniform grey (with dark bands across back when young) (Figs 2.2, 12.9, 12.70).
Tomistoma schlegelii False gharial, Malay gharial Males to 5 m, females to 4 m	Remnant populations in Peninsular Malaysia, Borneo, Sumatra, possibly Java (previously southern Thailand). May co-occur with C. porosus.	Forested freshwater lakes, slow-moving rivers and swamps, often with floating mats of vegetation. Reported to use burrows.	Affinities uncertain (Chapter 2). Diverse prey, from invertebrates to monkeys, small deer, birds and reptiles. One confirmed predation on human, others suspected.	Figs. 2.2, 3.20. Long narrow snout which blends gradually with base of head/skull (Figs 2.2, 3.20). Two rows of very small, barely distinct post-occipital scutes. Nuchal scales continuous with dorsal scutes and almost indistinguishable. Brown, with dark bands, including blotches and bands on snout and jaws.
Family Alligatoridae **(alligators and caimans)**				
USA and China				
Alligator mississippiensis American alligator Males to 5 m (uncommonly), females to 3 m	South-eastern United States: Alabama, Arkansas, North and South Carolina, Florida, Georgia, Louisiana, Mississippi, Oklahoma, Texas (Fig. 13.3).	Freshwater marshes, streams, lakes, sometimes brackish coastal areas, temporary foraging in ocean. Not equipped to tolerate sea water for long periods.	First crocodylian to attract serious scientific study. Burrows and 'gator holes' dug as retreats in winter/dry times benefit other wetland wildlife. Spectacular social behaviour. Commonly sympatric with humans but seldom aggressive unless provoked; surprisingly few attacks.	Fig. 3.1. Snout long and broad, with rounded tip (Figs 5.5, 9.4, 10.16). Upper surface of snout relatively smooth (Fig. 3.26). Nostrils separated by bony septum (Figs 3.23, 10.27). Mostly black (Fig. 6.20) with lighter stripes, especially when young (Figs 3.8, 9.2). No blotches on side of jaw (Figs 3.1, 9.2). Nuchal scutes in three rows of two scales (Fig. 9.4). Toes of front feet with extensive webbing (Fig. 3.1).
Alligator sinensis Chinese alligator Males 1.5–2 m	South-east China, lower Yangtse Valley.	Freshwater rivers, streams, ponds and swamps; survivors in wild are in farmlands.	The 'coldest' croc. Hibernates in burrows for 6–7 months in winter. Comprehensive book by Thorbjarnarson and Wang (2009). Distribution of the two *Alligator* species is puzzling.	Fig. 1.30. Snout short and broad. Nostrils separated by bony septum. No webbing on front toes. Black, with lighter stripes when young. Light bands on tail often wider than *A. mississippiensis*. Head often speckled.

TABLE 1.1 CONTINUED

Tropical America

Caiman crocodilus (several sub-species) Spectacled caiman, brown caiman, many others Males 2–2.5 m, females to 1.4 m	Northern two-thirds of South America and north to the southern tip of Mexico (Fig. 13.5); introduced to Florida.	Adaptable species, numerous in most freshwater wetland habitats; prefers slow moving water. Can occur with *C. acutus*, *C. intermedius*, *Melanosuchus niger* in different parts of its range.	Very numerous. Often likened to dogs in the way they carry their limbs and often hold their heads high. Source of huge harvests from the wild and much captive breeding too.	Fig. 13.6. Snout triangular, ranging from medium to narrow (Figs 3.20, 3.34). Curved ridge between eyes ('spectacles'), otherwise top of snout smooth. 1st and 4th teeth on lower jaw often protrude through holes in upper jaw (Fig. 3.20). Protrusions on eyelids (Figs 3.34, 11.30). Base colour varies, often olive to brown, but usually not heavily marked, especially on head (Figs 3.34, 13.6).
Caiman yacare Paraguayan caiman, jacaré, yacare Males 2–2.5 m, females to 1.4 m	Lowlands of Argentina, Bolivia and SW Brazil, and Paraguay (Fig. 13.5).	Freshwater wetland habitats, e.g. Pantanal. Geographically separate from *C. crocodilus* except for a small area towards the north of its range.	Previously a sub-species of *C. crocodilus*, judged separate in 1988. Harvests from wild and farming.	Fig. 13.6. Snout medium to broad, relatively smooth on top (Fig. 3.26). Curved ridge between eyes (Fig. 3.28). 1st and 4th teeth on lower jaw often protrude through holes in upper jaw (Fig. 3.28). Protrusions on eyelids (Fig. 3.26). Usually heavily marked dark spotting on lighter background, especially noticeable is a series of three to five dark blotches on lower jaw (Figs 3.20, 3.26, 4.3).
Caiman latirostris Broad-snouted caiman 1–2 m, males rarely to 3 m	Argentina, Paraguay, Bolivia, Uruguay and drainage of Brazil's SE coast (Fig. 13.5).	Highly aquatic; freshwater and brackish wetlands, and mangrove-lined estuaries on Brazilian coast. Extensive overlap with *Caiman yacare* in Bolivia and Paraguay.	Some populations in brackish and salt water (see Chapter 11). Commercial harvesting and farming.	Fig. 13.6. Snout very broad and short (Figs 3.13, 11.36). Curved ridge between eyes (Fig. 3.13). Raised ridges extending forward and laterally from front of eyes sometimes present. Protrusions on eyelids (Figs 3.13, 3.20). Dark markings on a lighter background, often heavily marked, especially when young (Fig. 13.6), including dark blotches on lower jaws (Figs 3.20, 11.36).
Melanosuchus niger Black caiman Males to 4 m, reported to 6 m	Northern South America; Amazon basin. Bolivia, Brazil, Colombia, Ecuador, French Guiana, Guyana, Peru, Venezuela (unconfirmed) (Fig. 13.5).	Freshwater rivers, streams, lakes, swamps, preferring slow-moving waters.	The largest of the caimans. Similar to *A. mississippiensis* in body shape and body colouration. Harvested, including illegally. Numbers very low. Threatened by habitat destruction and illegal hunting (Chapter 14).	Fig. 7.6 (juveniles). Snout broad, but with slightly pointed tip (Fig. 3.19). Curved ridge between eyes. Ridges on snout radiating forward and laterally from centre, starting in front of eyes (Fig. 4.12). Eye socket extends far forward of eyelid (Fig. 7.6). Paired column of raised dorsal scutes running down spine (Figs 4.12, 7.6). Black body with light markings (Figs 7.6, 13.6), head brownish with dark markings on jaws (Fig. 3.20).
Paleosuchus palpebrosus Cuvier's dwarf caiman	South America east of the Andes as far south as Paraguay.	Forest streams, rivers, flooded forest.	Recent discovery of larger individuals. Heavily armoured. Range over land at night, retreat to	Figs. 3.20, 10.19. Snout triangular, relatively high and square in cross section at base (in front of eyes). Eyelids prominent, bony (Fig. 10.19).

(Table continued)

TABLE 1.1 CONTINUED				
Reputation as smallest, to 1.6 m, but recent report at 2.1 m, larger than *Osteolaemus*			burrows by day. Similar ecology to Africa's dwarf crocodiles. , Nest against termite mounds for warmth (Chapters 10 and 13).	Minimum of four rows of dorsal scutes above hind limbs. Dark colouration, with darker markings. Head often lighter than body, brownish (Figs 9.1, 10.19). Iris brown.
Paleosuchus trigonatus Schneider's dwarf caiman, smooth-fronted caiman Males to 2.3 m	South America east of the Andes as far south as Bolivia. Orinoco and Amazon drainage basins.	Forest streams. Overlap with *P. palprebrosus*.	Heavily armoured, even more than *P. palpebrosus*. Nest against termite mounds for warmth (Chapters 10, 13).	Fig. 13.38. Snout longer than in *P. palpebrosus*, less square in cross section. Eyelids prominent, bony (Fig. 3.21). Minimum of two rows of dorsal scutes above hind limbs. Dark colouration, with darker markings. Head not noticeably lighter than body (Fig. 13.38). Iris dark (Fig. 3.14).
Family Crocodylidae **(crocodiles)**				
Tropical America				
Crocodylus acutus American crocodile Males to 5 m (old claims to 6 m)	Central America from Mexico to Colombia, Cuba, Jamaica, Dominican Republic, Southern Florida (Fig. 13.8).	Fresh, brackish and coastal waters, coastal lagoons and mangrove swamps. Burrows.	Florida population was critically endangered, now recovering. Likely to be as competent in salt water as *C. porosus* (Chapter 11).	Fig. 14.15. Relatively long snout with raised, rounded hump in front of eyes (Fig. 11.22). Dorsal scutes irregular, often asymmetrical due to 'missing' scutes (Fig. 14.15). Soft skin between some dorsal scutes. Olive to grey with indistinct darker markings when older (Figs 11.22, 14.15).
Crocodylus intermedius Orinoco crocodile Males to 5 m (old claims to 6 m)	Orinoco River basin, Colombia, Venezuela (Fig. 13.8).	Freshwater rivers and waterlogged llanos in wet season. In dry season may travel overland to find pools and also dig burrows.	Alexander von Humboldt famously reported a measurement of 6.78 m made in 1800 by the botanist travelling with him. (see below).	Fig. 1.31. Long narrow snout with slight or no raised area in front of eyes (Fig. 3.20). Dorsal scutes in slightly irregular rows, often extending to sides of body. Very light coloured, usually tan or yellowish to light grey or olive background colour, with darker markings (Fig. 1.31).
Crocodylus moreletii Morelet's crocodile Males to 3 m	Caribbean-facing watersheds of Belize, Guatamala, Mexico (Fig. 13.8).	Freshwater swamps and marshes in forested areas, also brackish water. Adults may aestivate during the dry season.	Extensive geographic overlap with *C. acutus* on Yucatan Peninsula and long standing hybridization between the two has been confirmed (see Chapter 13).	Short, broad snout (Fig. 13.9), often with raised area in front of eyes. Greenish-olive (Fig. 13.9) with darker markings to uniformly dark. Scale row incursions on base of tail (Fig. 3.7).
Crocodylus rhombifer Cuban crocodile Males to 3.5 m (5 m has been recorded)	Very restricted. In one swamp on Cuba and another on a nearby island; extinct on Cayman and Bahama Islands (Fig. 13.8).	Freshwater swamps, though tolerant of brackish water also.	Sympatric with *C. acutus*. Evidence of recent hybridisation has raised conservation concerns (Chapter 13).	Fig. 13.10. Snout medium width, triangular. Rugose on upper surface (Fig. 3.20). Pronounced ridges above ears, sometimes horn-like (Fig. 3.26, 13.10). Scales pronounced, heavily keeled, especially noticeable on limbs. Body robustly built, limbs heavy (Fig. 13.10), toes short and stubby. Brightly patterned, with dark spots on a yellow background, with blotches on lower

TABLE 1.1 CONTINUED

				jaw (Fig. 13.10). Adults often retain bright colours. Iris dark (unusual for *Crocodylus*) (Fig. 3.26).
South-East Asia				
Crocodylus palustris Mugger, marsh crocodile Males to 5 m	India, Sri Lanka, South Pakistan and South-East Iran, South Nepal, probably no longer in either Bhutan or Bangladesh.	Freshwater rivers, lakes and marshes, preferring slow-moving, shallower areas. Also in reservoirs, irrigation canals, and other man-made bodies of freshwater.	Extensive burrowing habits, some burrows used for nesting. In some places have accommodated to life in urban waterways (Chapter 13). Also in lagoons on salt pans; salt glands functional (Chapter 11).	Fig. 12.44. Snout medium to very broad, especially in older males (Fig. 3.20). Dorsal scutes irregular (Fig. 12.44), often with large, wide scutes replacing two smaller scutes, producing a row of four, rather than six, scutes in the middle of the back (compare with *C. niloticus*). Enlarged scutes on side of body, often forming a vague line angling downward from tail base forward (Fig. 12.44).
Crocodylus siamensis Siamese crocodile Males to 4 m (unusually)	Historically in Cambodia, Indonesia (including Borneo and possibly Java), Laos, Malaysia, Thailand, Vietnam. Range now reduced severely, possibly by 99%.	Little known with certainty. Thought to prefer slow-moving, sheltered parts of lakes, rivers, streams and swamps.	Critically endangered. Fragile breeding populations occur in Cambodia and Lao PDR. The Wildlife Conservation Society is coordinating conservation efforts by government, zoo, business and local community groups.	Fig. 12.49. Medium length, triangular snout. Slight ridges in front of eyes converging towards centre, forming vague triangle. Small longitudinal ridge sometimes present between eyes (not present in other species). Ridges above ears often lighter in colour than surrounding, giving the appearance of having been rubbed (Fig. 12.49). Olive with darker markings (Fig. 12.49), adults usually dark all over.
Crocodylus mindorensis Philippine freshwater crocodile Males to 3 m	A Philippines endemic (suitable habitat on islands of Busuanga, Luzon, Masbate, Mindanao, Mindoro, Negros and Samar).	Freshwater lakes and ponds, marshes.	Very similar in appearance to *C. novaeguineae* (was once thought a sub-species of that) and *C. porosus* Now generally regarded as warranting full specific status.	Fig. 14.7. Triangular snout. Usually six post occipital scutes (Fig. 14.7). Top of snout bumpy, sometimes with indistinct pre-orbital ridges (Fig. 14.7). Nuchal scutes meet in centre of neck (compare with *C. novaeguineae*). Yellowish brown to olive grey. Sides of head and neck very light coloured in young animals (Fig. 4.21).
Papua-New Guinea and Australia				
Crocodylus novaeguineae New Guinea freshwater crocodile Males to 3.5 m	Irian Jaya, Papua New Guinea. Two disjunct populations, north and south of the Central Highlands.	Freshwater swamps, marshes and lakes; rarely in coastal areas, and never with *Crocodylus porosus*.	Minor anatomical differences and some genetic distinctiveness between the two populations have raised the possibility of them being separate species.	Fig. 13.1. Long snout, somewhat narrow. Slight pre-orbital ridges, but generally smoother than *C. mindorensis*. Soft skin separating left and right nuchal scutes. Dark olive to grey, with darker markings.
Crocodylus johnstoni Australian freshwater or Johnston's crocodile	Northern Australia. Freshwater sections (mainly) of drainage systems flowing to the sea as far as 18°S, except watersheds	Freshwater river systems, lakes, billabongs and swamps. Some populations in brackish and even	Lose out in competition with *C. porosus*. Removal of *C. porosus* by hunting until the 1970s allowed downstream range expansion by *C. johnstoni*,	Figs 3.15, 13.42. Very long, narrow snout (Figs 3.15, 3.26, 9.9). Nuchal scutes in cluster of four large scutes, with smaller scutes on either side (typical *Crocodylus* pattern – compare with *Mecistops*). Dorsal scutes very

(Table continued)

TABLE 1.1 CONTINUED

Males to 2.5 m (max. 3 m)	draining Cape York eastwards into the Pacific Ocean.	hyperosmotic estuaries (Chapter 11).	now being reversed by recovery of *C. porosus* populations (Chapter 13).	even in size and arrangement (Figs 3.15, 12.43). Light yellowish brown to olive with darker markings (Figs 3.15, 13.42).
Crocodylus porosus Indo-Pacific, estuarine or saltwater crocodile Males to 6 m	East coast of India though tropical and subtropical coastlines and seas to Western Pacific.	Freshwater and brackish water, common in estuaries, capable of sea journeys.	Regarded as the largest of the crocodylians (discussion below), *C. porosus* also has the worst reputation as a man-eater. It also has the strongest bite (Chapter 3) and features extensively in this book.	Fig. 1.2. Snout medium, triangular (Figs 3.20, 10.9, 12.31). Raised ridges running from eyes forward on snout (pre-orbital ridges) (Figs 1.37, 4.33, 10.9, 14.6). No post occipital scales – neck between cranium and nuchal scutes smooth, covered in small scales only (Figs 3.2, 4.33, 9.5, 12.50, 14.6). Juveniles bright yellow with black spots (Figs 3.9, 11.1), adults often dark grey to black.
Africa				
Nile crocodile (*C. niloticus*) West African Nile crocodile, desert crocodile (*C. suchus*) Males to 5 m, old reports to 6 m (*C. suchus* smaller)	African countries south of ~15°S, with a 'tongue' going (historically) north to Egypt in the Nile Valley. Also west coast of Madagascar (Fig. 1.4).	Freshwater and brackish water, common in estuaries. Can tolerate sea water.	Molecular analysis shows that taxonomy has conflated two species. The proposed name for the western species, *C. suchus*, is in wide use but so far lacks formal description. See Figs 1.4–1.6 and Chapter 13.	Fig. 1.5 (*C. niloticus*) and Fig. 1.6 (*C. suchus*). Snout triangular, broad in large males. Upper surface relatively smooth (Figs 1.5, 1.6, 3.26, 3.37). Ridges above ears sometimes enlarged in older animals (Fig. 3.26). Dorsal scutes regularly spaced (Fig. 4.5). Colour variable, usually olive to grey with darker markings (Fig. 1.5).
Mecistops cataphractus. Two discrete taxa now recognised (Chapter 2) African slender-snouted crocodile Males to 4 m, commonly 2.5 m	The west African countries fronting the Atlantic ocean from Congo DR to Gambia, with the strongest populations in Congo DR and Gabon (Fig. 1.11).	A highly aquatic species found primarily in riverine habitat with dense vegetation cover and large lakes. Also in brackish water and an offshore island. Has operational salt glands but salinity tolerance not yet resolved (Chapter 11).	On the basis of molecular analysis and morphological attributes, this species now has separate generic status within Crocodylia and two species are recognised (Figs 1.11–1.13). Morphologically similar to *C. johnstoni* (see Box 2.1 Fig. 1).	Fig. 1.13. Long narrow snout (Figs 1.12, 1.13). Nuchal pattern of four rows of two scutes, with smallest pair adjacent to dorsal scutes (different from *Crocodylus*, which typically has six nuchal scutes arranged in a central cluster of four, with an additional scute on each side). Brownish colouration with indistinct markings (Central Africa) (Fig. 1.13, 3.20) or light background colouration with distinct darker markings, including strong spotting or banding on jaws (western species) (Figs 1.12, 12.75).
Osteolaemus tetraspis, *Osteolaemus osborni*, and a probable 3rd species. Dwarf crocodiles Males to 1.9 m, more typically 1.5 m	*O. osborni* in Congo DR, *O. tetraspis* in Gabon, Cameroon and Central African Republic. Situation westwards, from Nigeria to Gambia still uncertain. Genetic differences between drainage basins (Figs 1.7–1.9).	Mostly rainforest, but also in a variety of freshwater habitats, including cool streams under closed canopy rainforest, savannah pools and even mangrove swamps.	*Osteolaemus* is the most phylogenetically distant of the extant crocodylids, and the most ancient. They may use burrows during the day and in the dry season, some with an underwater entrance. Reported to forage nocturnally on land. Habits similar to *Paleosuchus*.	Figs 1.8, 1.9, 1.10. Short, broad snout (Figs 1.8, 3.20). Four nuchal scutes (Fig. 1.8), rarely six. Darkly coloured, with dark markings, including belly (which can be completely black) (Fig. 1.9). Iris brown (Figs 1.9, 3.20, 14.4).

Sources: Various, including crocodilian.com, and Crocodile Specialist Group and IUCN publications.

their modern representatives have had much more recent evolutionary radiations, most within the last 20 million years (My) and some species have differentiated within the last 1 My (Fig. 2.1). Their evolutionary history is explored in Chapter 2.

For the last few decades, and until quite recently, 23 species of modern crocodylians in eight genera were recognised (Table 1.1, Fig. 2.1). With more field work and a combination of molecular and morphological approaches, it has become clear that some of these species deserve subdivision or reassignment and the current score is 27 species in nine genera. Changes that are accepted already will be discussed below and some of the foreshadowed changes will be mentioned in Chapter 2. When the taxonomy is resolved, there are likely to be ~30 species recognised.

RECENT TAXONOMIC CHANGES IN AFRICAN CROCODILES

(We thank Matthew H. Shirley and Mitchell J. Eaton for their considerable contributions to this section).

Defining and naming new species has to be considered very carefully. Under the international rules of nomenclature, names used previously for a taxon are allowed to be resurrected, but new names should not be used until there has been a formal description that meets the mandatory requirements of the International Code of Zoological Nomenclature (e.g. published in an appropriate, peer-reviewed journal). For some of the newly recognised crocodylian species, names used previously are available and are coming into common use, but other species await formal description and the assignment of new scientific names. We have adopted the resurrected names where appropriate.

'Nile crocodiles'; Crocodylus niloticus *and* Crocodylus suchus

What was known until recently as the Nile crocodile has a long and diverse taxonomic history with many species and/or sub-species that have been synonymised with *Crocodylus niloticus*. However, *C. niloticus* has now been split in two. Evon Hekkala, Matthew H. Shirley, John Thorbjarnarson and

colleagues analysed DNA sequence and karyotype data representing the full extent of Nile crocodile distribution in Africa and found significant divergence between the West/Central African populations and those in East and southern Africa, including Madagascar (Hekkala *et al.* 2011) (Figs 1.4–1.6). Interestingly, they showed that the two lineages are not even sister species (Fig. 2.1). The West/Central populations represent a cryptic, older lineage, that is sister to a group containing the New World crocodiles *C. moreletii, C. rhombifer, C. acutus* and *C. intermedius* to which *C. niloticus* (Fig. 1.5) is more closely related (Hekkala *et al.* 2011; Meredith *et al.* 2011; Oaks 2011) (Fig. 2.1). The east–west segregation in Africa is additionally supported by both microsatellite divergence (Hekkala *et al.* 2010) and morphological analysis (Sadleir 2009; Nestler 2012). The analysis of Hekkala *et al.* 2011 included a series of mummified crocodiles from the Egyptian temples at Thebes and the Grottes de Samoun, all

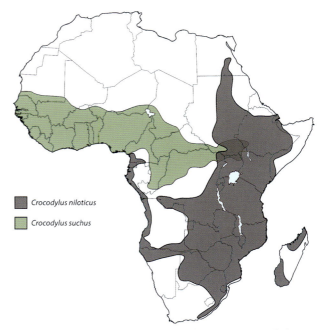

Crocodylus niloticus
Crocodylus suchus

Fig. 1.4. *Until recently Nile crocodiles were regarded as a single, widespread species,* Crocodylus niloticus, *but the west/central populations have now been recognised as a separate species,* Crocodylus suchus. *Morphological and molecular analyses revealed that they are not even sister species (Fig. 2.1). The ancient Egyptians also distinguished between two different crocodiles, preferentially keeping the smaller one in their temples (see Fig. 1.6). (Redrawn from Matthew H. Shirley pers. comm.)*

Fig. 1.5. *A Nile crocodile (*Crocodylus niloticus*) photographed in the Olifants River, South Africa. As the closest crocodiles to Europe, even being present in southern Europe until a few hundred thousand years ago, Nile crocodiles played a role in European folk stories, such as one of Rudyard Kipling's* Just So Stories, *about how the elephant got its trunk. Along with Australia's estuarine crocodiles,* C. porosus, *Nile crocs certainly match most closely the public image of a typical 'crocodile'. (Photo Louis Guillette)*

Fig. 1.6. Crocodylus suchus *at the sacred crocodile pool at Paga, Ghana. Although both* C. niloticus *and* C. suchus *have much folklore associated with them, the West African crocodile has by far the more extensive historical and cultural significance. For example, the ancient Egyptians regarded crocodiles as sacred, worshipping the crocodile-headed god Sobek and keeping and breeding the smaller, more docile* C. suchus *in temple pools, as his earthly representatives. Crocodile offspring were often mummified and given to parishioners who came to worship. Throughout West Africa, people associate* C. suchus *with the presence of water and, among other things, fertility, and sacred crocodile pools exist in most countries. (Photo Matthew H. Shirley)*

of which matched the West/Central African form. The taxon *Crocodylus suchus* was described from similar mummy crocodiles (Geoffroy Saint-Hilaire 1807) allowing Hekkala *et al.* 2011 to propose the resurrection of this previously existing name for the West/Central lineage (Fig. 1.6). That all the mummies analysed were *C. suchus* is consistent with the historical account by the ancient Greek Herodotus, quoted by Geoffroy Saint-Hilaire, that the ancient Egyptian priests recognised two types of crocodile and chose the smaller, more tractable crocodiles to keep in captivity and use in their religious ceremonies (Hekkala *et al.* 2011; Shirley *et al.* 2014b). These authors showed that the two species may have been sympatric in the Nile basin historically, but there is no evidence of contemporary co-distribution or hybridisation. This lack of hybridisation is also seen in zoos when the two species have been kept together (Shirley *et al.* 2014b) and may be due to their divergent karyotypes (Hekkala *et al.* 2011). Hekkala and colleagues pointed out that the lack of recognition hitherto for this cryptic western taxon has significant conservation implications (see Chapter 14).

Osteolaemus, *three species at least*

The African dwarf crocodiles, genus *Osteolaemus* (Crocodylidae), have had a rather fluid taxonomic history. Small differences in morphology across their combined range have long been recognised, and members of the genus have at different times been regarded as comprising an undifferentiated clinal species, *O. tetraspis*, two sub-species (*O. tetraspis tetraspis* and *O. tetraspis osborni*) or two full species (*O. tetraspis* and *O. osborni*). Ray *et al.* (2001) sequenced mitochondrial DNA from 10 captive *Osteolaemus* within US zoos, all assigned morphologically to *O. t. tetraspis,* and concluded that three of the individuals belonged to a separate clade. Ray and colleagues identified the need for more data, particularly of known provenance, because their study (and nearly all others) had relied on captive animals of unknown or uncertain origin. Brochu (2007) recognised two sub-species on the basis of comparative morphology, as did McAliley *et al.* (2006) using morphological and

limited molecular data. Pursuing the matter further, Mitchell Eaton and colleagues applied molecular phylogenetic methods to samples collected throughout West and Central Africa, as well as from zoological collections, to reveal three genetically distinct forms with equally divergent morphologies in three different biogeographic regions of western Africa. They noted the need for wider geographic sampling to delimit the fine-scale boundaries between these lineages and also that the third species will be recognised formally in due course (Eaton *et al.* 2009). From east to west (Fig. 1.7), the three *Osteolaemus* taxa are: *O. osborni* (Fig. 1.8) in the Congo Basin; *O. tetraspis*, mainly in Gabon, but extending into Cameroon, possibly Togo (N. Smolensky, F. Schmidt *pers. comm.* to M.H. Shirley) and south-west Congo (Fig. 1.9); and the new species in the Upper Guinean biome (Fig. 1.10).

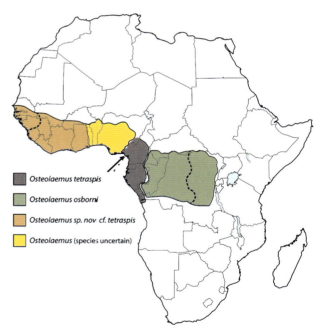

Fig. 1.7. *Proposed distribution of the three* Osteolaemus *species. Further sampling will be required before the exact distributional limit between* O. tetraspis *and the West African taxon is known. The arrow points to the Cameroon Volcanic Line (CVL), a volcanic mountain range in western Cameroon that extends as a chain of islands into the Gulf of Guinea. The broken lines within the* O. osborni *and West African species distributions represent potential delimitations between the proposed ESUs (Shirley* et al. *2014b). (Redrawn from Matthew H. Shirley* pers. comm.*)*

Fig. 1.8. Osteolaemus osborni. *This male (TL 77cm, 1.4 kg) was captured by a hunter/fisherman in June 2004, using a hook, the string of which has been used to tie the snout. It was caught in the Likouala River, Lac Télé Community Reserve, Republic of Congo. See also Fig. 14.4. (Photo Mitchell Eaton)*

In a further analysis, Shirley *et al.* (2014a) concluded that *O. osborni* split ~10–12 Mya, with the other two species splitting 7.5–8.0 Mya. Two of these taxa, *O. osborni* and the new West African taxon, may each comprise at least two distinct evolutionary units, likely as a result of Plio-Pleistocene forest

Fig. 1.9. Osteolaemus tetraspis. *Alain Mougoula weighs a large male captured in Loango National Park, Gabon. (Photo Mitchell Eaton)*

Fig. 1.10. Osteolaemus *sp. nov. cf.* tetraspis *(Côte d'Ivoire). No name has yet been associated with this 'new' Osteolaemus *sp.,* though two previously described names may be available pending further sampling: Osteolaemus (Crocodilus) frontatus *(Murray 1862) or* Osteolaemus (Halcrosia) afzelii *(Lilljeborg 1867). (Photo Matthew H. Shirley)*

refuge entrapment. Franke *et al.* (2013) uncovered the fourth lineage, somewhat divergent from the West African samples analysed by Eaton *et al.* (2009). However, as their study was on captive animals of unknown provenance, the geography of this fourth lineage was unknown until Shirley *et al.* (2014b), conducted similar analyses including samples from the Senegambia region of far western Africa and found that they matched the Franke *et al.* lineage. Shirley and colleagues also supported Eaton and colleagues' conclusion that eastern and western Congo Basin samples are likely evolving independently. Shirley and colleagues proposed that these isolated populations should be considered Evolutionarily Significant Units for future conservation planning. This makes a total of five genetically distinguishable entities within the three species (Fig. 1.7). The extant *Osteolaemus* are survivors of a diverse clade of African and Madagascan crocodylians that extend back ~20 Mya to the early Miocene (Brochu 2007; Shirley *et al.* 2014a).

Two African slender-snouted crocodiles

The African slender-snouted crocodile, long known as *Crocodylus cataphractus*, was recently recognised as a sister taxon to *Osteolaemus* and not a member of the genus *Crocodylus*. Because of this, resurrection of the monotypic genus *Mecistops* has been proposed (McAliley *et al.* 2006). This finding has been supported by all systematic studies of the Crocodylia since 2006 (Meredith *et al.* 2011; Hekkala *et al.* 2011; Oaks 2011; Shirley *et al.* 2014a) but, though in wide common usage, *Mecistops* awaits formal declaration. Recently, Matthew H. Shirley and colleagues analysed more than 200 specimens from across the range of *Mecistops*. They found clear molecular and morphological divergence warranting a split into two species, one on each side of the Cameroon Volcanic Line (Figs 1.11, 1.12, 1.13) (Shirley *et al.* 2014a). Formal redescription and recognition of the new species is in preparation (Matthew H. Shirley *pers. comm.* 2014).

THE GROWTH OF SCIENTIFIC KNOWLEDGE ABOUT CROCODYLIANS

BEGINNINGS

Humans have been observing crocodylians ever since there have been humans, mostly to learn how

Fig. 1.11. *Distribution of the two species of African slender-snouted crocodiles, Mecistops. The break shown at the Cameroon Volcanic Line (CVL) is not yet confirmed but is implied by skull samples and because the high country provided by the CVL is a known zoogeographic barrier. Species divergence occurred ~6.5–7.5 Mya, matching the final geological uplift forming the CVL and conforming quite well to the timing of the* Osteolaemus *split (Shirley* et al. *2014a). (Redrawn from Matthew H. Shirley pers. comm.)*

Legend:
- Mecistops cataphractus
- M. sp. nov. cf. cataphractus

and being four-footed, dangerous and not having a tongue. These authors were relying on travellers' tales, not first hand observations. Reports of the travels of Bartram and von Humboldt in the Americas in the late 1700s were written by the travellers themselves, recording both their own observations and information recounted by others. The American William Bartram (1729–1823) was a significant natural historian and he recorded many observations of American alligators in his travels in south-eastern USA in 1773–1774, recounted in the 1791 publication by James and Johnson; Philadelphia *Travels Through North and South Carolina, Georgia, East and West Florida, the Cherokee Country, the Extensive Territories of the Muscogulges, or Creek Confederacy, and the Country of the Chactaws; Containing An Account of the Soil and Natural Productions of Those Regions, Together with Observations on the Manners of the Indians* (Bartram 1791 in Bartram and Harper 1943; the orginal document can be found at: https://archive. org/stream/travelsthroughno00bart#page/n5/ mode/2up). Many of these observations, such as his account of a female taking care of several young, were ridiculed but have since proven to be correct. Wayne King, now Emeritus Curator at the Florida Museum of Natural History, has evaluated Bartram's behavioural reportage against modern knowledge, for the most part very favourably (King 2008). The Prussian Alexander von Humboldt

to avoid being eaten, and to eat them. Herodotus recorded information about crocodiles in his 'Histories', 5th Century BC. In the 1st century AD Pliny the Elder wrote of them living in the Nile

Fig. 1.12. Mecistops cataphractus *from West Africa. This juvenile was captured and photographed in the Gambia River, River Gambia National Park, The Gambia. Heavy jaw-spotting is characteristic of the West African species and is almost completely lacking in the Central African populations. (Photo Matthew H. Shirley)*

Fig. 1.13. Mecistops *sp. nov. cf.* cataphractus *from Central Africa. This animal was photographed in the N'Gowe River, Akaka, Loango National Park, Gabon. (Photo Mitchell Eaton)*

(1769–1859) travelled extensively in Venezuela, Colombia, Ecuador and Peru and recorded some observations about crocodylians, the most interesting being a measurement of 6.78 m in what was presumably an Orinoco crocodile (see below).

As in other scientific fields, most of what we know about crocodylians has accrued in the last 100 years. Choosing significant 'seminal' contributors from such a large field will inevitably be controversial but, for me, four stand out: A. M. Reese, E. A. McIlhenny, H. B. Cott and A. C. Pooley.

Albert Moore Reese (1872–1965) (Fig. 1.14) wrote *The Development of the Alligator* (Reese 1908) and *The Alligator and its Allies* (Reese 1915), which are still valuable after a century. He was Professor of Zoology at the University of West Virginia in Morgantown VA. His research reflected the times: anatomy and embryology. He also spent a lot of time in the field in Florida's Everglades and in the Okefenokee Swamp, which straddles the border between Florida and Georgia. He talked to hunters and alligator farmers and included a chapter on alligator biology, which is an interesting record of what was known or assumed at that time. He also described how alligators were hunted, using

Fig. 1.14. *Albert M. Reese 1923.* Embryo Project Encyclopedia *(1923). ISSN: 1940–5030 http://embryo.asu.edu/handle/10776/2634. (With permission of Marine Biological Laboratory Archives)*

Fig. 1.15. Edward McIlhenny and McIlhenny's Tabasco Sauce. The sales of this sauce must have funded the earliest well recorded observations of alligator behaviour, made by Edward (Mr. Ned) McIlhenny (left and right), who was President of the McIlhenny Co. for the first half of the 20th century. McIlhenny is well known to crocodylian biologists because he studied alligators around his home for much of his life and wrote the well known book The Alligator's Life History, *first published in 1935. The company is still based in Louisiana's bayou country. (Photos of Edward McIlhenny supplied by the company)*

an acetylene lamp, to pick up eye shines, and a shotgun.

Edward Avery McIlhenny (1872–1949) (Fig. 1.15) wrote *The Alligator's Life History* (1935), republished in 1976 by the Society for the Study of Amphibians and Reptiles. It is a classic. The McIlhenny family lived in Louisiana and the family made its fortune manufacturing and selling the legendary pepper sauce, Tabasco (Fig. 1.15). The sauce was invented by Edward's father in 1868 and Edward was president of the company from 1898 for more than 50 years. This allowed him to indulge his interests in natural history, which particularly focussed on birds and the alligators living in the bayous surrounding the family home and sauce factory on Avery Island. In the preface to his book, he refers to his lifelong interest in American alligators and his intention 'to record some of the things I have learned concerning their life history during more than three score years of living near to and observing them'. Crocodylian biologists should feel obliged to use their famous sauce!

Hugh Cott (1900–1987) (Fig. 1.16) was an English zoologist and an accomplished artist who put those two skills together, most notably in a 1940 book titled *Adaptive Coloration in Animals*. He was much in demand in the Second World War as an expert on camouflage. Crocodylian biologists appreciate him for his extraordinarily wide ranging and insightful

study of Nile crocodiles (Cott 1961), which is referred to frequently in this book. His research was conducted mostly in the cool seasons of 1952 (in Uganda), 1956 and 1957 in Northern Rhodesia (now Zambia) and at St Lucia Estuary in Zululand (now KwaZulu-Natal).

Tony Pooley (1938–2004) (Fig. 1.17) is revered among crocodylian biologists who knew him not only for his knowledge of Nile crocodile behaviour but for his passion for conservation and for educating

Fig. 1.16. Hugh Cott, photographed while looking at 'magnetic' termite mounds on the floodplain of the Tomkinson River near Maningrida, NT, 1976. The occasion was the Crocodile Specialist Group's 3rd working meeting, held at the University of Sydney's Crocodile Research Facility at Maningrida. (Photo GCG)

Fig. 1.17. *Tony Pooley at St Lucia Estuary, 1977 doing what he did so well and so usefully: sitting quietly and patiently watching crocodiles. (Photo GCG)*

people about nature. He was a wonderful raconteur. His 1982 book *Discoveries of a Crocodile Man* tells how his boyhood passion for birds and other wildlife led to a job with the then Natal Parks Board (now Ezemvelo KZN Wildlife) and a posting as a game ranger to Ndumu in the far north-east of South Africa where he became enthralled by crocodiles. His research tools were mainly curiosity and patience, combined with observational skills and what must have been an instinctive knack for experiment. If any one of his contributions were to be singled out among the myriad of behavioural and ecological observations, it would be his experimental approach to learning about crocodylian maternal care. As he was fond of demonstrating at his St Lucia Crocodile Research Centre in (now) KwaZulu-Natal, he could bring a Nile crocodile to the fence of the breeding pen by playing a hatchling call, hand over a squirming hatchling through the mesh, whereupon it would be taken in the jaws, carried down to the pond and

released (Fig. 12.71). Such experiments are described in Pooley (1977), one of his many published papers, and he also made and contributed to films and published recordings of wildlife sounds and the music of Tongaland (a region within KwaZulu-Natal).

MORE RECENT RESEARCH

Of the more than 1100 publications cited in this book, 90% have been published since 1970 and 45% of them since 2000. There has been an explosion of research into crocodylian biology in the last 40 years, with 25% of the citations having appeared since 2007, the year David and I began serious work on this book!

Here is a list of some of the significant recent advances:

- Several new species have been recognised, and more are likely (Chapters 1 and 3).
- Molecular studies have made great advances in unravelling the pattern and timing of the evolution of the living crocodylians (Chapter 2).
- Morphological studies employing modern descriptive and analytical approaches have contributed enormously to unravelling the evolutionary lineages and affinities within the extinct archosaurs and crocodylomorphs (Chapter 2).
- A new fossil was found in central Queensland, *Isisfordia duncani*, which may be a very early eusuchian, in a sister group to Crocodylia (Chapter 2).
- The bite forces generated by crocodylians have been measured as the greatest in the animal kingdom, with *Crocodylus porosus* at the top of the list (Chapter 3).
- Tracking data has shown coastal travel, homing ability and site fidelity by *C. porosus*, the latter confirmed by genetic data Chapters 4, 13).
- In addition to adjusting buoyancy by the state of lung inflation, crocodylians have been shown to control pitch and roll by muscular control over the fore and aft, left and right distribution of air in the lungs (Chapters 4, 7).
- The sacculus has emerged as a likely receptor for crocodylian infrasound, and the canal

Fig. 1.18. *Nile crocodiles,* C. niloticus *were the subject of research by both Hugh Cott and Tony Pooley. These two crocs were photographed at St Lucia Estuary in KwaZulu-Natal, an area in which both men carried out studies. (Photo Louis Guillette)*

joining left and right tympanic cavities (part of a strikingly complex Eustachian tube system) posited to allow sense of the direction a source of an infrasound source (Chapter 5).

- Provocative data from crocs, turtles and birds has suggested that the lagena, which mammals lack, may be a significant organ used in navigation (Chapter 5).

- Embryos have been shown to imprint on particular chemical stimuli, raising the possibility that pre-hatch learning may influence subsequent habitat choice, perhaps even the recognition of a 'home' river, facilitating site fidelity (Chapter 5).

- The functions of the conspicuous integumentary sense organs (ISOs), which had puzzled scientists for so long, have been explored. Neurophysiological studies have revealed that they are extremely sensitive touch receptors, useful for a diversity of functions including prey location (detection of water movement) and provision of tactile sensitivity in the jaws. ISOs dot every scale of all crocodylians but, paradoxically, alligatorids have none on the body scales (Chapter 5).

- Air has been shown to flow in the same direction past gas exchange elements of the lung during both inspiration and expiration, implying very efficient gas exchange. This supports the 'endothermic ancestry' proposition and provides another example of the affinity between crocodylians and birds (Chapter 7).

- New data has accumulated about one of the longest running puzzles in crocodylian physiology: the operation and role of the heart's unique Foramen of Panizza, which provides a conditional connection between the left and right aortas (Chapter 8).

- Data from freely ranging crocodiles has confirmed that they can stay submerged for hours and that long dives may be associated with rest. Physiological considerations imply that these long resting dives and almost all other dives are made aerobically (Chapter 9).

- Data gained from free ranging crocodiles and alligators have expanded our understanding of their thermal relations, revealed how different they are from other reptiles, shown how social status can influence thermoregulatory opportunities, provided a quantitative example of 'gigantothermy' and, ultimately, led to a new way to assess how 'thermoregulation' can be diagnosed and what this term actually means in reptiles (Chapter 10).

- A strong, but controversial, proposal has suggested that crocodylian ancestors were endothermic (Chapter 10).

- Increased recognition of the many similarities between crocodylians and birds has been much reinforced (highlighted in many chapters).

- Genetic studies of mating systems have shown that multiple paternity is commonplace among crocodylians (Chapter 12).

- A daily cycle in melatonin has been reported in *C. johnstoni*, suggesting that this may provide a mechanism for monitoring photoperiod and, thus cueing the start of the reproductive cycle (Chapter 12).

- Industrial water pollution has been shown to affect hormonal control of reproduction in American alligators, which can therefore be used as environmental indicators (Lou Guillette and colleagues pers. comm.).

- Concerning the near future, technical advances in genomics now enable whole genomes to be sequenced, which, with new mathematical analyses, promise myriads of exciting new understandings. Much more fine-grained analyses of genetic structure are now possible. Crocodylian research is well at the forefront of non-human studies, with the International Crocodilian Genomics Consortium delivering completed genomes of *C. porosus*, *G. gangeticus* and *A. mississippiensis*, and spin-off research. Genetic studies have contributed to many of the studies reported in this book, some listed above. Some of these will be fine-tuned by the new approaches, and new fields will be possible to explore. If this book were ever revised, contributions from crocodylian genomic studies would certainly be a large component, probably dictating many changes.

Some useful sources of information about crocodylians are listed in Box 1.1.

CROCODYLIANS AS RESEARCH SUBJECTS

Pros and cons

The interest humans have in crocodylians is biased towards the very large ones – the 'wow!' factor – and knowledge of crocodylian behaviour in the wild has come mostly from observations of large crocs. In contrast, and not surprisingly, detailed knowledge about their physiology and anatomy has come from studies of smaller, more manageable animals. Smaller animals can be kept in captivity more cheaply and more easily, and can be handled with less risk (lacerations rather than dismemberment). Small individuals are also more likely to be available for dissection. A second bias in crocodylian studies is that most work has been done on just a few species, whereas many species have been the subject of comparatively few studies. Luckily, their morphology and physiology have been shaped by similar lifestyles and, even though alligatorids and crocodylids have been separated since the Late Cretaceous, it is clear that they have been conservative and many generalisations from the few well-studied species apply to the group as a whole (with caution).

There are advantages and disadvantages in using crocodylians as research subjects. They can be very hard to study in the field because they are so wary. You could sit for a long time by a tropical river and see little or no interesting crocodylian behaviour. They do habituate to humans to some extent, so tourists on one of the three times daily cruises on Yellow Water in Kakadu National Park, Northern Territory, for example, may get a false impression about crocs' apparent lack of concern about boats or about the accompanying amplified commentary.

BOX 1.1 SOURCES OF INFORMATION ABOUT CROCODYLIANS

Books

Graham A, Beard P (1973) *Eyelids of Morning; the Mingled Destinies of Crocodiles and Men*. A&W Visual Library, New York.

Grenard S (1991) *Handbook of Alligators and Crocodiles*. Kruger Publishing Co., Malabar, Florida.

Grigg GC, Seebacher F, Franklin CE (2001) (Eds) *Crocodilian Biology and Evolution*. Surrey Beatty & Sons, Sydney.

Guggisberg CAW (1972) *Crocodiles; their Natural History, Folklore and Conservation*. Wren Publishing, Mt Eliza, Victoria.

McIllheny EA (1935) *The Alligator's Life History*. The Christopher Publishing House, Boston Massachusetts.

Neill WT (1971) *The Last of the Ruling Reptiles; Alligators, Crocodiles and their Kin*. Columbia University Press, New York.

Pooley AC (1982) *Discoveries of a Crocodile Man*. W Collins Sons & Co. Ltd, Johannesburg, South Africa.

Reese AM (1915) *The Alligator and its Allies*. GP Putnam's Sons, London.

Richardson KC, Webb GJW, Manolis SC (2002) *Crocodiles Inside Out*. Surrey Beatty & Sons, Sydney.

Schwimmer D (2002) *King of the Crocodylians; The Paleobiology of* Deinosuchus. Indiana University Press, Bloomington, Indiana.

Steel R (1989) *Crocodiles*. Christopher Helm Ltd, Beckenham, UK.

Thorbjarnarson J, Wang X (2010) *The Chinese Alligator; Ecology, Behavior, Conservation, and Culture*. The Johns Hopkins University Press, Baltimore, Maryland.

Trutnau L, Sommerland R (2006) *Crocodilians: Their Natural History and Captive Husbandry*. Edition Chimaira, Frankfurt, Germany.

Webb GJW, Manolis C (1989) *Crocodiles of Australia*. Reed, Sydney.

Webb GJW, Manolis SC, Whitehead PLJ (1987) *Wildlife Management, Crocodiles and Alligators*. Surrey Beatty & Sons, Sydney.

Websites

There are many websites that provide information about the diversity of crocodylians, their geographic distribution, how large they grow and when and where they nest. The following are likely to have some longevity and some provide links to many more sites.

Adam Britton's http://www.crocodilian.com/

(Supplementary material for this book is on Adam's website; please check out: http://crocodilian.com/books/grigg-kirshner.).

Bibliography of Crocodilian Biology http://utweb.ut.edu/hosted/faculty/mmeers/bcb/

(Created by Mason Meers and useful for publications up to 1996.)

Fauna of Australia; Amphibia and Reptilia http://www.environment.gov.au/biodiversity/abrs/publications/fauna-of-australia/fauna-2a.html

(Contains free downloads of more than 40 herpetological chapters, including four on crocodylians.)

Gharial information database http://www.gharial-info.com/

IUCN SSC Crocodile Specialist Group http://www.iucncsg.org/

(Provides access to much information and many downloadable crocodylian publications, such as CSG

BOX 1.1 CONTINUED

Newsletters, Action Plans, Proceedings.)

Kent Vliet's home page http://kvliet.crocodylia.com/

(Includes many links to other crocodylian and herpetological sites.)

Louisiana alligator program website: http://www.wlf.louisiana.gov/wildlife/alligator-program

Tomistoma Task Force http://tomistoma.org/pa/

Wikipedia http://www.wikipedia.org/

(Very useful for fossil history in particular, citing much original literature.)

Worldwide Crocodilian Attack Database http://www.crocodile-attack.info

Their large size and physical threat are also disadvantages for researchers. When monitoring body temperatures of large *C. porosus* in a very large pen made by fencing off natural habitat at Pormpuraaw, North Queensland (Fig. 10.8), we were much more comfortable making observations from the safety of a vehicle, the standard method for studying free-range carnivores from close by. But field work with large crocodiles in interesting places is also part of the appeal, and many crocodylian researchers whose work is heavily laboratory and analytically based also love to get out anywhere there are crocs (Fig. 1.19).

On the other hand, crocodylians offer many advantages. In field studies they can be marked in long-lasting ways by clipping tail scutes and web-tagging. The larger ones can carry radio-transmitters and other electronic monitoring equipment attached to the neck scutes. Their nests are often conspicuous and are a ready source of eggs (Fig. 1.20) (but keep a sharp lookout for their mother) and, after incubation, hatchlings for experiments or growth in captivity to found a captive colony. Another advantage is that sampling

Fig. 1.19. *Louis Guillette from the Medical University of Southern Carolina (on its head end) and Koos de Wet (on its tail end) restrain a large male* C. niloticus *captured in Loskop Dam on the Olifants River, South Africa. After getting blood and other samples for later analysis, the animal was released. (Photo Louis Guillette)*

Fig. 1.20. *David Kirshner in February 1981 at a grass nest of* Crocodylus porosus *in Myeeli Swamp near the mouth of the Liverpool River, Northern Territory. This was David's first field trip in Australia. The stick in David's left hand was used to prod the female's wallows (one of which is to David's right) to make sure she wasn't hanging around to defend the nest. This regularly used nest site was across the river from the University of Sydney's research base at Maningrida and was a convenient source of eggs. (Photo GCG)*

Fig. 1.21. *Hand capturing small* C. porosus *in the Liverpool River, Northern Territory, Australia (left to right: Gordon Grigg, Peter Harlow, Laurence Taplin). The spotlight person transfixes the croc (hopefully), the driver brings the boat in slowly and on target (hopefully) and the catcher dives onto the croc, grasping it firmly around the neck (hopefully). A wetsuit conveys a sense of invulnerability (sometimes). Such practices are ill advised anywhere there may be large crocs, may not comply with workplace safety regulations and are not recommended. They can, however, be very effective. (Left photo Peter Harlow, others GCG)*

blood is relatively easy (Fig. 11.37) and, compared with most reptiles, they can provide an adequate volume. In captivity, most crocodylians (except *Tomistoma schlegelii*) breed fairly readily once accustomed to their new surroundings, and this has been a source of much of the information on reproductive and other behaviour (Chapter 12).

First catch your croc

Capturing crocodylians alive and uninjured for research is not straightforward. Figures 1.21–1.29 show a range of approaches.

BODY SIZE AND AGE

CROCODYLIANS LARGE AND SMALL

Even the smallest crocodylian is very large compared with most reptiles. Indeed, 80% of reptile species weigh less than 20 g (Pough 1980), so even a hatchling croc weighs more than twice that of most adult reptiles. The largest crocs are at least 6 m and weigh more than a tonne. Only leatherback turtles (*Dermochelys coriacea*) come close: the largest leatherback reported so far weighed 916 kg, with a curved carapace length 256.5 cm, plastral length

Fig. 1.22. *Marcos Coutinho and Zilca Campos snare* Caiman yacare *at Nhumirin Farm in the Pantanal, Brazil. (Photos GCG)*

Fig. 1.23. *Large crocodylians can be captured in large mesh fishing nets. The main drawback is that the nets need to be checked frequently to avoid the animal struggling and drowning, as well as the difficulty of removing and securing a large animal safely, typically at night and from a small boat. Netting is also labour intensive, especially if a large number of fish are caught and need to be removed. (Photo Bill Green)*

134.6 cm (Eckert and Luginbuhl 1988). Next down the list are anacondas and Burmese pythons, which grow longer (to ~10 m) but weigh less than crocodylians of similar length (Fig. 1.36). Next down are Komodo dragons, ~3 m maximum and maybe 90 kg, falling pretty much on the same line as *C. porosus*.

Crocs are big, but they start out very small. A 6 m animal would weigh ~1000 kg but would have weighed as little as 50 g when it hatched: a 20,000-fold increase in body mass or four orders of magnitude! For comparison, a human's increase in body mass after birth is typically less than 30-fold.

Growth is usually described as indeterminate; that is, continuing throughout life, but data are lacking. Certainly, growth is variable in response to food supply. All species show a pronounced sexual dimorphism, with males growing larger than females and often at a more rapid rate. As a generalisation, the larger crocodylians can be said to start life at about 'a foot' long (25–30 cm) and grow 'about a foot

a year' until they reach sexual maturity (say 10 years for females, 15 for males) gradually slowing until, in their mature years, growth is slow. Smaller species are smaller at hatching and mature more rapidly (maybe 7 years for females and 10 for males).

Implications of body size

Biologists are always interested in body size and the implications of allometric scaling (see Chapter 7 and a recent interesting review with a crocodylian flavour by Seymour *et al.* 2013). A frog, a crocodile or a human twice the body mass of another of the same species requires much less than twice the amount of food. In biology, few things scale directly with body mass (but see lung volume, Chapter 9) and, when thinking about a species that grows in body mass through four orders of magnitude, one has always to keep size effects in mind. Also, one can predict that, even if a large croc will have a lower food requirement per kg of body mass than a small one, a

Fig. 1.24. *Handling a large croc caught in a net is not straightforward. Dean Stevens keeps its head above water. (Photo GCG)*

Fig. 1.25. *Collateral benefit. Grahame Webb and Gordon Grigg heft a nice barramundi netted accidentally in 1973 in the Liverpool River, NT. (Photographer not recorded)*

large croc has a capacity to manage much larger food items, which are taken much less frequently. Thus we see large male crocodiles eating wildebeests, whereas as hatchlings they ate bite-sized insects and small frogs. In parallel with their extraordinarily large increase in body size from when they hatch until when they become adult, their capabilities change dramatically along the way as well.

The smallest and largest of today's crocodylians

How small is the smallest and how large the largest croc, and what is the relationship between length and mass in different species? These are not easy questions and they do not have really satisfactory answers. Until recently, Cuvier's dwarf caiman, *Paleosuchus palpebrosus*, was regarded as the smallest among the extant crocodylians, with adult males

thought to grow to no more than 1.5 m in total length (TL) and 15 kg or so. However, in three Brazilian habitats, Campos *et al.* (2010) found several male *P. palpebrosus* greater than 1.8 m TL, with the largest 2.1 m TL (37 kg). This is on a par with Chinese alligators (Fig. 1.30) (largest wild-caught 2.03 m, Thorbjarnarson and Wang 2010) and African dwarf crocodiles, *Osteolaemus* spp., 2.0 m (Eaton 2009).

Mostly, the focus is on the large crocs and there is always great curiosity among people about just how big they grow. To put it into perspective, however, although we might wonder if the largest extant species grows to 6, 7 or even 8 m in length, it is also worth remembering that in the distant past there have been crocs that grew to at least 12 m (see Chapter 2). The estuarine or saltwater crocodile, *Crocodylus porosus*, is usually regarded as the largest species, although the famous naturalist and explorer Alexander von Humboldt reported the measurement of a 6.78 m crocodile (presumably the

Fig. 1.26. *Larger specimens are often captured by harpooning, using a detachable head that fits loosely into a hole in the end of the harpoon shaft (Webb and Messel 1977). At capture, the short barbs penetrate the skin and the animal becomes tethered by stout cord attached to the harpoon head. For large animals, attaching a second line is desirable in case one comes out. The idea is to 'play' the animal until it tires and can be brought alongside to have its jaws secured before being brought into the boat or secured alongside. This is not quite as easy as it sounds. (Photos GCG and Bill Green)*

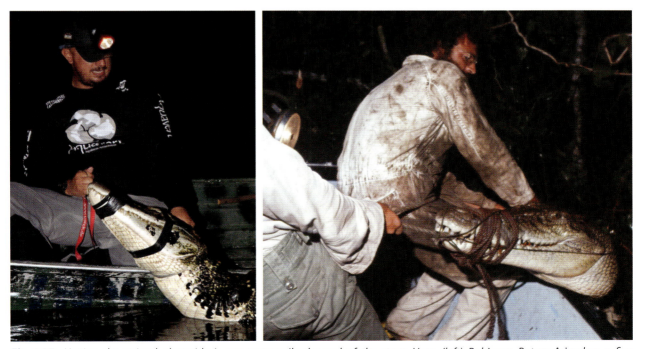

Fig. 1.27. *Having the animal alongside is not necessarily the end of the story. Here, (left) Robinson Botero-Arias has a fine* Melanosuchus niger *and (right) Dan Grace handles a large* C. porosus. *(Left photo R Botero-Arias, right Bill Green)*

Fig. 1.28. *Early on in the University of Sydney's Northern Territory crocodile research programme, traps made from rope cargo nets were developed and used successfully to catch* C. porosus *(Webb and Messel 1977). Set on a river bank, the net was held to shape with cords tied to vegetation and baited. Tugging on the bait by a lured crocodile released a weight (a sack of mud) from high in a tree. Its downward travel pulled the drawstring tight around the trap's throat. This was the forerunner of the floating metal traps now used routinely (Fig. 1.29). (Photo Bill Green)*

Fig. 1.29. *A floating metal crocodile trap. Such traps are effective and much easier to set and service than rope traps. (Left photo GCG, right Winston Kay*

Fig. 1.30. *A Chinese alligator,* Alligator sinensis, *in competition with species of* Osteolaemus *and* Paleosuchus *for the title of shortest extant crocodylian. (Photo John Thorbjarnarson, courtesy of his Estate)*

Fig. 1.31. *Orinoco crocodile,* Crocodylus intermedius. *Although Australia's estuarine crocodile,* C. porosus, *is reputed to be the largest living species,* C. niloticus, C. acutus *and* Gavialis gangeticus *are also claimed to reach 6 m and there may have been others. The Prussian naturalist/explorer Alexander von Humboldt reported a measurement of 6.78 m (22 feet 3 inches) made in 1800 from an Orinoco crocodile,* C. intermedius, *in the Orinoco River. Humboldt's reputation is such that it cannot be dismissed automatically as a 'traveller's tale'. (Photo César Luis Barrio Amorós)*

Orinoco crocodile, *Crocodylus intermedius*, Fig. 1.31) in the Orinoco River in 1800. The measurement was made by his co-traveller and, later, co-author the then 27 year old Aimé Bonpland, who was trained in medicine and botany and who later became a very well known botanist. Bonpland also has a lunar crater named after him. His measurement might, therefore, have been quite trustworthy but it would be useful to check the original French manuscript. The 1852 translation by Thomasina Ross records it as 22 feet 3 inches, but the original measurement would have been made and reported in metres. Ms Ross's mathematical reliability seems not to have been recorded. *C. porosus*, *C. niloticus*, *C. acutus* and *Gavialis gangeticus* are all known, or claimed, to reach 6 m. One very large *C. porosus* was shot near Darwin in 1945, but no length was apparently recorded (Fig. 1.32). Other species too get very large.

Fig. 1.32. *A large* Crocodylus porosus *shot among mangroves in the Howard River, a few km east of Darwin, Northern Territory, Australia, on 9 March 1945. This was before the end of the Second World War, with troops still stationed in defence of Darwin. (Photo courtesy of Selwyn Day and the Fortean Picture Library)*

Crocodylus palustris (the Mugger) and *Melanosuchus niger* (the Black Caiman) are reputed to grow to 5 m. Woodward *et al.* (1995) proposed 4.57 m as the likely maximum size for *Alligator mississippiensis*. We will never know with certainty the record for any of these species. Adam Britton has taken an interest in trying to unearth and evaluate stories about big crocs. He noted that old croc hunters in northern Australia told of crocs more than 8 m being shot (e.g. a 27 foot or 8.23 m one from the Staaten River in the 1970s, www.crocodilian.com) but accepting such an extraordinary claim would require the support of compelling documentation.

A very large *C. porosus* was captured in the Agusan Marsh near the village of Bunawan, Agusan del Sur, on the island of Mindanao in the Philippines in early September 2011. A weighbridge was available and a mass of 1075 kg or 2370 lbs was determined. Named Lolong, he was put on display and became quite a tourist attraction. His TL was measured by Adam Britton at 6.17 m (20.24 feet), with a head length of 700 mm (2.3 feet) (Britton *et al.* 2012) (head lengths are measured from the tip of the snout to the rear of the skull platform). Lolong was accepted into the Guinness Book of Records in June 2012 as the largest crocodile in captivity, supplanting Cassius, the 5.5 m 'saltie' living at Marineland Melanesia on Green Island, off Cairns in North Queensland. Unfortunately, Lolong died in January 2013, so the title has reverted to Cassius.

Britton *et al.* (2012) mentioned another *Crocodylus porosus* plausibly more than 20 feet (6.2 m), just a little longer than Lolong. The measurement was made of a skin with head still attached and is thought to be conservative (Whitaker and Whitaker 2008). This may be the largest crocodylian ever actually measured, unless of course von Humboldt's Orinoco crocodile could be accepted, but we'll never know about that. Bhitarkinika Wildlife Sanctuary in Orissa enjoys a reputation for having the largest *C. porosus*, but estimates are just that: it is hard to measure a free-living animal. These big crocs are very rare now, and perhaps there are not any really big crocs any more. It takes decades for them to reach such a size and, in a world in which human activity intrudes upon and modifies habitats ever increasingly, they face many hazards. On the other

Fig. 1.33. *George Craig feeds Gomek at Marineland Melanesia on Green Island, North Queensland in 1981. Gomek was once the largest crocodylian in captivity at 5.42 m (17.8 feet). He was purchased in 1985 and taken to Florida by Arthur Jones, a wealthy animal fancier and inventor, manufacturer and distributor of Nautilus Health Equipment who sent one of his personal aircraft (a Boeing 707) to Cairns to collect Gomek. There were delays over export paperwork so the aircraft returned without its precious cargo and Gomek subsequently rode on a Qantas flight. He probably needed more than one seat. Gomek died of heart disease in 1997 (Adam Britton pers. comm.). Arthur Jones died in 2007. The largest crocodylian in captivity is now Cassius, also living on Green Island. (Photos GCG)*

hand, with climate change underway, which will probably be advantageous for them (see Chapters 10, 14), and with the current human population's increasing unsustainably, there might come a day when they are once more in great abundance and have undisturbed opportunity to grow to their maximum size!

Reliable data about crocodile size are hard to obtain. Much of what is reported is based on estimation rather than actual measurement, often by comparison with the length of a boat, a truck or, perhaps, pacing out the length.

Skull length as an index of total length

In seeking 'world record' lengths, it is common to resort to skull size as an indication, because large skulls are often valued and kept. But the relationship between skull size and croc size is never sufficiently tight to substantiate a record, although a ratio of total length to head length (TL/HL) of 7:1 is sometimes claimed. Note that skull (or head) length is the measurement from the snout to the rear of the skull platform, not the posterior end of the mandible

(which would be much longer). Whitaker and Whitaker (2008) have written an entertaining review of their search for 'Who's got the biggest' croc skull, concluding that the largest known *C. porosus* skull is in the Paris Museum (MNHN PMP specimen #A11803). Its length is 760 mm. But that does not necessarily provide an accurate estimate of the TL of its original owner, because the ratio of HL to TL changes with total length and, anyway, is variable at any length. Fukuda *et al.* (2013) analysed data from 2755 *C. porosus* captured in Northern Territory waters. They found that the TL/HL ratio increased with body size, showing that the body grows proportionally more than the head. For the larger animals (120–420 cm TL), they reported a logistic increase in HL/TL ratio from 6.7 to 7.1. Few data were available for larger individuals but for animals longer than 5 m they recommended a ratio of 8.0. The ratio for Lolong, referred to above, is 8.8. If the Paris specimen were multiplied by 8.8 (as Lolong) a TL of 6.7 m would result, but there are no details for that specimen apart from its having been killed in Cambodia in the early 1800s. The conclusion

Fig. 1.34. *Size is important. Small animals are easy to manage. (Top left) Lou Pastro taking time out from her studies at the University of Sydney to volunteer on Winston Kay's crocodile study in the Ord River, in the Kimberley region of Western Australia, weighs a yearling* Crocodylus porosus. *(Top right) Winston Kay weighs a decent sized juvenile. Most crocs will tolerate suspension by a rope sling under their armpits, staying passive while a measurement is made. Sub-adults and adults of the larger species pose a bigger problem. (Bottom) Winston weighs a 2.92 m, 86.4 kg sub-adult male* C. porosus. *Weighing crocs becomes quite a task once they get heavier than is comfortable for a couple of people to lift. Longer than ~3.5 m (nearly 200 kg) a weighbridge is almost a necessity (remember to subtract the weight of the trailer). It is not surprising that the largest crocs that are captured or shot are seldom weighed. (Photos Winston Kay)*

by Fukuda *et al.* (2013) that TL/HL ratios appear to increase steeply above ~5 m lends credibility to some of the larger TLs that have been claimed, but reports of a record skull do not necessarily imply a record TL. The world record skull, however, is an 84 cm *Tomistoma* skull in the British Museum. The length of its original owner is unknown, but Whitaker and Whitaker (2008) reported a single data point from *Tomistoma* that would give a ratio of 6.4, implying a TL of ~5.38 m (18 feet) for that croc.

Part of the motivation behind the Fukuda *et al.* (2013) study was to determine a suitable ratio to use in converting estimates of head length made on spotlight counts of *C. porosus*. Their conclusion was that a ratio of 7 was appropriate for that purpose, 7.01 being the average ratio calculated for animals over a size range of 38–503 cm.

Stunted (dwarf) populations

There are some populations that are small because of food shortage in their habitat and *Crocodylus niloticus* and *Crocodylus johnstoni* (Webb 1985) both have stunted, sometime called 'dwarf' or 'pygmy', populations. Two dwarf *Crocodylus johnstoni* populations are known in northern Australia. Both recently had their habitats invaded by the ever-expanding population of Australia's introduced cane toads (*Rhinella marina*). Toads are known to be wreaking havoc on some populations of 'freshies' in the drier parts of their range (Letnic *et al.* 2008; Somaweera *et al.* 2013) and the dwarf populations, with their restricted distributions, are significantly affected so their future is uncertain (Britton *et al.* 2013 and further discussion in Chapter 13).

Body mass

Whereas reliable data on TL of very large crocodylians are in short supply, data on their body mass are much harder to get and therefore in even shorter supply. Small animals are easy to weigh, but, the larger they get, the more effort is required, until above a TL of ~3–3.5 kg a weighbridge is

Fig. 1.35. *A weighbridge would be necessary for this 4.22 m (~ 300 kg, from Fig. 1.36)* C. porosus *netted in the Goyder River in 1971. Harry Messel (behind) and (left to right) Gordon Grigg, Dean Stevens and Keith Brockelsby. (Photo Christine Lehmann)*

necessary (Figs 1.34, 1.35) and weights of the largest animals have rarely been taken, which is not at all surprising. Crocs usually die or get shot in remote locations where suitable machinery for handling and weighing their large carcasses is seldom available. Most locations are not only remote, but warm as well, so tissue rots (and stinks) rapidly. There are therefore very few data on body weight of the largest animals, and even many of the length measurements are open to question (crocs grow with each telling of the story). Despite the difficulty of measuring the mass of large individuals, if there is a good set of data for both TL and mass over a range of lengths (Fig. 1.36), a reasonably good relationship between these, for a species, will allow the mass of even very large individuals to be calculated, as exemplified in the next section.

RELATIONSHIPS BETWEEN LENGTH AND MASS

Length is much easier to measure than mass and relationships between the two are presented in Fig. 1.36 for *C. porosus*, *C. niloticus* and *A. mississippinesis*. Note that the increase in body mass increases in a positively allometric way: that is, doubling the length leads to far more than doubling the mass. Because the relationship between length and body mass is reasonably similar between species, these also allow a reasonable estimate of body mass from length in other species, at least as a guide. From Fig. 1.36, a *C. porosus* will get to 250 kg at ~4.1 m, 500 kg at ~5 m and 1000 kg at ~6.2 m. A 7 m individual, if one existed, would hit the scales at ~1500 kg and would need to grow to ~7.7 m to reach 2 tonnes. The relationships shown in Fig. 1.36 should be particularly useful for veterinarians and researchers needing to calculate dosages of drugs for immobilisation, analgesia or anaesthesia (see Box 1.2).

The graphs also allow cautious speculation about the mass of extinct species such as the extinct *Deinosuchus*: a probable 12 m alligatorid found in Late Cretaceous sediments in North America (see Chapter 2). Alligatorids are slightly heavier than crocodylids for the same length so, using the

A. mississippiensis equation, a body mass of ~11 tonnes could be expected.

Equations:

C. porosus:
Mass = 2.658*TL^3.242
i.e. ln Mass (kg) = 0.9776 + 3.242 (ln TL m) or
log Mass (kg) = 0.4246 + 3.242 (log TL m)

A. mississippiensis:
Mass = 2.264*TL^3.428
i.e. ln Mass (kg) = 0.8171 + 3.428 (ln TL m) or
log Mass (kg) = 0.3549 + 3.4248 (log TL m)

[Note by Laurence Taplin: Mass:length relationships are challenging to derive because the condition of crocodiles varies through space and time. The *C. porosus* data derive from five samples captured in the Northern Territory in 1974–75 and Queensland in the late 1980s. Four Northern Territory samples are subsets of animals captured and analysed by Webb and Messel (1978). Each subset consists of animals drawn from a single river (Tomkinson, Liverpool, Goomadeer or Andranangoo Creek) during mid dry-season (Jun–Aug) in either 1974 or 1975. The Queensland sample is heterogeneous, from animals captured between 1980 and 1988 across a range of rivers and at all times of year. The alligator data are from males taken in September–October, 1981–1990 from Orange Lake in Florida in a long-term harvesting study reported in Woodward *et al.* (1992). The regression for alligators is the weighted average of 10 estimates, one for each sampling year. The 95% confidence limits are slightly asymmetrical about the log:log curve but consistent throughout the size range, amounting to +23 and −18% of predicted mass. Analysed separately, female alligators averaged 2% heavier than males of the same length. Data from a series of large alligators reported by Woodward *et al.* (1995) are overlaid on the Orange Lake data and several of those lie outside the 95% confidence limits. Note that large crocodiles and alligators both tend to lie above the lines for mass predicted from length. That is, the simple allometric equations commonly employed can underestimate the mass of large animals, but the effect may be detectable only in large samples. For practical purposes the equations above can be regarded as sound approximations.]

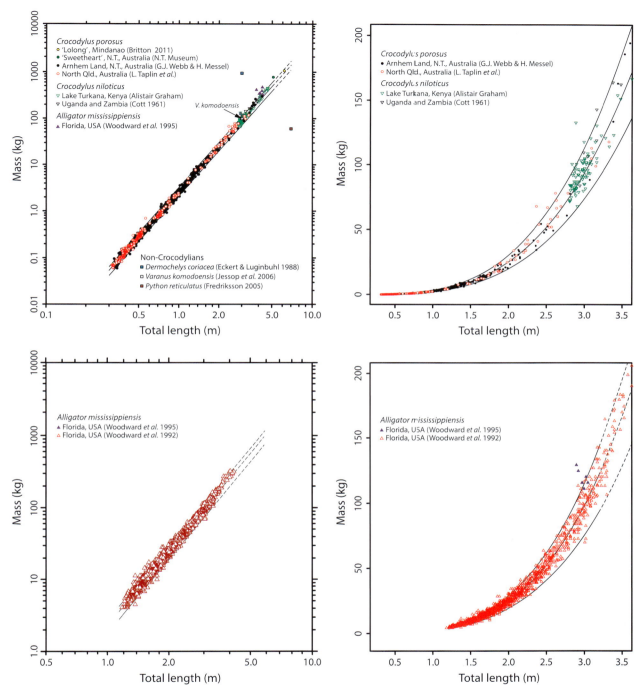

Fig. 1.36. *(Left) Double logarithmic and (right) linear plots of weight plotted against total length for a range of sizes of (upper)* Crocodylus porosus *and* C. niloticus *and (lower) male* Alligator mississippiensis. *(Top left) Some other iconic reptiles are shown for comparison. Males get much longer than females in all crocodylians. In these datasets, female alligators averaged 2% heavier, but data points for male and female crocodiles fell on the same line. On the log-log plots, the relationships look reasonably tight but in the arithmetic plots (right) it becomes clear that there is quite a lot of variability. Some variability will be from the amount of food in the stomach but most is probably due to differences in overall body condition. One could argue about which are the largest reptiles (upper left). Leatherback turtles are heavier per unit TL (all that globosity) but grow to only half the length of crocs. The largest pythons grow longer than crocs but are much lighter per unit length (all that elongation). Technically, crocodylians have a good claim to being the largest reptiles: a 6.0 m croc would weigh about a tonne, heavier than the largest known leatherback. (Data for* C. porosus *provided by Harry Messel and Grahame Webb, for* C. niloticus *by Alistair Graham and for* A. mississippiensis *by Alan Woodward. Figures and analyses by Laurence Taplin)*

BOX 1.2   IMMOBILISATION, ANALGESIA AND ANAESTHESIA

A practical application of the relationship between mass and length is its usefulness when the appropriate dosage of an immobilising, analgesic or anaesthetic drug needs to be determined, particularly for large individuals when weighing them is impractical.

Various procedures have been used to immobilise crocodylians for transport or measurement and for minor or major surgical interventions. The most recent review is the one by Olsson and Phalen (2012), with an earlier review by Loveridge and Blake (1987) helpfully including approaches that were unsuccessful.

Ketalar (ketamine hydrochloride) has sometimes been used for major surgical procedures and, in times past, cold-induced torpor was used, but no injectable anaesthetic agent has been found that is really well suited for crocodylians. Halothane anaesthesia is very effective when delivered via an endotracheal tube (Axelsson *et al.* 1996). For very large individuals, prior immobilisation may be desirable, and necessary.

For many years, Flaxedil (gallamine triethiodide), a muscle relaxant given by intramuscular injection, was the drug of choice to induce immobilisation. Its use in crocodylians was pioneered by Loveridge and Blake (1972) and operates by blocking transmission of nerve impulses across neuromuscular junctions. Its increasing cost and difficulty of acquisition led Bates *et al.* (2004) to explore Pavulon (pancuronium bromide) as an alternative, with very encouraging results. Its physiological action is very similar to Flaxedil and it is cheaper and easier to obtain. It is important to recognise that these muscle relaxants have no analgesic (pain killing) or sedative effect. If minor surgery is to be undertaken, such as drilling holes in neck scutes to attach a telemetry device, then infiltrating the area with a local anaesthetic such as lignocaine (lidocaine, xylocaine) is effective (e.g. Read *et al.* 2007). Sometimes it may be desirable to immobilise the animal with Pavulon before minor surgery.

Preliminary trials with Domitor (medetomidine) for deep muscle immobilisation have been undertaken on both *C. johnstoni* and *C. porosus* and the results are encouraging (Olsson and Phalen 2012). If its usefulness for crocodylians is confirmed by further work, its probable advantage over Pavulon is that recovery time is shorter and, additionally, its effect can be reversed by injection of Antisedan (atipamezole). Domitor is an α-two, adrenergic agonist, which also has some sedative and limited analgesic effect.

A strong word of caution is necessary about dose rates of any drug used systemically, particularly in crocodylians because of their large size range. Suitable doses have usually been determined experimentally on juveniles and they are typically reported in terms of mg/kg. This may be very misleading because the required dose can be relied upon to decrease with increasing body mass, possibly scaling in a manner similar to metabolic rate (see Chapter 7). The point is made well in a tragic account of the effect of being unaware of scaling effects. In *How Animals Work*, a 1972 book by the renowned Norwegian comparative physiologist Knut Schmidt-Nielsen (see Fig. 11.14), Knut described the death of an elephant as a result of it being given a human-sized dose of LSD scaled up arithmetically to the elephant's body mass (Schmidt-Nielsen 1972). In their seminal work with Flaxedil, Loveridge and Blake (1972) were well aware of this problem, advising decreasing dose rates for 10, 200 and 500 kg Nile crocodiles.

Adam Britton must have been faced with a dosage dilemma when called upon to immobilise Lolong, for a time the 'world record crocodylian in captivity', so that accurate measurements could be taken (Britton *et al.* 2012). He used Pavulon: two 2 mg doses were given into the base of the tail 40 min apart, with a single 10 mg dose of Valium (diazepam) to induce temporary mild sedation. Adam's aim was to be conservative, but also to avoid putting any handlers at risk. The dose he chose took 2 h to take effect and was sufficient to make Lolong 'lethargic', at which point he was restrained physically so that the many measurements could be made. No reversal agent was used and Lolong had resumed normal behaviour after about 6 h. The significance of recognising the effects of scaling can be demonstrated using Lolong as an example. If an 'average' dose

BOX 1.2   CONTINUED

recommended by Bates *et al.* (2004) from within the size range of their study animals for 'partial immobilisation (~0.5 mg for a 40 kg *C. porosus*; that is, 0.0125 mg/kg) had been applied to Lolong, at 1075 kg, that would have yielded a dose of more than 13 mg of Pavulon. Scaling according to the way metabolic rate scales with body mass – that is, at $M^{0.829}$ (Seymour *et al.* 2013) – yields a dose of ~2.4 mg, compared with the actual dose of 4 mg. Bates *et al.* (2004) suggested a dose of 2 mg for a 45 kg croc if longer immobilisation times were required (6–10 h), which would yield a dose of nearly 50 mg for Lolong if calculated arithmetically and ~8.4 mg if scaled according to $M^{0.829}$. Adam's chosen dose of 4 mg was between 2.4 and 8.4 mg, and Lolong 'woke up' after 2 h. Neatly done!

Bates and his colleagues, Adam Britton and, indeed, Loveridge and Blake (1972), all recognised that lower 'per kg' doses are required as body size increases, but scaling is not part of every undergraduate course and dosage is typically provided in publications as a bald 'per kg' rate, which could be misleading. Lolong could have been killed by a 50 mg dose and 13 mg might have kept him immobilised for a very long time and may even have led to his suffocation.

HOW LONG DO CROCODYLIANS LIVE?

Just as data about length and weight of really large crocs are scarce, so are data about their age. A lot is known about growth rates of smaller individuals up to sexual maturity (Chapter 13), but information about longevity is largely conjectural. Certainly, some of the largest crocodylians in captivity have been there for decades, and some of them seem to have changed very little in that time. One known well to me was Hymn (Fig. 1.37): a one-tonner living at the Edward River Crocodile Farm at Pormpuraaw on the western side of Cape York, Queensland. Hymn was a large *C. porosus* who stars in Chapters 10 (Thermal relations) and 12 (Reproduction). He was captured by Robbie Bredl in the Malaman River near Pormopuraaw in 1978 and was already fully grown and old looking when I met him soon after. He would have taken at least 30 years to achieve that size and appearance, but he could just as easily have already been 50. He died in December 2012 after 34 years in captivity. Gomek (Fig. 1.33) was captured as an already very large *C. porosus* in 1968 and died of heart disease 29 years later, in 1997, looking very much, according to George Craig, as he did when he was first captured. The folklore seems to imply that the larger species live longer than the smaller ones and, perhaps in line with this, that American alligators live shorter

lives than crocodiles, and caiman shorter still (perhaps 30–40 years?).

Anecdotal observations make it easy to believe that the large crocodiles have human-style lifespans, at least. To put the ageing of crocs onto a scientific basis, there has been reasonably good success from counting annual growth rings in the bony scutes (osteoderms) in the skin. This is because their growth is often seasonal, following temperature cycles (Chapter 10), so the osteoderms grow seasonally and this pattern can be seen by looking through a microscope at thin sections of the bony tissue. The best data so far come from a study on *Crocodylus johnstoni* by Anton Tucker (Chapter 13). If the seasonal temperature fluctuations are small, as in many places in the tropics, rings are likely to be less distinct, which is also the case when crocs get old and grow much more slowly. Tucker (1997) was able to conclude that *Crocodylus johnstoni* reach 50–60 years of age, so some of the 'guesstimates' of 100+ years for the much larger species are very plausible. In old animals, growth is too slow for rings to be detectable, so this approach will never reveal a maximum age. My suspicion is that, in general, humans have not yet come to terms with the realisation that many animals, and plants especially, have much longer life spans than humans, and longevity is a very under-studied area (for obvious reasons). For the time being, it seems sensible to take

Fig. 1.37. *Hymn, captured as a mature and very impressive animal in 1978, ruled the roost in the Edward River Crocodile farm for most of the 34 years he spent in captivity until his death in December 2012. Anecdotal though such observations are, they are the best we have so far for most of these long-lived beasts, and they point to life spans of human scale, at least. Hymn comes in for significant mention in Chapters 10 (Thermal Relations) and 12 (Reproduction). (Photo GCG)*

any unsubstantiated claim about age with a grain of salt, but also to keep an open mind. The bottom line summary for crocodylians is 'they live a long time'.

REFERENCES

Axelsson M, Franklin C, Loefman C, Nilsson S, Grigg G (1996) Dynamic anatomical study of cardiac shunting in crocodiles using high-resolution angioscopy. *The Journal of Experimental Biology* **199**, 359–365.

Bartram W, Harper F (1943) Travels in Georgia and Florida, 1773–1774. A report to Dr. John Fothergill. *Transactions of the American Philosophical Society. New Series* **33**(2), 121–242, http://www.jstor.org/stable/1005614

Bates L, Webb GJW, Richardson KC, Britton A, Bar-Lev J, Manolis SC (2004) "Pancuronium Bromide" – an immobilising agent for crocodiles. In *Crocodiles. Proceedings of the 17th Working Meeting of the IUCN-SSC Crocodile Specialist Group*. 24–29 May, Darwin, Northern Territory. pp. 447–451. IUCN, Gland, Switzerland.

Benton MJ, Clark JM (1988) Archosaur phylogeny and the relationships of the Crocodylia. In *The phylogeny and classification of the Tetrapods, volume 1*. (Ed. MJ Benton) pp. 295–338, Clarendon Press, Oxford.

Brazaitis P, Watanabe ME (2011) Crocodilian behaviour: a window to dinosaur behaviour? *Historical Biology* **23**(1), 73–90.

Britton ARC, Whitaker R, Whitaker N (2012) Here be a dragon: exceptional size in a Saltwater Crocodile (*Crocodylus porosus*) from the Philippines. *Herpetological Review* **43**, 541–546.

Britton ARC, Britton EK, McMahon CR (2013) Impact of a toxic invasive species on freshwater crocodile (*Crocodylus johnstoni*) populations in upstream escarpments. *Wildlife Research* **40**, 312–317.

Brochu CA (2007) Morphology, relationships, and biogeographical significance of an extinct horned crocodile (Crocodylia, Crocodylidae) from the Quaternary of Madagascar. *Zoological Journal of the Linnean Society* **150**(4), 835–863.

Brochu CA, Wagner JR, Jouvre S, Sumrall SD, Densmore LD (2009) A correction corrected: consensus over the meaning of Crocodylia and why It matters. *Systematic Biology* **58**(5), 537–543.

Campos Z, Sanaiotti T, Magnusson WE (2010) Maximum size of dwarf caiman, *Paleosuchus palpebrosus* (Cuvier, 1807) in the Amazon and habitats surrounding the Pantanal. *Amphibia-Reptilia* **31**, 439–442.

Cott HB (1961) Scientific results of an enquiry into the ecology and economic status of the Nile Crocodile (*Crocodylus niloticus*) in Uganda and Northern Rhodesia. *Transactions of the Zoological Society of London* **29**, 211–357.

Eaton MJ (2009) Systematics, population structure and demography of the African Dwarf Crocodile (*Osteolaemus* spp.): a perspective from multiple scales. PhD Thesis, University of Colorado.

Eaton MJ, Martin A, Thorbjarnarson JB, Amato GD (2009) Species-level diversification of African dwarf crocodiles (Genus *Osteolaemus*): a geographic and phylogenetic perspective. *Molecular Phylogenetics and Evolution* **50**, 496–506.

Eckert KL, Luginbuhl C (1988) Death of a giant. *Marine Turtle Newsletter* **43**, 2–3.

Franke FA, Schmidt F, Borgwardt C, Bernhard D, Bleidorn C, Engelmann WE, *et al.* (2013) Genetic differentiation of African Dwarf Crocodile *Osteolaemus tetraspis* Cope, 1861 within European zoological gardens. *Organisms, Diversity & Evolution* **13**(2), 255–266.

Fredriksson GM (2005) Predation on Sun Bears by Reticulated Python in East Kalimantan, Indonesian Borneo. *The Raffles Bulletin of Zoology* **53**(1), 165–168.

Fukuda Y, Saalfeld K, Lindner G, Nichols T (2013) Estimation of total length from head length of Saltwater Crocodiles (*Crocodylus porosus*) in the Northern Territory, Australia. *Journal of Herpetology* **47**(1), 34–40.

Geoffroy Saint-Hilaire E (1807) Description de deux crocodiles qui existent dans le Nil, compare´s au crocodile de Saint-Domingue. *Annales du Muséum d'Histoire Naturelle* **10**, 67–86.

Graham A, Beard P (1973) *Eyelids of Morning; the mingled destinies of crocodiles and men.* A&W Visual Library, New York.

Grenard S (1991) *Handbook of Alligators and Crocodiles.* Kruger Publishing Co., Malabar, Florida.

Grigg GC, Seebacher F, Franklin CE (Eds) (2001) *Crocodilian Biology and Evolution.* Surrey Beatty & Sons, Sydney.

Guggisberg CAW (1972) *Crocodiles; their Natural History, Folklore and Conservation.* Wren Publishing, Mt Eliza, Victoria.

Hekkala ER, Amato G, DeSalle R, Blum MJ (2010) Molecular assessment of population differentiation and individual assignment potential of Nile crocodile (*Crocodylus niloticus*) populations. *Conservation Genetics* **11**, 1435–1443.

Hekkala ER, Shirley MH, Amato G, Austin JD, Charter S, Thorbjarnarson J, *et al.* (2011) An ancient icon reveals new mysteries: mummy DNA resurrects a cryptic species within the Nile crocodile. *Molecular Ecology* **20**, 4199–4215.

Hopson JA (1977) Relative brain size and behavior in Archosaurian reptiles. *Annual Review of Ecology and Systematics* **8**, 429–448.

Jessop TS, Madsen T, Sumner J, Rudiharto H, Phillips JA, Ciofi C (2006) Maximum body size among insular Komodo dragon populations covaries with large prey density. *Oikos* **112**, 422–429.

King FW (2008) *Alligator Behavior: The Accuracy of William Bartram's Observations.* University of Florida, Gainesville, Florida.<http://web.uflib.ufl.edu/ufdc/?b=UF00088969>

Letnic M, Webb JK, Shine R (2008) Invasive cane toads (*Bufo marinus*) cause mass mortality of freshwater crocodiles (*Crocodylus johnstoni*) in tropical Australia. *Biological Conservation* **141**, 1773–1782.

Loveridge JP, Blake DK (1972) Techniques in the immobilization and handling of the Nile crocodile (*Crocodylus niloticus*). *Arnoldia* **40**, 1–14.

Loveridge JP, Blake DK (1987) Crocodile immobilization and anaesthesia. In *Wildlife Management, Crocodiles and Alligators.* (Eds GJW Webb, SC Manolis and PLJ Whitehead) pp. 301–317. Surrey Beatty & Sons, Sydney.

Martin JE, Benton MJ (2008) Crown clades in vertebrate nomenclature: correcting the definition of Crocodylia. *Systematic Biology* **57**(1), 173–181.

McAliley LR, Willis RE, Ray DA, White S, Brochu CA, Densmore LDIII (2006) Are crocodiles really monophyletic?—Evidence for subdivisions from sequence and morphological data. *Molecular Phylogenetics and Evolution* **39**, 16–32.

McIllheny EA (1935) *The Alligator's Life History.* The Christopher Publishing House, Boston, Massachusetts.

Meredith RW, Hekkala ER, Amato RG, Gatesy J (2011) A phylogenetic hypothesis for *Crocodylus* (Crocodylia) based on mitochondrial DNA: evidence for a trans-Atlantic voyage from Africa to the New World. *Molecular Phylogenetics and Evolution* **60**, 183–191.

Neill WT (1971) *The Last of the Ruling Reptiles; Alligators, Crocodiles and their Kin.* Columbia University Press, New York.

Nestler JH (2012) A geometric morphometric analysis of *Crocodylus niloticus*: evidence for a cryptic species complex. MSC. Thesis, University of Iowa, Ames.

Oaks JR (2011) A time-calibrated species tree of Crocodylia reveals a recent radiation of the true crocodiles. *Evolution* **65**, 3285–3297.

Olsson A, Phalen D (2012) Preliminary studies of chemical immobilization of captive juvenile estuarine (*Crocodylus porosus*) and Australian freshwater (*C. johnstoni* crocodiles with medetomidine and reversal with atipamezole. *Veterinary Anaesthesia and Analgesia* **39**, 345–356.

Pooley AC (1977) Nest opening response of the Nile crocodile *Crocodylus niloticus*. *Journal of Zoology* **182**, 17–26.

Pooley AC (1982) *Discoveries of a Crocodile Man.* William Collins, Sons, London.

Pough FH (1980) The advantages of ectothermy for tetrapods. *American Naturalist* **115**, 92–112.

Rachmawan D, Brend S (2009) Human-*Tomistoma* interactions in central Kalimantan, Indonesian Borneo. *CSG Newsletter* **28**(1), 9–11.

Ray DA, White PS, Duong HV, Cullen T, Densmore LD (2001) High levels of variation in the African dwarf crocodile *Osteolaemus tetraspis* In: Crocodilian Biology and Evolution. In *Crocodilian Biology and Evolution* (Eds G Grigg, F Seebacher and C Franklin) pp. 58–69. Surrey Beatty & Sons, Sydney.

Read MA, Grigg GC, Irwin SR, Shanahan D, Franklin CE (2007) Satellite tracking reveals long distance coastal travel and homing by translocated Estuarine Crocodiles, *Crocodylus porosus*. *PLoS ONE* **2**(9), e949.

Reese AM (1908) *The Development of the American Alligator.* Smithsonian Institute, Washington DC.

Reese AM (1915) *The Alligator and its Allies.* GP Putnam's Sons, London.

Richardson KC, Webb GJW, Manolis SC (2002) *Crocodiles Inside Out.* Surrey Beatty & Sons, Sydney.

Sadleir RW (2009) A morphometric study of crocodilian ecomorphology through ontogeny and phylogeny. PhD thesis. University of Chicago, Illinois.

Schmidt-Nielsen K (1972) *How Animals Work.* Cambridge University Press, Cambridge, UK.

Schwimmer D (2002) *King of the Crocodylians; The Paleobiology of* Deinosuchus. Indiana University Press, Bloomington, Indiana.

Senter P (2008) Homology between and antiquity of stereotyped communicatory behaviors of crocodilians. *Journal of Herpetology* **42**(2), 354–360.

Seymour RS, Gienger CM, Brien ML, Tracy CR, Manolis SC, Webb GJ, *et al.* (2013) Scaling of standard metabolic rate in estuarine crocodiles *Crocodylus porosus*. *Journal of Comparative Physiology. B, Biochemical, Systemic, and Environmental Physiology* **183**(4), 491–500.

Shirley MH, Vliet KA, Carr AN, Austin JD (2014a) Rigorous approaches to species delimitation have significant implications for African crocodilian systematics and conservation. *Proceedings. Biological Sciences* **281**, 20132483.

Shirley MH, Villanova V, Vliet KA, Austin JD (2014b) Genetic barcoding facilitates captive and wild management of three cryptic African crocodile species complexes. *Animal Conservation*. In press

Somaweera R, Shine R, Webb J, Dempster T, Letnic M (2012) Why does vulnerability to toxic invasive cane toads vary among populations of Australian freshwater crocodiles? *Animal Conservation* **16**, 86–96.

Steel R (1989) *Crocodiles*. Christopher Helm Ltd, Beckenham, UK.

Thorbjarnarson J, Wang X (2010) *The Chinese Alligator; Ecology, Behavior, Conservation, and Culture*. The Johns Hopkins University Press, Baltimore, Maryland.

Trutnau L, Sommerland R (2006) *Crocodilians: Their Natural History and Captive Husbandry*. Edition Chimaira, Frankfurt, Germany.

Tucker AD (1997) Skeletochronology of post-occipital osteoderms for age validation of Australian freshwater crocodiles (*Crocodylus johnstoni*). *Marine and Freshwater Research* **48**, 343–351.

Webb GJW (1985) Survey of a pristine population of freshwater crocodiles in the Liverpool River, Arnhem Land, Australia. *National Geographic Society Research Report* **1979**, 841–852.

Webb GJW, Manolis SC (1989) *Crocodiles of Australia*. Reed, Sydney.

Webb GJW, Messel H (1977) Crocodile capture techniques. *The Journal of Wildlife Management* **41**, 572–575.

Webb GJW, Messel H (1978) Morphometric analysis of *Crocodylus porosus* from the north coast of Arnhem Land, northern AustraliA. *Australian Journal of Zoology* **26**, 1–27.

Webb GJW, Manolis SC, Whitehead PLJ (1987) *Wildlife Management, Crocodiles and Alligators*. Surrey Beatty & Sons, Sydney.

Whitaker R, Whitaker N (2008) Who's got the biggest? *IUCN Crocodile Specialist Group Newsletter* **27**(4), 26–30.

Woodward A, Moore CT, Delany MF (1992) Experimental alligator harvest. Final report to the Florida Game and Freshwater Fish Commission, Gainesville, Florida.

Woodward AR, White JH, Linda SB (1995) Maximum size of the alligator (*Alligator mississippiensis*). *Journal of Herpetology* **29**, 507–513.

2

THE CROCODYLIAN FAMILY TREE

Impressive though the crocodylian survivors are, they are only a fragment (a 'biased sample') of a very diverse and geographically widespread group of crocodile-like reptiles that originated on land way back in the Triassic, 250 Mya. Some of them loped about on long limbs, some were bipedal, some were quite small and others were enormous. Most were carnivores but some were herbivorous. Most were terrestrial but there were marine groups too and some may even have been endothermic. It is an exciting history: the crocs were as diverse as the dinosaurs! So, those stories about crocodiles and alligators 'surviving unchanged since the age of dinosaurs' are misleading if taken to imply that crocs represent a group that didn't evolve much. The crocs we live with today are very different from their relatives living alongside dinosaurs. That's probably a good thing: some of the early critters grew to 13 m!

The fossil history of crocodylians and other archosaurs is a very active research field and much of the literature cited in this chapter has been published since 2000. New fossils are being discovered and described and analysed in new ways, and molecular analytical techniques are being applied to the extant species – all leading to reinterpretations of previous understandings about taxonomic affinities.

This chapter reviews the relationships between surviving species, as well as summarising high-lights of their ancient history – the crocodylian evolutionary tree.

THE MODERN CROCODYLIANS AND THEIR RELATIONSHIPS

EXTANT CROCODYLIA

Today's Crocodylia are survivors of three groups that were differentiated and widespread in the late Cretaceous: the Gavialoidea, Crocodyloidea and Alligatoroidea (see Fig. 2.3). Many reptile groups became extinct with the end of the Cretaceous (mosasaurs, plesiosaurs, pliosaurs, pterosaurs and, more famously, the non-avian dinosaurs). These extinctions are usually linked to a cataclysm that marked the end of the Cretaceous, at the well known K–Pg boundary (previously known as the K–T boundary). This was caused by a massive asteroid impact in what is now the Gulf of Mexico (Shulte *et al.* 2010). The ejecta would have darkened the skies, lowered temperatures, inhibited photosynthesis and are believed to have disrupted the entire planetary ecology. Nevertheless, the crocodylians were apparently little affected by the K–Pg extinction event and many snakes, lizards, tuataras and turtles also survived (Buffetaut 1990). Sebecids (see Fig. 2.25) also survived the end of the Cretaceous, until the Miocene.

Cast of type specimen (AMNH 3160) of Sebecus icaeorhinus, *an Eocene mesoeucrocodylian from Patagonia. Unlike crocodylians, which were mostly aquatic, sebecids were terrestrial predators with a laterally compressed, dog-like skull. A reconstruction of this specimen can be seen in Fig. 2.25. (Photo DSK at American Museum of Natural History)*

Brochu (2003 and *pers. comm.*) has estimated that around eight to nine times more fossil Crocodylia are known than extant species. There seem to have been two peaks of high crocodylian diversity – in the early Eocene and the Miocene – correlating with global mean temperature maxima. Perhaps our current global warming will prompt another crocodylian adaptive radiation? Scheyer *et al.* (2013) noted that in the late Miocene in Venezuela there were 14 species, with up to seven in sympatry. This compares with only two or three modern species being sympatric. But these all became extinct by the Pliocene. Members of the Crocodylia have commonly been referred to as 'crocodilians' but, as discussed in Chapter 1, that term is avoided here because of its ambiguity. Since the recognition and acceptance of the clade 'Crocodylia', it is more coherent to use the term 'crocodylians', reflecting the name of their clade. The phylogeny of the Crocodylia, including the known extinct species, has been reviewed by Brochu (2003).

Our focus in this section will be on the relationships between living species. More species are being recognised as genomic studies reveal diversity within previously accepted species (Chapter 1) and 26 can now be accepted, with 27 recognised (Table 1.1, Fig. 2.1). At least a couple more are foreshadowed: one is a possible sibling for *C. novaeguineae* (Chapter 13) and the taxonomy of South American caimans may not yet be settled.

Apart from taxonomic considerations, there has been a great deal of interest in unravelling the affinities between the extant crocodylians, particularly in the last 20 years, stimulated by the capacity for determining amino acid sequences in proteins and the availability of sophisticated computer-based methods for analysing both molecular and morphological datasets. With the newest techniques in genomics, the precision of such determinations can only increase. The best way to express affinities is through a phylogenetic tree: a cladogram. The field is a very active one and Fig. 2.1 reflects several recent studies, which, although differing a little, are broadly concordant with and confirm relationships published a couple of decades ago. This concordance is reassuring, as is the observation that (with one spectacular exception) morphological and molecular studies have yielded generally congruent results.

AFFINITIES BETWEEN SPECIES WITHIN CROCODYLIDAE

Unlike fossils, extant species provide molecular as well as morphological data, thus enriching greatly the databases that can be applied to determine the strength or otherwise of affinities between them. Morphological and molecular data have mostly been in harmony, but there is intriguing discord over the affinities of *Tomistoma* (see below).

Perhaps the most interesting finding from the early molecular approaches was the revelation that the surviving members of the genus *Crocodylus* – that is, the extant 'true crocodiles' –underwent their radiation as recently as the late Miocene, 5–6 Mya (Densmore 1983; Densmore and Owen 1989). Perhaps because of this, the differences between them are comparatively slight at a genomic level and different methodological approaches have yielded different resolutions of their interrelationships. In the last decade or so, there has been an 'evolution of cladograms' as techniques have improved and more loci from a wider diversity of species have been included. The most recent cladogram is that by Oaks (2011), which has been taken as the base for Fig. 2.1 and whose clade divergence ages have been used. His analyses confirmed the late Miocene radiation of extant *Crocodylus*. Despite newer methodology, the most recent cladograms are surprisingly similar to those developed more than a decade and a half ago. It is convenient, and not surprising, that some species of *Crocodylus* which group together are geographically correlated (Fig. 2.1). There is an identifiable 'New World' clade, which, curiously at first sight, includes both *Crocodylus niloticus* and *C. suchus* (Hekkala *et al.* 2010, 2011). However, dispersal between these two regions is quite likely, given their physiological capacities (Taplin and Grigg 1989; Chapter 11) and the presence of salt glands probably accounts for the circum-tropical distribution of the genus. It is interesting that *C. suchus* is not a sibling species to *C. niloticus*, but a hitherto cryptic, older lineage (Hekkala *et al.* 2011; Meredith *et al.* 2011). In earlier studies, *C. siamensis*, *C. palustrise* and *C. porosus*

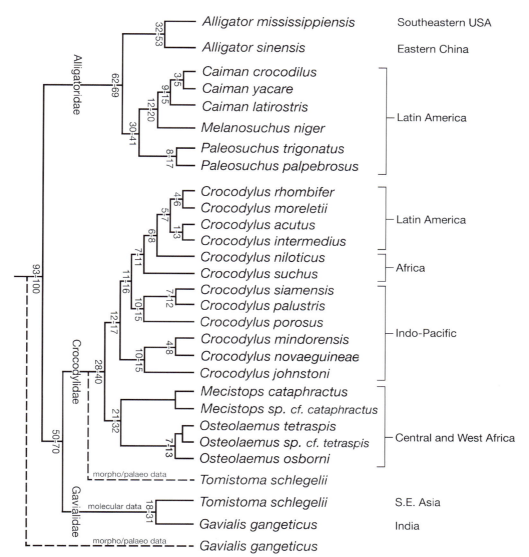

Fig. 2.1. *A cladogram representing the affinities between extant crocodylians (see Box 2.1 for notes about cladistics). This interpretation incorporates recent analyses by Oaks (2011), which provided conclusions about ages of divergence. A final consensus is probably still yet to come, but the latest versions (Meredith et al. 2011; Man et al. 2011; Oaks 2011) are quite similar to those reported a decade ago by Brochu and Densmore (2001) and Brochu (2003). One difference, however, is that the Oaks interpretation groups* C. siamensis, C. palustris *and* C. porosus *closer to the Latin American species, which seems zoogeographically discordant. The cladogram shown above includes the recognition that* C. cataphractus *has been judged worthy of generic distinction as* Mecistops (McAliley et al. 2006) *and also comprises two species (Shirley et al. 2014a; Shirley pers. comm. 2014). It also takes molecular research on* Gavialis *and* Tomistoma *into account (Janke et al. 2005; Willis et al. 2007; Roos et al. 2007; Man et al. 2011, Oaks 2011). For the most part, cladograms based on morphology and molecular analyses are concordant, but the position of* Tomistoma *is controversial. Morphology suggests* Tomistoma *groups with* Crocodylidae *but molecular studies imply grouping with* Gavialis. *Hence, both alternatives appear here. Note that the fossil record shows that all three lines were established before the end of the Cretaceous, yet the molecular clock suggests a more recent divergence between* Tomistoma *and* Gavialis. *Molecular analyses by Man et al. (2011) have confirmed the affinity of* Tomistoma *and* Gavialis, *but these authors have asserted that they both should be included within* Crocodylidae, *not in a separate family. The schema also includes the recognition that Nile crocodiles have been subdivided (Hekkala et al. 2011) and that full specific status is warranted for three species of* Osteolaemus *(Eaton et al. 2009; Shirley et al. 2014b) as discussed in Chapter 1.*

BOX 2.1   TAXONOMIC TOOLS, CLASSIFICATION AND CLADISTICS

Taxonomy has undergone a significant revolution since 1966 when Willi Hennig published his book *Phylogenetic Systematics* and launched cladistics: the science which makes it possible to bring evolutionary history and classification into harmony. If all were known, then the linear descent of all species throughout time would make a 'perfect' family tree. All descendants of a common ancestor form a clade, which can be said therefore to have a monophyletic origin, and the resultant 'family tree' is known as a cladogram. Figures 2.1 and 2.3 are examples of cladograms. As evolution occurs, lineages diverge, so clades come to be nested within clades, reflecting the evolutionary history. This way of looking at things, where only shared derived/advanced characters are considered, is known as cladistics, and it contrasts with the old Linnaean system, devised a hundred years before Darwin's *Origin of Species*. The Linnaean system focussed on a hierarchical system of classes, orders, families, and so on, based on the extent of morphological (usually) similarities and differences between groups. The problem with that system has been that organisms that have evolved independently and are not related often got lumped together in the same category (taxon) because their morphological characteristics look so similar (i.e. convergent evolution). In classifying animals, or plants, in the past, systematists (taxonomists) have had to rely on morphology for classification and to judge whether or not a particular character is analogous or homologous. Taking the example of crocs with long snouts (Fig. B2.1), has that character evolved independently (by convergent evolution, i.e. analogy) or does the presence of a long snout imply a close genealogical relationship (i.e. homology)?

Despite the pitfalls, many of the old taxonomists made good judgments and, because the similarities and differences that were the basis of categorisation under the Linnaean system resulted from their evolution, much of the old nomenclature transposes easily enough. Some of the implications of this system do, however, take some getting used to. For example, cladistically (and quite properly) the birds are part of the clade Reptilia (Fig. 2.3). Furthermore, if pelycosaurs and therapsids were still regarded as reptiles, so would we be!

In making taxonomic judgments, systematists have to juggle information about a myriad of attributes and decide which feature/s are more significant and which are likely to be 'ancestral' (plesiomorphic) and which are likely to be 'derived' (apomorphic). Within the last 30 years, systematists have been aided greatly by the availability of personal computers and numerical techniques. Sophisticated computer programmes such as

Fig. B2.1. *Long, narrow snouts have arisen independently in several crocodylian clades through convergent evolution, as demonstrated here by these strikingly similar crocs from different genera and different continents. (Left)* Mecistops *sp. was photographed in central Africa; (right)* Crocodylus johnstoni *was photographed in the Kimberley region of northern Australia. Classification based on morphological similarity can lead to categorisations that do not reflect evolutionary history; cladistics strives to bring phylogeny and classification into harmony. (Left photo Mitchell Eaton, right DSK)*

BOX 2.1   CONTINUED

PAUP (Phylogenetic Analysis Using Parsimony) allow the automatic construction of phylogenetic trees using hundreds of attributes, whether morphological or molecular or both combined, from a series of specimens, and quantifying the degree of similarity/dissimilarity with respect to an 'outgroup'. The geographic location of fossils and their geological age are also important considerations in such analyses because, after all, genetic relatedness does require there to have been proximity in both time and space. Data from historical zoogeography may provide important input into judgments about whether similarities are homologous or analogous and, hence, whether two species are closely related or just superficially similar.

Almost contemporaneously with the availability of sophisticated classificatory computer programmes came the capacity for easy determination of the amino acid sequences of proteins. Nowadays, while morphology is as important as ever – and that is all we have to go on when studying fossils – significant additional guidance about relationships between extant species comes from molecular evidence. This has brought the capacity to derive phylogenies based on evolutionary changes in nucleotide sequences, such as that in Fig. 2.1. As with morphological analyses, nucleotide sequences can be analysed from multiple sources, particularly from mitochondrial and nuclear DNA. Based on an assumption that rates of protein mutation are essentially constant, the probable age of divergences between clades can be assigned – a 'molecular clock', which can often be calibrated against dates from the fossil record to account for different groups having different rates of mutation. Extinct species, of course, cannot be subject to this sort of molecular analysis, but there is a compelling logic behind the idea that the 'real truth' of the genealogy of extant species should be encapsulated in their molecular evolution, unconfounded by questions about plesiomorphy and apomorphy. Not everyone would agree and, as we shall see, one of the more famous 'morphology versus molecules' conundrums appears in attempts to resolve the genealogy of the two most long-snouted crocodylians, *Tomistoma schlegelii* and *Gavialis gangeticus*.

Christopher Brochu at the University of Iowa and Llewellyn Densmore III at Texas Tech University, practitioners of the morphological and molecular approaches, respectively, have had an interest in crocodylians for a long time and got crocodylian cladistics off to an early start. In the last decade or so, the field has been joined by other researchers, particularly with the formation of the International Crocodilian Genomes Working Group (ICGWG), which is sequencing and assembling the genomes of the American alligator, the estuarine crocodile, the American alligator and the Indian gharial (St John *et al.* 2012). The growing wealth of crocodylian data and its analysis will provide many new insights and, in all likelihood, an abundance of vigorous discussion.

have been put with *C. mindorensis*, *C. novaeguineae* and *C. johnstoni* as an Indo-Pacific group, but whereas this was confirmed by Meredith *et al.* (2011), Oaks (2011) found them to be closer to the Latin American species (Fig. 2.1).

It is now accepted that the African slender-snouted crocodile, previously identified as *C. cataphractus*, is sufficiently different from *Crocodylus* to warrant a separate generic classification as *Mecistops* (McAliley *et al.* 2006) (reviving the name conferred by Gray in 1844). Both molecular and morphological data point to *Mecistops* and the African dwarf crocodiles, *Osteolaemus*, being closer to each other than

either is to *Crocodylus* (Man *et al.* 2011; Oaks 2011) (Fig. 2.1). *Osteolaemus* and *Mecistops* are much older than any of the extant *Crocodylus*, but still within Crocodylidae. Their recently increased species complements, discussed in Chapter 1 and shown in Fig. 2.1, make Africa the world's crocodylian diversity hotspot, with seven species.

AFFINITIES BETWEEN SPECIES WITHIN ALLIGATORIDAE

There seems to be general agreement that, within the Alligatoridae, the three genera of South American caimans form one clade and the two species of *Alligator* another (Fig. 2.1). Within the 'caiman' clade, the dwarf caimans, *Paleosuchus,* are more distant from *Melanosuchus* and *Caiman,* with robust support for a basal division between them (Brochu 2003) being confirmed by more recent work (Hrbek *et al.* 2008, Oaks 2011). Clearly there has been a significant radiation of alligatorids within the Americas, but one fascinating puzzle stands out: how did *Alligator sinensis* get to China? It is the only alligatorid outside the Americas. Various authors have speculated about this, including Taplin and Grigg (1989) and, more recently Brochu (2003) and Snyder (2007). None of the Alligatoridae has salt glands (Chapter 11) and *A. sinensis* is not large enough to have been buffered from hypernatremia for long enough to survive a long ocean voyage (see Chapter 11), so transoceanic dispersal is unlikely. Could they have gone overland during the Pleistocene (1.8M–12K years ago) via the Bering Land Bridge (Beringia), exposed periodically by falls in sea level during the ice ages? Taplin and Grigg (1989) favoured this explanation, as have both Brochu (2003) and Snyder (2007). The land bridge escaped glaciation and is thought to have sometimes been steppe-like grassland within Miocene temperate forests, which extended from Oregon to Japan. The alligators cope better with cooler climates than do other crocodylians (Chapter 10) and *A. sinensis* survives cold winters by taking to

burrows to hibernate (Chapters 10, 11). Snyder (2007) discussed the Beringia hypothesis comprehensively against other possible alternatives and noted that it was tempting to accept that explanation, almost by default. He also noted the extent of uncertainty about the Beringian climate at the relevant time. There are other vertebrate examples with a similar distribution. The freshwater restricted paddlefishes (Polyodontidae, Chondrostei) also occur only in both China and the USA. Cryptobranchid salamanders, which are also restricted to fresh water, are found in North America (the hellbender, *Cryptobranchus alleganiensis*), China (the Chinese giant salamander, *Andrias davidianus*) and Japan (the Japanese giant salamander, *Andrias japonicus*). For alligators, without oceanic crossings, the only alternative to a crossing via Beringia would seem to be dispersal across Laurasia, the northern half of the great continent Pangaea. That avenue was closed when what is now Greenland separated from Europe 55 Mya. Molecular data suggest that separation of the lineages yielding the two species of alligator occurred 32–52 Mya. However, as pointed out by Snyder, it seems unlikely that alligators could have lived in Europe for millions of years without leaving fossils. Until a better idea is put forward, dispersal via Beringia seems the most likely explanation. We need to find an *Alligator* fossil on Alaska's Seward Peninsula!

Fig. 2.2. *(Left) A juvenile Malay or false gharial,* Tomistoma schlegelii, *in the flooded forest in Mesangat Lake, east Kalimantan. (Right) A Gharial,* Gavialis gangeticus, *in the Kataraniaghat Wildlife Sanctuary on the Indo–Nepalese border. These two species are causing a lot of angst. Morphologically* Tomistoma schlegelii *appears to be one of the Crocodylidae but the molecular data suggest membership of the Gavialidae. Yet another view is that both species are crocodylids. (Left photo Agata Staniewicz, right Ruchira Somaweera)*

THE AFFINITY OF THE MALAY OR FALSE GHARIAL, *TOMISTOMA SCHLEGELII*

The main controversy about the relationships between extant croc species has been the status of the SE Asian long-snouted Malay or false gharial, *Tomistoma schlegelii* (Fig. 2.2). Is it a crocodylid or a gavialid? The uncertainty is expressed in Fig. 2.1 by putting *Tomistoma* in twice, once in the same clade as *Gavialis* and again (shaded) within Crocodylidae.

The matter is important, not just to have *T. schlegelii* recognised 'in the right place', but also because it is the survivor of a large group of at least a dozen genera of tomistomine crocodylians that had a substantial radiation 40–50 Mya. To quote Piras *et al.* (2007) in their substantial review, tomistomines had *'a glorious past with a world wide distribution documented by a conspicuous fossil record that starts at least in the early Eocene.'* A recent genomic comparison between populations of *T. schlegelii* on Peninsular Malaysia, Sarawak, Sumatra and East Kalimantan has revealed considerable genetic differentiation between them (Kaur *et al.* 2013). The samples from Peninsular Malaysia were sufficiently different that these authors suggested they may qualify that population as an evolutionarily significant unit (*sensu* Moritz 1994) or perhaps even a cryptic species. They noted a need for further work. *Tomistoma* is listed as Endangered on the IUCN's Red List, but it is in need of updating (Table 14.1). The desirability of conservation efforts having to take this new information into account will not make the job any easier.

The conundrum over whether *Tomistoma* is a gavialid or a crocodylid results from morphological and molecular datasets pointing in different directions. Morphologically, *T. schlegelii* has long been considered to be a member of the Crocodylidae (reviewed by Brochu 2003). Much of the analysis has focussed on skeletal attributes, often constrained that way to allow comparison with fossil material, but there is supporting evidence from soft anatomy as well. For example, Frey *et al.* (1989) found that the anatomy of the tail musculature in *Gavialis gangeticus* set it apart from *Tomistoma*, and Endo *et al.* (2002) drew a similar conclusion from differences between these two long-snouted species in the jaw musculature. More recently, Piras *et al.* (2010) compared skull growth patterns between *Tomistoma*, *Gavialis*, *Mecistops cataphractus* and *C. acutus* and concluded that, although *Tomistoma* shows some unique features: '*Gavialis* "starts" its post hatching morphology from a completely different region of morphospace'.

Despite all the apparently compelling morphological evidence, for more than a decade molecular data have been consistently grouping *Tomistoma schlegelii* with *Gavialis gangeticus* as a gavialid (Densmore 1983; Densmore and White 1991; White and Densmore 2001; Janke *et al.* 2005; McAliley *et al.* 2006, Roos *et al.* 2007; Man *et al.* 2011; Oaks 2011). Molecular studies have been criticised in the past for considering only a few species and/or a few protein sequences, but more recent studies have examined very comprehensive datasets over a wide diversity of species, which have pointed to strong similarity between *Tomistoma* and *Gavialis* (Janke *et al.* 2005; Willis *et al.* 2007; Oaks 2011). Willis *et al.* (2007) ended their paper with the observation that 'it is becoming increasingly difficult to ignore the consistent results of the molecular datasets that have examined the *Gavialis–Tomistoma* controversy'. The morphologists, of course, have the added benefit of many fossil species whereas the molecular biologists have only the smaller dataset from extant species.

The puzzle is more than just a discrepancy between morphological and molecular data: there is a big discrepancy in timing. The fossil record suggests a Cretaceous origin for Gavialoidea, but molecular evidence suggests that *Gavialis* diverged from other crocodylians much later, ~50 Mya, with *Tomistoma* and *Gavialis* diverging ~22–26 Mya (Roos *et al.* 2007) or, similarly, 18–31 Mya (Oaks 2011) (Fig. 2.1). But there are fossil 'tomistomines' and 'gavialoids' much earlier than that. The situation is still far from reconciled.

Also uncertain is whether or not Gavialidae are sufficiently different to be regarded as a separate family. Janke *et al.* (2005) concluded that the Gavialidae were close enough to Crocodylidae to be absorbed into that family and Man *et al.* (2011) came to the same conclusion. Willis *et al.* (2007) were of a different opinion and, so far, three 'families'

make a very good point: the whole of the vertebrate body plan – skeletal, muscular, nervous, digestive, cardiovascular and reproductive systems – is there in the fishes, and the fascination for biologists, especially comparative anatomists, comparative physiologists and evolutionary biologists, is how that basic theme has been reworked and extended during the ages.

The fossil history is particularly relevant to the functional theme of this book because many of the functional attributes we see among living crocs have their origins in the capabilities and habits expressed way back in deep time. One example of the relevance of extinct crocs to our functional story is that in Chapter 10 we will be discussing the proposal by Seymour *et al.* (2004) that some of the functional attributes of the modern crocs – their heart structure, for example (Chapter 8) – are consistent with the ancient ones having had a high metabolic rate and, possibly, being endothermic. Warm blooded crocs!

We will not start with the fishes but with the early 'reptiles'. Throughout the tour it will be helpful to refer to Fig. 2.3, which depicts a history of ~320 million years of reptilian evolution, showing where some of the most familiar groups sit in relation to our main focus, the few surviving Crocodylia.

AMNIOTES, SYNAPSIDS, REPTILES, ANAPSIDS AND DIAPSIDS

Amniotes are named because their embryonic development occurs within a membranous sac: the amnion. In Reptilia, the developing embryo is enclosed along with nutrient and water supplies within a porous and (usually) hard shell (see Chapter 12). Combined with internal fertilisation, this allowed tetrapods to be less dependent on water than their amphibian ancestors, which required free water in which fertilisation of the egg and subsequent development occurred. By the late Carboniferous, three clades had arisen within Amniota: Synapsida, Anapsida and Diapsida (Fig. 2.3). Synapsids have a single opening in the skull (a fenestra, = 'window') behind and below the eye socket, diapsids have two fenestrae and anapsids none.

Two important changes in classification have resulted from a combination of the shift to cladistics

and research conducted within the last decade or so. The first change is a redefinition of Reptilia as a clade, to exclude synapsids. The second is a realisation that turtles (Testudines) may not be anapsids, but diapsids that have secondarily lost their post-orbital fenestrae. Each of these revisions is shown in Fig. 2.3 and they both deserve comment.

First to the definition of Reptilia. Most of us have grown up with the idea that the pelycosaurs (the 'sail lizards' like *Dimetrodon*) and therapsids (the 'mammal-like reptiles') are reptiles. These were the earliest synapsids; pelecosaurs arose in the Carboniferous and therapsids in the Permian. Only therapsids survived the Permian–Triassic extinction event ~250 Mya. Their descendants survived further widespread extinctions at the end of the Triassic and again at the end of the Cretaceous (the K–Pg extinction event) and, luckily for us, a few survived even that cataclysmic event and live on as today's mammals – we are synapsids. According to cladistics, if pelecosaurs and therapsids were still regarded as reptiles, then we would be in the same clade – we would be reptiles! However, Modesto and Anderson (2004) mounted a well-structured argument based mainly on nomenclatural rules that the synapsids should be a clade on their own, separate from Reptilia. The detail of their argument need not concern us here and, although its goal was (presumably) not to 'rescue' us from membership of the clade Reptilia, it did.

That leaves two clades within Reptilia: Anapsida and Diapsida, which brings us to turtles. Turtles have always been regarded as typical anapsids: anyone familiar with the large skulls of marine turtles would realise that they have no post-orbital fenestrae. However, based on both molecular and morphological studies (Zardoya and Meyer 2001; Iwabe *et al.* 2005; Tsuji and Müller 2009), the view is strengthening that turtles are not anapsids at all, but diapsids that have lost the temporal fenestrae secondarily. Molecular data from extant representatives imply a sisterhood with crocodylians and birds – that is, with archosaurs – although it has not been without controversy (e.g. Harris *et al.* 2007). However, the association has been strengthened greatly by a substantial next-generation genomics study of living taxa by Chiari *et al.* (2012), which came to the same conclusion and implied a split between archosaurs and turtles ~250 Mya in the late Permian. Hence, in Fig. 2.3, turtles (Testudines) are shown (dotted) as a sister group to the Archosauriformes.

This leaves the Anapsida with only the small (40 cm to 2 m) marine carnivorous mesosaurs, the much larger (up to 3 m, 600 kg) terrestrial herbivorous pareiasaurs; and the very small (~30 cm) and possibly insectivorous procolophonids. The Permian–Triassic extinction event was progressive, rather than focussed in time, and may have had a series of causes, but only the procolophonids survived into the Triassic, during which they died out. Phylogeny within the Anapsida has been reviewed by Tsuji and Müller (2009).

THE EARLIEST 'ARCHOSAURS', ARCHOSAURIFORMES

Opinions about the relationships within and between early amniotes have not yet stabilised and Fig. 2.3 is an attempt to synthesise results from recent research. The two reptilian clades are sometimes called Parareptilia (= Anapsida) and Eureptilia (= Diapsida) and it is the latter that are relevant to crocodylian ancestry. The Diapsida includes the aquatic ichthyosaurs, plesiosaurs and pliosaurs, the lepidosaurians (snakes, lizards and tuatara) and the Archosauromorpha (including the Archosauria), but, in the Permian, included several distant, non-archosaurian terrestrial predatory diapsid lineages. Most did not survive past the Permian–Triassic extinction,

but a few of one sub-group, the Archosauriformes, did survive.

Euparkeria *and* Erythrosuchus

Some of the archosauriforms that survived past the end of the Permian prospered and, indeed, proliferated and diversified to become the dominant terrestrial vertebrates in the Triassic. The earliest archosauriforms ranged from small agile predators such as *Euparkeria* (Fig. 2.4), which may have been bipedal, to large quadrupedal animals such as *Erythrosuchus* (Fig. 2.5). Another group of archosauriforms were semi-aquatic and even looked superficially like modern crocodylians. These were the Phytosauria, long considered archosaurs, but now thought more likely to be archosauriforms.

Phytosaurs

Although they looked quite like crocodylians, these large, long-snouted, well armoured and semi-aquatic reptiles living in the late Triassic may not have been archosaurs at all, but a sister group (Nesbitt 2011) (Fig. 2.3). They were strikingly like modern Crocodylia in general body form (Fig. 2.6) and quite possibly in habits as well. Despite their name, they were anything but plant eaters! They grew to several metres and some authors have drawn parallels between them and the modern gharials. Fossils have been recovered from Eurasia, North America, North Africa and Madagascar. Unlike Crocodylia, their nostrils were well back on the snout, not far in front of the eyes and, like the early Crurotarsi, they had no bony secondary palate. In some, the nostrils were at the crest of a slightly raised area of the snout, suggesting an adaptation to lying low in the water as sit-and-wait predators like extant crocodylians (Chapter 3). Also similar to extant crocs, phytosaur mothers laid eggs and made a nest.

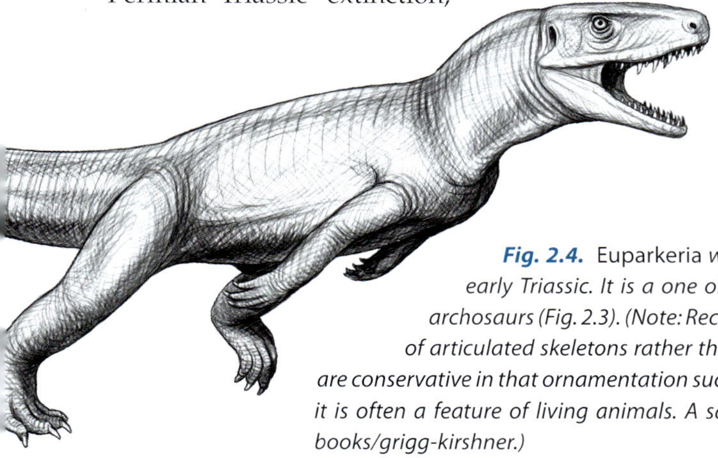

Fig. 2.4. Euparkeria *was a small (1 m) agile terrestrial (possibly bipedal) predator in the early Triassic. It is a one of the Archosauriformes, the group thought to have been basal to archosaurs (Fig. 2.3). (Note: Reconstructions in this chapter were, when possible, based on drawings of articulated skeletons rather than previous reconstructions. In most instances the reconstructions are conservative in that ornamentation such as pigment patterns, crests and frills was not added, even though it is often a feature of living animals. A sample of the process can be viewed here: http://crocodilian.com/books/grigg-kirshner.)*

One of these archosauriform lineages gave rise in the Triassic to the Archosauria, the group which became the dominant tetrapod fauna in the Mesozoic, the 'Age of Reptiles', and lives on today in the form of crocodylians and birds.

ARCHOSAURS; TWO MAJOR CLADES, CRUROTARSI (PSEUDOSUCHIA) AND AVEMETATARSALIA (ORNITHODIRA)

The Triassic was taxonomically crowded with reptiles, most of which were diapsids. It would have been a wonderful time for a herpetologist to visit! On land, there were Testudines (turtles), Lepidosauria (ancestors of today's snakes, lizards and tuataras) and Archosauria (ancestors of today's crocodylians and birds), the oceans were inhabited by Sauropterygia (plesiosaurs and pliosaurs), Ichthyopterygia (ichthyosaurs) and, in fresh water, different turtles.

Among this Triassic reptilian diversity, the archosaurs were the dominant land vertebrates. Archosaurs are diapsids with teeth (except for modern birds) that fit into sockets (thecodont dentition), a ridge on the femur that

Fig. 2.5. Erythrosuchus *was a member of a quite diverse family of mostly large (to 5 m) early to mid-Triassic terrestrial predatory archosauriforms (Fig. 2.3). The Erythrosuchidae were very widespread, with fossils having turned up in Europe, Africa and China.*

serves for muscle attachment and that may have facilitated the evolution of bipedality, and additional openings in the skull in front of their eyes (present early, and then lost, in crocodylians), as well as in the jaw bones, which may lighten the structures. They also have a heel bone – the calcaneum which, as in humans, serves for muscle attachment.

Archosaurs appeared in the fossil record in the early Triassic, although some late Permian fossils may also be better classified as archosaurs. They derived from two earlier 'non-archosaurian' clades, Archosauromorpha and Archosauriformes (Fig. 2.3), from which *Proterosuchus*, *Euparkeria* (Fig. 2.4) and *Erythrosuchus* (Fig. 2.5) are among the best known. By the mid-Triassic, two major clades of Archosauria were apparent: the Avemetatarsalia (Benton 1999) (known also as Ornithodira because of the upright, 'bird-like' stance in many); and the Crurotarsi (= cross-ankle bones (Sereno and Arcucci 1990) and known also as Pseudosuchia). All the archosaurs nest in either one of

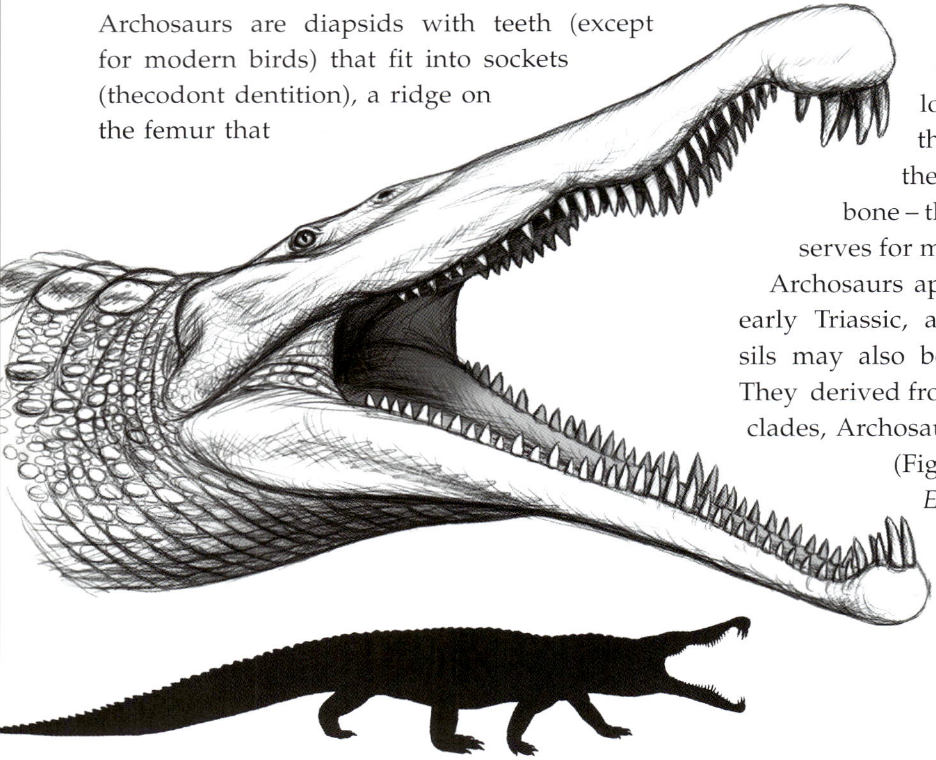

Fig. 2.6. *A phytosaur. These superficially crocodylian-like reptiles were large (2–12 m, mostly 3–4 m), semi-aquatic reptiles that must have had similar lifestyles to today's crocodylians; they are also known to have made a nest and laid eggs. Although long regarded as archosaurs, this has been thrown into doubt by a very comprehensive analysis of their morphology (Nesbitt 2011), which concluded that they are not within Archosauria at all, but are a sister group within Archosauromorpha (Fig. 2.3).*

these two clades. The Ornithdira (Avemetatarsalia) include those most celebrated ancient reptiles, the dinosaurs, which, in turn, gave rise to the birds. It also includes what must have been equally spectacular creatures: the flying pterosaurs, some of which had wingspans of more than 10 m. The other track led to another very diverse collection: the Pseudosuchia, or Crurotarsi, the 'crocodile-like reptiles', of which only the Crocodylia have survived into modern times (Fig. 2.3). The difference between the two clades relates to the structure of the foot skeleton (Fig. 2.7): the ankle bones of the Crurotarsi confer more flexibility, with a capacity for rotation, which allows them to walk with either a sprawling gait or with an erect stance.

CRUROTARSAN DIVERSIFICATION IN THE EARLY TRIASSIC: THE EARLIEST CROCODILE-LIKE REPTILES

By the end of the Triassic there were many widely distributed groups of Crurotarsi: ornithosuchids,

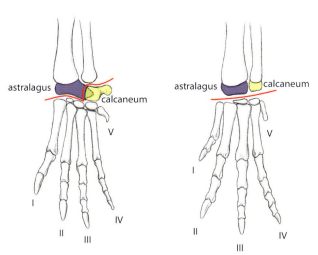

Fig. 2.7. *(Left) Top views of the left hind foot of an American alligator and (right) a generalised dinosaur. Archosaurs fall into two major groups on the basis of the skeletal structure of their ankle joints: the Crurotarsi (= Pseudosuchia) (crocodile-like reptiles) and the Avemetatarsalia (dinosaurs and pterosaurs). In Crurotarsi, a peg on the astralagus fits into a socket in the calcaneum, conferring more flexibility, which enables both sprawling gaits as well as the crocodylian high walk (Chapter 4). Dinosaurs have a simpler hinge joint, conducive to a more erect carriage and non-sprawling gaits.*

poposaurids, rauisuchians, aetosaurs, *Gracilisuchus* and sphenosuchians and the group from which the extant Crocodylia are derived, the Crocodylomorpha (Fig. 2.3). These groups will be discussed below. They were very diverse in appearance, body size, habitat and diet. Traditionally included within them have been the semi-aquatic Phytosauria (Fig. 2.6), but, as discussed above, it may be more appropriate to acknowledge them outside the archosaur clade, as a sister group. Others were very different in body form and lifestyle: the rauisuchians (4–6 m) were active, agile terrestrial predators; and the aetosaurs (1–5 m) were heavily armoured terrestrial herbivores. They all differed from Crocodylia in having no secondary bony palate. What the taxonomic relationships were between the Triassic Crurotarsi is not yet fully resolved, but reviews by Brusatte *et al.* (2010a) and Nesbitt (2011) have drawn attention to the issues and clarified many of them.

Most of the Triassic reptiles, especially the large ones, were wiped out by the End-Triassic Extinction Event, 201 Mya. This may have been caused by an enormous volcanic explosion in what is now the mid-Atlantic Ocean (Whiteside *et al.* 2010). From the Crurotarsi, only some crocodylomorphs survived past the Triassic and the extinction of the remainder, particularly the large carnivorous rauisuchians, may have opened the way for the evolution of large predatory dinosaurs in the Jurassic and Cretaceous.

Some of the most conspicuous examples of Triassic non-crocodylomorph crocodile-like reptiles are listed below.

Gracilisuchus

Gracilisuchus is a small (30 cm) bipedal mid-Triassic carnivore known from fossils in Argentina (Fig. 2.8). Its affinity has been the subject of much debate. At various times it has been classified as a dinosaur; at others, within Crocodylomorpha and also as a sphenosuchian. Brusatte *et al.* (2010a) placed it more basally (Fig. 2.3). Part of the interest it has generated is probably due to its agile body form.

Ornithosuchidae

These late Triassic archosaurs were quite like dinosaurs and were judged that way for quite some

Fig. 2.8. Gracilisuchus, *a small (30 cm) mid-Triassic crurotarsan that has been variously classified, most recently as more basal than the crocodylomorphs.*

time. However, they are now recognised as basal Crurotarsi (= Pseudosuchia) and characterised by two genera: *Ornithosuchus* from Scotland and *Riojasuchus* from Argentina (Figs 2.3, 2.9). The confusion with dinosaurs came from them having larger hind legs than forelegs, implying at least facultative bipedality, and large teeth implying predatory habits. Braincase morphology and ankle structure, however, show their affinity with the crocodylian line of evolution, not the dinosaur line. The limbs were under the body, rather than to the sides. Their position in relation to other crurotarsans has been uncertain, but in his recent, and very comprehensive, analysis Nesbitt (2011) judged them to be a sister group to the Suchia and the basal-most of the Crurotarsi.

Aetosaurs

Also occurring in the late Triassic, but not beyond, the aetosaurs (Fig. 2.10) could be claimed to be the most extraordinary of the early crurotarsans. For a start, they were herbivorous. They were also extraordinarily heavily armoured, sometimes with the osteoderms extended into long spines, presumably for defence against predators. They had an upright stance with the limbs well under the body and small heads with upturned snouts. Perhaps they used their snouts for digging up roots and tubers, like modern pigs. More than 20 genera have been described, ranging from 1 to 4 m. Bowl-shaped nests with eggs have been found, apparently dug out, raising the possibility that the maternal care seen in extant crocodylians is extremely old. Indeed, many dinosaur nests are known too, so nesting and maternal care by reptiles are very likely to be older than the separation of the archosaurs into two clades.

Poposauroidea

A diverse collection of large carnivorous terrestrial Triassic Crurotarsi has for some time been grouped together as 'Rauisuchia'. Many were large (4–6 m) and had an erect stance and large skulls well armed with teeth. Many were dinosaur-like and some were sail-backed. Some were bipedal, providing an example of convergent evolution with theropod dinosaurs. The affinities of the 'rauisuchians' and whether or not they were monophyletic has been a controversial issue. Brusatte *et al.* (2010a)

Fig. 2.9. Ornithosuchus, *a late Triassic crurotarsan. Growing to 4 m it was a widespread terrestrial predator. Somewhat dinosaur-like, one conspicuous difference is that it had five toes, not four.*

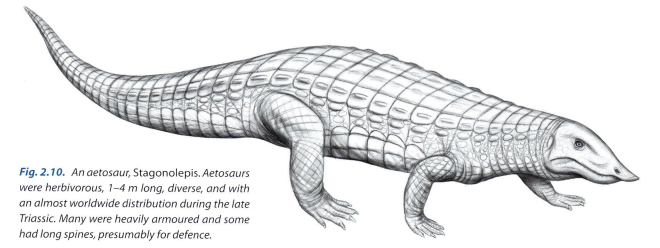

Fig. 2.10. *An aetosaur,* Stagonolepis. *Aetosaurs were herbivorous, 1–4 m long, diverse, and with an almost worldwide distribution during the late Triassic. Many were heavily armoured and some had long spines, presumably for defence.*

referred to the collection of fossil genera listed as rauisuchians as a 'nebulous assemblage', which was 'the most problematic issue in crurotarsan phylogeny'. On the basis of analysis of a very large dataset, Nesbitt (2011) concluded that the genera usually included as 'rauisuchians' are paraphyletic and that the term Rauisuchia should be abandoned. He recognised two sub-sets of genera: the Poposauroidea forming a monophyletic group outside the Loricata and the Rauisuchidae forming a sister group to Crocodylomorpha (Fig. 2.3). Poposaurs included the sail-backed *Arizonasaurus* and *Ctenosauriscus,* the superficially dinosaur-like bipedal runners *Effigia, Poposaurus* and *Shuvosaurus* and a possibly semi-aquatic quadruped, *Qianosuchus.* The Rauisuchidae included the large-bodied quadrupedal predators *Postosuchus, Polonosuchus* and *Rauisuchus.*

A re-evaluation of the sail-backed *Ctenosauriscus koeneni* concluded that it was from the early Triassic (Butler *et al.* 2011). Palaeontology presents an ever-changing story, but, putting this with the comprehensive analysis by Nesbitt (2011), the Poposauroidea can be regarded as a monophyletic clade of early Triassic Crurotarsi (Pseudosuchia) and the first global radiation of archosaurs. Many of them were very similar to dinosaurs and some were mis-diagnosed that way. The nine or so genera were large (2–4 m) and somewhat diverse; some were sail-backed and some lacked teeth.

Two genera are of particular interest: *Qianosuchus mixtus,* from a magnificent specimen in China (Fig. 2.11), was almost certainly aquatic (Li *et al.*

2006); and the bipedal *Effigia okeeffeae* (Fig. 2.12) (Nesbitt and Norell 2006; Nesbitt 2007). *Effigia* had so many characters convergent with dinosaurs that Nesbitt (2007) suggested that a theropod dinosaur body plan developed in the crocodile-like archosaurs before it appeared in the dinosaurs. Given the likelihood that the crocodylians have endothermic ancestry (Seymour *et al.* 2004; Chapter 10), it is tempting to posit that *Effigia* was endothermic.

'Rauisuchians' (Rauisuchidae)

Nesbitt (2011) provided a substantial review of the history of the use of the term Rauisuchia, recognising it as paraphyletic and separating from it the Poposauroidea (see above). He recommended that the term Rauisuchia should be abandoned. The Rauisuchidae includes a range of large (310 m) terrestrial carnivorous Crurotarsi, such as *Postosuchus,* with an erect stance and large skulls well armed with teeth (Fig. 2.13). Unlike poposaurs, there seem to have been no bipedal genera and most had a sturdy quadrupedal build, with the forelimbs shorter than the hindlimbs. Some had two or more rows of osteoderms dorsally. They are thought to have been agile predators, despite their sturdy build. Their fossils have been found in Europe and North and South America and they are usually regarded as having become extinct at the end of the Triassic.

BASAL CROCODYLOMORPHS

In the late Triassic, there were several groups of small, lightly built terrestrial archosaurs whose affinities

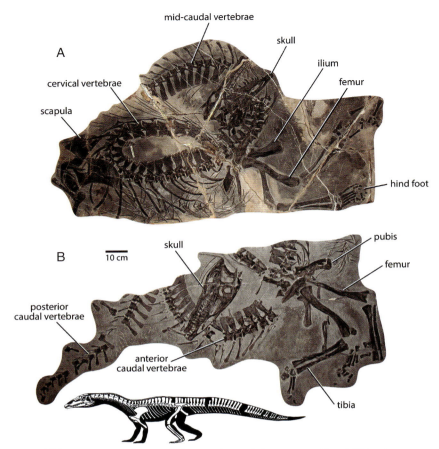

Fig. 2.11. *(A) Holotype and (B) paratype specimens of* Qianosuchus mixtus, *an aquatic mid-Triassic poposaur from China. (Modified from a photograph supplied by Xiao-chun Wu, Institute of Vertebrate Paleontology and Paleoanthropology, Academia Sinica, Beijing, People's Republic of China)*

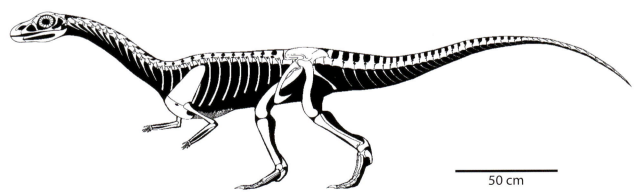

Fig. 2.12. *A skeletal reconstruction of the late Triassic poposaur* Effigia okeefeae *(from Nesbitt (2007) with permission). Nesbitt noted that the reconstruction is based on all available specimens and that details of distal tail, the number of vertebrae and proportions of the femur are uncertain. Two things are of particular interest about* Effigia: *its similarity to ornithomimid dinosaurs, which did not appear until the Cretaceous and that, after its collection in the late 1940s, it lay undiscovered and still in its plaster jacket in the American Museum of Natural History for ~60 years before Sterling Nesbitt opened it up. It was collected by the famous zoologist and palaeontologist Edwin H Colbert at Ghost Ranch Quarry in New Mexico, and its generic name means 'ghost'.*

Fig. 2.13. *This late Triassic member of the Rauisuchidae, Postosuchus, is named for Post Quarry in Texas where it was first found. It was a large (3–4 m) quadrupedal carnivore, with the limbs well in under the body. The Rauisuchidae may have been a sister clade to Crocodylomorpha (Nesbitt 2011) (Fig. 2.3).*

with each other are somewhat uncertain, but which can be regarded as basal crocodylomorphs. All were apparently carnivorous. The smallest were ~30 cm long, the largest ~1.5 m and some may have been bipedal.

'Sphenosuchians'

About 20 genera of small (30 cm to ~1.5 m), lightly built terrestrial archosaurs that have been found in North and South America, Eurasia and Africa, and whose affinities with each other are somewhat uncertain, are commonly grouped as the Sphenosuchia. They were around from the late Triassic to late Jurassic. They had long limbs, an erect stance with the limbs held under the body, procoelus vertebrae and other anatomy suggesting a capacity for flexing the spine, reduced osteoderms and reduced digits. These attributes all suggest running ability, and they were probably fast, agile predators. They were somewhat reminiscent of today's varanid lizards

but with longer limbs. Some may have been bipedal. It is probable that Sphenosuchia is not a monophyletic group (Clark and Sues 2002; Clark *et al.* 2004) and, like so many of the fossil archosaur groups, the positions of 'sphenosuchians' such as *Terrestrisuchus* (Fig. 2.14), *Sphenosuchus* and *Saltoposuchus* are uncertain. But they are accepted as 'basal crocodylomorphs', despite being so apparently different from the typical crocs. This does not imply that today's crocodylians are descended from them, but they were a group that branched away early from what became the main stem.

CROCODYLIFORMES (EXCLUDING EUSUCHIA)

The clade Crocodyliformes was erected by Benton and Clark (1988) and subsequently extended by Sereno *et al.* (2001) who defined it more specifically. It includes the Protosuchia and Mesoeucrocodylia, so the surviving crocs are all within this clade (Fig. 2.3).

Protosuchians

The name literally means 'first crocodiles' and they occurred from the late Triassic to

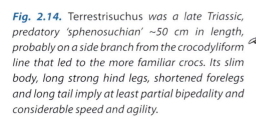

Fig. 2.14. *Terrestrisuchus was a late Triassic, predatory 'sphenosuchian' ~50 cm in length, probably on a side branch from the crocodyliform line that led to the more familiar crocs. Its slim body, long strong hind legs, shortened forelegs and long tail imply at least partial bipedality and considerable speed and agility.*

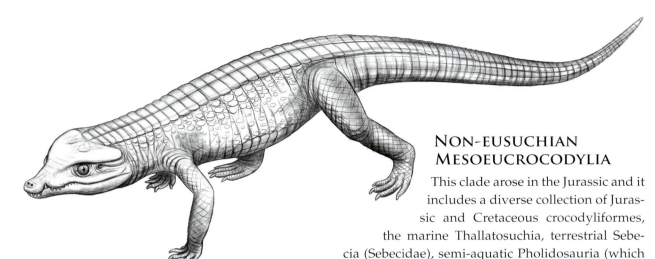

Fig. 2.15. *The early Jurassic* Protosuchus *(1 m) is one of several genera of protosuchians: a sister group to the Mesoeucrocodylians (Fig. 2.3).*

early Jurassic. The skeletons and skulls were very crocodile like but, like the 'sphenosuchians', they were terrestrial, had long limbs and walked with the limbs beneath them (Fig. 2.15). Their skeletal features and dentition suggest they were terrestrial carnivores. They were widespread in what are now the Americas, Africa and China.

So, through most of the Triassic, the 'crocs' were mostly terrestrial, with a great range of body sizes from small to very large, including some which were obviously very active predators. Most of them were wiped out by the end-Triassic extinction but many crocodylomorphs survived, with the small, long-limbed sphenosuchians and the larger (to 1 m) protosuchians on land, and the spectacular thallatosuchians, such as the Metriorhynchidae, in the oceans.

NON-EUSUCHIAN MESOEUCROCODYLIA

This clade arose in the Jurassic and it includes a diverse collection of Jurassic and Cretaceous crocodyliformes, the marine Thallatosuchia, terrestrial Sebecia (Sebecidae), semi-aquatic Pholidosauria (which includes *Sarcosuchus imperator*, the much celebrated 'Supercroc'), the strange, almost 'mammal-like' Notosuchia and the earliest Eusuchia (which includes the Crocodylia). The mesoeucrocodylians were known previously as the Mesosuchia. Only the sebecids and eusuchians survived the K–Pg extinction, with the former dying out in the Eocene.

Thallatosuchians: a marine invasion

These were marine crocodyliformes such as *Pelagosaurus* (Teleosauridae) and *Metriorhynchus* (Metriorhynchidae). Teleosaurs retained limbs and would have been capable of venturing onto land, but metriorhynchids had limbs modified as paddles and presumably were entirely aquatic. Metriorhynchids were well adapted for aquatic life, with a streamlined, elongated

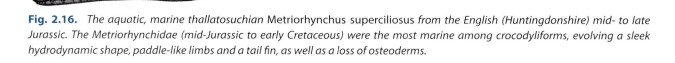

Fig. 2.16. *The aquatic, marine thallatosuchian* Metriorhynchus superciliosus *from the English (Huntingdonshire) mid- to late Jurassic. The Metriorhynchidae (mid-Jurassic to early Cretaceous) were the most marine among crocodyliforms, evolving a sleek hydrodynamic shape, paddle-like limbs and a tail fin, as well as a loss of osteoderms.*

body form (Fig. 2.16). The forelimbs were paddle-like and the hindlimbs appear to have been evolving in that direction. Pelvic and pectoral girdles were reduced, presumably because the paddles were not weight bearing, and they had a well-developed tail fin. The tail fin was supported on a downward-turned vertebral column; that is, a hypocercal tail similar to that of the ichthyosaurs and some of the Mosasaurs (late Cretaceous squamates). Metriorhynchids were carnivores, feeding on fish and belemnites (extinct straight-shelled cephalopods that were very numerous in the Jurassic and Cretaceous). There is a comprehensive review of Metriorhynchidae by Young *et al.* (2010) and a detailed study of their dentition by de Andrade *et al.* (2010).

We can be fairly sure that Metriorhynchids had salt glands (Fernández and Gasparini 2000, 2008; Gandola *et al.* 2006), although these were not on their tongues as in Crocodylidae (Figs 11.15, 11.16) but associated with the orbits (Fig. 2.17).

Notosuchians: very different beasts

In the late Cretaceous of Gondwana, there was a substantial adaptive radiation of small (up to ~1 m) and very different, short-snouted mesoeucrocodylians that can be regarded as the most unusual of all the Crurotarsi. Their body

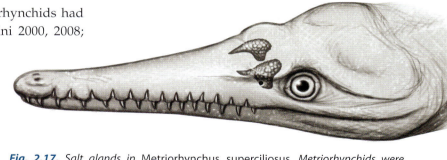
cavity

Fig. 2.17. *Salt glands in* Metriorhynchus superciliosus. *Metriorhynchids were entirely marine and paired cavities in front of the eyes with associated ducts have been interpreted as salt glands, probably draining to the exterior via openings on each side of the snout (Fernández and Herrera 2009). One of the cavities is visible in the palatal view of this skull (top) from a late Jurassic* Metriorhynchus superciliosus *from England, held in the National Museum of Ireland–Natural history (NMING:F16892 and NMING:F21731) (Gandola* et al. *2006). The position of the glands is based on* M. superciliosus *and the related late Jurassic* Cricosaurus *(formerly* Geosaurus*)* araucanensis *(Fernández and Gasparini 2000, 2008; Gandola* et al. *2006; Fernández and Herrera 2009). (Photo National Museum of Ireland and Rob Gandola)*

forms and dentitions suggest lifestyles more like mammals than reptiles. Two genera, *Simosuchus* from Madagascar and *Pakasuchus* from Tanzania, have attracted considerable interest. Their posture was erect, not sprawling, and the post-cranial skeletons were similar to those of other terrestrial mesoeucrocodylia, but their fossilised skulls are dramatically different. In *Simosuchus* (Figs 2.18, 2.19) the skull was high-domed, described by Buckley *et al.* (2000) as 'pug-nosed', with a most uncroc-like squarish jawline. Just as striking, the teeth were multi-cusped and 'clove-shaped', suggesting a herbivorous diet. *Simosuchus* also had a substantial covering of osteoderms, presumably supporting scutes, even on the limbs (Fig. 2.19). It was recently the subject of an extensive supplement to the *Journal of Vertebrate Paleontology* (Krause and Kley 2010). *Pakasuchus*, described by O'Connor

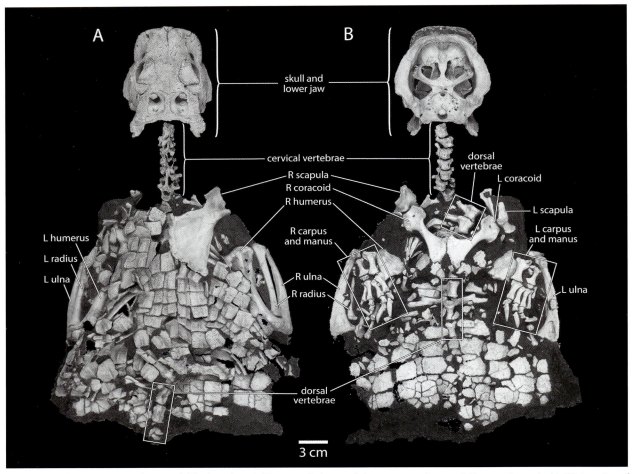

Fig. 2.18. *This is a painted cast of (A) dorsal and (B) ventral views of the type specimen of* Simosuchus clarki *(University of Antananarivo specimen UA 8679) discovered in 1998 by L. L. Randriamiaramanana in NW Madagascar (from Krause* et al. *2010). The authors noted that the specimen is a virtually complete and articulated skull and lower jaw, with much of the post-cranial skeleton as well. (Photo David Krause)*

et al. (2010), had a short broad skull, with a robust jaw holding a reduced number of highly differentiated teeth (Fig. 2.20). The resting position of the head was angled down, the neck provided mobility and the dentition looks more mammalian than reptilian. These features are reflected in the name *Pakasuchus,* where *Paka* is Kiswahili for 'cat', and indeed it was about the size of a domesticated cat. 'Canines', 'pre-molars' and 'molars' are identifiable, and the dentition suggests a capacity for tearing and shearing flesh. The jaw articulation may have allowed more mobility than in extant croc species, and even a capacity to chew and process food more like mammals than crocs or, indeed, any extant reptile. The osteoderms were reduced in *Pakasuchus,* except on the tail. Understandably,

descriptions of this fossil caught the attention of the world's media.

Two additional notosuchian genera are of particular interest: *Anatosuchus* and *Araripesuchus.* *Anatosuchus* (Figs 2.21, 2.22) is a really extraordinary, duck-billed beast from the early Cretaceous of Niger (Sereno *et al.* 2003) growing to ~70 cm. These authors observed that it had 'one of the most specialised snouts among crocodylomorphs', probably having a fleshy area around the nares that may have been elaborated for olfaction and tactile senses. The number of teeth on the upper jaw increased to suit the broadened upper jaw and medially in the lower jaw was a bony projection that would engage behind the tooth row with the front of the upper jaw and be suitable for crushing prey. The fingers were

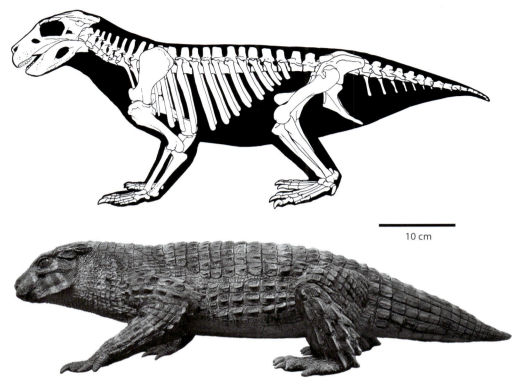

10 cm

Fig. 2.19. *(Top) Reconstruction of the skeleton of* Simosuchus clarki, *illustration by L Betti-Nash, based primarily on the type specimen (UA 8679) and three in the Florida Museum of Natural History (FMNH PR 2596, FMNH PR 2597, and FMNH PR 2598. (Bottom) Reconstruction of* Simosuchis clarki, *based on a photograph of sculpture by B. Filipović, which was then adjusted digitally by L. Betti-Nash. (Both images from Krause* et al. *2010, provided by David Krause)*

elongated. Sereno and colleagues suggested that *Anatosuchus* may have captured invertebrates, frogs and small fish by scratching and digging through vegetation in shallow water.

There are at least six species of *Araripesuchus* and their fossilised remains have been found at widespread Gondwanan locations including Brazil, Argentina, Niger (Figs 2.23, 2.24) and Madagascar. The largest grew to ~1 m and they were probably omnivores. There has been uncertainty about whether they are notosuchians. Turner (2006), in a detailed study describing a new species from Madagascar, concluded otherwise, but separate analyses by Pol and Apesteguia (2005) and Sereno and Larsson (2009) placed them within Notosuchia. For our purposes, what is more important is that they represent another geographically widespread, but Gondwanan, small terrestrial crocodylomorph that was possibly omnivorous, thus broadening the known diversity of the Cretaceous mesoeucrocodylia.

Although it is clear that the notosuchians had a substantial adaptive radiation in Gondwana, whether or not they were restricted to that landmass is uncertain because of fossils found in China. Wu *et al.* (1995) described *Chimaerasuchus paradoxus* from Cretaceous rocks in the province of Hubei and assigned it to Notosuchia on the basis of many similarities, including heterodont dentition suggestive of herbivory. Further analyses by both Wu and Sues (1996) and O'Connor *et al.* (2010) confirmed its proximity to notosuchians. The puzzle about distribution may not be resolved until more fossil material is known.

Sebecia (Sebecidae, sometimes called Sebecosuchia)

These were found in European and South American deposits from Cretaceous through to the mid-Miocene. They were a clade of large terrestrial carnivorous mesoeucrocodylia that grew to 3–4 m (Fig. 2.25, p. 66). They had laterally compressed

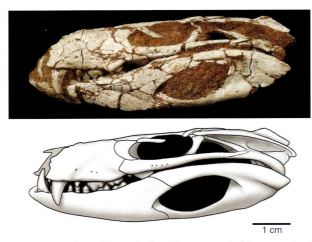

Fig. 2.20. Pakasuchus *skull. This remarkable, cat-sized notosuchian had highly differentiated, mammal-like teeth, with 'canines', 'pre-molars' and 'molars' identifiable. The pre-maxilla (grey) is conjectural.* Pakasuchus *may have been able to move the lower jaw fore and aft, implying a capacity for shearing between upper and lower 'molars': a facility unknown in any other crocodyliform. (Photo Patrick O'Connor)*

heads with laterally compressed, pointed and serrated teeth (ziphodont dentition). Some of them, such as *Iberosuchus* from Spain, may have been able to gallop, as do some of the modern crocodiles (see Chapter 4). To my perception, sebecids are candidates for having been endothermic (see Chapter 10). Turner and Calvo (2005) and Larsson and Sues (2007) provide recent reviews about sebecosuchians and their phylogenetic affinities (and uncertainties).

Kaprosuchus: *the boar croc*

Another very extraordinary crocodyliform skull was found in Upper Cretaceous rocks in the Sahara Desert in the Republic of Niger. Its discovery led to the erection of a new genus and species, *Kaprosuchus saharicus*, meaning 'boar crocodile of the Sahara Desert' (Sereno and Larsson 2009) (Fig. 2.3). The name recognises the superficial similarity to a boar's head (Greek *kapros* = boar; *souchos* = crocodile) and

Fig. 2.21. *Lateral, dorsal and ventral views of the skull of* Anatosuchus minor *(from Sereno and Larsson 2009). Scale bar = 5 cm. The pink tone indicates restored jaw margin. (Photos Carol Abraczinskas and Paul Sereno)*

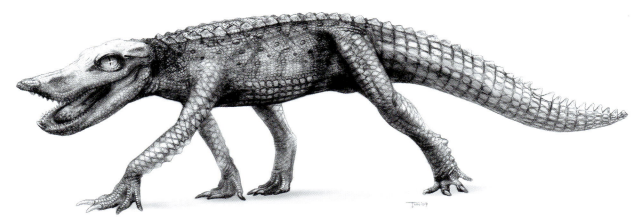

Fig. 2.22. *Artist's reconstruction of* Anatosuchus minor *(Sereno and Larsson 2009). (Artwork Todd Marshall)*

where it was discovered. The similarity to a boar is conveyed to some extent by the overall skull shape, but most particularly by the hyperdeveloped, tusk-like dentition (Fig. 2.26). These were large animals. The skull was ~51 cm long, the lower jaw a shade over 60 cm and Sereno and Larsson projected a total length of ~6 m. Finding that skull, almost complete, must have been an enormous thrill! Not only are the teeth large, the broadened anterior of the pre-maxillae is rough looking and rugose – like a well-trimmed moustache according to Sereno and Larsson – and suggestive of a raised, keratinous shield. The authors considered that it is more likely

20 cm

Fig. 2.23. *Block containing multiple specimens of* Araripesuchus wegeneri *from the Lower Cretaceous of Niger (from Sereno and Larsson 2009). Three partly articulated specimens lie parallel and there is part of a skull to the top right. (Photo Carol Abraczinskas and Paul Sereno)*

to have been a terrestrial than an aquatic predator. No post-cranial remains have yet been found.

Pholidosaurs (Sarcosuchus)

The pholidosaur that stands out and deserves a special mention is *Sarcosuchus*, nicknamed 'Supercroc' (Sloan and Sereno 2002) (Fig. 2.27). It was discovered in Cretaceous sediments in the Nigerian Sahara by Albert-Félix de Lapparent, a French palaeontologist and priest, and was described by de Broin and Taquet (1966). de Lapparent found only fragments, mostly teeth and dermal bones, but significant quantities of skeletal material were recovered by Paul Sereno in 1997. *Sarcosuchus* was at least as big as the eusuchian *Deinosuchus* (see below), reaching ~12 m and 8–11 000 kg, depending on whether it scales like an alligator or a crocodile (Fig. 1.36). That is much more than 100 times my bodyweight and its skull (Fig. 2.28, p. 68) is about as long as I am tall! The jaw is quite narrow, suggestive of a fish-eating habit, and it was large enough to eat very large prey. Maybe it was an ancient equivalent of today's gharials, except that it was much larger. The back was armour-plated with huge osteoderms and Steve Salisbury (*pers. comm.* 2010) suggested it may not have had even the mobility of *Gavialis* on land (see Chapter 4). The end of the upper jaw contained a large circular depression, suggesting that the soft tissue was elaborated into some sort of a chamber. The function of this is unknown and suggestions have been the obvious ones: perhaps used for olfaction or for resonating sounds used in communication. Is this enigmatic

Fig. 2.24. *A reconstruction of* Araripesuchus wegeneri, *a small late Cretaceous notosuchian from Niger (Sereno and Larsson 2009). (Artwork Todd Marshall)*

anatomy of the *Sarcosuchus* snout a further parallel with modern gharials, with their bladder-like ornamentation on the snout of males?

BERNISSARTIA AND EXTINCT EUSUCHIA (INCLUDING EXTINCT CROCODYLIA)

The Crocodylia are all in a broader clade, the Eusuchia (*eu* = 'well', *such* = Egyptian word for crocodiles so, by this derivation, these are the 'proper' crocodylians). The eusuchians evolved during the Cretaceous more than 100 Mya within the clade Mesoeucrocodylia (Fig. 2.3 and see below).

Fig. 2.25. Sebecus *(a sebecid).*

The Eusuchians had and have 'ball and socket' joints between the vertebrae (i.e. 'procoelus' vertebrae) (Fig. 2.29) and internal nostrils well back on the secondary palate, presumably behind a palatal valve as in the extant species, ducting air directly into the glottis (Fig. 5.18 and Chapter 7). These characteristics are indicative, but not diagnostic, of both extinct and extant eusuchians.

As the major theme of this book is crocodylian function, a review of the fossil Eusuchians is beyond its scope. However, I have chosen several to highlight eusuchian diversity: the mekosuchines (Crocodyloidea) and *Pristichampsus* in the Caenozoic, and *Deinosuchus* (Alligatoroidea) and *Isisfordia* in the Cretaceous. Another famous croc known only from fossils is *Bernissartia*, which is sometimes included with the eusuchians.

Bernissartia

A small crocodyliform reptile was found fossilised in early Cretaceous sediments in a quarry in Belgium and described by Dollo (1883) as *Bernissartia fagesii* (Figs 2.30, 2.31, p. 69). The genus may have been quite widespread. A nearly complete skull was found in Spain and characteristically rounded teeth from the Isle of Wight have been attributed to *Bernissartia*, as have fragments from North America (Buffetaut and Ford 1979). It has been classified as a neosuchian (Fig. 2.3), and

Fig. 2.26. Kaprosuchus saharicus, *the 'boar crocodile from the Sahara' (from Sereno and Larsson 2009). Growing to ~6 m, this late Cretaceous neosuchian (Fig. 2.3) was probably a terrestrial predator. (Photo Carol Abraczinskas and Paul Sereno)*

is considered to be more primitive than early eusuchians such as *Isisfordia* (Fig. 2.32 and see below). It was probably restricted to fresh water. It grew to ~60 cm and would have weighed less than 1 kg so is among the smallest of crocs.

Despite not being a eusuchian, the similarity between *Bernissartia* from 130 Mya and today's crocodylians is striking. If one were released today, it could probably pass as just another small croc in a North Queensland or Florida wetland. This could be said about most of the early eusuchians as well, except for the very large ones, and it would be their large size not an unusual body form that would draw comment.

Isisfordia duncani: *a eusuchian in Gondwana*

Until recently, the most primitive known eusuchian was *Hylaeochampsa vectiana* from early Cretaceous sediments on the Isle of Wight. It had a secondary palate and posteriorly situated internal nares.

However, it may have been supplanted by spectacularly good fossil material that turned up in Queensland in the mid-1990s near the small town of Isisford, and which has been identified by Steve Salisbury as another early eusuchian, from the early Cretaceous ~100 Mya. It is apparently the earliest eusuchian found so far and Salisbury *et al.* (2006) have suggested that the Eusuchia may, in fact, have evolved in Gondwana.

The original specimen, articulated and almost complete except for the skull, is shown in Fig. 2.32. After painstaking preparation by Kerry Geddes and Steve Salisbury at the University of Queensland, most of the skeleton of a 1 m croc can be seen. Most crocodylian fossils are far less complete than this one. Palaeontologists are usually working from skull fragments, hopefully including teeth and maybe skull and jaw articulations, and any identifiable post-cranial material is a real bonus. This one is essentially complete, except for the skull. However, a lucky hammer blow on a boulder at the same site in April 2005 (Fig. 2.33) produced a skull which, after preparation (Fig. 2.34, p. 71), confirmed the early diagnosis of *I. duncani* as a very early eusuchian. A reconstruction by David Kirshner is shown in Fig. 2.35 (p. 71).

Although *Isisfordia* is disjunct from the main line that gave rise to the three major Crocodylian clades, Gavialoidea, Crocodyloidea and Alligatoroidea, it appears to be a relic close to that stem (Fig. 2.3). The features that make *Isisfordia* exciting are the anatomy of its vertebrae and the position of the internal nares and bony structure surrounding them (Fig. 2.29). In the Crocodylia, the vertebrae articulate with each other via ball-and-socket joints, which give the spine flexibility. In the earlier crocodylomorphs

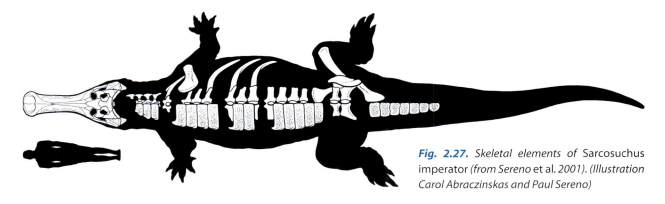

Fig. 2.27. *Skeletal elements of* Sarcosuchus imperator *(from Sereno* et al. *2001). (Illustration Carol Abraczinskas and Paul Sereno)*

Fig. 2.28. *The skull of* Sarcosuchus imperator. *(Photo DSK, courtesy of Museum and Art Gallery of the Northern Territory)*

intermediate. Also intermediate were the position of the internal nares and the extent of development and fusion of the bony plates through which the nares pass. Salisbury (2006) considered that these two attributes may have enabled the famous 'death roll' (Chapter 6), making possible the capture and overpowering of large prey.

What excites me about *Isisfordia* almost as much as the phylogenetic information it conveys and its discovery in Australia, is how similar it is to the crocs that we are all familiar with, even though it lived 100 million years ago. The physiology, anatomy and probable lifestyles of the three extant groups, all of which had evolved by the late Cretaceous, are so similar that we can reasonably assume that these attributes are shared with their common ancestor, whatever that was.

Pristichampsus

This Palaeocene–Eocene croc (a pristichampsine, Brochu 2003) (Fig. 2.36, p. 72) can be classified within Crocodylia. It may be the most unusual of

the vertebrae were slightly concave at the ends, allowing less angular movement and which, with the osteodermal armour (Chapter 3), restricted flexibility. The vertebrae of *Isisfordia* were

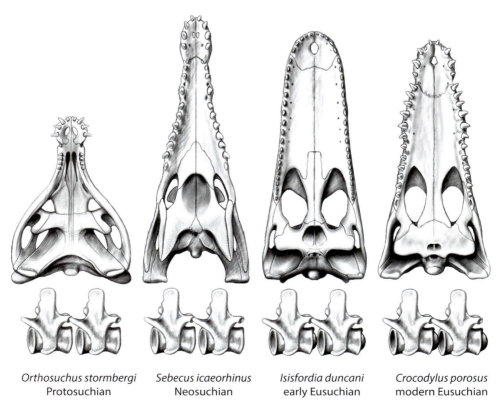

| *Orthosuchus stormbergi* | *Sebecus icaeorhinus* | *Isisfordia duncani* | *Crocodylus porosus* |
| Protosuchian | Neosuchian | early Eusuchian | modern Eusuchian |

Fig. 2.29. *Rearward movement of the internal nares and evolution of procoelus vertebrae. Eusuchians have 'ball and socket' joints between the vertebrae, conferring more spinal flexibility. Earlier crocodile-like reptiles had vertebrae with flatter ends (the anterior of vertebrae is on the left).*

Fig. 2.30. Bernissartia fagesii, *a very small Cretaceous neosuchian. Although considered more primitive than eusuchians, it probably looked very similar to modern crocodylids. It was described from a quarry in Belgium in 1883. The genus was reasonably widespread in Europe and, possibly, North America.*

the eusuchians, a reminder that not all eusuchians looked like the species living today. They likely arose in the late Cretaceous, survived the K–Pg extinction to diversify and become widespread in Europe, Asia and North America. Several species are known, growing to 2–3 m . They are thought to have been more terrestrial than other eusuchians, with limb structure suggesting a capacity for running on land, and an unusual foot structure often described as 'hoof-like'. Their snouts were laterally compressed, holding laterally compressed, and sometimes serrated, teeth.

Deinosuchus, *a late Cretaceous alligatoroid*

Its name means 'terrible crocodile' and it has come to be regarded as one of the really 'iconic' fossil crocodylians, becoming famous because of its very large size and obviously predatory habits, and a substantial review of its palaeobiology has been written (Schwimmer 2002). It abounded widely in North America during the late Cretaceous. In general body form *Deinosuchus* was much like modern alligators (Fig. 2.37). The broad snout was somewhat distended at the tip. Fossils have been found over a wide geographic area with a range of sizes. It remains uncertain whether there are several species or only one (Brochu 1999) and the specimens are usually referred to by just the generic name. Their skulls were ~2 m long, with very impressive dentition. Their maximum length is somewhat uncertain because of insufficient skeletal material, but 10 m seems a safe bet and Schwimmer (2002) made a good case for 12 m. The question of the affinity of *Deinosuchus* has been taken up comprehensively by Brochu (1999) who concluded that it is not 'the largest crocodile', but 'one of the largest alligators', as shown in Fig. 2.3. Brochu (1999) pointed out that gigantism is a recurring theme within crocodylians, with the 11–12 m north African early Cretaceous pholidosaur *Sarcosuchus* (see above) and the 10–12 m South American early Miocene alligatorid *Purussaurus* (a giant caiman) conspicuous examples.

Fossil *Deinosuchus* have been found in freshwater and marine/estuarine deposits. As an alligatoroid, *Deinosuchus* is unlikely to have had lingual salt

Fig. 2.31. *The mounted lectotype skeleton of* Bernissartia fagesii, *IRScNB R46. Total length ~600 mm. (Photo S. W. Salisbury and the Institut Royal des Sciences Naturelles de Belgique)*

tail vertebrae

left hind limb

left forelimb

base of skull

base of tail

right hind limb

dorsal osteoderms

10 cm

Fig. 2.32. (Left) In this assembled collection of rock is a wonderful fossil found in the mid-1990s by Ian Duncan on a property near the town of Isisford, Queensland. (Right) After its careful preparation by Kerry Geddes, the skeleton of Isisfordia duncani, named for Mr Duncan, is revealed. (Photos Steve Salisbury)

glands, such as are found in Crocodylidae (Chapter 11). However, the presence of fossils in marine sediments need be no surprise. *Alligator mississippiensis* and *Caiman latirostris* are both known to forage in estuarine and marine environments, depending upon periodic access to freshwater to re-establish osmotic homeostasis (Chapter 11). The length of time they can go between drinks depends on body size; larger animals can last longer. Adult *Deinosuchus* would have been able to travel quite large oceanic distances without difficulty because their small surface area:mass ratio would have

Fig. 2.33. (Left) Field work near Isisford, Queensland and (right) Steve Salisbury from the University of Queensland surveys a boulder in a creek bed nearby, which, when cracked open, revealed a skull which could be identified as the first specimen of an Isisfordia duncani skull. The prepared skull, released from the matrix, is shown in Figs 2.34, 2.35. (Photos Steve Salisbury)

Fig. 2.34. *Skull of* Isisfordia duncani *being extracted from the matrix. (Photo Steve Salisbury)*

Alligatoridae, Gavialidae and Crocodylidae

Crocs identifiable as belonging to one of these still extant groups evolved during the Cretaceous. Many of them survived the K–Pg extinction event and sallied forth more than 60 Mya into the early Palaeocene. Their broad similarity to each other, mostly 'crocodylian' in appearance, some with long snouts and some with short snouts, might convey the impression that there has not been much crocodylian evolution going on since then. However, there was a large crocodylian radiation in the Palaeocene and the fossil record shows many other genera and species in all three families (Brochu 2003; Scheyer *et al.* 2013). Many fossils have been found in Europe, northern Asia, North America and other places where the climate has changed (e.g. the final stages of continental drift, occurrence of ice

exchanged water and salts with the surroundings proportionally more slowly than in small individuals.

Fig. 2.35. Isisfordia duncani, *a 1.1 m early non-crocodylian eusuchian that lived between 98 and 95 Mya in the mid-Cretaceous swamps of what is now central Queensland. Despite some fundamental differences in its skeleton, this animal could pass unnoticed on a North Queensland or a Florida beach as a modern croc, with no trouble at all. (Reconstruction by DSK). Inset:* Isisfordia *skull after extraction from the matrix. (Photo Steve Salisbury)*

Fig. 2.36. *Unlike typical Palaeocene and Eocene eusuchians, species of* Pristichampsus *grew to 2–3 m and are thought to have been mostly terrestrial. They were also unusual in body form. They had extensive body armour, which may have compromised their lateral flexibility. The muzzle was laterally compressed, housing blade-like teeth, some with serrations. In this reconstruction by John Sibbick, made with input from Dino Frey and Steve Salisbury, a pristichampsid preys on small Eocene horses. It is easy for me to imagine these large, terrestrial predators being endothermic (see Chapter 10). (Image John Sibbick)*

Fig. 2.37. *Deinosuchus (= terrible crocodile) from the North American late Cretaceous, one of the largest known eusuchians, is thought to have grown to 10–12 m which, extrapolating from* Alligator *data (Fig. 1.36), implies a body weight up to 11 tonnes. (Reconstruction photographed at WILD LIFE Sydney Zoo)*

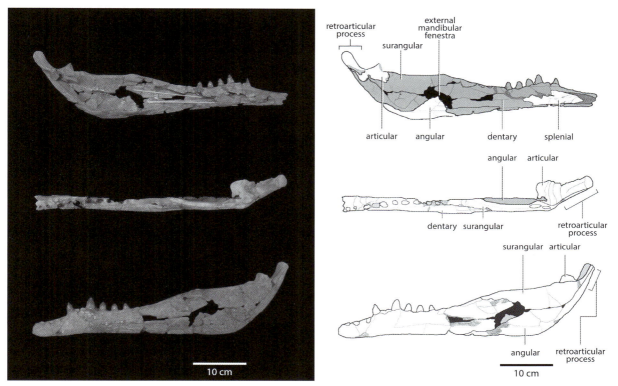

Fig. 2.38. *Palaeontologists usually have to work from fragments. This 60 million years old fossilised left lower jaw of a 3–4 m crocodylian came from Palaeocene deposits near Gladstone. It was described as a new species of mekosuchine,* Kambara molnari, *named for Ralph Molnar in honour of his contributions to palaeontology in Australia (Holt* et al. *2005). (Modified from images supplied by Steve Salisbury)*

ages) in ways that make them no longer suitable for crocs. Most, of course, are now extinct, but an example of a group that survived throughout most of the Caenozoic is provided by the mekosuchines.

The South Pacific mekosuchines

About nine genera of extinct Crocodylia have been designated as Mekosuchinae (see Fig. 2.3), identified from Eocene–Pleistocene deposits over the eastern two-thirds of the Australian mainland and islands of the South Pacific (Salisbury and Willis 1996; Willis 1997; Willis 2001). The earliest ones (*Kambara* spp.) inhabited the Australian continent while it was still joined to Antarctica, and one species (*Mekosuchus inexpectatus*) may have survived almost to modern times. It has been suggested that its extinction from New Caledonia, Fiji and Vanuatu may have resulted from predation following the arrival of human inhabitants. Paul Willis, previously at the University of New South Wales, has taken a particular interest in them. They are known mainly from skull and jaw

fragments such as in Fig. 2.38 (*Kambara molnari* from Palaeogene sediments near Gladstone, Queensland). In size they ranged from 1 m or less (*Mekosuchus*) to ~5 m (*Pallimnarchus*) and form an apparently monophyletic clade – a sub-group of Crocodyloidea, possibly as a separate family (Mekosuchidae) or sub-family (Mekosuchinae) within Crocodylidae. They showed a great diversity of head shapes, presumably reflecting dietary specialisations. Some mekosuchines had differentiation of the posterior teeth in a way that suggested usage for crushing molluscs, slightly reminiscent of the much more extreme tooth differentiation in Notosuchia (above). Some were terrestrial and, based on post-cranial skeletal material, Willis (1997) has suggested that *Mekosuchus inexpectatus* may have been arboreal. The mekosuchines apparently became extinct on the Australian mainland in the Pliocene, at the same time as did the megafauna and also coinciding with the first appearance of fossils attributed to *Crocodylus porosus* (Willis and Archer 1990). Their occurrence on Pacific Islands and their probable evolutionary proximity to Crocodylidae suggests that they may have had salt glands.

SUMMARY

The evolutionary pattern of the Amniota since the emergence of the clade is one of adaptive radiation and diversification, into both terrestrial and aquatic habitats, punctuated by dramatic extinction events at the end of the Permian, Triassic and Cretaceous. Given the number of reptile groups that did not make it through these extinction events, we are lucky to have any of the crocs left at all! As it has worked out, we have living representatives of each of three major Crocodylian lines that survived the K–Pg extinction event. Sadly, the sebecids did not make it past the Eocene. The survival of crocodylians to modern times has given us opportunities not just to study them but also to speculate constructively about the biology and physiology of many of the extinct taxa, even though the survivors are not necessarily wholly representative of the group as a whole.

Their history has been spectacularly diverse. The earliest crocs were terrestrial, and terrestrial examples are found throughout much of their history. Aquatic and semiaquatic lifestyles were invented and reinvented many times, as were excursions into gigantism and different locomotor styles, from a quadrupedal sprawling gait to erect stance with the limbs held under the body, galloping (technically bounding, which we see in some of today's crocs, Chapter 4) and swimming. Some were aquatic to the extent that they developed paddles instead of limbs. Aquatic lifestyles were facilitated by the evolution of locomotion driven by a laterally compressed tail and, in the Eusuchia, which were (and are) all aquatic or semi-aquatic, additional flexibility of the trunk. A more sinuous locomotion became possible with the reduction of the bony osteoderms. There have been far more carnivores than herbivores, and dentition has evolved to reflect diet. Skull shapes have varied between dorsally flattened and laterally compressed, long-snouted and short-snouted. They have evolved to be very large several times, to 12–13 m but, even the smallest, at 25 cm or so, is large compared with most of today's non-croc reptiles. There have been some presumed to be restricted to fresh waters, and some to salt, and many will have lived through times when the sea was both fresher and saltier than now.

What implications does their history have for interpreting and understanding functional aspects in modern species? Like the ancient crocs, the modern ones too have a range of adult sizes. The crocodylian physiology can obviously support either very large or quite small animals. However, the three most interesting functional aspects of the modern crocs, to my thinking, are those that relate to their salt and water balance, their thermal physiology and the unusual and very specialised structure of their heart. These are the three areas that have always captured my attention. The first, tolerance for salt water, or lack thereof, has been a recurring theme in their evolutionary history for a couple of hundred million years. The last two, thermal physiology and heart structure, are probably interrelated, and the heart structure in turn has implications for their full adaptation to aquatic life, to allow long submergences. Many ancient crocs had body forms that suggest they were just as likely as dinosaurs to have been endothermic. It is not at all hard to imagine that early crocodile-like reptiles such as some of the 'Sphenosuchia', Sebecidae and Protosuchia, with their erect stance and active predatory habits, were endothermic. Unlike in the debate about dinosaur endothermy, we do at least know with certainty that the living crocs have a four-chambered heart, and we can guess that the ancient ones probably did too. It is a provocative thought, and one will be taken up in Chapters 8 and 10.

We can now look at the biology of living crocs in the context of 320 million years of reptilian evolution, nearly 250 million of them since the evolution of the first 'crocodile-like' reptiles.

REFERENCES

Benton MJ (1999) *Scleromochlus taylori* and the origin of dinosaurs and pterosaurs. *Philosophical Transactions of the Royal Society of London* **354**, 1423–1446.

Benton MJ, Clark JM (1988). Archosaur phylogeny and the relationships of the Crocodylia. In *The Phylogeny and Classification of the Tetrapods. Volume 1.* (Ed. MJ Benton) pp. 295–338. Clarendon Press, Oxford.

Brochu CA (1999) Phylogenetics, taxonomy, and historical biogeography of Alligatoroidea *Society of Vertebrate Paleontology Memoir* **6**, 9–100.

Brochu CA (2003) Phylogenetic approaches toward crocodylian history. *Annual Review of Earth and Planetary Sciences* **31**, 357–397.

Brochu CA, Densmore LD (2001) Crocodile phylogenetics: a review of current progress. In *Crocodilian Biology and Evolution.* (Eds G Grigg, F Seebacher and CE Franklin), pp. 3–8. Surrey Beatty & Sons, Sydney.

Brochu CA, Wagner JR, Jouvre S, Sumrall SD, Densmore LD (2009) A correction corrected: consensus over the meaning of Crocodylia and why it matters. *Systematic Biology* **58**(5), 537–543.

Brusatte SL, Benton MJ, Desojo JB, Langer MC (2010a) The higher level phylogeny of Archosauria (Tetrapoda: Diapsida). *Journal of Systematic Palaeontology* **8**, 3–47.

Brusatte SL, Nesbitt SJ, Irmis RB, Butler RJ, Benton MJ, Norell MA (2010b) The origin and early radiation of dinosaurs. *Earth-Science Reviews* **101**, 68–100.

Brusatte SL, Benton MJ, Lloyd GT, Ruta M, Wang SC (2011) Macroevolutionary patterns in the evolutionary radiation of archosaurs (Tetrapoda: Diapsida). *Earth and Environmental Science Transactions of the Royal Society of Edinburgh* **101**, 367–382.

Buckley GA, Brochu CA, Krause DW, Pol D (2000) A pug-nosed crocodyliform from the Late Cretaceous of Madagascar. *Nature* **405**, 941–944.

Buffetaut E (1990) Vertebrate extinctions and survival across the Cretaceous-Tertiary boundary. *Tectonophysics* **171**, 337–345.

Buffetaut E, Ford RLE (1979) The crocodilian *Bernissartia* in the Wealden of the Isle of Wight. *Palaeontology* **22**(4), 905–012.

Butler RJ, Brusatte SL, Reich M, Nesbitt SJ, Schoch RR, Hornung JJ (2011) The sail-backed reptile *Ctenosauriscus* from the Latest Early Triassic of Germany and the timing and biogeography of the early Archosaur radiation. *PLoS ONE* **6**(10), e25693.

Chiari Y, Cahais V, Galtier N, Delsuc F (2012) Phylogenomic analyses support the position of turtles as sister group of birds and crocodiles (Archosauria). *BMC Biology* **10**, 65–79.

Clark JM, Sues H-D (2002) Two new basal crocodylomorph archosaurs from the Lower Jurassic and the monophyly of the Sphenosuchia. *Zoological Journal of the Linnean Society* **136**, 77–95.

Clark JM, Xu X, Forster CA, Wang Y (2004) A Middle Jurassic 'sphenosuchian' from China and the origin of the crocodylian skull. *Nature* **430**, 1021–1024.

de Andrade MB, Young MT, Desojo JB, Brusatte SL (2010) The evolution of extreme hypercarnivory in Metriorhynchidae (Mesoeucrocodylia: Thalattosuchia) based on evidence from microscopic denticle morphology. *Journal of Vertebrate Paleontology* **30**(5), 1451–1465.

de Broin F, Taquet P (1966) Découverte d'une crocodile nouveau dans la Crétacé inférieur du Sahara. *Comptes Rendus de l'Académie des Sciences Paris* **262**, 2326–2329.

Densmore LD 1983. Biochemical and immunological systematics of the order Crocodilia. In *Evolutionary Biology* (Eds MK Hecht, B Wallace and GH Prance) pp. 397–465. Plenum, New York.

Densmore LD, Owen RD (1989) Molecular systematics of the order Crocodilia. *American Zoologist* **29**, 831–841.

Densmore LD, White PS (1991) The systematics and evolution of the Crocodilia as suggested by restriction endonuclease analysis of mitochondrial and nuclear ribosomal DNA. *Copeia* **1991**, 602–615.

Dollo L (1883) Premiere note sur les Crocodiliens de Bernissart. *Bulletin du Musée Royal d'Histoire Naturelle de Belgique* **2**, 309–338.

Eaton MJ, Martin AP, Thorbjarnarson J, Amato G (2009) Species-level diversification of African dwarf crocodiles (Genus Osteolaemus): a geographic and phylogenetic perspective. *Molecular Phylogenetics and Evolution* **50**, 496–506.

Endo H, Aoki R, Taru H, Kimura J, Sasaki M, Yamamoto M, *et al.* (2002) Comparative functional morphology of the masticatory apparatus in the long–snouted Crocodiles. *Anatomia, Histologia, Embryologia* **31**, 206–213.

Fernández M, Gasparini Z (2000) Salt glands in a Tithonian metriorhynchid crocodyliform and their physiological significance. *Lethaia* **33**, 269–276.

Fernández M, Gasparini Z (2008) Salt glands in the Jurassic metriorhynchid *Geosaurus*: implications for the evolution of osmoregulation in Mesozoic crocodyliforms. *Naturwissenschaften* **95**, 79–84.

Fernández M, Herrera Y (2009) Paranasal sinus system of *Geosaurus araucanensis* and the homology of the antorbital fenestra of metriorhynchids (Thalattosuchia: Crocodylomorpha). *Journal of Vertebrate Paleontology* **29**, 702–714.

Frey E, Riess J, Tarsitano SF (1989) The axial tail musculature of recent crocodiles and its phyletic implication. *American Zoologist* **29**, 857–862.

Gandola R, Buffetaut E, Monaghan N, Dyke G (2006) Salt glands in the fossil crocodile *Metriorhynchus*. *Journal of Vertebrate Paleontology* **26**, 1009–1010.

Harris SR, Pisani D, Gower DJ, Wilkinson M (2007) Investigating stagnation in morphological phylogenies using consensus data. *Systematic Biology* **56**, 125–129.

Hekkala ER, Amato G, DeSalle R, Blum MJ (2010) Molecular assessment of population differentiation and individual assignment potential of Nile crocodile (*Crocodylus niloticus*) populations. *Conservation Genetics* **11**, 1435–1443.

Hekkala ER, Shirley MH, Amato G, Austin JD, Charter S, Thorbjarnason J, *et al.* (2011) An ancient icon reveals new mysteries: mummy DNA resurrects a cryptic species within the Nile crocodile. *Molecular Ecology* **20**, 4199–4215.

Holt T, Salisbury SW, Willis PMA (2005) A new species of mekosuchine crocodilian from the middle Palaeogene Rundle formation, central Queensland. *Memoirs of the Queensland Museum* **50**(2), 207–218.

Hrbek T, Vasconcelos WR, Rebelo G, Farias IP (2008) Phylogenetic relationships of South American Alligatorids and the Caiman of Madeira River. *Journal of Experimental Zoology* **309A**, 588–599.

Iwabe N, Hara Y, Kumazawa Y, Shibamoto K, Saito Y, Miyata T, *et al.* (2005) 2005 Sister group relationship of turtles to the bird–crocodilian clade revealed by nuclear DNA–coded proteins. *Molecular Biology and Evolution* **22**(4), 810–813.

Janke A, Gullberg A, Hughes S, Aggarwal RK, Arnason U (2005) Mitogenomic analyses place the gharial (Gavialis gangeticus) on the crocodile tree and provide pre-K/T divergence times for most crocodilians. *Journal of Molecular Evolution* **61**, 620–626.

Kaur T, Japning JR, Sabki MS, Sidik I, Chong LK, Ong AH (2013) Genetic diversity of *Tomistoma schlegelii* inferred from mtDNA markers. *Biochemical Genetics* **51**(3–4), 275–295.

Krause DW, Kley NJ (Eds) (2010) *Simosuchus clarki* (Crocodyliformes: Notosuchia) from the Late Cretaceous of Madagascar. Society of Vertebrate Paleontology Memoir 10. *Journal of Vertebrate Paleontology* **30**(6, Supplement).

Krause DW, Sertich JJW, Rogers RR, Kast SC, Rasoamiaramanana AH, Buckley GA (2010) Overview of the discovery, distribution, and geological context of *Simosuchus clarki* Crocodyliformes: Notosuchia) from the Late Cretaceous of Madagascar. *Journal of Vertebrate Paleontology* **30**, 4–12.

Larsson HCE, Sues H-D (2007) 2007 Cranial osteology and phylogenetic relationships of *Hamadasuchus rebouli* (Crocodyliformes: Mesoeucrocodylia) from the Cretaceous of Morocco. *Zoological Journal of the Linnean Society* **149**, 533–567.

Li C, Wu X-C, Cheng Y-N, Sato T, Wang L (2006) An unusual archosaurian from the marine Triassic of China. *Naturwissenschaften* **93**, 200–206.

Man Z, Yishu W, Peng Y, Xiaobing W (2011) Crocodilian phylogeny inferred from twelve mitochondrial protein-coding genes, with new complete mitochondrial genomic sequences for *Crocodylus acutus* and *Crocodylus novaeguineae*. *Molecular Phylogenetics and Evolution*.

McAliley LR, Willis RE, Ray DA, White PS, Brochu CA, Densmore LD (2006) Are crocodiles really monophyletic? Evidence for subdivisions from sequence and morphological data. *Molecular Phylogenetics and Evolution* **39**, 16–32.

Meredith RW, Hekkala ER, Amato RG, Gatesy J (2011) A phylogenetic hypothesis for *Crocodylus* (Crocodylia) based on mitochondrial DNA: Evidence for a trans-Atlantic voyage from Africa to the New World. *Molecular Phylogenetics and Evolution* **60**, 183–191.

Modesto SP, Anderson JS (2004) The phylogenetic definition of Reptilia. *Systematic Biology* **53**(5), 815–821.

Moritz C (1994) Defining "evolutionary significant units" for conservation. *Trends in Ecology & Evolution* **9**, 373–375.

Nesbitt SJ (2007) The anatomy of *Effigia okeeffeae* (Archosauria, Suchia), theropod convergence, and the distribution of related taxa. *Bulletin of the American Museum of Natural History* **302**, 1–84.

Nesbitt SJ (2011) The early evolution of archosaurs: relationships and the origin of major clades. *Bulletin of the American Museum of Natural History* **352**, 1–292.

Nesbitt SJ, Norell MA (2006) Extreme convergence in the body plans of an early suchian (Archosauria) and ornithomimid dinosaurs (Theropoda). *Proceedings. Biological Sciences* **273**, 1045–1048.

O'Connor PM, Sertich JJW, Stevens NJ, Roberts EM, Gottfried MD, Hieronymus TL, *et al.* (2010) The evolution of mammal – like crocodyliforms in the Cretaceous Period of Gondwana. *Nature* **466**, 748–751.

Oaks JR (2011) A time-calibrated species tree of Crocodylia reveals a recent radiation of the true crocodiles. *Evolution* **65**, 3285–3297.

Piras P, Delfino M, Del Favero L, Kotsakis T (2007) Phylogenetic position of the crocodylian *Megadontosuchus arduini* (de Zigno, 1880) and tomistomine palaeobiogeography. *Acta Palaeontologica Polonica* **52**, 315–328.

Piras P, Colangelo P, Adams DC, Buscalioni A, Cubo J, Kotsakis T, *et al.* (2010) The *Gavialis–Tomistoma* debate: the contribution of skull ontogenetic allometry and growth trajectories to the study of crocodylian relationships. *Evolution & Development* **12**, 568–579.

Pol D, Apesteguia S (2005) New *Araripesuchus* remains from the early Late Cretaceous (Cenomanian–Turonian) of Patagonia. *American Museum Novitates* **3490**, 1–38.

Roos J, Aggarwal RK, Janke A (2007) Extended mitogenomic phylogenetic analyses yield new insight into crocodylian evolution and their survival of the Cretaceous–Tertiary boundary. *Molecular Phylogenetics and Evolution* **45**, 663–673.

Salisbury SW (2006) Dawn of a crocodilian dynasty. *The Australian Geographer* **83**, 52–53.

Salisbury SW, Willis PMA (1996) A new crocodylian from the early Eocene of southeastern Queensland and a preliminary investigation into the phylogenetic relationships of crocodyloids. *Alcheringa* **20**, 179–226.

Salisbury SW, Molnar RE, Frey E, Willis PMA (2006) The origin of modern crocodyliforms: new evidence from the Cretaceous of Australia. *Proceedings. Biological Sciences* **273**, 2439–2448.

Scheyer TM, Aguilera OA, Delfino M, Fortier DC, Carlini AA, *et al.* (2013) Crocodylian diversity peak and extinction in the late Cenozoic of the northern Neotropics. *Nature Communications* **4**, 1907.

Shulte P, Alegret L, Arenillas I, Arz AJ, Barton PJ, Bown PR, *et al.* (2010) The Chicxulub Asteroid impact and mass extinction at the Cretaceous–Paleogene Boundary. *Science* **327**, 1214– 1218.

Schwimmer DR (2002) *King of the Crocodylians: The Paleobiology of Deinosuchus.* Indiana University Press, Bloomington, Indiana.

Sereno PC, Arcucci AB (1990) The monophyly of crurotarsal archosaurs and the origin of bird and crocodile ankle joints. *Neues Jahrbuch für Geologie und Palaontologie. Abhandlungen* **180**, 21–52.

Sereno PC, Larsson HCE (2009) Cretaceous crocodyliforms from the Sahara. *ZooKeys* **28**, 1–143.

Sereno PC, Larsson HCE, Sidor CA, Gado B (2001) The giant crocodyliform *Sarcosuchus* from the Cretaceous of Africa. *Science* **294**, 1516–1519.

Sereno PC, Sidor CA, Larsson HCE, Gado B (2003) A new notosuchian from the Early Cretaceous of Niger. *Journal of Vertebrate Paleontology* **23**, 477–482.

Seymour RS, Bennett–Stamper CL, Johnston SD, Carrier DR, Grigg GC (2004) Evidence for endothermic ancestors of crocodiles at the stem of archosaur evolution. *Physiological and Biochemical Zoology* **77**, 1051–1067.

Shirley MH, Vliet KA, Carr AN, Austin JD (2014a) Rigorous approaches to species delimitation have significant implications for African crocodilian systematics and conservation. *Proceedings. Biological Sciences* **281**, 20132483.

Shirley MH, Villanova V, Vliet KA, Austin JD (2014b) Genetic barcoding facilitates captive and wild management of three cryptic African crocodile species complexes. *Animal Conservation*. in press

Sloan C, Sereno PC (2002) *Supercroc and the Origin of Crocodiles*. National Geographic Society, Washington DC.

Snyder D (2007) Morphology and systematics of two Miocene alligators from Florida, with a discussion of *Alligator* biogeography. *Journal of Paleontology* **81**, 917–928.

St John JA, Braun EL, Isberg SR, Miles LG, Chong AY, Gongora J, *et al.* (2012) Sequencing three crocodilian genomes to illuminate the evolution of archosaurs and amniotes. *Genome Biology* **13**, 415.

Taplin LE, Grigg GC (1989) Historical zoogeography of the eusuchian crocodilians: a physiological perspective. *American Zoologist* **29**, 885–901.

Tsuji LA, Müller J (2009) Assembling the history of the Parareptilia: phylogeny, diversification, and a new definition of the clade. *Fossil Record* **12**(1), 71–81.

Turner AH (2006) Osteology and phylogeny of a new species of *Araripesuchus* (Crocodyliformes: Mesoeucrocodylia) from the Late Cretaceous of Madagascar. *Historical Biology* **18**, 255–369.

Turner AH, Calvo JO (2005) A new sebecosuchian crocodyliform from the Late Cretaceous of Patagonia. *Journal of Vertebrate Paleontology* **25**, 87–98.

White PS, Densmore LD III (2001) DNA sequence alignment and data analysis methods: their effect on the recovery of crocodylian relationships. In *Crocodilian Biology and Evolution* (Eds GC Grigg, F Seebacher and CE Franklin) pp. 29–37. Surrey Beatty & Sons, Sydney.

Whiteside JH, Olsen PE, Eglinton T, Brookfield ME, Sambrotto RN (2010) Compound-specific carbon isotopes from Earth's largest flood basalt eruptions directly linked to the end-Triassic mass extinction. *Proceedings of the National Academy of Sciences of the United States of America* **107**(15), 6721–6725.

Willis PMA (1997) Review of fossil crocodilians from Australasia. *Australian Zoologist* **30**, 287–298.

Willis PMA 2001. New crocodilian material from the Miocene of Riversleigh (northwestern Queensland, Australia). In *Crocodilian Biology and Evolution* (Eds GC Grigg, F Seebacher and CE Franklin). pp. 64–74. Surrey Beatty & Sons, Sydney.

Willis PMA, Archer MA (1990) A Pleistocene longirostrine crocodilian from Riversleigh: first fossil occurrence of Crocodylus johnstoni Krefft. *Memoirs of the Queensland Museum* **28**, 159–163.

Willis PMA, Stilwell JD (2000) A probable piscivorous crocodile from Eocene deposits of McMurdo Sound, East Antarctica. *Antarctic Research Series* **76**, 355–358.

Willis RE, McAliley RL, Neeley ED, Densmore LD (2007) 2007 Evidence for placing the false gharial (*Tomistoma schlegelii*) into the family Gavialidae: Inferences from nuclear gene sequences. *Molecular Phylogenetics and Evolution* **43**, 787–794.

Wu XC, Sues HD (1996) Anatomy and phylogenetic relationships of *Chimaerasuchus paradoxus*, an Unusual Crocodyliform Reptile from the Lower

Cretaceous of Hubei, China. *Journal of Vertebrate Paleontology* **16**(4), 688–702.

Wu XC, Sues HD, Sun A (1995) A plant-eating crocodyliform reptile from the Cretaceous of China. *Nature* **376**, 678–680.

Young MT, Brusatte SL, Ruta M, de Andrade MB (2010) 2010 The evolution of Metriorhynchoidea (mesoeucrocodylia, thalattosuchia): an integrated approach using geometric morphometrics, analysis of disparity, and biomechanics. *Zoological Journal of the Linnean Society* **158**, 801–859.

Zardoya R, Meyer A (2001) The evolutionary position of turtles revised. *Naturwissenschaften* **88**(5), 193–200.

3

CROCODYLIANS CLOSER UP

Today's crocodylians are the product of a very long evolutionary history (Chapter 2). Deconstructing them, they can be thought of as sophisticated lever-systems (skeleton) covered by skin for protection and operated by muscles (this chapter), moving within their habitat (Chapter 4), including underwater (Chapter 9), under the control of coordinating sense organs and brain (Chapter 5). These components all require an energy supply to fuel their operation, including the respiratory and blood systems (Chapters 7, 8), and we can think of the remaining organ systems – that is, the alimentary system and other viscera (Chapter 6), the kidneys and salt glands (Chapter 11) and so on – as support systems for the supply and distribution of energy to the muscles, skin, sense organs and brain, and the maintenance of a suitable cellular environment for their operation, all within an appropriate temperature range (Chapter 10). All of these systems, indeed the whole life of an individual crocodylian, as in all living individuals, support reproduction for the successful production of another generation of individuals (Chapter 12) and for maintaining the population and the species (Chapter 13).

This Chapter is a brief introduction to the external features of the living crocodylians and the skeletal and muscular elements that make up the lever system that supports the body and causes its movement. More detailed anatomical information is included as appropriate in the relevant chapters.

THE EXTERNAL FEATURES OF CROCODYLIANS

Crocodylians are sturdy lizard-shaped reptiles, with a short and not very flexible neck (Fig. 3.1). They are specialists for a life in and near the interface between water and land. The head, trunk and limbs are well armoured by horny skin and scales and the tail has a distinctively jagged upper edge. They are well adapted for aquatic life, with a laterally compressed tail and webbed hind feet, and eyes, ears and nostrils mounted high on the head so they can continue to see, hear and smell even while the bulk of their body is almost completely hidden under the water's surface. The nostrils and ears (imperfectly) can be sealed to keep water out when the animal is completely submerged and a third eyelid, the nictitating membrane, protects the eye under water. There is a rigid plate of tissue at the rear of the oral cavity, the palatal flap, which closes the throat against the entry of water.

There is no obvious external sign of their sex apart from adult males growing much larger than females: the penis and clitoris are discreetly hidden. The alimentary, urinary and reproductive systems empty into a single cavity, the cloaca, with only one external opening, visible as a longitudinal mid-ventral slit a little posterior to the origin of the hind legs (vent in Fig. 3.7). This, too, is under muscular control to keep

---

An estuarine crocodile, Crocodylus porosus, *in the early morning sun, Kakadu National Park, Northern Territory. (Photo DSK)*

Fig. 3.1. *A representative crocodylian, the American alligator,* Alligator mississippiensis, *basking in January in Everglades National Park, Florida. (Photo DSK)*

water out when the crocodylian is submerged. If a croc has been captured, its gender can be identified even in quite small individuals by spreading the cloacal slit open to determine whether a clitoris or a penis is present. Unlike lizards, which have two penises or two clitorises (called hemipenes and hemiclitori) crocodylians have only one (Fig. 12.16).

BODY AND LIMBS

The trunk is cylindrical and usually wider than high, with a short neck and a long tapering tail, laterally compressed into a vertical blade for much of its length. The whole of the trunk, tail and limbs are well protected by scales. The tail has two rows of elongated triangular scales (scutes) on the upper edges of its base, converging to a single row posteriorly (Figs 3.2, 3.3). Elongated scutes also occur on the trailing edges of the limbs in some species (Fig. 3.4). These flattened scutes increase the surface area and so facilitate swimming and manoeuvring in water (Chapter 4).

If the tail is taken as that part of the body posterior to the body cavity, from approximately the posterior end of the cloaca, it is about half the total length and about one-third of the body mass. The limbs are short and project laterally. The position of the limbs suggests a sprawling gait, typical of salamanders and lizards, which crocodylians sometimes use, but they can also carry the body in a semi-erect stance in a high walk or gallop (Chapter 4). Because the ancestral crocodylomorphs were terrestrial and had an erect stance (Chapter 2), the sprawling gait (e.g. belly run) and semi-erect high walk of modern crocodylians and, in some, a gallop, can be seen as secondary adaptations and not representatives of intermediates in a reptile to mammal transition from sprawling to erect (Parrish 1987; Reilly and Elias 1998, Reilly *et al.* 2005). That transition was, anyway, on a completely different evolutionary line (Fig. 2.3).

The hindlimbs are larger than the forelimbs: they have more weight to support on land. The front foot has five fingers, the inner three of which have

Fig. 3.2. (Left) Crocodylus porosus *and (right)* Alligator mississippiensis *showing off their scaly dorsal armour. (Photos DSK)*

claws, while the others end in blind points (Fig. 3.4). The hind foot has four toes, with just the skeletal rudiments of a fifth, with only the three inner toes clawed (Fig. 3.4). Although the front foot may be partially webbed or not webbed at all, depending on the species, the hind foot is usually significantly webbed (Fig. 3.5).

SKIN AND SCALES

The main advantages reptiles have over amphibians, both being terrestrial tetrapods, are their cleidoic eggs (Chapter 12) and keratinised integument, which is almost impervious to water loss. Crocodylian skin (Fig. 3.6.) has non-overlapping scales, or scutes – epidermal plates covered with hard wear-resistant β keratin – yet the skin is also surprisingly soft to the touch, particularly on the

flanks where higher densities of smaller scales are embedded in the more finely arranged and flexible intermediate skin (Figs 3.1, 3.31). Many scales have bony plates (osteoderms) embedded (see below). The neck and posterior flanks are covered by more flexible skin, whereas the ventral surfaces, as well as the sides of the tail, are sheathed in large flat squarish scutes arranged neatly in rows and columns. The osteoderms and robust, collagenous dermis provide tough protection against serious mechanical damage and a barrier to the movement of water and ions (Chapter 11), yet the thin epidermis abrades and bleeds relatively easily. This poses a challenge to capturing them uninjured and adds double meaning to the statement that crocodylians need to be handled with care!

Scale patterns are extraordinarily similar between individuals of the same species. Side by side they can look as though they came from the same mould. The intricacies of the scale patterns are what make crocodylian leathers so attractive for handbags, belts, boots and other apparel or accessories (Chapter 14). The scale patterns are similar between individuals of the same species, but different across species. This is useful, because it allows Customs officials to identify incorrectly labelled crocodylian hides, making it harder for a poacher to export an endangered species by passing it off as a

Fig. 3.3. Lateral view of the single row of vertical scutes on the tail of C. porosus. *(Photo DSK)*

Fig. 3.4. *Front right and back right feet of a juvenile* Crocodylus porosus. *The front foot has five fingers, three of which have claws. The hind foot has four prominent toes and only the rudiments of a fifth; three of the toes are clawed with webbing between. Note the elongated scutes on the trailing edge of the hind foot. Such scutes are lacking in Alligatoridae. (Photos DSK)*

Fig. 3.5. *Sole of the right hind foot of 13 foot 10 inch (4.2 m)* C. porosus, Goyder River NT 1972. (Inset) *Underside of right front foot of large captive* C. porosus *photographed at Crocosaurus Cove, Darwin. Callouses may be a consequence of captivity. (Photo GCG, inset DSK)*

common, and therefore legally tradeable, species. While at the Bronx Zoo in New York, and since, Peter Brazaitis – a world expert on identifying crocodylian hides – provided a service identifying questionable imports, and has written guides and conducted training programmes for Customs Officers (Brazaitis 1987, 2001). There is a comprehensive hide identification guide published by CITES (1995) and available on their website: http://www.cites.org/eng/resources/publications.php

The scalation is different on different parts of the body in a characteristic way across all species (Fig. 3.7). On the back are thicker, larger and more regularly arranged scales, making what is called the 'horny back'. These enlarged scutes are supported by bone, the osteoderms, and in some species bone has invaded most of the scales, forming almost impenetrable armour (Figs 3.6, 3.12, 3.13 and below). The 'horny back' finds novelty use in the commercial trade, whereas most handbags and shoes are made from the flank and belly skin (Chapter 14). On the back of the neck, larger, thickened, bone-strengthened scutes, the nuchal scutes, form a species-specific pattern of armour, (Fig. 3.7). In *C. porosus* and some other species, these provide a convenient platform to which monitoring devices such as dive recorders and satellite transmitters can be attached (Chapter 4); the scales on the sides of

the tail and the belly are rectangular and regular, like graph paper, and arranged in transverse rows. The vertical scutes on the upper edges of the tail offer field biologists a convenient way to mark crocodylians with a permanent number, removing a combination of scutes numbered fore and aft from the bifurcation to represent single digits, 10s or 100s. Unfortunately, as if assuming crocodylians were sedentary, researchers in different jurisdictions have not followed any standardised protocol, raising the possibility of confusion in some species whose individuals travel long distances (see Chapter 4).

SKIN COLOUR

The colour derives from pigment in the epidermis, particularly melanin (Fig. 3.6), with the colour pattern in a particular species depending upon the relative proportions of the different pigments and their distribution. The dorsal surface and the flanks are usually well pigmented, often patterned, which presumably breaks up the outline and provides some camouflage. The ventral surface is pale – cream in most species, pale grey in others – but often dark in the smaller genera (*Paleosuchus*, *Osteolaemus*). The lower jaws are often paler than the rest of the head, broken in some species of both alligatorids and crocodylids with dark blotches

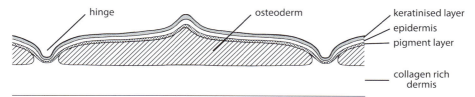

Fig. 3.6. *Schematic transverse section of the dorsal skin of a crocodylian.*

(e.g. Figs 3.20, 4.3). Alibardi (2011) described the histology of skin from several species, identifying a range of chromatophores in the epidermis and dermis: melanophores, which produce and hold black pigment; xanthophores, which are yellow; and, infrequently, iridophores, which reflect light from plates of guanine. The colour of the dorsum and flanks varies between species, and in captivity the skin often becomes dull and unrepresentative of free-living wild animals. Alligatorids tend to be greyish/brown with the paler regions such as lips and underbelly cream with a pinkish hue. *Crocodylus* species tend towards yellow/olive-green with black splotches and *Osteolaemus* are dark brown-grey to black.

Within a species, colour varies between individuals and also changes with age. Younger individuals are usually more brightly coloured and show more patterning, but they darken and lose much of the juvenile pattern as they age. The

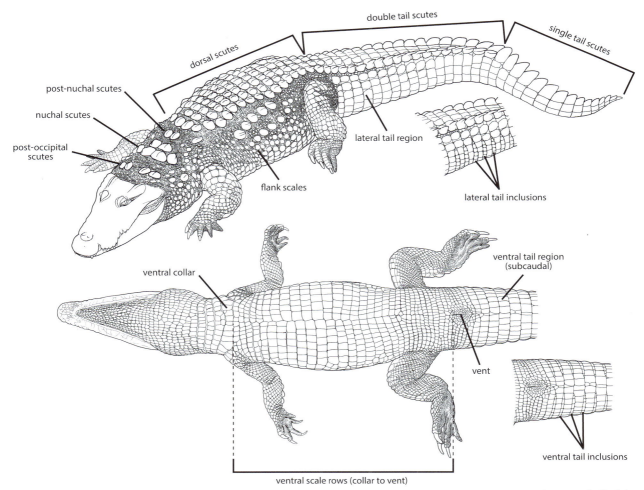

Fig. 3.7. *Crocodylians have different but characteristic patterns of scalation on different parts of their bodies. The upper half of the figure represents a generic crocodylian, based on* Crocodylus palustris, *modified by adding post nuchal scales and altering some of the dorsal scale rows. The lower half is a direct tracing from a photograph of a juvenile* Crocodylus porosus.

banded prettiness of hatchling American alligators (Fig. 3.8) gradually shows less contrast and a darker more monochromatic hue as they grow. In adult *C. porosus*, their overall colour is a yellow or olive green to greenish brown background with black spotting, the degree conveying the overall impression of lightness or darkness. Adults living in estuaries and the sea seem to retain more of their yellow background colour than adults living in fresh water and, indeed, experimental work has shown that skin colour in this species responds to the light characteristics of the habitat. Kirshner (1985) raised groups of sibling hatchling 'salties' separately in black tanks and white tanks for a year. Those kept in the black tanks were much darker in colour, first by darkening of the background pigmentation but, in time, the black spotting became more extensive as well (Fig. 3.9). The difference between the two groups was apparent after 1 month. When 'white-tank' hatchlings were moved to the black tank, their background colour darkened quite quickly: within 3 weeks. In contrast, although 'black-tank' individuals did lighten when moved to white tanks, the change occurred much more slowly.

In both alligatorids and crocodylids, colour variants occur, particularly very pale individuals, which are much sought after for display (Figs 3.10, 3.11).

BONY ARMOUR

Like the ostracoderm ('shell-skinned', from Greek) fishes back in the Devonian, crocodylians are encased to varying degrees in armour of dermal bone, not only on their heads but on their bodies as well.

Dermal bone

Bone in vertebrates arises either from the ossification of cartilaginous elements (chondral bone) deep within the embryonic tissue (endoskeleton) or by ossification within the skin (dermal bone, sometimes called membrane bone), forming an exoskeleton. Most of the elements we think of as being part of a tetrapod skeleton, including our own, start out in embryonic development as cartilage, except for some of the bones of the skull roof and jaws, the clavicle and, in humans, our knee caps, which are sesamoid bones (bones formed within tendons). The dermal bone exoskeleton is evolutionarily older than the chondral skeleton: think of the bony exoskeletons of the Devonian ostracoderm fishes.

In most tetrapods, there are few remnants of this ancient exoskeleton, but in crocodylians dermal bone features much more significantly. Indeed, the living crocodylians have a more diverse array of dermal bone than any other extant group. Dermal bone defines the shape of the crocodylian 'face' (the robust dermatocranium), makes the bony gastralia ('stomach ribs'), which form a rib-like layer of protection in the ventral abdominal body wall (Fig. 3.55), and also the osteoderms: bony elements embedded within many of the scutes, particularly dorsally (Figs 3.12–3.18). Many species also have palpebral ossicles in the anterior upper

Fig. 3.8. *Hatchling American alligators are beautifully marked and although their disruptive colouration occasionally makes them quite visible (left), in some habitats it usually renders them well camouflaged (right). (Photos DSK)*

Fig. 3.9. *Three-month-old* Crocodylus porosus, *one kept since hatching in a black tank (left), the other in a white tank. (Photo DSK)*

part of the orbits: flattened elements of dermal bone articulating with the pre-frontal bones and embedded in the upper eyelids, where they probably have a protective function (Fig. 3.44). Their evolutionary history and homology have been reviewed by Nesbitt *et al.* (2013). Crocodylians lack a clavicle but have a dermal bone interclavicle (Fig. 3.55). Vickaryous and Hall (2008) provided a comprehensive analysis and review of the dermal skeleton in alligators. The dermatocranium and gastralia ossify early in development, earlier than the cartilaginous chondral bone, but ossification of the osteoderms is delayed until well after hatching. The dermatocranium will be discussed in the section on the skull, but we will now discuss the osteoderms and, in a later section, the gastralia. Crocodylians lack scleral ossicles: bony plates of dermal origin that are found within the eyes of many fishes, turtles, ichthyosaurs and birds.

Osteoderms

Osteoderms ('bony skin', from Greek) (Fig. 3.12) are most typical of the dorsal scutes, which commonly bear conspicuous keels, but some species have them in almost all of their large scutes, encasing them in

Fig. 3.10. *An albino American alligator at the St Augustine Alligator Farm. Although popular as zoo exhibits, albino crocodylians present the unusual problem (for a crocodylian) of being at risk of sunburn. Consequently, exposure to the sun is minimised and the diet must be supplemented with vitamin D3. (Photo DSK)*

Fig. 3.11. *A leucistic* Crocodylus porosus *resident at Crocosaurus Cove, Darwin, NT. Unlike albinos, leucistic crocodylians usually have normally coloured eyes, not pink. (Photo DSK)*

Fig. 3.12. *Osteoderms. These are from dorsal scutes of unknown origin, possibly from* Paleosuchus trigonatus. *Note the raised keel and the anterior flange, which fits under the next-most anterior osteoderm, contributing to an imbricating bony pavement. (Photo GCG)*

bony armour (Fig. 3.13). The most heavily armoured among the extant crocodylians is *Paleosuchus trigonatus* (Fig. 3.14). It is not a straightforward matter to judge ossification from external appearance. On casual inspection, *C. johnstoni* does not look to have many more osteoderms than *C. porosus*, but it is far better protected, with bony scales surrounding the trunk and down the legs (Fig. 3.15) whereas *C. porosus* has them mainly dorsally (Figs 3.16, 3.17, pp. 92, 93). Species with more ossification are less sought for hides by the tanning industry and the difficulty of dissolving out the bone is one of the reasons *C. porosus* is preferred over *C. johnstoni*, although techniques for processing bony hides have now improved. Alligators too have significant development of osteoderms in their dorsal scutes (Figs 3.18, 3.42, pp. 94, 110).

Osteoderms are well vascularised, particularly the dorsal ones (Seidel 1979). Control over their blood flow probably lets them act as heat absorbers or radiators and they are assumed to be an important part of the crocodylian heat exchange, although Farlow *et al.* (2010) have pointed out that conclusive data about that are lacking so far (see Chapter 10). Apart from serving as armour, the osteoderms provide attachment sites for muscles, which, inserting on adjacent ribs, stiffen the dorsum (Fig. 3.53) and also sequester lactate (Jackson *et al.* 2003) as well as serving as a source of calcium for seasonal production of eggs (Chapter 12).

HEAD

This is probably the most interesting part of a crocodylian: it houses most of the sensory system and its brain, as well as the 'gleaming teeth' and strong jaws that provoke such comment. Head shape and its diversity, as well as the eyes, ears and nostrils will be discussed here, with the jaws and their functional aspects explored later, in the section on the skull.

Shape: a balance between feeding habits and hydrodynamic realities

Biologists have long noted the differences between species in head shape and dentition and wondered to what extent those differences are 'fit-for-purpose' adaptations to different lifestyles or simply expressions of phylogenetic relatedness. This is an im-

Fig. 3.13. *A broad-snouted caiman,* Caiman latirostris, *showing its well armoured dorsal surface. Most of its scales have embedded osteoderms and the dorsal ones are conspicuously ridged. Those on the neck are the nuchal scutes and the pattern they form is different in different species and, therefore, useful for identification (e.g. Brazaitis 1987). In some species, nuchal scutes are also useful for attaching measurement devices such as satellite tracking transmitters and dive recorders (see Figs 4.33, 9.12). (Photo GCG)*

Fig. 3.14. *The smooth-fronted caiman (or Schneiders dwarf caiman),* Paleosuchus trigonatus, *is regarded as the most heavily armoured crocodylian. (Photo DSK)*

portant question, particularly for palaeontologists. It is often said that alligatorids are broad snouted, crocodylids have narrower, more triangular snouts and gavialids have the longest and narrowest snouts of all (Fig. 1.2), but this is an oversimplification (Fig. 3.19). The African *Osteolaemus* species, Indian mugger crocodile, *C. palustris* and Central American *C. moreletii* are crocodylids with broad snouts, whereas among the alligatorids *P. trigonatus* and some sub-species of the spectacled caiman, *Caiman crocodilus*, have narrower, triangular shaped snouts (Fig. 3.20, p. 96). The gharial of India, *Gavialis gangeticus* certainly has the longest, narrowest snout, followed closely by the false gharial (*Tomistoma schlegelii*), but the species of *Mecistops* (Africa) and *C. johnstoni* (Australia's freshwater crocodile) are also very long snouted. Greg Erickson at Florida State University and his colleagues quantified snout shape in captive adults of almost all extant species as part of a broader study (discussed below). They defined rostral proportion (RP) as the ratio of the width of the snout at the anterior margin of the orbits to the length from there to the tip of the snout (Erickson *et al.* 2012) and these values are shown in Table 3.1. The data emphasise that the common generalisation made about snout shape and familial category has many exceptions. Note also that snout shape is not constant once an animal becomes adult, so the values of RP in Table 3.1 should be taken as indicative, rather than fully representative, for a species.

Broader, flatter snouts have been seen as sturdier and thus suitable for sturdier prey, whereas long snouts are seen to be more suitable for eating fish and thought to be weaker. McHenry *et al.* (2006) took a numerical approach to this issue, using beam and finite element modelling techniques (arising from structural mechanics) to compare skull shape and strength in seven species, covering a range of head shapes. They noted that most prey captured by crocodylians involves a sideways swipe with the head, and that lengthening the snout provides a more rapid strike at the tip for the same angular acceleration. A disadvantage would be that lengthening a broad snout leads to more drag in the water, whereas narrower and flatter snouts would have less water resistance. They concluded that, rather than being optimised only for the strength to manage prey, head shape of a particular species may be a balance between feeding optimisation and the hydrodynamic constraints of catching agile aquatic prey. In a follow up study of the same species Walmsley *et al.* (2013) compared computer simulations of the different skull shapes in their resistance to the biting, twisting and shaking loads associated with dealing with large prey. They found that the lower jaws of longirostrines were more likely to break under these loads, limiting their ability to deal with large prey. The lower jaws of short-snouted species fared better. They noted that the strength to cope with biting and twisting loads was associated with shorter symphyses (junctions) between the left and right rami of the lower jaw; longirostrine crocodylians have a longer symphysis and are thus less suited to feed on large prey. Part of the impetus for this study was palaeontological, and Walmsley and colleagues noted that this correlation may help in reconstructions of the feeding ecology of extinct marine reptiles such as icthyosaurs, pliosaurs and plesiosaurs and, of course, extinct crocodylomorphs.

Pierce *et al.* (2008) conducted a broadly similar analysis and came to similar conclusions. The shorter heads are stronger, whereas the long, thinner skulls are less strong, and this has implications for diet. In particular, these authors emphasised that head (and skull) shape is determined by ecological and behavioural factors and is not an expression of taxonomic grouping. They also showed that head shape is not a correlate of geography. Indeed, biogeographic regions had more similarities in their diversity of head shapes than taxonomic affinity, implying perhaps some form of niche specialisation. Pierce *et al.* (2008) then applied these conclusions to an analysis of feeding ecology in two groups of crocodylomorphs within the Thallatosuchia (Fig. 2.3), the metrirhynchids and teleosaurids (Chapter 2), which went extinct in the early Cretaceous. Teleosaurids were very long snouted and, with reference to their earlier study, Pierce and colleagues concluded that these and the long-snouted metriorhynchids were probably restricted to making lateral, swiping attacks, like the modern longirostrines. Shorter snouted metriorhynchids, however, may have been able to

Fig. 3.15. *Australia's freshwater crocodile,* Crocodylus johnstoni, *and its external skeleton. The armour provided by osteoderms is presumed to have evolved for protection, but it is not effective against* C. porosus *(Fig. 13.7). It did, however, provide protection against hunters for quite a while, because it took a long time for tanneries to find a way to deal with the sturdy embedded osteoderms. The same issue arose in many other species too. In* C. johnstoni *that problem is largely solved, but 'freshie' skins are still not nearly as valuable as skins of* C. porosus. *(Upper photo DSK, lower photo Darwin Museum and Art Gallery)*

grasp prey and shake it and possibly even make 'death roll' manoeuvres.

What emerges from these studies of extant species is that crocodylian skulls are a functional compromise between hydrodynamics and structural mechanics. That is, head shape in crocodylians has been quite labile in response to evolutionary selection pressures, reflecting feeding habits and not their cladistic affinity. Questions about bite force and tooth pressure are discussed below.

One feature of head structure that does reflect cladistic affinity in living species is the location of the tooth rows, but that breaks down if extinct forms are included. The extant Alligatoridae and Crocodylidae differ conspicuously in the position of the tooth rows (Figs 1.3, 3.20). The alligatorids have an 'overbite', with the teeth of the lower jaw lying inside the teeth of the upper jaw when the mouth is closed, the largest of these teeth fitting into pits in the palate. Typically in alligators and caimans, only

Fig. 3.16. *The dorsum of* Crocodylus porosus *showing the six rows of large keeled scutes. The nuchal (neck) scutes are just visible in the foreground. (Photo DSK)*

the top teeth are visible when the animal closes its mouth. In crocodylids most upper and lower teeth are visible, the lower teeth interlocking with, and in line with, the upper teeth, fitting into notches between them. This is most noticeable with the large 4th mandibular tooth. Put another way, more of the dentition of crocodiles (and gavials) is visible when the jaws are closed. The location of the tooth row in gavialids is crocodylid-like.

Eyes and ears

Vision and hearing are dealt with in Chapter 5, so here there will only be a description of the external appearance of eyes and ears.

The eyes of crocodylians are located high enough on the head to provide a good view with almost all of the rest of the head submerged. They are highest in gharial, *Gavialis gangeticus* (Fig. 3.21, p. 97). The eyes are closed by movement of both eyelids but mostly the upper, which is armoured in some species by a bony plate, the tarsus (Underwood 1970). The tarsus is more conspicuous in some species than in others and can be quite elaborate. When the eye is closed, the tarsus protects the eye. Presumably this is useful during the struggles associated with the life of an active predator.

Crocodylians also have a nictitating membrane or 'third eyelid': a cartilage-supported semi-transparent membrane that sweeps horizontally backwards from the anterior corner of the eye when the animal submerges, as if protecting it (Figs 3.22 on p. 98 and 5.6, 9.6, 9.9). Many other vertebrates also have nictitating membranes and in humans it is vestigial. The muscle that sweeps the nictitating membrane across the eye is the same one that pulls the eyes of frogs down during swallowing. It is activated by the abducens cranial nerve (Fig. 5.16, Table 5.1). Lachrymal (tear) glands lubricate the upper and lower lids, while Harderian glands discharge lubricating mucus onto the inside of the nictitating membrane.

The ears are visible externally as curved longitudinal slits running posteriorly from behind

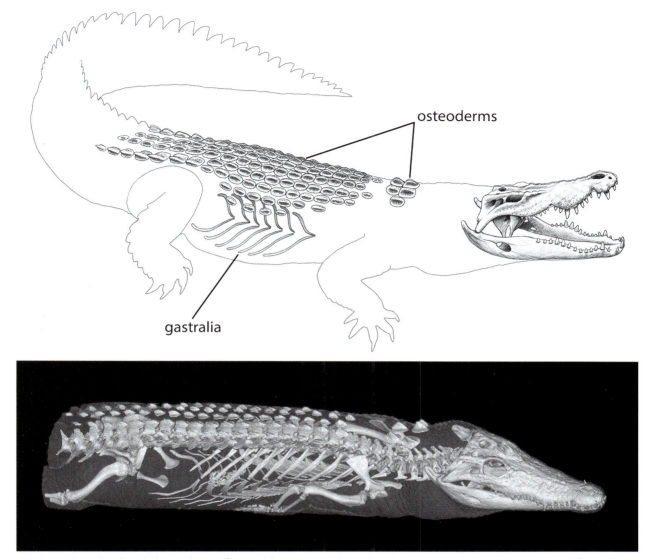

osteoderms

gastralia

Fig. 3.17. *Dorsal and nuchal osteoderms of* Crocodylus porosus, *shown here in a 3D computer model reconstructed from CT scan data (lower) and schematic illustration (upper). The gastralia (see Fig. 3.55) are also made of dermal bone, as is the visible surface of the skull (the dermatocranium), which defines the shape of a crocodylian's face. This species has less of an external skeleton than most crocodylians. The bony dorsal elements are important in what Steve Salisbury refers to as the 'the self-carrying system' (Fig. 3.53). (3D image Simon Collins and Steve Johnston)*

the eyes (Fig. 3.21). The anatomy and musculature of the ears of *A. mississippiensis*, *Caiman crocodilus* and *Crocodylus niloticus* have been described by Shute and Bellairs (1955). They found all three species to be similar. Each slit is covered dorsally by a moveable leaf-like flap, the earlid. Less obvious is the smaller muscular flap ventrally and at the anterior end of the slit. The slit is the external auditory meatus (= opening) and, when above the water line, its anterior end is usually slightly agape, letting sound waves in to contact the tympanic membrane lying in the floor of the ear cavity (Fig. 5.10). The opening and closing of the anterior end of the meatus is brought about by muscular control of the ventral ear flap. The meatus is closed when a crocodylian submerges. Occasionally crocodylians flap the upper ear flaps rapidly, either together or separately, briefly opening the auditory meatus along its whole length. What prompts this is unknown. Anecdotally, some observers have attributed this as a sign of stress or inner tension and another suggestion is that water can be flicked away.

Fig. 3.18. *An American alligator floats in Florida's Everglades National Park, well protected from above by its dorsal armour. (Photo DSK)*

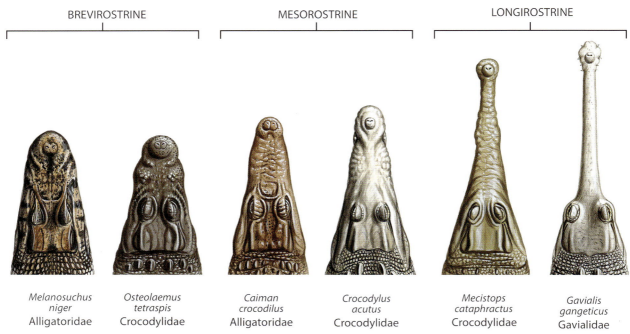

Fig. 3.19. *A rough visual guide to crocodylian snout shape categories, as defined by Erickson et al. (2012), exemplified by a few sample species. The occurrence of crocodylids and alligatorids in both brevirostrine and mesorostrine categories shows that the widespread notion of snout shape being the best way to tell 'alligators' from 'crocodiles' is an oversimplification. (Illustration by DSK, courtesy of Weldon Owen Publishing)*

TABLE 3.1 SNOUT SHAPE OF CROCODYLIANS LISTED IN ORDER OF ROSTRAL PROPORTION (RP, THE RATIO OF THE WIDTH OF THE SNOUT AT THE ANTERIOR MARGIN OF THE ORBITS TO THE LENGTH FROM THERE TO THE TIP OF THE SNOUT)

The terms brevirostrine and longirostrine are convenient, but RP grades almost continuously from the *C. latirostris* to *G. gangeticus*, so the break between brevirostrine and mesorostrine is somewhat arbitrary. The notion that alligatorids are broad snouted and crocodylids have narrower more triangular snouts has significant exceptions and, as discussed in the text, head shape in crocodylians has evolved in response to feeding habits and is not an indication of cladistic affinity. (Data from Erickson *et al.* 2012).

	Rostral proportion (RP)	Species
Brevirostrine ('short snouted')	0.89	*Caiman latirostris*
	0.83	*Alligator sinensis*
	0.75	*Crocodylus moreletii*
	0.73	*Melanosuchus niger*
	0.69	*Alligator mississippiensis*
	0.69	*Paleosuchus palpebrosus*
	0.69	*Caiman yacare*
	0.66	*Osteolaemus* species
	0.66	*Crocodylus palustris*
	0.65	*Crocodylus mindorensis*
	0.64	*Crocodylus siamensis*
Mesorostrine ('middle snouted')	0.58	*Caiman crocodilus*
	0.55	*Crocodylus novaeguineae*
	0.51	*Paleosuchus trigonatus*
	0.51	*Crocodylus rhombifer*
	0.47	*Crocodylus niloticus (and C. suchus)*
	0.41	*Crocodylus porosus*
	0.40	*Crocodylus acutus*
	0.35	*Crocodylus intermedius*
Longirostrine ('long snouted')	0.25	*Mecistops* species
	0.24	*Crocodylus johnstoni*
	0.18	*Tomistoma schlegelii*
	0.10	*Gavialis gangeticus*

Nostrils

The openings of the external nostrils (nares) are paired, crescent-shaped and situated on a nasal disk ('button') high at the front of the snout (Figs 3.23, 3.24 on pp. 99, 100, and 5.18). The nostrils are usually half open in a resting animal, but can be closed tightly or flared wide open by opposing muscles that are uniquely crocodylian (Bellairs and Shute

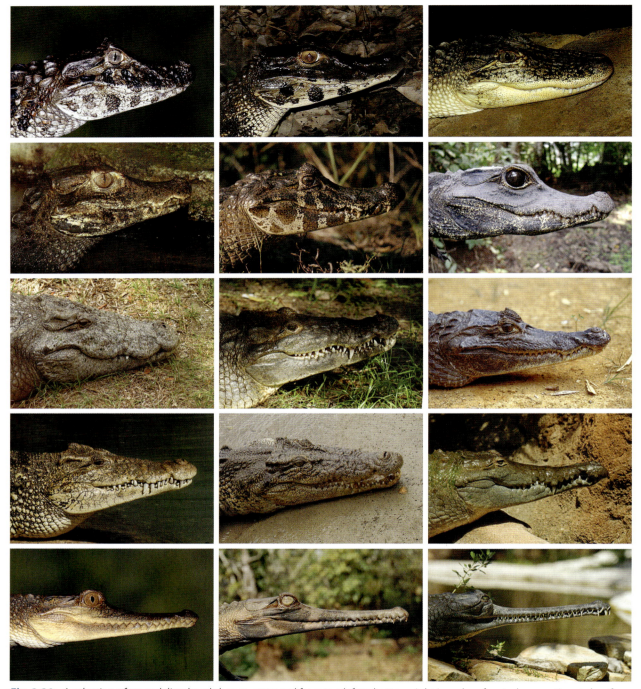

Fig. 3.20. *A selection of crocodylian head shapes, arranged from top left to bottom right in order of rostral proportions taken from* Erickson *et al. (2012). The lone exception is the image of* Crocodylus moreletii, *as the photograph is of a sub-adult that had not yet attained the brevirostrine rostral proportions characteristic of a mature individual. Identifications and photo credits:* Caiman latirostris *(GCG),* Melanosuchus niger *(Robinson Botero-Arias),* Alligator mississippiensis *(DSK),* Paleosuchus palpebrosus *(DSK),* Caiman yacare *(DSK),* Osteolaemus *(Mitchell Eaton),* Crocodylus palustris *(DSK),* Crocodylus moreletii *(DSK),* Caiman crocodilus *(DSK),* Crocodylus rhombifer *(DSK),* Crocodylus porosus *(DSK),* Crocodylus intermedius *(DSK),* Mecistops cataphractus *(Mitchell Eaton),* Tomistoma schlegelii *(Agata Staniewicz),* Gavialis gangeticus *(DSK)*

Fig. 3.21. *Eyes and ears in: (top) an alligatorid,* Paleosuchus trigonatus; *(middle) a crocodylid,* Crocodylus johnstoni; *and (bottom) a gavialid,* Gavialis gangeticus. *The vertical pupil is characteristic of crocodylians. The longish earflaps (earlids) are under muscular control and are usually open only at the anterior end. (Photos top Zilca Campos, others DSK)*

1953) (Fig. 3.24). The epithelium lining the anterior and posterior walls of the nostrils is greatly thickened, forming pads that prevent water entering when they are pressed together. Anterior to the nostrils is the mass of cavernous tissue, well supplied with blood vessels and which swells when engorged with blood. Its function is unknown.

TEETH

The teeth seize and hold prey, enabling it to be crushed and killed. All living crocodylians have many teeth, which are pointed, hollow and conical and set along the jaw margins in sockets (thecodont

dentition) in pre-maxillary, maxillary and dentary bones (Figs 3.25, 3.26, 3.43 on pp. 101, 102, 111). Structurally and developmentally, they are similar in all living crocodylians but the number, size and disposition of them varies between species, and the combination of skull structure and tooth complement and arrangement relates to diet, altering as the animal grows and its diet changes (Fig. 3.45, p. 113).

Structure and shedding

The teeth do not have roots and are wide open at the base. Their embryological development was described for American alligators by Westergaard and Ferguson (1990) and crocodylians hatch from the egg with a full set of functional teeth. Teeth have a hard time, frequently becoming broken during prey capture, and they are replaced throughout life (= polyphyodont, an endless succession of teeth). Poole (1961) described the successional process and estimated that each tooth in a 13 foot (3.9 m) Nile crocodile would have been replaced at least 40 times, and probably 45–50 times. The pattern of tooth replacement in *C. niloticus* was examined in more detail by Kieser *et al.* (1993) who described how the new teeth form within the same socket, the new tooth pushing up (or down) and replacing the old one. The process slows down and stops in older individuals, so some of the largest individuals are likely to be without teeth (edentulous) (Fig. 12.12). A hatchling has the same number of tooth sockets as an adult, and the sockets and the teeth grow progressively with the growth of the animal.

The teeth are made of dentine, covered with enamel and have cementum around the base. In crocodiles, but not alligators, vertical striations, the striae of Retzius, are visible on the body of the tooth as brown striae, which are thought in mammals to reflect nutritional status, like growth rings in a tree (Schour and Hoffman 1939). This interpretation may apply to crocodylians.

In hatchlings, the teeth are sharp and needlelike (e.g. Fig. 13.39) and are all of similar size, but as growth proceeds they differentiate, thicken and become sturdier, equipped to deal with sturdier prey. Growth is accompanied in the mesorostrine and brevirostrine species by the teeth becoming of uneven size along the jawline (heterodont condition).

Fig. 3.22. *Eye of* Crocodylus porosus *open and closed, with two stages of closure of the nictitating membrane in between.* *(Photos DSK)*

In these species, the jawline is vertically undulating rather than straight when viewed from the side. The upper convexities coincide with the presence of the larger teeth (Fig. 3.26, p. 102). Mammalian dental terminology does not fit well, but in some species a 'canine-like' tooth at the anterior convexity (good for seizing prey) and a 'molar-like' tooth at a posterior convexity (often blunter, suitable for crushing) can be identified. The sinuosity increases with body size, along with the visible extent of heterodonty, reflecting the changing diet during growth from hatchling to adulthood (e.g. Fig. 3.45, p. 113). In the long-snouted (longirostrine), 'fish-eating' crocodylians (but see Chapter 6), the teeth are far more equal in size, approaching a homodont condition, and the jawline is straight rather than undulating (Fig. 3.26).

Dental formulae

Sometimes a 'dental formula' is used to aid description (Iordansky 1973). Although there have been attempts to define incisors, canines and even molars (e.g. Kieser *et al.* 1993) and derive dental formulae in a mammalian way, it is more effective to identify the number of teeth arising from sockets in each of the bones of the upper and lower jaw. Thus, the dental formula for *C. porosus* is 4(5) + 13 – 14/15, which translates to 4 (5 sometimes) in the premaxilla and 13–14 in the maxilla, making up the tooth row in one side of the upper jaw, and 15 in the dentary (lower jaw) – that is, a total of 64–68 teeth. *C. johnstoni* has a formula of 5+14–16/15, *Alligator mississippiensis* is 5+13–15/19–20 and *Gavialis* gangeticus 5+23–24/25–26 (106–110 teeth). The formulae are useful, but do not take into account size differences between the teeth or the shape of the jawline. The upper jawline and tooth row in crocodylids shows a pronounced notch on each side, close to the end of the snout and behind the nares, into which fit the large anterior teeth in the lower jaw. In some crocodylids and alligatorids the most anterior pair of teeth in the lower jaw may

Fig. 3.23. *An American alligator surfaces and inhales deeply, its nares wide open. (Photo DSK)*

protrude through sockets in front of and on either side of the nares (Figs 3.27, 3.28, 3.29, p. 103). There can be no doubting the capacity of a croc to get a good grip!

INTEGUMENTARY SENSE ORGANS (ISOS)

Every scale on the head of crocodylians bears at least one slightly raised, circular pigmented 'spot' that is sensory in function (Figs 3.30–3.35, pp. 104–106). These are the multifunctional integumentary sense organs (ISOs). They are discussed in further detail in Chapter 5, with a focus on their structure and function. They occur at higher densities on the upper and lower jaws, concentrating this unique sensory apparatus anteriorly. They occur on all body scales of all the Crocodylidae and most body scales of *Gavialis*, but are completely lacking on the body scales of alligators and caimans, revealing a striking and very interesting difference between alligatorids and crocodylids. Given that both groups have such similar lifestyles this poses a fascinating puzzle. What are alligators missing out on?

SKIN GLANDS

In general, chemicals produced from integumentary glands in reptiles are important in protecting the animals from pathogenic microorganisms, from ectoparasites (including vectors that spread disease), from predators, and also in social communication such as for the attraction of mates and other pheromone-driven behaviour: individual recognition, alarm signalling and territoriality (reviews by Weldon and Ferguson 1993; Weldon *et al.* 2008; Mason and Parker 2010).

All crocodylians possess three types of glands in the skin: prominent pairs of so-called gular glands (at the angle of the jaw, so mandibular might be a better term), which are often everted and become visible in response to capture; paracloacal (cloacal) glands; and, much less conspicuously, a series of small dorsal integumentary glands embedded under scales, in two rows close to the dorsal midline. These are all exocrine glands (as opposed to endocrine), formed by infolding of the epidermis to form a flask or sac, and their secretions result from the proliferation of modified epidermal cells

Fig. 3.24. *(Upper panels) Nasal disk and nostrils of a large male* Crocodylus porosus *showing how the nostrils are opened and closed by opposing musculature (after Bellairs and Shute 1953). Opening occurs by contraction of the dilator muscle, drawing the posterior wall of the nostril posteriorly. (Lower panels) Closure is brought about by the constrictor muscle, a cuff of muscle fibres encircling the dilator muscle. When the constrictor muscle contracts, the (relaxed) dilator muscle expands forwards, bulging the posterior wall of the nostril forwards, to plug the opening. When both sets of muscles are relaxed the nostrils are half open, the typical situation in a crocodylian at rest. (Photos DSK)*

that break down to release the secretion (that is, their secretion is of the holocrine type).

Both mandibular and cloacal glands produce oily secretions. Vliet (1989) noted in American alligators that the secretion is pale brown and floats on water. Paul Weldon of the Smithsonian Institution has made a comprehensive study of reptilian epidermal secretions over many years and has published extensive reviews (Weldon and Wheeler 2001 with a crocodylian focus and Weldon *et al.* 1990, 2008 with a broader focus). The diversity of chemicals produced by these two pairs of glands is both bewildering and puzzling. Both pairs secrete sterols, free fatty acids, triglycerides, steryl esters and aliphatic alcohols and, commonly, a range of hydrocarbons. The proportions of different components differ

Fig. 3.25. *A juvenile black caiman,* Melanosuchus niger, *shows a toothy smile. (Photo Cássia Camillo* per *Robinson Botero-Arias)*

between sexes, glands and individuals and between mandibular and paracloacal glands, suggesting different functions.

There are still many uncertainties about the glands' functions and what is known so far about all three glands raises more questions than answers. Mason and Parker (2010) noted the need for experiments conducted under controlled conditions to resolve the roles played by these glandular secretions.

Mandibular (or gular or chin) glands

The mandibular glands, unique to crocodylians (Weldon and Sampson 1988), lie hidden, folded into short longitudinal slits in the skin just anterior to the angles of the jaw (Figs 3.36, 3.37, p. 107). Pooley (1962) noted in Nile crocodiles that they are the same size in both sexes and secrete a dark green greasy substance with a fishy smell. They can be everted at will and frequently do so in response to stress, such as when a small individual is picked up. The histology of the glands has been described in both *A. mississippiensis* (Reese 1921; Weldon and Sampson 1988) and *C. porosus* (Wright and Moffat 1985) and it is similar. The glands are ovoid, contained within a capsule of well-vascularised dense collagenous connective tissue and housed within a cup of striated muscle. They are divided by septa into lobules, within which the modified epithelial cells proliferate and move towards the centre and degenerate, releasing their secretions, so that the gland 'lumen' is filled with disintegrated cells and secretory products. There is no duct and the everted glands are heavily pigmented with

Fig. 3.26. *Heterdont dentition in: (top left)* Crocodylus niloticus *and (top right)* C. rhombifer, *(centre left)* Caiman yacare *and (centre right)* A. mississippiensis; *and homodont dentition in (bottom left)* C. johnstoni *and (bottom right)* Gavialis gangeticus. *(Photos* C. niloticus *Louis Guillette,* C. yacare *GCG, others DSK)*

melanin. Their ultrastructure has been described by Wright and Moffat (1985). The holocrine secretions are primarily lipid and evidence suggests that at least some of their function is as pheromones, a category of semiochemical chemicals (that is, 'signalling' chemicals, from the same Greek root as 'semaphore').

When everted (Fig. 3.38, p. 107), the mandibular glands resemble small brushes and perhaps that is at least one of the ways they are used, because there are numerous reports of males or females rubbing their chins over the ground, a nest or each other. Tony Pooley (*pers. comm.* 1977) described female *C. niloticus* rubbing their mandibular glands over males before nesting and Kofron (1991) reported that during courtship males and females investigate each other's heads and tail bases and rub their mandibular and gular areas over each other's heads (Fig. 5.30). Similar behaviour was reported in American alligators by McIlhenny (1935) and Garrick (1975) and has been described in detail by Vliet (2001) (Chapter 12). The sexual connection is emphasised by evidence for increased glandular activity during the breeding season, particularly in males (Mason and Parker 2010).

Immature American alligators respond to exudates of both mandibular and paracloacal glands by increased rates of gular pumping, presumably 'sniffing' the air (Johnsen and Wellington 1982) (see

Fig. 3.27. *A young adult* Crocodylus porosus *watching an approaching boat with interest, its lower teeth protruding visibly through holes in its upper jaw. Photographed on Yellow Water in Kakadu National Park, where the frequent passage of tourist boats has habituated the resident crocs. (Photo GCG)*

Chapter 5) so a pheromonal role in intraspecific communication is quite likely.

The diversity of chemical components found in secretions from the mandibular glands of American alligators has been reviewed by Weldon and Sampson (1988), Weldon and Wheeler (2001) and Weldon *et al.* (2008). Among other things they secrete squalene, cholesterol, free fatty acids and α-tocopherol (vitamin E).

Fig. 3.28. Caiman yacare *showing lower teeth protruding impressively through holes in its premaxillae. Rio Negro, Pantanal, Brazil, 2003. (Photo GCG)*

Fig. 3.29. *Teeth near the tip of the snout of a* Crocodylus johnstoni. *Note the extent to which the lower 'incisor' teeth (one visible here) protrude through holes in the upper jaw in this specimen. Note also the typical crocodylid pattern, with upper and lower teeth alternating and fitting neatly between each other, and the large fourth tooth of the lower jaw fitting into a notch in the upper jaw, despite the homodont condition. (Photo GCG)*

Fig. 3.30. *Integumentary sense organs (ISOs) on the head of a juvenile* Crocodylus porosus. *All crocodylians have ISOs on their head scales, typically more than one per scale, often many more than one, with particular concentration around the jawline and teeth. (Photo Winston Kay)*

Fig. 3.31. *Crocodylidae and Gavialidae have ISOs on their post-cranial scales as well as on the head, typically one per scale and including the limbs and the belly. This is the flank of a juvenile* C. porosus. *(Photo DSK)*

Fig. 3.32. *Dorsal view of the neck region of a juvenile* C. porosus *showing ISOs and also a nice view of the nuchal scute pattern typical of this species. (Photo DSK)*

Fig. 3.33. *Flank ISOs on a gharial,* Gavialis gangeticus. *(Photo Ruchira Somaweera)*

Fig. 3.34. *ISOs on the head of a spectacled caiman,* Caiman crocodilus, *photographed in Parrot Bay Lagoon, Puerto Jimenez, Costa Rica. Unlike the other two clades, Alligatoridae have ISOs only on their cranial scales. This is a striking difference. The cranial ISOs are surely homologous across all three clades, suggesting that the lack of body scale ISOs in alligatorids is a derived condition. With such similar lifestyles, the lack in alligatorids presents an intriguing puzzle. (Photo Brandon Sideleau)*

Fig. 3.35. *ISOs concentrated along the jawline of an American alligator. Note also, in contrast to crocodylids (Fig. 3.29), the alligatorid 'overbite', with lower teeth mostly concealed when the jaws are closed. (Photo DSK)*

Fig. 3.36. *Mandibular (or chin or gular) glands in Australia's freshwater crocodile,* Crocodylus johnstoni, *everted through the slits behind which they normally 'hide'. (Photo Paul Weldon)*

Paracloacal (or cloacal) glands

If the vent is spread open the paracloacal glands can be seen, embedded in the lateral walls of the proctodaeum (Figs 3.39, 3.40). Their histology has been described in *A. mississippiensis* (Reese 1921; Weldon and Sampson 1987) and in *C. porosus* by Wright and Moffat (1985). They are formed by

Fig. 3.38. *A mandibular gland everted from a 3 m* Crocodylus porosus *captured in 1984 in the Limmen Bight River, Northern Territory. The snout is to the left foreground. The darkly pigmented surface of the gland has a brush-like appearance. (Photo GCG)*

Fig. 3.37. *A Nile crocodile,* Crocodylus niloticus, *photographed on the Olifonts River, South Africa with an everted mandibular gland visible. Nice teeth too! (Photo Louis Guillette)*

infolding of the proctodael wall and are therefore lined with epidermal cells. They are larger than the mandibular glands and have a single duct opening into a spheroidal sac-like chamber partially occluded by cellular debris and secretion. They are enclosed by dense collagenous connective tissue and striated muscle. Like the mandibular glands, their holocrine secretions are primarily lipid and evidence suggests that at least some of the contents act as pheromones.

Musky smelling secretions that are thought to come from the cloacal glands are reported from adult male American alligators, particularly in the breeding season. Kent Vliet noted that secretions could be detected in the air after a head-slap (Chapter 12), when an oily sheen appeared on the water surface, positioned as if coming from the paracloacal glands (Vliet 1989). It has been suggested that the secretions may mark territories and Weldon *et al.* (2008) reported a previously unpublished observation by Paul Weldon that a group of free-ranging juvenile alligators dispersed after thawed secretions from adult males were poured into the water where they had aggregated to feed. He suggested that the secretions may signal the presence of aggressive adults. A little surprisingly, however, no differences in composition were found between males and females in their paracloacal gland secretions, although limited gender-related differences were found in the Chinese alligator (Weldon *et al.* 2008).

Chemical constituents of the secretions from the paracloacal glands of American alligators and several other species include hydrocarbons, sterols, steroids, free fatty acids, alcohols, ketones, triacylglycerols, phospholipids and an array of esters (Weldon and Wheeler 2001; Weldon *et al.* 2008). It is not hard to believe that the secretions stink!

Although most work has been done on American alligators, there are enough data from other crocodylians to show that they also have a numerous and diverse array of chemicals in the secretions. Weldon and Wheeler (2001) noted that citronellol and some other compounds isolated from caiman paracloacal glands are unknown in other vertebrates. The aromatic ketone dianeackerone has been found in *Osteolaemus* (Whyte *et al.* 1999), which is unknown from any other natural source. Among paracloacal

secretions from a range of caimans, Kruckert *et al.* (2006) found 'a new family of 43 aliphatic carbonyl compounds that includes aldehydes, ketones, and β-diketones with an ethyl branch adjacent to the carbonyl group'.

Much remains to be found about the functions of these glands and, with secretions from both mandibular and paracloacal glands apparently influencing mating and nesting activities, a focus on mature animals in particular is desirable.

Dorsal integumentary glands

These small structures (1–3 mm) were originally called dorsal organs, but now that their glandular structure is recognised it is appropriate to call them glands. They have an odd distribution, embedded in the dorsal musculature in two rows, close in under the dermis, one row on either side, scale by scale (mesoanteriorly), on the second scale row out from the mid-line between the neck and the pelvis, including some nuchal scutes. They are known from both American (Fig. 3.41) and Chinese alligators and from *C. niloticus*, *C. johnstoni* and *C. porosus*, so they are probably a feature of all crocodylians. They have been known since 1899, at least and their embryonic development in American alligators was described in detail by Reese (1921). Cannon *et al.* (1996) looked at their histology and histochemistry in 1.9–2.3 m alligators. Embryonic and hatchling *C. porosus* and *C. johnstoni* were studied by Linda Heaphy (1985) in an unpublished thesis and by Richardson and Park (2001) who also reviewed the earlier literature. These last authors reported 19 pairs of dorsal glands in *C. porosus* and 24 pairs in *C. johnstoni*, but all studies agree that numbers may vary between individuals and even between left and right rows, although it is difficult to make an accurate count. The glands themselves have been described as sac-like and ovoid in shape except that, anteriorly, the glands in *C. johnstoni* are lobed. Richardson and Park (2001) reported that in *C. porosus* they begin to develop as epidermal invaginations ~23 days into embryonic life and increase dramatically in size during the last third. At this stage, and in hatchlings, a short straight duct opens from each gland between the second and third rows of scales. However, Heaphy (1985)

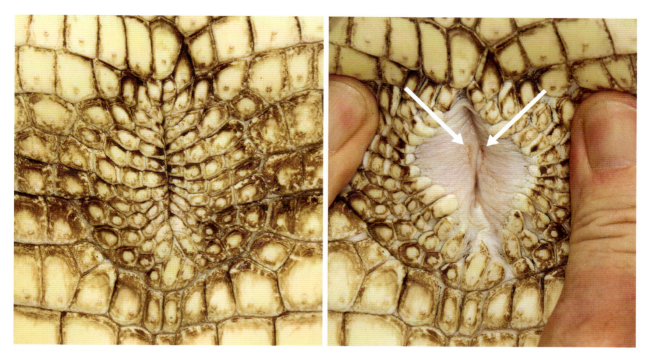

Fig. 3.39. *(Left) Cloaca of a* Crocodylus porosus *and (right) spread to show the openings of the cloacal glands in the walls of the proctodaeum, anterior to the proctodael opening. (Photos DSK)*

reported, differently, for both species, that the upper region of the gland forms 'a relatively long, thin plug, which connects the end of the organ to the scale hinge region'. Indeed, she reported that at no stage was a duct observed, 'nor a break in the stratum corneum overlying the plug'. Moreover, electron micrographs in the possession of Michael Beal at the University of New South Wales (*pers. comm.*) suggest that the plug may be porous. In American alligators also, the apparent duct is occluded by epithelial cells (Reese 1921) and there is uncertainty about whether or not there is an open duct (Cannon *et al.* 1996). On the basis of histological preparations, it appears that the glands are most active late in embryonic life and, by implication, early post-hatching. The detailed post-hatchling life of these glands is not described but, in the *Crocodylus* species, at 2 years and older their remains are embedded entirely within the epaxial muscles and they appear to be no longer functional. Other puzzles remain, for example Heaphy (1985), Cannon *et al.* (1996) and Richardson and Park (2001) all described rods of 'crystalline material' within the glands. Additionally, the nature of the secretions remains unresolved, with contradictory results

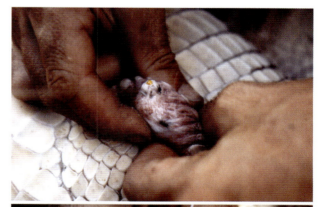

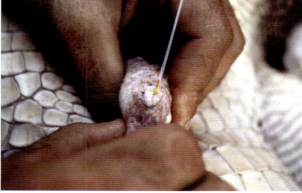

Fig. 3.40. *Exudate being collected from the paracloacal glands of a 3 m* Crocodylus porosus *captured in 1984 in the Limmen Bight River, Northern Territory. (Photo GCG)*

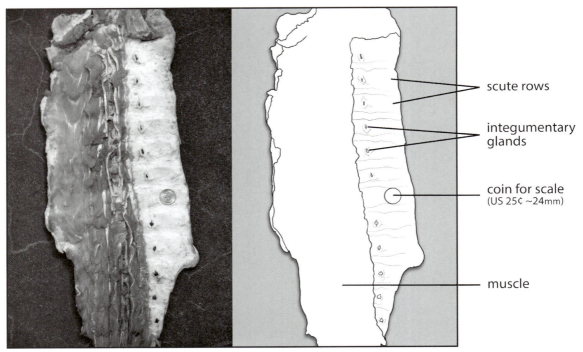

Fig. 3.41. *The under surface of the dorsal skin of an American alligator showing the dorsal integumentary glands. (Photo Paul Weldon)*

about the presence of lipids. Energy-dispersive X-ray analysis of the rods revealed calcium, copper, iron, lead, potassium and zinc (Cannon *et al.* 1996).

Their function, too, is so far a mystery. The glands are holocrine, their epithelial cells releasing secretory product through lysing after being shed into the lumen of the gland. This is functionally analogous to the operation of mammalian sebaceous glands, which may have prompted Reese to suggest that their function could be to keep the skin in good condition,

Fig. 3.42. *Complete skeleton of an American alligator, including osteoderms. (Specimen prepared by and photo supplied by the Northern Territory Museum and Art Gallery, Darwin)*

like mammalian oil glands. The clean appearance of crocodylians in the wild is certainly impressive. Mud seems to flush right off them, suggesting the presence of a detergent-like or oily substance. However, if the dorsal glands were the source of such an agent, far more of them would be expected, and they would be expected to grow proportionally with the animals, not regress and become completely embedded within the epaxial musculature.

Innervation is too slight to imply any sensory function (Cannon *et al.* 1996) and Reese (1921) noted that their secretion had no odour. However, later authors connected function with the development of the glands in mid- to late-term embryos and their subsequent regression and Heaphy (1985) and Richardson and Park (2001) apparently came independently to the hypothesis that the glands may secrete a pheromone that assists in crèche formation (see Chapter 12), keeping a pod of hatchlings together and, perhaps, reminding their mother not to eat them.

Further research into these glands would make a good PhD research topic.

SKULL AND MUSCULOSKELETAL SYSTEM

The musculoskeletal system of Crocodylia is typical of tetrapods in general and of archosaurs in particular, with the skull and pelvis particularly specialised. It is similar in the three extant groups and this section will summarise some crocodylian specialties: the diversity of skull shape in different species, the elaborate lever system for closing the jaws, the skeleton and musculature of the neck, and the role of osteoderms in the structural bracing of the trunk.

General descriptions of the musculoskeletal systems in crocodylians have been published by Reese (1915), Chiasson (1962) and Richardson *et al.* (2002). Research papers on specific topics will be referred to where appropriate.

SKULL AND JAWS

Large and robust skulls and jaws are characteristic of adult crocodylians and they are very impressive (Figs 3.43, 3.44). These beautifully sculptured and

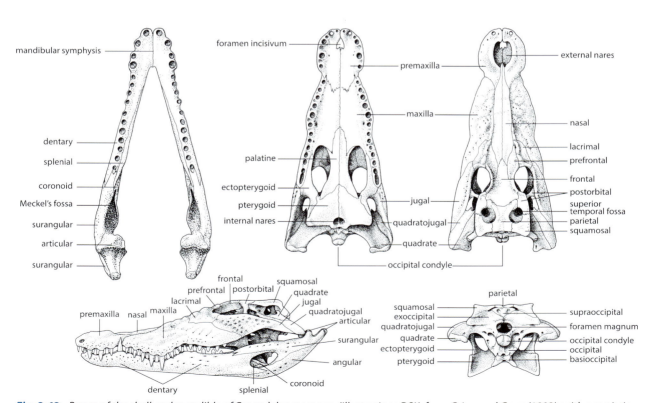

Fig. 3.43. *Bones of the skull and mandible of* Crocodylus porosus. *(Illustrations DSK, from Grigg and Gans (1993), with permission from the Australian Biological Resources Study, Canberra)*

111

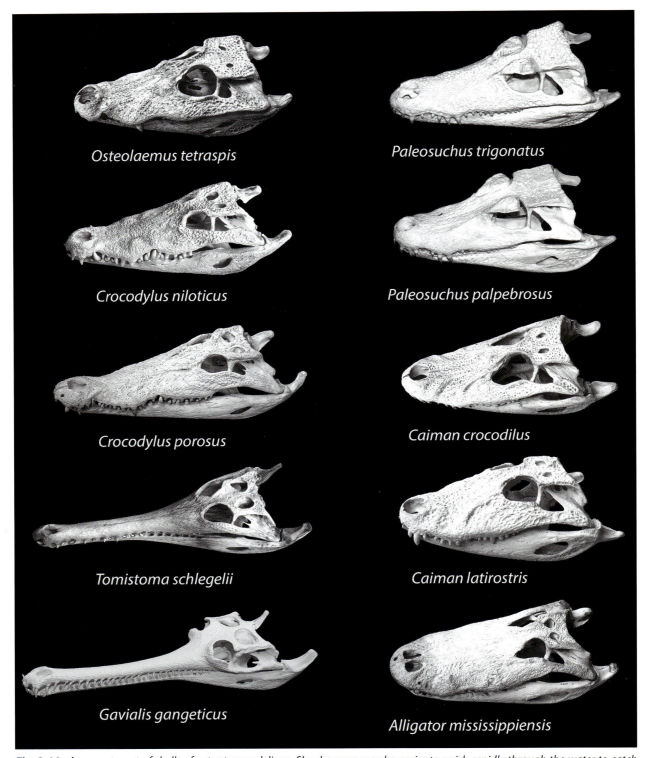

Osteolaemus tetraspis

Paleosuchus trigonatus

Crocodylus niloticus

Paleosuchus palpebrosus

Crocodylus porosus

Caiman crocodilus

Tomistoma schlegelii

Caiman latirostris

Gavialis gangeticus

Alligator mississippiensis

Fig. 3.44. *An assortment of skulls of extant crocodylians. Slender ones may be easier to swish rapidly through the water to catch fish but are less robust. Shorter thicker ones are more suitable where food needs crushing, such as eating turtles or having to kill and subdue mammals. Specimens from the University of Florida-Florida Museum of Natural History Herpetology Collection as follows:* Osteolaemus tetraspis – *UF33749,* Crocodylus niloticus–*UF54812,* Tomistoma schlegelii–*UF54210,* Caiman latirostris–*UF62649,* Alligator mississippiensis–*UF 35129;* Paleosuchus trigonatus *and* P. palpebrosus *are from Kent Vliet, private collection;* C. porosus *from Gordon Grigg private collection; and* Gavialis gangeticus *from the WCS Bronx Zoo. The large palpebral ossicles that give the species its name are still attached in the specimen of* P. *palpebrosus. (Photos GCG, DSK, Kent Vliet)*

Fig. 3.45. Skulls of 1 m, 3 m and 5 m Crocodylus porosus, scaled to the same length to show ontogenetic change in morphology. Note that the head and also the teeth become proportionally more massive, reflecting a capacity to deal with much larger prey. (Photos DSK, skulls from DSK, GCG and lowest courtesy Northern Territory Museum and Art Gallery, Darwin)

often massive bony structures house the brain, most of the sensory systems and, with the lower jaw (which is technically not part of the skull), the teeth. Muscles attached to the skull operate the jaws with forces either strong enough to crush bone, or with a delicacy sufficient to ease a hatchling from an egg and carry it gently to the water.

Reviews of the structure of crocodylian skulls may be found in Iordansky (1973), Langston (1973), Schumacher (1973), Brochu (1999) and Cleuren and De Vree (2000). More recently, there has been a comprehensive and very well illustrated description of the broad (brevirostrine) skull of *Caiman latirostris* (Bona and Desojo 2011). This includes a description of the musculature, a comparison with *Caiman yacare* and a comprehensive literature review. As Archosaurs, crocodylians are diapsid, although the post-temporal fenestrae are reduced. The skulls are dorsoventrally flattened (much less

so in *Paleosuchus*) and have a distinct cranial table perforated with the temporal openings through which some of the closing muscles bulge (Fig. 3.44). They have a solid secondary palate and (usually) long jaws bearing numerous teeth. The secondary (hard) palate is made up of shelf-like processes from the premaxillae, maxillae, palatines and pterygoids (Fig. 3.43). It permits ventilation from nostril to glottis, bypassing the mouth, via the secondary choanae opening through the pterygoids (Chapter 7).

As in all tetrapods, the skull is the result of a coherent merger between dermal bones (which arise within the dermis and evolved initially forming a protective shield over the bodies and heads of some of the earliest fishes) and the chondral bones of the braincase, which, along with the vertebrae and limbs, started out as cartilage and later became ossified (see above). In the skull of crocodylians, most of the bones that form the visible surface of the skull are dermal bones. The only chondral bones visible are the quadrates, supra-, ex- and basi-occipitals (the last with the occipital condyle) and, in the lower jaw, the articular. The premaxillae, maxillae and dentaries bear peg-like conical thecodont teeth. Out of sight are the chondral bones surrounding the braincase and auditory capsules. Pterygoids and quadrates are tightly plastered onto the lateral walls of the braincase. The middle ear region is modified extensively, and the quadrate inclined so that the hinge of the jaw is displaced far posteriorly. All species lack a parietal (pineal) foramen.

The surface of the skull is ornamented by bony sculpturing formed by secondary dermal ossifications. These adhere firmly to much of the bone on the dorsal surface, becoming more complex with age, and very likely strengthening the skull structure. Some species (e.g. *Alligator mississippiensis*, *Osteolaemus* spp. and *Paleosuchus* spp.) have palpebral ossicles in the upper eyelids, articulating with the pre-frontal bones of the skull (Nesbitt *et al.* 2013) and which, presumably, offer protection.

Many cranial bones are pneumatised, having gas-filled cavities connected to the Eustachian tubes of the middle ear and the nasal passages. These may equalise pressure in the inner ear (Colbert 1946) or

Fig. 3.46. *Greg Erickson from Florida State University measures bite force in a large American alligator. The force was measured by placing a leather-padded sensor, comprising piezoelectric transducers sandwiched between steel plates, below either the left or right most prominent maxillary molariform tooth; the relevant place to measure force used to crush prey. (Photo Greg Erickson)*

isolate the inner ear from underwater sounds when listening above water (Iordansky 1973). However, pneumatisation also reduces the cranial mass (and inertia), while maintaining its strength, which may be an important factor in both buoyancy and feeding.

The lower jaw (mandible) is made up of dermal bones, except for the articular (with which it articulates with the skull). The dentary is the largest bone, splinted anteriorly by the splenial and the coronoid and extended posteriorly by the surangular (Fig. 3.43). Jaw-closing muscles insert through the pre-articular (Meckel's) fossa (Figs 3.43, 3.48).

Among extinct crocs, there is a much greater diversity of skull and jaw shapes than in extant forms and, because they are the most frequently fossilised body part, palaeontologists are always interested in skulls and teeth. The solid skulls of the modern crocs with their enormous jaw-closing muscles seem a far cry from the lighter models

possessed by the terrestrial crocodylomorphs such as the sphenosuchians (Chapter 2). Solidification of the skull elements and increased jaw musculature are among a suite of features defining the Crocodyliformes, compared with the rest of the Crocodylomorpha (see Fig. 2.3). However, Clark *et al.* (2004) described a sphenosuchian-like and highly cursorial crocodylomorph from the middle Jurassic of China that showed similarly sturdy attributes. They concluded that 'the consolidation of the crocodylian skull thus began well before crocodylians entered the water'.

OPENING AND CLOSING THE JAWS

Force and pressure at impact

Among the many features crocodylians are famous for, snapping the jaws shut rapidly and with very great force would be near the top of anyone's list. Although the jaws can be held closed by an adult

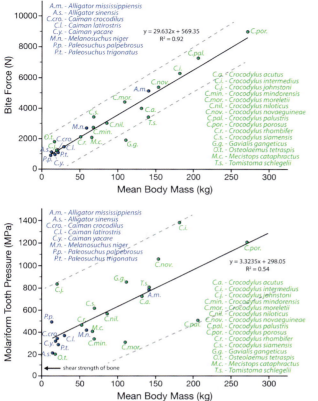

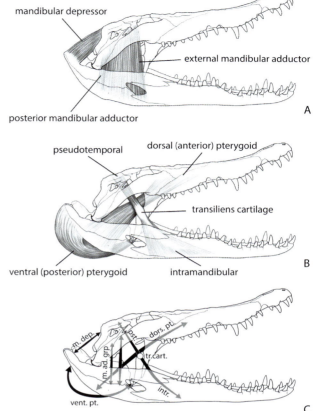

Fig. 3.47. *These graphs show the relationship between body mass and (top) bite force and (bottom) tooth pressure measured on 23 species of crocodylian at the large molar-like teeth. No difference in either bite force or tooth pressure emerged between crocodylids and alligatorids, and bite force was not determined significantly by snout shape. In bite force, only the two most longirostrine species,* Gavialis gangeticus *and* Tomistoma schlegelii, *emerged as markedly weaker than the throng. In tooth pressure, the fish-eating* Crocodylus johnstoni *and* Crocodylus intermedius *showed higher. The dotted lines represent ± 95% confidence limits around the regression equation for the line. (From Erickson* et al. *2012)*

Fig. 3.48. *Diagrammatic view of the major jaw muscles of* Crocodylus porosus. *(A) Adductors and depressors. (B) Deeper muscles. (C) Schematic view of the skull with arrows representing the lines of action of the muscles. Whereas opening the jaws involves only one small set of muscles, the mandibular depressors, several large, powerful muscle groups are involved in closure. Accordingly, although the jaws can be held closed relatively easily, they can be snapped closed with stunning speed and force and, once clenched, are nearly impossible to open. (Illustrated with reference to Schumacher 1973; Busbey 1989; Bona and Desojo 2011; Cleuren and De Vree 2000)*

human relatively easily, it is almost impossible for them to be opened, once clenched. In fact, crocodylians can snap them shut and hold them together with a speed and a force that outclasses any other jaws in the animal kingdom. This has been shown by Greg Erickson at Florida State University in Tallahassee, and colleagues. They used precision electronic force transducers to measure bite force in American alligators (Erickson *et al.* 2003, 2004) (Fig. 3.46) and in a subsequent study across 23 species of living crocodylians (Erickson *et al.* 2012). In their 2003 study, they reported a force of 13 172 N (2961 pounds) in a 3.8 m alligator, noting that

it was the 'highest bite force ever measured for a living animal'. But this was eclipsed when they measured 16 414 N (3690 pounds) in a 531 kg *C. porosus*. (Note: a Newton, N, is a unit of force: the force required to accelerate a mass of 1 kg/s², which is hard to relate to. One way to help put this into perspective is that the force of gravity acting on a 70 kg human is 686 Newtons, so a croc bite of 16 400 N is equivalent to the force of gravity exerted on a mass of 1673 kg.)

Erickson and colleagues found a strong correspondence between bite force and body mass

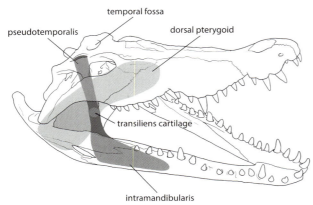

Fig. 3.49. *Deep jaw muscles of* Crocodylus porosus, *simplified. Much of the speed and power of a croc's snap is powered by a scissoring action performed by two muscle groups set more or less at right angles to each other. Both the pseudotemporalis muscle and the huge dorsal pterygoid muscle originate deep within the skull, but run in opposite directions to their points of insertion, crossing at the transiliens cartilage. The pseudotemporalis inserts into the lower jaw via the transiliens cartilage and intramandibular muscle. These two may actually be a single muscle (Tsai and Holliday 2011).*

(Fig. 3.47), with no sexual dimorphism. *Gavialis* and *Tomistoma* stood out from the others. No differences emerged between crocodylids and alligatorids. It has commonly been held that long snouted forms would generate lower bite forces than short snouted forms. To test this, Erickson and colleagues quantified jaw shape as RP (rostral proportion), the width of the snout at the anterior margin of the orbits divided by the length from there to the tip of the snout (Table 3.1). Values ranged from 0.1 in *Gavialis* to 0.89 in *Caiman latirostris*. The effect of this on bite force was, however, far outweighed by body mass in having an influence, so jaw shape 'demonstrates surprisingly little correlation to bite force'. Extrapolating, they calculated that a 6.7 m *C. porosus* would have a bite force of 27 531 N–34 424 N (6187 to 7736 lbs) and it would be a little over 100 000 N in an 11 m (3450 kg) *Deinosuchus*. Walmsley *et al.* (2013), in their analysis of seven species using finite element analysis, predicted lower values for bite force than the measurements made *in vivo,* but they too found that bite force was heavily dependent upon body size rather than skull morphology.

Pressure is force per unit area, and Erickson's team took that into account by measuring the relevant area of contact made by a tooth's first impact upon a target prey item. They did this by making a 1 mm indent in modelling clay with epoxy replicas of relevant teeth and translating photographs of the result into an area of contact. Combining that with the relevant measured force, they were able to calculate pressure at impact. The area depends upon the shape of the tooth tip, which differs in different species. Like bite force, pressures exerted by both the large canine-like teeth (anteriorly) and molar-like teeth (further back) increased with body mass, but the relationship was less tight. The plot of a molar-like tooth pressure is shown (Fig. 3.47). These pressures ranged from 203 MPa (29 443 psi) in *Osteolaemus tetraspis* (RP 0.66) to 1388 MPa (201 312 psi) in *C. intermedius* (RP 0.35). The slender-snouted and commonly fish-eating *C. johnstoni* (RP 0.24) and *C. intermedius* (RP 0.35) are comparatively high and the broader-snouted, more generalist feeders *C. palustris* (RP 0.66) and *C. moreletii* (RP 0.75) are comparatively low.

The study refutes the idea that bite force is a correlate of snout shape. For example, similarly sized *C. johnstoni* (RP 0.24) and *Caiman latirostris* (RP 0.88) are vastly different in rostral morphology yet have comparable molariform bite force. Their molar-like tooth pressures are, however, very different (372 and 832 MPa, respectively.) Clearly, tooth morphology and snout shape have evolved independently.

Taken together, the studies by McHenry *et al.* (2006), Walmsley *et al.* (2013) and Erickson *et al.* (2003, 2012) have made it abundantly clear that the diversity of crocodylian skull shapes and dentition are essentially unrelated to phylogeny and very much the product of specialisation to particular dietary niches.

How can such huge forces be generated? The answer is by a combination of the molecular structure of the muscles themselves and their anatomical arrangement.

Speed: superfast myosin

The muscles that mammals use for posture and locomotion are made up of four different types of fibres, one 'slow-twitch' and three 'fast-twitch', each with different types of myosin, contraction speeds, tolerance for fatigue and so on. Their jaw muscles, however, have an additional, different form of myosin: 'superfast myosin', sometimes called 'masticatory myosin' (Hoh *et al.* 2001), which confers powerful and high-speed contraction. Superfast myosin is found in mammalian carnivores, both marsupial and eutherian, and also in primates and some reptiles and fishes. Superfast myosin is found in the jaw-closing muscles of *C. porosus* (Hoh *et al.* 2001; Hoh 2002)) and a caiman (species not reported) (Rowlerson 1994) and, presumably, other crocodylians as well. It was not found by these authors in either the jaw-opening or locomotory muscles, which have only slow and fast myosins, but only in the jaw adductor muscles. It seems safe to agree with the conclusion by Hoh *et al.* (2001) that 'the speed and power of jaw closure in *C. porosus* may be attributed to the unique properties of this molecular motor': superfast myosin.

The 'speed and power' depends also upon adequate mechanical leverage, conveyed by the arrangement and functioning of the jaw muscles. Crocodylians do not chew their food, so there is no food preparation in the way that it occurs in mammals and in some herbivorous reptiles where plant material is ground before swallowing. Accordingly, the jaws operate in only one plane, with no capacity for lateral or forward-aft movement. Therefore the muscle movements that are involved are comparatively straightforward: they open and close about the hinge joint between the quadrate bones of the skull and the articular bones of the mandible (Figs 3.43).

Muscles to open the jaws

The jaws can be opened by contraction of the mandibular depressor muscles, which run from the back of the skull to the articular and angular bones on the jaw (Figs 3.43, 3.48A). In practice, opening the jaw often involves raising the head, as it does when a croc is on land, with a whole suite of muscles to assist (Fig. 3.52). Of course, as pointed out by Iordanksy (2000), both may occur simultaneously. Most people have heard, and some have experienced, that a crocodylian is unable to open its mouth against a firmly held grip on the snout. The reason is clear. Even with the lower jaw bones extended posteriorly, well behind their hinges

with the quadrates (Figs 3.43, 3.48), it is obvious that the leverage these muscles can exert will be minimal compared with the leverage one can get by holding the jaws together at the end of the snout.

Muscles to close the jaws

Positioning the jaw-closing muscles has posed a particular evolutionary problem. The famously wide crocodylian gape has been made possible by the articular hinges being so far posterior, even behind the foramen magnum (Fig. 3.43). However, the gape would be compromised if the muscles were attached to the sides of the skull and jaws reasonably well forward, as in humans. Crocodylians maintain the wide gape by having the muscles set well back in the gape, but achieving a rapid and forceful closure from that position is a bit like trying to use chopsticks while holding them at their very ends.

How do they manage it? Part of the solution is the sheer bulk of muscle devoted to jaw closure. Also, with the 'hinge' so far back, there is room for considerable muscular development between it and the gape (Fig. 3.48A, B). Two of the major muscles between skull and jaw, anterior to the hinge, attach posteriorly but reach well forwards at shallow angles, crossing posterior to the gape (Figs 3.48C, 3.49). Finally, the 'post-hinge' extension of the lower jaws allows muscles to reach around and exert additional leverage.

Looking in more detail, closure depends upon six (or five, see below) major muscle pairs (Fig. 3.48). The most superficial are the external and posterior mandibular adductor muscles, which are attached to the sides of the skull and run behind the jugal bones and insert into the lower jaw via Meckel's fossa (Figs 3.48B, C, 3.43). It is these muscles that hold jaws closed while holding prey (Busbey 1989). Deeper in are the dorsal and ventral pterygoids, and the pseudotemporalis and intramandibularis, which link in the middle at the transiliens cartilage (see below) and which may, actually, be a single muscle (Tsai and Holliday 2011). These are set more or less at right angles to each other, attaching deep into the skull (Figs 3.48, 3.49). The pseudotemporalis muscles bulge up through the temporal fossae on the cranial platform during contraction. The pseudotemporalis muscle inserts into the lower jaw via the transiliens cartilage and intramandibular muscle, which enters the lower jaw via Meckel's fossa and runs forward into the hollow dentary, further than the external mandibular adductor. The transiliens cartilage is an interesting fibrocartilaginous structure, usually regarded as being unique to crocodylians (including all crocodyliforms) and, possibly turtles, which have a similar structure of uncertain homology. Its name comes from the Latin *transilio*, meaning to jump over. Tsai and Holliday (2011) examined its histology using several techniques and came to the conclusion that it is actually a sesamoid bone formed within a single muscle, the pseudotemporalis. Sesamoids often form in mammalian tendons where they pass over a joint. The human patella is an example. They are common in birds too, but less common in reptiles (reviewed by Haines 1969). In this case, the sesamoid has formed where it passes the pterygoid buttress. This reinterpretation implies a homology with similar structures in the jaws of birds and turtles. The transiliens cartilage also has attachments to the dorsal pterygoid, which seems odd until it is realised that all these muscles contract simultaneously during closure. The last 'closing' muscle to draw attention to is the ventral pterygoid: the large muscle that reaches from the ventrally extended pterygoid buttress around to the ventral and internal surfaces of the posteriorly extended articular and surangular bones of the jaw, providing additional leverage via the posterior elongation of the jaw past the hinge.

Thus the jaw is a lever system with short and long arms disposed about a fulcrum, the articular hinge, with five (or six) muscles involved in closure, two of which hold the closed position, and one muscle for lowering the jaw and opening the mouth, often in tandem with muscles that raise the head (Fig. 3.52).

VERTEBRAE AND VERTEBRAL COLUMN

There are eight or nine cervical vertebrae (including the atlas and axis elements, which form the joint

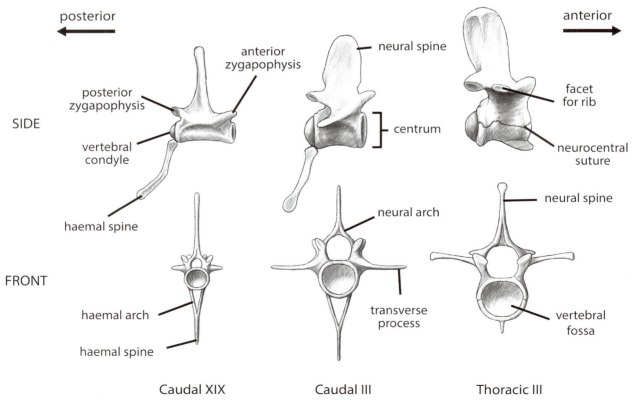

Fig. 3.50. *Selected vertebrae. Note the neurocentral suture, a remnant from ontogeny, in which the centrum and neural arch start off as separate elements that fuse. The fusion occurs from the tail to the head, so the sutures are no longer visible in these caudal vertebrae but are still present in the cervical vertebrae (Fig. 3.51) (lateral views are based on the same skeleton).*

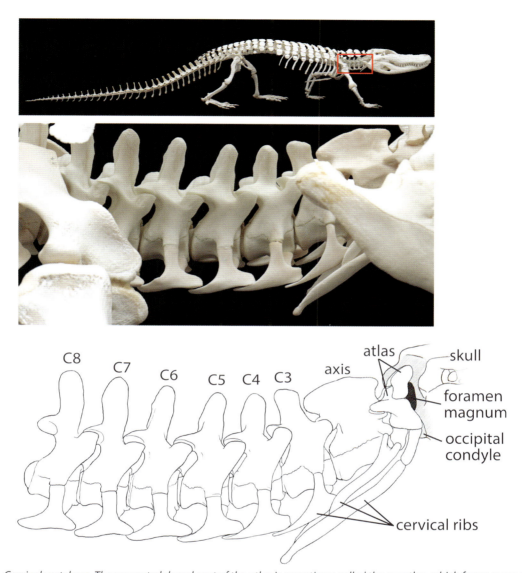

Fig. 3.51. *Cervical vertebrae. The separated dorsal part of the atlas is sometimes called the proatlas, which forms a cap sitting over the gap to the foramen magnum, protecting the spinal cord. The neural spines and the broad lateral cervical ribs provide attachment points for the muscles that move the head (Fig. 3.52). Note the neurocentral sutures. (Photos Northern Territory Museum and Art Gallery, Darwin).*

with the skull), 15–16 dorsal (thoracic and lumbar are not much differentiated, but the former bear ribs), two sacral (to which the ilium of the pelvis is attached) and 30–40 caudal vertebrae (Hoffstetter and Gasc 1969) (Fig. 3.42). Most vertebrae are procoelous – that is, concave anteriorly (Figs 2.29, 3.50) – the exceptions being the atlas and axis, the two sacrals, which are flat anteriorly and posteriorly, and the first caudal, which is also flat to match the sacral. Procoelous vertebrae allow a degree of flexure of the spine. Salisbury and Frey (2001) and

Salisbury *et al.* (2006) have discussed the evolution of this type of vertebral structure in the context of the evolution of the Eusuchia (see Chapter 2). The paper by Hoffstetter and Gasc (1969) provides a detailed description of the vertebral column.

NECK: SUPPORTING A HEAVY HEAD

The heavy skull and striking dentition are major contributors to crocodylian success, and they require the support of a sturdy, muscular neck. Yet the head is not wielded only with brute force,

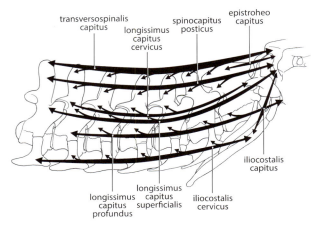

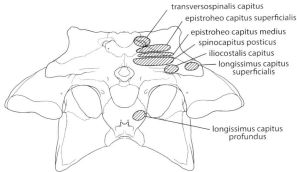

Fig. 3.52. *Muscles alongside and attaching to the cervical vertebrae and the rear of the skull allow controlled movement of the heavy head, lifting it as the mouth is opened by contraction of the mandibular depressor (Fig. 3.48A), and important in 'inertial feeding' (Chapter 6). These muscles must also be very strong for a croc to be able to drag a large prey item into the water to drown. This diagram, adapted from Cleuren and De Vree (2000) describing* Caiman crocodilus, *shows the lines of action of the cervical muscles (above) and their sites of attachment at the back of the skull (below).*

The cervicals articulate on their procoelous 'peg and socket' joints, held together by ligaments and the enveloping longitudinal tracts of muscle, most of which insert onto the back of the skull (Fig. 3.52). The neck muscles are structurally segmented (see below) and attach to each adjacent vertebra. Muscle insertions on the blade-like neural spines and short cervical ribs provide leverage. Reference to Fig. 3.52 will explain how the operation and coordination support and move the head. The first two cervical ribs are notably longer than the others.

TRUNK MUSCULATURE: ANALOGY WITH AN I-BEAM

Groups of epaxial muscles lie dorsal to the vertebral transverse processes, while the hypaxial muscles make the lateral and abdominal body walls (Fig. 3.53). The viscera are enveloped within a flexible cage of bone and muscle, suspended from the vertebral column. This is typical of tetrapods but crocodylians have an interesting addition, with the building of the dorsal osteoderms into the support structure, as described below.

Several groups of muscles make up the epaxials: the spinalis group (articulospinalis, spinalis and multifidus); tendinoarticularis; longissimus dorsi; and the ilio-costals (Fig. 3.53 lower panel). The last two muscle groups are segmented by fascia into myomeres, each of which is cone shaped and inserts posteriorly into the next cone, reaching across several adjacent vertebrae (Murakami *et al.* 1991). This metameric (serially repeated) structure is not seen in the hypaxial musculature (external and internal oblique and transverse abdominus muscles). Bilateral reciprocal contraction and relaxation of epaxial and hypaxial musculature flexes the trunk dorsally or ventrally, although in crocodylians this is obviously limited. It probably reaches its most impressive expression in species that bound or gallop (Figs 4.8, 4.9). Contraction of musculature on one side with relaxation on the other curves the body laterally.

The mass of the viscera, slung below the vertebral column, must weigh heavily when crocs are on land, particularly when high walking (Chapter 4). Salisbury and Frey (2001) have taken an interest

but also with fine, subtle control. Collectively, the bones and muscles of the neck form an externally supported flexible strut supporting the head. The skull articulates with the first two cervical vertebrae, the atlas and axis (Fig. 3.51). There is also a small bone, sometimes called the pro-atlas, whose homology is uncertain, that covers the area between the front of the atlas and the back of the skull. The atlas and axis allow limited rotation, left–right and up and down movement of the skull about the odontoid peg, which protrudes forward from the axis vertebra. Developmentally, the odontoid peg is formed from the centrum of the atlas. The remaining six or seven cervical vertebrae bear short ribs attached to transverse processes.

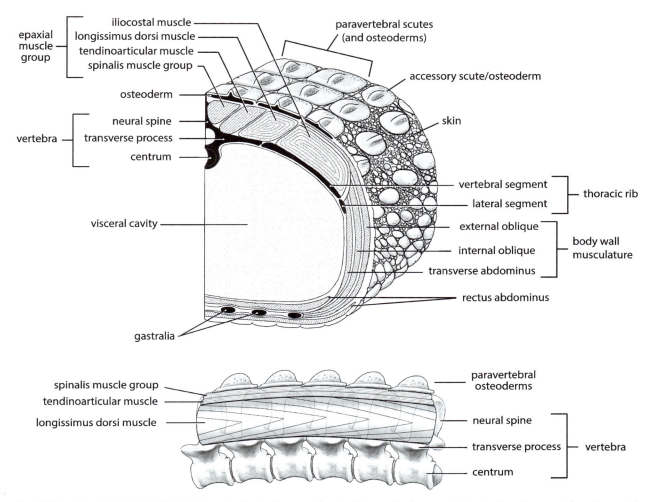

Fig. 3.53. *Anterior to the left. (Top) Schematic block diagram of part of the trunk of a crocodylian to show the epaxial and hypaxial musculature and other structures, including the transverse processes of the vertebrae and the paravertebral scutes and their underlying osteoderms, which form the bracing system. (Bottom) Side view of the epaxial musculature between the row of transverse processes below and osteoderms above. The segmental elements of the longissimus dorsi muscle (myotomes) are interdigitating cones, adhering by fascia to each other and the skeletal elements above and below, to make a flexible but sturdy braced structure that has been likened to the I-beams used in the construction of buildings (see text). (Adapted from Salisbury and Frey 2001)*

in the structural implications. They likened the combination of osteoderms above and vertebrae below to the flanges on the I-beams used in building construction and referred to it as a 'hydraulically stabilised, longitudinally cross-braced, segmented I-beam'. To give a simplified version, the epaxial musculature is attached between fascia of the osteoderms and vertebral transverse processes and ribs, across a distance of several vertebrae, ensuring head–tail compressive forces, while their bulge during isometric contraction provides lateral stability. They refer to this as a 'eusuchian-type bracing system' because it differs from other Crocodylomorpha. It is worth noting that

the bracing system in *Gavialis* differs a little, but is still of the eusuchian type (Salisbury and Frey 2004).

TRUNK: RIBS AND GASTRALIA

The musculoskeletal elements of the trunk include the vertebrae, ribs, sternum, gastralia and relevant musculature. There are 15–16 vertebrae, which are not much differentiated into thoracic and lumbar and two sacral vertebrae to which the ilium of the pelvis is attached.

All but the last few vertebrae bear ribs that curve ventrally (Figs 3.42, 3.54) to articulate with the (predominantly) cartilaginous sternum. The ribs are

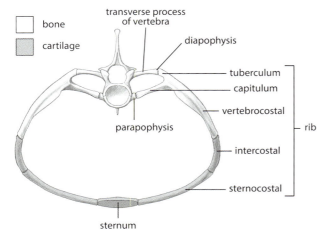

Fig. 3.54. *Ribs articulating with a thoracic vertebra, front view. The jointed and slightly w-shaped nature of the ribs allows their anterior/posterior and folding movement to accommodate the thorax when expanding and contracting during respiration (Chapter 7), described by Claessens (2009) using cineradiography.*

jointed, allowing expansion and contraction of the thorax during respiration (Claessens 2009). Their distal segments are cartilaginous.

The gastralia, sometimes called abdominal ribs, are rib-like bony elements posterior to the sternum and reaching posteriorly as far as the pelvis, embedded above the large ventral rectus abdominus muscles (Figs 3.17, 3.55). These are slender dermal bones, not ribs (which are chondral bone; that is, having their origin as cartilage that becomes bony as development proceeds). The gastralia serve for muscle attachment and for protection of the abdomen. They also appear to stiffen it against collapse during inhalation (see Chapter 7). They are joined along the midline by a tough membrane that runs from the posterior of the sternum to the pubic bones of the pelvic girdle.

TAIL: THE CROCODYLIAN PROPELLER

Proportionally to the rest of the body, crocodylians have very large, muscular tails (Fig. 3.56). The caudal vertebrae taper gradually to the tail tip. Near the sacrum they are similar to the dorsal vertebrae, with well developed neural spines and transverse processes and, ventrally, haemal arches (Figs 3.42, 3.50). Towards the tip of the tail these processes diminish to the extent that the last few are simply elongated centra.

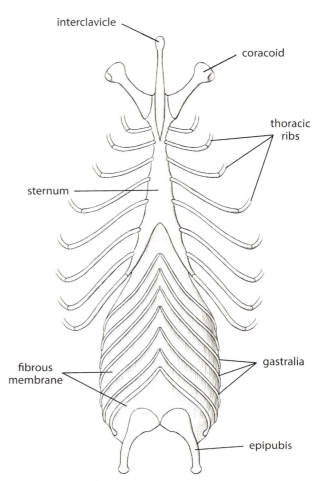

Fig. 3.55. *Ventral view of the gastralia (literally 'stomach ribs') and their interconnections via a membrane with the sternum anteriorly and the pelvic girdle posteriorly. The interclavicle and coracoids are part of the pectoral girdle.*

The dorsal musculature of the tail is basically an extension of the anterior axial musculature, with the main epaxial muscle a posterior extension of the latissimus dorsi, renamed the longissimus caudalis. Running parallel, more medially, is the much smaller tendinoarticularis muscle group (Fig. 3.57). Below the transverse processes, two large hypaxial muscles dominate: the caudofemoralis muscles (caudal femorals) medially, wrapped ventrolaterally by the ilio-ischiocaudalis muscles, which run from the pelvic girdle and insert along the shaft of the tail. The caudal femorals insert via long tendons onto the hind limb. The caudal epaxials and the ilio-ischiocaudals are noticeably segmental, comprising a series of inter-fitted cones traversing the length of several vertebrae (Frey *et al.* 1989). The apices of

Fig. 3.56. *The tail of a large* Crocodylus porosus *drowned in an illegally set fisherman's net is carted off, headed for dinner. Maningrida, Northern Territory, c. 1979. The tail has much of the body mass (and most of the meat) and is a heavy weight to manage on land (Chapter 4), but a wonderful propulsive organ in the water. The common belief that a crocodile or alligator will use the tail to sweep a victim off its feet is more myth than reality. (Photo GCG)*

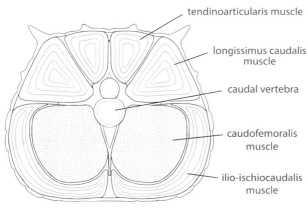

tendinoarticularis muscle

longissimus caudalis muscle

caudal vertebra

caudofemoralis muscle

ilio-ischiocaudalis muscle

Fig. 3.57. *Cross section of the tail of a typical crocodylian. Not only are these muscles important to crocs for swimming, they also stabilise the tail and the caudofemoralis muscles attached to the femora and are therefore important for walking. Tail musculature is also the main source of crocodylian meat for human consumption.*

the cones point anteriorly in the tendino-articularis muscles and posteriorly in the longissimus caudalis. Interestingly, these authors dissected tails of a good representation of crocodylians (*Alligator mississippiensis*, *Osteolaemus tetraspis*, *Paleosuchus trigonatus*, *Caiman crocodilus*, *Crocodylus acutus* and *Tomistoma schlegelii*) and noted structural differences in caudal musculature that set *Gavialis gangeticus* apart from the others, including *Tomistoma*. This is relevant to ideas about the taxonomic position of *Tomistoma* (Chapter 2).

FORELIMBS AND PECTORAL GIRDLE

Meers (2003) made the interesting observation that the forelimb of archosaurs may be one of the most functionally diverse of all tetrapod structures.

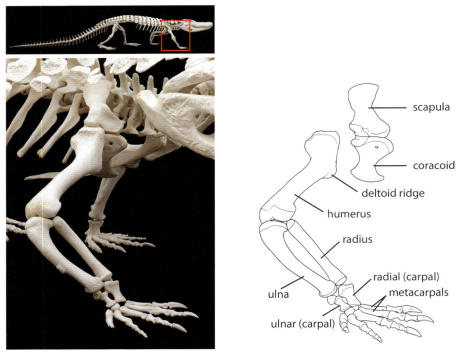

Fig. 3.58. *Pectoral girdle and forelimb of* A. mississippiensis. *(Photo Northern Territory Museum and Art Gallery, Darwin)*

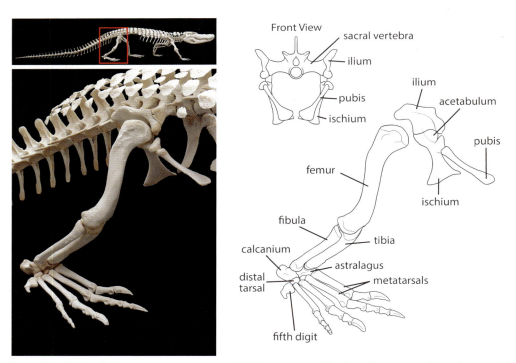

Fig. 3.59. *Pelvis and hind limb of* Alligator mississippiensis. *The ilium is secured firmly to the two sacral vertebrae and the two ischia are connected ventrally by a cartilage, providing a firm platform for the articulation of the femur. The femur fits into the acetabulum: a 'notch' made from the conjunction of the ilium and the ischium. Unlike the pubes (the third pelvic element) in most tetrapods, the crocodylian pubes are moveable and do not contribute to the acetabulum. They project ventrally and are expanded towards the tips. Muscles draw the pubes ventrally during inspiration (Fig. 7.9). (Photo Northern Territory Museum and Art Gallery, Darwin)*

It has evolved for running (in both early and modern crocodylomorphs), swimming (paddles in thallattosuchians), as pillars to support sauropods in their graviportal locomotion and, twice, for flight, in pterosaurs and birds. Nevertheless, in the modern crocodylians the forelimbs are of the fairly standard pentadactyl type (Fig. 3.58) without substantial specialisation, apart from a reduction in claws.

The humerus of each forelimb articulates with a socket in the pectoral girdle, which is a sturdy bony 'shoulder' platform, set in muscle and not connected to the vertebral column. The girdle has, on each side, a blade-like scapula (homologous to our own) and a coracoid (which we lack), curving ventrally around the anterior thorax to meet the mid-line interclavicle. The connection with the interclavicle is loose and the girdle is not fused to the axial skeleton, but held in place by muscles and fascia. The socket into which the humerus fits, the glenoid fossa, is formed where the scapula and coracoid join and is made more effective by part of the scapula arching over it (Fig. 3.58). The pectoral girdle was described by Reese (1915), Kälin (1929) and Nicholls and Russell (1985).

The musculature of the forearm and pectoral girdle has been described by Chiasson (1962) and also Richardson *et al.* (2002). The most detailed description of the forearm muscles is by Mason Meers (Meers 2003). He examined a phylogenetically diverse sample of extant species (*Alligator mississippiensis*, *Crocodylus siamensis*, *Crocodylus acutus*, *Osteolaemus tetraspis* and *Gavialis gangeticus*) and found the musculature to be highly conserved, with subtle differences in the position of origin and insertion of various muscles. Perhaps the most striking difference was found in *Gavialis*, in which one triceps was enlarged, perhaps in relation to its unusual habit of reaching out with all four limbs and then drawing itself forwards (Bustard and Singh 1977 and Chapter 4).

HINDLIMBS AND PELVIC GIRDLE

Unlike the pectoral girdle, the pelvic girdle is connected firmly to the vertebral column, its dorsal blade fused to the two sacral vertebrae (Fig. 3.59).

This is characteristic of tetrapods, as is a 'tri-radiate' structure of its three elements – the ilium, ischium and pubis – with the socket-like acetabulum at their junction forming an articulation point for the head of the femur. The elements of the pelvic girdle are sites for attachment of the muscles that move the hind limbs. The basic tri-radiate pattern has evolved differently in different tetrapod groups: a good example is provided by the characteristically different pelvic girdles in the two major groups of dinosaurs, saurischia (lizard-hipped) and ornithischia (bird-hipped). An unusual feature of crocodylians is that the pubis (plural pubes) does not form part of the acetabulum. A minor controversy that dates back to 1887 is an opinion that the knob on the ischium that forms the ventral side of the acetabular cup is actually a fused pubis and that the bone usually identified as the pubis should properly be recognised as a pre-pubis. This has been refuted by Claessens and Vickaryous (2012), based on their study of the embryological development of the pelvic girdle in American alligators. The ischia are expanded ventrally, left and right elements joined by a cartilaginous connection, so forming a stable platform with which the limbs articulate. Pubes too expand ventrally and, apart from being important sites for muscle attachment, are moveable and have a role in respiration, rotating ventrally to assist inspiration (Chapter 7).

Like the forelimbs, the hindlimbs also show a standard pentadactyl pattern, but with a reduction to four digits and the absence of a claw on the fourth digit. They bear most of the weight of the body when a crocodylian is high walking and they function over a range of different postures, from sprawling in the 'belly run' to being fully extended during high walking, and providing the driving force in galloping (Chapter 4). This functional flexibility has provoked interest in their muscle anatomy and, because of its relevance to the evolution of locomotion in archosaurs, there has been comprehensive study of the musculoskeletal system of the pelvis and hind limbs (Gatesy 1997; Reilly and Elias 1998; Reilly and Blob 2003; Hutchinson and Gatesy 2000; Reilly *et al.* 2005; Allen *et al.* 2010). These studies have focussed on

American alligators, but Otero *et al.* (2010) provided a detailed description of the pelvic musculature of *Caiman latirostris*, reporting comparatively minor differences from alligators and it is likely that results obtained from alligators will apply to other species as well. One point of interest has been in comparing the mechanics of locomotion in sprawling and semi-erect walking, because it is unusual to find these two types practised by the same animal. It is often considered that in the evolution of tetrapod locomotion there has been a progression from sprawlers (amphibians and squamates) to animals that move with an erect stance (mammals and dinosaurs). However, in a very comprehensive study of hindlimb kinematics, patterns of muscle activity and femoral loading patterns in alligators, Reilly *et al.* (2005) concluded that their characteristics are derived and are not representative of any ancestral semi-erect condition.

REFERENCES

Colbert EH (1946) The Eustachian tubes in the Crocodilia. *Copeia* **1946**, 12–14.

Alibardi L (2011) Histology, ultrastructure, and pigmentation in the horny scales of growing crocodilians. *Acta Zoologica (Stockholm, Sweden)* **92**, 187–200.

Allen V, Ellsey R, Jones N, Wright J, Hutchinson JR (2010) Functional specialisation and ontogenetic scaling of limb anatomy in *Alligator mississippiensis. Journal of Anatomy* **216**, 423–445.

Bellairs A d'A, Shute CCD (1953) Observations on the narial musculature of Crocodilia and its innervation form the sympathetic system. *Journal of Anatomy* **87**, 367–378.

Bona P, Desojo JB (2011) Osteology and cranial musculature of *Caiman latirostris* (Crocodylia: Alligatoridae). *Journal of Morphology* **272**, 780–795.

Brazaitis P (1987) Identification of crocodilian skins and products. In *Wildlife Management: Crocodiles and Alligators.* (Eds GJW Webb, SC Manolis and PJ Whitehead) pp. 373–386. Surrey Beatty & Sons, Sydney.

Brazaitis P (2001) *A Guide to the Identification of the Living Species of Crocodilians.* Science Resource Center, Wildlife Conservation Society, New York.

Brochu CA (1999) Phylogeny, systematics, and historical biogeography of Alligatoroidea. Society of Vertebrate Paleontology Memoir **6**, 69–100.

Busbey AB III (1989) Form and function of the feeding apparatus of *Alligator mississippiensis. Journal of Morphology* **202**, 99–127.

Bustard HR, Singh LAK (1977) Studies on the Indian gharial, *Gavialis gangeticus* (Gmelin) (Reptilia, Crocodilia). Change in terrestrial locomotory pattern with age. *Journal of the Bombay Natural History Society* **74**, 534–536.

Cannon MS, Davis RW, Weldon PJ (1996) Dorsal glands of *Alligator mississippiensis*; a histological and histochemical study. *Journal of Zoology (London, England)* **239**, 625–631.

Chiasson RB (1962) *Laboratory Anatomy of the Alligator.* WC Brown, Dubuque, Iowa.

CITES (1995) *CITES Identification Guide – Crocodilians.* Environment Canada, Ottawa, Canada. <http://www.ec.gc.ca/Publications/default.asp?lang=En&xml=839625F8-BF8A-4169-BFB0-399233745A49>

Claessens LPAM (2009) A cineradiographic study of lung ventilation in Alligator mississippiensis. *Journal of Experimental Zoology* **311A**, 563–585.

Claessens LPAM, Vickaryous MK (2012) The evolution, development and skeletal identity of the crocodylian pelvis: revisiting a forgotten scientific debate. *Journal of Morphology* **273**, 1185–1198.

Clark JM, Xu X, Forster CA, Wang Y (2004) A Middle Jurassic 'spenosuchian' from China and the origin of the crocodylian skull. *Nature* **430**, 1021–1024.

Cleuren J, De Vree F (2000) Feeding in crocodilians. In *Feeding.* (Ed. K Schwenk) pp. 337–358. Academic Press, San Diego.

Erickson GM, Lappin AK, Vliet KA (2003) The Ontogeny of bite-force performance in

American alligator (*Alligator mississippiensis*). *Journal of Zoology* **260**, 317–327.

Erickson GM, Lappin AK, Parker T, Vliet KA (2004) Comparison of bite-force performance between long-term captive and wild captured American alligators (*Alligator mississippiensis*). *Journal of Zoology* **262**, 21–28.

Erickson GM, Gignac PM, Steppan SJ, Lappin AK, Vliet KA (2012) Insights into the ecology and *evolutionary success of crocodilians revealed through bite-force and tooth-pressure experimentation*. *PLoS ONE* **7**(3), e31781.

Farlow JO, Hayashi S, Tattersall GJ (2010) Internal vascularity of the dermal plates of *Stegosaurus* (Ornithischia, Thyreophora). *Swiss Journal of Geosciences* **103**, 173–185.

Frey E, Riess J, Tarsitano SF (1989) The axial tail musculature of recent crocodiles and its phyletic implication. *American Zoologist* **29**, 857–862.

Garrick LD (1975) Love among the alligators. *Animal Kingdom* **78**, 2–8.

Gatesy SM (1997) An electromyographic analysis of hindlimb function in *Alligator* during terrestrial locomotion. *Journal of Morphology* **234**, 197–212.

Grigg GC, Gans C (1993) Crocodilia: morphology and physiology. In *Fauna of Australia. Volume 2A. Amphibia and Reptilia*. (Eds CJ Glasby, GJB Ross and BL Beesley) pp. 326–336. Australian Government Publishing Service, Canberra.

Haines RW (1969) Epiphyses and sesamoids. In *Biology of the Reptilia Volume 1*. (Eds C Gans, AD Bellairs and TS Parsons) pp. 81–116. Academic Press, New York.

Heaphy L (1985) A study of the dorsal organs and cutaneous papillae of the Australian crocodiles. BSc (Hons) thesis, University of New South Wales, Sydney.

Hoffstetter R, Gasc J-P (1969) Vertebrae and ribs of modern reptiles. In *Biology of the Reptilia Volume 1*. (Eds C Gans, AD Bellairs and TS Parsons). pp. 201–310. Academic Press, New York.

Hoh JFY (2002) 'Superfast' or masticatory myosin and the evolution of jaw-closing muscles of vertebrates. *Journal of Experimental Biology* **205**, 2203–2210.

Hoh JFY, Lim JHY, Kang LDH, Lucas CA (2001) Expression of superfast myosin in the jaw-closing muscles of *Crocodylus porosus*. In *Crocodilian Biology and Evolution* (Eds GC Grigg, F Seebacher and CE Franklin) pp. 156–164. Surrey Beatty & Sons, Sydney.

Hutchinson JR, Gatesy SM (2000) Adductors, abductors, and the evolution of archosaur locomotion. *Paleobiology* **26**, 734–751.

Iordansky NN (1973) The skull of the Crocodilia. In *Biology of the Reptilia Volume 4*. (Eds C Gans and TS Parsons) pp. 201–260. Academic Press, New York.

Iordansky NN (2000) Jaw muscles of the crocodiles: structure, synonymy, and some implications on homology and functions. *Russian Journal of Herpetology* **7**, 41–50.

Johnsen PB, Wellington JL (1982) Detection of glandular secretions by yearling alligators. *Copeia* **1982**, 705–708.

Jackson DC, Andrade DV, Abe AS (2003) Lactate sequestration by osteoderms of the broad-nose caiman, *Caiman latirostris*, following capture and forced submergence. *Journal of Experimental Biology* **206**, 3601–3606.

Kälin JA (1929) Uber den Brustschulterapparat der Krokodile. *Acta Zoologica Stockholm* **10**, 343–399.

Kieser JA, Klapsidis C, Law L, Marion M (1993) Heterodonty and patterns of tooth replacement in *Crocodylus niloticus*. *Journal of Morphology* **218**, 195–201.

Kirshner D (1985) Buoyancy control in the estuarine crocodile, *Crocodylus porosus* Schnieder. PhD thesis. University of Sydney.

Kofron CP (1991) Courtship and mating of the Nile crocodile (*Crocodylus niloticus*). *Amphibia-Reptilia* **12**, 39–48.

Krückert K, Flachsbarth B, Schulz S, Hentschel U, Weldon PJ (2006) Ethyl-branched aldehydes, ketones, and diketones from caimans (*Caiman* and *Palaeosuchus*; Crocodylia). *Journal of Natural Products* **69**, 863–870.

Langston W (1973) The crocodilian skull in historical perspective. In *Biology of the Reptilia. Volume 4*

(Eds C Gans and TS Parsons) pp. 263 – 289. Academic Press, New York.

Mason RT, Parker MR (2010) Social behaviour and pheromonal communication in reptiles. *Journal of Comparative Physiology. A, Neuroethology, Sensory, Neural, and Behavioral Physiology* **196**, 729–749.

McHenry CR, Clausen PD, Daniel WJT, Meers MB, Pendharkar A (2006) Biomechanics of the rostrum in crocodilians: A comparative analysis using finite element modeling. *Anatomical Record* **288A**, 827–849.

McIlhenny EA (1935) *The Alligator's Life History.* Christopher Publishing House, Boston, Massachusetts.

Meers MB (2003) Crocodylian forelimb musculature and its relevance to Archosauria. *Anatomical Record Part A* **274**, 892–916.

Murakami G, Akita K, Sato, T (1991) Arrangement and innervation of the iliocostalis and longissimus muscles of the brown caiman (*Caiman crocodilus fuscus*: Alligatoridae, Crocodilia) American Journal of Anatomy 192, 241–256.

Nesbitt SJ, Turner AH, Weinbaum JC (2013) A survey of skeletal elements in the orbit of Pseudosuchia and the origin of the crocodylian palpebral. *Earth and Environmental Science Transactions of the Royal Society of Edinburgh* **103**, 365–381.

Nicholls EL, Russell AP (1985) Structure and function of the pectoral girdle and forelimb of *Struthiomimus altus* (Theropoda: Ornithomimidae). *Palaeontology* **28**, 643–677.

Otero A, Gallina P, Herrera Y (2010) Pelvic musculature and function of *Caiman latirostris. The Herpetological Journal* **20**, 173–184.

Parrish JM (1987) The origin of crocodilian locomotion. *Paleobiology* **13**, 396–414.

Pierce SE, Angielczyk KD, Rayfield EJ (2008) Patterns of morphospace occupation and mechanical performance in extant crocodilian skulls: a combined geometric morphometric and finite element modelling approach. *Journal of Morphology* **269**, 840–864.

Poole DFG (1961) Notes on tooth replacement in the Nile crocodile *Crocodylus niloticus. Proceedings of the Zoological Society of London* **236**, 131–140.

Pooley AC (1962) The Nile Crocodile, *Crocodylus niloticus. The Lammergeyer* **2**, 1–5.

Reilly SM, Blob RW (2003) Motor control locomotor hindlimb posture in the American alligator (*Alligator mississippiensis*). *Journal of Experimental Biology* **206**, 4327–4340.

Reese AM (1915) *The Alligator and its Allies.* GP Putnam's Sons, New York.

Reese AM (1921) The structure and development of the integumental glands of crocodilian. *Journal of Morphology* **35**, 581–611.

Reilly SM, Elias JA (1998) Locomotion in *Alligator mississippiensis* kinematic effects of speed and posture and their relevance to the sprawling-to-erect paradigm. *Journal of Experimental Biology* **201**, 2559–2574.

Reilly SM, Willey JS, Biknevicius AR, Blob RW (2005) Hindlimb function in the alligator: integrating movements, motor patterns, ground reaction forces and bone strain of terrestrial locomotion. *The Journal of Experimental Biology* **208**, 993–1009.

Rowlerson A (1994) An outline of fibre types in vertebrate skeletal muscle: histochemical identification and myosin isoforms. *Basic and Applied Myology* **4**, 333–352.

Richardson KC, Park JY (2001) The histology of the dorsal integumentary glands in embryonic and young Estuarine crocodiles, *Crocodylus porosus* and Australian Freshwater crocodiles, *Crocodylus johnstoni*. In *Crocodilian Biology and Evolution* (Eds GC Grigg, F Seebacher and CE Franklin) pp. 180–187. Surrey Beatty & Sons, Sydney.

Richardson KC, Webb GJW, Manolis SC (2002) *Crocodiles Inside Out.* Surrey Beatty & Sons, Sydney.

Salisbury SW, Frey E (2001) A biomechanical transformation model for the evolution of the semi-spheroidal articulations between the adjoining vertebral bodies in crocodilians. In *Crocodilian Biology and Evolution.* (Eds GC Grigg,

F Seebacher F and CE Franklin) pp. 85–134. Surrey Beatty & Sons, Sydney.

Salisbury SW, Frey E (2004) Anatomical correlates associated with the bracing system of extant crocodilians: addressing the locomotor inadequacies of the Indian gharial. *Proceedings of the 17th Working Meeting of the IUCN–SSC Crocodile Specialist Group. 24–29 May, Darwin. p. 394.* IUCN Gland, Switzerland.

Salisbury SW, Molnar RE, Frey E, Willis P M. A. (2006) The origin of modern crocodyliforms: new evidence from the Cretaceous of Australia. Proceedings of the Royal Society of London, Series B, **273**, 2439–2448.

Schour I, Hoffman MH (1939) Studies in tooth development: I. the 16 microns calcification rhythm in the enamel and dentin from fish to man. *Journal of Dental Research* **18**, 91–102.

Schumacher GH (1973) The head muscles and hyolaryngeal skeleton of turtles and crocodilians. In *Biology of the Reptilia. Volume 4* (Eds C Gans and TS Parsons) pp. 101–200. Academic Press, New York.

Seidel MR (1979) The osteoderms of the American alligator and their functional significance. *Herpetologica* **35**, 375–380.

Shute CCD Bellairs A d'A (1955) The external ear in crocodilia. Proceedings of the Zoological Society of London **124**, 741–749.

Tsai HP, Holliday CM (2011) Ontogeny of the alligator cartilago transiliens and its significance for sauropsid jaw muscle evolution. *PLoS ONE* **6**(9), e24935.

Underwood G (1970) The eye. In *Biology of the Reptilia Volume 2.* (Eds C Gans and TS Parsons) pp. 1–97. Academic Press, New York.

Vickaryous MK, Hall BK (2008) Development of the dermal skeleton in *Alligator mississippiensis* (Archosauria, Crocodylia) with comments on the homology of osteoderms. *Journal of Morphology* **269**, 398–422.

Vliet KA (1989) Social displays of the American Alligator (*A. mississippiensis*). *American Zoologist* **29**, 1019–1031.

Vliet KA (2001) Courtship behaviour of American Alligators, *Alligator mississippiensis*. In *Crocodilian Biology and Evolution* (Eds GC Grigg, F Seebacher and CE Franklin) pp. 383–408. Surrey Beatty & Sons, Sydney.

Walmsley CW, Smits PD, Quayle MR, McCurry MR, Richards HS, Oldfield CC, *et al.* (2013) Why the long face? The mechanics of mandibular symphysis proportions in crocodiles. *PLoS ONE* **8**(1), e53873.

Weldon PJ, Flachsbarth B, Schulz S (2008) Natural products from the integument of nonavian reptiles. *Natural Product Reports* **25**, 738–756.

Weldon PJ, Ferguson MWJ (1993) Chemoreception in crocodilians: anatomy, natural history, and empirical results. *Brain, Behavior and Evolution* **41**, 239–245.

Weldon PJ, Sampson HW (1987) Paracloacal glands of *Alligator mississippiensis*: a histological and histochemical study. *Journal of Zoology* **212**, 109–115.

Weldon PJ, Sampson HW (1988) The gular glands of *Alligator mississippiensis*: histology and preliminary analysis of lipoidal secretions. *Copeia* **1988**, 80–86.

Weldon PJ, Wheeler JW (2001) The chemistry of crocodilian skin glands. In *Crocodilian Biology and Evolution* (Eds GC Grigg, F Seebacher and CE Franklin), pp. 286–296. Surrey Beatty & Sons, Sydney.

Weldon PJ, Scott TP, Tanner MJ (1990) Analysis of gular and paracloacal gland secretions of the American Alligator (*Alligator mississippiensis*) by thin layer chromatography: gland, sex, and individual differences in lipid components. *Journal of Chemical Ecology* **16**, 3–12.

Westergaard B, Ferguson MWJ (1990) Development of the dentition in *Alligator mississippiensis*: Upper jaw dental and craniofacial development in embryos, hatchlings, and young juveniles, with a comparison to lower jaw development. *The American Journal of Anatomy* **187**, 393–421.

Whyte A, Yang Z-C, Tiyanont K, Weldon PJ, Eisner T, Meinwald J (1999) Reptilian chemistry:

characterization of dianeackerone, a secretory product from a crocodile. *Proceedings of the National Academy of Sciences of the United States of America* **96**, 12246–12250.

Wright DE, Moffat LA (1985) Morphology and ultrastructure of the chin and cloacal glands of juvenile Crocodylus porosus (Reptilia, Crocodilia). In *Biology of Australian Frogs and Reptiles* (Eds GC Grigg, R Shine and H Ehmann) pp. 411–422. Surrey Beatty & Sons, Sydney.

4

LOCOMOTION, BUOYANCY AND TRAVEL

It is an impressive and memorable sight to see a very large crocodile travelling so fast that it is 'up on the plane', surfing on its chest, driven by the powerful tail and leaving a wake of roiled water! If it is surfing straight at you when you are thigh deep in tannin-stained water, it is even more impressive! The occasion was in 1977. Bill Magnusson, Janet Taylor and I were on an aerial survey for crocodile nesting habitat and were staying with the legendary Northern Territory wildlife ranger Dave Lindner on Coburg Peninsula. Dave wanted to capture and mark a 4 m C. porosus *whose tracks he had seen crossing back and forth between the sea and this shallow freshwater lagoon behind the dunes. His plan was to put a net across the seaward end and we, being naïve in those days, would walk down the lagoon in a line, prodding the bottom with sticks and spears in the hope of driving the croc ahead of us towards possible escape into the sea, so entangling itself in the net.*

The plan was put into action. We walked in a line. We prodded. The dark water hid anything it might contain. Bubbles rose from the soft bottom as we walked. We were now well down the lagoon. I doubted there was any croc there at all, until suddenly – there was! Dave threw his spear. But instead of escaping to the sea, the croc headed for us, specifically at me! I had a couple of seconds' view of an enormous head and teeth coming straight at me, right up out of the water, surfing on a froth-laden, hissing bow wave, the front third of his enormous body being powered over the surface by his tail. And then he veered slightly,

sped past me, and sank into the dark water behind us, out of sight.

A more colourful account of this has been written (Magnusson 1990), but I cannot agree with the description that I 'appeared to be running on the surface of the water'. My recollection is that I was frozen to the spot, and it all happened far too quickly for me to even think about running!

Good locomotor skills are very important for crocodylians. They affect the capacity to escape predation, to find and capture food, to choose a place to live and to find a mate. They are most at home in the water, yet they also have a wide range of terrestrial gaits. Their original body design evolved for life on land (Parrish 1987; Reilly and Elias 1998), one sign of which is the way the vertebral column is braced in a way that is better suited for locomotion on land (Chapter 3). Their ancestors were long legged, loping, cursorial (running) land animals (Chapter 2), which might even have been endothermic (Chapter 10) and from which they evolved into ectothermic, aquatic vertebrates whose body shape is much more efficient for swimming. In adapting to aquatic life, they came to rely on an eel-like tail instead of limbs to provide the main propulsive force, and the limbs became much reduced. The shift involved a large change in musculature too: much aquatic propulsion depends on the axial musculature, whereas terrestrial

---

Nile crocodiles, Crocodylus niloticus, *demonstrating the control crocodylians have over their buoyancy. (Illustration DSK, courtesy Weldon Owen Publishing)*

locomotion depends on the paraxial musculature of the limbs and associated pectoral and pelvic girdles.

The movements of crocodylians on land are often said to be clumsy or ungainly, whereas in the water crocs are superbly elegant and look beautifully at home. In the water, they can float, sink, swim several styles, jump, roll, dive and remain underwater for long periods (see Chapter 9).

CROCODYLIANS ON LAND

Despite their evolution into the water, all crocodylians go ashore to lay eggs and many spend time there guarding their nests (Chapter 12). Most also leave the water to warm the body by basking (Chapter 10). Most feed in water, at the water's edge, sometimes with a brief chase onto land (Chapter 6). Some species travel overland routinely and sometimes extensively. *Crocodylus johnstoni* often travel between pools in the dry season (Webb and Gans 1982) and *C. niloticus* make journeys well away from water to feed (Cott 1961;

Graham 1968). *C. palustris*, too, hunt on land and make long seasonal trips on land between water bodies (Whitaker and Andrews 1989). In some seasonally drying habitats, crocodylians may migrate considerable distances overland to seek pools that still contain water. One species famous for its overland travel is *Caiman yacare*, which, in the enormous wetland of the Pantanal in South America, often congregates in enormous numbers during the dry season around surviving water bodies (Fig. 4.2).

Campos *et al.* (2003) have described groups of caiman moving overland between pools, sometimes even travelling in a line! Some journeys are apparently prompted by human disturbance, such as may be caused by a group of hunters or researchers, and the largest group Zilca Campos reported had 50 individuals! However, she has also found solitary travellers (Fig. 4.3).

The least terrestrial crocodylian is the gharial, which, proportionally, has the smallest limbs. Although juvenile gharials can 'high walk', once

Fig. 4.1. Crocodylus porosus *moving off the draining floodplain of the South Alligator River, Northern Territory, to a tidal gutter and thus to the estuary. They are sliding on their bellies and using their feet for propulsion. This sprawling gait could be called a sprawling walk or a belly run, depending upon speed. (Photo Buck Salau)*

Fig. 4.2. *Numerous* Caiman yacare *around a pond in the Brazilian Pantanal. Zilca Campos and colleagues in Brazil have written about caiman travelling overland to find pools still holding water as the country dries up after the wet season (Campos* et al. *2003). (Photo DSK)*

they have grown to more than ~2 m, they seem capable only of a sprawling belly slide, pushing forward with four limbs in synchrony in a manner reminiscent of marine turtles (Bustard and Singh 1977). Nevertheless, a 2.3 m gharial climbed out of its pen at the Madras Crocodile Bank in Tamil Nadu, India and slid 650 m, apparently surprising everybody (Whitaker and Andrews 1989).

Apart from excursions on land for particular reasons, crocs obviously regard the water as 'home'. They seek refuge there, hide there, relax there and, possibly, sleep there. They also feed there and mate and have most of their social life there.

GAITS ON LAND

On land, crocodylians have a wide variety of gaits. Allen *et al.* (2010) noted that they are the only extant tetrapods known to use such a diversity of locomotory patterns on land, from laterally undulating 'sprawling' gaits to more erect walking, plus bounding and galloping. They can propel themselves on their bellies using their feet (the sprawling walk or, if faster, the belly run). They can lift themselves off the substrate and walk or trot (the high walk) and, most spectacularly, several species can bound or gallop (they are the only reptiles that do). They can also transition between these gaits. Among the first authors to categorise gaits in crocodylians were Cott (1961), Zug (1974), Webb and Gans (1982) and Frey (1984). Let's look at each gait in turn.

High walk

The normal gait on land is a 'high walk', seen when a croc hauls itself up out of the water to bask, or to visit its nest or, indeed, whenever they travel on land (Figs 4.4, 4.5). It is a slow mode of travel. Large individuals may travel comfortably at 2–4 km/h but they stop frequently to rest. For a short distance they can speed up to ~12–14 km/h (Adam Britton *pers. comm.*). The increase in speed is achieved by increasing both step rate and stride length (Reilly

Fig. 4.3. *A solitary* Caiman yacare *out for an early evening walk in the Brazilian Pantanal, crossing from one water body to another. Its skin has become polished from pushing through the dry grass. The animal has stopped and lowered itself, so it is not showing the high walk it had been displaying when we chanced upon it. (Photos GCG)*

and Elias 1998). Technically, the high walk is a symmetrical walking trot, a gait in which alternate diagonal limbs step nearly in synchrony. The limbs are extended, held almost in under the trunk and hold the body well up off the ground, except for the tail, which drags. There are always at least two feet on the ground (Willey *et al.* 2004). The right front and hind left move forward simultaneously, then the alternate pair, with only a little lateral flexing of the trunk. The hind feet touch the ground just a little ahead of the front feet in each step cycle (Willey *et al.* 2004). The vertebral column and its associated

Fig. 4.4. *High walking in (top left to bottom right)* Crocodylus johnstoni, C. porosus, C. palustris *and* Alligator mississippiensis. *This is the usual method of travel on land if there is no need to hurry. The heavy tail is a bit of a liability on land, throwing additional weight onto the hind legs and providing drag. In the water, the tail is the main source of propulsion. (Photos Ruchira Somaweera, Bill Green, Brandon Sideleau, DSK)*

muscles and tendons brace the trunk against the tendency to flex axially (see Chapter 3). The tail is heavy and the centre of gravity is just in front of the hind limbs (Fig. 4.6), so the hind limbs carry most of the weight of the body and generate the propulsive force.

The tail – such a wonderful propeller in the water – is a dead weight on land. Willey *et al.* (2004) measured force dynamics of 2–4 kg alligators walking over a force platform and found that the drag of the tail is substantial, giving the hind legs (mostly) and also the front legs extra work. The hindlimbs also have to cope with lifting some of the weight of the tail, which Willey and colleagues found to be nearly 28% of bodyweight in small alligators.

Belly run (or belly slide, sprawling walk)

The belly run, or belly slide (Fig. 4.7), technically a sprawling gait, is usually seen on mud or some other wet substrate (e.g. Fig. 4.1). It can be slow and measured or very fast, such as when a croc is disturbed high on a river bank and rushes headlong downslope into the water, seeking refuge. A fast belly run is used over only short distances, usually when escaping a threat. The belly and tail remain on the substrate and propulsion comes from the toes gripping the substrate as they pull themselves along, assisted a little by lateral flexion of the trunk. This is often accompanied by the tail thrashing from side to side as if ready to start swimming the moment the croc hits the water. Speeds of 8–10 km/h are common in a belly run.

A frequently asked question is whether a human can run faster than a croc – we can, but the commonly held misconception that zig-zagging is helpful is not correct. In reality, it is most unlikely that a human would be actually chased by a croc: that is not their normal habit. In hunting, crocodylians usually rely on surprise, ambushing prey at the water's edge and most humans that are attacked are caught there or in the water, not on land (see Chapter 14). Maximum speeds reached by a crocodylian on land are likely to be when it is heading for the water in escape.

Fig. 4.5. *A Nile crocodile at St Lucia Estuary, Kwa Zulu Natal, turning in a high walk. The red colouration in the teeth is due to minerals in the water. (Photo Louis Guillette)*

Bounding/galloping

The fastest gait employed by crocs is usually referred to as a gallop, although technically it is bounding. In bounding the two front feet strike the ground simultaneously, as do the hind feet. Galloping can be thought of as asymmetrical bounding, with the two forefeet striking the ground sequentially, and the hind feet likewise. For convenience, the term galloping will be used here. It is a gait that only some species are known to use and it is usually seen in nature when they are escaping from a threat, but it could be effective in a chase too. The term is essentially self-explanatory (Figs 4.8, 4.9): a galloping croc bounds along, alternating between weight being on its hind feet and front feet, each pair hitting the ground nearly, but not necessarily in perfect synchrony, with the tail appearing to act as a counterbalance (which seems not yet to have been analysed biomechanically). When the hind feet are drawn forward for the next step cycle, there is a small dorsoventral flexure to arch the vertebral

column and the hind feet reach forward to almost meet or even precede the front feet. The movement has been likened to the way a hare gallops. A croc can commence this gait from a standing start, such as in response to being startled, or can make the transition from a high walk. Galloping was described originally by Hugh Cott (1961) in small (< 2 m) *Crocodylus niloticus* and George Zug (1974) subsequently described it in a 45 cm *C. porosus* at a crocodile farm outside Port Moresby (although it occurs uncommonly in this species). Whitaker and Andrews (1989) reported it in *C. novaeguineae*, *C. palustris* and *Osteolaemus tetraspis* and refer to its occurrence also in juvenile *Gavialis gangeticus*. The only report I could find of galloping in an alligatorid was by Reilly and Elias (1998) who noted that *Alligator mississippiensis* on a treadmill attempted to gallop, for part of a stride, when their treadmill was at speed. The most galloping-prone by far is *Crocodylus johnstoni*. If one is released 20–30 m from, and in sight of, the water it is highly likely to oblige by

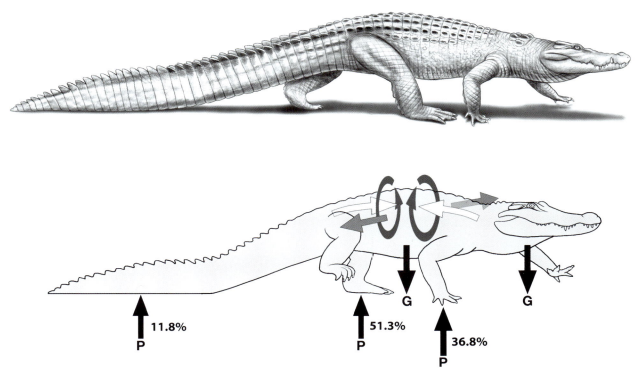

Fig. 4.6. *With such a heavy tail, the centre of gravity is well back towards the hind legs in the lumbar region and these take the 'lion's share' of the weight and have to contribute disproportionally to thrust as well, in order to drag the tail. Data from 2–4 kg alligators show that the hindlimbs support about half the body weight, the front legs 37% and the tail the rest (Willey* et al. *2004). Similar data are unavailable for any large crocodylian, but can be expected to be similar. The vertical components of the propulsive forces (P) balance gravitational forces (G) and generate alternating shear forces with each step, which are countered by muscular bracing of the vertebral column (Chapter 3). (Lower panel modified from Salisbury and Frey 2001, with permission).*

demonstrating its gallop. Speeds are size dependent: a 1 m individual might travel at ~18 km/h (Webb and Gans 1982), but they cannot keep it up for more than 30–50 m. Most sightings of crocodylians galloping have been of relatively small species or small individuals. Allen *et al.* (2010) have probably explained why this is so, by measuring ontogenetic changes in alligator muscle mass and architecture, which imply a decrease in terrestrial athleticism as they grow. Although galloping is usually seen only in small or immature individuals, I once surprised an adult female *C. porosus* on her nest as I was walking alongside the East Alligator River, Northern

Territory (the nest had been spotted from the air, see Fig. 12.38) and she bolted down to the water along her track through the grass to the water. The grass was nearly 1 m high, so she was mostly obscured, but she looked me right in the eye as she passed and the pitching movement of her front end with each stride convinced me she was galloping.

The original work on this gait by Zug (1974) and subsequent work by Renous *et al.* (2002) reveal that

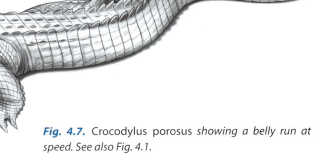

Fig. 4.7. Crocodylus porosus *showing a belly run at speed. See also Fig. 4.1.*

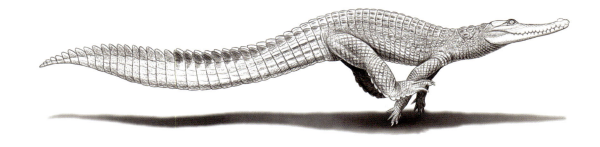

Fig. 4.8. *Galloping (technically bounding) is the fastest gait employed by crocodylians and is most often employed by juveniles escaping to water, as demonstrated here by (top and lower left)* Crocodylus johnstoni *and (lower right)* C. niloticus. *(Photos GCG, Matthew H. Shirley)*

it is rather more complicated than it looks, and the order in which the feet strike the ground is variable. A detailed analysis on seven captive *Crocodylus johnstoni* in the Frankfurt Zoo by the latter authors showed that as well as front and hind pairs of feet striking the ground simultaneously (bounding), they also showed a 'transverse gallop', which is often seen in horses. In this, the lead hind foot and lead front foot in each step cycle are on the same side (e.g. hind LR, front LR). The crocs also showed 'rotary galloping', which is common in dogs, where the lead hind foot and lead front foot in each step cycle are on the opposite sides (e.g. hind LR, front RL, so the foot order 'rotates'). Zug (1974) reported the same thing in *C. porosus*. It remains to be determined whether or not these variants are biologically meaningful.

My suspicion is that, as time goes by, other species will be found to show this type of gait. More and more people carry compact video cameras with them and many love to put croc images onto the worldwide web. Maybe one day someone will validate my 'almost' observation of galloping in an adult *Crocodylus porosus*!

Several things are certain: such a gait in crocodylians is rapid, exhausting, and also gives them good capacity to jump small obstructions and make their way over rough ground much more quickly than they could in either a high walk or a belly run. As a general rule, a crocodylian on land can go about twice as fast in a belly run as in a high walk, and twice as fast again in a gallop.

Evolutionary history of walking in crocs

Parrish (1987) concluded, from morphological studies of the foot and limb structure that the earliest crocodylians, the crocodylomorphs (see Chapter 2), were adapted for an erect posture, which he considered to be pleisiomorphic (see Box 2.1) for the archosaurs. The implication he drew was that the modern crocodylians are aquatic secondarily and had ancestors that walked erect. An analysis of the mechanics of walking in alligators by Reilly and Elias (1998) has confirmed this conclusion and revealed some additional evolutionary implications. Whereas mammals (and dinosaurs) are considered to be 'erect' in their walking gait, with their limbs underneath,

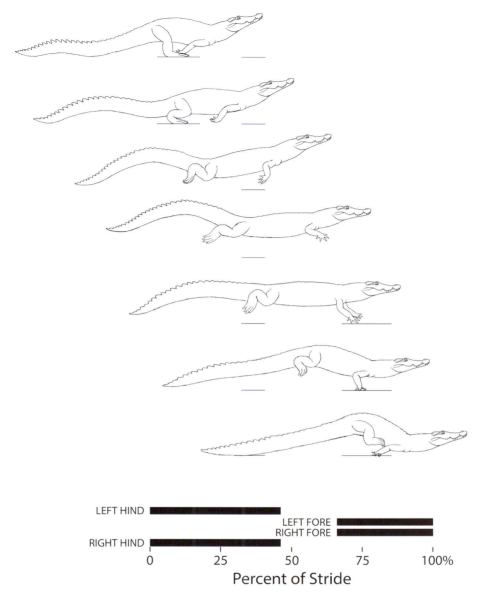

Fig. 4.9. *A single stride made by a small crocodile at full gallop. The most habitual galloper is* Crocodylus johnstoni, *but its first analysis, shown here, was of a juvenile* C. porosus. *(Modified from tracings by George Zug with his and the publisher's permission, from a cine sequence shot by him in Papua New Guinea; Zug 1974).*

and lizards and amphibians (such as salamanders) are considered to be 'sprawlers', with their limbs out to the sides, crocs are sometimes referred to as 'semi-erect', with their high walk gait being thought of as intermediate between sprawling and erect. Because they use a sprawling gait as well, and as mammals have evolved from sprawlers, Reilly and Elias (1998) studied the kinematics of walking in alligators to see if crocodylians are, indeed, intermediate. If so, this could provide a model for investigating the evolutionary transition from the sprawling gait typical of reptiles to the erect stance and gait of mammals. However, they found that the crocodylian sprawl is functionally quite different from the primitive sprawl locomotion shown by lizards and salamanders and is mechanically not very different from the high walk. It might be better termed a 'low walk'.

That many of the crocodylian ancestors were once long-legged terrestrial cursorial beasts with an erect stance was discussed in Chapter 2 and is, of course, quite exciting to those of us who suspect that ancient crocodylians were endotherms (see Chapter 10).

CROCODYLIANS IN THE WATER

Cott (1961) was one of the earliest scientists in modern times to wax lyrical about a crocodylian's mastery of aquatic locomotion and there have been many since. This section will review their skills in aquatic locomotion, before discussing one of the most intriguing puzzles: their buoyancy and the role of the stomach stones.

TYPICAL POSTURES AT REST IN WATER

At rest in water, crocs typically have only the eyes, cranial platform, ears and nostrils protruding (Figs 4.10, 4.11 and Chapter 5). As water deepens, the front legs lose contact with the bottom but the waterline around the head is maintained. If the water deepens further, the hindlimbs extend slightly to maintain contact as long as possible. In water too deep to allow bottom contact, the body angle steepens, the tail droops, the hind legs spread wide for stability and make rowing movements to counteract disturbances, and there are only small ventilatory movements of flanks and abdomen. They can also float with the body parallel to the surface, maintaining the same waterline around the head. If they want to send a message to other crocs, or for some other reason, they can choose to float higher in the water, and they do that with equal skill (Figs 4.12, 4.13). In rough water they float higher, as if to keep the nostrils clear of the water. Their management of buoyancy is spectacularly good and its control will be discussed later in the chapter.

SWIMMING

Crocodylians are more at home in water than on land, and wonderful to watch (Figs 4.14, 4.15). Much of their general aquatic slow manoeuvring is brought about by paddling with the limbs, often in combination with flexure of the body and neck. Davenport and Sayer (1989) analysed some of these movements by hatchling *C. porosus*, in addition to swimming *per se*, and provided charming diagrams of them 'treading water' as they caught and

Fig. 4.10. *An alligator in Big Cypress National Preserve, Florida floats with just its eyes, ears (and skull platform) and nostrils out, with the rest of its body underwater, out of sight from potential prey. The 'iceberg effect' is evident, demonstrating what a small proportion of the animal needs to show above the surface for it to keep tabs on its surroundings. Crocodylian waterline management is spectacularly good and they are almost impossible to see among floating or emergent vegetation. They can do this even in water too deep to feel bottom, maintaining the pose for long periods. (Photo DSK)*

swallowed insects thrown to them as food. But they also swim to travel, tucking the limbs at their sides and propelling themselves with the tail.

The kinematics of swimming has been studied in *Alligator mississippiensis* (Manter 1940; Fish 1984), in *Osteolaemus tetraspis* (Frey and Salisbury 2001) and in *Crocodylus porosus* (Davenport and Sayer 1989; Seebacher *et al.* 2003). All studies have been on either hatchlings or juveniles in captivity.

Paraxial swimming

For slow swimming (< 0.5 body lengths/second in a hatchling; Davenport and Sayer 1989), particularly

Fig. 4.11. *A large* Crocodylus porosus *watches and listens from the surface of a billabong off the South Alligator River in northern Australia, its reflection visible in the almost still water. They can stay motionless like this for long periods, their respiration proceeding imperceptibly. Meanwhile, under the surface, the integumentary sense organs on the submerged scales monitor things underwater (Chapter 5). Note the horse fly (Family Tabanidae) on the nostrils of the crocodile in this photo. These flies often suck blood from the nasal button and around the eyes and they remain attached even when the animals submerge or jump for food (Fig. 4.20). (Photo DSK)*

Fig. 4.12. *This* Melanosuchus niger *is floating higher, taking an interest in what is going on and not shy about being seen, indicated by the raised scutes along its spine (a characteristic of this species) being visible above the surface. It is in Mamirauá Reserve near Tefé on the Solimões river ~500 km upstream from Manaus, Brazil. (Photo William E. Quatman)*

Fig. 4.13. *An estuarine crocodile,* Crocodylus porosus, *in Yellow Water, Kakadu National Park, Northern Territory, in a 'high float', with its dorsum raised well above the water surface. This is often used as a visual signal to advertise status or intimidate conspecifics. Adult males may show the entire dorsal surface from the snout to the base of the tail. (Photo DSK)*

when manoeuvring, crocs use all four limbs alternately for propulsion in the normal tetrapod pattern, much the way one would expect. This is called paraxial swimming (Fig. 4.16). The front feet are only partially webbed, or not webbed at all, but the hind feet are completely webbed and able to generate greater forces. In tight manoeuvring, the body, too, can be turned.

Hybrid swimming

For slightly faster swimming (0.5–1.6 body lengths/sec in a hatchling), limb thrust is supplemented by sinusoidal movement of the tail. Frey and Salisbury (2001) named this motion 'hybrid' swimming, because it is a mixture of paraxial and axial swimming (Fig. 4.17).

Axial swimming

When swimming faster still, at 1–2 body lengths/second (hatchling), the tail is the main driving force and the limbs are held by the sides and have only a steering role. This is called axial swimming (Fig. 4.18). Most of the propulsive force comes from that part of the tail blade with a single scale crest: this has the larger surface area and its angle of attack is optimum. The body is streamlined and its surface smooth and slick, which minimises drag. Most of the weight is in the front half of the body's length, which is good hydrodynamically.

Increased speed in axial swimming derives from an increase in the frequency of the tail beat, not from an increase in amplitude (Fish 1984; Davenport and Sayer 1989; Seebacher *et al.* 2003). The head as well as the limbs are used for steering by a crocodylian under way. A slight change in the direction of the head will initiate a turn and it must also form a significant 'diving plane': small movements up or down leading to the croc angling up or down in the water column. A sudden turn can be effected by the crocodile extending only one hind foot,

Fig. 4.14. *An Everglades American alligator cruises slowly, its buoyancy control precise enough to maintain the waterline at the preferred level. (Photo DSK)*

with digits and webbing spread, in the direction to which it wishes to turn. To stop, both hind feet, and occasionally the front feet, are splayed out to the sides with the digits and webbing spread, at times facing slightly forwards. As in slow swimming, use of the limbs and body flexion can increase manoeuvrability.

Transitions between one type of swimming and another can see limbs and tail employed together, as when a rapid start up and acceleration are generated, the limbs then being folded as axial swimming takes over.

There are not a lot of data on top speeds that a swimming croc can reach. Clearly that will depend upon body size and the fastest speeds will be supported by anaerobic metabolism (Chapter 7) so it is of interest to know the fastest burst speed as well as the fastest speed that can be sustained. There are surprisingly few data, and almost none from crocodylians in the wild. Some measurements have been made on captive animals swimming in a

tank or a flume and, naturally enough, these focus on small animals. Elsworth *et al.* (2003) found in *C. porosus* (300–110 cm total length) that maximum sustainable swimming speed increased with increased length, but that length-specific speed decreased. Speed increased also with temperature between 15 and 23°C, was constant over 23–33°C and decreased at temperatures higher than that; there was a broad plateau of thermal independence. A hatchling could swim at ~1 km/h and a 1 m animal at about twice that. Campbell *et al.* (2013) exercised *C. porosus* to exhaustion while measuring oxygen consumption and found a broad plateau of performance over a similar range of temperature. It is worth noting that Emshwiller and Gleeson (1997) found that *A. mississippiensis* showed a plateau of maximal oxygen consumption and terrestrial locomotor performance between ~25 and 35°C, surprisingly similar to *C. porosus*, given that *A. mississippiensis* can be regarded as a less thermophilic species (see Chapter 10).

Fig. 4.15. *A large* Crocodylus porosus *slowly under way. It is always impressive to see a large crocodylian swimming at the surface in clear water, with its snout, eyes, skull platform and parallel rows of dorsal scutes visible, with the tail moving in lazy sweeps from side to side and leaving a shallow triangular wake. See http://crocodilian.com/books/grigg-kirshner. They can, of course, also swim very well underwater, often making no disturbance of the water's surface. (Photo GCG)*

Fig. 4.16. *Paraxial swimming. At slow speeds, a crocodylian uses all four limbs, as if 'walking' in the water, with little to no involvement of the tail. (Modified from Frey and Salisbury 2001, with permission)*

Fig. 4.17. *Hybrid swimming. At intermediate speeds, a crocodylian combines the limb movements of paraxial swimming with tail undulations, a hybrid between paraxial and axial swimming. (Modified from Kirshner 1985, with reference to Frey and Salisbury 2001)*

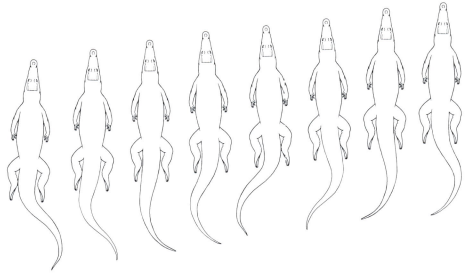

Fig. 4.18. In axial swimming, the limbs trail or, when travelling fast, are held close to the sides and propulsion comes from the sinuous movement of the muscular tail. Steering is accomplished by turning the head and/or using the limbs. (Modified from Frey and Salisbury 2001 with permission and input from Kirshner's material)

Fig. 4.19. Jumps with a 'tail walk'. These are two of many C. porosus in the Adelaide River, Northern Territory, which have learned to jump for food. If they hold position high for a few seconds, they demonstrate 'tail walking'. Rapid tail undulations propel the croc up and out of the water (left) as far as its hind legs and (right) sometimes even further, especially with smaller individuals. (Photos GCG, DSK)

Fig. 4.20. *The things one discovers while photographing leaping crocodiles! Horse flies (Family Tabanidae) in general lack substantial claws on their feet, or any elaborate wherewithal for clinging on to a crocodile, yet these are clearly very tenacious, (inset left) even hanging on as an estuarine crocodile powers up out of the Adelaide River, northern Australia. (Right top) Tabanids on a large* C. porosus, *photographed from a bridge overlooking Adelaide River using a 500 mm lens. (Bottom right) A black caiman,* Melanosuchus niger, *with tabanids on its snout and behind its eye, photographed in the Mamiraua Reserve, Brazil. (Photos DSK, Ruchira Somaweera and William E. Quatman)*

These attributes allow crocs to move with great speed and efficiency when that is desirable. However, crocodylians are slow in comparison with fish, equally as efficient as aquatic mammals and more efficient than semi-aquatic mammals (Seebacher *et al*. 2003).

JUMPING AND 'TAIL WALKING'

This has been made famous by the Northern Territory's 'jumping crocodiles' tourist attractions on the Adelaide River where free-ranging, but habituated, *C. porosus* have learned to jump high out of the water to take food offered on a string suspended from a pole (Figs 4.19, 4.20). Because they can hold the position high out of the water for a few seconds, this behaviour has also been referred to as tail walking (the 'cauda walk'). Crocodylians undoubtedly make practical use of these skills in a natural context by powering up out of the water to seize prey from overhanging branches: birds from nests or, perhaps, low roosting flying foxes. Mechanically, the movement is essentially the same as the familiar ambush rush from the water, *de rigueur* on wildlife TV documentaries, as a gazelle, wildebeest, bullock or kangaroo, is ambushed, seized, and dragged back into the water to drown. The initial acceleration for the jump is probably aided by thrusting with the hind limbs, but the main propulsive force comes from the tail whose rapid undulations continue conspicuously as the animal rises far out of the water exposing the hind limbs and much of the base of the tail, and appears

Fig. 4.21. *A juvenile Philippine crocodile,* Crocodylus mindorensis, *underwater in Dunoy Lake II, San Mariano, on the island of Luzon, ~400 km north of Manila. (Photo Merlijn van Weerd)*

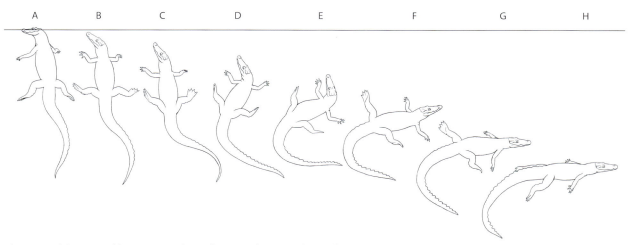

Fig. 4.22. *(A) A crocodile resting at the surface can depart with very little ripple and no splash. (B–D) Its departure is initiated by a scooping movement of the hindlimbs with the head tilted upwards as it leaves the surface. In this example, movement of the tail is passive, but in a very rapid departure the downward movement can be assisted by lateral scooping with the tail (Fig. 4.23). (E) Once submerged, the croc turns its head forwards and begins to turn, (F–H) paddling with the hind feet. In this example the upper (left) foot has the webbing spread for a thrust stroke while the other foot acts as a rudder (F–G) and the animal then swims away, with the limbs against the flanks and using the tail for thrust. (From Kirshner 1985)*

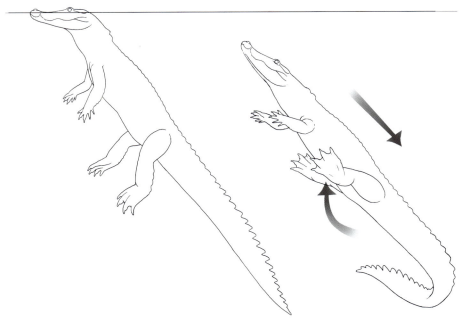

Fig. 4.23. *In this rapid departure from the surface, not only do the hind feet scoop forwards in synchrony, but additional thrust backwards is provided by active lateral flexure of the tail, providing an additional active surface to 'row' the animal backwards. The movement is often accompanied by an abrupt exhalation to air to decrease buoyancy. (After Kirshner 1985; Frey and Salisbury 2001)*

to hang there briefly before falling back with a mighty splash.

It is not only estuarine crocodiles that jump: it is probably characteristic of all species (perhaps except *G. gangeticus*?) and from hatching through to adulthood. Davenport and Sayer (1989) described captive hatchling *C. porosus* jumping to catch food (cockroaches) and provided useful cinematic analysis. Some individuals jumped almost completely out of the water. They prepared for a jump by bringing both hind limbs far forward and extending the webbing, and bending the tail at right angles to the body-axis. The jump resulted from simultaneously driving with the hind limbs and tail in 'a propulsive action lasting ~0.08 s'– a reversal of the actions taken in a rapid dive (Fig. 4.23). They noted that as the animal leaves the water, the lower viscosity of the air allows some acceleration towards the apogee of the jump. Davenport and Sayer observed that lunges to capture insects on the water surface were driven in a similar fashion. It is probable that the same actions described for the hatchlings accompany jumps by large individuals at tourist attractions, but the water is usually too turbid to see.

DIVING

The small amount of data available so far suggest that crocodylians spend quite a lot of their time underwater, some foraging, but most resting on the bottom or taking refuge (Fig. 4.21). Patterns of submergence behaviour and its physiology are discussed in Chapter 9. The mechanics of diving have been described in *C. porosus* by Kirshner (1985) and Davenport and Sayer (1989) and in *O. tetraspis* by Frey and Salisbury (2001) working on small individuals. It is likely that the same patterns occur in other species and in larger individuals.

Dives may be made from a resting position or while swimming, and may be passive or active. A swimming croc may submerge gently and smoothly by angling the head downwards a bit, trailing a stream of bubbles as buoyancy is adjusted to assist the dive. Recent work by Uriona and Farmer (2008) has shown that such dives are helped by muscular contraction shifting lung air posteriorly or laterally to adjust the centre of buoyancy (discussed further below). From a resting position at the surface, an undisturbed croc may submerge quietly and unobtrusively by simply releasing air slowly from the lungs and sinking slowly. Usually the

as the animal submerges backwards and the head tips up to be in line with the body for smoother movement through the water. The tip of the snout seems always to be the last to leave the surface while submerging, and sometimes the forelimbs are held high above the head as the animal sinks.

In a more active dive (Fig. 4.23), for example when a croc resting at the surface is disturbed suddenly, the forward movement of the hind limbs is preceded slightly by a quick curling of the tail to one side, producing a synchronised unidirectional scooping. This pulls the croc underwater rapidly, after which it will roll or turn and swim downwards to safety.

BOTTOM WALKING

Crocs can move along the bottom in deep water or in water too shallow for swimming by moving the limbs in a step cycle similar to walking on land (Fig. 4.24). This has been termed 'bottom walking'. As in walking on land, the forelimbs are used for grip or, in very shallow water, to provide support. The webbing of the hindlimbs is often extended so that the thrust may come from a combination of grip on the substrate with the claws and paddling. The tail simply trails, having no role in bottom walking unless there is sufficient depth of water and speed is required.

SURFACING TO BREATHE

Crocodylians are slightly negatively buoyant when they dive, and become more so as the submergence progresses (more on this below). Consequently, they can remain motionless on the bottom. However, unless frightened by a predator, they visit the surface periodically to breathe

Fig. 4.24. *Bottom walking in* C. porosus. *(A) The crocodile grips the bottom with the claws of the left foot and right hand, as the left forelimb and right hindlimb are swung forward in a ventrolateral arc. The left hindlimb is protracted and the right forelimb retracted, pulling the crocodile forward. The webbing of the left foot is extended and the foot rotated so that the plantar surface lies in a vertical plane, aiding in propulsion. (B) The left forelimb and right hindlimb continue their forward motions, the digits folded back to reduce drag. The left hand and right foot gain purchase on the substrate as the right forelimb and left hindlimb reach the end of their rearward movement. (C) At this stage the left foot acts mainly as a paddle and the claws barely grip the substrate. (From Kirshner 1985)*

submergence is accomplished backwards, aided by the hindlimbs scooping forward in synchrony, drawing the animal under water (Fig. 4.22). By simply spreading the digits and drawing the hindlimbs forwards, a croc can submerge quickly and with little effort, leaving minimum disturbance at the surface. The tail may curve forward passively

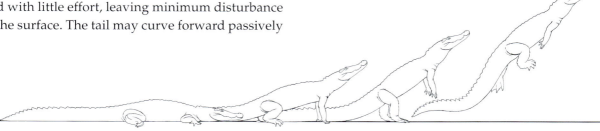

Fig. 4.25. *A juvenile* C. porosus *surfaces to breathe in water too deep for a simple head lift to suffice. The animal lifts its forequarters until only the tail and hind feet are in contact with the substrate, then commences paddling with the hindlimbs and, once it is clear, strokes with the tail until the surface is reached. (From Kirshner 1985)*

Fig. 4.26. *A saltwater crocodile (above) and an American alligator (below) lift their heads while swallowing food (a fish, Tilapia in the case of the alligator). This is made possible by arching their backs and lifting their tails clear of the water as a counterpoise. (Photos GCG, Louis Guillette)*

(see Chapter 9). The first sign of this is usually a slight upward lift of the head, then a raising of the forequarters, pivoting on the hindlimbs (Fig. 4.25). This can occur even if the forelimbs are not touching the substrate, perhaps by a combination of flexing the spine and/or muscular redistribution of air within the lungs (Uriona and Farmer 2008, see below). If the depth is such that the croc needs to swim to the surface, it will paddle with the hindlimbs, rotating them at each stroke and presenting minimum drag through the water on the forward stroke (the way a human paddles a canoe). If necessary, undulations of the tail begin once it is free of the substrate, driving the croc the rest of the way to the surface while the hindlimbs continue to paddle alternately. The forelimbs have little or no role in this.

HEAD AND TAIL LIFT

Crocodylians usually swallow food above water, presumably at least partly to avoid taking in water (particularly if it is salt water; see Chapter 11). Hatchlings do this by 'treading water' to hold the head out of the water, as described by Davenport and Sayer (1989). Larger individuals, however, raise the head and tail simultaneously so that the weight of the tail acts as a counterpoise (Fig. 4.26) and by raising the head they probably get a bit of gravity assistance acting on the food bolus. Similar head and tail lift behaviour is seen when crocodylians bellow, the most well publicised being the bellowing of *A. mississippiensis* in the breeding season (Chapter 12). This is often referred to as the HOTA position (head oblique tail arched).

BUOYANCY

BEHAVIOURS THAT DEPEND ON GOOD BUOYANCY CONTROL

It will have become clear by now that crocs have excellent control over their buoyancy. They can float motionless and parallel to the surface or at a steep angle to it, or anything in between. They can sink silently and unnoticeably without any apparent movement and they can lie motionless on the bottom for long periods (see Chapter 9). Typically, they have long periods without taking a breath (apnoea, see Chapter 7), even when they are on land. When floating in water, they manage to exhale and inhale with very little vertical movement of the exposed head, although the posterior of the body tends to move up and down a little. Being cryptic is an important characteristic for sit-and-wait predators, which is one of a croc's strategies, and ripples associated with breathing could give the animal's presence away.

It has been known for a long time that the lungs are the buoyancy organ of crocs. By changing the volume of air, they can float high or low in the water, and they use this skill routinely, for feeding, in their reproductive and social behaviour, and even in thermoregulation (Chapter 10).

Uriona and Farmer (2008, see below) have recently added a whole new dimension by showing

that alligators are even able to use their lungs to control pitch and roll, using an array of muscles to contract and expand different areas of the lung, adjusting its centre of buoyancy.

Several behaviours in particular show a precise control over buoyancy, as described below.

Carrying a load of mud or waterweed

In summer, I have seen *C. porosus* swimming around at the surface carrying a load of mud on the dorsal surface, and other crocodylians do this too (see Fig. 10.21). They are likely to risk exposure to an excess of solar radiation in the summer and it is probable that the load of mud acts like an umbrella by reducing radiant heat load or UV. Similarly, both *C. porosus* and *C. niloticus* are known to acquire large loads of the waterweed *Salvinia molesta*, perhaps for the same purpose. To be floating or travelling at what appears to be the 'normal' level in the water, they must be carrying a larger lung volume to accommodate the load.

Bellowing and other social signalling

The HOTA posture is seen also in male alligators when they make their spectacular bellowing display with water made to dance on the back (Vliet 1989, 2001) (Chapter 12). Similar behaviour is seen in other crocodylians, but it seems to be most pronounced, and is certainly better recognised, in alligators. The lungs must be full of air at the start of the sequence, lifting the dorsum clear of the water and, as bellowing proceeds and air is exhaled, the animal sinks deeper into the water. As the dorsum submerges, the low frequency vibrations of the bellow cause droplets of water to dance up off the scutes of the dorsum. Here, their capacity to control flotation is being used in a very spectacular manner!

Floating high in the water is also used as a signal to other crocs, conveying information about status and intentions (Figs 4.12, 4.13, Chapter 12).

MECHANISMS OF BUOYANCY REGULATION

Let's look more deeply at some of the buoyancy issues. Watching a croc resting at the surface, cryptic with just its nares, eyes, ears and skull platform exposed, the animal must be just slightly positively buoyant and, because it is in equilibrium, just enough so to equal the weight of the exposed parts in air. Likewise, a croc can submerge and lie motionless on the bottom, so it must then be negatively buoyant. That they can leave the bottom and rise to the surface without much effort suggests that this negative buoyancy underwater is slight. In other words, even casual observations suggest that they are never too far from neutral buoyancy: too light and they'd struggle to submerge, too heavy and they'd expend too much energy returning to the surface. These casual observations have been backed up by experimental work by David Kirshner (1985) who showed that they maintain control over their buoyancy within very narrow limits. This is apparently crucial for them and they manage it very effectively.

Crocodylian density and specific gravity

Animal tissue in general is heavier than water. Different tissues have different densities: bone is denser, lipids less dense and muscle in between, so that overall densities will be different in different species, depending particularly on the amount of lipids and bone. Hugh Cott (1961) measured the density of nine *C. niloticus* (24.4–44.0 kg), finding an average of 1.09 g/mL. Kirshner (1985) measured overall tissue densities of ~1.08 and 1.09 g/mL respectively in juvenile *C. porosus* and *C. johnstoni*. The slightly higher density of *C. johnstoni* over *C. porosus* reflects it having more dermal ossification (Fig. 3.15) and there are many crocs with much more ossification in their scutes, such as *Paleosuchus*, which would be denser still. These values can be compared with pure fresh water having a density very close to 1.0 g/mL (actually 0.9982 g/mL for water at 20°C and 0.9957 at 30°C) and sea water ~1.025 g/mL. So, without any air in its lungs, a croc would sink.

The old croc shooters knew this. They knew that they had to get a line onto a croc (via a harpoon or spear) as soon as it was shot because otherwise it was likely to sink to the bottom. There it would remain until gas accumulated as a result of the decomposition and it would float at last to the surface; the skin, of course, would then be useless.

The relevant measure of buoyancy is the relative density compared with water (= specific gravity) of the whole animal, including whatever air space it contains; that is, the combination of the density of the body tissue and the volume of the lungs. Because specific gravity is a ratio of two densities, it is dimensionless and, therefore, has no units.

Control over lung inflation adjusts buoyancy within narrow limits

Simple observations imply that crocs must be slightly positively buoyant when resting at the surface with just the top of their head exposed, and slightly negatively buoyant when they submerge, but never very far from neutral. This has been confirmed quantitatively by Kirshner (1985), studying juvenile *C. porosus* (0.25–3.76 kg). He found that they set up their negative buoyancy just before voluntary submergences by adjusting the volume of air in the lungs, usually making a small exhalation. Their measured specific gravities at submergence averaged 1.028 with almost no variation, and he found exactly the same value in *C. johnstoni*, despite their slightly greater absolute density. This implies that *C. johnstoni* dive with proportionally larger lung volumes and this is what was found: 57.3 mL in a 1 kg *C. johnstoni* compared with 47.8 mL in a 1 kg *C. porosus*. According to Kirshner 'this (specific gravity) seems to suit the needs of negative buoyancy yet allows the croc ease of movement in the water'. It presumably minimises the energetic costs of underwater activity. In sea water, a little less air is taken down: in sea water David measured a reduction of specific gravity at submergence to 1.014.

Compensation for added weight

It has always been assumed in discussions about the role of stomach stones (see next section) that, because stone is typically much heavier than tissue, the presence of stones will add to their density and thus reduce buoyancy. This turns out not to be the case. Kirshner (1985) loaded five of his juvenile *C. porosus* (0.891–3.711 kg) with steel ball bearings equating to ~1% of body weight, the typical gastrolith load of wild crocodylians found carrying them. He placed them into the stomach (to mimic natural gastrolith location) and found that

Fig. 4.27. C. porosus, *slightly positively buoyant, floating at the surface either parallel to it or at a steep angle, controlled by the degree of inflation of the lungs (Kirshner 1985). Part of the explanation is that the lungs reach further posteriorly as they inflate, shifting the centre of buoyancy caudally; in this example, the degree of inflation has been enough to expose the dorsum.*

buoyancy under water for ease of movement and minimal energy expenditure – a 'buoyancy comfort zone'. It seems reasonable to predict that when measurements are made on other species, there will be similar results.

Buoyancy control gets priority over oxygen storage

An important implication from this stout defence of a particular specific gravity (buoyancy state) by adjusting lung volume is that crocs obviously give priority to buoyancy considerations ahead of respiratory needs, at least within certain limits (see above). This means that they must be able to ventilate their lungs from a wide range of 'starting points' (residual volumes) (Chapter 7), allowing them to adjust buoyancy appropriately but still accommodate an appropriate tidal volume of air during breathing. A large meal of bone, for example, or an accumulation of gas in the gut from the digestive processes would change their buoyancy, to which the croc accommodates by changing residual lung volume and then adding the appropriate ventilatory cycle on top. Interestingly, when crocodiles in David's experiments had experimentally induced larger or smaller lung volumes, their voluntary submergence times were unmodified. Most dives were short in duration, so it appears that time spent submerged is not a direct consequence of lung volume and must depend upon more than just oxygen levels in the gas or blood. Indeed, juvenile *C. porosus* are known to surface from short dives with most of the oxygen store still available (see Chapter 9).

Buoyancy decreases during submergence

Buoyancy does not remain constant once a croc submerges. They get heavier for two reasons. First, the pressure of the water compresses the air in the lung, so a croc will have a higher specific gravity on the bottom than at the surface. Second, the volume of gas in the lungs decreases slowly as the

the crocs compensated almost completely (85%) for the added weight by increasing the volume of air in the lungs by ~20%. In contrast, attaching polystyrene flotation (at volumes displacing the equivalent of 2% of body weight) had the opposite effect: the crocs reduced lung volume, again by ~20%. However, compensation in this direction was limited by the crocodiles and they seemed reluctant to sacrifice oxygen storage beyond a certain point, even though that point was at a lung volume well above the oxygen storage needs for an average dive. Presumably crocs adjust lung volume in the same way to accommodate a load of mud, or *Salvinia molesta*, when using it as a solar umbrella in summer (Chapter 10, Fig. 10.21, 10.22).

These measurements and experiments backed up the conclusion implied by casual observations: crocs are rarely far from being neutrally buoyant. *C. porosus* and *C. johnstoni* obviously defend a particular buoyancy state very successfully: one that apparently gives them just enough negative dive proceeds and consumed oxygen is replaced by a lesser volume of CO_2 (i.e. respiratory quotient, RQ < 1, see Chapter 7). David Kirshner confirmed this by weighing crocs underwater on a wire mesh platform suspended in a large tank and found that, as expected, they became less buoyant (heavier) over time as they rested at the bottom. As we shall see (Chapter 8), some observations he made incidentally at the same time have been recognised subsequently to have very significant implications for an understanding about how the heart functions during submergence.

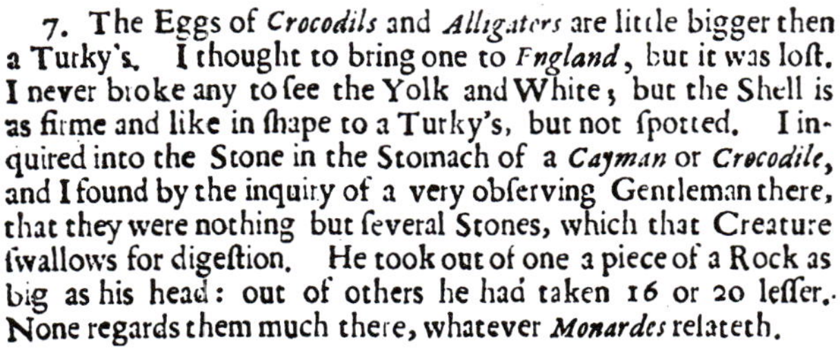

Postural control using the lungs

Apart from adjusting the total volume of air in the lungs to achieve appropriate overall buoyancy, postural adjustments can be made as well. As noted above, the angle at which a croc rests at the surface (in water too deep for a footing) can change a little with inhalation and exhalation, and a croc can choose to float parallel to the surface or at a steep angle (Fig. 4.27). Kirshner (1985) showed that the angle correlates nicely with the volume of air in the lungs. A croc can control the angle at which it floats by adjusting the amount of air: the lungs elongate as they fill and the centre of buoyancy moves posteriorly. This finding has been confirmed in a modelling study (Henderson 2003), which will be discussed later.

The centre of buoyancy, however, can be modified by more than just inflation and deflation of the lungs: it is under muscular control as well. In a quite spectacular recent addition to our understanding about how the lungs can contribute to postural adjustments, TJ Uriona and Colleen Farmer at the University of Utah (2008) monitored the electrical activity of the respiratory muscles of *Alligator mississippiensis* diving in an experimental tank. They found a pattern of contraction of the respiratory muscles (see Chapter 7) during a dive cycle that implied movement of the liver, and thus the lung gas, either fore or aft, thereby changing the centre of buoyancy in relation to the centre of gravity. Relative movement between these pivot points will cause pitch up or down. Contraction of the diaphragmaticus, ischiopubis, rectus abdominis and internal intercostal muscles would shift lung

7. The Eggs of *Crocodils* and *Alligators* are little bigger then a Turky's. I thought to bring one to *England*, but it was loft. I never broke any to fee the Yolk and White; but the Shell is as firme and like in fhape to a Turky's, but not fpotted. I inquired into the Stone in the Stomach of a *Cayman* or *Crocodile*, and I found by the inquiry of a very obferving Gentleman there, that they were nothing but feveral Stones, which that Creature fwallows for digeftion. He took out of one a piece of a Rock as big as his head: out of others he had taken 16 or 20 leffer. None regards them much there, whatever *Monardes* relateth.

Fig. 4.28. *The original description of stomach stones in a crocodylian, possibly* C. acutus. *(Stubbes H. 1668)*

gas posteriorly, and the measured electrical activity suggested contraction of these muscles during a head-down dive from a horizontal posture. Adding weights experimentally to either the head or the tail enhanced the activity, implying stronger muscle contractions. The researchers also found that by differential use of left and right diaphragmaticus and rectus abdominis muscles, an alligator can squeeze air from one side to the other and, thus, exerting a stabilising effect, counteracting any tendency to roll. This is an exciting finding. I have often wondered, with the stomach lying not in the mid-line but to the left side, whether stomach stones might provoke a list to that side. But crocs seem always to float symmetrically, without a list. That must be how! Perhaps crocs even make small cyclical adjustments to left and right lung volumes while swimming, to offset any tendency to roll.

Fig. 4.29. *A collection of stones from the stomach of a 214 cm (total length)* C. johnstoni, *Lake Argyle, Kimberley District, Western Australia. (Photo Ruchira Somaweera)*

A myth busted

There has been a myth that crocs can control their buoyancy while underwater by contracting thoracic musculature, compressing the gas in the lung and reducing its volume and, thus, its buoyancy. According to this idea, a croc could then reduce the muscular tension, increase its buoyancy and rise to the surface without any need to swim. The role of muscles described by Uriona and Farmer (2008) does not involve compression or expansion of the gas enclosed within the lung and, indeed, Kirshner (1985) found no pressure changes within the lung which would be consistent with that idea.

A very significant implication that follows from the discovery that crocodiles give so much priority to defending a particular relative density is that explanations for the presence of stomach stones cannot be based on assumptions that ingested stones alter buoyancy. And yet it is exactly that assumption upon which most current theories about gastrolith function rely. Read on …

STOMACH STONES (GASTROLITHS)

Few observations about crocodylians have prompted more intrigue and speculation than the stones they often have in their stomachs (Fig. 4.29). The term gastrolith seems to have been used first by Mayne (1854), quoted in Baker (1956), and published information about stones in croc stomachs goes back at least to 1668 when the English physician Henry Stubbes described stomach stones in a crocodylian in an article published by the Royal Society, (Stubbes 1668) (Fig. 4.28). His opportunity to learn about them presumably came while he was His Majesty's Physician for Jamaica from 1661 to 1665.

Wings (2007) noted the many different types of 'stones' and other inert objects that may be found in the stomachs of a wide variety of animals, including extinct taxa and suggested that the term gastrolith should apply to 'a hard object of no caloric value (e.g. a stone, natural or pathological concretion) which is, or was, retained in the digestive tract of an animal.' He suggested adjunct nomenclature to specify the origin of the 'stone'. Thus, calcareous or other 'stony' concretions formed pathologically

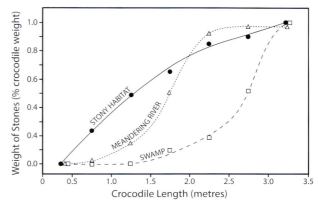

Fig. 4.30. *Eminent early crocodylian scientist Hugh Cott collected body weights and gastrolith weights from 507* C. niloticus *in Uganda and what is now Zambia. Adult individuals ended up with a stone load of ~1% body weight, but the rate of acquisition was slower in habitats having fewer stones (Cott 1961). He proposed that crocs must have made excursions to collect stones. Regardless of how they acquired them, this dataset is probably the most convincing that, in this species anyway, gastroliths do serve some useful function and that they may be more important in adult animals. (Modified from Cott 1961, © 1961 The Zoological Society of London)*

would be termed patho-gastroliths, balled up vegetable matter or fur balls such as can occur in crocodylians (Fig. 6.23) and many mammals could be called bio-gastroliths, whereas those of 'true' stone could be called geo-gastroliths. It is the last that are our present focus.

THE SEARCH FOR A FUNCTION

Many reptiles, birds and mammals, both terrestrial and aquatic, sequester stones or stony material in their alimentary canal. Wings (2007) provided a comprehensive review. The most familiar examples are seen in seed-eating birds, which retain grit in their muscular crop, presumably for grinding seeds (birds lack teeth). Some herbivorous or omnivorous terrestrial reptiles also may use grit or sand for the same purpose. It is probably logical, therefore, that when many crocodylians were found to have stomach stones there was speculation that they had some digestive role. However, a careful review of the evidence by Taylor (1993), which included a comparative study of what is known about the occurrence of gastroliths in many of the fossil marine reptiles, has put that idea to rest. After all, crocodylians are carnivorous and have excellent digestive capabilities (Chapter 6). Taylor concluded that the role of gastroliths in aquatic tetrapods (in general) is for buoyancy control, not for food processing. As we shall shortly see, it is not quite that simple, but some useful speculation can be made.

Other suggestions too have been made about their possible function. Hugh Cott (1961) reported body lengths and weights, and stomach stone weights from 507 *C. niloticus* collected in the 1950s in Uganda and 'Northern Rhodesia' (now Zambia). There are no other datasets as comprehensive as Cott's and, in presenting an analysis of the results, he reviewed suggestions about function. His data and his comprehensive and thoughtful discussion are contributions which are still relevant today.

Cott's data showed that, in adults, the stones are typically only ~1% of body weight, – so little that it seems hard to see how it could be important, – yet his data suggested very strongly that the presence of stones was not the result of accidental ingestion. Some crocs, particularly larger ones, had stones that had clearly been collected large distances away from where the crocs lived and, he assumed, were the result of a collecting trip. Smaller, younger individuals had fewer or no stones, depending on the ease of availability of stones in their habitat (very few in swamps). The rate of their acquisition depended upon habitat (Fig. 4.30), but the proportion of *C. niloticus* with stones gradually increased with body size until almost all crocs had them as adults. Two implications could be drawn: gastroliths are important, somehow, and they are more important in adult animals.

Taylor (1979) examined stomach contents of hatchling, juvenile and sub-adult *C. porosus* in the Liverpool and Tomkinson Rivers in Australia's Northern Territory and found that only one out of 289 juvenile *C. porosus* had stones. These rivers are particularly muddy, sinuous and mangrove-lined, with stones in very short supply. However, and very tellingly, there were stomach stones in both of the adults Taylor had an opportunity to examine (Taylor 1977), one at 0.8% of the body weight. Webb *et al.* (1982) found stones in the stomachs of 88.8% of the 153 *C. johnstoni* they sampled (~0.3–2.6 m total length). They found that larger individuals had larger gastrolith loads, which were made up of larger stones rather than an increase in number of small ones, and that loads tended to be greater in the dry season. On average, ~79 stones were carried, which was uncorrelated with body size. Typical loads were, however, considerably lighter than those found by Cott (1961). These results are broadly consistent with the data for *C. niloticus*: presence is not accidental, and fewer stones in smaller animals and heavier loads with increasing body size. Does a seasonal change imply a capacity for adjustment? After all, crocs are good at regurgitating fur balls and other indigestible items.

But why have them? Cott discussed and dismissed the digestive (food trituration) hypothesis, and the idea that they swallow stones to satisfy hunger pangs (!), and came firmly to the conclusion that their function is hydrostatic: for adjusting buoyancy, to make crocs 'heavier'. He suggested also that they may act as an 'anterior counterpoise' and as

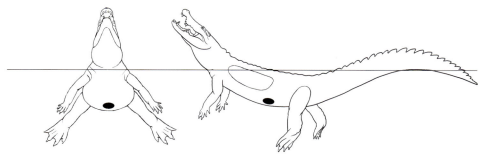

Fig. 4.31. *In the head-up tail-up posture adopted by large crocodylians when swallowing food or by bellowing alligators the gastroliths will be at or near the part of the animal that is lowest in the water, with the potential to impart the greatest turning moment to resist roll, and thus assist stability. However, modelling shows the effect of 1% body weight of stones to be insignificant: another good idea bites the dust!*

'a ventral stabilising force', because the stomach is anterior to the centre of gravity and the stones lie close to the ventral body surface. These 'buoyancy-related' possibilities for gastrolith function in crocodylians are now discussed in more detail.

WHAT FUNCTION COULD THE STONES SERVE?

Not buoyancy adjustment

The most long-standing and resonant hypothesis about the function of gastroliths has been that they provide a benefit by giving crocs extra weight in the water; that is, that they increase specific gravity, making them less buoyant. The idea goes back at least to the middle of the 19th century, with von Humboldt (1852) reporting that native tribes in South America believed that crocodiles 'like to augment their weight'. Cott (1961) was able to evaluate the idea with his new data. He, like several authors before and since, drew attention to two benefits that might accrue from increased weight in the water as a result of having stomach stones; it would be easier to stay still on the bottom, particularly in a current, and additional effective weight might be useful to hold a buffalo or a water buck under water to drown it.

However, as reviewed above, Kirshner (1985, reported by Grigg and Gans 1993) found experimentally that *C. porosus* simply compensate for added weight at typical gastrolith loads by increasing lung volume: they defend their 'buoyancy comfort zone'. So the presence of the stomach stones does not increase their effective specific gravity and Kirshner (1985) drew the conclusion that buoyancy modification cannot be a satisfactory explanation for their function.

The same conclusion was drawn independently by Henderson (2003) from a comprehensive mathematical/computational modelling analysis. He developed a mathematical model based on *A. mississippiensis* and considered the implications of various states of lung inflation. He found that, with the lungs fully inflated, addition of stones equal to 6% of the body weight would be required to make his model animal negatively buoyant. Without stones, deflation of the lungs by 40–50% would be required to attain negative buoyancy. In all situations, the model was resistant to capsizing. He came to the conclusion that 'The relatively small amounts of gastroliths (< 2% body mass) found in aquatic tetrapods are considered to be inconsequential for buoyancy and stability, and the lungs are the principle agent for hydrostatic buoyancy control.'

If sequestering stones does not decrease buoyancy there must be some other explanation.

Stability? Anterior counterpoise?

Another common theory for the presence of gastroliths in crocodylians is that they aid stability in water. It is an old idea. In a letter to *The Field, The Country Gentleman's Newspaper* in 1925, Dunbar Brander (probably the big game hunter AA Dunbar Brander who wrote a book 'Wild Animals in Central India') attributed buoyancy (postural?) adjustment by crocodiles to having stones on board, noting 'the crocodile is a top heavy animal. He requires the finest adjustment and poise in the water. He is specially equipped and has marvellous powers of adjusting his equilibrium' (Brander 1925).

Cott (1961) likened the presence of stones in the stomach to the stabilising effect of the cargo in a ship's hold, with the dorsal position of the lungs adding further value. He speculated also that the weight of the stones in front of the centre of gravity would counteract the anterior position of the lungs and noted that young *C. niloticus*, lacking stones, were top heavy and tail heavy and 'shun open water'. Henderson's study (2003) found no convincing support for the role of stones as ballast. However, his mathematical model was based on a crocodylian having uniform tissue density, with the exception of the lungs. Brander and Cott both noted that Nile crocodiles are top heavy, which can be expected from the concentration of denser skeletal tissue, particularly the vertebrae and osteoderms, in the upper half of the body.

Although Kirshner (1985) refuted the idea of gastroliths being used to increase a crocodylian's specific gravity, he did not dismiss the idea of gastroliths possibly aiding stability. He noted that juvenile crocodiles, when floating at an angle at the surface, rest with outstretched limbs for balance and frequently use paddling motions to counteract roll. Uriona and Farmer (2008) have shown that crocodylians can modify lung position by differential contraction of respiratory muscles, both antero-posteriorly and laterally, thereby influencing the position of the centre of buoyancy in relation to the centre of gravity and therefore suggesting that gastroliths would not be necessary to stabilise the animal's posture in the water. However, Kirshner (1985) highlighted one of the dilemmas of exploring gastrolith function in crocodylians, which is that, although only large individuals seem to accumulate them in the wild, for obvious logistical reasons, laboratory studies on crocodylians concentrate on small individuals. The increase in relative gastrolith load with size in wild crocodylians, as noted by Cott (1961) and others, suggests that gastrolith ingestion may be related to morphological changes associated with growth, so it is possible that the benefits of gastrolith ingestion can only be noticed in crocodylians of a size that would ingest them naturally and that extrapolation from studies of animals smaller than this may be prone to error.

The allometric changes that could theoretically affect buoyancy and postural control are relative increases in skull size (Dodson 1975; Webb and Messel 1978; Chentanez *et al*. 1983) and dermal ossification (P. Brazaitis *pers. comm*.), combined with a reduction in relative limb size (Dodson 1975). If crocodylians are top and front heavy, they would appear to become more so with increased size. Is it possible that gastroliths help counteract the increasing energetic costs of limb movements and muscular adjustment of the lungs, to maintain a stable posture at the surface, as they grow? Could they be useful in more extreme situations, such as the feeding and bellowing (HOTA) posture?

Can they stabilise the HOTA posture?

Henderson's (2003) study concluded that the stones were not important for stability in the water. As stated above, his study was based on a model of uniform density. However, his model was also of an alligator in a prone position. What if the back were arched, the way it is when large crocs adopt the 'head oblique, tail arched' (HOTA) posture

(Fig. 4.26), such as in alligators when they bellow and in many crocodylians when they swallow large food items? The stomach is well posterior (Fig. 6.22), close to the centre of gravity (Kirshner 1985) and it is also ventral in the body cavity. In such a posture, the lungs will be high in the water, and the stones in the stomach will be at the deepest part of the croc (Fig. 4.31). Together, the lungs and the stones are where they could provide the greatest turning moment to counteract roll, no doubt in concert with widely spread hind feet.

If the stones confer benefit by stabilising the head-up tail-up feeding or bellowing posture, that would also be consistent with the finding that it is large individuals that accrue stones, as it is the large crocs which are most likely to show the head-up tail-up posture.

This suggestion was put to Donald Henderson at the Royal Tyrrell Museum of Palaeontology, Alberta, and he kindly explored it using a modification of the model he had developed for alligators, with reference to the criteria shipbuilders use to quantify stability. The answer is that a crocodylian in the HOTA position will have ~15% of the body mass above the water surface at peak lung capacity, and it will be self-righting with or without the addition of gastroliths. Adding stones to 1% of the bodyweight does not change the existing stability in any significant way. This extended and confirmed the results he obtained for an alligator in the prone position (Henderson 2003) and seems to be the last nail in the coffin to bury proposals that gastroliths modify buoyancy or stabilise posture.

But one idea remains standing, and it is a good one …

Prolongation of dive time

Roger Seymour at the University of Adelaide suggested that the presence of gastroliths in a crocodylian could benefit the animal by increasing the volume of its lungs and thus its oxygen store, extending dive time (Seymour 1982). He calculated that stones of 1% body weight would increase lung volume by 12%.

Could that be the benefit from having gastroliths – to lengthen dive time? Kirshner's (1985) finding that *C. porosus* defends a particular specific gravity showed quite clearly that, to be in its 'buoyancy comfort zone', a crocodylian diving with stomach stones will have a larger lung volume than one without, and he was able to measure that experimentally. From measurements of tissue density of individual crocodiles and their submerged weight, he was able to determine lung volumes at submergence in five individuals (890– 3700 g) over several days of voluntary dives, before and after the addition of a load of ball bearings, as pseudogastroliths, equal to 1% of body weight. The resultant average increase in lung volume at voluntary submergence was just under a substantial 20%. He found no increase in voluntary dive time as a result of loading them with ball bearings, but this is not surprising because crocodylians typically surface long before their oxygen store is used (see Chapter 9).

Henderson (2003) took up the same question, with remarkably similar results. Data from his model implied that adding stones equal to 1% of the body weight would increase the volume of air in the lungs by 20% (from 50% inflation to 60% inflation) to achieve the same specific gravity (neutral buoyancy in his case).

It is reasonable to propose that a 20% increase in lung volume will result in an addition to both its 'comfortable dive time' and its 'maximum dive time'. Although it would be quite a step to conclude from these observations that the reason large crocs sequester stones is to extend dive time, from time to time it must be advantageous to a croc to remain underwater as long as it can, and presumably a larger lung volume would then become an advantage. There is surprisingly little information about diving behaviour and its purpose, and the longest voluntary submergences recorded so far seem to be spent at rest. It is almost certain that such dives are conducted aerobically (see Chapter 9), so a gastrolith-induced increase in lung volume at submergence can only be advantageous.

Finally, what of the implication that gastroliths are more important for larger crocodylians? If a benefit of carrying stones in the stomach is to increase oxygen store, does that imply that a capacity for long dives is more important in adult crocs than in juveniles? This is an important question, because studies of diving physiology so far have, very reasonably, focussed on juveniles. Perhaps the physiological mechanisms by which long dives are supported, and the controversy over the significance or otherwise of the pulmonary by-pass shunt (Chapters 8 and 9) will not be resolved until studies are undertaken on adults.

In the absence of certainty about how gastroliths assist crocodylians, we can be sure that one benefit of the stone-carrying habit is that it facilitates retention of temperature-sensitive radio-transmitters (see Chapter 10).

CAPACITY FOR LONG DISTANCE TRAVEL

Much of Chapters 3 and 4 dealt with the structural and functional aspects of crocodylian locomotion, on land and in the water. What they use locomotion for is mostly taken up in other chapters: to feed (Chapter 6), to bask or seek shade (Chapter 10), to vacate seasonally drying habitat (this chapter), to patrol territory, to mate and to nest (Chapter 12), and also because population pressures or some innate drive in sub-adult males (mostly) prompts dispersal (Chapter 13). Most of the travel associated with these behaviours is comparatively short, at least over short timeframes, but there are some striking exceptions. VHF radio-tracking and, more recently, satellite/GPS and acoustic tracking have provided new capabilities for monitoring long-distance travel, and the results are interesting, provocative, and the subject of this section.

There have always been suspicions, anecdotal accounts or folklore about long-distance travels, such as a rumoured annual migration of estuarine crocodiles around the coastline of Papua New Guinea and also reports of large crocodiles being seen way out in the ocean. The distribution of *C. porosus*, from India to the Solomon Islands could be taken to imply a capacity for travel, or, at least, range extension across chains of land masses and islands by a sort of survival 'sweepstakes', even as far as the Solomon Islands (perhaps even Madagascar!). Campbell *et al.* (2010) have suggested that they may undertake directed oceanic travel, taking advantage of surface currents, and their homing ability implies good navigational skills (see below and Chapter 5). Certainly their salt glands give them a marine capability (Chapter 11) for long journeys.

Mark–recapture studies and radio-telemetry have yielded much information of movements at 'study area' scale. In the former category, Campos *et al.* (2006) and Lance *et al.* (2011) recaptured more than 500 marked *Caiman yacare* and *A. mississippiensis* in wetland habitats over 5–15 years and 6 years, recording maximum displacements of 18 km and 90 km, respectively. Brien *et al.* (2008) radio-tracked 13 adult *C. porosus* in a non-tidal waterhole in northern Australia, calculated home ranges and reported that adult males apparently moved freely through each others' territories (Chapter 12). All of these movements were over comparatively small distances, particularly over short timeframes. Aerial tracking enables a larger area to be monitored.

For example, Winston Kay monitored radio-tagged *C. porosus* in the Ord and King Rivers and Cambridge Gulf, northern Western Australia, by boat and from the air (Kay 2004). He found females that travelled seasonally between their core dry season habitat and their nesting sites, 15–62 km distant (Fig. 12.46). Using satellite transmitters that uploaded stored GPS locations periodically (described by Franklin *et al.* 2009), Hamish Campbell and colleagues reported the same phenomenon in *C. porosus* in the Wenlock and Ducie Rivers on the western side of Cape York, northern Queensland (Campbell *et al.* 2013). This study reported females travelling to presumed nesting locations 28–55 km from their core areas and, most curiously, visiting the sites twice in the season, early, then returning to the core area (to mate?) and then spent long periods at the probable nesting sites. Some of the males too were surprisingly mobile, cruising 100 km or more

Fig. 4.32. *Mark Read (green cap) and others untying the lashing before releasing a 4.3 m* C. porosus *into the Endeavour River, near Cooktown, Queensland, in May 2003. Satellite and VHF transmitters are in a package secured to the nuchal scutes. This large male was tracked for 8 months, during which time he travelled only within 10 km of river and 7 km of side creeks, without visiting the ocean ~10 km downstream (Fig. 4.34). This animal was christened Charles in honour of Charles Tanner (see text). (Photo GCG)*

Fig. 4.33. *Saltwater crocodile, C. porosus, about to be released with a satellite transmitter attached between its medial nuchal scutes. One antenna sends signals to the satellite, the other sends out a VHF signal, which can be used for ground based location if desired. The transmitter is held in place by stainless steel wire laced through fine holes drilled through the bony scutes: a method that has proven very effective and causes no harm to the animal. (Photo Mark Read)*

of the rivers repeatedly in the breeding season, as if on the lookout for an undefended female. These observations are discussed further in Chapter 12. Similarly long-distance riverine journeys by *C. porosus* were reported in the North Kennedy River, eastern Cape York (Campbell *et al*. 2010), using sonic tags and receivers placed at intervals over 63 km of river. Eight out of 20 tagged animals made long journeys, leaving the river for a period, beyond the limit of the receiver array and returning subsequently. In these studies, the long-distance movements were recorded somewhat incidentally, in the course of considering other questions.

Satellite telemetry is not geographically constrained and several studies, including some ongoing, and so far unpublished, have gained information about long-term movements in this way. The first was the result of a bequest to the Queensland National Parks and Wildlife Service by Charles Tanner when he died in 1996. Charles was a well known herpetologist who lived for many years in Cooktown where he kept a diverse array of venomous snakes and milked them for research and the production of anti-venoms. The bequest was to be used for wildlife research, particularly reptiles, and when the time came to select an appropriate project Mark Read was the QNPWS crocodile biologist, based in North Queensland. Mark was charged with management of issues related to human–crocodile conflict and he and I had often lamented the paucity of information about how widely these large and dangerous animals travel, considering that to be a serious management issue. Setting aside protected areas 'where crocodiles could be crocodiles' was futile if it should turn out that they travelled far beyond its limits (as we

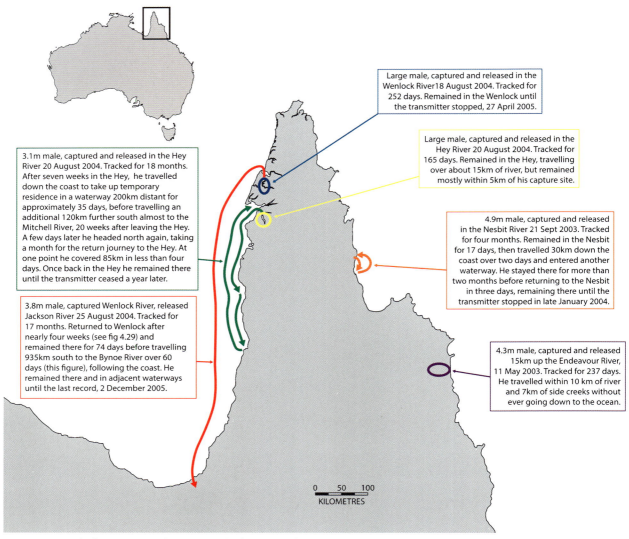

Large male, captured and released in the Wenlock River 18 August 2004. Tracked for 252 days. Remained in the Wenlock until the transmitter stopped, 27 April 2005.

Large male, captured and released in the Hey River 20 August 2004. Tracked for 165 days. Remained in the Hey, travelling over about 15km of river, but remained mostly within 5km of his capture site.

3.1m male, captured and released in the Hey River 20 August 2004. Tracked for 18 months. After seven weeks in the Hey, he travelled down the coast to take up temporary residence in a waterway 200km distant for approximately 35 days, before travelling an additional 120km further south almost to the Mitchell River, 20 weeks after leaving the Hey. A few days later he headed north again, taking a month for the return journey to the Hey. At one point he covered 85km in less than four days. Once back in the Hey he remained there until the transmitter ceased a year later.

4.9m male, captured and released in the Nesbit River 21 Sept 2003. Tracked for four months. Remained in the Nesbit for 17 days, then travelled 30km down the coast over two days and entered another waterway. He stayed there for more than two months before returning to the Nesbit in three days, remaining there until the transmitter stopped in late January 2004.

3.8m male, captured Wenlock River, released Jackson River 25 August 2004. Tracked for 17 months. Returned to Wenlock after nearly four weeks (see fig 4.29) and remained there for 74 days before travelling 935km south to the Bynoe River over 60 days (this figure), following the coast. He remained there and in adjacent waterways until the last record, 2 December 2005.

4.3m male, captured and released 15km up the Endeavour River, 11 May 2003. Tracked for 237 days. He travelled within 10 km of river and 7km of side creeks without ever going down to the ocean.

0   50   100
KILOMETRES

Fig. 4.34. *Results from six animals tagged in the first study of movement by a crocodylian using satellite technology. Whereas most of the 12 males tagged in 2003–2004 remained where they were captured, three voluntarily made substantial coastal journeys. Newer technology now combines GPS data and returns more accurate locations. The data show not only that C. porosus may remain in a smallish area for a long time, they also make substantial journeys and can be quite seafaring, well supported by physiology that enables them to live indefinitely in salt water (Chapter 11). (Data from an unpublished study by Mark Read, Craig Franklin, Steve Irwin and Gordon Grigg)*

both suspected they would) and some people were recommending translocating 'problem' animals away from human habitation, even though studies had already shown their proclivity for swimming home (Walsh and Whitehead 1993). The suggestion to use Charles' legacy for using satellite telemetry to track crocodiles met with approval and the study got underway in 2003 (Figs 4.32, 4.33). Apart from Queensland National Parks and Wildlife Service (Mark and his staff), other participating institutions

were the University of Queensland (Craig Franklin and myself) and Australia Zoo (Steve Irwin).

Figures 4.34 and 4.35 show some of the results from the first phase of the study in which only males were studied. For the realised lifetimes of the transmitters, 3–18 months, most of the dozen animals left to themselves remained in the rivers where they were captured, but three moved out and travelled along the coast, two for very considerable distances (Fig. 4.34). The latter

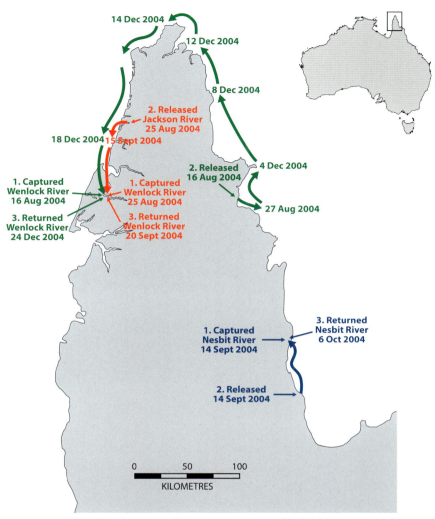

Fig. 4.35. *Despite some previous data showing it is ineffective, some people still talk about moving large 'problem' crocodiles away from habitation to remote sites, for safety. These three* C. porosus *were captured, fitted with satellite transmitters and moved by helicopter, two of them for some distance along their coastline and one right across Cape York. All of them returned to their capture sites, one making the spectacular > 400 km journey in 20 days. (Modified from Read* et al. *2007)*

sometimes showed rates of movement of ~20 km/day. In the life of a crocodylian, these periods of monitoring are brief, but the proportion that moved long distances suggests that long coastal journeys are probably routine. Additionally, three animals were translocated (Fig. 4.35) (Read *et al.* 2007). This was prompted partly by concern about human–crocodile conflict and partly to extend a previous study (Walsh and Whitehead. 1993 and Chapter 14), which had concluded that translocating large crocodiles would be ineffective as a safety measure; they were likely to simply return 'home'. In this part of the study, three large males (3.1,

3.8 and 4.5 m) were captured, fitted with satellite transmitters, translocated a substantial distance (56, 77 and 411 km coastline distance, respectively) by helicopter and released.

All returned to their capture sites 21, 26 and 130 days later, but none started out immediately; they spent 13, 15 and 108 days in the vicinity of their release site and then travelled 'home' as if purposefully. The animal translocated the farthest had been shifted from the west to the east side of Cape York. I wondered if he might return overland, more than 120 km! Mark Read pointed out to me that he could have gone upstream to the headwaters of

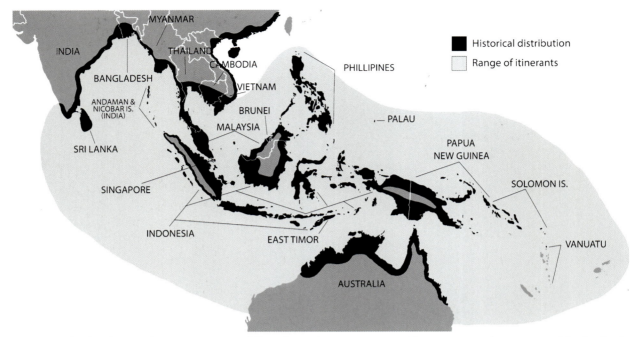

Fig. 4.36. *The distribution of C. porosus as it would have been about a century ago. There are no very large gaps and the distribution can be explained easily enough by journeys such as those reported in the two previous figures, combined of course with their physiological tolerance for living in salt water (Chapter 11). (Modified from a figure by Brandon Sideleau, courtesy of Adam Britton)*

the Olive River, made a 1.5 km overland journey to the headwaters of the Wenlock River and followed that downstream and home. But he took the coastal route, through Torres Strait and then south to his capture site on the Wenlock River. This journey of 411 km (minimum) took 20 days: an average of more than 20 km/day and on one day he travelled 30.4 km. He arrived at the Wenlock on 24 December (home for Christmas) and remained there as long as the transmitter lasted, until January 22.

This study and the long journey home by the 4.5 m animal in particular, raise a diverse range of interesting questions. What metabolic support provided for swimming an average of 20 km/day? To what extent may the journeys have been current assisted? How did the translocated animals navigate their way back from a (presumably) unfamiliar location? Or were the two shifted 56 and 77 km along the coast already familiar with their release sites? Does the delay by all three animals before they started for home imply some 'working out' which way to travel home, like a GPS recalculating after a long relocation? (They all travelled in the right direction.) Were they waiting for a favourable current? Did the long journeys

shown in Figs 4.34 and 4.35 depend on ocean currents? Campbell *et al.* (2010) observed that an ability to sustain sustained swimming of 20 km/day (0.2315 m/s) 'seems astonishing', and so it is, particularly given the crocodylian reputation for having only limited aerobic capacity, which they drew attention to. They attributed much of the movement reported by Read *et al.* (2007) to riding surface water currents travelling at 0.3–0.5 m/s, a rate fast enough to account for the observed rate of travel and which they said 'negates the need for active swimming'. Campbell *et al.* (2010) found that for short distances, less than 10 km, there was no association with current direction, but the data clearly showed how well *C. porosus* choose tidal flows to move up and down rivers and there can be little doubt that they take good advantage of currents when available.

Taking this a step further, and noting the lack of diversification within *C. porosus* compared with freshwater-restricted species, Campbell and colleagues hypothesised 'that sea-voyages by *C. porosus* are a frequent occurrence, and should not be viewed as occasional mishaps of navigation but as a successful dispersal strategy'.

Fig. 4.37. *Visitors report seeing* C. porosus *of various sizes on Adele Island, ~80 km off the north-west coast of Western Australia. The crocs have been seen feeding on the young of nesting seabirds and have also been reported visiting other islands distant from the shore in the same general area. (Photos Kevin Coate, Shana Nerenberg) (Note: In April 2014 a >2 m* C. porosus *was photographed at Browse Island, about 80 km further out than Adele Is.; Jarrod Hodgson pers. comm.)*

A consideration of the historic distribution of *C. porosus* (Fig. 4.36), now much reduced of course, shows that the distances between many of the occupied land masses are not extravagantly longer than coastal journeys undertaken by the three translocated individuals described by Read *et al.* (2007). Gratten (2004) analysed mitochondrial DNA and microsatellites from more than 200 *C. porosus* sampled from the western end of Sumatra through to Bougainville and the Buka Islands at the north-western end of the Solomon Islands Archipelago. He concluded that the populations he sampled comprise a single 'evolutionarily significant unit (ESU, *sensu* Moritz 1994)', but with individual populations 'strongly structured on the basis of allele frequency differences and at least seven regional populations qualify as management units'. He identified 'several episodes of range expansion coupled with recurrent gene flow', 'a pattern of isolation by distance', 'no evidence of male-biased gene flow' and concluded that, compared with the freshwater species *C. novaeguineae* and *C. siamensis*, the lack of phylogeographic structure in *C. porosus* 'may reflect a greater capacity for long-distance marine migration'. Gratten interpreted this in the context of *C. porosus* being accomplished at living in salt water (Chapter 11). All three species have lingual salt glands (Taplin 1988) but the salt glands in the other two species, restricted to fresh water, are likely to be less competent.

Whether the lack of phylogenetic structure within *C. porosus* is a consequence of their marine-adapted physiology combined with an aerobic swimming capability, or is dependent upon a capacity to use oceanic surface currents for 'frequent invasion of island populations' (Campbell *et al.* 2010) is not known, but poses a fascinating question. More data from satellite tracking, matched with contemporaneous data on water currents could answer the question. In the meantime, we can only speculate. It could be argued that a particular skill at making use of oceanic currents for very long distance travel might have led to them having a pan-tropical distribution, rather than being restricted to the Indo-pacific. Marine turtles, for example, well known exploiters of ocean currents, are found in all the world's oceans.

Could the journeys home (Fig. 4.35) have been made without a current to ride? Few relevant data are available, but Elsworth *et al.* (2003) swam juvenile *C. porosus* in a flume, seeking the maximum swimming speed that could be supported aerobically, over a range of temperatures. Admittedly it is a wild extrapolation from animals 30–110 cm in total length, but, from their data a 4.5 m *C. porosus* could be expected to sustain aerobically a swim speed of 1.75 m/s at 28°C. This is ~6 km/h: a bit faster than a comfortable human walking speed, but believable for an animal that size, and more than seven times the average speed required to swim the 411 km (minimum distance) from Temple Bay to the Wenlock River in 20 days if there were no current. Swimming it without relying necessarily on a current does not seem impossible, but knowledgeable use of currents is obviously part of their skills base.

There are other examples of significant, purpose-directed swims, some over long distances. *C. porosus* congregate on king tides at a barrage on the Mary River, Northern Territory, to feed on mullet and other fish (Britton *pers. comm.* and Chapter 6). Some come from a distance. One that happened to be wearing a satellite transmitter travelled more than 45 km up the Mary River from the mouth in less than 24 h. Likewise, many large *C. porosus* congregate at Crab Island, off the western tip of Cape York, north Queensland, to coincide with the hatching of flatback turtles (Limpus 2007) and, though hard data are lacking, the implication is that some will have travelled a substantial distance.

Even quite small *C. porosus* can make substantial oceanic journeys. Recently I learned that people visiting islands 70–80 km out to sea from the Kimberley Coast of Western Australia often see estuarine crocodiles or their tracks (Fig. 4.37). Predation on seabird chicks has been observed and there are sleeping birds at night and also turtles and fish. It would be fascinating to have tracking data from some of these animals, to see where they have come from and whether or not such journeys are not only routine but also seasonal to match turtle or seabird activity. Even modest-sized sub-adults can apparently make the trip (e.g. Figure 11.21). This is interesting not only because of its implications for oceanic travelling and navigation, but also for its evidence of capability in salt water for long periods, even by animals of modest size (Chapter 11).

These long journeys imply a capacity for navigation beyond a simple familiarity with the area, although that is undoubtedly very important. Much evidence of 'homing' behaviour over short distances has been reported as well. Yet little is known about the navigational mechanisms. However, some exciting new discoveries about homing pigeons may well be relevant to crocodylians too, and these will be discussed in Chapter 5.

References

Allen V, Elsey RM, Jones N, Wright J, Hutchinson JR (2010) Functional specialization and ontogenetic scaling of limb anatomy in *Alligator mississippiensis*. *Journal of Anatomy* **216**, 423–445.

Baker AA (1956) The swallowing of stones by animals. *Victorian Naturalist* **73**(6), 882–895.

Brander D (1925) Stones in crocodile's stomach. *Field* **146**, 537

Brien ML, Read MA, McCallum HJ, Grigg GC (2008) Home range and movements of radio-tracked estuarine crocodiles (*Crocodylus porosus*) within a non-tidal waterhole. *Wildlife Research* **35**, 140–149.

Bustard HR, Singh LAK (1977) Studies on the Indian gharial *Gavialis gangeticus* (Gmelin) (Reptilia, Crocodilia). Change in terrestrial locomotory pattern with age. *Journal of the Bombay Natural History Society* **74**, 534–536.

Campbell HA, Watts ME, Sullivan S, Read MA, Choukroun S, Irwin SR, *et al.* (2010) Estuarine crocodiles ride surface currents to facilitate long-distance travel. *Journal of Animal Ecology* **79**(5), 955–964.

Campbell HA, Sissa O, Dwyer RG, Franklin CE (2013) Hatchling crocodiles maintain a plateau of thermal independence for activity, but at what cost? *Journal of Herpetology* **47**(1), 11–14.

Campos Z, Coutinho M, Magnusson WE (2003) Terrestrial activity of caiman in the Pantanal, Brazil. *Copeia* **2003**, 628–634.

Campos Z, Coutinho M, Mourão G, Bayliss P, Magnusson WE (2006) Long distance movements by caiman. *The Herpetological Journal* **16**, 123–132.

Chentanez T, Huggins SE, Chentanez V (1983) Allometric relationships of the Siamese crocodile, *Crocodylus siamensis*. *Journal of the Science Society of Thailand* **9**, 5–26.

Cott HB (1961) Scientific results of an inquiry into the ecology and economic status of the Nile crocodile (*Crocodilus niloticus*) in Uganda and Northern Rhodesia. *Transactions of the Zoological Society of London* **133**, 561–572.

Davenport J, Sayer MDJ (1989) Observations on the aquatic locomotion of young salt-water crocodiles (*Crocodylus porosus* Schnieder). *The Herpetological Journal* **1**, 356–361.

Dodson P (1975) Functional and ecological significance of relative growth in *Alligator. Journal of Zoology* **175**, 315–355.

Elsworth P, Seebacher F, Franklin CE (2003) Sustained swimming performance in crocodiles (*Crocodylus porosus*): effects of body size and temperature. *Journal of Herpetology* **37**, 363–368.

Emshwiller MG, Gleeson TT (1997) Temperature effects on aerobic metabolism and terrestrial locomotion in American alligators. *Journal of Herpetology* **31**(1), 142–147.

Fish FE (1984) Kinematics of undulatory swimming in the American Alligator. *Copeia* **1984**(4), 839–843.

Franklin CE, Read MA, Kraft PG, Liebsch N, Irwin SR, Campbell HA (2009) Remote monitoring of crocodilians: implantation, attachment and release methods for transmitters and data-loggers. *Marine and Freshwater Research* **60**, 284–292.

Frey E (1984) Aspects of the biomechanics of crocodilian terrestrial locomotion. In *Third Symposium on Mesozoic Terrestrial Ecosystems*. (Eds WE Reif and F Westphal) pp. 93–97. Attempto Verlag, Tübingen, Germany

Frey E, Salisbury SW (2001) The kinematics of aquatic locomotion in *Osteolaemis tetraspis* Cope. In *Crocodilian Biology and Evolution*. (Eds GC Grigg, F Seebacher and CE Franklin) pp. 165–179. Surrey Beatty & Sons, Sydney.

Graham A (1968) 'The Lake Rudolf crocodile (*Crocodylus niloticus* Laurenti) population'. Report to the Kenya Game Department by Wildlife Services Limited. Nairobi, Kenya, <http://ufdcimages.uflib.ufl.edu/aa/00/00/75/90/00001/lakerudolfcrocodilesalistairgraham.pdf>

Gratten J (2004) The molecular systematics, phylogeography and population genetics of the Indo-Pacific *Crocodylus*. PhD thesis, The University of Queensland.

Grigg GC, Gans C (1993) Crocodilia: morphology and physiology. In *Fauna of Australia Volume 2A. Amphibia and Reptilia*. (Eds CJ Glasby, GJB Ross and PL Beesley) pp. 326–336. Australian Government Publishing Service, Canberra.

Henderson DM (2003) Effects of stomach stones on the buoyancy and equilibrium of a floating crocodilian: a computational analysis. *Canadian Journal of Zoology* **81**, 1346–1357.

Kay WR (2004) Movements and home ranges of radio-tracked *Crocodylus porosus* in the Cambridge gulf region of Western Australia. *Wildlife Research* **31**, 495–508.

Kirshner DS (1985) Buoyancy control in the estuarine crocodile, *Crocodylus porosus* Schneider. PhD thesis, The University of Sydney.

Lance VA, Elsey RM, Trosclair PL, III Nunez LA (2011) Long-distance movement by American alligators in southwest Louisiana. *Southeastern Naturalist* **10**(3), 389–398.

Limpus CJ (2007) *A Biological Review of Australian Marine Turtle Species. 5. Flatback Turtle, Natator depressus (Garman).* Queensland Environmental Protection Agency.

Magnusson WE (1990) Crocs, cows and colleagues on Coburg Peninsula. *Bulletin of the Chicago Herpetological Society* **25**(6), 97–100.

Manter JT (1940) The mechanics of swimming in the alligator. *Journal of Experimental Zoology* **83**, 345–358.

Moritz C (1994) Defining "evolutionary significant units" for conservation. *Trends in Ecology & Evolution* **9**, 373–375.

Parrish JM (1987) The origin of crocodilian locomotion. *Palaeobiology* **13**(4), 396–414.

Read MA, Grigg GC, Irwin SR, Shanahan D, Franklin CE (2007) Satellite tracking reveals long distance coastal travel and homing by translocated Estuarine Crocodiles, *Crocodylus porosus*. *PLoS ONE* **2**(9), e949.

Reilly SM, Elias JA (1998) Locomotion in *Alligator mississippiensis*: kinematic effects of speed and posture and their relevance to the sprawling-erect paradigm. *Journal of Experimental Biology* **201**, 2559–2574.

Renous S, Gasc J-P, Bels VL, Wicker R (2002) Asymmetrical gaits of juvenile *Crocodylus johnstoni*, galloping Australian crocodiles. *Journal of Zoology* **256**, 311–325.

Salisbury SW, Frey E (2001) A biomechanical transformation model for the evolution of the semi-spheroidal articulations between the adjoining vertebral bodies in crocodilians. In *Crocodilian Biology and Evolution*. (Eds GC Grigg, F Seebacher and CE Franklin) pp. 85–134. Surrey Beatty & Sons, Sydney.

Seebacher F, Elsworth PG, Franklin CE (2003) Ontogenetic changes of swimming kinetics in a semi-aquatic reptile (*Crocodylus porosus*). *Australian Journal of Zoology* **51**, 15–24.

Seymour RS (1982) Physiological adaptions to aquatic life. In *Biology of the Reptilia* (Eds C Gans and FH Pough) pp. 1–41. Academic Press, London.

Stubbes H (1668) An enlargement of the observations, formerly publisht Numb. 27, made and generously imparted by that learn'd and inquisitive physitian, Dr. Stubbes. *Philosophical Transactions* **3**, 699–709.

Taplin LE (1988) Osmoregulation in crocodilians. *Biological Reviews of the Cambridge Philosophical Society* **63**, 333–377.

Taylor JA (1977) The foods and feeding habits of sub-adult *Crocodylus porosus* Schneider, in northern Australia. MSc Thesis, The University of Sydney.

Taylor JA (1979) The foods and feeding habits of subadult *Crocodylus porosus* Schneider in Northern Australia. *Australian Wildlife Research* **6**, 347–359.

Taylor MA (1993) Stomach stones for feeding or buoyancy? The occurrence and function of gastroliths in marine tetrapods. *Philosophical Transactions of the Royal Society of London. Series B, Biological Sciences* **341**, 163–175.

Uriona TJ, Farmer CG (2008) Recruitment of the diaphragmaticus, ischiopubis and other respiratory muscles to control pitch and roll in the American alligator (*Alligator mississippiensis*) *Journal of Experimental Biology* **211**, 1141–1147.

Vliet K (1989) Social displays of the American alligator (*Alligator mississippiensis*). *American Zoologist* **29**, 1019–1031.

Vliet K (2001) Courtship behaviour of American alligators, *Alligator mississippiensis*. In *Crocodilian Biology and Evolution*. (Eds GC Grigg, F Seebacher and CE Franklin) pp. 383–408. Surrey Beatty & Sons, Sydney.

von Humboldt A (1852) *Personal Narrative of Travels to the Equinoctial Regions of America. Volume 2.* Henry G. Bohn, London.

Walsh B, Whitehead PJ (1993) Problem crocodiles, *Crocodylus porosus*, at Nhulunbuy, Northern-Territory – an assessment of relocation as a management strategy. *Wildlife Research* **20**, 127–135.

Webb GJW, Gans C (1982) Galloping in *Crocodylus johnstoni* – a reflection of terrestrial activity? *Records of the Australian Museum* **34**(14), 607–618.

Webb GJW, Messel H (1978) Morphometric analysis of *Crocodylus porosus* from the north coast of Arnhem Land, northern Australia. *Australian Journal of Zoology* **26**, 1–27.

Webb GJW, Manolis C, Buckworth R (1982) *Crocodylus johnstoni* in the McKinlay River area, N.T. I. Variation in the diet, and a new method of assessing the relative importance of prey. *Australian Journal of Zoology* **30**, 877–899.

Whitaker R, Andrews H (1989) Notes on crocodilian locomotion. *Journal of the Bombay Natural History Society* **85**(3), 621–622.

Willey JS, Biknevicius AR, Reilly SM, Earls KD (2004) The tale of the tail: limb function and locomotor mechanics in *Alligator mississippiensis*. *Journal of Experimental Biology* **207**, 553–563.

Wings O (2007) A review of gastrolith function with implications for fossil vertebrates and a revised classification. *Acta Palaeontologica Polonica* **52**(1), 1–16.

Zug GR (1974) Crocodilian galloping: a unique gait for reptiles. *Copeia* **1974**(2), 550–552.

5

SENSORY SKILLS AND BRAIN

In 1973 Grahame Webb and I were travelling at speed up a stretch of the Liverpool River in Arnhem Land, Northern Territory, in an open 5 m aluminium boat when we spotted a pygmy goose settled on the water up ahead. Almost simultaneously, a movement caught our eyes as a 2.5 m C. porosus ran down a high bank opposite the goose, dived into the water and disappeared. Was it after the goose? We pulled to a halt. The tidal current was driving eddies of floating leaves. Spits of rain bounced droplets from the surface of the turbid river. The goose paddled on. Suddenly the croc surfaced midstream, looking straight at the goose now about 10 m away and submerged again. Within a few more seconds, the goose was pulled down and disappeared under the muddy water. We saw no further sign of either goose or croc.

Think about the sensory skills required for this and similar acts. The goose was originally about 40 m away, so the croc showed good visual acuity. It swam underwater, following a planned course across the current, probably with some knowledge of its depth. Its brief surfacing allowed a check on direction and distance. But it could not rely on vision to travel the last 10 m, the water was too muddy. Yet it located the goose and pulled it under without even breaking the surface. I was impressed! For some time, we've known that crocodylians have good visual acuity and binocular vision, and any creature with a lung might be able to sense hydrostatic pressure and so assess depth. But how did it find the goose? Could it sense the feet, presumably paddling away in the murky stream? The answer to that has become clearer only

recently: they use their integumentary sensory organs (ISOs) and these and other crocodylian sensory organs and skills will now be explored.

Crocodylians need to sense two worlds: air and water. The different physical properties of these two media in transmitting light, sound (pressure waves and vibrations) and chemicals have 'design' implications for crocodylian sensory systems. Crocs often float at the surface almost completely submerged, with just the eyes, nostrils and (usually) ears visible, well placed for sensing the world above the waterline (Figs 5.1, 5.2). They have the same senses that we do: vision, touch, hearing, smell and taste (and balance and sensitivity to temperature). They also have some very interesting sense organs on their skin (integumentary sense organs, ISOs) that are a bit reminiscent of the lateral line organs in fishes but are in fact unique to crocodylians. Alligatoridae have ISOs only on the head, but other crocodylians have them all over, presumably enabling more input from underwater sources (Fig. 5.3). New research is generating exciting results about ISOs and it has become clear that they are multifunctional, sensing action below the waterline as well as around the teeth and jaws, all useful in feeding, and even sensing changes in pH. ISOs are particularly dense on the head, and crocs pay a lot of attention to each others heads during courtship, stroking and rubbing (Fig. 5.30, Chapter 12) so they almost certainly play an important role in this behaviour.

The eye of Crocodylus porosus, *with part of the ear opening and a number of integumentary sense organs also visible. (Photo DSK)*

Fig. 5.1. *The monitoring and command centre is mainly above water. Crocodylians can lurk hidden in water with just their eyes and nostrils breaking the surface (vision, olfaction and respiration). In this photo, the ears are above water too, but their underwater hearing is also good anyway via bone conduction. Sense organs are distributed over the body surface, and there are specialised sensory receptors on the head scales of Alligatoridae, and* **all** *the scales of Crocodylidae, which respond to a range of inputs from the surrounding water, including touch, water movement and even pH. (Photo DSK at WILD LIFE Sydney Zoo)*

Crocodylians may also have magnetoreceptors, giving them navigational skills. The last chapter reported that at least some crocodylians make long journeys and seem to be good navigators. New research posits the possibility of navigation by magnetoreception, highlighting a sense organ that is lacking in mammals but characteristic of all other amniotes – the lagena – which is part of the vestibular system. As far as we know, crocodylians lack electrosensory perception, which is found in many fishes and in echidnas and platypus.

This chapter reviews crocodylian sense organs and briefly discusses brain, learning and intelligence.

VISION

Crocodylians rely heavily on vision. They hunt actively by night and by day and in most species their eyes have to cope with a wide range of light intensity, from bright beaches to palest moonlight, and under water. Anatomically their eyes are not very different from our own, except the pupil is a vertical slit (Fig. 5.4) and they have a reflective tapetum that enhances night vision. They have good visual acuity, are equipped to see colours and they have binocular vision. There is still some uncertainty about their visual acuity underwater.

OPERATIONAL ASPECTS, BINOCULARITY, VISION UNDERWATER

The external appearance of the eyes and their closure, including the operation of the nictitating membrane were discussed in Chapter 3.

The eyes are set at the highest points of the skull in most crocodylians and, even when almost

Fig. 5.2. *If the water were not so clear and still, or if there were even a small amount of flotsam or floating vegetation, this animal would be nearly impossible to see, yet it is well able to keep a good surveillance over the surroundings both above and below water and could act in a flash with a strong thrash of the tail should a wallaby or a goose chance along. Intermittent ventilation is an advantage, but their control over buoyancy is so good that their breathing can be almost imperceptible. (Photo Garry Lindner)*

Fig. 5.3. *An underwater view of a large* C. porosus *to complement Fig. 5.2. Every scale of this animal, as well as all crocodylids and gavialids, has at least one integumentary sense organ (ISO), unique to crocodylians. Alligatorids have them too, but only on their heads. It is only in the last couple of years that the depth and breadth of the capabilities of ISOs are being revealed. (Photo DSK)*

all the animal is below the water surface, they command a good view of surroundings. This view is wide angle: ~260° compared with 200° in humans. Underwood (1970) reported that the cornea subtends an angle of ~128°, with the fields of view overlapping anteriorly by 25° (Fig. 5.5). This, with their eyes being set sufficiently wide apart, implies a capacity for at least some binocular vision (stereopsis) and, thus, depth perception. The species on which those measurements were made seems not to have been recorded. However, Pettigrew (*pers. comm.*) has confirmed overlapping fields of view in *C. porosus* and, further, Pettigrew and Grigg (1990), by labelling and tracing retinal connections histochemically to the forebrain in *C. johnstoni*, confirmed that crocodylians have the neural substrate for binocular vision. Combined with behaviour that implies excellent depth perception, these observations collectively make a strong case for good binocular vision. Interestingly, the neural connections between the eye and the brain in *C. johnstoni* (Pettigrew and Grigg 1990) and, presumably other crocodylians, follow the avian (archosaur) pattern. That is, their stereopsis is achieved in the same way as birds and differently from other reptiles and from mammals. This is another example of the similarity of crocodylians to birds, and their distinctness from other reptiles. In addition to the obvious advantages to a predator of being able to judge distance, binocular vision is also thought to aid the detection of low luminance objects at night and under water, and the perception of direction of movement (Stevens 2006).

When submerged, the eye is open but covered by the nictitating membrane, sweeping across from in front (Figs 5.4, 5.6, 3.22, 9.5, 9.6, 9.9). Because the nictitating membrane is semi-transparent, there has been speculation that it functions as a correcting lens, adapting the eye to underwater vision. This is an appealing idea, but the membrane is translucent, rather than transparent, and unlikely to permit sharp underwater vision. Fleishman *et al.* (1988) made direct observations on the optics of six species – *Caiman crocodilus*, *Crocodylus johnstoni*, *C. acutus*, *C. rhombifer*, *Paleosuchus palpebrosus* and *Gavialis gangeticus* – representing all three crocodylian families (Alligatoridae, Crocodylidae

and Gavialidae). All showed a good ability to focus in air, but were severely farsighted under water, as are humans. Comparing the refractive properties of the eye with and without the cover of the nictitating membrane, they concluded that the membrane does not modify refraction so, presumably, it does not function as a correcting lens. Furthermore, unlike in predatory birds, accommodation by the lens was found insufficient to adapt the eye for underwater vision. They also observed that, contrary to what one might expect in lower light, the pupil size diminished in submerged animals and the size of the pupil in Fig. 9.9 seems to confirm this. This could increase the depth of field, like using a higher f-stop on a camera, and raises the possibility of at least some limited improvement in vision while submerged. However, the Fleishman team's overall conclusion was that crocodylian vision is well adapted for vision in air but poorly so for vision under water, and that 'sensory systems other than vision must play an important role in prey capture underwater'. Their comment was prescient, for we now know about the integumentary sense organs (ISOs), discussed below.

Despite the conclusion that underwater vision in crocodylians is much less acute than above water, it is good enough to be useful and the question of underwater vision cannot yet be regarded as completely settled. At the very least, enough light passes through the nictitating membrane to provide information about gradients in light intensity and thus the direction to the bottom or to the water surface. However, the extent to which they can form images begs further experimentation. Observations of Smaug, a large captive male *C. porosus* (Figs 9.5, 9.6), have shown that crocs submerged in clear water can see an object moving above water (Adam Britton *pers. comm.* 2012). Accustomed to being fed chickens, Smaug can see one from below the water surface, moving his head to track the movement of the white carcass before launching himself out of the water at great speed. Likewise, if he misses it he is well able to see and seize the chicken under water as it drifts within striking range of his head. Crocodylians probably do have blurry underwater vision, as we do, with low acuity and not much capacity for focus, but still enough to be useful.

Fig. 5.4. *The eye of a juvenile saltwater crocodile,* Crocodylus porosus. *Note the vertical slit pupil and the edge of the nictating membrane anteriorly. (Photo DSK)*

Many (most?) of them spend most of their time in turbid water anyway, where eyes may be of little use beyond detecting light intensity gradients.

Fleishman and Rand (1989) showed that crocodylians hunt very well under water, despite having poor vision there. They went on to show that *Caiman crocodilus* could capture prey under water in total darkness, as long as there was some movement of the prey. From recent work, it can be assumed that the main sensory systems operating in this situation are almost certainly the specialised integumentary sense organs (ISOs), in combination with touch receptors and the ears (see below).

If it seems surprising that crocodylians have comparatively poor underwater vision, it is worth remembering that their evolutionary history is of descent from a long lineage of terrestrial forebears (Chapter 2). Besides, somewhere in their adaptation to aquatic life they acquired ISOs, a supplementary sense with 360° input whose evolution may have

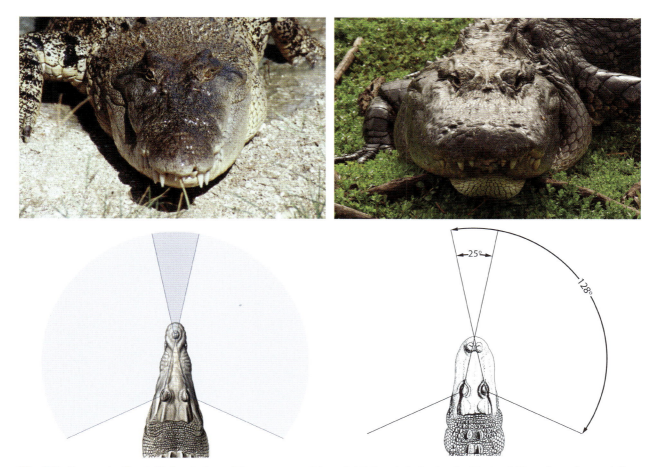

Fig. 5.5. *Stereopsis. (Top left) Front view of* C. porosus *and (top right)* A. mississippiensis. *(Bottom) Plan view of crocodylians showing fields of view as reported by Underwood (1970). We have assumed, for the sake of illustration, that all crocodylians have a similar field of view. (Photos GCG, DSK)*

Fig. 5.6. *Underwater photograph of the eye of the estuarine crocodile,* C. porosus, *covered by its translucent nictitating membrane. Speculation that the membrane acts as a correcting lens to aid underwater vision is not confirmed by research, but anecdotal observations show that* C. porosus, *anyway, has at least blurry vision underwater. While submerged, captives seem well able to see their white chicken dinner arriving (see text). More research is needed. (Photo Ruchira Somaweera)*

reduced selection pressures for improvement to underwater vision.

ANATOMY OF THE EYE

The crocodylian eye has the usual vertebrate structure (Fig. 5.7) and is moved within the orbit by the usual array of muscles controlled by cranial nerves (Fig. 5.16, Table 5.1). The cornea is the main refractile element focussing light onto the retina. It is highly curved, suggesting specialisation for terrestrial vision. Its shape is able to be modified to some extent by ciliary muscles (Fleishman *et al.* 1988).

Crocodylians have pupils that close to a slim vertical slit in bright light and open to a wide circle in the dark. Such pupils are known as stenopeic (narrow) and are typical of animals that are active over a wide range of light levels. It could be that during the day, when the pupil is a narrow slit, the depth of field is improved, offsetting the limited capacity of the lens for accommodation. Additionally, Malmström and Kröger (2006) have found that slit pupils are commonly associated with multi-focus lenses, and this was true in the two species they studied, *C. niloticus* and *Osteolaemus tetraspis*. Such lenses have multiple focal points – that is, distinct concentric zones of different refractive powers – each tuned to a

different wavelength of light and, so, compensating for chromatic aberration caused by different colours having different wavelengths. Multifocal optical systems lead to sharpened colour images.

Many fishes, reptiles and birds have plate-like bony scleral ossicles surrounding the iris, but they are lacking in crocodylians. They occurred, however, in many fossil crocodylians, and it has been suggested that their evolutionary loss has accompanied the adoption of nocturnal habits. Without ossicles, the aperture of the pupil is able to open more widely and let in more light.

Crocodylians, probably more than most animals, have to operate over a great range of light intensities. In a single day, they may have to cope with intensity ranging from the daytime glare and brightness of a sandy beach to the darkness of a moonless night. Opening and closing the pupil is only part of the mechanism by which the sensitivity of the eye is adjusted. The eye can also adapt to dark or light by facilitating exposure of the appropriate light receptors. This is typical of many vertebrates, and is achieved by the migration of pigments within the retinal cells and/or lengthening or shortening of the cells themselves.

RETINAL CELLS, PIGMENTS, COLOUR VISION

Vertebrate retinas consist of millions of cellular photoreceptors (rods and cones, named for the shape of their outer segments; Fig. 5.7), which contain photosensitive pigments. The pigments within the photoreceptors are sensitive to light of a particular wavelength and transform it into a minute electrical signal. This is carried by the retinal interneurons (horizontal, bipolar and amacrine cells) (Fig. 5.7) to the ganglion cells and conveyed along the optic nerve to the brain where the electrical activity of the myriad nerve fibres is integrated and interpreted in the optic lobes. Functional aspects of the interneuron cells and their interconnections in birds, likely to be very similar to crocodylians, can be found in Husband and Shimizu (2001). The spectral sensitivity of the receptors determines the nature of the resultant image that the animal has of its surrounding world. The retinas of crocodylians have both rods and

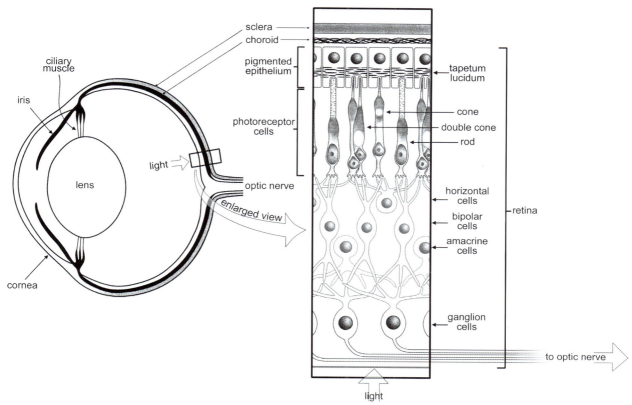

Fig. 5.7. *Gross anatomy of the crocodylian eye, with a cross section of the retina. Note the tapetum lucidum 'behind' the retina, which reflects incoming photons of light, increasing the likelihood of photon absorption in photoreceptors and thus enhancing night vision. It is reflection from the tapetum that causes crocodylian eyes to shine red so spectacularly at night when a torch or spotlight is used (Fig. 5.9). The rods are very sensitive and are used in dim light, while the cones are comparatively less sensitive and require bright light conditions to function. Colour vision arises from different classes of cones, each of which contains a different type of pigment with peak absorption at different wavelengths (Fig. 5.8). The double cones consist of a 'principal' cone and an 'accessory' cone. (Drawn from various sources, including Walls (1963), Laurens and Detwiler (1921) and Kalloniatis and Tomisich (1999))*

cones. In general, rod photoreceptors are sensitive in dim light, cones in bright light, and colour vision results from cones with several types of pigment, each with a sensitivity maximised at a different wavelength.

The photoreceptor cells, their pigments and their spectral sensitivities have been examined in *Alligator mississippiensis* (Sillman *et al.* 1991) and enough work has been done on *Caiman crocodilus* and *C. niloticus* to indicate general similarities across the group. Birds and most reptiles have oil droplets in their photoreceptors, but they seem to be lacking in crocodylians (Underwood 1970).

The alligator retina is made up mainly of rods, with ~28% cones (Sillman *et al.* 1991), and so is adapted for both scotopic (low light) and photopic (well-lit) vision. The highest densities of rods and cones occur in the central area of the retina (*area centralis*) and in crocodylians this is in the form of a horizontal band, which results in the greatest visual acuity around their visual horizon, at and a bit above the plane of the water surface. This must be a very important for crocodylians: a retinal specialisation that enhances their ability to detect prey at the water land interface. Like mammals, birds and many other vertebrates, crocodylian rods contain rhodopsin (Wald *et al.* 1957), a pigment based on Vitamin A1. *Caiman crocodilus* and *Crocodylus niloticus* have rods with rhodopsin sensitivity at around 500 nm (Fig. 5.8) and when Sillman and colleagues measured rhodopsin of *A. mississippiensis* they found a similar sensitivity at 501 nm. Rhodopsin has been shown to have a capacity for rapid dark adaptation and in *A. mississippiensis* the rods expand in the dark and contract in the light (Laurens and Detwiler 1921). Cones in *A. mississippiensis* are either single or double, and Sillman *et al.* (1991) found them to contain four pigments with overlapping spectral sensitivities, providing for good colour vision. They found that one type of single cone has a pigment sensitive at 535 nm and another at 444 nm (Fig. 5.8). Each double cone has a principal and an accessory cone (Fig. 5.7). The principal double cone is sensitive at 566 nm and the accessory cone at 503 nm, similar to the rods. As well as offering colour vision, this multiple cone pigment system allows for good wavelength discrimination: a mechanism for enhancing contrast and so improving visual acuity.

Double cones are found in all four classes of tetrapod vertebrates, but not in placental mammals. Fish have twin cones, which may be similar. Underwood (1968) made the intriguing suggestion that double cones may allow the detection of polarised light: a proposition put forward independently by Cameron and Pugh (1991). Hart and Hunt (2007), reviewing visual pigments of birds, observed that this idea remains controversial. A capacity for crocodylians to perceive polarised light would be interesting in relation to their (now) well-documented navigational abilities (Chapter 4). Other theories are that double cones have a role in detecting luminance and/or movement, but their actual function remains unresolved.

During the day, therefore, the eyes of crocodylians are well set up for their daytime lives when visual acuity and colour discrimination are vital. As predators, and with their complex social signalling (Chapter 12), it is clear that they need a good visual system. They see also well adapted for good night vision.

NIGHT VISION, THE TAPETUM LUCIDUM

For night vision, the rods adapt to dark (see above) and the slit pupil widens to round, letting in more light. Crocodylians also have an additional 'string to their bow': the retinal tapetum lucidum (literally 'shining carpet' in Latin). This is composed of crystalline platelets containing guanine in a layer within the tips of the retinal pigment cells, on the choroid side (Fig. 5.7). The parallel arrangement of the crystalline platelets has been likened to fibre optics (Dieterich and Dieterich 1978, working on *Caiman crocodilus*), whereby incident light is guided, rather than scattered, within the retina and, by reflection, maximises the chance of photons activating the breakdown of photosensitive pigment and generating a signal. In this way, the tapetum reflects photons that were not absorbed on their first pass through the layer of photoreceptors, giving them a second chance. Thus, visual sensitivity in low light is enhanced and the reflection from the tapetum is what gives crocodylians their 'eye shine' in response to a light shone at them at night. Among reptiles, only the crocodylians have a tapetum (Schwab *et al.* 2002) but tapeta also occur in many nocturnal birds, mammals and fishes, with structure differing a little between different groups. Most people will be familiar with the tapetum 'eye shine' in cats, dogs and foxes. Humans lack a tapetum: the familiar 'red eye' in humans photographed using a flash occurs when the pupils are dilated, with the red being a reflection from the blood vessels of the choroid.

The 'eye shine' of crocodylians, seen when the eyes reflect light back to an observer shining a light (Fig. 5.9), is well known, is exploited by hunters and by researchers and has given crocodylians an unfortunate 'vulnerability' in modern times because hunters can find them so much more easily. On a dark night, even a dim flashlight is often enough to disclose a lurking crocodylian that would otherwise be well hidden behind vegetation at the water's edge.

If the retina is inspected through the lens, the tapetum is found to be visible only in the upper half of the retina, which, because of the way lenses work, inverting and reversing the image, 'sees' the lower half of a scene. This differentiation between upper and lower is apparently because in the light-adapted eye the reflective tapetal crystals in the lower half are masked by pigment in melanosomes within the retinal cells. That is, in bright light, the tapetum is occluded in the lower half of the retina, turning down the light sensitivity in the upper half of the field of vision, while leaving the upper half of the retina sensitive to lower light in the lower visual

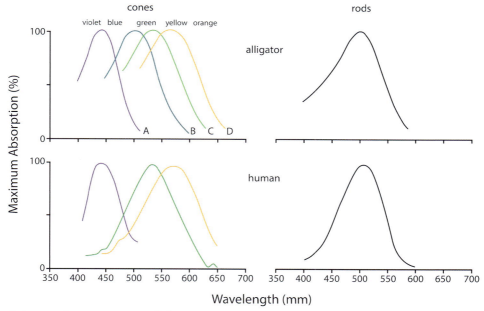

Fig. 5.8. *Retinal pigment absorption curves of* Alligator mississippiensis *and humans. In crocodylians, as in humans, the rods are specialised for dim light and different cones contain different photosensitive pigments, four types in total, which implies colour vision. A typical rod had a maximum absorbance at 501 nm, single cones were measured in the violet at 443 nm (A) and green, 535 nm (C), with an accessory cone at 503 nm (blue, B) and a principal cone at 566 nm (yellow, D). (Redrawn after Sillman* et al. *1991 and other sources.)*

field, such as for hunting on the bottom (Dieterich and Dieterich 1978). In adapting to darkness, it is presumed that the pigment migrates to expose the tapetum in the lower half of the retina as well. These attributes underline nicely the way that crocodylian eyes are adapted to cope with the visual challenges posed by life at the water–air interface and by being both diurnal and nocturnal.

Modern crocodylian families have apparently changed little since the late Cretaceous in the Mesozoic Era, many having broadly similar lifestyles. The multiple-cone pigment system is probably as old as that or older and may well have been typical of archosaurs in general. This supports the common assumption that dinosaurs, too, had good visual acuity and colour vision.

HEARING, MOVEMENT AND BALANCE

Very unusually among reptiles, crocodylians vocalise and communicate with each other by sound. They vocalise as adults, as hatchlings and even from within the egg. Sound production and the

use of vocalisation for intraspecific communication will be discussed in Chapter 12.

Good hearing is a pre-requisite for effective communication and it is also very important for monitoring sounds in the environment, such as those made by animals that could be sources of food or danger. Crocodylians' acute sensitivity to airborne sound is easy to observe and has been confirmed by studies on several species. They have good hearing under water as well, both in the human audible and in the infrasound part of the spectrum. Vergne *et al.* (2009) provided a substantial review of sound production and detection in crocodylians, noting that the functional anatomy and sensitivity of the ears are similar to those in birds. Crocodylians undoubtedly have good skill at sensing acceleration, position and balance. These diverse but anatomically linked capabilities are now discussed.

FUNCTIONAL ANATOMY OF THE EARS AND ASSOCIATED SENSE ORGANS

The external location and appearance of the ears have been discussed in Chapter 3 (Fig. 3.21). The

Fig. 5.9. *Nocturnal cities of crocodylian eye shines. (Left) A multitude of juvenile* C. porosus *in a crocodile farm in Madang, Papua-New Guinea and (right) a night-time feeding aggregation of* Melanosuchus niger *in Mamiraua Reserve near Tefé on the Solimões river ~500 km upstream from Manaus, Brazil. (Photos Mark Read and William E Quatman)*

visible flaps, together with the short flattish cavity in which the eardrum lies, comprise the outer ear (Fig. 5.10). The middle ear consists of a tympanum (eardrum), which overlies the tympanic cavity and a bony rod, the columella, which transmits vibrations through the tympanic cavity to the sensory inner ear where these sound waves are sensed in the cochlea (Fig. 5.11). The cochlea in crocodylians and birds is an elongated bent tube through bone, subdivided into three canals: the central one (cochlear duct) containing the basilar papilla, which bears sensory hairs through which sound is transduced to an electrical signal. The basilar papilla is the homologue of the mammalian Organ of Corti. As in other vertebrates, a system of semi-circular canals containing fluid and fine sensory hairs is associated with each ear and, by analogy with other vertebrates, conveys information about movement and position in space to maintain balance and posture (Fig. 5.11).

Although there are significant differences at the level of the sensory cells (Manley 2000), the ears and associated canal systems of crocodylians are sufficiently similar to human ears (and tetrapods in general) that similar terminology can be used. The main differences at a superficial level are that: the cochlea is only a short, bent tube terminating in a conspicuous lagena, which mammals lack; and the connection between the eardrum and the cochlear structures lacks the more complicated, multi-element and lever action system of mammals, into which lower jaw elements have become

incorporated. Crocodylian ears and bird ears are very similar both structurally and functionally, and authors tend to discuss the inner ears of birds and crocodylians together as the 'archosaur' ear. This means that some likely operational details for crocodylians can be inferred from what is known for birds. Gleich and Manley (2000) and Vergne *et al.* (2009) provide comprehensive reviews.

The inner ear of crocodylians consists of three separate elements (Fig. 5.11) – the semicircular canals, the otolith organs (utricle, sacculus and lagena) and the cochlea – each providing important and different situational information to the brain. The semicircular canals are mutually at right angles, so they can respond to rotational motion in three dimensions. Head rotation gives rise to movement of fluid within the semicircular canals, through its inertia, so stimulating the sensory cells in the cristae within the ampullae. The semicircular canals communicate with the three otolith organs – the utricle at their base, the adjacent sacculus and the third, the lagena, at the apex of the cochlea. These three otolith organs respond to gravity acting on the relatively more dense mineralised otoliths (literally 'earstones'), which, in each organ, sit on sensory patches called maculae. They provide information about the angle of the head and also linear accelerations, so conveying information about movement and position in space in order to maintain balance and posture. The cochlea houses the basilar papilla, which is the organ of hearing. Arriving soundwaves are translated into tiny movements of the eardrum, which are transmitted to the cochlea via the columella and generate corresponding bulk movements of fluid within the scala vestibuli and the scala tympani, connected fluid-filled canals on either side of the basilar papilla. These canals connect at the apex of the cochlea. Fluid displacement is facilitated by elasticity of the fenestra rotunda (round window) (Fig. 5.11). Movement of the fluid on its way around the cochlea stimulates specific sensory hair cells in the basilar papilla, depending on sound

frequency and intensity, and the resultant electrical signals travel to the brain via the 8th (auditory and vestibular) cranial nerve (Fig. 5.16). These signals are then processed and interpreted in the medulla and cerebral cortex. Vergne *et al.* (2009) have discussed the neurological pathways and events, drawing attention to the overall similarity between crocodylians and birds.

The sacculus and the lagena may each have sensory roles in addition to orientation and balance. The sacculus may be the focus of infrasound reception and the lagena may have specialisations for sensing the Earth's magnetic field. Each of these possibilities will be discussed below.

The tympanic cavities, which are air-filled, are connected to the pharynx by the Eustachian tubes which, presumably, as in mammals, allow pressure equalisation (Fig. 5.15). Also, the tympanic cavities are interconnected by an air-filled passage running transversely across the head (Wever 1971). This provides an interconnection between left and right Eustachian tubes and its possible significance for hearing is discussed below.

HEARING SENSITIVITY IN AIR AND WATER IN THE HUMAN AUDIBLE RANGE

Crocodylians have excellent hearing in air. Their acute sensitivity to airborne sound is easy to observe and has been confirmed in several studies on several species. Audiograms of three species were measured by Wever (1971) by monitoring cochlear potentials in response to generated sounds. All were most sensitive in the mid-range: *A. mississippiensis* at 100–1000 Hz, *Caiman crocodilus* at 300–2000 Hz and *C. acutus* at somewhat higher frequencies, 700–2000 or 3000 Hz (two different specimens. Wever rated the sensitivity as 'excellent in comparison with most birds and with many of the mammals'. Klinke and Pause (1980) made recordings from single fibres of the auditory nerve of *Caiman crocodilus* and reported broadly similar results.

Crocodylians have good hearing under water as well. Higgs *et al.* (2002) measured the brain stem responses of *Alligator mississippiensis* to stimulation over a range of frequencies and concluded that they compared very favourably with below water

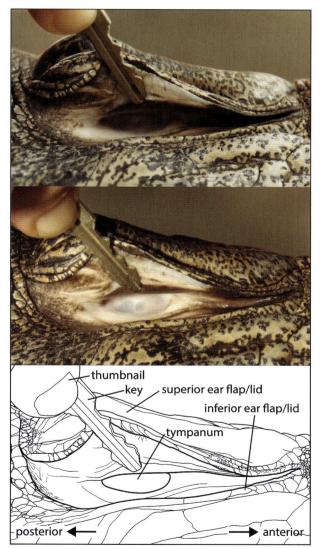

Fig. 5.10. *The ears are behind the eyes. The earlids are moveable by muscles and typically only the anterior third is held slightly open as a result of holding the lower lid depressed. (With reference to Shute and Bellairs 1955) (Photos DSK)*

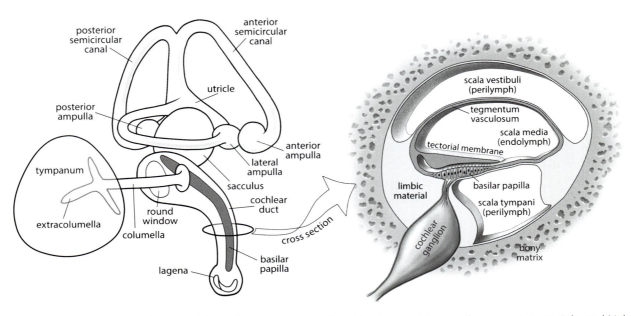

Fig. 5.11. *(Left) Schematic diagram of the auditory apparatus and semicircular canals in an archosaur, anterior to right, and (right) diagrammatic cross section through the cochlear duct. Sound waves stimulate movements of the tympanum, which are transmitted via the extra-columella and columella to the base of the cochlear duct, next to the round window. The cochlea is much shorter and straighter than in mammals, but its length is similar between crocodylians and birds. It is connected to the sacculus and houses the sensory basilar papilla along its length and the lagena at its tip (after Baird 1970; Khan et al. 1982; Lewis et al. 1985; Gleich and Manley 2000; Witmer et al. 2008).*

hearing in goldfish and above water hearing in budgerigars (Higgs *et al.* 2002) (Fig. 5.12). Higgs and colleagues found that a bubble of air trapped under the earlid in the external ear canal did not affect auditory sensitivity under water. They concluded that, under water, hearing in the human audible range occurs via sound waves transmitted through the bones of the skull, not through the middle ear. This is true of human underwater hearing also.

Understandably, these studies have been on juveniles. Because the nature of vocalisations and social communications change as individuals mature, data from adults would be of considerable interest. Significant steps in that direction have been taken recently by Neil Todd from the University of Manchester, working on mature *Alligator mississippiensis*, as described below.

SUB-AUDIBLE VIBRATIONS, INFRASOUND: POSSIBLE ROLE OF THE SACCULUS

Crocodylians make and also hear sounds at low frequencies, below 100 Hz, and even in the 'infrasound' below the limit of human hearing at ~30 Hz. Indeed, this low frequency is a major channel of communication for them, which Kent Vliet recognised while studying courtship in *Alligator mississippiensis* in Florida (Vliet 1989, 2001). Research by Neil Todd at the Australian Reptile Park, just north of Sydney, also working on *A. mississippiensis*, has confirmed it (Todd 2007 and *pers. comm.*). *A. mississippiensis* is regarded as the most vocal of all the crocodylians (Chapter 12). Adult males and females frequently bellow soon after dawn, and often in chorus. Todd recorded very loud (up to 140 dB) low frequency bellows well below 100 Hz, typically 30–50 Hz and as low as 20 Hz. In adult males, the infrasound components are referred to as sub-audible vibrations (SAVs) because they cannot be detected by the human ear. The low frequency sounds produce a very visible 'water dance' from the flanks and over the alligator's back and this is significant in courtship (Fig. 5.13 and Chapter 12).

It seems likely that alligators can detect airborne sounds from other alligators from more than 150 m and water borne SAVs from 1.5 km (Fig. 5.14)

(Todd 2007). What part of the alligator ear is picking up these SAVs? The shape of the 100–10 000 Hz audiogram suggests that the basilar papilla is unlikely to be stimulated and Todd (2007) has suggested that the sacculus is the much more likely candidate. The sacculus is the primary hearing organ in fish, and plays a role in seismic sound detection in frogs. It has been thought to have no auditory function in mammals and to function only as an otolith organ, as discussed above. However, it is easy enough to see how the sacculus could respond to the low frequency SAVs. Todd made the argument, based on work in fishes and amphibians, that hearing via the sacculus is related to reproductive behaviour, and if the sacculus is an organ of hearing in alligators, this would mean they (and, by implication, other crocodylians

also) use two distinct sensory mechanisms: the cochlear apparatus for 'environmental' sounds in the (human) audible range, and the sacculus for reproductive signals delivered by infrasound. Todd suggested that the sacculus may have an auditory function in all vertebrates, hitherto overlooked in the amniotes, and he has gone so far as suggesting (in a seminar at the University of Queensland in 2007) that the appeal of the thumping bass percussion in rock music may lie in its stimulus to the sacculus, with ancient sexual overtones. More will be discussed about the role of sound, including sub-audible vibrations, in Chapter 12.

MAGNETORECEPTION? POSSIBLE ROLE OF THE LAGENA

Crocodylians have well developed, but so far unexplained, navigational skills. There is much evidence of 'homing' behaviour over short distances, and some of the long journeys discussed in Chapter 4 suggest navigational skills beyond learned familiarity with an area. Explanations for the outstanding navigational skills shown by so many animals, birds and sea turtles in particular, have been, and are still, an enduring puzzle for biologists. Ideas about magnetoreception have been at the forefront among hypotheses, with a sensor based on a magnetite-like iron mineralisation located … somewhere, and such mineralisations are not uncommon in the animal kingdom.

Magnetoreception is the sensory ability to perceive magnetic cues, transduce them and transfer them to the nervous system and to the brain, where processing and interpretation occurs (Wajnberg *et al.* 2010). Behavioural experiments by Kenneth and Catherine Lohmann and colleagues at the University of North Carolina have demonstrated magnetoreception by hatchling loggerhead sea turtles (Lohmann and Lohmann 1996; Putman *et al.* 2011; Lohmann *et al.* 2012). They have presented convincing evidence that the hatchlings can perceive both latitudinal and longitudinal information and interpret it against an inherited 'magnetic map'.

Such a skill depends on being able to sense and integrate magnetic field polarity, inclination

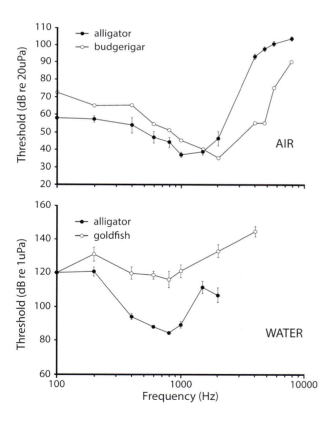

Fig. 5.12. *Audiograms of* A. mississippiensis *compared with (top) budgerigars and (bottom) goldfish; the lower the threshold value, the greater the sensitivity. The presence or absence of an air bubble in the outer ear, under the earlid (Fig. 5.10), made no difference to the under water sensitivity of alligators to sound, suggesting that sound reception under water occurs via the skull rather than via the tympanum. (Adapted from Higgs* et al. *2002)*

Fig. 5.13. *Water dancing from the flanks of a male American alligator early in a bellow. (Photo Kent Vliet)*

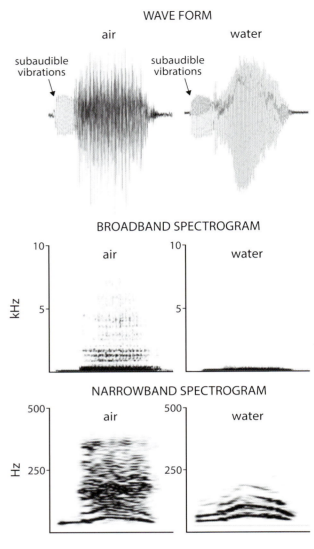

WAVE FORM

air water

subaudible vibrations subaudible vibrations

BROADBAND SPECTROGRAM

air water

NARROWBAND SPECTROGRAM

air water

1 second

Fig. 5.14. *(Top) Waveforms and (middle and bottom) spectrograms made from recordings in air (left) and water (right) of a male American alligator bellowing. The lower pair of spectrograms show the frequencies below 500 Hz where most of the sound energy is concentrated. Note the sub-audible vibrations preceding the vocalisations. Neil Todd has suggested that the sacculus, which is the primary acoustic receptor in fishes, is the most likely candidate in crocodylians for the reception of infrasound. (Reprinted and adapted with permission from Todd 2007, © 2007 Acoustical Society of America) (Recordings of these sounds are available at http://www.acoustics.org/press/151st/1012_male1_mic.wav (in air) and http://www.acoustics.org/press/151st/1012_male1_hyd.wav (in water))*

and intensity. The first two are enough to provide latitude; the challenge is to derive longitude: a challenge that defeated human navigators for centuries! Measurement of intensity, however, can provide information about longitude, albeit with different precision at different places. This is because, although inclination and intensity both grade with latitude, lines of equal inclination (isoclinics) diverge sufficiently across many wide areas from lines of equal intensity (isodynamics) to provide, in combination with the other magnetic field attributes, unique geopositional information.

How can a sea turtle sense and interpret these tiny forces (and is that relevant to navigation by crocodylians)? Most of the 'mechanism' studies have been done on birds and with so many homologies between turtles, crocodylians and birds, any explanation for the navigational skills of turtles and birds could be relevant to crocodylians

too. That the homing behaviour of pigeons can be disrupted by attaching magnets to them was shown long ago (Keeton 1971) but their ability to use the

Earth's magnetic field for navigation, along with other cues, seems to be less well established than it is in sea turtles.

The answer may lie in the lagena: a somewhat enigmatic organ at the tip of the cochlea. All vertebrate groups except the mammals have a lagena. Harada *et al.* (2001) reported iron-rich particles in the lagenal otoliths of some fishes and pigeons and speculated that these may have magnetic properties that could convey geomagnetic information useable for navigation. Subsequently, Harada (2008) implanted tiny magnets near the semicircular canals of pigeons or severed the lagenal nerves and found that this disrupted their return to the loft.

So the focus for a mechanism to explain navigation is now on the lagena as the sense organ and the brainstem for its resolution. Wu and Dickman (2012) recorded electrical activity from vestibular neurons in the brainstem of alert pigeons subjected to stimulation by changes in magnetic field. They concluded that these cells, which receive sensory input from the lagena, are able to encode magnetic field direction, intensity and polarity, all of the attributes identified as necessary for computing navigational information. Commenting on this discovery in an article titled '*An avian magnetometer*', Winklhofer (2012) suggested a model in which chains of magneto-receptor cells sit among hair cells in the macula of the lagena, on which the otolith sits, and sense geomagnetism as tension in the membrane, which is then translated to an electrical potential. This raises a puzzle, however, because it is in the otolith from which the iron mineralisation has been reported (Zhao *et al.* 2009).

Crocodylians have a lagena (Fig. 5.11). Do they also have a capacity for magnetoreception? Whether any of them ever make transoceanic journeys like marine turtles and birds, as implied by Campbell *et al.* (2010) (Chapter 4), is unknown. However, the 400 km return journey by the translocated *C. porosus* from Temple Bay on the east coast of Cape York, north Queensland, to its capture site in the Wenlock River on the west coast (Fig. 4.35) implies some capacity for 'global positioning'. Could they be using geomagnetic information? All three of the translocated animals in the Read *et al.*

(2007) study spent time at the release point and then headed home with purpose. Could they have been somehow 'working out' a geolocation before they left?

There are some data that suggest a capacity for crocodylian magnetoreception. An intriguing experiment that deserves more attention was reported by Dominguez-Laso (2007) in which magnets were attached with duct tape to the heads of 20 *C. acutus*, *C. moreletii* and *Caiman crocodilus* (1.4–4.0 m) in Mexico. This was an attempt to prevent translocated animals (moved 1.3–120 km) from returning home, and was prompted initially following repeated returns by a translocated 'nuisance' 1.4 m *C. acutus*. None of the 'treated' animals returned, even though the magnets were removed before release!

As mammals, humans have trouble coming to terms with magnetoreception. I think this may be because we find it almost unbelievable that such small forces could be 'read' and because we ourselves lack that sense (and a lagena!). Harada *et al.* (2001) found that the otoliths in the lagenas of several species of fish contained iron mineralisation and suggested that it might be used for navigation in them too. Who knows, in a few decades it may have become accepted that the lagenas of non-mammalian vertebrates comprise a sophisticated magnetosensory organ.

EUSTACHIAN TUBES: FOR DIVING AND/OR SENSING DIRECTION OF SOUND, OR INFRASOUND?

Like other tetrapods, crocodylians have Eustachian tubes that provide an airway and thus an opportunity for pressure equalisation between the middle ear and the pharynx (throat) (Fig. 5.15). The Eustachian tubes of crocs were first described by the eminent British comparative anatomist and vertebrate palaeontologist Sir Richard Owen (Owen 1850) and later attracted the attention of the American Edwin H. Colbert: another eminent vertebrate palaeontologist (Colbert 1946). Both commented on the tubes' unique complexity. Whereas most amniotes have a single tube on each side, crocodylians have lateral tubes that join ventrally and have a mid-line connection with the throat just behind the internal nares. As if that were not already sufficiently complicated, the median tube divides in the mid-line into Y-shaped anterior and posterior branches. The three branches on each side connect dorsally into chambers that connect with the tympanic cavities of the middle ear. These chambers were referred to by Owen (1850) as rhomboidal sinuses, formed as dorsal enlargements of the lateral Eustachian tubes.

Although both Owen and Colbert speculated (94 years apart) about the role of these tubes in equalising air pressure between the pharynx and the middle ear, neither of them discussed possible function in the context of a diving animal. Owen speculated, rather hesitatingly, that the tubing provided access for sound via the nasal passages when a croc had only its nostrils and eyes above the surface, but this unlikely interpretation was dismissed by Colbert. Pressure equalisation between the throat and the ears is important for a diving animal, but that does not easily explain their bizarre complexity. Jack Pettigrew (*pers. comm.* 2008) drew my attention to a much more intriguing possibility: that the inter-connection between the tympanic cavities may enable crocodylians to localise low frequency sounds. As a result of recent work, this now seems almost certainly so, and again there is similarity with birds. It has been known for more than 100 years that humans and other mammals localise the direction of a sound source by assessing the delay between its arrival at one ear compared with the other: the interaural delay. The smaller an animal's head, the shorter the delay and the more challenging it is to measure the delay. Birds have small heads, and they also have an air-filled canal connecting between their tympanic cavities – the interaural canal – which is lacking in mammals. Coles *et al.* (1980) drew attention to the role of this canal in providing acoustic coupling between the two ears and concluded that birds' ears functioned as pressure difference receivers (PDRs) in receiving directional information. Calford (1988) and Calford and Piddington (1988) modelled the effect of the interaural canal in magnifying the interaural delay, and confirmed it by direct measurement in several bird species. They found that the effect was greater at low frequencies. Low frequency sounds are the most difficult to locate, because the wavelength is long compared with the width of the head, so the interaural canal would be particularly useful to birds communicating at low frequencies. The plains wanderer (*Pedionomus torquatus*) is such a bird: a ground-dwelling quail-like bird of inland south-eastern Australia that calls with a low dominant frequency around 360Hz. It also has a large interaural canal. Pettigrew and Larsen (1990) combined behavioural data and neurophysiological measurements to explain how a plains wanderer's large interaural canal could enable the location of a source of sound. The role of acoustic coupling provided by the interaural canal to increase the range of interaural time differences in birds was confirmed experimentally by Hyson *et al.* (1994) and Hyson (2005), using chickens. Like plains wanderers, crocodylians communicate by low frequency sound and Pettigrew (*pers. comm.*), noting that the junction between Eustachian tubes in crocodylians provides an interaural connection, suggested its use for sound localisation, as has been suggested for birds.

The case that this occurs in crocs is now much strengthened by two recent studies. Using data from *in vivo* preparations of *Alligator mississippiensis* and *Caiman crocodilus*, Carr *et al.* (2009) demonstrated large interaural time delays and were also able to describe the neural circuits for detecting them, showing they were similar to those in birds. To take the matter further, Bierman *et al.* (2014) examined whether acoustic coupling between the eardrums of alligators enhances their ability to detect the direction of a sound. In addition to the ventral interaural canal, they found another potential dorsal route for acoustic connection. By monitoring responses to directional sound stimuli in both the movement of the eardrums and within the auditory brainstem, they were able to show that the interaural time delay is enhanced by the eardrums being coupled. Whether the interaural canal is the most significant connection is not yet known. The response was most effective at low frequencies, up to 1.5 kHz but the study did not extend to infrasound. The authors concluded that, as in birds, coupled ears increase the time delay between the arrival of sounds at each eardrum.

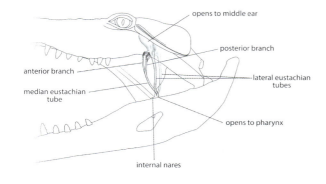

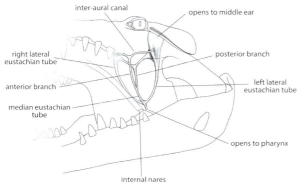

Fig. 5.15. *Two views of the Eustachian tubes, showing their extraordinary complexity. Note that in the upper diagram the skull is rotated to its right, so that view is from slightly posterior. Simpler tubes such as most amniotes have would afford pressure equalisation between the ears and the throat. The functional benefit/s conveyed by such additional complexity is intriguing. From recent work, it seems almost certain that acoustic coupling of the eardrums via the interaural canal and/or another continuous passage within the head allows localisation of a source of low frequency sound, but whether that includes infrasound remains unknown. Another, not mutually exclusive potential function of such complexity is some connection with a croc's diving capabilities. (Illustration developed with reference to Owen 1850, Colbert 1946 and a skull of* C. porosus*)*

This system could work well in both air and water, and Dinets (2013) has recently reported data that imply that American alligators can detect the direction of water-borne sounds. Colbert (1946) was clearly puzzled about the function of such elaborate Eustachian tube architecture. He could not see why pressure equalisation would be better served by having three tubes rather than a single large one on each side. His only speculation was that the 'branched passages may allow direct pressure equalisation between the ears', rather than via the throat. He added 'Perhaps it will be possible for someone to devise and carry out

experiments that will afford a clue as to the reason for the complexity of the eustachian tubes in the crocodiles'. Good progress is being now made on this front, but there are still questions. Whether or not the interaural canal is directly involved remains unproven, and whether or not the PDR allows the source of infrasound to be located is also unknown. Elsewhere in this Chapter the possibility that infrasound is detected via the sacculus is discussed. If so, perhaps that also takes advantage of the PDR. Yet another possibility is that integumentary sense organs (ISOs) are involved in detecting the source of infrasound, as well as all their other senses. After 160 years, the crocodylian Eustachian tube story is still being unravelled, but I am sure that Owen and Colbert, were they still alive, would be delighted to learn of the track that recent ideas are following. Clearly this intriguing topic is worthy of further exploration.

BRAIN AND CRANIAL NERVES

The crocodylian brain follows the usual vertebrate pattern, with three major regions: forebrain, midbrain and hindbrain (Fig. 5.16). Anteriorly, the forebrain comprises olfactory lobes and cerebral hemispheres. The cerebral hemispheres in mammals are dominated by the grey matter, or cerebral cortex, which forms the main bulk of the brain and is the main site for cognition, memory, learning, thought and consciousness. The cerebral hemispheres are comparatively larger in crocodylians than in other reptiles, but are much smaller than in mammals or birds. They are dominated by the corpus striatum: a solid mass of tissue that is thought to be a centre for instinctive behaviour. Among the reptiles, only crocodylians have what can be described as a cerebral cortex, but it is comparatively small. The diencephalon is tucked behind and between the cerebral hemispheres. It acts as a major conduit between sensory centres and the cerebral hemispheres. In typical reptiles, the diencephalon also bears the pineal body: the primary source of melatonin whose changing levels are influential in regulating seasonal physiology and behaviour. Crocodylians lack a pineal body, but Firth *et al.* (2010) recently observed rhythmicity in melatonin

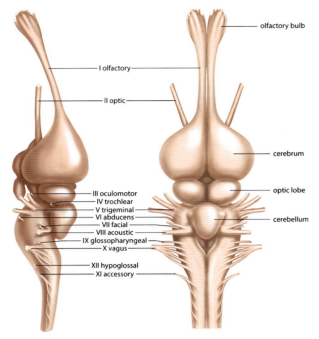

olfactory bulb

I olfactory

II optic

cerebrum

III oculomotor
IV trochlear
V trigeminal
VI abducens
VII facial
VIII acoustic
IX glossopharyngeal
X vagus

optic lobe

cerebellum

XII hypoglossal
XI accessory

Fig. 5.16. *Crocodylian brain, side and dorsal views. Anterior towards the top. (Drawn from various sources, including Chiasson 1962)*

levels in *Crocodylus johnstoni*, suggesting an alternate source or location. The diencephalon also contains the thalamus and hypothalamus, which are important in coordinating many metabolic activities. The pituitary gland, source of so many regulatory hormones by which homeostasis is maintained, is attached by the infundibulum below the hypothalamus. The optic lobes are the most conspicuous part of the midbrain, functioning in the integration of both visual and auditory information, unlike mammals where much of the processing of visual information occurs in the forebrain. The roof of the midbrain, the tectum, is also an important coordinating centre. The hindbrain comprises the cerebellum. The cerebellum is larger in crocodylians than in other reptiles and, as in higher vertebrates, is a site for processing sensory information and the coordination of muscular activity for locomotion. The medulla oblongata controls respiration and some of the auditory information. It extends seamlessly into the spinal cord which exits the skull through the foramen magnum (Fig. 3.43). Compared with both mammals and birds, the crocodylian brain is very small: ~0.004% of body

weight in *Alligator mississippiensis* (Hurlburt 2001). This is similar to other reptiles, but birds have brains ~10 times heavier (Jerison 1969), making a departure from the similarity between crocodylians and birds in so many other areas.

Much of the sensory information gathered by crocodylians from their environment enters the brain via the cranial nerves. Crocodylians and other reptiles have 12 pairs of cranial nerves, as do mammals and birds, whereas fishes and amphibians have 10. The paired nerves reflect the typical vertebrate pattern of primitive segmentation and subsequent cephalisation (evolutionary modification of the anterior segments to form the head). Table 5.1 lists the cranial nerves and their major function and location. The most striking feature of the crocodylian cranial nerves is the huge enlargement of the Vth pair, the trigeminal nerves, with their three main branches serving sense organs on the surface of the face and jaws (Fig. 5.17), as well as the integumentary sense organs to be discussed shortly.

OLFACTION AND GUSTATION (CHEMOSENSATION)

There is good evidence that crocodylians have a keen sense of smell (olfaction) and presumably good taste too (gustation). Along with birds, they lack a third chemosensory organ, the vomeronasal organ, often called Jacobsen's organ, which is present in snakes and lizards. Actually, the rudiments of a vomeronasal system do appear during early embryogenesis but are soon lost. Schwenk (2008) provides a useful review of the anatomy and physiology of chemical senses of aquatic reptiles, including crocodylians.

OLFACTION (SMELL)

Crocodylians undoubtedly use smell to find food (Chapter 6) and, presumably, taste (and texture) to discriminate the suitability or otherwise of food in the mouth. Weldon and Ferguson (1993) reviewed observations and studies that show that crocodylians use chemical cues to locate food on both land and under water. These include reports

TABLE 5.1   CRANIAL NERVES OF CROCODYLIA (AFTER CHIASSON 1962 AND OTHER SOURCES)

Cranial nerve	Action	Target area
I Olfactory	Sensory	Nasal epithelium
II Optic	Sensory	Retina
III Oculomotor	Motor	Eye muscles
IV Trochlear	Motor	Eye muscles
V Trigeminal Ophthalmic branch Maxillary branch Mandibular branch	Sensory and motor Sensory and motor Sensory	Skin, eye muscles, eyelids Eye muscles, palate, upper jaw Lower jaw
VI Abducens	Motor	Eye muscles
VII Facial	Motor and sensory	Masticatory muscles, palate, skin
VIII Auditory (vestibulocochlear)	Sensory	Acoustic and vestibular apparatus of ear
IX Glossopharyngeal	Motor and sensory	Larynx, oesophageal, tongue, face
X Vagus	Motor and sensory	Pharynx, larynx, trachea, lungs, oesophagus, stomach, heart
XI Accessory	Motor	Neck muscles
XII Hypoglossal	Motor	Tongue muscles

of mugger crocodiles, *Crocodylus palustris*, leaving the water to locate a tiger carcass 450 m from their pool, and *Crocodylus niloticus* around an elephant carcass 700 m from the river. Presumably these were located by olfaction. Scott and Weldon (1990) took an experimental approach and found that, at night, captive *Alligator mississippiensis* removed paper bags containing food more frequently than control bags containing only paper. Neonates can apparently imprint on particular chemical stimuli, with hatchlings showing clutch-specific food preferences (Webb *et al.* 1990). Sneddon *et al.* (2000) showed that chemosensory learning can occur even before hatching, and that the learning can accommodate stimuli that are anomalous, in this case strawberry essence. They wiped essence onto the surface of eggs of *C. porosus* from 65–87 days post-laying and thereby 'trained' hatchlings to prefer food flavoured with strawberry essence. Olfaction is also important in crocodylian social interactions, and this will be explored further in

Chapter 12. Suffice it to say here that experiments on juvenile *A. mississippiensis* showed that their gular (throat) pumping rate increased in response to exposure to airstreams containing scents from either the gular or paracloacal glands (Weldon and Ferguson 1993) (see Chapters 3 and 12).

Each nostril (Figs 3.23, 3.24) opens via a short vertical duct, the vestibulum, to a long, broad nasal cavity (*cavum nasi proprium* = nasal cavity proper, often called the cavum) (Fig. 5.18). This cavity has opening from it a series of diverticulae, or paranasal sinuses (see below). Then, about half the head's length back, the cavum opens posteriorly into the nasopharyngeal duct. The paired nasopharyngeal ducts open posteriorly through the internal nares (choanae) close to the glottis, through which air enters the trachea to ventilate the lungs. Thus, the paired air supply ducts can each be divided into three sections: the vestibulum, the nasal cavity proper (cavum) and the nasopharyngeal duct (Parsons 1970).

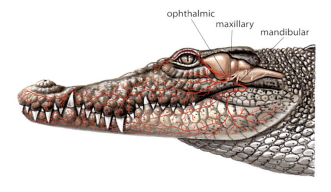

ophthalmic
maxillary
mandibular

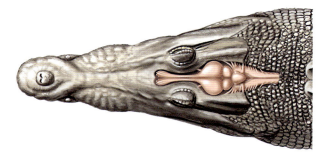

Fig. 5.17. *Top and side views of the brain in situ, to show in particular the extravagant elaboration of the three branches of the trigeminal cranial nerve, reflecting the high density of integumentary sense organs (ISOs) on the head (Figs 3.30, 3.34). (Illustration developed with reference to Chiasson 1962, Romer and Parsons 1977 and Leitch and Catania 2012)*

Respiration is interrupted by swallowing, when food passes down the oesophagus between the internal nares and the glottis. However, crocodylians can breathe uninterrupted with their mouth full because, like mammals, they have a secondary palate: a bony plate that forms the roof of the mouth and separates the respiratory and nasal passage from the mouth (buccal cavity). In mammals, the secondary palate allows chewing and other food preparation in the mouth without interrupting respiratory air flow. Although crocodylians do not chew their food, the secondary palate allows them to continue to breathe while holding struggling prey for a long period (Chapter 6), while the gular and palatal flaps close off the back of the mouth and give them a watertight, as well as airtight, front end (Figs 5.18, 9.3). Most other reptiles lack a secondary palate, but it has also evolved in the lizard family Scincidae (Greer 1989). The early crocodylians, the protosuchians, lacked a secondary palate,

and its evolution, with the progressive rearwards movement of the choanae, forms an interesting series that has been useful to palaeontologists (Fig. 2.29).

The anatomy of the olfactory chambers is the most complex among reptiles, but it is essentially similar within all Crocodylia. Early literature was reviewed by Parsons (1970) and more recent reviews put the anatomy into an evolutionary context (Witmer 1995, 1999). In crocodylians, each cavum has three thin, curled, shell-like bony projections called conchae (from 'shell-like') that project from its lateral walls and increase the surface area on which olfactory epithelium is exposed. In general, tissue lining the nasal chambers of air breathing vertebrates is well supplied with blood and it humidifies the respiratory air during inhalation. In mammals, the nasal conchae are elaborated into spongy bones, which are usually called turbinals and are an important site for heat exchange (see Chapter 10). Adding to the complexity in crocodylians are a series of sinuses, sacs and blind ducts opening laterally from each cavum. A reader seeking detailed description is referred to Parsons (1970) and Witmer (1995, 1999). Nasal glands also open into the cavum, secreting mucus to moisten the delicate olfactory epithelium. The air spaces within the skull cause it to be much lighter than it would be otherwise: a phenomenon known as pneumatisation, which reaches its greatest development in birds and the now extinct (unfortunately) flying reptiles, the pterosaurs.

The olfactory epithelium is ventilated during normal respiration and, because crocodylians are intermittent breathers (Chapter 7), particularly by gular pumping. In gular pumping, the floor of the pharynx is raised and lowered in a rhythmic way while the palatal and hyal flaps are held closed. Gular pumping is sometimes misidentified as respiratory movements, but it occurs quite independently of respiration and without air entering or leaving the lungs. The olfactory bulbs of the forebrain are well developed, consistent with the importance of olfaction (Fig. 5.16). Gular pumping increases in the presence of meat odours (Weldon *et al.* 1992; Weldon and Ferguson 1993) and is a visible sign of air sampling by olfaction (Fig. 5.19).

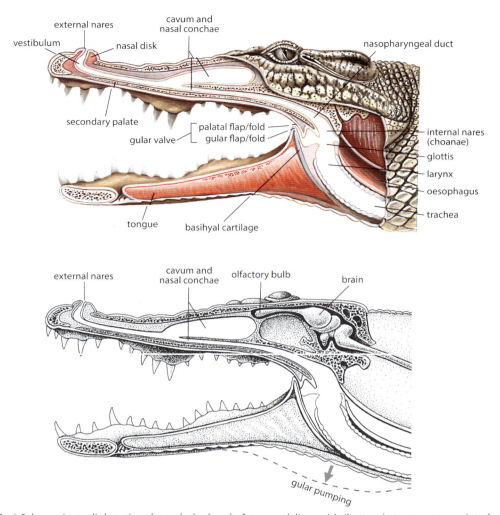

Fig. 5.18. (Top) Schematic medial section through the head of a crocodylian, with (bottom) structures associated with olfaction highlighted. In olfaction, the glottis and gular valve are closed and air is flushed in and out over the olfactory epithelium in the nasal conchae by expansion and contraction of the gular region: that is, gular pumping. (Modified from Grigg and Gans 1993 and adopting the terminology of Putterill and Soley 2006). (Illustrations DSK, upper figure courtesy Weldon Owen Pty Ltd, lower courtesy ABRS)

GUSTATION (TASTE)

Taste, or gustation, is another chemosensation. Crocodylians have taste buds on the tongue, palate and on the walls of the buccal cavity and pharynx (Weldon and Ferguson 1993) (Fig. 5.20). Franklin *et al.* (2005) reported structures resembling taste receptors around the sides of the excretory pores of the salt glands (Chapter 11). Taste helps in the evaluation of the quality and suitability of a food item once in the mouth. It is known that crocodylians prefer fresh, rather than putrid, food, though they will eat carrion. There are no data to support the commonly held belief that crocs routinely stash a

carcass to let it rot before they eat it. In general, even inexperienced crocodylids avoid drinking salt water, whereas naïve alligatorids drink it indiscriminately (Chapter 11). One would assume they use buccal taste buds to assess salinity. However, Jackson and Brooks (2007) found that yearling *Crocodylus porosus* with the body coated with petroleum jelly drank fresh water and salt water indiscriminately, whereas controls without petroleum jelly avoided the salt. They concluded that the more heavily keratinised buccal epithelium in Crocodylidae prevented them tasting salt water and proposed instead that salinity is sensed via the integumentary sense organs

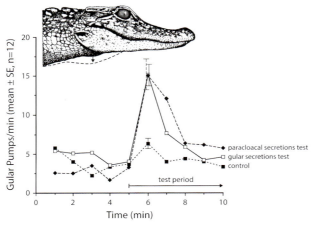

Fig. 5.19. *Gular pumping increases with interesting things to smell. In this experiment with juvenile alligators, rates of gular pumping increased with the introduction (test period) of methylene chloride extracts of secretions from the mandibular or paracloacal glands of adults into an airstream with methylene chloride only. The control confirms that the increase is not an artefact of the method of introducing the test airstream. (Adapted from Weldon and Ferguson 1993 with permission)*

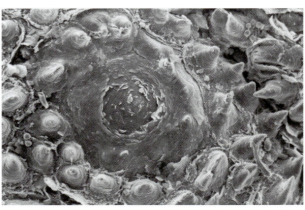

Fig. 5.20. *Scanning electron micrograph of a taste bud on the tongue of an American alligator. A fungiform (mushroom-like) papilla (gustatory) in the centre is surrounded by smaller papillae analogous to the filiform (thread-like) non-gustatory papillae on the human tongue. (Photo Paul Weldon)*

(ISOs, see below). However, *Crocodylus porosus* kept captive in salt water, such as Cassius on Green Island (see Chapter 1), learn quickly to drink fresh water from a hose. The question will be discussed further in Chapter 11 and there is a need for further experimental work.

SENSE ORGANS OF THE SKIN

Despite having good vision and a wide angle of view horizontally, a crocodylian's field of view is limited ventrally because the eyes in most are set pretty much on top of the head. Also, the neck and trunk are not very flexible (Chapters 3, 7). So, even in daylight, much of a crocodylian's awareness of its immediate surroundings must come from other senses, particularly the ears and sensors on the skin. At night or under water, when vision is much reduced, these receptors are even more crucial. The skin has a range of sense organs, both visible and invisible to the naked eye (von Düring and Miller 1979). They have simple touch receptors (mechanoreceptors) over the general body surface, as well as many more specialised, tiny dome-shaped sense organs, the 'ISOs' which have often been described as 'pits' because they show as pits on tanned hides. The ISOs are easily visible, but

there are also many touch receptors that are too small to see.

TOUCH RECEPTORS

Nobody who has watched an adult female crocodile, with her eyes on top of her head, rolling an egg gently in her jaws between the tongue and the palate and assisting the wriggling hatchling to emerge, and then carrying it, oh so carefully to the water, could be any doubt about the sensitivity and fine motor control of those jaws. The sensation of touch is provided by a diversity of mechanoreceptors, specialised nerve endings located under slight thinnings in the stratum corneum, the outer layer of epidermis (von Düring and Miller 1979). Deformation of the surface of the sense organ stimulates nerve endings within the receptor, generating a nerve impulse.

INTEGUMENTARY SENSE ORGANS, THE MULTIPURPOSE ISOs

Observers were puzzled for years about the function of small dark pigmented spots or pits on crocodylian scales. Quite 'off beam', I once suggested that they might secrete some sort of detergent-like substance (Grigg and Gans 1993) because crocs in the wild are always so clean! But a sensory function has always been most favoured and that is now known to be correct. (I still think there may be some sort of surfactant on the skin though.) There has also been uncertainty about what they should be called, but it is now settled that they can be called 'integumentary sense organs', ISOs, a name coined by Brazaitis (1987). In a long career working with the reptile collection at New York's Bronx Zoo, Peter Brazaitis noted how crocodylians used their body regions with the highest ISO profusion in courtship and feeding and he was in no doubt that they are sensory in function. He was also well aware of the usefulness of these 'pits' in the scales for species identification of tanned crocodylian hides and in leather goods such as handbags: a service he often provided for the US Fish and Wildlife Service as a 'forensic herpetologist' (see Chapter 14). 'I was never comfortable with the terms pit-glands, pits, apical pits, follicle glands, touch papilla that were in common use, and had been long used by the leather trade. So I decided to create a standard term that might fit for what I saw – a sensory organ in the skin (integument) – that others might use to talk about their function and morphology...' (Brazaitis *pers. comm.* 2012). His name has stuck, but there have been others. Histological study on *Caiman crocodilus* by von Düring (1973, 1974) led to their diagnosis as being mechanosensory, leading to her calling them 'touch papillae'. Their putative sensory capacity was confirmed experimentally by Soares (2002) (see below) who re-named them 'dome pressure receptors' (DPRs). However, Jackson *et al.* (1996) and Leitch and Catania (2012), in papers providing further evidence of their sensory role, preferred the ISO terminology. Retaining this as an umbrella term for these superficially similar structures on the head and body scales of the various crocodylians seems preferable, at least until much more is known about a possible diversity of sensory modalities.

Unlike touch receptors, ISOs are quite conspicuous (Figs 5.21–5.24). Although they show as pits in a tanned hide, in life the 'pit' of an ISO is actually a shallow dome-shaped button. ISOs occur on scales of the head, body and limbs of all Crocodylidae (including *T. schlegelii*) and also on *Gavialis gangeticus*, but, quite paradoxically, only on the head scales of Alligatoridae. To a biologist more familiar with species of *Crocodylus*, alligators look somewhat naked! So here is another difference between Alligatoridae and other crocodylians, and it

makes it easy to tell if your crocodile handbag really is crocodile, or made from less valuable alligator or caiman hide (Brazaitis 1987). Most scales of the body and limbs have a single ISO (Figs 5.23–5.26), but the scales on the head have more, commonly up to seven or eight, so they are much more numerous at the anterior end (Figs 5.21, 5.22). Despite being so numerous on the jaws and snout and around the eyes, sides of the head and the snout, there are few or none on the cranial platform; it is as if their function is to receive only water-borne information. But the lack of ISOs post-cranially in Alligatoridae is a mystery: a crocodylid or a gharial (or false gharial) might well wonder how they get by without them!

Histological and ultrastructural work leaves no doubt about their sensory nature. Studies of ISOs on the head scales of *Caiman crocodilus* (von Düring 1973, 1974) and on the post-cranial scales of *Crocodylus porosus* (Jackson *et al.* 1996) described a thinning in the stiff stratum corneum to expose the underlying epidermis above a presumably fluid-filled pocket in the dermis, which houses nerve terminals (Fig. 5.27). Von Düring described

Fig. 5.21. *(Top)* Alligator mississippiensis *(upper) and (bottom)* Crocodylus porosus *showing the distribution of ISOs on the mouth and jaws. (Photos DSK)*

Fig. 5.22. *ISOs on the snout of* C. acutus. *Note also the bottom 'incisor' teeth protruding through the upper jaw. (Photo Louis Guillette)*

Fig. 5.24. *Fingers and toes of Crocodylidae are well supplied with ISOs. Sensitive fingers may be particularly useful in 'cross-posture' feeding (Chapter 6) except that behaviour has been described also in* Caiman yacare, *which lack post-cranial ISOs, raising again the puzzle posed by that apparent evolutionary loss in the alligatorids. (Photo DSK)*

Merkel cells and lamellated receptors, both of which are commonly found in vertebrate skin and associated with mechanoreception and referred to the structures as 'touch papillae' deducing a mechanosensory function. Jackson *et al*. (1996) acknowledged their likeness to touch receptors, but indicated a need for further work to determine function, referring to the possibility of salinity sensing (Jackson and Brooks 2007), a proposal later discounted by experiment (Leitch and Catania (2012). Structurally, ISOs in the three species studied so far are broadly similar, except that Jackson *et al*. (1996) noted some microanatomical differences between the post-cranial ISOs of *Crocodylus porosus* and the cranial ISOs of *Caiman crocodilus*: enough to hesitate about assuming a commonality of function throughout. Soares (2002) looked at

A. mississippiensis and referred to thinning of the skin at the dome, with a dermal fold at the margin and a highly branched nerve bundle innervated by a branch of the trigeminal nerve. Soares noted that the ISOs in alligators (which she called dome pressure receptors, DPRs) are distributed directly over the many foramina in the bones of the upper and lower jaws. With such a profusion of ISOs cranially and the enormous development of the trigeminal cranial nerves in crocodylians (Fig. 5.17), a major functional role was implied, and so it has turned out.

A detailed microstructural study by Leitch and Catania (2012) extended von Düring's seminal

Fig. 5.23. *ISOs on the flank of* C. porosus *(head to the right), one per scale, as on the whole of the body and limbs. (Photo DSK)*

Fig. 5.25. *One ISO posteriorly on every dorsal scute (C. porosus): well placed to keep a good lookout behind! (Photo DSK)*

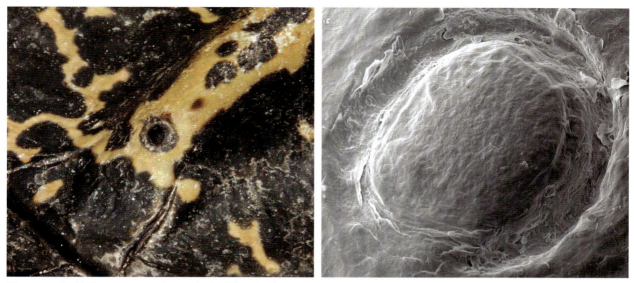

Fig. 5.26. *(Left) Close up view of a single dorsal scale of* C. porosus *with a single ISO posteriorly on the keel and (right) a scanning electron micrograph of a single ISO from an American alligator. (Photos DSK and EM from Leitch and Catania 2012, with permission)*

studies, taking advantage of the improved resolution provided by confocal microscopy. The results are shown schematically in Fig. 5.28. The most prominent sensory nerve endings were those associated with a column of flattened Merkel cells, providing a receptor centrally in the dome. These were only found associated with ISOs. Within the connective tissue below the dome, they also found numerous nerve endings ensheathed in lamellated Schwann cell processes, reminiscent of Pacinian corpuscles found in mammalian skin and sensitive to rapid changes in pressure or vibrations. Additionally, there are free nerve endings in the

dermis and 'discoid receptors' between the stratum lucidum and stratum spinosum. Structurally, Leitch and Catania found that crocodylian ISOs have many features in common with known mechanoreceptors. They also noted the vast network of nerves just below the epidermis, servicing the ISOs and sheathed in bone so that they are protected from mechanical injuries they could well acquire in feeding or fighting.

Electrophysiological recordings were made from the trigeminal ganglion and while stimulating large and small regions of ISOs (Leitch and Catania 2012). These showed that the ISOs are extremely sensitive,

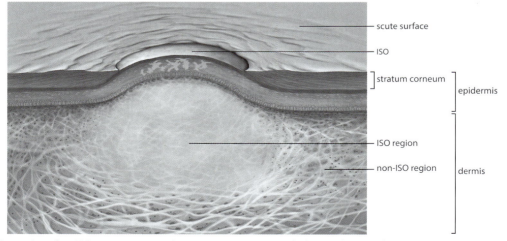

Fig. 5.27. *Illustration of an ISO in cross section. (From various sources, including Jackson et al. 1996)*

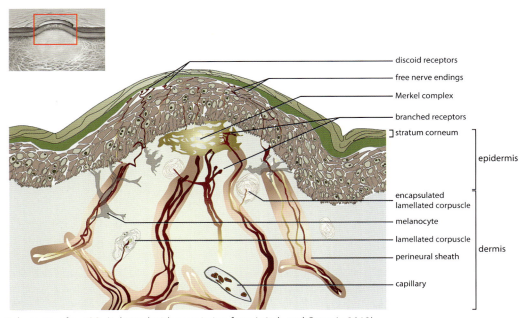

discoid receptors
free nerve endings
Merkel complex
branched receptors
stratum corneum
epidermis
encapsulated lamellated corpuscle
melanocyte
lamellated corpuscle
perineural sheath
capillary
dermis

Fig. 5.28. *Schematic of an ISO. (Adapted with permission from Leitch and Catania 2012)*

leading to the observation that 'the ISOs impart a mechanical sensitivity exceeding that of primate fingertips'. They also allow a great deal of spatial resolution. Some nerves in the trigeminal complex were stimulated by only a few, or even only one, ISO. Recordings made from spinal afferent nerves in *C. niloticus* showed that the post-cranial ISOs are less sensitive than the cranial ISOs of alligators and provide less spatial resolution. The webbing between the third and fourth digits of the forelimbs is, however, more sensitive than elsewhere post-cranially. Di-Poï and Milinkovitch (2013) confirmed the mechanosensory function of ISOs and found them also to be sensitive to changes in pH and, in a less than straightforward way, to changes in temperature, and emphasised their multi-sensory nature. There can be no doubt that sensation gained via ISOs contributes in a really major way to crocodylian lives. What is more, they also have a long evolutionary history within crocodile-like reptiles. Soares (2007) noted that, because the innervation of ISOs is via numerous foramina in the bones of the jaws, the likely occurrence ISOs in extinct groups could be judged. She found examples of crocodylomorphs with similar foraminal distributions as far back as the Triassic, but only

in groups interpreted as having been semi-aquatic (Soares 2007).

Logically, one might propose that these structures would sense water movement or vibrations, similar perhaps to the lateral line system in fishes and some amphibians and, indeed, work by Daphne Soares (2002) at the University of Maryland showed that this is the case, at least for ISOs on the head scales of *Alligator mississippiensis*. She took a behavioural approach and found that *A. mississippiensis* can turn or lunge towards a surface disturbance even in total darkness. Covering the ISOs experimentally with an elastomer obliterated the response and she was also able to record from the trigeminal nerve ganglion to show that there was a graded response to a series of graded surface wave stimuli. The response did not occur if the alligator was completely submerged or if its head was completely out of the water, so she concluded that the frequency response must be tuned to pick up surface wave activity. Similar experiments were conducted by Leitch and Catania (2012) on both alligators and *Crocodylus niloticus*. It is highly likely that ISOs on the heads of all crocodylians function similarly. This later study, however, found that ISOs are not stimulated only by surface waves: a fish swimming beside the tail of a juvenile *C. niloticus* commanded immediate attention.

So what is their function? Clearly crocodylians deprived of vision can respond to, and snap at, water-borne disturbances such as surface vibrations and ripples. With ISOs distributed post-cranially as well, crocodylids (and presumably gharials) can also respond to subsurface water movements such as those made by fish in close proximity. This would be very useful in turbid water. On the basis of the very high density of ISOs around the jaws, including particularly the gingivae ('gums') (Fig. 5.29), Leitch and Catania (2012) suggested that the ISOs also help in the discrimination of objects seized by the jaws and in the manipulation of prey once it is held. It is easy to see that much of the information about what crocodylians capture in the sideways sweeping of the head they make when feeding in the surface waters is conveyed by the ISOs, not the eyes, because their eyes are too high on the head and the water is often turbid. Information ISOs provide must assist also in the so-called 'inertial feeding' behaviour (Gans 1969, Chapter 6). Kent Vliet has described the extensive touching of the sides of the head and around the eyes that alligators perform in their courtship behaviour (Vliet 2001) (Fig. 5.30). The areas that attract most attention have very high densities of ISOs. Similar behaviour is seen in the courtship behaviour of other crocodylians.

It seems that prey capture and feeding are almost certainly benefited by sensory information from the ISOs, both from direct touch and indirectly by the receipt of water movements nearby. It is interesting to speculate about other potential benefits. The finding by Di-Poï and Milinkovitch (2013) that they can respond to changes in pH implies some capacity to sense some aspect of water quality. Leitch and Catania (2012) developed a tuning curve for cranial ISOs of alligators, finding them to be at their most sensitive at low frequencies with the minimum at ~20 Hz, so it seems certain that ISOs are stimulated by infrasound (see above and Chapter 12) and that has been suggested by both Vergne *et al.* (2009) and Brazaitis and Watanabe (2011). Bellowing by alligators is frequently preceded by a burst of infrasound (sub-audible vibrations, SAVs,

Fig. 5.29. *Concentration of ISOs around the jaws and gingivae of* C. porosus. *(Photo DSK)*

see Chapter 12) and the bellowing individual is usually in a HOTA posture at that stage, with the ISOs above water. But alligators with the head at water level could receive that. Infrasound is generated by crocodylids as well (Chapter 12) and perhaps their post-cranial ISOs provide its major sensation. Vergne *et al.* (2009) noted that there were no measurements in crocodylians of their auditory sensitivity to very low frequencies and that is still the case. It would be surprising if their ears are insensitive to infrasound, but it is easy to imagine that a burst of infrasound could give a crocodile (or an alligator with its head partly submerged) quite a frisson of sensation, particularly in courtship or from an aggressive animal nearby (see Figs 12.2, 2.6). Todd (2007) and Dinets (2011a, b) have shown that infrasound can provide long range signalling, but could ISOs detect the direction of a distant source of infrasound? Dinets (2013) has shown that American alligators are able to detect the direction of water-borne sound, but its frequency was undetermined. Theoretical work on the lateral line system of African clawed frogs by Franosch *et al.* (2003) has shown that it has the capacity to determine the direction of a wave source and they suggested that ISOs too would have that capability. If crocodylians can sense the direction of a distant source of infrasound, which seems very likely, the explanation will probably lie in either the ISOs or the interaural canal which interconnects left and right tympanic cavities, as discussed previously in the section on Eustachian tubes.

Fig. 5.30. *Courting American alligators at the St Augustine Alligator Farm. In this case, it is the female who is taking the initiative, bumping and rubbing her head on the male's and pressing down. Kent Vliet has drawn particular attention to these tactile interactions and quantified touches to different parts of the head and neck (Vliet 2001). With such a density on the heads of all crocodylians, the ISOs must have particular significance in courtship. See also Chapter 12. (Photos DSK)*

Apart from the possibility of infrasound reception and the detection of potential prey swimming nearby, the ubiquity of ISOs on the post-cranial scales of crocodylids raises all sorts of other possibilities to explore. Certainly their operation is monitored by a range of different nerves and it is likely that the post-cranial ISOs respond to differently conveyed information. The positioning of ISOs on the dorsal scutes (Fig. 5.25) might provide information about whether or not the dorsum is submerged or exposed. They could provide information about swimming speed and water flow over the body surface. Their responsiveness to changes in temperature needs further exploration. The dorsal scutes are often said to be crocodylian solar panels (Chapter 10), and Brazaitis and Watanabe (2011) noted that crocodylians push their heads into prospective nest sites as if checking temperature and suggested that the ISOs on the head could be providing thermal information. Obviously more work is needed. It could be that the term 'integumentary sense organ', ISO, may end up being a generic term for a range of superficially similar skin sensors with different *modus operandi* and distribution.

Despite the gaps in current information, it is fairly clear that these many sense organs over the surface of crocodylians, both visible and too small to see, are providing crocodylians with valuable and diverse information about their immediate surroundings and activities, including feeding and courting and, almost certainly, swimming. And crocodylids have quite high densities of them around their cloacas too.

COGNITIVE CAPACITIES AND LEARNING

Before dismissing the mental capabilities of crocodylians, it is as well to remember the complexity of their predatory and reproductive behaviour (see Chapters 6, 12) and their homing capacities (Walsh and Whitehead 1993; Read *et al.* 2007 and Chapter 4). Reptilian brains are about one tenth the mass of avian or mammalian brains at comparable body size, and crocodylians appear to be no exception (Jerison 1969). Nevertheless, brain size can be very misleading. A large crocodylian

has a brain that is larger than the whole body of one of the smaller insectivorous bats, yet these show very elaborate behaviour, with a knowledge of large geographic areas and the food resources they hold, as well as the time of year to tap them. The knowledge that crocodylians show of the geography surrounding them is not trivial: females use the same nest sites repeatedly, even though these may be far distant from the places where they live for most of the year (Chapter 12).

Various experiments on crocodylians in captivity show a capacity for learning. Northcutt and Heath (1971) tested spatial learning in a simple maze (one choice point) and found that *Caiman crocodilus* learned to overcome a natural turning preference and return successfully to their home pool. Somaweera *et al.* (2011) found that naïve hatchling *C. johnstoni* learned quickly to avoid eating metamorphs of the introduced cane toads, *Rhinella* (formerly *Bufo) marina*. This offers hope that as adults they will avoid eating the large, more toxic toads that are at present causing severe reductions in some *C. johnstoni* populations in northern Australia (see Chapter 13). Sneddon *et al.* (2001) demonstrated embryonic chemosensory learning, exposing embryos of *C. porosus* to strawberry essence by wiping it on the egg shells. Compared with a control group wiped with water, strawberry-exposed hatchlings preferred strawberry flavoured food items over the control group, and also over food flavoured with a novel, orange flavour.

Very strikingly, there is an intriguing report suggesting tool use (Dinets *et al.* 2013). These authors reported the apparent use of sticks balanced across the snout as bait to entice nest-building egrets within snapping distance. This was observed in captive muggers, *C. palustris*, at the Madras Crocodile Bank in Tamil Nadu, India, and in American alligators in captivity at the St Augustine Alligator Farm, Florida and in the wild in Louisiana (Fig. 5.31). Stick-displaying behaviour was noted to be strongly associated with egret colonies, an observation confirmed by the finding that in Louisiana it was more frequent in alligators near egret rookeries than in areas lacking such a rookery. Also, it was more prevalent during the early part of the breeding season, when the egrets were building nests, than later, even though the egrets were still present. One (unsuccessful) capture attempt by a mugger was reported, but the Brueggens have observed successful captures at St Augustines by stick-displaying captive alligators and this behaviour is reported to be common during the egrets' nest-building season.

This is the first claim of tool use by a reptile. Whether it is a learned behaviour, an evolved instinct or the result of individual insight is unknown. The report is likely to stimulate some healthy discussion and also to provoke reports of similar behaviour and further speculations about crocodylian tool use. Stick display might also enhance crypsis, providing a useful adjunct to baiting. Can the acquisition of mud or the waterweed *Salvinia*, as an apparent parasol (Chapter 10), also be regarded as tool use? That depends upon whether the acquisition is accidental or deliberate. And do crocs use a covering of *Salvinia* as camouflage to enhance

a stalking approach to prey? These could be good study questions.

In the wild, the predatory behaviour of crocodylians conveys every sense of a capacity for anticipation based on experience. Many *C. porosus* show up at a particular causeway on the Mary River in Australia's Northern Territory when king tides and the breeding runs of mullet coincide to provide a good feeding opportunity. Satellite tracking data have shown that the crocs travel many kilometres in anticipation. Such apparently regulated patterns of behaviour are usually assumed to have arisen over millennia and become 'hard wired'. But the causeway has only been there for ~50 years, only a few crocodylian generations. Varanid lizards are said to be able to count, up to six anyway (King and Green 1999). Maybe crocodylians can too. The South African biologist Tony Pooley, one of the giants upon whose shoulders we crocodylian aficionados stand, had a charming story to explain that Nile crocodiles

Fig. 5.31. *In what has been interpreted as tool use by a reptile, a mugger,* C. palustris, *at an egret colony at the Madras Crocodile Bank, Tamil Nadu, India, balances sticks on its snout. The sticks are apparently used as bait to lure a nest building egret within snapping range. The same behaviour has been seen in American alligators too, both in captivity and in the wild and it is most prevalent during the nest-building phase of the egret's nesting season (Dinets* et al. *2013). (Photo Vladimir Dinets)*

too can count. 'On just one day of the week', he said, 'Offal from an abattoir was discharged into the river'. Every 7 days the crocodiles would swim up the river in anticipation, arriving in time for the discharge. 'It can't be smell', said Tony, because this occurred even on bank holidays when there was no discharge. 'They can count,' he said, 'at least up to seven!' Awaiting more definitive experimental results, and considering other observations reported in this book, you can form your own opinion about their learning and cognitive skills!

References

Baird IL (1970) The anatomy of the reptilian ear. In *Biology of the Reptilia. Volume 2.* (Eds C Gans and TS Parsons) pp. 193–275. Academic Press, New York.

Bierman HS, Thornton JL, Jones HG, Koka K, Young BA, Brandt C, *et al.* (2014) Biophysics of directional hearing in the American alligator (*Alligator mississippiensis*). *The Journal of Experimental Biology* **217**, 1094–1107.

Brazaitis P (1987) Identification of crocodilian skins and products. In *Wildlife Management: Crocodiles and Alligators*. (Eds GJ Webb, SC Manolis and PJ Whitehead) pp. 373–386. Surrey Beatty & Sons, Sydney.

Brazaitis P, Watanabe ME (2011) Crocodilian behaviour: a window to dinosaur behaviour? *Historical Biology* **23**(1), 73–90.

Calford MB (1988) Constraints on the coding of sound frequency imposed by the avian interaural canal. *Journal of Comparative Physiology. A, Neuroethology, Sensory, Neural, and Behavioral Physiology* **162**, 491–502.

Calford MB, Piddington RW (1988) Avian interneural canal enhances interaural delay. *Journal of Comparative Physiology. A, Neuroethology, Sensory, Neural, and Behavioral Physiology* **162**, 503–510.

Cameron DA, Pugh EN (1991) Double cones as a basis for a new type of polarization vision in vertebrates. *Nature* **353**, 161–164.

Campbell HA, Watts ME, Sullivan S, Read MA, Choukroun S, Irwin SR *et al.* (2010) Estuarine crocodiles ride surface currents to facilitate long-distance travel. *Journal of Animal Ecology* **79**(5), 955–964.

Carr CE, Soares D, Smolders J, Simon JZ (2009) Detection of interaural time differences in the alligator. *The Journal of Neuroscience* **29**(25), 7978–7982.

Chiasson RB (1962) *Laboratory anatomy of the alligator.* WC Brown, Dubuque, Iowa.

Colbert EH (1946) The Eustachian tubes in the Crocodilia. *Copeia*(1), 12–14.

Coles RB, Lewis DB, Hill KG, Hutchings ME, Gower DM (1980) Directional hearing in the Japanese Quail (*Coturnix coturnix japonica*) II Cochlear physiology. *The Journal of Experimental Biology* **86**, 153–170.

Di-Poï N, Milinkovitch MC (2013) Crocodylians evolved scattered multi-sensory micro-organs. *EvoDevo* **4**, 19.

Dieterich CE, Dieterich HJ (1978) Electron microscopy of retinal tapetum (*Caiman crocodilus*) *Graefe's Archive for Clinical and Experimental Ophthalmology* **208**, 159–168.

Dinets V (2011a) The role of habitat in crocodilian communication. PhD thesis, University of Miami. *Open Access Dissertations.* Paper 570. <http://scholarlyrepository.miami.edu/oa_dissertations/570>

Dinets V (2011b) Effects of aquatic habitat continuity on signal composition in crocodilians. *Animal Behavior* **82**(2), 191–201.

Dinets V (2013) Underwater sound locating capability in the American alligator (*Alligator mississippiensis*). *Journal of Herpetology* **47**(4), 521–523.

Dinets V, Brueggen JC, Brueggen JD (2013) Crocodilians use tools for hunting. *Ethology Ecology and Evolution*.

Dominguez-Laso J (2007) Relocation of crocodilians using magnets. *Crocodile Specialist Group Newsletter* **27**(3), 5–6.

Firth BT, Christian KA, Belan I, Kennaway DJ (2010) Melatonin rhythms in the Australian freshwater crocodile (Crocodylus johnstoni): a reptile lacking a pineal complex? *Journal of*

Comparative Physiology. B, Biochemical, Systemic, and Environmental Physiology **180**, 67–72.

Fleishman LJ, Rand AS (1989) *Caiman crocodilus* does not require vision for underwater prey capture. *Journal of Herpetology* **23**(3), 296.

Fleishman LJ, Howland HC, Howland MJ, Rand AS, Davenport ML (1988) Crocodiles don't focus underwater. *Journal of Comparative Physiology. A, Neuroethology, Sensory, Neural, and Behavioral Physiology* **163**, 441–443.

Franklin CE, Taylor G, Cramp RL (2005) Cholinergic and adrenergic innervation of lingual salt glands of the estuarine crocodile, Crocodylus porosus. *Australian Journal of Zoology* **53**, 345–351.

Franosch J-MP, Sobotka MC, Elepfandt A, van Hemmen JL (2003) Minimal model of prey localisation through the lateral line system. *Physical Review Letters* **91**(15), 1–4.

Gans C (1969) Comments on inertial feeding. *Copeia* 855–857.

Gleich O, Manley GA (2000) The hearing organ of birds and reptiles. In *Comparative Hearing: Birds and Reptiles* (Eds RJ Dooling, RR Fay RR and AN Popper) pp. 70–138. Springer, London.

Greer AE (1989) *The Biology and Evolution of Australian Lizards.* Surrey Beatty & Sons, Sydney.

Grigg GC, Gans C 1993 Crocodilia: morphology and physiology. In *Fauna of Australia Volume 2A. Amphibia and Reptilia.* (Eds CJ Glasby, GJB Ross and PL Beesley) pp. 326–336. Australian Government Publishing Service, Canberra.

Harada Y, Taniguchi M, Namatame H, Lida A (2001) Magnetic materials in otoliths of bird and fish lagena and their function. *Acta Oto-Laryngologica* **121**(5), 590–595.

Harada Y (2008) The relation between the migration function of birds and fishes and their lagenal function. *Acta Oto-Laryngologica* **128**(4), 432–439.

Hart NS, Hunt DM (2007) Avian visual pigments: characteristics, spectral tuning, and evolution. *American Naturalist* **169**, S7–S26.

Higgs DM, Brittan-Powell EF, Soares D, Souza MJ, Carr CE, Dooling RJ, *et al.* (2002) Amphibious auditory responses of the American Alligator (*Alligator mississippiensis*). *Journal of Comparative Physiology* **188**, 217–223.

Hurlburt GR (2001) Ontogeny of relative brain size, maximum brain size, and body size measures in the American alligator (*Alligator mississippiensis*). *Journal of Vertebrate Paleontology* **21**(3 Supplement), 64A.

Husband S, Shimizu T (2001) Evolution of the avian visual system. In *Avian visual cognition* (Ed. RG Cook) Published by RG Cook in cooperation with Comparative Cognition Press, Tufts University, Boston, Massachusetts,. <http// www.pigeon.psy.tufts.edu/avc/husband/>

Hyson RL (2005) The analysis of interaural time differences in the chick brain stem. *Physiology & Behavior* **86**, 297–305.

Hyson RL, Overholt EM, Lippe WR (1994) Cochlear microphonic measurements of interaural time differences in the chick. *Hearing Research* **81**, 109–118.

Jackson K, Brooks DR (2007) Do crocodiles co-opt their sense of 'touch' to 'taste'? A possible new type of vertebrate sense organ. *Amphibia-Reptilia* **28**, 277–285.

Jackson K, Butler DG, Youson JH (1996) Morphology and ultrustructure of possible integumentary sense organs in the estuarine crocodile (*Crocodylus porosus*). *Journal of Morphology* **229**, 315–324.

Jerison HJ (1969) Brain evolution and dinosaur brains. *American Naturalist* **103**, 575–588.

Kalloniatis M, Tomisich G (1999) Amino acid neurochemistry of the vertebrate retina. *Progress in Retinal and Eye Research* **18**(6), 811–866.

Keeton WT (1971) Magnets interfere with pigeon homing. *Proceedings of the National Academy of Sciences* **68**(1), 102–106.

Khan NS, Shwab U, Trinker DEW (1982) Sensory transduction and neuronal transmission as related to ultrastructure and encoding of information in different labyrinthine receptor systems of vertebrates. *Archives of Oto-Rhino-Laryngology* **236**, 27–39.

King D, Green B (1999) *Goannas; The Biology of Varanid Lizards.* 2nd edn. University of New South Wales Press, Sydney.

Klinke R, Pause M (1980) Discharge properties of primary auditory fibres in *Caiman crocodilus Experimental Brain Research* **38**, 137–150.

Laurens H, Detwiler SR (1921) Studies on the retina. The structure of the retina of *A. mississippiensis* and its photomechanical changes. *The Journal of Experimental Zoology* **32**, 207–234.

Leitch DB, Catania KC (2012) Structure, innervation and response properties of integumentary sensory organs in crocodilians. *The Journal of Experimental Biology* **215**, 4217–4230.

Lewis ER, Leverenz EL, Bialek W (1985) *The Vertebrate Inner Ear.* CRC Press, Boca Raton, Florida.

Lohmann KJ, Lohmann CMF (1996) Detection of magnetic field intensity by sea turtles. *Nature* **380**, 59–61.

Lohmann KJ, Putman NF, Lohmann CMF (2012) The magnetic map of hatchling loggerhead sea turtles. *Current Opinion in Neurobiology* **22**, 336–342.

Malmström T, Kröger RHH (2006) Pupil shapes and lens optics in the eyes of terrestrial vertebrates. *The Journal of Experimental Biology* **209**, 18–25.

Manley GA (2000) Cochlear mechanisms from a phylogenetic viewpoint. *Proceedings of the National Academy of Sciences of the United States of America* **97**, 11736–11743.

Northcutt RG, Heath JE (1971) Performance of caimans in a T-maze. *Copeia* **1971**, 557–560.

Owen R (1850) On the communication between the cavity of the tympanum and the palate in the Crocodilia (gavials, alligators and crocodiles). *Philosophical Transactions of the Royal Society of London* **140**, 521–527.

Parsons TS (1970) The nose and Jacobson's organ. In *Biology of the Reptilia, Volume 2.* (Eds C Gans and TS Parsons) pp. 99–191. Academic Press, New York).

Pettigrew JD, Grigg GC (1990) Avian pattern of connections for binocular vision in crocodiles. *Proceedings of the Australian Neuroscience Society* **1**, 114

Pettigrew JD, Larsen ON (1990) Directional hearing in the Plains-wanderer, *Pedionomus torquatus*. In *Information Processing in the Mammalian Auditory and Tactile Systems*, (Eds LM Aitkin and MJ Rowe) pp 179–190. Alan R Liss, New York.

Putman NF, Endres CS, Lohmann CMF, Lohmann KJ (2011) Longitude perception and bicoordinate magnetic maps in sea turtles. *Current Biology* **21**, 463–466.

Putterill JF, Soley JT (2006) Morphology of the gular valve of the Nile crocodile, *Crocodylus niloticus* (Laurenti, 1768). *Journal of Morphology* **267**(8), 924–939.

Read MA, Grigg GC, Irwin SR, Shanahan D, Franklin CE (2007) Satellite tracking reveals long distance coastal travel and homing by translocated Estuarine Crocodiles, *Crocodylus porosus*. *PLoS ONE* **2**(9), e949

Romer AS, Parsons TS (1977) *The Vertebrate Body.* 5th edn. WB Saunders, London.

Schwab IR, Yuen CK, Buyukmihci NC, Blankenship TN, Fitzgerald PG (2002) Evolution of the tapetum. *Transactions of the American Ophthalmological Society* **100**, 187–200.

Schwenk K (2008) Comparative anatomy and physiology of chemical sense in non-avian aquatic reptiles. In *Sensory Evolution on the Threshold. Adaptations in Secondarily Aquatic Vertebrates.* (Eds JGM Thewissen and S Nummels). University of California Press, Berkeley.

Scott TP, Weldon PJ (1990) Chemoreception in the feeding behaviour of adult American alligators (Alligator mississippiensis) to meat scents. *Animal Behaviour* **39**, 398–400.

Shute CCD, Bellairs A, d'A (1955) The external ear in crocodilia. *Proceedings of the Zoological Society of London* **124**, 741–749.

Sillman AJ, Ronan SJ, Loew ER (1991) Histology and microspectrophotometry of the photoreceptors of a crocodilian, *Alligator mississippiensis*. *Proceedings. Biological Sciences* **243**(1306), 93–98.

Sneddon H, Hepper PG, Manolis C (2001) Embryonic chemosensory learning in the saltwater crocodile *Crocodylus porosus*. In *Crocodilian Biology and Evolution* (Eds GC Grigg, CE Franklin and F Seebacher) pp. 378–382. Surrey Beatty & Sons, Sydney.

Soares D (2002) Neurology: an ancient sensory organ in crocodilians. *Nature* **417**, 241–242.

Soares D (2007) The evolution of dome pressure receptors on crocodiles. *In Evolution of Nervous*

Systems: A Comprehensive Reference. Volume 2. Non-mammalian vertebrates. (Eds JH Kaas and TH Bullock) pp. 157–162. Elsevier, Amsterdam, Netherlands and Academic Press, Boston, Massachusetts)

Somaweera R, Webb JK, Brown GP, Shine R (2011) Hatchling Australian freshwater crocodiles rapidly learn to avoid toxic invasive cane toads. *Behaviour* **148**, 501–517.

Stevens KA (2006) Binocular vision in theropod dinosaurs. *Journal of Vertebrate Palaeontology* **26**(2), 321–330.

Todd NPM (2007) Estimated source intensity and active space of the American Alligator (*A. mississippiensis*) vocal display. *The Journal of the Acoustical Society of America* **122**(5), 2906–2915.

Underwood G (1968) Some suggestions concerning vertebrate visual cells. *Vision Research* **8**, 483–488.

Underwood G (1970) The eye. In *Biology of the Reptilia Volume 2.* (Eds C Gans and TS Parsons) pp. 1–97 Academic Press, New York.

Vergne AL, Pritz MB, Mathevon N (2009) Acoustic communication in crocodilians: from behaviour to brain. *Biological Reviews of the Cambridge Philosophical Society* **84**(3), 391–411.

Vliet KA (1989) Social displays of the American Alligator (*A. mississippiensis*). *American Zoologist* **29**, 1019–1031.

Vliet KA (2001) Courtship behaviours of American alligators, *Alligator mississippiensis*. In *Crocodilian Biology and Evolution* (Eds GC Grigg, CE Franklin and F Seebacher) pp. 383–408. Surrey Beatty & Sons, Sydney.

von Düring M (1973) The ultrastructure of lamellated mechanoreceptors in the skin of reptiles. *Zeitschrift fur Anatomie und Entwicklungsgeschichte* **143**, 81–94.

von Düring M (1974) The ultrastructure of the cutaneous receptors in the skin of *Caiman crocodilus*. *Abhandlungen der Rheinisch - Westfaelischen Akademie der Wissenschaften.* **53**, 123–134.

von Düring M, Miller MR (1979) Sensory nerve endings of the skin and deeper structures. In *Biology of the Reptilia Volume 9.* (Eds C Gans, RG Northcutt and P Ulinsky) pp. 407–442. Academic Press, New York.

Wajnberg E, Acosta-Avalos D, Alves OC, de Oliveira JF, Srygley RB, Esquivel DMS (2010) Magnetoreception in eusocial insects: an update. *Journal of the Royal Society, Interface* **7**, S207–S225.

Wald G, Brown PK, Kennedy D (1957) The visual system of the alligator. *The Journal of General Physiology* **40**(5), 703–713.

Walls GL (1963) *The Vertebrate Eye and its Adaptive Radiation.* Hafner, New York.

Walsh B, Whitehead PJ (1993) Problem crocodiles, *Crocodylus porosus*, at Nhulunbuy, Northern Territory: an assessment of relocation as a management strategy. *Wildlife Research* **20**(1), 127–135.

Webb GJW, Manolis SC, Cooper-Preston H (1990) Crocodile management and research in the Northern Territory: 1988–90. In *Proceedings of the 10th Working meeting IUCN–SSC Crocodile Specialist Group.* April 1990, Gainesville, Florida. pp. 253–273. IUCN, Gland, Switzerland.

Weldon PJ, Ferguson MWJ (1993) Chemoreception in crocodilians: anatomy, natural history and empirical results. *Brain, Behavior and Evolution* **41**, 239–245.

Weldon PJ, Brinkmeier WG, Fortunato H (1992) Gular pumping responses by juvenile American alligators (*Alligator mississippiensis*) to meat scents. *Chemical Senses* **17**, 79–83.

Wever EG (1971) Hearing in the Crocodilia. *Proceedings of the National Academy of Sciences of the United States of America* **68**(7), 1498–1500.

Winklhofer M (2012) An avian magnetometer. *Science* **336**, 991–992.

Witmer LM (1995) Homology of facial structures in extant archosaurs (birds and Crocodilians), with special reference to paranasal pneumaticity and nasal conchae. *Journal of Morphology* **225**, 269–327.

Witmer LM (1999) The phylogenetic history of paranasal air sinuses. In *The Paranasal Sinuses of Higher Primates: Development, Function and Evolution* (Eds T Koppe, H Nagai and KW Alt) pp. 21–34). Quintessence, Chicago.

Witmer LM, Ridgely RC, Dufeau DL, Semones MC (2008) Using CT to peer into the past: 3D visualization of the brain and ear regions of birds, crocodiles, and nonavian dinosaurs. In *Anatomical Imaging: Towards a New Morphology.*

(Eds H Endo and R Frey) pp. 67–87. Springer, Tokyo.

Wu LQ, Dickman JD (2012) Neural correlates of a magnetic sense. *Science* **336**, 1054–1057.

Zhao Y, Huang Y, Shi L, Chen L (2009) Analysis of magnetic elements in otoliths of the macula lagena in homing pigeons with inductively coupled plasma mass spectrometry. *Neuroscience Bulletin* **25**(3), 101–108.

FEEDING, DIGESTION AND NUTRITION

Late one night on the Cox River, which flows into the Limmen (see Chapter 11), I was leaning out over the bow of our small flat bottomed dinghy, gazing down my head torch beam into the deep clear water as we slid smoothly over forests of aquatic plants growing up from the bottom. At the edge of my vision, I suddenly became aware of a large dead wallaby on the water surface about a metre in front of me and, about a nanosecond later, I realised that it was in the jaws of a very large crocodile, now looking straight up at me. A fraction later as the bow was almost over the wallaby, the croc slid backwards and out of sight, taking the wallaby down with him. The image persists in my memory: the clear smooth water; the dead wallaby at the water surface and the sudden dark mass of a huge head and two bright pale yellow eyes. It was a magic moment.

WHAT DO CROCODYLIANS EAT?

All crocodylians are carnivorous (Figs 6.1–6.3) and, although not all attacks are feeding related, some of the large species certainly take humans for food, part of the reason they stir such emotion (Chapter 14). But, of course, large crocs start off small, so their diet when small must be appropriately small. As a generalisation, crocodylians eat a wide variety of invertebrate and vertebrate prey, depending on what is available in their habitat, with the diet changing from insects and other invertebrates when they are hatchlings to include, as they grow, larger and larger items, typically vertebrates.

In most tropical fresh and brackish water habitats, they are the top predator in the food chain. They are skilled predators, even taking birds and bats skimming along a water surface as they drink. They will often 'sit and wait', but they also hunt actively, and will feed on carrion, too, often lured by the smell (Chapter 5).

There is an enormous body of literature reporting numerous dietary studies of crocodylians over the last 50 or so years. In this chapter, I will refer particularly to papers which could be helpful to anyone beginning a new study and papers from which generalisations about crocodylian diet emerge.

STUDY METHODS

Determining diet by observing foraging behaviour of crocs is not practical because most feeding occurs at night, they are notoriously wary and they often live in turbid waters. Studies of diet therefore rely almost entirely on examining stomach contents, which is also not straightforward, either to collect the samples or to analyse the observations.

Stomach contents from harvested individuals

Much information has been gathered from biologists fitting in with hunters. For example, Cott (1961) examined 857 stomachs from *C. niloticus* in Uganda and Northern Rhodesia (now Zambia) between 1952 and 1957, mostly shot for their skins. Another

An Australian freshwater crocodile, Crocodylus johnstoni, *with an eel-tailed catfish,* Neosilurus hyrtlii. *(Photo Ruchira Somaweera)*

Fig. 6.1. *A saltwater crocodile,* C. porosus, *lifting its head to swallow a magpie goose. (Photo Buck Salau) (See video clips of feeding at http://crocodilian.com/books/grigg-kirshner)*

seminal study that provided valuable data from stomach contents early on was by Alistair Graham (1968) on the population ecology of Nile crocodiles (*C. niloticus*) in Lake Rudolf (now Lake Turkana) in Kenya. This study, financed by him selling the skins of his 500 study animals, features in Chapter 13.

Early last century, biologists often killed animals just to get stomach contents and it still occurred when I was a student in the early 1960s. Thankfully, it became unacceptable soon after that and ingenious new ways of getting stomach contents from crocodylians had to be found. However, there are still opportunities for collecting stomachs or other samples by coordinating with hunters engaged in wild harvests, notably from *A. mississippiensis* in the USA and *Caiman crocodilus* in Venezuela and elsewhere.

Stomach flushing

Stomach scooping and/or flushing with water has become a routine technique. Janet Taylor's study

of *C. porosus* in 1975–76 at Maningrida, northern Australia, was one of the first. With the mouth propped open by a short length of padded pipe she used a lubricated scoop on a long handle (Fig. 6.4), which could be manoeuvred carefully down into the stomach, rotated and withdrawn, hopefully with most of the contents. She also flushed food items out, using a large bore, lubricated, clear PVC tube inserted carefully into the stomach, which was then filled with water while the head was held high. Squishing the stomach mixed food items up into the water in the tube and the tube and crocodile were then upended to empty into a container (Taylor *et al.* 1978). The process was repeated several times until no more food items were produced.

A simplified example of this process is shown in Fig. 6.5 and some product in Fig. 6.6 (p. 214). In a study of *C. johnstoni*, Webb *et al.* (1982) devised an improvement to this method in which the stomach is filled with water (croc 30° head up) before the scoop is inserted, and material scooped out with the

Fig. 6.2. *A 71 cm* Osteolaemus tetraspis *tucking into a crab in a small stream in Petite Loango region of Loango National Park, Gabon. (Photo Mitchell Eaton)*

head 60° down. A further variant was developed by Ayarzagüenesan in 1983 and described by Fitzgerald (1989), in which flushing was conducted using a garden hose at a high flow rate, combined with repeated 'Heimlich manoeuvres' (squishing) to lavage the stomach and wash out the contents from a croc tied to a plank and inclined head down at ~20° (e.g. Fig. 6.7, p. 215). Fitzgerald (1989) used the opportunity of having access to harvested *Caiman crocodilus* to compare the effectiveness of the three methods, concluding that the garden hose plus squishing technique evacuated essentially everything, whereas scooping and flushing removed on average ~70% of the food diversity and ~50% of all the total contents. Rice (2004) and Rice *et al.* (2005) confirmed the garden hose method as the most effective in their comparison of *A. mississippiensis* diets in three Florida lakes. A garden hose on a tap is seldom available in the field, but a battery operated bilge pump in a bucket can be used instead. If a researcher is interested mainly

in what is being eaten, without needing to know how much, the simpler scoop and flush technique (or flush alone) is simple to apply and less likely to damage the animal. If a full quantitative answer is required, Fitzgerald's and Rice's analyses suggest that the garden hose technique may be necessary.

Sampling stomach contents from large, live crocodylians poses problems of course, and it is always useful to take advantage of opportunities provided by accidental deaths (Fig. 6.8, p. 216).

Analysis of stomach contents

This is not nearly as straightforward as might be thought. Identification of items from fragments can be challenging and soft bodied food items may be digested rapidly, leaving no trace. Some studies have tried to avoid this by counting only 'freshly ingested' prey, introducing a degree of subjectivity. Relative importance of prey may be expressed in terms of the proportion of stomachs with a particular type of food, the relative numbers of a type of prey,

Fig. 6.3. *A large, nearly 3 m female American alligator with a bowfin (Amia calva). She stayed like this on the bank for about half an hour. (Photo Louis Guillette)*

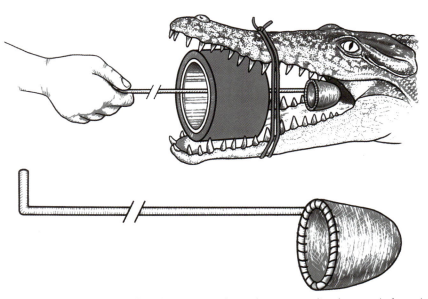

Fig. 6.4. *A simple stomach scoop can be made from bent wire and attaching a sampling bag made from the thumb of a rubber glove or a suitable mesh bag. The croc's mouth is propped open with a short piece of metal pipe coated with something to soften the surface and prevent tooth damage. Several sizes of scoop and pipe allow a range of different sized animals to be sampled. (Drawing by Fred Duncan, from Taylor et al. 1978, copyrighted material included courtesy of SSAR)*

Fig. 6.5. *Stomach flushing a* C. johnstoni, *Limmen Bight River, Northern Territory, late dry season 1984; (left to right) Lyn Beard, Laurence Taplin and Tim Pulsford (see Chapter 11). (Left) The croc is held vertically and a wide bore plastic tube inserted carefully into the stomach, followed by an aliquot of water. (Right) Squishing (surely the correct technical term) the stomach mixes the water with the stomach contents and the croc is then upended into a bag, so catching the contents (right). The process is repeated until no more is recovered. Had we been seeking a quantitative answer, this technique would have fallen short, but we wanted to know only what they had been eating (crabs). This method is more challenging in 4 m individuals, but see Fig. 6.7. (Photos GCG)*

or relative mass or volume, either as found or as 'reconstituted'. Webb *et al.* (1982) reviewed different ways to measure 'importance' and introduced the concept of 'target size', concluding that the size of the prey was more important than its taxonomic group. In any quantitative analysis of stomach contents, it is well to be mindful that some food items are digested more rapidly than others (Jackson *et al.* 1974). Hair and chitinous remains are not digested at all (Garnett 1985) and may be retained in the stomach, leading to an overestimate of the importance of their previous owners in the diet. To further complicate matters, the retention of these indigestible remains appears to depend on the overall size of the meal. Janes and Gutzke (2002) explored this experimentally in alligators and found that turtle scutes were retained

longer if they had been part of a larger meal. Nifong *et al.* (2012) determined rates of digestion of crabs by alligators, providing a visual guide that could be used to assess time since ingestion. Papers by Webb *et al.* (1982), Magnusson *et al.* (1987) and Wallace and Leslie (2008) provide useful reviews of different approaches.

DIET

As mentioned previously, all of the extant crocodylians are carnivorous, typically taking live prey but carrion feeding is also reported. There have been many studies across a wide range of species and habitats, so a comprehensive review is beyond our present scope, but some examples drawn from the literature will be helpful and some

Fig. 6.6. *Stomach contents flushed from a large broad-snouted caiman,* Caiman latirostris, *harpooned by Tereca Queirol Melo in highly saline water in the estuary of the Rio Jacariú, Ilha do Cardoso, southern Brazil, in May 1992 (see Chapter 11). We saw the same species of crab clambering around in the mangrove roots where the caiman were living. (Photo GCG)*

generalisations can be made. Informative recent reviews are those by Platt *et al.* (2006) reporting a study of more than 400 *C. moreletii* in northern Belize and Wallace and Leslie (2008) on the diets of Nile crocodiles *(C. niloticus)* in the Okavango Delta, Botswana.

Crocodylians emerge as opportunistic carnivorous predators that eat a wide variety of animal taxa across a wide range of prey sizes, larger individuals including a higher proportion of larger species, which are often vertebrates, and fewer smaller prey such as insects and spiders as they grow. They feed primarily at the water's edge but also on land. They will take carrion and may go onto land to retrieve it, even to some distance. They do not show much preference, and stomach contents typically reflect the animal biota that is available in their habitat, targeting animals large enough to stimulate interest yet not too large to manage.

Dietary changes through life

Their changing diet as they grow was shown very nicely by Cott (1961), while analysing stomach contents from a very large sample of *C. niloticus* (Fig. 6.9) and has been confirmed recently in a comprehensive study of the same species, in Botswana, by Wallace and Leslie (2008). The shift, from insects and other invertebrates as hatchlings to small vertebrates such as frogs and small fishes and then larger vertebrates, via increasingly large species of fishes, birds, reptiles and mammals, characterises many crocodylians. Of course, the diet also depends on what is available in the habitat and crocs' adaptability to survival on a wide variety of animal prey is a major reason for their success. For example, in contrast to the wide diversity of biota found by Cott, Graham (1968) found that large *C. niloticus* in Lake Rudolf (now Lake Turkana) lived almost entirely on fish: 223 stomachs out

Fig. 6.7. *Sampling from the stomach of large living crocodylians is very challenging. Here, Amanda Rice and John White are using the garden hose technique on a large American alligator in Florida (Rice et al. 2007). Try this with a big 'saltie'! (Photo Amanda Rice Waddle)*

of 254 with food present had only fish. Graham pointed out that Lake Turkana and its surroundings are rather barren and opportunities for feeding on mammals and birds are much fewer. An additional 239 stomachs were empty, of which more later. The implications from these early studies on *C. niloticus* are that crocodylians include larger prey items as they grow (Fig. 6.9) and the same species can subsist on very different diets in different habitats depending upon what is available. Neither of these conclusions is surprising, and they have been reinforced by studies on other members of the Crocodylidae, such as *C. porosus* (Taylor 1979), *C. johnstoni* (Webb *et al.* 1982; Tucker *et al.* 1996), *C. moreletii* (Platt *et al.* 2006) and in Alligatoridae: *A. mississippiensis* (Delany and Abercrombie 1986), several species of Amazonian caimans (Magnusson *et al.* 1987) and *Caiman yacare* in the Pantanal, Brazil (Coutinho 2000). This last study was able to examine stomachs from slaughtered individuals.

Janet Taylor (1979) examined stomach contents of 289 juvenile *C. porosus* up to 180 cm total length, caught in a variety of habitats in tidal creeks and rivers in Arnhem Land, Northern Territory, Australia. Diet differed between habitats and correlated with salinity, as did the fauna available as potential prey at each site. Fish, aquatic and terrestrial insects, crabs and shrimps were caught across all size classes and animals > 120 cm had also captured birds, rodents and bats. Some samples had snakes and skinks. Overall, crabs and shrimps predominated, caught at the land–water interface.

In a much more recent study, Steven Platt and colleagues (Platt *et al.* 2006) examined stomach contents from 420 *C. moreletii* in northern Belize over a size range from 23 to 255 cm total length (i.e. hatchling through to adult). Mostly they used stomach flushing. The smallest individuals fed mainly on insects and spiders, graduating to aquatic gastropods, crustaceans, fish and other vertebrates as they grew. Larger animals took progressively larger prey, but continued to take smaller prey as well. Insects and spiders were present even in adults, although their main diet comprised aquatic gastropods, crustaceans and fish. Interestingly, they found a small number of *Rhinella marina* (formerly

Fig. 6.8. *Most information from very large individuals has resulted either from hunted individuals or opportunistically from accidental deaths. This* C. porosus *was drowned in a fisherman's net set illegally in the Liverpool River, Northern Territory. Laurence Taplin found that it had eaten a magpie goose, a wallaby and several barramundi … similar to food selections that we might make, actually! (Photo GCG)*

Bufo marinus), a toad native to Belize but introduced to Australia where it is known to be fatal to the Australian *C. johnstoni* (Smith and Phillips 2006 and see Chapter 13). They also found a few reptiles (iguanas, anoles, turtles and, snakes), a few birds (blackbirds, cormorants, herons and egrets) and mammals (opossums, and a porcupine (!) and several other species of rodent).

Empty stomachs

In any study, a proportion of stomachs are found to be empty. Cott (1961) found 52% of nearly 600 *C. niloticus* with stomachs either empty or containing only indigestible fragments such as beetle elytra, molluscan opercula, fish scales, turtle scutes, feathers, hair and/or claws. Studying the same species in another area, where the diet was almost exclusively of fish, Graham (1968) found 48% with empty stomachs. Taylor (1979) found 15% of the juvenile *C. porosus* captured in a tidal habitat had empty stomachs. Cott and Graham both had many large individuals, whereas Taylor's study was on juveniles, and it is now recognised that the proportion of empty or nearly empty stomachs increases with body size. This has been shown particularly nicely for *C. niloticus* by Wallace and Leslie (2008) who reported fresh food in 70–80% of yearlings (30 cm SVL), declining to 10–20% in animals with SVL > 100cm. The occurrence of so many empty stomachs is likely a consequence of crocodylians not needing to eat every day. In larger individuals, relying more on terrestrial vertebrates, the higher proportion of empty stomachs may relate to their lower abundance than invertebrates and larger size.

Food preferences

As a generalisation, crocodylians typically eat insects, molluscs, crustaceans and small fishes and frogs when they are hatchlings and juveniles, adding larger fishes, birds and mammals as they grow, and the species recorded are a good reflection of what is available to them. However, Taylor (1979) found in juvenile *C. porosus* that prey taken correlated quite well with what was available in the different habitats, except that she found fewer fish than might have been expected. This raises the question of food preference. How selective might crocodylians be in a habitat rich in both diversity and abundance of suitable prey items? To demonstrate a preference for or against a particular food item, a study would need to demonstrate either superabundance or shortfall of that item in the diet, compared with its proportional availability in the habitat. This is hard to do because of the great difficulty in assessing the proportional quantities of available suitable prey. Webb *et al.* (1982) observed that there were 'virtually no data showing marked preferences for specific prey which cannot be interpreted as simply the availability of that prey' and that it was 'hard to challenge' the conclusion reached by most authors that 'crocodiles eat anything available to them

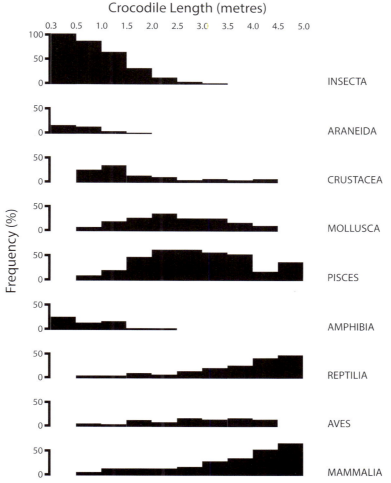

Fig. 6.9. *Hugh Cott analysed stomach contents from 857* C. niloticus *shot (mostly) for their skins in Uganda and Northern Rhodesia (now Zambia) from 1952 to 1957. This histogram shows the proportional occurrence of each of nine animal taxa in 10 crocodile size classes. Although some of most of the taxa are eaten throughout life, proportional trends with increasing size are marked and, within the fish, for example, larger crocodylians eat progressively larger fish as they grow. (Adapted from Cott 1961, with permission. © 1961 The Zoological Society of London)*

within acceptable size limits'. On the face of it, this seems an unlikely conclusion. Surely a croc would prefer a succulent rat or duck over a catfish? But, in the decades since 1982, despite various hints to the contrary (e.g. Fig. 6.8) and, although it seems to be unlikely, there still seem to be no data supporting preferences for particular food items.

Indeed, with such an abundance of data on crocodylians and so much of it showing that stomach contents depend primarily on what is available to them within the right size range, a cynic might say that the main use of a study on crocodylian stomach contents these days is to provide an assessment of the animal biota where they live!

The long-snouted 'fish eaters'

Lack of a particular preference seems to be true even for the four long-snouted supposed 'fish eater' species as well: *Gavialis gangeticus, Tomistoma schlegelii, Mecistops cataphractus* and *Crocodylus johnstoni*. Although their elongated snouts are interpreted as adaptations for predation on fish, this does not necessarily indicate a preference for fish, or that they eat fish exclusively. It may reflect only the preponderance of fish among suitable prey in the habitats in which they have evolved. For example, although *C. johnstoni* certainly eat fish, they also eat a wide variety of other prey as well and their stomach contents provide a good sampling of what is available in their habitat, with the usual progression in prey size (and taxa) as they grow. Tucker *et al.* (1996) made a comprehensive dietary study of *C. johnstoni* in the Lynd River in Queensland (see also Chapter 13). He found they eat spiders, aquatic and terrestrial insects, shrimps, fish, frogs, turtles, snakes, birds and mammals, with a strong correlation between diet and size, larger animals taking larger prey. Above ~60cm SVL they took more fish, turtles and snakes. Birds and mammals were taken infrequently and, presumably, opportunistically. Pauwels *et al.* (2003) reviewed information about prey sought by *M. cataphractus* and reported that, as well as fish, the list includes grasshoppers, shrimps, freshwater crabs, water snakes and one specimen even had a chevrotain (mouse deer) in its stomach. Detailed dietary information about gharials seems to be lacking and, because they are Critically Endangered, the last thing we need is a study on wild gharial stomach contents. However, there are opportunistic observations and they are widely reported to be specialist fish eaters, as their skull shape and dentition imply: very similar to dolphins (particularly riverine species) and some piscivorous fish. On the face of it, they may be the most 'fish specialised', but there is a risk of a circular argument here, and they also eat reptiles and mammals when they get an opportunity, as well as insects, spiders and frogs when they are juvenile (Whitaker and Basu 1983). Concerning the diet of *Tomistoma schlegelii*, the Tomistoma Task Force (reported by Rachmawan and Brend 2009) has found that they take a wide range of prey, from small invertebrates to birds, reptiles, monkeys and small deer. These authors also reported one probable and one confirmed attack on humans by *Tomistoma*, in different rivers, both late in 2008. The confirmed attack was by a 4 m (approx.) female and the motivation for the attack was not discussed directly but feeding rather than nest defence was implied; remains of the victim were found in the stomach after the croc was captured. The authors commented that part of the reason that *Tomistoma* are thought not to prey on humans is because of their reputation as fish eaters.

Despite this caution, the skull morphology of the long-snouted species is well suited to fish eating, as discussed in Chapter 3. However, the conventional wisdom that the long-snouted crocs are different in their dietary habits through preference needs to be treated with caution, and data.

Seasonal changes in diet

Many species of crocodylians show seasonal changes in food intake correlating with seasonal temperature changes. As ectotherms (Chapter 10), crocodylians are likely to be less active during winter months and take less food. Webb *et al.* (1982) found that *C. johnstoni* in the McKinlay River, Northern Territory ate more in the wet season, which is also the hot season there. Hutton (1987) reported higher food intake in the hot season in *C. niloticus* also and was able to relate this to faster growth rates and better body condition. Wallace and Leslie (2008) found that both juvenile and adult *C. niloticus* had a higher proportion of empty stomachs in the winter months, reflecting a lower need for food in cooler conditions. On the other hand, Taylor (1979) found no correlation with season in her study: juvenile *C. porosus* fed throughout the dry (cool) season in her study area. More temperate species such as *A. mississippiensis* and other species living in areas subject to low winter temperatures feed less, and alligators at higher latitudes cease or reduce feeding for longer in winter than their lower latitude cousins.

Vegetable matter, fruit consumption

The presence of vegetable matter in crocodylian stomach contents has drawn frequent comment. It has usually been attributed to accidental ingestion or, perhaps, secondary ingestion from having eaten a herbivore with a full gut. However, sometimes there is a large amount of vegetable material (e.g. Taylor 1979 in *C. porosus*, Webb *et al.* 1982 in *C. johnstoni,* McNease and Joanen 1977 and Rice 2004 in *A. mississippiensis*), even though it mostly does not look particularly nutritious for an animal without the means to deal with cellulose. A ball of indigestible vegetable material has sometimes been reported accumulated in a croc's stomach, forming what Wings (2007) would refer to as a 'bio-gastrolith'. Nobody seems to know how or why this occurs and whether it is injurious or, perhaps, can be regurgitated at will.

Fruit is a different matter. Many carnivores also eat fruit and captive *Caiman latirostris* (Brito *et al.* 2002) and both *A. mississippiensis* and *A. sinensis* (Brueggen 2002) have been reported eating fruit and other vegetable matter. Recently, Platt *et al.* (2013) reviewed the literature and found fruits or seeds reported from the stomach contents or faeces of 13 out of 18 species of crocodylian. They also found several convincing reports of crocodylians feeding on fruit in the wild and concluded that fruit consumption (frugivory) is both deliberate and widespread. They suggested that, although fruit and seeds may make only a small contribution to nutrition, there was no need to assume that it was not beneficial. At a more ecological scale, there could be a role in seed dispersal. In the last decade in particular the extensive journeys made by at least some crocodylians have become better recognised and it is not difficult to envisage large crocodylians as plant dispersal agents over long distances.

Cannibalism

Large crocodylians sometimes feed on smaller ones and it has been suggested that this is a significant factor in regulating populations. Accordingly, this topic is discussed in Chapter 13.

Do crocs feed their young?

McIlhenny (1935) reported seeing mother American alligators feeding their young on eight occasions. He wrote (p. 54) 'I have seen mother alligators catch large fish, large snakes, and turtles in their jaws and crunch them to a pulp, holding them at the surface of the water between their jaws, so that the young could gather bits of food from the crushed flesh'. John Brueggen at the St Augustine Alligator Farm in Florida described seeing a female Siamese crocodile twice feeding young by holding a piece of meat in her mouth: http://www.alligatorfarm.us/Reseach%20-%20croc%20feeding%20young.html. A video is posted at http://www.youtube.com/watch?v = wZdWRR5tRmc and it is convincing.

Tomas Blohm observed behaviour in a captive female Orinoco crocodile (*Crocodylus intermedius*) and, less often, a captive male, that was very similar to what McIlhenny described in alligators, but the hatchlings did not take advantage of it by feeding. Before hatching was due, the female showed odd behaviour, which became more common after her young hatched. To quote Blohm, she 'began to hiss softly, sink, and surface, then to suddenly crunch and chew the prey and slam it on the water's edge'. This was done in the vicinity of the hatchlings and produced bits of flesh and fat, floating on the water and some on land. The male sometimes did likewise. This might sound like convincing evidence for parental provisioning of the young, except Blohm reported that 'the hatchlings never paid attention nor fed from the remnants of the disintegrated chickens' Blohm (1982). Perhaps, being in captivity, they were already full.

To put this topic further into context, the embryos of an early Jurassic prosauropod dinosaur *Massospondylus carinatus* lacked teeth and had body proportions that imply that a hatchling could not have foraged for itself (Reisz *et al.* 2005). The authors speculated therefore that it was altricial, like most birds, and needed to be fed by its parent/s. Brazaitis and Watanabe (2011) have presented a good case that crocodylian behaviour provides good insight into the likely behaviour of dinosaurs. Given the extent of their maternal care, it would not be surprising to hear more observations of maternal provisioning among crocodylians.

PREY CAPTURE AND HANDLING

Feeding has several phases: the initial capture of prey, holding, killing and subduing it if necessary; breaking the food item up and manipulating it within the mouth; and, finally, the swallowing process in which the food item is moved between the palatal and gular flaps and into the pharynx and oesophagus. It is convenient to discuss separately the behaviour by which the food is acquired and the mechanics by which it is handled subsequently. Cleuren and De Vree (2000) have provided a valuable review of these topics.

FEEDING BEHAVIOUR

Crocodylians prefer their food fresh, and the well-known story that they stash food somewhere to rot lacks confirmatory evidence. It has probably arisen because they will also eat carrion, so the putrefying remains of an earlier victim may well be revisited, by the same or a different crocodylian. They are known to find rotting carcasses even at a distance from a waterway, presumably drawn by the smell (see Chapter 5), but they usually catch their prey and consume it straight away. Mostly prey is swallowed whole, but larger items may be broken up by being lifted from the water and 'whiplashed' (Figs 6.16, 13.20) in a pattern also used by monitor lizards and some birds.

Active hunters and 'sit-and-wait predators'

Most feeding occurs in the water (e.g. Fig. 6.10) and at the water surface, much of it at night. Although often referred to as 'sit-and-wait predators', this is only one of the strategies crocodylians employ, because they are also active hunters, such as in the underwater attack we saw on the goose, described in the introduction to Chapter 5. Juvenile *C. porosus* in northern Australian estuaries can often be seen working the shallows at the river edge, snapping at prawns flushed by their movement. A sideways strike of the head is the typical prey capture technique for this, often curving the tail as if corralling the prey towards the bank. In the dark, the snapping is presumably directed by the many integumentary sense organs (ISOs) on the jaws (Chapters 3, 5), relaying information to the brain via the massive trigeminal nerves. In daylight, and in moonlight, the eyes are important. I have seen *C. porosus* in northern Australian rivers lurking below cormorants nesting communally among the branches of trees overhanging the water, waiting for a chick to fall. This is probably visual, combined with sensing the splash through their ISOs (see Chapter 5). *C. porosus* and other species have been described catching birds and bats flying low to the water to skim up a drink, which must be visual. Juveniles take insects and frogs among the reeds with a sideways strike or a forward lunge driven by the tail. Often the insects are taken at emergence, so a lot of terrestrial insects with an aquatic life stage are captured.

A recent paper by Dinets *et al.* (2013) described a novel feeding mechanism, balancing sticks across the snout as a lure to attract nest building egrets (Fig. 5.31). This may be the first report of tool use in a reptile (see Chapter 5 for discussion).

An interesting fishing tactic (Fig. 6.11) has been described in both *Caiman yacare* (Olmos and Sazima 1990) and *C. porosus* (Britton and Britton 2013). In both species, the distinctive posture involves stretching the forelimbs at right angles to the trunk with the palms and fingers raised and facing outwards, cruising or floating at the surface with the lower jaw ajar. Fish swimming close to the head or

Fig. 6.10. *A large male American alligator with a Florida gar,* Lepisosteus platyrhincus. *The fish too is interesting. Like* Amia calva, *the bowfin (Fig. 6.3),* Lepisosteus *has a vascularised air sac connected to the oesophagus, which acts as a lung in poorly oxygenated water. (Photo Louis Guillette)*

Fig. 6.11. *A sub-adult* C. porosus *fishes at Cahill's Crossing on the East Alligator River, Northern Territory, Australia, using the distinctive 'cross posture' tactic. With arms at right angles and the lower jaw ajar, the presence of a fish stimulates ISOs on the head and limbs and triggers a sideways snapping swipe. (Photo Kelly Clark)*

outstretched arms are snapped at with a sideways swipe of the head. The particular circumstances in which this technique is employed are not yet described, nor how widespread it may be among crocodylians. It certainly seems to be a way in which crocs are able to hunt by touch, spreading their touch receptors across a wider span and although Olmos and Sazima reported an observation at night, whether its prevalence may increase then is so far unknown. Apart from the behaviour itself, two facets are of particular additional interest. First, unless the behaviour has evolved convergently it must be very old: alligatorids and crocodylids have been separated since the Mesozoic. Second, although the behaviour seems designed for taking advantage of the high density of integumentary sense organs on the fingers (Fig. 5.24) this can apply only in crocodylids: *Caiman yacare* and other alligatorids do not have ISOs here. Presumably they rely on their less conspicuous touch sensors.

The prodigious strength and speed displayed by large crocodylians capturing large animals drinking at the water's edge is probably their most spectacular feeding behaviour. Such a scene is *de rigueur* in African natural history documentaries, typically of Nile crocodiles attacking migrating wildebeest and zebras as they cross the Mara River in Tanzania. A peaceful scene with a few impala drinking in the still water is suddenly shattered by the rush of a huge crocodile, powered by its tail, and a wildebeest or a zebra is dragged under and drowned, often combined with the famous 'death roll'. A big Nile crocodile can swallow an impala whole, or several may cooperate to tear a larger prey item apart. Not only are the jaws strong, but the shortened neck could well be an adaptation to enable seizing and subduing large prey. The atlas, axis and cervical vertebrae, whose length and extent of rotational capacity dictates the length and flexibility of the neck of vertebrates, are shortened and sturdy in

Crocodylia, with extensive musculature inserting onto the back of the skull, and with posterior bony projections for muscle attachment (Fig. 3.52). The neck anatomy is well suited for dragging an animal as powerful as a buffalo into the water against its will.

Their death roll is useful also for tearing chunks from a large carcass. A grip is taken with the anterior teeth and, with its legs held against the flanks, the croc rotates rapidly, driven by the tail and with the head canted initially at an angle to the body-axis, and pulled away, tearing off a mouthful. Occasionally workers at crocodile farms lose an arm in this way when hand feeding. There have been some remarkable escapes, in which the 'attackee' has the presence of mind to roll with the croc until it realises its mistake or loosens its grip for some other reason. The mechanics of the death roll have been studied quantitatively by Fish *et al.* (2007) using hatchling alligators averaging 66 g as a model. Keeping the limbs against the body reduces drag and so increases the spin rate. The authors calculated that a shear force of 0.015 N was generated at the snout, which increases disproportionately with size, so that a 3 m alligator generates a shear force of 138 N.

The death roll may be a way for crocs to get around their lack of rotational flexibility, which has implications also for picking food off the ground. Being low to the ground and having very long jaws and a short, fairly inflexible neck, picking up food on land requires crocodylians to turn the head sideways, partly rotating the body (Fig. 6.12). Pooley (1962) pointed out that turning the head would also be useful to get a better sighting on a small food item on land.

Terrestrial hunting

Although crocodylians are specialist hunters at the land–water interface, many also hunt on land, particularly at night. A review by Dinets (2011) records that adults of the larger species conduct ambush hunting away from water. This is known from observations on *Alligator mississippiensis, Crocodylus niloticus* and *C. porosus,* while *C. palustris, C. acutus, C. intermedius* and *Melanosuchus niger* have been observed away from the water in places well chosen for an ambush and apparently lurking with intent. Although caiman are well known to make journeys over land (Chapter 4), this is to relocate between water bodies as they dry up. Hunting away from water seems not to have been reported so far. Terrestrial hunting seems to be less common in smaller species or juveniles, perhaps because they would be more at risk from predation themselves, away from the safety of the water. *Paleosuchus trigonatus,* however, may be an exception, with small terrestrial mammals and snakes appearing frequently in stomach contents (Magnusson *et al.* 1987).

Feeding aggregations and cooperative feeding

A food source can draw in several crocs from the surrounding area. For example, *C. porosus* have learned that flatback turtles, *Natator depressus,* congregate to nest on Crab Island, west of the tip of Cape York, Australia (Limpus 2007) and at Cape Dommets, northern Western Australia and Melville Island, off Darwin (Whiting and Whiting 2011). The crocodiles feed on the turtles, their hatchlings and their eggs. This is of conservation concern, particularly since the recovery of *C. porosus* in northern Australia (Chapter 13), because these are major nesting beaches used by this endemic Australian sea turtle, which is listed in Australia as Vulnerable ('Data Deficient' on IUCN Red List). At the Melville Island rookery, most of the turtles are olive ridleys (*Lepidochelys olivacea*), which are also listed as Vulnerable on the IUCN Red List.

Another aggregation of *C. porosus* forms at a barrage across the Mary River, Northern Territory when fish take advantage of king tides causing a flow of water over the concrete causeway. The high numbers of fish crossing in the shallow water are easy prey and attract crocodiles, which concentrate and catch the fish as they cross. What makes this particularly interesting is that the crocodiles travel long distances to these feeding opportunities. One croc with a satellite transmitter attached moved purposefully along the coast from the west and up the Mary River to the barrage, a distance of more than 45 km (Adam Britton *pers. comm.*). It arrived at the site 2 days before the king tide and left soon after! It

Fig. 6.12. *Crocodylians lack a lot of rotational flexibility in the neck, so pick up food items from the ground by rotating the whole anterior body. With the eyes so high on the head, this may also give them a better sighting of a small close food item at ground level. (Photos GCG, Adam Britton)*

is common to think of such behaviour by animals as being 'hard wired', embedded almost as an instinct and acquired over many generations. However, the barrage is comparatively recent, having been built in the 1980s and completed in 1988. How do they know? Of course, so far nobody knows the answer to that, which bears on questions about learning and intelligence, discussed briefly in Chapter 5.

Some aggregations seem to have an almost social dimension. With his characteristically engaging word pictures, Tony Pooley described a feeding aggregation by *C. niloticus*. From a hide on Lake Inyamiti adjoining the Pongola River in Ndumu Game reserve, Zululand, Tony watched 27 *C. niloticus* feed in an orderly fashion from the bloated carcass of a hippopotamus (Pooley 1982). He described the crocs going in one or two at a time from a semi-circle of individuals, biting and twisting off large chunks of flesh then submerging to exit and surfacing a short distance away to swallow while another croc went in for a bite. There were some aggressive growls and hisses if they got in the way of each other but 'the degree of tolerance and orderliness was remarkable'. This might seem fanciful – an over-interpretation by a person who really loved crocs – but Tony was also an astute observer. Re-reading his description recently reminded me that on Christmas Island in 2009 I had seen frigate birds flying in a wide circle around a Parks ranger who was throwing up morsels of food one at a time, with the birds taking turns in a very orderly fashion. Tony referred to his crocodile observation as 'social feeding', but as my notes from conversations with him in 1977 record, not all group feeding is harmonious. When, on another occasion, he saw 30–40 crocodiles around a buffalo carcass 'it was accompanied by a large amount of threatening hisses and growls and occasional spouting of water as individuals moved in to tear a piece off and then retreat to the outside of the throng to deal with their prize before going in to the melee again'.

Another type of aggregation has been described by Dinets (2010), observing 20 or so alligators circling within a pond and feeding on a variety of fish (Fig. 6.13). He saw this at five different locations and many times. In some cases the ponds were receding, which may have led to a concentration of fish. Dinets described this as cooperative feeding. His impression was that the alligators were 'in such a dense crowd because they somehow benefited from each others presence, probably by snatching fish that others flushed from the bottom mud' (Vladimir Dinets *pers. comm.* 2011). He referred also to having seen similar 'cooperative feeding' in muggers, *C. palustris*, and to anecdotal accounts, including TV documentaries and videos of it in all species of *Caiman* and in *Melanosuchus niger* (Dinets 2010). A similar aggregation is seen in Fig. 6.14.

A somewhat different type of feeding aggregation, also referred to as cooperative, has been described in *A. mississippiensis*, *C. niloticus*, *Caiman crocodilus* and *Caiman yacare* (reviewed by King *et al.* 1998 and King 2008), in which individuals line up across the width of a stream flowing into or out of a water body and harvest the fish washing through (Fig. 6.15). A typical example of this was described in *C. niloticus* by Pooley and Gans (1976); 'Another example of cooperation may be seen in the early spring, when rivers rise and the water flows into channels leading to pans, or natural depressions, along the river. Sub-adult crocodiles often form a semicircle where a channel enters a pan, facing the inrushing water and snapping up the fish that emerge from the river. Each crocodile stays in place and there is no fighting over prey. Any shift in position, of course, would leave a gap in the crocodiles' ranks through which the fish could escape, so that what might be a momentary advantage for one crocodile would be a net loss for the group'.

Readers will note that these examples are certainly 'feeding aggregations', but they may have wondered if any were actually 'cooperative'. In the feeding aggregations described so far there is obviously some cooperation, in the sense that individuals are tolerating others close by. This might be a 'passive cooperation' or, as Schaller and Crawshaw (1982) described it, 'proto-cooperation'. They noted that the semicircles of crocs across an inflow or outflow that they saw in *Caiman crocodilus* and those described by Pooley and Gans (1976) in *C. niloticus* were 'proto-cooperation' rather than 'cooperation in the sense of a joint action in the performance of a defined task', adding that 'any

Fig. 6.13. *American alligators congregated in a shallow slough in the Florida Everglades. They swam in circles, making quick sidewise lunges and snaps at disturbed fish. One alligator is shown lifting its head to swallow a fish. Whether or not such behaviour meets the formal definition of 'cooperative feeding' is open to question. (Photo Vladimir Dinets)*

help fishing crocodilians gave each other seemed inadvertent'. Additionally, Gans (1989) wondered if 'the few reports on cooperation represent accidental events or part of a more general phenomenon'.

There seems to be room for thought about what terminology is appropriate. It may often, of course, be hard to judge whether or not behaviour is actively cooperative, but the main difference between the behaviour of *C. porosus* at the Mary River barrage and that of the *Caiman yacare* in Brazil (Fig. 6.15) is the number of individuals in relation to the width of the water flow; that is, the proximity of animals to each other. The tolerance of close neighbours could be judged as cooperative at a low level, but where to draw the line? Cooperative behaviour in general has received much discussion by animal behaviourists, so there is guidance. Dugatkin and Mesterton-Gibbons (1996) defined cooperation (in general) as 'an outcome that – despite potential individual costs – is 'good' in some appropriate sense for the

members of a group of two or more individuals, and whose achievement requires collective action'. The *Encyclopaedia Brittanica* defines cooperative foraging as 'the process by which individuals in groups benefit by working together to gain access to food'.

It is not hard to see that tolerance for other individuals in a feeding aggregation could evolve into more active cooperation, and in some cases there can be no doubt about the cooperation being active and deliberate. Two examples follow. There must be many more, but several authors have referred to the lack of reporting or proper study and documentation so far.

I had the good fortune to see a very good example of active cooperation when visiting Tony Pooley at St Lucia Estuary, Zululand, South Africa in October 1977. Tony decided we would feed a pelican from his freezer to a male Nile crocodile that lived with two females in a large pen. He was curious to see how a croc would get on with this difficult meal: all

Fig. 6.14. *An aggregation of black caiman,* Melanosuchus niger, *at Mamiraua Reserve near Tefé on the Solimões river ~500 km upstream from Manaus, Brazil. The birds are large-billed terns and yellow-billed terns, plus a great egret. (Photo William E Quatman)*

beak, legs and wings. The male was very large and at Tony's call he came to the high fence to collect the thawed pelican carcass. He took it back to the water in his mouth but found it difficult to manage and recruited one of the females to assist, as told in Fig. 6.16.

Tony described another compelling example. At the Ndumu game reserve in northern Zululand near the Mozambique border, where his interest in crocodiles was first stimulated (see Pooley 1982), he had seen two Nile crocodiles carry a large nyala (a type of antelope) to the water where they collaborated in dismembering it, just as we saw with the pelican. Whether the animals had killed the animal was not recorded, but there was apparently no doubt about their collaborative effort in taking

Fig. 6.15. *A dozen* Caiman yacare *lined up across a channel in Brazil, facing the current and side by side, tolerating each others' proximity, with their mouths open ready to snap at any fish. (Photos by Carlos Yamashita per Wayne King)*

One of Tony Pooley's male Nile crocodiles is fed a pelican carcass at one of the St Lucia Estuary crocodile breeding pens in October 1977.

He tried to dismember it by raising his head high, drawing the pelican up and then flicking it with a powerful, swift sideways and downwards thrash of the head. He tried repeatedly, but the pelican remained intact.

After about 15 unsuccessful minutes, he swam across to one of his females with the carcass in his mouth and stopped in front of her, as if offering it.

She took hold and, while the male held firm, rolled a couple of times (left), twisting the carcass. Then they tugged and the pelican came apart.

The male took away his bit and tried to break it up further while the female ate what she had torn off. His bit was still too unmanageable so he got more help from the female and eventually their cooperation tore the pelican into swallowable pieces for both of them.

Fig. 6.16. *A large Nile crocodile,* C. niloticus, *at St Lucia, Zululand, South Africa, having difficulty dismembering a pelican carcass, invited and got cooperation from one of the two females in their breeding pond. (Photos GCG)*

it to the water. A third example, which is probable rather than convincing, is Pooley's observation of half a dozen crocs swimming and feeding together in a shallow river, their combined movements in the shallow water scared up fish so that there was mutual benefit from the group behaviour. This could have been similar to the aggregation described in alligators by Dinets (2010), and the movements of the crocs implied collaboration. And if a dozen crocs surround a floating, bloated buffalo and thus stabilise it somewhat as pieces are torn and twisted from it (Britton *pers. comm.* 2011), the participating individuals benefit from the presence of the others because their work is made easier, but are they working cooperatively? Or is the benefit occurring accidentally?

Many readers will have their own examples; hopefully the few recounted here will showcase some of the diversity of feeding in aggregations, whether cooperative or not, and underline the need for more research and discussion. But before getting carried away about crocodylians helping each other out, remember too that a lot of disputation arises over food ownership (Fig. 6.17).

FEEDING MECHANICS

Whereas most mammals start the breakdown of food by chewing and grinding it with saliva containing digestive enzymes, crocodylians use the mouth only for prey capture, repeated biting to kill it if necessary, and breaking up the food item to pieces that can be swallowed into the wide oesophagus and stomach. A paper by Cleuren and De Vree (2000) provides a comprehensive review.

Mouth and gular region

The elongation of the jaws and the formation of the secondary palate are significant in this process because, with the throat closed by the gular valve (Figs 6.18, 5.18), breathing can continue while food is held in the mouth, even while holding an animal under water to drown it. The gular valve is made up of the palatal and gular 'flaps', from above and below, which are pressed together and, in conjunction with smaller lateral folds, make a good seal. The gular flap is stiffened by cartilage and both flaps are under muscular control. The anatomy and histology of the gular valve have been described by Putterill and Soley (2006). The effective mouth opening is, therefore, not at lips or the jawline the way it is in so many vertebrates, but at the back of the throat, at the entrance to the pharynx (Fig. 6.20). It has been suggested that the evolution of a secondary palate was favoured because of the

Fig. 6.17. *They don't always cooperate! Two large* C. porosus *in a Northern Territory billabong in dispute over a magpie goose.* (Photo Buck Salau)

Fig. 6.18. *Crocodylian gape demonstrated by* Caiman latirostris. *The jaws, teeth and tongue are used for prey capture and management, while the gular valve, made by gular and palatal folds or 'flaps', pressed together at the angle of the jaw, guard entry to the pharynx. Substantial mucus production towards the rear of the buccal cavity and on the valve lubricates the passage of food during swallowing. (Photo Lyn Beard)*

additional strength it brought to the upper jaw and that the respiratory benefit it brings is secondary.

Having the 'mouth' opening at the rear of the buccal cavity answers the question I am often asked, 'What is the sense in having salt glands that excrete onto the surface of the tongue, still in the mouth and at risk of being swallowed?' (Figs 11.15, 11.16). The answer of course is that the animal can rinse its buccal cavity (mouth) by moving the head side-to-side while keeping the gular and palatal flaps firmly closed. These same structures prevent water entering the gullet during the feeding process.

The mouth is, therefore, mostly about prey capture and killing, with its contribution to food preparation limited to breaking up and crushing large items.

Inertial feeding

A food item is often seized between the tips of the jaws. It can then be repositioned and shifted towards the back of the mouth by a technique known as inertial feeding, described originally for snakes by Carl Gans (1961) and then broadened to a generalisation (Gans 1969). In this, the jaws quickly release the object and, relying on its inertia, shift the grip to a new position. This is obviously more effective in air than in water and may be another reason why ingestion usually occurs with the head above the water. Quite commonly, a croc will be seen to rotate a food item in its jaws in this way while the head is lifted clear of the water, with swallowing following rapidly once the chunk is positioned optimally. Although the tongue is not mobile fore and aft or left and right, it is very muscular (Fig. 5.18) and can play a role in inertial feeding, throwing a food bolus vertically within the mouth just as the head is rotated, thus facilitating the repositioning (Cleuren and De Vree 1992).

Inertial feeding requires a certain mobility of the head and neck, and there is a suite of muscles from the cervical (neck) vertebrae to the rear of the skull (Fig. 3.52) that allow controlled movements of the head. These are very strong, for the head of most crocodylians is very heavy, but it can be moved with very fine control.

Crushing power of the jaws

Crocodylians are renowned for the strength and speed of jaw closure, and most people know too that the muscles that open them are much weaker, so the jaws can be held together with comparatively little force. The behaviour has prompted biomechanical studies into what forces are generated and the musculoskeletal system that contrives them is discussed in Chapter 3. There is no doubt about the crushing power. Erickson *et al.* (2012) measured bite force and pressure in 23 species of crocodylians (Fig. 3.47). Bite force correlates with body size and the measurements made on *C. porosus* exceeded those from any other living animal. The jaws and teeth are the main weapons for killing prey. Particularly when dealing with large prey, long strong teeth are useful to provide a good grip to prevent escape, and also for stabbing and killing. It need come as no surprise that the capture of lively struggling prey is followed by repeated and rapid biting, inflicting further damage and doing so with great force until the struggling stops.

Swallowing

Having held prey under water to drown it (if necessary), a crocodylian will usually lift it clear of the water to swallow it, raising the tail as a counterpoise (Figs 6.19, 4.26). This makes it less likely that water will be ingested along with the food and, also, makes use of gravity to assist with swallowing. This behaviour has led to some colourful interpretations, including one from New Guinea about a missionary who was taken by a large *C. porosus* and then displayed, held high in its jaws for the horrified villagers to see as the croc swam slowly away.

Crocodylians often lift the head to swallow on land as well, standing up as best they can on their short front legs, but they seem to prefer to eat in the water, often taking food there if it has been acquired on land, perhaps because it is easier to achieve the head lift posture. Note that the stomach is pretty much at the lowest part of the body when this 'head oblique tail arched' (HOTA) posture is adopted.

To swallow a food item, the pharynx opens by muscular relaxation of the gular and palatal flaps and the lateral folds of the gular valve and the tongue is lowered, opening a large sack behind the palatal valves (Figs 6.20, 6.21). The whole gullet area is quite distensible. Swift lifting and thrusting of the head uses both gravity and the inertia of the food item to propel it towards the pharynx. Closing the mouth, combined with movements of the hyoid apparatus and tongue, squeezes the bolus of food further back, over the glottis and into the long, distensible oesophagus, where it is presumably assisted by muscular peristalsis into the stomach. Busbey (1989) and Cleuren and De Vree (1992) provide more detailed descriptions.

DIGESTION

Digestion is the process by which food items are broken down both mechanically and then chemically, the complex large molecules of protein, carbohydrate and lipid being reduced to their small component parts, namely amino acids, simple sugars, glycerol and free fatty acids, which are small enough to be absorbed across the walls of the small intestine.

Crocodylians do not chew their food, so most of the initial breakdown is chemical, under the action of the low pH and churning in the muscular stomach. The presence of stones (gastroliths) in the muscular gut has prompted some authors to liken this to the grit-containing gizzard of some birds, which acts in the trituration of food. Other authors have rejected this, favouring hypotheses related to ballast and buoyancy control. These ideas are not mutually exclusive, but neither survives under scrutiny. The topic is discussed in Chapter 4.

ANATOMY

It takes less anatomical specialisation and processing to get the nutrients from animal food than from most plant food, so crocodylians, like other carnivores, have a comparatively short and simple gut (Fig. 6.22). General descriptions of the anatomy in several species may be found in Chiasson (1962), Van der Merwe and Kotze (1993) and Romão *et al.* (2011). Large food items, either whole prey or pieces torn off, are swallowed and carried to the stomach via a longish, muscular and very distensible oesophagus (Uriona *et al.* 2005). The stomach is surprisingly far back, posterior to the heart, which is itself positioned halfway between the origins of front and rear limbs (Chapter 8). The stomach is large and muscular and the oesophagus delivers food to the cardiac portion via the cardiac sphincter at the (animal's) left anterior corner. The cardiac sac is further divided into left and right halves by a thick collar of muscle and spongy tissue, which may operate as a gizzard. The duodenum originates from the much smaller pyloric region, which opens to the duodenum via a pyloric sphincter.

PHYSIOLOGY AND BIOCHEMISTRY OF DIGESTION

Crocodylians digest flesh fully, including the bones. There is no truth in the belief that they need to store food until it rots. Contractions of the muscular stomach walls churn and mix the food bolus with the stomach acid and enzymes during digestion. Not surprisingly, crocs are well equipped to digest proteins. The cardiac (or fundic) part of the stomach has a glandular epithelium that secretes hydrochloric acid (HCl) and pepsinogen, the precursor to pepsin, which has been identified in *Caiman crocodilus* (Diefenbach 1975) and *C. porosus* (Read and Anderson 2001) and has a pH optimum in the range of 2.2–3.0. In pancreatic and duodenal tissue,

Fig. 6.19. *A large* C. porosus *seizes and then swallows a magpie goose, lifting his head as he does so and raising his tail as a counterpoise. This is Roughnut, a free-living but well known Boss Croc (see Chapter 12.) See http://crocodilian.com/books/grigg-kirshner. (Photos Garry Lindner)*

Fig. 6.20. *The open gullet of an American alligator, with the palatal and gular flaps drawn apart, giving a view into the distensible oesophagus. (Photo Michael Stern http://www.sternphotos.com/-/sternphotos/)*

Read and Anderson also identified the proteolytic enzymes trypsin, chymotrypsin, carboxypeptidase A, carboxipeptidase B and aminopeptidase, with pH optima of 8.0, 8.5, 7.0, 7.0 and 7.5, respectively. From this it may be concluded that digestion in crocodylians is like that in other carnivorous vertebrates. That is to say, food in the stomach will stimulate the release of pepsinogen and HCl, which converts the pepsinogen into the active enzyme pepsin, which, in the acid environment, hydrolyses proteins into the component polypeptides. Pepsin is also the most effective enzyme for breaking collagen into polypeptides: an important task because collagen is a major structural protein, present in connective tissue, and comprising 20–25% of the protein in vertebrates. Trypsin, chymotrypsin and carboxypeptidases, released into the duodenum via the pancreatic duct, operate in an alkaline environment, and reduce polypeptides to smaller peptides and, ultimately, to amino acids, which are small enough to be absorbed across the walls of the small intestine.

The secretion of a large volume of HCl into the stomach relies on chloride ions being taken up by cells in the stomach wall, causing a drop in plasma chloride levels and, to maintain the electrical balance, bicarbonate is released into the blood plasma, resulting in a rise in pH. This is known as the 'alkaline tide' because of the reciprocal ebb and flow of chloride and bicarbonate ions out of and into the plasma, and the accompanying changes in plasma pH. The same phenomenon accompanies digestion in humans, but in crocs it has been said to be much more pronounced. Coulson *et al.* (1950) fed whole rats to alligators (1.5–7.0 kg) and described large falls in plasma chloride during alligator digestion, even to 15% of normal levels, accompanied by pH changes from the normal pH 7.3 or so to values approaching pH 8.0. There was considerable individual variation, which could have been related

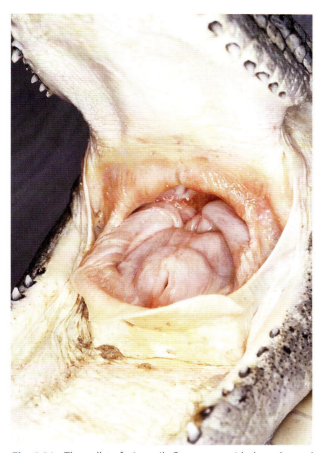

Fig. 6.21. *The gullet of a juvenile* C. porosus *with the gular and palatal flaps relaxed, showing the slit-like glottis in the floor of the pharynx. The glottis has a cartilaginous skeleton and opens into the lungs via the larynx. Food being swallowed slides over the glottis and into the stomach via the long and very distensible oesophagus. (Photo GCG)*

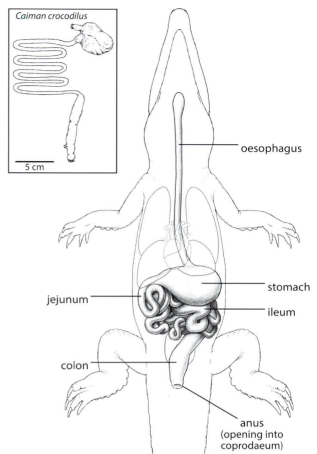

Fig. 6.22. *Anatomy of the crocodylian gut and thoracic cavity, based on Van der Merwe and Kotze (1993) and our own material. (Inset) Carnivores typically have short digestive tracts and crocodylians are no exception, as shown by this dissected out digestive tract from* Caiman crocodilus. *(Inset redrawn from Stevens and Hume 1995)*

to body size (individual sizes were not reported). Coulson and Hernandez (1983) expanded on their 1950 publication, saying 'no other known animal is subject by nature to the enormous changes in blood pH and plasma bicarbonate that occur almost daily in the crocodilian. They go into ... a grave acidaemia every time they exert themselves, and the simple act of eating forces them into an equally pronounced alkalosis'.

How common such large departures from normal values actually are under natural circumstances is unknown. They may be less reliant on anaerobic metabolism to support routine activity than is usually assumed (Chapters 7, 9) and claims about an enormous alkaline tide should be viewed with

caution. In field studies of *C. porosus* in northern Australia and *Caiman latirostris* in coastal Brazil, we found no evidence of very low plasma chloride in any of more than 100 individuals sampled shortly after capture. Rather, our data suggested very tight regulation of plasma chloride (Grigg 1981: 118 ± 2 mmol/L; Grigg *et al.* 1998 103 ± 1.2 mmol/L) (Chapter 11). Most of the animals in both studies had food in their stomachs at capture. In fact, analysis of stomach contents of the *C. porosus* captured in 1975 provided much of Janet Taylor's 1977 MSc study of diet (Taylor 1979). She found that ~15% of the crocs had empty stomachs, and no fewer in the dry season than the wet season. It is possible that a really significant 'alkaline tide' occurs in crocs only after very large meals. Coulson and Hernandez described feeding their captive alligators 25% of their body mass and, while such large meals may be taken naturally, that is probably uncommon (see below). The 'huge alkaline tide' in crocodylians has achieved wide acceptance, but further study seems to be warranted.

Crocodylians can also digest carbohydrates. Read and Anderson (2001) found amylase activity in the pancreatic and duodenal tissue, with a pH optimum of 7.5. Amylase breaks down starches, which are non-structural polysaccharides, to disaccharides, which are then split into monosaccharides (i.e. simple sugars), which can be absorbed. Amylase does not digest cellulose and there is no suggestion that crocodylians have any capacity for digesting the structural polysaccharides of plant cell walls.

Lipid digestion in other vertebrates occurs in the small intestine by lipases secreted by the pancreas acting on triglyceride and other fats that have been emulsified by detergent-like components of bile secreted from the gall bladder. The breakdown results in free fatty acids and glycerol, which are small enough to be absorbed. It seems that lipases in crocodylians have not yet been studied, but they have a gall bladder so it seems reasonable by analogy to assume that they have a typical vertebrate capability to digest and absorb lipids.

Chitinases are present in many reptiles (Jeuniaux 1963), but have not been reported in crocodylians. Steve Garnett (1985) noted the persistence of chitinous beetle elytra in hatchling *C. porosus* and suggested that crocodylians may lack chitinase. Chitinous and keratinous components, such as parts of turtle shell, snail opercula and hair, remain undigested. I recall opening a *C. porosus* stomach containing the disarticulated keratinous plates from a hawksbill turtle, the bone all digested away to leave only the horny exterior. Turtle scutes remain undigested in *A. mississippiensis* as well (Janes and Gutzke 2002). Indigestible remnants are likely to be cast out via the mouth, as in many birds and Pooley (1962) reported that *C. niloticus* readily regurgitate food. Crocodiles in captivity fed feral pig routinely regurgitate fur balls (trichobezoars) (Fig. 6.23). Regurgitation is under

Fig. 6.23. *Fur balls of pig hair, disgorged by a crocodile fed feral pig at the Edward River Crocodile Farm, Cape York, Queensland. According to terminology suggested by Wings (2007), this could be termed a 'bio-gastrolith' to distinguish it from stone 'geo-gastroliths' which are discussed in Chapter 4. (Photo GCG)*

voluntary control in those species that practise it, whereas vomiting (emesis) is usually involuntary: a protective mechanism against toxins. However, Andrews *et al.* (2000) suggested that in crocodylians a similar mechanism may account for both. They noted that crocodylians lack a diaphragm (see Chapter 7) and may achieve emesis by contracting abdominal muscles and elevating intra-abdominal pressure, thus compressing the stomach to eject material.

Taking blood samples from numerous juvenile *C. porosus* in the Liverpool and Tomkinson Rivers in the Northern Territory in 1975, we came across a small number of samples in which the plasma, when the red cells were spun down by centrifugation, was opaque and fatty looking (lipemic). This is the result of fatty acids taken up across the small intestine following digestion. Coulson and Hernandez (1983) measured the progress of digestion in alligators, concluding that in the summer months food remains in the stomach for 24–36 h, the plasma becomes lipemic on the second day, during which presumably the nutrients are being transported to the liver via the hepatic portal vein, and the lipaemia disappears after ~60 h.

It is likely that digestion in crocodylians is controlled hormonally in the same way as in other vertebrates, and the presence of gastrin-like and cholecystokinin-like hormones (Rawdon *et al.* 1981; Dimaline *et al.* 1982) provides supporting evidence. Gastrin is secreted by glands within the stomach mucosa and, in the presence of food or an olfactory signal, stimulates the gastric glands to release HCl and pepsinogen. Cholecystokinin is secreted by the duodenum in response to the presence of partly digested food entering from the stomach and it stimulates the pancreas to release digestive enzymes via the pancreatic duct.

Nutrition

Nutritional requirements

Despite all of their sophisticated and impressive feeding skills and adaptations, crocodylians in general have a very low-energy lifestyle. Like other reptiles, they have a very high rate of conversion of food into body tissue. Many also go routinely for long periods without feeding at all, such as when ambient temperatures are low or, sometimes, during a dry season when the water dries up. They survive then on energy stored in fatty tissue (Chapters 7, 11).

Although much is known about diet from stomach content studies, little is known about the nutritional composition of food taken by crocodylians living in the wild. In a study of feeding and growth of juvenile *C. porosus* in the Adelaide River, Northern Territory, Webb *et al.* (1991) found that their diet comprised either few large animals (mainly native rats) or many small ones (mainly crustaceans). It averaged 70–71% water. Of the remainder, protein made up 42–49% and lipids 10–15% with 15–17% 'unknown', possibly carbohydrate which was at the time of the study thought not to be assimilated. The remainder was ash. These figures reflect the high protein diet one would expect in an aquatic carnivore.

The main interest in crocodylian nutrition has arisen from a need to provide farmed animals suitable diets at affordable cost. Ideal foods in farms are those which, as well as providing sufficient nutrition, are also available locally, are relatively cheap and are easy to store and to present in the pen. For that reason crocodile and alligator farms are often near chicken farms or sources of second rate meat and fish and much of the research has been into the development of artificially formulated pelletised food. Experimental approaches have involved measuring growth rates on a range of diets with different proportions of protein, carbohydrates and lipids, with a range of supplements such as calcium and phosphorus. Some experimental diets have tried casein, a comparatively cheap but stable protein derived from milk, and others have assessed whether animal protein can be replaced by cheaper vegetable protein. A summary is outside our scope, but Read (2000) has reviewed the relevant literature. Protein-only diets are less satisfactory than those including some fat and carbohydrates (Staton *et al.* 1987, 1990) and the best artificial diets seem to be those which are similar to those reported in the limited field data. For example, a comprehensive experimental comparison of a range of different diets force fed at known volumes by Mark Read to *C. porosus* at a crocodile farm in Lae, Papua New Guinea, concluded that fastest growth rates were obtained from diets with protein and lipids of ~60% and 8–14% lipid (dry matter), respectively. This is quite similar to the composition of the natural dietary intake of juvenile *C. porosus* in the field (Webb *et al.* 1991). Read found that growth rates were similar on diets in which soybean meal replaced animal protein.

The statistic by which comparisons are made between growth rates on different diets is usually the food conversion ratio.

Food conversion ratio

The concept of the food (feed) conversion ratio (FCR) comes out of agriculture: a measure of the rate at which stock feed supply is converted into growth. Maximising FCR in commercial animal production systems is economically desirable. FCR is calculated by dividing the mass of food consumed by the gain in body mass. A low number speaks to efficient conversion. Carnivores can be expected to be more efficient than herbivores and ectotherms more efficient than endotherms, through not spending energy to produce body heat. Typical values for farmed fish are less than 2, compared with 2 for poultry, 4 for pork and 7 for sheep and cattle (Rosegrant *et al.* 1999). These numbers are ratios and have no units. They are also somewhat idiosyncratic because 'feed' is counted as dry mass, whereas the growth increment is counted as wet mass. The idiosyncrasy reflects the history of the measure, originating in herbivorous stock fed grain. Biologically, expressing conversion as a percentage of the food incorporated as flesh would be more sensible. Despite its clumsiness, FCR is a useful concept, with practical application for crocodile and alligator farmers. There have been several experimental determinations on captive crocodylians, mostly as hatchlings, yielding a range of results. Interpretation is not straightforward because some authors have not indicated whether food intake was quantified as wet or dry mass.

Table 6.1 shows several results for crocodylians, expressed in different ways. The high conversion rates from the two field studies are of particular interest: more data are needed but on the face of it so far diets or feeding regimens in captivity may be less than optimal, despite growth in captivity usually exceeding growth in the wild.

Measuring growth rate is one way to determine whether farmed crocodylians are getting sufficient nutrition. In the field, a useful 'snapshot' can be gained by determining 'condition factor'.

Condition factor

One way to assess how well animals are doing in a particular habitat or a particular season is to calculate some measure of body condition, usually called a 'condition factor' (CF). Typically, some measure of 'plumpness' is related to a linear dimension – usually body mass at a particular length, such as in the calculation of body mass index in humans. An early crocodylian application was that by Taylor (1979) when she compared relative condition factors among juvenile and sub-adult *C. porosus* captured in different habitats. She included a useful description of the method. The calculation relies on the relationship between body mass and length for the population, $M = aL^b$ (where M is body mass in grams, L is total length in centimetres, a is a constant and b is an exponent that expresses mean growth characteristics of the individuals in the population under consideration). The values of a and b are calculated for individuals in the population by linear regression of ln mass on ln length. The constant a is an expression of the average condition factor for that population, and

the CF for any individual in it can be calculated as: $CF = M.L^{-b}$ (inserting data relevant to the individual) and multiplied by 1000 (or some other multiple of 10) to make it a whole number between 1 and 10, for convenience. Calculating CFs in this way provides a measure of the extent of an individual's deviation from the 'average' for that population. In Taylor's study, she found significant differences between average CF values in different habitats, ranging from ~5.5 to 7.

One drawback of the method is posed by the difficulty of weighing large crocodylians. In such cases, tail girth, which is where fat is stored, can be used as a surrogate for body mass (Zweig 2003).

The biggest drawback of this method of assessing body condition is that the values of the constants of a and b are not characteristic of a species, nor can they be assumed to be stable through time. They may be different in different populations, so values of a and b calculated in one part of the range of a species may be inappropriate for application elsewhere. That is to say, there is nothing 'biological' about these values, or the CFs calculated from them, but they do provide a useful way to make quantitative comparisons as long as the constraints are understood.

Zweig (2003) provided a useful discussion about various ways to assess body condition of alligators and suggested that one way to get around this difficulty would be to use 3 as the value for b, which would be the case if growth were not allometric but isometric. In some circumstances, this could have merit. Likewise, other measures of length and 'plumpness' could be used in the calculation. Zweig (2003) compared different combinations of body measurements as indicators of body condition in alligators. She considered snout–vent length (SVL), total length (TL), head length (HL) and hind-foot length (HFL) as linear dimensions representing body 'size' (skeletal length) and neck girth (NG), tail girth (TG), chest girth (CG) and body mass as indicators of body volume. Her results showed that within population comparisons are probably best done as described by Taylor (1979) but using SVL and body mass, whereas there was merit in using HL/L^3 where comparisons between populations are required. Referring to Green

TABLE 6.1 FOOD CONVERSION RATIOS EXPRESSED IN DIFFERENT WAYS FOR SEVERAL SPECIES OF CROCODYLIANS

The figures in bold are drawn from the data as published; the other values are calculated using the assumption of a water content in food of 70% by mass (except *Caiman latirostris* in which food water content was taken to be 42% and Read's diet for *C. porosus* where the published value was used). For carnivores, the most intuitive measure is probably the proportion of the wet food intake converted to flesh. The highest conversion rates found so far have come from animals in the field. Captive studies show large variation depending upon type of food, feeding frequency, density of captives, ambient temperature and other factors.

Species	Wet intake/wet growth		Dry intake/ wet growth	Source
	Ratio	Percentage absorbed	Ratio	
Field animals				
C. porosus	1.2	**83%**	0.36	Webb *et al.* (1991)
C. niloticus	**1.56–1.61**	64–62%	0.47–0.48	Games (1990) in Read (2000)
Captive animals				
C. johnstoni (mean of various trials)				
	4.5	**22%**	1.35	Webb *et al.* (1983)
C. porosus				
(fed daily)	2.5	**40%**	0.75	Webb *et al.* (1990)
(fed every 2nd day)	3.6	**28%**	1.1	
C. porosus	2.5–3.1	**40–32%**	0.75–0.93	Manolis (1993)
Fed pork	2.8	36%	**0.83**	Garnett and Murray (1986)
Fed fish	5	20%	**1.5**	
Fed beef	3.2	31%	**0.97**	
A range of diets	**2.2**	45%	**1.27**	Read (2000)
A. mississippiensis				
Hatchling year	2.5	**40%**	0.75	Coulson *et al.* (1973)
Fed fish	6.7	15%	**2.0**	Joanen and McNease (1987)
	4.0	25%	1.2	Staton and Edwards (1987) Read 2000
Caiman latirostris (fed 60% chicken heads, 40% dry pellets)				
29ºC	6.7–9.3		**3.9–5.4**	Parachú Marcó *et al.* (2009)
33ºC	4.0–4.5		**2.3–2.6**	

(2001), who evaluated a different way to assess body condition (mass/length residuals), Zweig reminded practitioners that calculations of CF by the method described above assume that body length is not affected by condition, that length is an accurate measure of structural size and that the relationship between the logarithms of body size and mass is linear.

It is also worth noting that it should not be assumed automatically that animals with the highest CF are the healthiest, because they may be obese. Indeed, Rice (2004) found that her 'fattest' alligators had lower reproductive rates. Additionally, gravid females will have an artificially, but temporarily, high CF.

Although calculation of CF in crocodylians is usually aimed at comparing nutritional state or general health, perhaps related to food availability in a particular habitat, it can be used to pursue other questions. For example, Grigg *et al.* (1986) compared CFs in juvenile *C. porosus* in hyperosmotic and hypo-osmotic sections of an estuary, seeking a measure of how well they were coping in the more saline conditions. Using the same mathematics, they were also able to compare 'hydration factors' and 'sodium factors' between these two habitats, having determined total body water and exchangeable body sodium using tritiated water and radioactive [22]Na, respectively (Chapter 11).

This Chapter has discussed how energy is captured from the environment and taken up into the body (feeding and digestion), the nature of the food that is required and measurements of feeding success (nutrition status). The next Chapter looks at the processes by which that food is metabolised to produce energy for biological work.

REFERENCES

Andrews PLR, Axelsson M, Franklin CE, Holmgren S (2000) The emetic reflex in a reptile (*Crocodylus porosus*). *Journal of Experimental Biology* **203**, 1625–1632.

Britton A, Britton E (2013) Saltwater crocodile (*Crocodylus porosus*). Fishing behavior. *Herpetological Review* **44**(2), 312

Blohm T (1982) Husbandry of Orinoco crocodiles (*Crocodylus intermedius*) in Venezuela. In *Crocodiles. Proceedings of the 5th Working Meeting of the Crocodile Specialist Group*. Gainesville, Florida, August 1980. pp. 267–285. IUCN– The World Conservation Union, Gland, Switzerland.

Brazaitis P, Watanabe ME (2011) Crocodilian behaviour: a window to dinosaur behaviour? *Historical Biology* **23**(1), 73–90.

Brito SP, Andrade DV, Abe AS (2002) Do caimans eat fruit? *Herpetological Natural History* **9**(1), 95–96.

Brueggen J (2002) Crocodilians: fact vs. fiction. In *Crocodiles. Proceedings of the 16th Working Meeting of the Crocodile Specialist Group*. Gainesville, Florida. pp. 204–210. IUCN–The World Conservation Union, Gland, Switzerland.

Busbey AB (1989) Form and function of the feeding apparatus of *Alligator mississippiensis*. *Journal of Morphology* **202**, 99–127.

Chiasson RB (1962) *Laboratory Anatomy of the Alligator*. WC Brown, Dubuque, Iowa.

Cleuren J, De Vree F (1992) Kinematics of the jaw and hyolingual apparatus during feeding in *Caiman crocodilus*. *Journal of Morphology* **212**, 141–154.

Cleuren J, De Vree F (2000). Feeding in crocodilians. In *Feeding: Form, Function, and Evolution in Tetrapod Vertebrates*. (Ed. K Schwenk) pp. 337–358. Academic Press, San Diego.

Cott HB (1961) Scientific results of an enquiry into the ecology and economic status of the Nile Crocodile (*Crocodylus niloticus*) in Uganda and Northern Rhodesia. *Transactions of the Zoological Society of London* **29**, 211–356.

Coulson RA, Hernandez T (1983) *Alligator Metabolism: Studies on Chemical Reactions* in vivo. Pergamon Press, New York.

Coulson RA, Hernandez T, Dessauer HC (1950) Alkaline tide in the alligator. *Proceedings of the Society for Experimental Biology and Medicine* **74**, 866–869.

Coulson TD, Coulson RA, Hernandez T (1973) Some observations on the growth of captive alligators. *Zoologica (New York)* **58**, 45–52.

Coutinho ME (2000) Population ecology and the conservation and management of *Caiman yacare* in the Pantanal, Brazil. PhD thesis, University of Queensland, Brisbane.

Delany MF, Abercrombie CL (1986) American Alligator food habits in north-central Florida. *Journal of Wildlife Management* **50**, 348–353.

Diefenbach COda C (1975) Gastric function in *Caiman crocodilus* (Crocodylia: Reptilia)–II. Effects of temperature on pH and proteolysis.

Comparative Biochemistry and Physiology **51**, 267–274.

Dimaline R, Rawdon BB, Brandes S, Andrew A, Loveridge JP (1982) Biologically active gastrin/CCK-related peptides in the stomach of a reptile, *Crocodylus niloticus*; identified and characterized by immunochemical methods. *Peptides* **3**, 977–984.

Dinets V (2010) Nocturnal behavior of the American alligator (*Alligator mississippiensis*) in the wild during the mating season. *Herpetological Bulletin* **111**, 4–11.

Dinets V (2011) On terrestrial hunting by crocodilians. *Herpetological Bulletin* **114**, 15–18.

Dinets V, Brueggen JC, Brueggen JD (2013) Crocodilians use tools for hunting. *Ethology Ecology and Evolution*.

Dugatkin LA, Mesterton-Gibbons M (1996) Cooperation among unrelated individuals: reciprocal altruism, by-product mutualism and group selection in fishes. *Bio Systems* **37**, 19–30.

Erickson GM, Gignac PM, Steppan SJ, Lappin AK, Vliet KA (2012) Insights into the ecology and evolutionary success of crocodilians revealed through bite-force and tooth-pressure experimentation. *PLoS ONE* **7**(3), e31781.

Fish FE, Bostic SA, Nicastro AJ, Beneski JT (2007) Death roll of the alligator: mechanics of twist feeding in water. *Journal of Experimental Biology* **210**, 2811–2818.

Fitzgerald LA (1989) An evaluation of stomach-flushing techniques for crocodilians. *Journal of Herpetology* **23**, 170–172.

Games I (1990) The feeding ecology of two Nile crocodile populations in the Zambezi Valley. PhD thesis, University of Zimbabwe, Harare, Zimbabwe.

Gans C (1961) The feeding mechanism of snakes and its possible evolution. *American Zoologist* **1**(2), 217–227.

Gans C (1969) Comments on inertial feeding. *Copeia* **1969**, 855–857.

Gans C (1989) Crocodilians in perspective. *American Zoologist* **29**, 1051–1054.

Garnett ST (1985) The consequences of slow chitin digestion on crocodile diet analysis. *Journal of Herpetology* **19**, 303–304.

Garnett ST, Murray RM (1986) Parameters affecting the growth of the Estuarine Crocodile *Crocodylus porosus* in captivity. *Australian Journal of Zoology* **34**, 211–223.

Graham A (1968) 'The Lake Rudolf crocodile (*Crocodylus niloticus* Laurenti) population'. Report to the Kenya Game Department by Wildlife Services Limited. Nairobi, Kenya, <http://ufdcimages.uflib.ufl.edu/aa/00/00/75/90/00001/lakerudolfcrocodilesalistairgraham.pdf>

Green AJ (2001) Mass/length residuals: measures of body condition or generators of spurious results? *Ecology* **82**, 1473–1483.

Grigg GC (1981) Plasma homeostasis and cloacal urine composition in *Crocodylus porosus* caught along a salinity gradient. *Journal of Comparative Physiology* **144**, 261–270.

Grigg GC, Taplin LE, Green B, Harlow P (1986) Sodium and water fluxes in free-living *Crocodylus porosus* in marine and brackish conditions. *Physiological Zoology* **59**(2), 240–253.

Grigg GC, Beard LA, Moulton T, Queirol Melo MT, Taplin LE (1998) Osmoregulation by the broad-snouted caiman, Caiman latirostris, in estuarine habitat in southern Brazil. *Journal of Comparative Physiology. B, Biochemical, Systemic, and Environmental Physiology* **168**, 445–452.

Hutton JM (1987) Growth and feeding ecology of the Nile Crocodile, *Crocodylus niloticus*. *Journal of Animal Ecology* **56**, 25–35.

Jackson JF, Campbell HW, Campbell KE (1974) The feeding habits of crocodilians: validity of the evidence from stomach contents. *Journal of Herpetology* **8**, 378–381.

Janes D, Gutzke WHN (2002) Factors affecting retention time of turtle scutes in stomachs of American Alligators, *Alligator mississippiensis*. *American Midland Naturalist* **148**(1), 115–119.

Jeuniaux C (1963) *Chitine et Chitinolyse, un Chapitre de la Biologie Moléculaire*. Masson, Paris.

Joanen T, McNease L (1987) Alligator farming research in Louisiana In *Wildlife Management: Crocodiles and Alligators* (Eds GJW Webb, SC Manolis and PJ Whitehead) pp. 329–340. Surrey Beatty & Sons, Sydney and the Conservation Commission of the Northern Territory, Darwin.

King FW (2008) *Alligator Behavior: The Accuracy of William Bartram's Observations*. University of Florida, Gainesville, Florida, <http://web.uflib.ufl.edu/ufdc/?b=UF00088969>

King FW, Thorbjarnarson J, Yamashita C (1998) *Cooperative Feeding, A Misinterpreted and Under-Reported Behavior of Crocodilians*. University of Florida, Gainesville, Florida, <http://www.flmnh.ufl.edu/herpetology/herpbiology/bartram.htm >

Limpus CJ (2007) *A Biological Review of Australian Marine Turtle Species. 5. Flatback Turtle,* Natator depressus *(Garman)*. Queensland Environmental Protection Agency, Brisbane.

Magnusson WE, da Silva EV, Lima AP (1987) Diets of Amazonian crocodilians. *Journal of Herpetology* **21**, 85–95.

Manolis SC (1993) Nutrition of crocodiles. In *Proceedings of the 2nd Regional (eastern Asia, Oceania, Australasia) Meeting of the Crocodile Specialist Group. 12–19 March 1993. Darwin.* IUCN, the World Conservation Union, Gland, Switzerland.

McNease L, Joanen T (1977) Alligator diet in relation to marsh salinity. Proceedings of the Annual Conference of Southeastern Association of Fish and Wildlife Agencies 31, 36–40.

McIlhenny EA (1935) *The Alligator's Life History*. Christopher, Boston Massachusetts.

Nifong JC, Rosenblatt AE, Johnson NA, Barichivich W, Silliman BR, Heithaus MR (2012) American alligator digestion rate of blue crabs and its implications for stomach contents analysis. *Copeia* **2012**(3), 419–423.

Olmos F, Sazima I (1990) A fishing tactic in floating Paraguayan caiman: the cross-posture. *Copeia* **1990**, 875–877.

Parachú Marcó MV, Piña CI, Larriera A (2009) Food conversion rate (FCR) in *Caiman latirostris* resulted more efficient at higher temperatures. *Interciencia* **34**(6), 428–431.

Pauwels OSG, Mamonekene V, Dumont P, Branch WR, Burger M, Lavoué S (2003) Diet records for *Crocodylus cataphractus* (Reptilia: Crocodylidae) at Lake Divangui, Ogooué-Maritime province, southwestern Gabon. *Hamadryad* **27**(2), 200–204.

Platt SG, Rainwater TR, Finger AG, Thorbjarnarson JB, Anderson TA, McMurry ST (2006) Food habits, ontogenic dietary partitioning and observations on foraging behavior of Morelet's crocodile *Crocodylus moreletii* in northern Belize. *Journal of Herpetology* **16**, 281–290.

Platt SG, Thorbjarnarson JB, Rainwater TR, Martin DR (2013) Diet of the American crocodile (*Crocodylus acutus*) in marine environments of coastal Belize. *Journal of Herpetology* **47**, 1–10.

Pooley AC (1962) The Nile Crocodile, *Crocodylus niloticus. The Lammergeyer* **2**, 1–55.

Pooley AC (1982) *Discoveries of a Crocodile Man*. William Collins & Sons, London.

Pooley AC, Gans C (1976) The Nile crocodile. *Scientific American* **234**(4), 114–125.

Putterill JF, Soley JT (2006) Morphology of the gular valve of the Nile crocodile, *Crocodylus niloticus* (Laurenti, 1768). *Journal of Morphology* **267**(8), 924–939.

Rachmawan D, Brend S (2009) Human-*Tomistoma* interactions in central Kalimantan, Indonesian Borneo. *CSG Newsletter* **28**(1), 9–11.

Rawdon BB, Brandes S, Andrew A, Loveridge JP (1981) Gastrin-immunoreactive cells in the crocodile stomach. *South African Journal of Science* **77**, 92

Read MA (2000) Aspects of protein utilisation and metabolism by post-hatchling estuarine crocodiles (*Crocodylus porosus* Schneider). PhD thesis, University of Queensland, Brisbane.

Read MA, Anderson AJ (2001) Proteolytic and starch-digesting enzymes of the stomach, duodenum and pancreas of post-hatchling estuarine crocodiles (*Crocodylus porosus*). In *Crocodile Biology and Evolution*, (Eds GC Grigg, F Seebacher and CE Franklin). pp. 317–326. Surrey Beatty & Sons, Sydney.

Reisz RR, Scott D, Sues HD, Evans DC, Raath MA (2005) Embryos of an early Jurassic prosauropod dinosaur and their evolutionary significance. Science 309(5735), 761–764.

Rice AN (2004) Diet and condition of American Alligators (*Alligator mississippiensis*) in three central Florida lakes. MSc thesis. University of Florida. Gainesville, Florida, <http://myfwc.com/media/310266/Alligator_Rice_A.pdf>

Rice AN, Ross JP, Finger AG, Owen R (2005) Application and evaluation of a stomach flushing technique for alligators. *Herpetological Review* **36**, 400–401.

Rice AN, Perran Ross J, Woodward AR, Carbonneau DA, Percival HF (2007) Alligator diet in relation to alligator mortality on Lake Griffin, FL. *Southeastern Naturalist (Steuben, ME)* **6**(1), 97–110.

Romão MF, Santos ALQ, Lima FC, De Simone SS, Silva JMM, Hirano LQ, *et al.* (2011) Anatomical and topographical description of the digestive system of *Caiman crocodilus* (Linnaeus 1758), *Melanosuchus niger* (Spix 1825) and *Paleosuchus palpebrosus* (Cuvier 1807). *International Journal of Morphology* **29**(1), 94–99.

Rosegrant MW, Leach N, Gerpacio RV (1999) Meat or wheat for the next millennium? Alternative futures for world cereal and meat consumption. *The Proceedings of the Nutrition Society* **58**, 219–234.

Schaller GB, Crawshaw PGJr (1982) Feeding behavior of Paraguayan caiman (*Caiman crocodilus*). *Copeia* **1982**(1), 66–72.

Smith JG, Phillips BL (2006) Toxic tucker: the potential impact of cane toads on Australian reptiles. *Pacific Conservation Biology* **12**, 40–49.

Staton MA, Edwards HM, Brisbin IL (1987) Growth responses of alligators (*Alligator mississippiensis*) to diets varying in fat and carbohydrate content. *American Zoologist* **27**, 137A

Staton MA, Edwards HM, Brisbin IL, Joanen T, McNease L (1990) Protein and energy relationships in the diet of the American Alligator (*Alligator mississippiensis*). *Journal of Nutrition* **120**, 775–785.

Stevens CE, Hume ID (1995) *Comparative Physiology of the Vertebrate Digestive System*. 2nd edn. Cambridge University Press, Cambridge, UK.

Taylor JA (1979) The foods and feeding habits of sub-adult *Crocodylus porosus* Scheider in northern Australia. *Australian Wildlife Research* **6**, 347–359.

Taylor JA, Webb GJW, Magnusson WE (1978) Methods of obtaining stomach contents from live crocodilians (Reptilia, Crocodilidae). *Journal of Herpetology* **12**, 415–417.

Tucker AD, Limpus CJ, McCallum HI, McDonald KR (1996) Ontogenetic dietary partitioning by *Crocodylus johnstoni* during the dry season. *Copeia* **1996**, 978–988.

Uriona TJ, Farmer CG, Dazely J, Clayton F, Moore J (2005) Structure and function of the esophagus of the American alligator (*Alligator mississippiensis*). *Journal of Experimental Biology* **208**, 3047–3053.

Van der Merwe NJ, Kotze SH (1993) The topography of the thoracic and abdominal organs of the Nile crocodile (*Crocodylus niloticus*). *Onderstepoort Journal of Veterinary Science* **60**, 219–222.

Wallace KM, Leslie AJ (2008) The diet of the Nile crocodile (*Crocodylus niloticus*) in the Okavango Delta, Botswana. *Journal of Herpetology* **42**(2), 361–368.

Webb GJW, Manolis SC, Buckworth R (1982) Crocodylus johnstoni in the McKinlay River area, N.T. I. Variation in the diet, and a new method of assessing the relative importance of prey. *Australian Journal of Zoology* **30**, 877–899.

Webb GJW, Buckworth R, Manolis C (1983) *Crocodylus johnstoni* in a controlled-environment chamber: a raising trial. *Australian Wildlife Research* **10**, 421–432.

Webb GJW, Manolis SC, Cooper-Preston H (1990) Crocodile research and management in the Northern Territory: 1988–90. In *Proceedings of the 10th Working Meeting IUCN–SSC Crocodile Specialist Group*. April 1990, Gainesville, Florida pp. 253–273. IUCN, Gland, Switzerland.

Webb GJW, Hollis GJ, Manolis SC (1991) Feeding, growth, and food conversion rates of wild juvenile Saltwater Crocodiles (*Crocodylus porosus*). *Journal of Herpetology* **25**, 462–473.

Whitaker R, Basu D (1983) The Gharial (*Gavialis gangeticus*): a review. *Journal of the Bombay Natural History Society* **79**, 531–548.

Whiting SD, Whiting AU (2011) Predation by the saltwater crocodile (*Crocodylus porosus*) on sea turtle adults, eggs, and hatchlings. *Chelonian Conservation and Biology* **10**(2), 198–205.

Wings O (2007) A review of gastrolith function with implications for fossil vertebrates and a revised classification. *Acta Palaeontologica Polonica* **52**(1), 1–16.

Zweig C (2003) Body condition index analysis for the American alligator (*Alligator mississippiensis*). MSc thesis. University of Florida, Gainesville, Florida, <http://etd.fcla.edu/UF/UFE0000836/zweig_c.pdf >

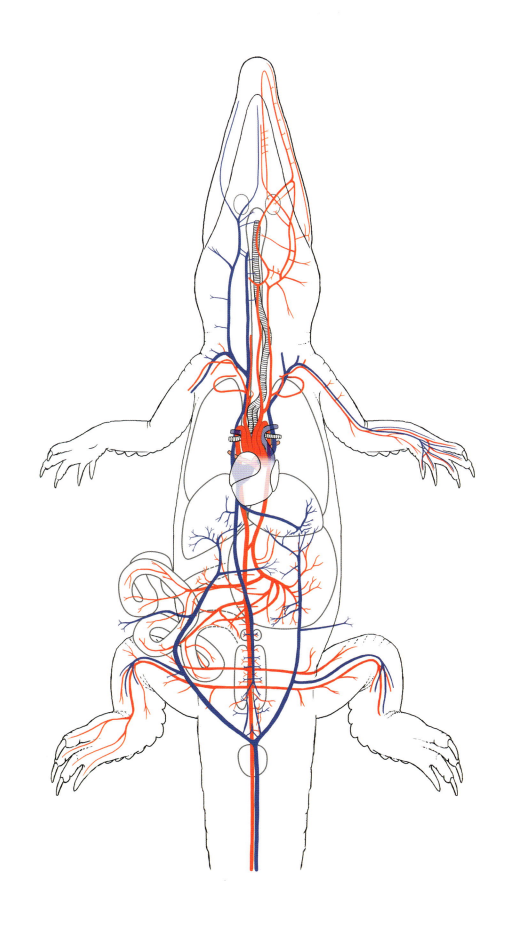

7

ENERGY SUPPLY AND DELIVERY

I came to crocodylian biology via interests in the respiratory and metabolic physiology of fishes (Queensland lungfish, Antarctic fish, a North American catfish and Port Jackson sharks), so it seemed natural to start wondering about similar topics in crocodiles. Most particularly, I wondered about crocodylian diving capacities. That wonderful English gentleman the late Hugh Cott had written a marvellous paper about crocodiles in Uganda: one of the first wide ranging studies on crocodile biology and ecology, and he had written about the length of time they could stay under water (i.e. a long time!). Long dives prompt questions about their metabolism, their lung capacity, their blood and, of course, their extraordinary heart (Chapter 8). Forty years ago people thought that diving by vertebrates involved anaerobic metabolism, with lactate accumulation resulting in a big 'oxygen debt' to 'pay back' after surfacing. But crocs could obviously dive repeatedly, with only a short breath in between and no time for such a luxury. There were so many questions! We will talk about diving in Chapter 9 after reviewing the heart. This Chapter will be about crocodylian metabolic requirements, respiration and circulation, and their capacity for both aerobic and anaerobic metabolism. It will provide a good background for understanding how crocodylians can remain submerged for so long (Chapter 9).

Different behaviours require different amounts of energy: resting, walking, running, galloping, swimming and diving all require the regulation and management of energy supply. Increased physical activity relies on up-regulation of oxygen usage at the mitochondria to produce more energy-delivering ATP, accompanied by an increase in the supply of oxygen to the tissues by increased blood supply and, in turn, increased ventilation to oxygenate the blood being circulated. As in humans, when metabolic requirements exceed what can be supplied aerobically, glycogen is metabolised locally and activity is supported by anaerobic metabolism.

This Chapter will discuss aerobic and anaerobic metabolism, the lungs, ventilation, and oxygen supply and delivery to the working tissues during rest and during physical activity. Submergence inhibits breathing and, because diving is such a significant behaviour in crocs, we can expect special adaptations for the prolonged dives that they make. The main discussion of diving and its metabolic support will be postponed until Chapter 9, after a discussion of the structure and function of the heart (Chapter 8), but this Chapter provides a useful background.

THE METABOLIC ENGINES: CROCODYLIAN BIOCHEMISTRY AND METABOLISM

(This is the shortest biochemistry course you'll ever have!)

From substantial research on alligators in the 1960s through to 1980s by Roland Coulson and Thomas Hernandez at Louisiana State University

---

The circulatory system plays an important role in the supply of energy to a crocodylian's body.

in New Orleans (Coulson and Hernandez 1983), we know that the biochemistry of crocodylians is generally similar to that in other vertebrates, including humans. Most of the enzymes we ourselves rely on, and know from standard biochemistry courses, are present in crocs, so we can infer broad similarities. A crucial difference from humans, of course, is that body temperatures are lower in crocs, varying both daily and seasonally (Chapter 10), and metabolic rates are about one-fifth (or less) of human rates. A very simplified overview is all that is needed here, with some focus on aspects of particular crocodylian interest.

THE PRODUCTS OF DIGESTION

The digestion of food results in its breakdown into molecules small enough to be taken up into the blood across the gut wall, mostly into the hepatic portal vein, which transports it from the gut to the liver (Fig. 7.19). These are glucose and other simple sugars derived from carbohydrates, fatty acids and monoglycerides from lipids, and amino acids from protein digestion. A simplified schematic of the fates of each of these energy substrates is shown in Fig. 7.1.

Amino acids

There is little or no storage of amino acids: they are used in protein synthesis, particularly during growth, but, even in adults whose growth is very slow, amino acids are incorporated into enzymes and haemoglobin, which are being turned over all the time, and for the maintenance of muscles. Excess amino acids are modified biochemically into carbohydrates and lipids, which can be stored (e.g. glycogen in muscle tissue): a process which produces ammonia as a by-product, which is excreted either as it is or converted to urea or uric acid (see Chapter 11).

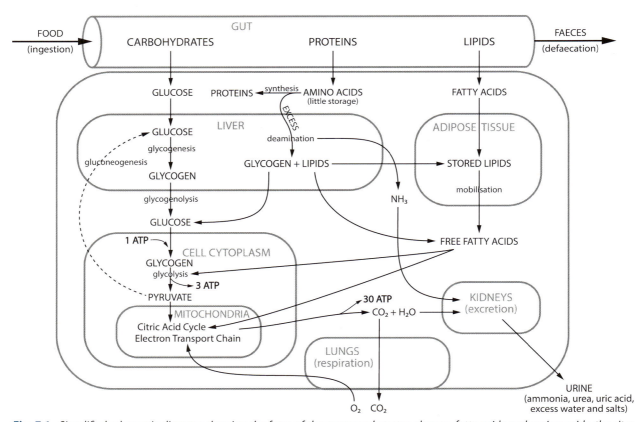

Fig. 7.1. *Simplified schematic diagram showing the fates of the energy substrates glucose, fatty acids and amino acids, the sites and processes of ATP production, storage sites and exchanges with the environment (exchange of gases, ingestion and defaecation, excretion of nitrogen and excess water and salts).*

Fatty acids

Lipids (fats) are a major component of cell membranes and steroid hormones, and are an important source of stored fuel. Excess lipids are stored as an energy source. In crocodylians, lipids are not stored in distinct fat bodies as they are in many reptiles, but dispersed as adipose tissue in muscles and in the skin, and even as a core to the tongue. Most cells can metabolise fatty acids just as easily as glucose, reacting with coenzyme A and thus entering the citrate cycle.

Carbohydrates

Glucose and other simple sugars taken up across the gut wall are stored in the liver as glycogen and reconverted to glucose when an increase in blood sugar is needed. In crocodylians, this is under the control of insulin secreted by the pancreas, just as in mammals (Lance *et al*. 1993). The other major site for glucose storage is in the muscles as glycogen, where it powers physical work by a process known as anaerobic glycolysis (see below).

HARVESTING ENERGY FROM NUTRIENTS FOR PRODUCTIVE BIOLOGICAL WORK

Blood glucose and other nutrients

The biological work of the cells and tissues depends upon a continuous supply of energy. This is delivered by the circulating blood, mostly as glucose and free fatty acids. Blood glucose is well regulated in crocodylians. Michael Cairncross and I measured blood glucose at capture in ~100 *C. porosus* in the Liverpool and Tomkinson Rivers and found little variation: typically 4–6 mmol/L. Similar levels were found in American alligators during summer (Coulson and Hernandez 1983) and a range of other captive crocodylians (reviewed by Stacy and Whitaker 2000). These values compare with 4–8 mmol/L in humans, depending upon the time elapsed since a meal. If this seems surprisingly similar, it is worth reflecting that, because of a much higher blood flow in mammals, probably 4–5 times higher than in a crocodile, the delivery rate of glucose in mammals and birds is much higher.

Blood glucose levels in alligators show seasonal variation, falling in cold winters (Coulson and Hernandez 1983). To counter seasonally low levels or any long periods without food, free fatty acids are released from their storage as lipids in fatty tissues, maintaining the availability of an energy supply. Alligator insulin is very similar to chicken insulin, with only three amino acid substitutions, and plasma glucose and amino acids become depressed after injection of alligator, chicken and even bovine insulin (Lance *et al*. 1993).

Cellular respiration

Glucose and/or free fatty acids are taken up across the cell membranes from the circulating blood. The release of energy is achieved stepwise in a series of complex biochemical processes: anaerobic glycolysis, which proceeds without oxygen, and the citric acid cycle (= Krebs cycle = tricarboxylic acid cycle) in combination with the electron transport system, in which oxygen is consumed (Fig. 7.1).

Glycolysis occurs in the cytoplasm of the cell and, for each mole of glucose, two moles of adenosine triphosphate (ATP) are released, with the still energy-rich pyruvate as a by-product. If oxygen is available, the pyruvate is transported from the cytoplasm into the mitochondria and metabolised further in the linked stages of the citric acid cycle and electron transport system, consuming oxygen and producing ATP, carbon dioxide and water. Earlier estimates of ATP production per mole of glucose have been revised downwards, from 38 to ~30 moles of ATP. The combined steps harvest the energy from glucose into a useable form, ATP, whose subsequent conversion to adenosine diphosphate (ADP) releases energy, which is the direct driver of cellular work.

Respiratory quotient and respiratory exchange ratio

The ratio of the volumes of carbon dioxide (CO_2) produced to oxygen consumed is known as the respiratory quotient, RQ. Strictly, this refers to the exchange going on at the tissue level but it is usually measured on the whole animal. RQ is usually 1 or less, and differs with the food type being metabolised: indicative values are fat 0.7, protein

0.9, carbohydrate 1.0. In anaerobic metabolism, more CO_2 is produced at the tissue level, resulting in RQ > 1.0. A related term is 'respiratory exchange ratio', RER or sometimes just R, referring to gas exchange of the whole animal and calculated in the same way. In an animal in steady-state, RER may equal, or be very close, to RQ. However, during exercise an animal is commonly not in steady-state and CO_2 and oxygen may be exchanged at rates that do not immediately reflect tissue RQ, so a separate term is useful. In crocodylians, RER is often much less than 0.7 because in fresh water much CO_2 is excreted in combination with ammonia in the urine, as ammonium bicarbonate (Grigg 1978; Chapter 11).

Glycolysis and the 'oxygen debt' or EPOC

If oxygen is not available, perhaps because its delivery is not keeping up with the need, as when skeletal muscle is exercised heavily, only the glycolytic step occurs, releasing 3 moles ATP (Fig. 7.1) and producing pyruvate, which is then catalysed by the enzyme lactate dehydrogenase (LDH) and converted to lactate. Lactate is still rich in energy and is processed further when oxygen is again available. This may occur either in the skeletal muscle or, after diffusion into the blood, in the liver where it is transformed by a different form of LDH to pyruvate, which then enters the citric acid cycle, producing more ATP. An alternative pathway is for the pyruvate to be converted to glucose by gluconeogenesis in the liver and stored as glycogen, although this is energetically costly.

Typically, heavy muscular work is fuelled by ATP produced by glycolysis within the muscle cells, leading to a build up of lactate and protons, which appear in the blood and reduce pH. Lactate is a useful proxy for the extent of acidification, although it should be noted that the source of excess protons arises from ATP hydrolysis and several other components of the anaerobic pathway (Hochachka and Mommsen 1983, Robergs *et al.* 2004). The subsequent metabolism of lactate via the citrate cycle and electron transport chain at the cessation of exercise leads to an increase in oxygen consumption and produces CO_2 and water: a process often referred to as 'paying back

the oxygen debt'. Anaerobic metabolism may also occur during crocodylian diving, although much less often than previously thought, because it is now clear that most voluntary dives are conducted aerobically (Chapter 9). However, 'forced' dives, dives involving a lot of exercise and some very long dives do rely on anaerobic metabolism, with a surge of lactate into the blood at the end of the dive when ventilation of the lungs is resumed. The resultant 'oxygen debt' is then 'paid back', as at the end of a bout of exercise (see Chapter 9). 'Oxygen debt' is more frequently referred to now as 'excess post-exercise oxygen consumption' (EPOC). This is oxygen consumption in addition to uptake needed to replenish the oxygen store.

The physiology of exercise to exhaustion and recovery from lactic acidosis is discussed below.

WHOLE BODY METABOLISM (ANAEROBIC AND AEROBIC) AND THE EFFECTS OF BODY MASS, TEMPERATURE, DIGESTION AND ACTIVITY

MEASUREMENT OF METABOLIC RATE

The total metabolic rate of an organism is the sum of all of the energy exchanges going on in its body. The most accurate measurement of this would be heat production, but this is technically difficult. Oxygen consumption is much easier to measure, so metabolic rate is usually measured as oxygen consumption. If there has been anaerobic metabolism, any 'postponed' oxygen consumption, the EPOC, needs to be taken into account (Hancock and Gleeson 2008). If required, heat production equivalents can be calculated from measurements of oxygen consumption. Some studies have measured CO_2 production instead of oxygen consumption, with the two related by the respiratory quotient (RQ).

Knowledge of an animal's metabolic rate can allow calculation of its food requirements and thus its quantitative exchanges with the environment and how these may change daily, seasonally and throughout life. At a practical level, knowledge of metabolic rate can be useful to a crocodile farmer, for example, needing to assess food requirements for different sized animals and even calculate likely waste production in order to determine water flow requirements to ensure a safe level of effluent.

Ideally, we would like to know metabolic rates of crocs over the complete size range, at rest and post-absorptive, akin to basal metabolic rate (BMR) in humans, at different body temperatures and at different types of activity, up to the maximum. Clearly that would be a major challenge! Even measuring metabolic rate of a crocodylian 'at rest' (RMR) poses problems: how do you tell a croc to rest? Indeed, just being in captivity and in a respirometer is a source of stimulation to most crocs, which are typically flighty as well as feisty, so getting a 'minimal' rate is not easy. Many studies have been compromised because the measurements were made on animals that were not properly relaxed and, further, after too little time had elapsed since they had fed. And, of course, temperature is important because crocs are ectothermic (Chapter 10) and metabolic rate varies with body temperature. Measuring the metabolic rate of active crocs also poses particular challenges, but some data have been gained from small crocs walking on a treadmill and/or swimming (Farmer and Carrier 2000a; Owerkowicz and Baudinette 2008; Eme *et al.* 2009). Getting similar data from very large crocs would be extremely difficult, except with very expensive apparatus. Despite these difficulties, quite a lot is known, although the knowledge is based almost entirely on juveniles and, in many cases, quite small juveniles, with only a couple of exceptions.

Because of the difficulty of persuading non-human animals to rest and be 'mentally at ease', which are requirements for measuring BMR, zoologists refer to 'resting metabolic rate' (RMR) or 'standard metabolic rate' (SMR), which can be defined as the rate measured on a resting, post-absorptive, 'non-stressed' animal either in a thermally neutral environment (endotherms) or in thermal equilibrium with the measurement temperature.

It is convenient to discuss aerobic and anaerobic metabolism separately.

AEROBIC METABOLISM

Effect of body mass on standard metabolic rate (SMR)

Oxygen consumption is more difficult to measure in large crocodylians than in small ones, so most measurements have been on juveniles, usually 1–2 kg. One exception is the 53 kg American alligator whose resting CO_2 production was measured by the American physiologist Francis Gano Benedict (Benedict 1932). Benedict became a bit of a legend in reptile biology because of his measurements of metabolism of large pythons. He was also a performing magician! See http://www.whonamedit.com/doctor.cfm/3319.html. The other exception is a recent study by Seymour *et al.* (2013) and that is much more valuable because these authors measured standard metabolic rate (SMR) of 44 *C. porosus* over a size range from 190 g to 389 kg. Making measurements over a size range is very important because animal size and animal metabolism are not related in a directly proportional way (see Box 7.1). The German physiologist Max Rubner published a paper in 1883 which recognised that a larger animal requires proportionally less oxygen than a smaller one. He proposed that metabolic rate is proportional to surface area; that is, would scale with body mass with an exponent of two thirds. This is attractive as an explanation for mammalian and avian metabolic scaling, because that would parallel the way body surface area, and heat loss, scale with mass. However, Kleiber (1932) reported a larger scaling exponent of three-quarters from the mammals he measured; 'quarter power scaling'. Benedict (1932) published a log–log plot, including more species, which subsequently became famous as the 'mouse to elephant curve'. Collectively, these papers introduced the realisation that in biology almost everything scales with body size in a less than 1:1 fashion (see Box 7.1). The plot thickened as the realisation grew that similar exponents applied to ectotherms and even to individual cells, and debate has raged ever since about what explains an exponent less than 1, about what exponent is 'correct', about whether or not one exponent can fit all, and if not why not, and so on. A consensus

BOX 7.1 SCALING STANDARD METABOLIC RATE AND BODY MASS; THE FAMOUS 'MOUSE TO ELEPHANT CURVE'

Typically, standard metabolic rate (SMR) of vertebrates scales to body size with an exponent of ~0.75 (although there is ongoing controversy about this and many exceptions); that is. MR α mass$^{0.75}$ or, alternatively, MR = a × mass$^{0.75}$ (a = a constant whose value will differ between different species or animal groups) and, more usefully, log MR = log a + 0.75 (log mass) and, even though mammals and birds have higher metabolic rates (larger 'a' values than reptiles, or fishes), the effect of body size on MR (as expressed by the exponent) is broadly similar within each group.

The easiest way to visualise this is to look at a graph. Figure 7.2 includes a log–log plot of the metabolic rate of mammals and birds against body mass over a very large size range: a 'mouse to elephant curve'. A small mammal weighing 100 g, perhaps one of Australia's native rats *Melomys*, can be expected to have an SMR of ~116 mL/h (= 1.16 mL/(g.h). If SMR increased proportionally, a 10 kg mammal, perhaps a small dingo, *Canis lupis dingo*, would have an SMR of 100 times that: 11 600 mL/h. Because of the way SMR scales with body weight, however, the curve predicts a value of much less: 3483 mL/h (= 0.3483 mL/(g.h) in this example). That is, the resting oxygen requirement per gram of tissue gets smaller as body weight increases. An elephant weighing 1000 kg can be expected to have a 'mass specific' SMR of only 0.1043 mL/(g.h); 1 g of a 100 g rat requires ~11 times more oxygen than 1 g of a 1 tonne elephant.

An important practical implication of this is that the very common habit of expressing oxygen consumption (whether resting or active, or anywhere in between) in terms of mL/(g.h), mL/(min.kg) or some other weight-based expression) is very limited because it is valid only at the size at which the measurement was made. It is not by itself a useful generalisation that allows easy comparison between animals of different sizes or application to larger or smaller animals. That manner of expression is even less meaningful if the mass of the animals measured is not disclosed, because if that is known a researcher at least has an opportunity to calculate an approximate MR at a different mass using an assumed exponent in the equation, such as 0.75. The correct way to express metabolic rate is to include the appropriate exponent, as in mL/ (mass$^{0.75}$.h).

Another very serious implication of these scaling effects is in calculating dosages of drugs. Knut Schmidt-Nielsen (1972) reported a study (not by him!) of the effect of the drug LSD on elephants. The researchers had scaled up the dose from a human dose in direct proportion to body weight and injected 297 mg. The elephant died and the researchers concluded that elephants are particularly sensitive to LSD. Scaling according to metabolic rate would have suggested an appropriate dose may have been 3.9 mg. Knowledge about the way metabolic rate scales with body mass can be very important!

Wildlife veterinarians facing a need to calculate a suitable dose of a drug to immobilise a crocodylian have to be aware of this (Box 1.2), and if the animal is large they are probably unable to weigh it. Length can usually be measured, and this can be used to make a reasonable estimate of body mass (Fig. 1.36).

---

seems to have been reached that exponents differ and White *et al.* (2006) have suggested a range in vertebrates from ~0.64 to 0.88, with reptiles at 0.76. An interested reader can explore the debate by consulting Glazier (2010) and White (2011), because it is beyond the scope of this book.

Knowledge about the effects of body size is particularly important in crocodylian studies (and for croc farmers calculating food requirements) because a croc's body mass increases so dramatically during

its life (see Chapter 1). Even small crocodylian species go through several orders of magnitude as they grow and mature. The largest, a male *C. porosus*, may weigh less than 100 g at hatching yet grow to more than 1 000 000 g (1000 kg); that is, four orders of magnitude.

Results from several studies measuring metabolic rate of crocodylians are shown in Fig. 7.2. It would be good to know the relationships between body mass and resting and active metabolic rates,

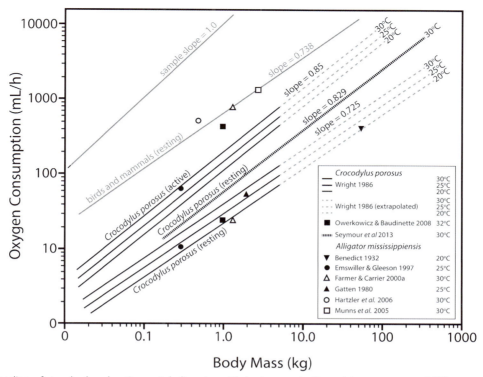

Fig. 7.2. *The scaling of standard and active metabolic rates with body mass in* Crocodylus porosus *and* Alligator mississippiensis. *Discrepancies between studies are more likely to stem from the way the measurements were made than from differences between species. Wright (1986) focussed particularly on the effects of body size and temperature, but his size range was not large (49 g to 4.078 kg); dotted lines show extrapolations of his data beyond this range. The most comprehensive study has been that by Seymour* et al. *(2013) who measured animals over a much larger size range, from 0.19 to 389 kg.*

but discrepancies between studies make this difficult (Fig. 7.2). These probably reflect the ways the measurements were made more than differences between species. The equation given by Seymour *et al.* (2013) for *C. porosus* is SMR = 1.01 M$^{0.829}$ (i.e. log SMR in mL/min = 0.004321 + 0.829 (log mass in Kg)). Seymour *et al.* discuss this surprisingly high exponent, values closer to 0.75 being more common across the animal kingdom.

Effect of body temperature

Metabolic rate in ectotherms is affected exponentially by body temperature (Tb). The effect is expressed as the ratio of the rates for a 10°C increase in temperature, known as Q_{10}. This is actually an archaic way to describe it, but it has become embedded. It would be better expressed as the slope of the line in a plot of metabolic rate (MR) on temperature, or the exponent in the relationship MR = aTb^b. In most biological systems, Q_{10} is between 2 and 3 (i.e. a doubling or tripling of MR for a 10°C rise in temperature). The equation $Q_{10} = (k_2/k_1)^{10/T_2-T_1}$ (where k_2,

k_1 are rates at T_2, T_1 respectively) is useful for calculations, including correcting for a temperature effect when Q_{10} is known. Q_{10} is known to vary across a temperature range and, further, is useful only within the normal range of body temperatures that the animal experiences and can cope with.

Combining the effects of body mass and temperature on SMR

It would be useful to have an equation or a series of equations that describe resting metabolic rate in crocs as a function of body size and temperature, and at a range of activity levels. A comprehensive series of measurements by Jon Wright at the University of Sydney (Wright 1986) dealt with the many difficulties posed by such a study and produced a generalised equation for SMR in juvenile *C. porosus*, which includes the effect of temperature, as follows:

log Vo$_2$ = −1.364 + 0.725 log M + 0.026 Tb (r^2 = 0.93, Vo$_2$ = resting oxygen consumption in mL/h, M =

mass in grams, Tb = body temperature in °C). This relationship in plotted in Fig. 7.2. The limitations of Wright's data are that the relationships were drawn from just one species, *C. porosus*, and over a very small size range compared with the total range in body size: 24 individuals over a size range from 49 g to 4 kg. The larger scaling exponent for metabolic rate with increasing mass reported by Seymour *et al.* (2013) is based on data over a much wider size range, so the value of 0.829 is likely to be more reliable.

Wright measured a Q_{10} of ~1.8 in the range 25–30°C, typical of Tb values of animals spending much of their time in water. Body temperatures of crocs with an opportunity to bask often raise Tb to higher levels, typically peaking at around 32°C (Chapter 10), and over the 30–33°C temperature step the effect of temperature steepened, with a Q_{10} of 2.6. Although it may be tempting to interpret this pattern as the lower Q_{10} minimising the effect of cooler winter temperatures on metabolism and the higher Q_{10} conveying significant benefit when the animal has an opportunity to bask and select warmer Tb, physiologists should be cautious about developing 'just so' stories (see Chapter 8 for discussion and Hicks 2002).

Increased oxygen consumption associated with digestion and absorption

Because of the oxygen cost of the digestive process, SMR needs to be measured well after feeding: several hours after in the case of humans, days after in the case of crocs. The increase in oxygen consumption associated with digestion is often referred to as the 'specific dynamic action' (SDA) of food and the whole topic was reviewed by Secor (2009). The costs include the mechanical and chemical breakdown of food and the processes of its absorption and protein re-synthesis. When you or I eat a meal, our oxygen consumption goes up and our heat production increases, and so it is with crocodylians although, of course, we are endotherms whereas crocs are ectotherms (Chapter 10) and the absolute increases are proportionally smaller in crocs. Not surprisingly, different increases have been reported, probably depending on meal size. Benedict's (1932) large snakes increased oxygen consumption 3–7 times over resting rates after feeding. McCue *et al.* (2005)

found similar values in Burmese pythons. Gatten (1980) reported an increase of only 1.6 times in *Caiman crocodilus,* whereas Busk *et al.* (2000) found an approximately 4-fold increase in *A. mississippiensis* (but they pooled data across a 10-fold size range to 9 kg without taking the effect of body size into account; see Box 7.1). More recently, Gienger *et al.* (2011) measured SDA in *C. johnstoni* and *C. porosus* and recorded a gradual rise in MR over 24 h to 1.6–2 times SMR, with a decline to SMR taking 3–4 days. The effect scaled with body mass with an exponent of 0.85 over a size range from 0.19 to 25.96 kg. They noted that the SDA effect was smaller in response to homogenised meals than 'intact' meals. This may have implications for improving production efficiency on farms. As an aside, it is worth noting that a feeding-induced increase in metabolic rate is more typical of carnivores, perhaps because herbivores tend to feed and process food almost continuously, whereas carnivores usually eat large meals periodically.

Apart from the direct effect of increased metabolic rate associated with processing food, from which some heat will be produced, crocs given the opportunity to bask and absorb heat will likely choose to bask longer and select a warmer Tb during the digestive phase of feeding. This has an additive effect on oxygen consumption and thereby accelerates digestion and the rate of passage of food through the gut (Chapter 10).

Aerobic support for activity; absolute and factorial aerobic scope

Crocodylians have a reputation for only very limited aerobically fuelled activity, but the long-distance travel they undertake (Chapter 4) suggests that they have greater capability than is usually assumed. Measuring oxygen consumption in an active crocodylian is not easy, but researchers have managed it in both alligators and crocodiles by training them to walk on a treadmill or swim in a flume. The subject wears a mask and measurements are made by analysing the depletion of oxygen as air flows through the mask at a known rate. An additional challenge is to measure a sustainable maximum rate, and so far it is not certain that this has been achieved. The extent to which aerobic metabolism can increase above resting values is called metabolic scope, which can be expressed as the actual increase (absolute aerobic scope, AAS) or the proportional increase (factorial aerobic scope, FAS).

A wide range of values has been reported for active oxygen consumption, making it difficult to draw a consensus about maximum rates (Fig. 7.2) or about factorial aerobic scope. Eme *et al.* (2009) recognised the difficulty of identifying maximum rates and reported 'peak' rather than 'maximum' Vo_2 values from a study on American alligators. This has merit and the wide spread in reported values probably results from the difficulty of contriving performance to be at a 'maximum'. In a study focussed particularly on sustainable rates, Owerkowicz and Baudinette (2008) measured 1 kg *C. porosus* on a treadmill. Untrained animals at 32°C showed a peak Vo_2 of 414 mL/h, which increased to

510 mL/h after substantial training. They measured SMR of 23.4 mL/h, yielding a FAS of ~18 (22 when trained). Farmer and Carrier (2000a) measured average maximum rates of 9.3 mL/min/kg (748 mL/h) in alligators averaging 1.34 kg walking on a treadmill at 25°C, 17 times higher than the pre-exercise values and 31 times their measured SMR values (!). Emshwiller and Gleeson (1997) measured Vo_2 max of 12.6 mL/h in 300 g alligators, on a treadmill at 25°C, with a FAS of ~6. All of these results are plotted in Fig. 7.2.

In addition to his measurements on crocs at rest, reported above, Wright (1986) attempted to establish the effect of body size and temperature on 'Vo_2 max'. He exercised *C. porosus* (49– 4078 g) to exhaustion at 20, 25, 30 and 33°C while measuring their oxygen consumption and lactate production. The differences between his study and the four that reported markedly higher values (Fig. 7.2)

Fig. 7.3. *Aggressive activity between two male* C. porosus. *Explosive activity such as this is powered, as in humans, by anaerobic metabolism of glycogen in the cytoplasm of the muscle cells. Following the activity, lactate floods into the blood, lowering the pH and, over a period of time, the surplus is metabolised through the citrate cycle and electron transport chain ('paying back the oxygen debt', or 'excess post exercise oxygen consumption' EPOC). (Photo GCG)*

could lie in the different way he induced activity, holding them by the tail while they ran vigorously on a slippery floor until they appeared exhausted. Despite this difference, it is worth reporting his results because they point to effects of body size and temperature, including higher FAS values in larger animals. His results were generalised in the following equation (and see Fig. 7.2).

$$\log V_{O_2} = -1.031 + 0.85 \log M + 0.025\ T_b\ (r^2 = 0.83,$$

V_{O_2} = active oxygen consumption in mL/h, M = mass in grams, T_b = body temperature in °C).

Wright's results imply an increase in FAS with increasing body size and this has significant implications. Although his active values are probably significant underestimates of maximum oxygen consumption, his experiments were conducted over three orders of magnitude in body size and comparable activity protocols were followed throughout. He found that FAS (as measured) was scarcely influenced by temperature but increased with body mass. Increasing FAS with increasing body mass has been reported in a wide range of birds and mammals (Bishop 1999) and

Killen *et al.* (2007) found similar increases in FAS in a comprehensive study of three species of sedentary marine fish over six orders of magnitude in body mass from larval to adult sizes. The phenomenon appears to be a general pattern among vertebrates and Weibel *et al.* (2004) sought a mechanistic explanation to account for it. They suggested that it follows from active oxygen consumption occurring mostly within the locomotor musculature (which, in mammals, receives 90% of the pulmonary oxygen during activity) and the way that the components of the oxygen transport system themselves scale with body mass. In the case of crocodylians, it is interesting that larger crocs have proportionally larger lungs than smaller crocs (Chapter 4 and below) and that may be additionally significant.

Whatever the mechanism, the recognition of this relationship is important because, as we shall see, the capacity of crocodylians to undergo activity, such as struggling during capture, is strongly size-dependent (Fig. 7.4) and their increasing factorial aerobic scope may be one of the explanatory factors.

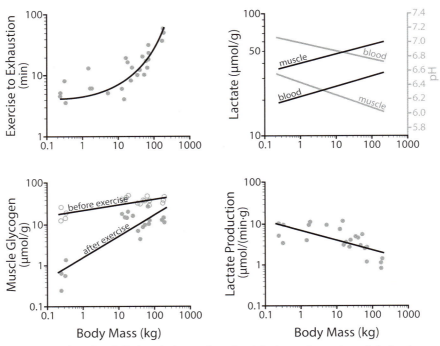

Fig. 7.4. *Data from* C. porosus *show (top left) that the larger the animal the longer it can struggle before becoming exhausted. (Top right) At exhaustion, large individuals show lower pH and higher lactate concentrations than small ones, in both blood and muscle coming closer to levels which put them at significant risk. (Bottom left) They have higher muscle glycogen stores than small crocs, yet use proportionally less of it, and (bottom right) have a lower rate of lactate production (and thus a lower anaerobic scope) than small crocs. (Adapted from Baldwin* et al. *1995)*

ANAEROBIC METABOLISM

It has long been held that diving and activity are both powered mostly by anaerobic metabolism. As we shall see in Chapter 9, it is probably unusual for a crocodylian to employ anaerobic metabolism during dives, unless these are associated with heavy activity, such as prey capture. How much of their unhurried peaceful activity, walking and swimming at a leisurely pace and minding their own business is anaerobic is unknown but my strong suspicion is that crocs, like humans, become anaerobic only to support substantial exercise. We may have gained a biased view of crocodylians' reliance on anaerobiosis because our engagement with them often promotes an explosive response as they attempt to escape and are finally subdued. Swimming is an efficient way to travel, but to me the 20 or more kilometres per day made by our translocated crocs returning 'home' (Read *et al.* 2007) implies a capacity for more than trivial aerobic capacity, even if their journey was to some extent current assisted (Campbell *et al.* 2010, discussed in Chapter 4). Crocodylians clearly have at least moderate capacity for aerobic work.

However, we can be sure that much of the high energy activity involved in subduing large prey or in escape, such as when struggling to escape capture, is powered by anaerobic metabolism.

Powering explosive activity

Every book about crocs will tell how they have limited aerobic capacity and tire quickly. Anyone who has harpooned big crocs and played them on the end of a line for a few minutes knows that their capacity for explosive struggling gradually flags, and waits for that so they can be drawn close enough to the boat to slip a noose on their jaws (Fig. 1.26).

That explosive power for which crocs are so famous (e.g. Fig. 7.3) is powered anaerobically, by glycolysis, using glycogen stored 'on-site' in the muscle cells, producing pyruvic acid and, after another biochemical step, lactate as a (still energy rich) by-product (Fig. 7.1). This is the same process by which the metabolism of glycogen in muscles powers the performance of human sprinters. The energy can be tapped immediately, limited only by the amount of the stored glycogen and the rate-limiting enzymes, until the glycogen runs out, or glycolysis or the mechanics of muscular contraction are inhibited by the accumulation of by-products and their effect, such as a fall in pH. Because anaerobic glycolysis is not much affected by temperature (Bennett 1982), it can produce ATP in a short time, regardless of the body temperature. One downside of this, as noted above, is that anaerobic (glycolytic) metabolism yields much less energy than aerobic metabolism, 1 mole of glucose yielding only 2 moles of ATP, compared with 30 with aerobic metabolism. Another downside is that the stored energy is used quickly and so is a short-lived source. Exhaustion is accompanied by high plasma and muscle lactate and low pH (Fig. 7.4) from which a croc may take a long time to recover. Meanwhile, the by-products are converted to (lesser amounts) of glucose and the glycogen stores in the liver and muscles are rebuilt.

Accordingly, it must be advantageous for a croc to avoid relying on anaerobic metabolism whenever possible and most crocodylian activity is 'slow and peaceful' (Seymour *et al.* 1985), and powered aerobically. I think it is almost certain that the swimming associated with their journeys to feeding and mating sites (Chapter 4) is powered aerobically.

Anaerobic capacity in large and small crocodylians

There have been some interesting studies on the capacity for anaerobic metabolism of crocs over a size range, prompted initially by unexpected deaths of very large crocs after capture. In northern Australia, when some very large *C. porosus* is harassing people or dogs, the normal practice is to move it, rather than kill it, usually to a croc farm. Before effective trapping was developed and became routine (Chapter 14), crocs were usually captured by harpooning from a boat. Once the harpoon/s is in place, tethering the animal, it will struggle to exhaustion and can be brought alongside to secure the jaws before hauling it into the boat or alongside, or onto a truck or trailer for transport. Recovery from this takes time and, sometimes, if the croc has become over-stressed, it will die a few hours later. Grahame Webb and others noticed that crocs larger than ~700 kg were much more likely to die than smaller ones. Suspecting that this was the result of an acute metabolic acidosis and other biochemical disturbances, and that large crocs may be more sensitive than small ones. Roger Seymour from the University of Adelaide and Al Bennett from the University of California and several colleagues (Bennett *et al.* 1985; Seymour *et al.* 1985; Baldwin *et al.* 1995) exercised *C. porosus* to exhaustion over a size range from 400 g to 180 kg. They analysed blood samples taken from the exhausted animals and found that the larger the individual, the longer it took to become exhausted, 30 min in their largest, and the greater the accumulation of lactate in the blood (Fig. 7.4).

The highest values of plasma lactate, 40–50 mmol/L, were the highest ever recorded as a result of activity in any animal, compared with values close to zero in *C. porosus* at rest. They also measured the lowest recorded values for plasma pH, 6.49, compared with a normal value of 7.4 at 30°C (Seymour *et al.* 1985). Recovery was partial after 2 h and one animal was still unresponsive after 4 h, although the next day it was 'active and aggressive'. The results pointed to larger individuals taking much longer to become exhausted and having an extraordinary capacity for tolerating high levels of blood lactate and low pH. Extrapolation of the data suggests that very large individuals may be 'perilously close to their physiological limits' when forced into activity leading to exhaustion. This is undoubtedly the explanation for the apparent 'fragility' of very large *C. porosus* when captured. They may need many hours to recover and if the stressful situation of capture persists they may not survive. Jackson *et al.* (2003) showed in force-dived *Caiman latirostris* that lactate is sequestered in the osteoderms, as in the bones of turtles, and that it may be a vertebrate phenomenon for the bony skeleton to buffer lactate production in those species whose reliance of anaerobic energy production is substantial.

Most comparatively short bursts of activity are probably supported by anaerobic metabolism, as in humans. The slow recovery from exhaustion suggests that, although large crocodiles have outstanding capacity for anaerobic metabolism

when they need it – for example, to subdue a buffalo or a wildebeest – it is unlikely to be a routine feature of their lives. It is most unlikely that a croc would exercise voluntarily to the point where it becomes comatose through low pH and high lactate. More will be said about this in Chapter 9 in a discussion about diving behaviour and physiology.

Because anaerobic scope (a measure of the rate of lactate production, as an index of ATP production) decreases with increasing body size, it seems odd that larger crocs take much longer to become exhausted (Fig. 7.4). Baldwin *et al.* (1995) realised that this would be explained if larger crocs had greater aerobic capacity than smaller ones; that is, a greater FAS, which was exactly what Wright (1986) had found, as in other vertebrates. From exercising juvenile *C. porosus* to exhaustion, he was able to calculate that at 30–33°C, aerobic metabolism contributed 30–40% of the total increased energy supply, the rest being anaerobic. Higher FAS in larger animals implies that the aerobic contribution to activity will be proportionally larger in larger animals.

Recovery from metabolic acidosis

Hartzler *et al.* (2006) looked in more detail at the mechanisms driving recovery from exhaustion by alligators. They found that 480 g (average) alligators, after forced exercise, became exhausted in 3–4 min at 30°C, similar to what was been found in small *C. porosus* (Baldwin *et al.* 1995). They described physiological events accompanying recovery from exhaustion, monitoring a range of respiratory and metabolic data. At exhaustion, there was a significant increase in oxygen consumption, arterial oxygen partial pressure (Pa_{O_2}) (see Box 7.2), plasma lactate, metabolic rate and ventilation, and a decrease in arterial Pco_2, pH and bicarbonate. Oxygen consumption returned to resting levels within 15 min, but CO_2 excretion remained elevated for 30 min, as did arterial Pa_{O_2}. Plasma pH recovered from 7.0 to a more normal 7.4 in about 2 h, and bicarbonate recovered from ~6 to 23 mmol/L, more or less in parallel with pH but taking about 6 h. Respiratory ventilation was high for about an hour. Plasma lactate fell in ~8 h from 16 mmol/L to the normal level of 1 mmol/L or less.

Carbon dioxide production was much higher than would be expected during the recovery phase. With plasma pH and most of the respiratory parameters recovering much more quickly than lactate levels, Hartzler and colleagues concluded that much of the metabolic acidosis resulting from exhausting activity by crocodylians is handled by respiratory mechanisms rather than the gluconeogenesis of lactate and other metabolic pathways (Fig. 7.1). Seymour *et al.* (1985) came to the same conclusion in *C. porosus*, noting a two stage recovery process in which respiratory compensation is important in the early phase of the recovery, with ventilation reducing CO_2 to lower than normal levels, reducing the acidosis, and removal of lactate metabolically over a longer time frame. Indeed, excretion of lactate

to effect recovery would be would be inefficient and maladaptive: the efficient course is to recycle the lactate *in situ*.

THE RESPIRATORY SYSTEM: LUNGS, GAS EXCHANGE AND TRANSPORT BY THE BLOOD

Oxygen is supplied to the working tissues and CO_2 excreted via the lungs and the blood system. The exchange of these gases occurs via diffusion at both the tissue end and in the lungs, with respiratory muscles ventilating the gas exchange surfaces in the lungs and the heart muscle driving the circulation of blood. The loading and unloading of the blood with oxygen and CO_2 is governed by the blood's

---

BOX 7.2   WHY PARTIAL PRESSURE IS USED IN DESCRIBING AND UNDERSTANDING GAS EXCHANGE

Gases diffuse across membranes down partial pressure gradients and, furthermore, partial pressures are easy to measure using appropriate electrodes, so gas exchange is usually evaluated using partial pressure data. Furthermore, partial pressure data can be translated to oxygen content of gases – of blood or of solutions (in mmol/L, mL/L, vol% or whatever) – as required. The convention is to express partial pressure as an upper case P and the gas species identified as a subscript, e.g. P_{O_2}, with units of mmHg, torr (named for Evangelista Torricelli, a colleague of Galileo who invented the mercury barometer) or, more properly in the SI system, kilopascals (kPa) where 1 mmHg = 1 torr = 0.1333 kPa (named after Blaise Pascal, a French mathematician, physicist and theologian who also invented the mechanical calculator).

Some people find the partial pressure concept difficult. The easiest way to understand gas partial pressures is to follow an actual example: the fraction of oxygen in air is ~20.9% by volume, so dry air at an atmospheric pressure of 760 mmHg (101 kPa) will have an oxygen partial pressure of 0.209 × 760 = 159 mmHg (21.2 kPa). Air usually has some water vapour, and in saturated air at 25°C the water vapour pressure is very close to 24 mmHg (3.2 kPa). The partial pressures of all the gaseous components of a gas mixture add up to the total pressure (by Dalton's law), so if the saturated air is at an atmospheric pressure of 760 mmHg (101 kPa), the partial pressure of the oxygen in it will be (760–24) × 0.209 = 154 mmHg (20.5 kPa).

Partial pressure measurement also provides a useful way to quantify oxygen content in solution and also in blood. Think of water stirred and in equilibrium with air. Oxygen partial pressures will be the same in the air above and in the water, because any difference in partial pressures between the two will drive diffusion until there is equilibrium. Partial pressure in distilled water or in a solution (e.g. blood plasma, sea water) can be translated to oxygen content (e.g. mL/100 mL, mmol/L), if the solubility coefficient (different in different solutions) and temperature are known. Partial pressure measurements of oxygen made in whole blood can be converted to oxygen content using the oxygen equilibrium curve (sometime called the oxygen dissociation curve), if the curve is known for the appropriate temperature, CO_2 and pH values (see Fig. 7.15). Typically, P_{O_2} is translated using the curve to a percentage saturation of the haemoglobin and thus to oxygen content from knowledge of the blood's haemoglobin content.

respiratory properties. Each component is discussed separately below.

LUNGS AND BREATHING

Gross anatomy of the respiratory system

Crocodylians are particularly well adapted to breathing without taking water into either the nasal passages or the trachea when they are in water. The nostrils and their musculature, the position of the internal nares in proximity to the glottis, and the gular and palatal valves in the throat have been described elsewhere (Chapters 5, 6). Crocodylians can open their mouths under water (Fig. 9.3) without risk of water flooding into either the trachea or the oesophagus. Although they almost always surface to swallow food, it is not obligatory and I have watched Adam Britton's very large male Smaug (Figs 9.5, 9.6) snatch a chicken at the surface, sink to the bottom of his pool and swallow it while remaining submerged. Presumably at least some water must get in to the oesophagus and stomach when crocs swallow under water.

The glottis is the gateway to the trachea and it lies at the back of the throat, on the floor of the posterior pharyngeal cavity (Figs 5.18, 6.21), supported by the cartilaginous hyoid plate. The paired lungs are in the thorax, each in a separate pleural space. They are narrower anteriorly and extend well posterior to the heart, overlying it.

During breathing, the muscular glottis is opened and pressed up to the socket of the internal nares, minimising the respiratory dead space, and air can pass from the closable external nares, through the nasal passages above the secondary palate and into the trachea. The trachea is stiffened by annular cartilaginous supports, as in most air breathing vertebrates, and divides to a pair of bronchi that deliver air to and from the lungs. Details of the histology of the trachea of *Caiman crocodilus* have been described recently by Santos *et al.* (2011). Interestingly, in large *C. porosus* and *C. niloticus* (and perhaps in all crocodylids except when small,) the trachea is thrown into a large loop in the upper thorax (Fig. 7.5), reminiscent of the water-trap under a kitchen sink, but whose function is sometimes said to be to allow the airway to lengthen and be moved aside when large food items are swallowed (Chapter 6). Reports are few but alligators and caimans apparently lack this tracheal loop. In the black caiman, *Melanosuchus niger* (Fig. 7.6), the trachea is wide anteriorly and tapers towards the bronchial bifurcation (Fig. 7.7). The function of this widening is unknown; could it function somehow to amplify bellowing? The lungs extend anteriorly well forward of the hilus (the point where the extrapulmonary bronchi enter the lung).

Lung ventilation is normally intermittent

The repetitive movement of the throat (gular) region seen in crocodylians is not respiration, but 'sniffing', drawing air in and out through the nostrils to ventilate the nasal sinuses for olfaction (Chapter 5), with the glottis remaining closed (Naifeh *et al.* 1970, working on *Caiman crocodilus*). Breathing – that is, ventilation of the lungs – is seen when the body wall inflates and deflates and this is not continuous in crocodylians except when they are active. At rest their ventilation pattern is periodic: a series of breaths followed by a period of apnoea with the glottis closed and the lungs held full of air, and then another series of breaths which begins with an exhalation (Fig. 7.8). The pattern is quite variable between individuals. It also depends upon oxygen requirements. When more oxygen is required, such as when walking on a treadmill (Farmer and Carrier 2000a), at higher temperatures or during exposure to artificially lowered ambient oxygen levels (Munns *et al.* 1998), ventilatory frequency and tidal volume increase and periods of apnoea shorten, raising the volume of air processed. Ventilation can also be continuous during locomotion (Farmer and Carrier 2000a), which may at first seem obvious, except that ventilation is compromised in many reptiles by walking or running. Furthermore, Munns *et al.* (2005) have shown that venous return to the heart is not reduced by the increased abdominal pressure that accompanies locomotion, unlike in Varanidae and some other lizards. Crocodylians seem well equipped to maintain oxygen delivery during locomotion.

Mechanics of lung ventilation

There was early interest in how crocodylian lungs are ventilated because, like other reptiles, they have

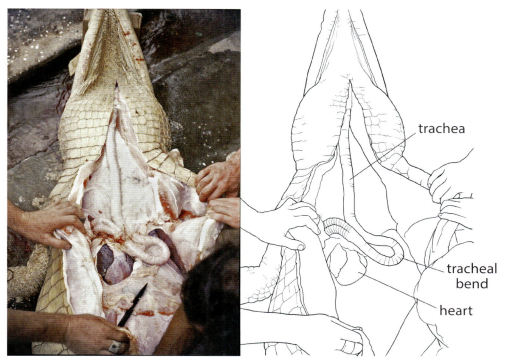

trachea

tracheal bend

heart

Fig. 7.5. *Trachea in a large* C. porosus, *showing its division into two extrapulmonary bronchi and the way the trachea is thrown into a loop reminiscent of a water trap under a sink. Small* C. porosus *and* C. niloticus *lack this tracheal loop: it develops as they mature. Other Crocodylidae show the same pattern, but the tracheal loop is apparently lacking in Alligatoridae. It has been suggested that the loop allows the trachea to lengthen to accommodate distension of the oesophagus when swallowing large items. Another possibility is some connection with sound production (Colleen Farmer pers. comm.). But why such a conspicuous difference between crocodylids and alligatorids? (Photo GCG)*

no diaphragm. It turns out that their ventilatory mechanics are extraordinarily complex, with about six contributory elements, of which the first two do most of the work: forward and aft movement of the viscera (the 'hepatic piston'); expansion and contraction of the rib cage (costosternal movement); ribs in three jointed sections to allow extended expansion of the thorax; rotation of the pubes which increases abdominal volume; movement of the gastralia; and vertebral flexion and extension. The most recent work is by Claessens (2009) who studied juvenile alligators by cineradiography (movie X-rays). He was able to clarify, quantify and add to previous understanding, and he also provided a comprehensive and constructive review of earlier work.

Although crocodylians lack a diaphragm, they do have a sheet of connective tissue anterior to the liver that divides the pleural cavities in which the lungs lie from the rest of the visceral cavity (Fig. 7.9). This is sometimes referred to as a pseudodiaphragm. Gans

and Clark (1976) explored ventilatory mechanics in *Caiman crocodilus* by electromyography (diagnosing electrically which muscles are at work and in what sequence). They concluded that the main driver of inspiration was contraction of the thin sheets of the paired diaphragmaticus muscles attached to the pelvis, which pull the liver and other viscera posteriorly and so draw air into the lungs (Fig. 7.9). Functionally, then, the pseudodiaphragm is analogous to the mammalian diaphragm, but its operation is driven by different muscles and it is innervated by spinal nerves, not the vagus (Claessens 2009). The liver can be thought of as a 'hepatic piston', driven by the diaphragmaticus muscles. Gans and Clark (1976) drew attention additionally to the supplementary role of the intercostal musculature, whose contraction causes the rib cage to expand, as in humans. It was suggested quite early that the gastralia (see Chapter 3) and associated musculature might stiffen and so help resist collapse of the ventral body wall

Fig. 7.6. *Juvenile black caiman,* Melanosuchus niger, *near Mamirauá, ~500 km up the Solimões River from Manaus, Brazil. (Photo Cássia Camillo* per *Robinson Botero-Arias)*

during inspiration: a 'floppy' belly could confound the development of negative pressure in the thorax. At the end of inspiration, the glottis is closed and air in the lungs is held under small positive pressure, in the vicinity of 8–12 mmHg (1.0–1.6 kPa) (Claessens 2009). With the glottis closed, the diaphragmaticus can be relaxed during the breath hold (Claessens 2009).

Slight positive pressure in the lungs means that expiration can occur partly passively. There is also contraction of the superficial intercostal muscles to decrease the volume of the rib cage, and contraction of the rectus abdominus muscles. This increases hydrostatic pressure within the abdomen, driving the viscera forward and helping to drive the air out (Gans and Clark 1976; Farmer and Carrier 2000b). In what I found to be an interesting historical note, Claessens pointed out that the unique role of the diaphragmaticus muscle in crocodylian respiration was recognised as early as the 17th century in a 1674 publication by the Danish scientist Olaus Borrichius

(cited in Claessens 2009), 300 years ahead of Gans and Clark (1976) and there were references to it through the 1800s.

So far, that is two of the six components: the 'hepatic piston' and muscular expansion and contraction of the rib cage by intercostal muscles. In the late 1990s, Colleen Farmer and David Carrier suspected that there was more to this story. They noted that when holding a young alligator 'in the palm of one's hand, a ventral expansion of the caudal abdomen is felt during inspiration' (Farmer and Carrier 2000b). They noted also that the pelvis in crocs is unusual because, as well as being moveable, the pubes are widened and flattened anteriorly so that they lie in the posterior ventral wall of the abdomen (Fig. 3.59) and they wondered why. Using electromyography, they recorded and correlated the activity of eight different pairs of muscles with ventilation and abdominal pressure in alligators at rest and while walking, and identified the muscles involved in both inhalation and exhalation.

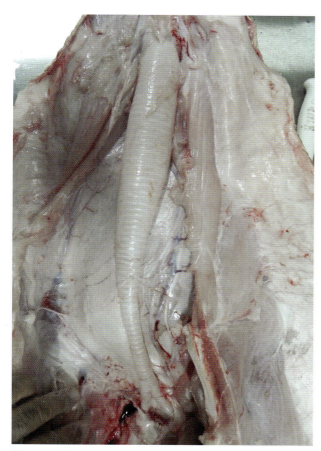

Fig. 7.7. *Black caiman,* Melanosuchus niger, *have an unusual trachea, widened considerably for its anterior two-thirds before tapering to its bifurcation to the bronchi. This photograph is of an adult animal, but the same widening is evident in a photograph of a juvenile in Romão et al. (2011). The function of this is so far unknown, but hypotheses about sound production are appealing. (Photo Marcos Coutinho)*

They found that contraction of the ischiopubic, truncocaudalis and ischiotruncus muscles rotated the pubes ventrally during inspiration, increasing the abdominal volume and facilitating ventilation of the lungs. Using high speed cineradiography, Claessens (2004) was able to observe rotation of the pubes during inspiration and, in another historical note dug up by Claessens (2004), the German embryologist Martin Rathke at the University of Königsberg had previously hypothesised that rotation of the pubic bones was a component of the aspiration pump (Rathke 1866).

Subsequently, Claessens (2009) identified three additional features and quantified the relative contribution of the major ones. He noted that flexion and extension of the vertebral column accompanies the respiratory cycle and that the gastralia also move in synchrony with breathing. It is unclear to me whether the gastralia movements are an active, muscle-driven contribution to ventilation or a stiffening response. Any stiffening of the ventral body wall would seem to be advantageous in facilitating the viscera to be drawn towards the rear by the diaphragmaticus muscles. Claessens also pointed out that the ribs are in three jointed sections (Fig. 3.54), which facilitate expansion and contraction of the body wall. He attributed up to 60% of the tidal volume to forward and aft movement of the viscera and up to 40% to costosternal movement. He scored pubic rotation at an average of 4%, and 3% to vertebral flexion.

An adaptable ventilatory system

Despite these quantifications, the respiratory system has considerable flexibility, as was highlighted in a study of small *A. mississippiensis* by Uriona and Farmer (2006). They found that disabling the diaphragmaticus muscles did not compromise ventilation in resting, fasted animals and did not even reduce their vital capacity (lung volume at maximum inspiration). Presumably the alligators managed sufficiently well at rest with the costosternal muscles, perhaps with enhanced pelvic rotation. After alligators had taken a large meal, however, which would take up significant space within the body cavity and could compromise ventilatory volume, Uriona and Farmer found that alligators with the diaphragmaticus muscles disabled were less able to compensate and showed reduced vital capacity. This would surely reduce aerobic scope and shorten submergence time (see Chapter 9). The diaphragmaticus muscles clearly become more important in some situations than others. They are probably recruited also during increased aerobic activity, and are likely to be important for breathing in water, although hydrostatic pressure is likely to assist exhalation, inhalation will require more effort and this may call for more work by the diaphragmaticus muscles (Fig. 7.10). If you have noticed the posterior flanks of a crocodile or alligator in water sucking in while the thorax is expanding during an inhalation, the extra work and extra

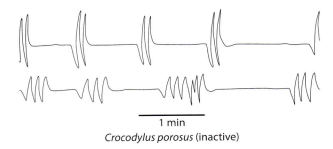

Crocodylus porosus (inactive)

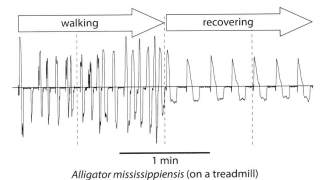

Alligator mississippiensis (on a treadmill)

Fig. 7.8. *Ventilatory patterns in crocodylians. Like most reptiles, crocs at rest ventilate their lungs intermittently, with several consecutive expirations and inspirations separated by non-ventilatory periods. With increased oxygen consumption, the non-ventilatory periods shorten until respiration becomes continuous. (Upper trace adapted from Wright 1985; lower trace adapted from Farmer and Carrier 2000a, with permission)*

negative pressure required to cope with hydrostatic pressure probably provides the explanation.

Using some of the same muscles which ventilate the lungs, alligators were shown by Uriona and Farmer (2008) to be able to move air fore and aft and from side to side, so adjusting their centre of buoyancy and thus their pitch when they initiate a dive, or to counteract roll. This has been discussed in Chapter 4.

Crocodylian lungs have a great capacity for combining the dual functions of buoyancy management and respiration. Crocs adjust lung volume before they dive, often needing a small exhalation, to establish a particular level of buoyancy (Chapters 4 and 9) which, typically, requires well below maximum lung inflation (Kirshner 1985). While floating at the surface, crocs can breathe over a wide range of residual volume states, depending on what is required to adjust buoyancy (Perry 1988). This means that they can achieve an appropriate respiratory tidal volume for ventilating the lungs

from a wide range of inflation starting points. In a study of small (3 kg) Nile crocs by Perry (1988), the maximum possible lung volume (determined by inflating the lungs of dead crocs from −10 to +10 cm H_2O pressure) was ~12% of body volume (using SG = 1.0772, Kirshner 1985), suggesting a maximum possible change in volume of ~100 mL in a 1 kg croc. Residual lung volume (determined by the volume of air that could be drawn from the lungs of dead crocs to −10 cm H_2O pressure) was ~13% of that. These would seem to be upper and lower limits of what might happen naturally in individuals this size. David Kirshner (1985) measured lung volumes of ~5% of body volume at the onset of voluntary submergence in a series of *C. porosus* up to 4 kg. So, at the start of a dive, a croc usually has its lungs about half inflated, which leaves plenty of room to compensate for taking on board a heavy food item, a collection of stones, or a load of mud as a sun screen and still superimpose a ventilatory cycle (see Chapter 4, Chapter 10).

The anatomy and functioning of the respiratory system have been described in a variety of crocodylians by Gans and Clark (1976), Duncker (1978) and Perry (1988, 1989, 1990 and a review of reptilian lungs in 1998) and only minor differences have emerged between the different species. Their anatomy is strikingly complex (Figs 7.11, 7.12). Until recently, the lungs were interpreted as being multicameral (many chambered), with bidirectional air flow into and out of many blind ending, well perfused sacs. However, work by Farmer and Sanders (2010) on American alligators and Nile crocodiles (Schachner *et al.* 2013) has revealed an entirely different reality, as follows.

Unidirectional airflow in the lungs

In one of the most exciting recent discoveries about crocodylians, Colleen Farmer and Kent Sanders at the University of Utah found that air flow through parts of the lungs of juvenile *A. mississippiensis* is unidirectional, flowing in the same direction during both inspiration and expiration (Fig. 7.13) (Farmer and Sanders 2010). A follow up study on *C. niloticus* has led to the same conclusion (Schachner *et al.* 2013). This is the case in birds too and the finding is striking for two reasons. First, anatomical similarities

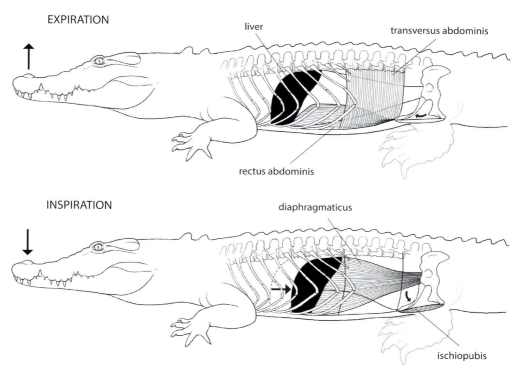

Fig. 7.9. *Inspiration is always an active process and is driven by the intercostals and the diaphragmaticus, expanding the rib cage and drawing the liver backwards (so-called 'hepatic piston'), with a contribution from posterior rotation of the pubes with contraction of the ischiopubic and ischiotrunchus (not shown) muscles. Between breaths, the lungs are held inflated and at a slight positive pressure and in resting animals expiration is largely passive upon opening the nares. With more rapid and more vigorous respiration during activity, expiration is driven by contraction of the transverse abdominals, which squeeze the abdomen, and the rectus abdominus, which rotates the pubes forward, reducing the volume of the abdominal cavity. (Adapted from Farmer and Carrier 2000b, with permission).*

Fig. 7.10. C. niloticus *surfacing to fill its lungs … inhaling against hydrostatic pressure. (Illustration DSK, courtesy of Weldon Owen Publishing)*

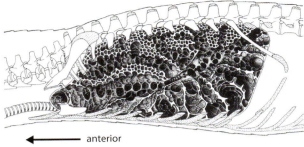

Fig. 7.11. *A longitudinal section through an inflated lung of* Caiman crocodilus *showing its complexity. (From Duncker 1978, image supplied by HR Duncker)*

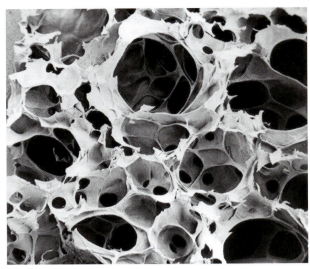

Fig. 7.12. *Scanning electron micrograph of a section taken from the dorsal portion of the lungs of* Caiman crocodilus. *(From Duncker 1978, image supplied by HR Duncker)*

between crocodylian and bird lungs have been recognised since the late 19th century (Huxley 1882) and, if there is functional similarity as well, that emphasises the likelihood of homology with birds and that such a respiratory system is plesiomorphic for archosaurs. Second, unidirectional flow of air across a respiratory surface provides the possibility for a cross-current or counter-current gas exchange system, markedly increasing the efficiency of gas exchange.

In their study, Farmer and Sanders (2010) measured flows in different parts of the major bronchi in excised lungs and *in vivo*, using thermistor flow probes during induced as well as natural ventilation. They also visualised flows within an excised lung by using fluorescent microspheres suspended in physiological saline.

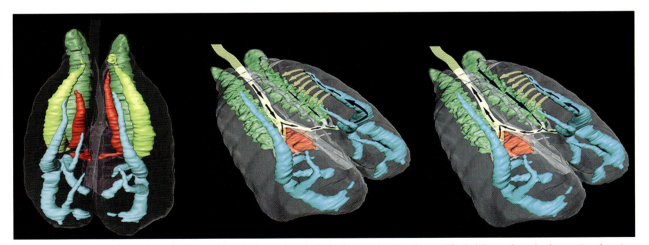

Fig. 7.13. *The major bronchi in the lung of* A. mississippiensis *(left, dorsal view) and simplified oblique dorsal schematics showing airflow during (centre) inspiration and (right) exhalation, as unravelled by Farmer and Sanders (2010). The trachea is shown in yellow, dividing into left and right secondary bronchi. Similarities abound between the anatomy within the lung and that of birds, with the cervical ventral bronchi (green) similar to the avian ventral bronchi, the three caudal bronchi (blue) similar to the avian dorsal bronchi, and these and the dorsolateral bronchi (chartreuse) and medial paracardiac bronchi (red) originating from a common chamber and spiralling anteriorly like the dorsal bronchi of birds. Air flow is shown by arrows. The dorsal bronchi (blue) connect to each other and to the cervical ventral bronchi (green) through numerous anastomosing passages 1–1.5 mm in diameter (brown arrows), which have been interpreted as parabronchi, perhaps homologous with the sites of gas exchange in avian lungs. Note that the direction of airflow through the parabronchi is the same during inhalation and exhalation. In birds, unidirectional airflow through the parabronchi is known to promote efficient gas exchange and it is likely to function similarly in crocodylians. (Photos Colleen Farmer)*

All three approaches yielded the same results – the direction of air flow through numerous, tiny branching tubes connecting the dorsal bronchi and the cervical ventral bronchus was the same during both inspiration and expiration (Fig. 7.13). They found the anatomical similarity to avian lungs compelling. They likened the cervical bronchus to the ventral bronchus of birds and the tubules to the parabronchi, which are the sites of gas exchange in birds, and adopted the same terminology, envisaging the same function; that is, as the sites of gas exchange. Their initial interpretations about the anatomical similarities between alligator and bird lungs were confirmed by a further study using computed tomography (Sanders and Farmer 2012).

In bird lungs, the functional units of gas exchange are a series of parallel tubes called parabronchi, which have an increased surface area and are very well vascularised. The rest of their respiratory system is made up of air sacs that act like bellows during ventilation and whose operation maintains the flow of air through the parabronchi. The maintenance of a unidirectional flow through the parabronchi means that the directions of air flow and blood flow remain consistent throughout the respiratory cycle, without the reversal that would accompany tidal flow. Gas exchange in birds occurs according to a 'serial multicapillary' or 'cross current' model, which results in high arterial oxygen partial pressures: higher than in mammals. The crux is that in birds the parabronchi are exposed to air at closer to atmospheric partial pressures than are the alveoli in mammalian lungs. Scheid (1979) and Maina (2000) reviewed bird lungs and their operation, including the bellows-like role of the air sacs. Peter Scheid played a significant role in explaining the mechanism that engineers flow to be in only one direction in birds. Because there are no valves, this was mysterious for a long time, but, using soot suspended in air as well as glass models, he was able to show that the unidirectionality is contrived by the internal configuration of the airways, relying on a Venturi effect.

In a subsequent study, Schachner et al. (2013) used computer tomography to examine the internal anatomy of the excised lungs of both alligators and Nile crocodiles and used heated thermistor probes implanted into secondary bronchi to measure airflow. Anatomically, primary and secondary bronchi are similar in both species, with variation emerging at the level of the tertiary bronchi. Ventilating the lungs artificially they found that air flows towards the head in the dorsal bronchi and posteriorly in the central ventral bronchus, as in alligators (Fig. 7.13), and in birds. They proposed that, also as in alligators and birds, the tidal flow of air in the trachea is converted passively to a unidirectional flow through the parabronchi by the geometry at junctions between the primary and secondary bronchi providing aerodynamic valves.

This implies that crocodylian lungs may be very effective at gas exchange, which could explain why they can get by with smaller lungs in comparison with other reptiles; with an efficient gas exchange organ they would have been able to allow lung volume to be dictated by buoyancy requirements rather than by metabolic demands. The 'proof of the pudding', so to speak, will be in the performance of the lungs. In humans, there is a large ventilatory dead space and no special arrangements for alveoli to be exposed to particularly fresh air. Unidirectional airflow across the respiratory surface avoids this constraint. As we shall see in the section on gas exchange, the lungs of crocodylians apparently outperform human lungs in blood oxygenation.

Are crocs' respiratory systems over-engineered: a legacy from their cursorial ancestry?

All of this adds up to an extraordinarily competent ventilatory system for a reptile. Perry (1988) was puzzled that crocodylians, as comparatively inactive ectotherms whose activity is always regarded as being mainly supported anaerobically, have lungs that seem to be more complex than they need. Farmer and Carrier (2000b) observed how surprising it is that such a large proportion of the alligator musculo-skeletal system is devoted to maintaining ventilation, and that this highly derived, complex and competent respiratory system seems out of place in a primarily sit-and-wait predator with comparatively low metabolic rate. They wondered (Farmer and Carrier 2000a) if an 'explanation for this 'paradoxical assemblage of characters is that extant crocodylians may have inherited a cardio-pulmonary system from ancestors that were specialised for the ability to sustain vigorous locomotion', and drew attention to the early crocodylomorphs having been active terrestrial predators. Claessens (2004) noted that crocodylians 'seem to be able to use only a fraction of their gas exchange potential'. Additionally, their capacity for galloping (Chapter 4), a completely divided and highly evolved heart, and pressure differential between systemic and pulmonary circuits (Chapter 8), and highly specialised ventilatory and gas exchange systems all seem out of place in animals with 'a poor capacity for sustained vigorous terrestrial locomotion' (Farmer and Carrier 2000a). Farmer's and Sanders' (2010) recent discovery of a unidirectional flow of air through apparent parabronchi in alligator lungs only heightens the seeming paradox. They speculated that these highly evolved systems were 'a legacy from cursorial ancestors rather than an adaptation to a lifestyle as amphibious sit-and-wait predators'. Farmer (2010) uncovered an additional and intriguing piece of information about the functioning of alligator lungs. She found that with each beat of the heart there is a unidirectional net flow of air within the lung during periods of apnoea. Putting that together with the other information she developed a hypothesis that 'internal structures requisite for unidirectional flow were present in the common ancestors of birds and crocodylians and may have pre-adapted the lungs of archosaurs to function advantageously during the oxygen poor period of the early Mesozoic'.

Not mentioned by Farmer and Sanders but, as discussed in Chapter 10, Seymour *et al.* (2004) proposed that today's crocodylians have endothermic ancestry and that could well have depended upon the evolution of a highly efficient respiratory system.

Certainly there is a strong body of evidence that crocodylian ancestors were more capable of sustained vigorous activity than current crocs (Chapter 2) and, by extrapolation from other reptiles, there may have been too easy an acceptance of crocodylians being severely limited aerobically. Certainly one of their feeding strategies

is to sit and wait, but they are also active predators (Chapter 6) and it is becoming clearer that they are very mobile, making considerable journeys in surprisingly short times (Chapter 4). If 30–40% of the active metabolism of a 2–4 kg animal is aerobic and factorial aerobic scope (FAS) increases with body size (Wright 1986, see above), the implication is that the larger they become the more aerobic they are likely to be. It does seem as though the more we learn about crocodylians, the more they are revealed as being more competent aerobically than their current reputation allows.

GAS EXCHANGE; OXYGEN UPTAKE AND CARBON DIOXIDE EXCRETION

The first part of this section will review typical oxygen and CO_2 partial pressures at different steps of the 'oxygen cascade' from outside air to the tissues in different situations and, in less detail, the CO_2 cascade from the tissues out. The second part will discuss the way partial pressure dictates the binding of oxygen to haemoglobin, and the factors influencing it.

Does lung performance in crocodylians reflect a benefit from unidirectional air flow in lung parabronchi?

The answer seems to be 'yes' as far as equilibrating blood is concerned. In humans, blood oxygen comes into partial pressure equilibrium with the air in the alveoli at ~100 mmHg (see Box 7.2) compared with ~150 mmHg (20 kPa) in the outside air. This is about as high a level of oxygenation as can be expected in alveolar lungs with a tidal, bi-directional flow of air. In birds, unidirectional airflow through the parabronchi means that the perfusing blood 'sees' air at a higher oxygen partial pressure than it does in mammals, allowing more efficient oxygen uptake via a 'cross-current' arrangement between capillaries and airflow (see review by Maina 2000). It might be thought that this would show up as higher values of arterial Pa_{O_2}. However, Pa_{O_2} in birds is typically similar to mammals, a consequence of the relative flows of air and blood. Because of the shape of the oxygen equilibrium curve (see below), higher Pa_{O_2} would make little difference to the level of oxygenation of the blood, and the high blood flow can take advantage of the efficient cross-current arrangement of the anatomy.

What is the situation in crocodylians? Several studies have measured arterial oxygen partial pressures (Pa_{O_2}) very much higher than 100 mmHg (13.3 kPa), commonly 120 mmHg (16 kPa) and even higher, quite high enough to imply the operation of a cross-current or similar arrangement between air flow and blood flow. For example, Seymour *et al.* (1985) sampled blood from the carotid artery of freshly caught *C. porosus* (43 g to 7 kg) and measured blood gases at rest and following exercise to exhaustion (5 min period). In resting, undisturbed animals they found values of arterial oxygen (Pa_{O_2}) averaging ~102 mmHg (13.6 kPa), range 58–27 mmHg (at Pco_2 = 33 mmHg, 4.4 kPa). They noted that Pa_{O_2} did not change significantly with exercise, but there was considerable variability and their graph shows that most values in the first 30 min post exercise exceeded 100 mmHg (13.3 kPa) with several higher than 120 mmHg (16 kPa). Also, Hicks and White (1992), working on 500 g resting alligators, sampled blood from the left atrium (equivalent to blood leaving the lungs) and reported Pa_{O_2} values averaging 100 ± 10 mmHg (13 ± 1 kPa) but, from their graph, ranging from ~90 to 140 mmHg (12–19 kPa). More recently, Hartzler *et al.* (2006) monitored the recovery of alligators from being exercised to exhaustion on a treadmill. At the cessation of exercise, oxygen consumption and CO_2 excretion were elevated immediately, accompanied by an increase in both respiratory rate and tidal volume. Pa_{O_2} rose to more than 120 mmHg (16 kPa) and returned to lower values more typical of resting values (≈80 mmHg, 10.7 kPa) within ~30 min. None of the authors drew attention to these surprisingly high values of Pa_{O_2}. Averaging data often draws one's attention away from things that are really interesting! Clearly and in sharp contrast to mammals, the level of oxygenation in crocodylian blood varies over a wide range, with the highest values much higher than in humans.

Two things emerge. First, the higher values are comparable to arterial oxygen partial pressures in fishes with their well publicised (in every first year biology textbook) countercurrent gas exchange at the gills. This is strong evidence that the peak performance of crocodylian lungs is enhanced by the architecture of airflow direction in relation to blood flow. Second, the level of oxygenation of the arterial blood is not only very high, but it is variable and responsive to demand, such as during activity (Hartzler *et al.* 2006). Birds also show an increase in Pa_{O_2} when active (e.g. Millard *et al.* 1973, penguins; Butler *et al.* 1977, pigeons), although the increases are smaller than in alligators and some birds do not show it (e.g. Grubb *et al.* 1983, emus). Fish too increase Pa_{O_2} during activity. I once watched in fascination as Pa_{O_2} in the dorsal aorta of a cannulated Port Jackson shark, resting quietly in its experimental tank, started to rise conspicuously for no apparent reason until suddenly, with some muscular contractions of its flanks, a large defaecation occurred, after which Pa_{O_2} slowly returned to the previous level. In fish gills, there is an osmoregulatory cost to respiration because of water and ion movements between the blood and the water, so it makes sense to reduce oxygen uptake and operate at a lower Pa_{O_2} when oxygen demand is low. The compromise between oxygen uptake and osmoregulation in fishes was pointed out by Randall *et al.* (1972) and has been explored further by Gonzalez and Mcdonald (1994). Further research is needed to explain the range of Pa_{O_2} in crocodylians; it could be simply a consequence of intermittent ventilation at times of low oxygen demand. However, the possibility should not be overlooked that crocodylians are able to generate airflow through the parabronchi during periods of apnoea: perhaps this provides the explanation for movements of the flanks when they are submerged. Farmer's and Sanders' (2010) discovery prompts many questions and hints at new and exciting understanding. Despite the capacity of the lungs to provide high arterial oxygenation, it should be noted that high rates of oxygen uptake have not been recorded. The essential enigma remains: crocodylians seem to be over-engineered for their current lifestyles in many ways, relicts quite possibly of a cursorial and endothermic ancestry.

Gas exchange at rest

The crocodylian cyclical breathing (Fig. 7.8) shows up in the pattern of resting gas exchange. In both

C. porosus (Wright 1985) and *A. mississippiensis* (Hicks and White 1992) the highest levels of oxygen in both the lungs and the arterial blood occur at the end of the ventilatory period, then fall slowly during the ensuing apnoea. Apnoeic periods are quite variable in length, so reporting average values conveys only part of the story. Accordingly, Fig. 7.14 shows the approximate ranges of Po_2 and Pco_2 values that accompany the cyclical pattern of respiration in a resting crocodylian. The figure depicts the 'oxygen cascade' of a crocodylian at rest, visualising the diffusion gradients that drive gas flows between the separate compartments of the gas transport system: that is, the movement of oxygen into and CO_2 out of the cells.

Both Wright (1985) and Hicks and White (1992) noted a surprisingly large gradient of ~20 mmHg (2.7 kPa) between lung oxygen (Pa_{O_2}) and arterial oxygen (Pa_{O_2}), which seems at odds with efficient oxygen uptake. This is probably explained by these being resting values of Pa_{O_2}, which are lower than values typical of activity (see below). Another curious finding was that Pco_2 varies so little between arterial and venous blood (see Fig. 7.14). However, Jensen *et al.* (1998), apparently unaware of the earlier reports, predicted perceptively that there would be minimal change between arterial and venous Pco_2 in crocodylians because of the very strong Haldane effect (Grigg and Cairncross 1980) (Fig. 7.15). The strong Haldane effect also explains another curious observation about Pco_2 levels. One might expect that, as the period of apnoea progresses, Pco_2 would remain higher in the pulmonary artery than in the pulmonary vein and lungs, but the reverse is true (Wright 1985; Hicks and White 1992): the CO_2 remains bound to haemoglobin instead of dissolving in the plasma to be seen as a partial pressure. A similar 'reverse gradient' is seen during a diving apnoea (see below) in the turtle, *Chrysemys picta* (White *et al.* 1989). During ventilation, re-oxygenation of the haemoglobin drives off the CO_2 so that CO_2 excretion from the blood and the lungs is periodic. The Haldane effect will be discussed further in the section below on oxygen and CO_2 transport.

Gas exchange during digestion

Other activities in crocodylians are also accompanied by an increase in arterial oxygenation, to support increased oxygen consumption. Busk *et al.* (2000) reported that oxygen consumption in alligators increased approximately 4-fold during digestion of a large meal, with an increase in arterial oxygen (Pa_{O_2}) from an average of 60 (8 kPa), to ≈80 mmHg (10.7 kPa), the higher levels being maintained for several days during digestion.

Gas exchange during diving

When submerged, crocodylians have only the oxygen stores they take down with them (Chapter 9), so their 'exchange' of gases is with the gases in their lungs. The earliest studies of diving involved holding them under water for an hour, and sometimes more. Alligators forced to dive used up most of the oxygen in the lungs and blood, dropped heart rate to a few beats a minute and developed a

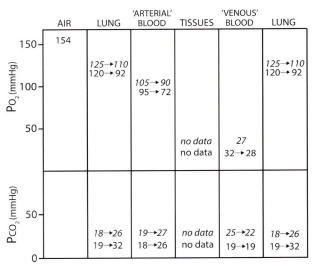

Fig. 7.14. (Upper panels) Indicative 'oxygen cascade' and (lower panels) CO_2 levels in the cyclical breathing pattern in a crocodylian, based on data from (upper values) A. mississippiensis (Hicks and White 1992) and (lower values) C. porosus (Wright 1985). Data are shown at the end of ventilation (i.e. start of apnoea) and at the end of a 5 min apnoea (end of arrow). Features to note are: the large Po_2 gradient of ~20 mmHg (2.7 kPa) from lung gas to 'arterial' blood; relative stability of 'venous' Po_2 during the cycle; and reversal during apnoea of the Pco_2 gradient between blood and lungs (see text). The 'arterial' and 'venous' values from alligators were sampled from the left and right atria, respectively. In C. porosus, femoral artery and jugular vein were sampled. (Note: 1 mmHg = 0.1333 kPa.)

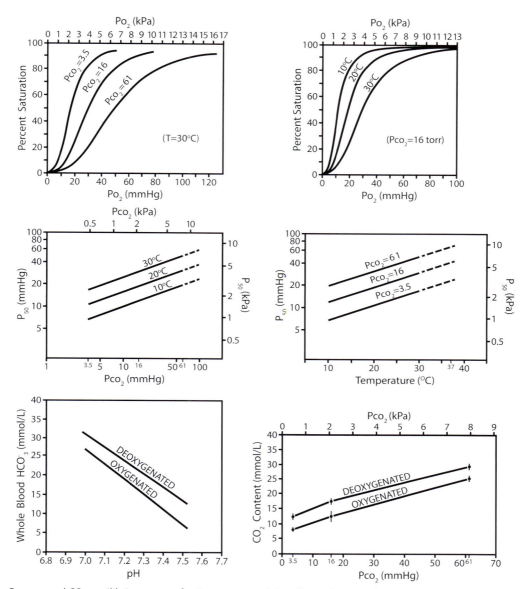

Fig. 7.15. *Oxygen and CO$_2$ equilibrium curves for* C. porosus *and the effects of Pco$_2$ and temperature (top), the effects of Pco$_2$ and temperature on oxygen affinity (P$_{50}$) (middle), buffering capacity of whole blood (bottom left) and CO$_2$ equilibrium curve (bottom right) showing the large Haldane effect (increased capacity of deoxygenated blood to bind CO$_2$). (From Grigg and Cairncross 1980, copyright © 1980, Elsevier).*

profound peripheral vasoconstriction and acidosis from build up of high CO$_2$ levels and high blood lactate (Andersen 1961). During the 1970s, it became clear that voluntary dives were typically much shorter and involved only modest cardiovascular adjustments and, subsequently, that most dives are conducted aerobically (reviewed in Chapter 9).

Wright (1985) monitored oxygen and CO$_2$ partial pressures in the lungs, femoral artery and jugular

vein of juvenile *C. porosus* (1.1–4.5 kg) at 25°C diving undisturbed in an experimental tank screened from the investigator. Most dives were short, the longest 13 min, and all were aerobic with no increase in plasma lactate. After submergence, oxygen partial pressures declined quite rapidly in both the lung and the femoral artery (from 100 down to 70 mmHg in 5 min) (13.3 to 9.3 kPa) and more slowly in the jugular vein (30 down to 23 mmHg) (4 to 3.1 kPa), a consequence of the shape of the oxygen equilibrium curve (see below and Fig. 7.15).

Wright found that partial pressures of CO_2 rose very slowly in the lungs with little or no change in P_{CO_2} in the artery or the vein as the dive progressed. Accordingly, the gradient from lung to blood was reversed compared with what might be expected. This was puzzling at the time, but is now recognised to be a consequence of the strong Haldane effect, associated with a lack of sensitivity to red cell organic phosphates and enhanced capacity to bind bicarbonate (see below). This means that, as blood oxygen contents falls, more CO_2 is taken up by haemoglobin. Loading properties of the blood of *C. porosus* are minimally decreased by falling pH, but sharply modified by increased P_{CO_2} (see below), so the small increases in P_{CO_2} in the blood and lungs during a dive leave the blood well able to load oxygen from the lungs. This is explored further below.

TRANSPORT OF GASES BY THE BLOOD

Just as in humans, oxygen and CO_2 are transported in the blood between the lungs and the working tissues. A conspicuous difference is that in crocs, as in other reptiles (and birds), the red cells (erythrocytes) containing the oxygen-transporting pigment haemoglobin are oval, not round, and are nucleated.

Oxygen binds reversibly to haemoglobin, loading oxygen at the lungs and giving it up to the working tissues.

Oxygen-carrying capacity

A small amount of oxygen is carried in solution – only ~2% in *C. porosus* – the remainder being bound reversibly to haemoglobin (1 g of Hb binds ~1.35 mL of oxygen). The average haemoglobin concentration of 96 juvenile and sub-adult *C. porosus* captured in the wild was 7.74 g/100 mL blood, which equates to 10.4 mL/100 mL = 10.4 vol% or 4.63 mmol/L (Grigg, unpublished observations). Because it is so easy to measure, haematocrit is sometimes used as a proxy for oxygen-carrying capacity. Haematocrit is the volume percentage of red cells in the blood, determined in a centrifuge, and often called 'packed cell volume', PCV. Grigg and Cairncross (1980) reported an average haematocrit of 24.8% (range 19.1–31.3) in the 96 wild caught *C. porosus*.

Such values are not unusual among crocodylians, although the database is not large. Jelkmann and Bauer (1980) found juvenile *Caiman crocodilus* with a haematocrit of 23.7% ± s.d. 5.7, with haemoglobin averaging 7.5 g/100mL and an oxygen capacity of 9.7 mL O_2/100mL.

Oxygen-carrying capacity is not necessarily fixed and can vary in response to conditions. For example, haematocrits and haemoglobins may decline in captivity, presumably because crocs in close captivity are less active. Eme *et al.* (2009) found that haematocrit increases as a response to training in alligators (see above) and embryonic alligators exposed to 17% oxygen increased haematocrit from 19% to 27% (and haemoglobin from 7.54 to 10.26 g/100 mL blood) (Warburton *et al.* 1995).

In passing, and now of mainly historical interest, we should note that it was mistakenly thought for a time that reptiles typically had large proportions of their haemoglobin in the oxidised form (methaemoglobin) and therefore non-functional. This would be somewhat bizarre. However, this misunderstanding is likely to have arisen as an artefact of applying the standard mammalian analytical assay to species with nucleated red cells, because cell debris may distort the results. High levels could be an indication of stress, however, as can occur in fish (Rufus Wells *pers. comm.*). Certainly methaemoglobin levels in healthy *C. porosus* are negligibly low (Gruca and Grigg 1980) and Jelkmann and Bauer (1980) found the same in *Caiman crocodilus*.

Oxygen equilibrium curves: effects of temperature, carbon dioxide and pH

The loading of oxygen onto haemoglobin is not directly proportional to its availability, but is described by the oxygen equilibrium curve (sometime called the oxygen dissociation curve).

The shape and position of this curve influence the uptake of oxygen at the lungs and its release to the tissues. The position of the curve is an expression of the affinity of the haemoglobin for oxygen, modified a bit by the cellular environment, and it is influenced by temperature, pH and CO_2. Data for sub-adult *C. porosus* are shown graphically in Fig. 7.15.

The oxygen affinity of blood is commonly represented in respiratory physiology by a single value, the P_{50}, which is the oxygen partial pressure (in mmHg) at which the blood is half saturated with oxygen. In *C. porosus* the P_{50} is ~24 mmHg (3.20 kPa) at 25°C and a Pco_2 of 20 mmHg (2.7 kPa), quite similar to human blood at human body temperature. The equilibrium curve can be linearised by the Hill equation, named for a famous British physiologist A.V. Hill, by plotting log (Y/(100–Y)) against log Po_2 (where Y = % saturation), in which case the slope of the line at P_{50}, designated '*n*' provides a measure of the extent to which the equilibrium curve is sigmoid. In *C. porosus* blood, *n* depends to some extent on temperature and Pco_2 but in the physiological range '*n*' approximates 2.7, essentially the same as in humans. The functional significance of sigmoidicity is that in the steep part of the curve, a large proportion haemoglobin-bound oxygen can be loaded or unloaded for a relatively small change in blood Po_2. In alligator blood, Weber and White (1986) found a lower Hill '*n*' value, 2.2, indicating a slightly less sigmoidal curve.

In most vertebrate bloods, higher body temperature shifts the curve to the right (i.e. to a lower affinity for oxygen, a higher P_{50}). This is particularly significant in ectotherms, where body temperatures typically vary daily and seasonally (Chapter 10): oxygen is more readily unloaded to meet increased respiratory demand at warmer temperatures. In *C. porosus* the effect of temperature is quite marked (Fig. 7.15). Weber and White (1986) found similar sensitivity in alligators.

In general, CO_2 and pH affect oxygen affinity independently and these can be differentiated experimentally into the 'CO_2 Bohr effect' and the 'fixed acid Bohr effect'. However, because CO_2 affects pH, their combined effect is normally referred to as the 'Bohr effect' (described in 1904 by Christian Bohr, the famous Danish physiologist and father of physicist and bomb maker Niels Bohr). In Crocodylia, the fixed acid Bohr effect is strikingly low and the CO_2 Bohr effect is strong (Grigg and Cairncross 1980) and these can be seen together as a 'crocodylian speciality' that is functionally significant. Crocs build up significant protons in their blood when very active, so insensitivity to low pH makes sense. Enhanced sensitivity to CO_2 is also adaptive. Accumulating CO_2 during a period

of apnoea, whether on land or associated with an aerobic dive, will right-shift the equilibrium curve and help unload oxygen to the tissues. When breathing resumes, the CO_2 can be blown off quickly and the rapid left-shift of the curve will increase the affinity of the blood for oxygen, facilitating reloading. If a dive is an active, hunting dive, which might require anaerobic support (see Chapter 9), rapid blow-off of CO_2 at the end of a dive would still be advantageous while, at the same time, insensitivity to low pH is advantageous because of the slow metabolic re-oxidation of accumulated lactate (Grigg and Gruca 1979; Seymour *et al.* 1985). There will be more on this topic below.

The effects of temperature and Pco_2 are shown graphically for *C. porosus* in Fig. 7.15. Combining them, the oxygen equilibrium curves for *C. porosus* can be described by a general descriptive equation:

$$\log P_{50} = 0.4163 + 0.02\ T°C + 0.3763 \log Pco_2$$

The same equation matches the alligator data from Weber and White (1986) quite well in the physiological range of Pco_2 20–30 mmHg (2.7–4.0 kPa) and at 25°C, predicting P_{50} to within ~1 mmHg (0.133 kPa).

Red cell organic phosphates

Red cell organic phosphates provide the most interesting oxygen transport story in crocodylians, because of the unusual way they modulate oxygen affinity. Haemoglobin has a very high intrinsic oxygen affinity, which, in most vertebrates, is lowered into the functional range by one of the red cell organic phosphates (RCOP). In humans and most other mammals, the relevant RCOP is 2,3 diphosphoglycerate (2,3 DPG). Without 2,3 DPG, the P_{50} of human blood would be ~16 mmHg (2.1 kPa), compared with 26 mmHg (3.5 kPa) at normal physiological conditions in a resting human. In birds, the active RCOP is inositol hexaphosphate (IHP) and in fishes it is ATP (ATP). Alligators, however, were found by Sullivan (1974) to have very low levels of RCOP and he suggested that they must have some other way to regulate blood oxygen affinity. Following that up, Bauer and Jelkman (1977) looked at blood from a specimen of

C. porosus from the Munich Zoo. They found that P_{50} of a solution of the haemoglobin (haemolysate) was insensitive to a range of RCOPs, but very sensitive to CO_2. In solution, P_{50} was right-shifted from 3.9 to 28.8 mmHg (0.5–3.8 kPa) under the influence of 40 mmHg (5.3 kPa) CO_2, approximating the situation in whole blood. They proposed that, in crocodiles, CO_2 plays a similar role to RCOP in right-shifting the oxygen equilibrium curve. Their report attracted the attention of *New Scientist* (15 Dec 1977) under the headline 'Crocodiles get their oxygen by a metabolic trick'. It is quite a neat trick, really, because CO_2 is a product of oxidative metabolism, and is then harnessed to lower the blood's oxygen affinity, helping to unload oxygen to the working tissues. Jelkmann and Bauer (1980) confirmed their finding in a study of *Caiman crocodilus* and, furthermore, found that much of the CO_2 is bound to haemoglobin as bicarbonate; thus bicarbonate parallels the action of RCOPs in crocodylians (Bauer *et al*. 1981). Low RCOP has been reported also in *C. johnstoni, C. novaeguineae, C. moreleti* and *C. niloticus* (Grigg and Gruca 1979), so it is likely to be crocodylian-wide and older than the split between crocodylids and alligatorids back in the Cretaceous. It is not a typical archosaur trait, however, because most birds rely on inositol pentaphosphate (IPP) to modulate oxygen affinity (Weber and Jensen 1988).

We can be sure that crocodylians lack their RCOP as a result of evolutionary specialisation because they have two different ones during embryonic development and lose them both by the time they hatch (Grigg *et al*. 1993) (Fig. 7.16). As embryos, they also have two distinct haemoglobins. Early in development, they have an embryonic haemoglobin that is sensitive to ATP, the RCOP which is present in high levels initially and declines throughout embryonic life. Embryonic haemoglobin is replaced progressively by adult haemoglobin, and there is a surge of 2,3-DPG produced not long before hatching, then lost. This is all puzzling because the oxygen affinity of the adult haemoglobin is unaffected by either ATP or 2,3-DPG, so the significance of this pre-hatching surge remains unknown.

Some mammals too have strikingly low levels of RCOP, specifically Felidae (cat, lion), Lemuridae (lemurs), Cervidae (deer, elk), Antilocapridae

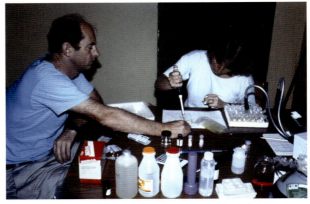

Fig. 7.16. *Rufus Wells and Lyn Beard in a makeshift laboratory at the Edward River Crocodile Farm, Pormpuraaw, North Queensland, assaying red cell organic phosphates in embryonic blood from* C. porosus. *We also took an Aminco Haemoscan to determine oxygen equilibrium curves. (Photo GCG)*

(pronghorn), some Bovidae (goats, sheep, cattle) and Giraffidae (giraffe) (Bunn *et al*. 1974; Kay 1977). Kay observed that these low-RCOP mammals typically expend energy in short bursts: the sprinters, dashers and pouncers, in contrast to endurance performers which have higher, 'normal' levels of 2,3 DPG. Crocodylians fit that pattern. Furthermore, Duhm (1976) found that stripping the 2,3 DPG from human haemoglobin enhances the CO_2 Bohr effect and reduces the 'fixed-acid' Bohr effect. Accordingly, Grigg and Gruca (1979) proposed that the loss of red cell organic phosphates in crocodylians at hatching may be an adaptation to cope with metabolic acidosis, which arises after a burst of activity and could compromise oxygen uptake.

Linking oxygen consumption and oxygen transport

These two functions are linked by a variant of the Fick Equation, i.e.

Oxygen consumption = cardiac output × (arterial oxygen content – mixed venous content)

or

Oxygen consumption = (heart rate × stroke volume) × (A-V difference)

Cardiac output can be calculated from a knowledge of oxygen consumption, the respiratory properties of blood for a species, the haemoglobin concentration of the individual in question and the oxygen and CO_2 partial pressures (or pH) of arterial and mixed venous blood. Stroke volume may be calculated from heart rate. A graphic representation of this for a 1 kg *C. porosus* at rest is shown in Fig. 7.17. Note that the large oxygen reserve can be tapped to increase arterio-venous difference in oxygen content (A-V difference). To cope with the need for increased oxygen supply during activity, an increase in aerobic metabolism by the mitochondria draws oxygen from the capillaries, lowering Pv_{O_2} and increasing the A-V difference.

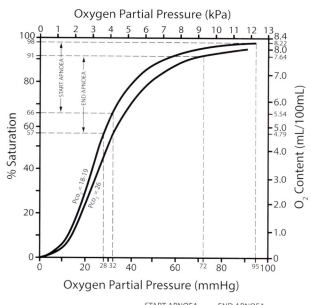

	START APNOEA	END APNOEA
$C_{a_{O_2}}$ (mL/100 mL)	8.22	7.64
$C_{v_{O_2}}$ (mL/100 mL)	5.54	4.79
A-V difference (mL/100mL)	2.68	2.85
Cardiac output (mL/min)	24.6	22.4
Heart rate (bpm)	20	17
Stroke volume (mL)	1.2	1.3

Fig. 7.17. *Quantification of the link between gas exchange and oxygen transport exemplified by a respiratory cycle with an apnoeic period of 5 min of a 1 kg* C. porosus *resting at 25°C with an oxygen-carrying capacity at 100% saturation of 8.39 vol% (data drawn from Wright 1985). Oxygen and CO_2 partial pressures are drawn from Fig. 7.14 and oxygen capacities from the relevant oxygen equilibrium curve. In this short apnoeic interval, oxygen consumption can be assumed to be maintained, 38.4 mL/h in this example, heart rate can be expected to fall by ~14%, from perhaps 20 to 17 bpm and pulmonary by-pass shunting, though possible, is unlikely. For the sake of the exercise, the small amount of oxygen carried in solution in the plasma is ignored. If the apnoea were to be accompanied by submergence, the pattern would be essentially identical (see Chapter 9).*

The combined effect of an increased A-V difference and an increased heart rate (which increases cardiac output) provides a greater rate of oxygen supply.

In Fig. 7.17, showing a 1 kg *C. porosus* at rest, the cardiac output is calculated to be 22–25 mL/min. This can be compared with averages of 18.4–26.6 mL/kg/min reported in three small alligators by Powell and Gray (1989) and an estimate by Robertson and Smith (1979) of 70 mL/min in a 3 kg alligator. To put this into context, using the resting oxygen consumption extrapolation in Fig. 7.2 and assuming similar gas exchange and oxygen transport values, an 85 kg croc can be expected to have a cardiac output of ~450 mL/min (at 25°C), compared with an 85 kg human of something like 6000 mL/min (at 37°C, but the difference is due to much more than temperature of course). A 1000 kg croc might have a resting cardiac output of ~2700 mL/min (~160 L/h), still much less than a human weighing 8% of the croc's body mass.

Transport of carbon dioxide

Carbon dioxide is carried in the plasma in solution (oxygen is ~30 times more soluble than CO_2), in the red cells reversibly bound with water to form bicarbonate ions and also in combination with haemoglobin as carbamino compounds. In crocodylians, however, there is a unique and very significant add-on (Jensen *et al.* 1998): because of a few amino acid substitutions, bicarbonate ions are bound to the deoxygenated haemoglobin molecules at the site normally occupied by RCOPs. Thus, there is oxygenation-linked bicarbonate binding which explains the strong effect that CO_2 has on oxygen affinity (see above). It also explains the large effect of oxygenation state on bicarbonate ion binding (the Haldane effect) (Fig. 7.15). Jensen *et al.* (1998) reported a Haldane effect of 0.72 mmol CO_2/mmol Hb in alligator blood, which is about twice that of human blood but a little less than the 0.93 mmol CO_2/mmol Hb we found in *C. porosus* (Grigg and Cairncross 1980).

THE CIRCULATORY SYSTEM

The cellular biochemistry that supplies energy for biological work depends upon the cells having a good blood supply to deliver nutrients and hormones and remove CO_2 and ammonia. Apart from the heart, which is unique (Chapter 8), the crocodylian circulatory system is fairly typical of reptiles, including the presence of a renal portal system. Useful descriptions are to be found in Reese (1914, 1915) and Chiasson (1962), but each has shortcomings and there are some inconsistencies between them. Because the anatomy of the heart and the central arterial system is very different from typical reptiles, it has received an enormous amount of interest and attention from researchers and is the focus of Chapter 8. A major difference is that in typical reptiles left and right aortas both receive blood from the left ventricle and join symmetrically above the lungs to form the main (dorsal) aorta. In crocodylians, however, only the right systemic arch arises from the left ventricle. It emerges from the heart as a very large vessel, sometimes called the innominate artery, which normally carries oxygenated blood and branches early to three distinct vessels (Fig. 7.18): the common carotid; the right subclavian; and the vessel that sweeps laterally upwards around the lungs to become the dorsal aorta, which supplies the posterior body and viscera. The left systemic arch is present but much reduced and, paradoxically, arises from the right ventricle alongside the pulmonary arch. It too sweeps dorsally but sends a large vessel to supply the gut before joining the right aorta asymmetrically (Fig. 7.18 and Chapter 8). It is convenient to describe separately the blood supply to the lungs and the anterior and posterior parts of the body.

PULMONARY CIRCULATION

Deoxygenated blood flows to the lungs via the left and right pulmonary arteries. These branch from the pulmonary arch, which exits from the right ventricle alongside the left aorta. The entry to the pulmonary arch is guarded by one of the most extraordinary structures in any blood vascular system in any animal: the so called cog-wheel valve, which is under muscular and nervous control to meter the flow of blood into the lungs. The structure and significance of this will be discussed in more detail in Chapter 8. Past the cog-wheel valve, the pulmonary arch widens considerably into a thick-

walled triangular vessel, which, presumably, helps to smooth the flow of blood into the pulmonary arteries and protects the delicate capillaries in the lungs against any surge in pressure. This function is similar to the bulbus arteriosus in the ventral aorta of teleost fish: a distensible and elastic reservoir that smooths the flow of blood into the gills. This unusual feature of the crocodylian pulmonary blood supply and possible function seem to have attracted little attention by physiologists so far.

ANTERIOR SYSTEMIC CIRCULATION (FIG. 7.18)

Anterior arterial circulation is supplied by two subclavian arteries and the common carotid. Each subclavian has a different origin. The right subclavian branches directly from the right aorta (innominate) and, early on, gives branches to the oesophagus, thyroid, rib cage, thoracic vertebrae and a large vessel, the collateralis colli (right side) (see below), which runs forwards to the lower side of the head. Past the origin of the collateralis colli, the subclavian enters the shoulder region, gives off the subscapular (to shoulder skin and muscles) and the thoracic (to shoulder muscles and posterior forelimb) and continues into the forearm as its major blood supply, the brachial artery. The left subclavian branches from the common carotid (Fig. 7.18), but otherwise branches similarly to the right subclavian, including the collateralis colli (left side), which runs forward beside the oesophagus and trachea, interconnects with the common carotid in a complex way and, like the right collateralis colli, sends branches to the lower side of the head: that is, the jaw muscles, larynx, lower jaw (mandibular artery) and tongue (lingual artery).

The common carotid artery is unpaired. It runs forward as a separate branch from the innominate and, past where it gives off the left subclavian, interconnects in a complex way with the collateralis colli and provides blood to the cervical vertebrae and much of the upper parts of the head, including some of the jaw muscles, the eyes and eye muscles, ears, nasal cavity, upper jaw and, via the internal carotid, the brain.

The anterior venous drainage is as shown in Fig. 7.18. Little more needs to be said beyond what

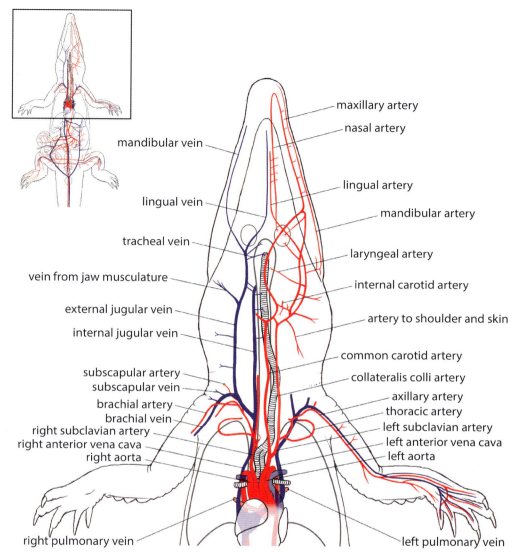

Fig. 7.18. *Anterior circulatory system of a crocodylian, from sketches by June Jeffery of a dissection of* C. porosus *by Christine Lehmann and Gordon Grigg and with reference to Reese (1914, 1915), Chiasson (1962) and Franklin and Grigg (1993).*

is obvious from the figure. The anatomy of the veins anteriorly is essentially symmetrical. The skin and muscles of the shoulder, neck, tongue, lower jaw and the massive jaw muscles are drained by the paired external jugulars, which, at the level of the forelimbs, receive blood from the shoulder and upper arm (subscapular vein) and the rest of the forearm (brachial vein). The upper part of the head, including the cranium and the brain, are drained by the internal carotids, which run parallel to the oesophagus and trachea, receive several smaller inflows and then join posteriorly with the external jugular. These vessels empty into left and right anterior vena cavae, which unite in the sinus venosus from which venous blood enters the right atrium.

POSTERIOR SYSTEMIC CIRCULATION (FIG. 7.19)

The right aorta is larger in diameter than the left and, at their junction, dorsal to the lungs, the anastomosis is muscular and under nervous and pharmacological control (Chapter 8). The left aorta appears well placed to be a major source of blood for the gut, via the coeliac artery, during digestion (Fig. 7.19) and, indeed, Axelsson *et al.* (1991) and Farmer *et al.* (2008) have shown that blood flow into the gut via

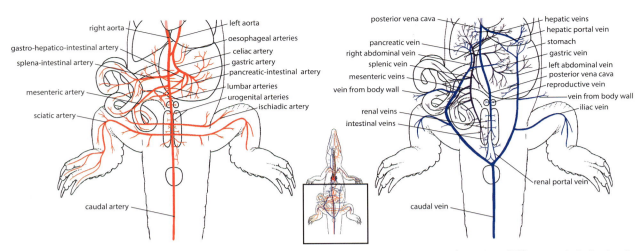

Fig. 7.19. *(Left) Posterior arterial and (right) venous system of a crocodylian, based on Reese's dissection of* Alligator mississippiensis *(after Reese 1914, 1915).*

the left aorta increases markedly during digestion. Whether the source of that blood is the right aorta, as suggested by Farmer and colleagues, or the left ventricle via the foramen of Panizza (Axelsson) is so far unresolved (discussed in Chapter 8). The coeliac artery supplies blood to the posterior oesophagus, stomach, spleen, pancreas, liver and anterior part of the small intestine. The right aorta, known as the dorsal aorta past its convergence with the left dorsal to the lungs, carries most of the blood that perfuses the rest of the posterior body: skin, musculature and the posterior length of the gut, plus other organs within the body cavity (Fig. 7.19). Distal to the confluence, a series of short lateral branches supply the body wall in the thoracic and lumbar region, while most of both the small and large intestines are perfused by the mesenteric artery. Further posteriorly, several pairs of short urogenital arteries supply the reproductive organs and kidneys. Next is a pair of ischiadic arteries, which supply blood to the ventral body wall, anterior of the thigh and pelvis (hence the name: relevant to the ischium). The hindlimbs receive blood from a pair of iliac arteries each of which divides into femoral and external iliacs to supply the anterior and posterior of the thigh, respectively. Past the knee, the femoral artery divides to tibial and fibular arteries supplying the anterior and posterior of the distal limb.

The posterior venous system is more interesting, having a renal portal system in addition to the hepatic portal system, as is typical of non-mammalian vertebrates (Figs 7.19, 11.7). Portal veins are veins that begin and end in capillary beds, neither arising from nor discharging directly to the heart. Hepatic portal veins drain blood from the absorptive regions of the gut, the small and large intestine particularly (Chapter 6), collecting the products of digestion, mostly sugars, fatty acids and amino acids, and transporting them to the liver. The hepatic portal vein also receives blood from the spleen and pancreas. The liver is drained by a pair of hepatic veins that join the sinus venosus next to the posterior vena cava. Renal portal veins drain venous blood from the posterior of the body and divide into capillary beds bathing the kidney tubules (Chapter 11). They are characteristic of fishes, amphibians, reptiles and birds, but are lacking in mammals, whose kidneys receive blood only from the dorsal aorta. Blood leaving the kidneys returns to the heart via a major posterior vein, the posterior vena cava (or postcaval vein), which traverses ventrally across the right lobe of the liver to deliver blood to the sinus venosus and, so, into the right atrium. Paired ventral abdominal veins (called epigastric veins by Reese) complete the posterior venous drainage. Anteriorly they join the hepatic portal vein. Posteriorly they join the renal portal veins and the iliac veins and, in between, they receive small branches from the stomach, the body wall and skin, the hindlimbs, the pelvis and the tail. They are assumed to be homologous with the unpaired ventral abdominal vein in amphibians, which provides an alternative pathway to the renal portal veins for venous return from the posterior body, emptying anteriorly into the hepatic portal system.

Venous systems are more variable between individuals than arterial systems, and are likely to be variable between crocodylian species as well.

LYMPHATIC SYSTEM

Crocodylians have an extensive lymphatic system, described in detail by Ottaviani and Tazzi (1977) in an extensive review of reptilian lymphatics. The lymphatic system serves to return to the blood system the plasma that accumulates in the tissues' interstitial fluid by filtration across the capillaries. It also has a role as a component of the immune system. The crocodylian lymphatic system is the most complex and modified among the reptiles. It lacks the great sinuses that are typical of other reptiles and has finer, more plexiform trunks, which are generally similar to those of mammals. The flow of lymph is driven largely by hydrostatic pressure.

REFERENCES

Andersen HT (1961) Physiological adjustments to prolong diving in the American alligator, *Alligator mississippiensis*. *Acta Physiologica Scandinavica* **53**, 23–45.

Axelsson M, Fritsche R, Holmgren S, Grove DJ, Nilsson S (1991) Gut blood flow in the estuarine crocodile, *Crocodylus porosus*. *Acta Physiologica Scandinavica* **142**, 509–516.

Baldwin J, Seymour RS, Webb GJW (1995) Scaling of anaerobic metabolism during exercise in the estuarine crocodile (*Crocodylus porosus*). *Comparative Biochemistry and Physiology. A. Comparative Physiology* **112**, 285–293.

Bauer C, Jelkmann W (1977) Carbon dioxide governs the oxygen affinity of crocodile blood. *Nature* **269**, 825–827.

Bauer C, Forster M, Gros C, Mosca A, Perrella M, Rollema SR, *et al.* (1981) Analysis of bicarbonate binding to crocodilian hemoglobin. *Journal of Biological Chemistry* **256**, 8429–8435.

Benedict FG (1932) *The Physiology of Large Reptiles: with Special Reference to the Heat Production of Snakes, Tortoises, Lizards and Alligators*. Carnegie Institute, Washington DC.

Bennett AE (1982) The energetics of reptilian activity. In *Biology of the Reptilia. Volume 13.* (Eds C Gans and FH Pough), pp. 155–201. Academic Press, London.

Bennett AF, Seymour RS, Bradford DF, Webb GJW (1985) Mass dependence of anaerobic metabolism and acid-base disturbance during activity in the salt-water crocodile, *Crocodylus porosus*. *Journal of Experimental Biology* **118**, 161–171.

Bishop CM (1999) The maximum oxygen consumption and aerobic scope of birds and mammals: getting to the heart of the matter. *Proceedings. Biological Sciences* **266**, 2275–2281.

Bunn FH, Seal US, Scott AF (1974) The role of 2,3-diphosphoglycerate in mediating hemoglobin function of mammalian red cells. *Annals of the New York Academy of Sciences* **241**, 498–512.

Busk M, Overgaard J, Hicks JW, Bennett AF, Wang T (2000) Effects of feeding on arterial blood gases in the American alligator *Alligator mississippiensis*. *Journal of Experimental Biology* **203**, 3117–3124.

Butler PJ, West NH, Jones DR (1977) Respiratory and cardiovascular responses of the pigeon to sustained level flight in a wind tunnel. *Journal of Experimental Biology* **71**, 7–26.

Campbell HA, Watts ME, Sullivan S, Read MA, Choukroun S, Irwin SR, *et al.* (2010) Estuarine crocodiles ride surface currents to facilitate long-distance travel. *Journal of Animal Ecology* **79**, 955–964.

Chiasson RB (1962) *Laboratory Anatomy of the Alligator*. WC Brown, Dubuque, Iowa.

Claessens LPAM (2004) Archosaurian respiration and the pelvic girdle aspiration breathing of crocodyliforms. *Proceedings of the Royal Society of London. Series B, Biological Sciences* **271**, 1461–1465.

Claessens LPAM (2009) A cineradiographic study of lung ventilation in *Alligator mississippiensis*. *Journal of Experimental Zoology* **311A**, 563–585.

Coulson RA, Hernandez T (1983) *Alligator Metabolism: Studies on Chemical Reactions* in vivo. Pergamon Press, New York.

Duhm J (1976) Dual effect of 2,3-diphosphoglycerate on the Bohr effects of human blood. *Pflügers Archive – European Journal of Physiology* **363**, 55–60.

Duncker H-R (1978) General morphological principles of amniotic lungs. In *Respiratory Function in Birds, Adult and Embryonic.* (Ed. J Piiper) pp. 2–15. Springer-Verlag, Berlin.

Eme J, Owerkowicz T, Gwalthney J, Blank JM, Rourke BC, Hicks JW (2009) Exhaustive exercise training enhances aerobic capacity in American alligator (*Alligator mississippiensis*). *Journal of Comparative Physiology. B, Biochemical, Systemic, and Environmental Physiology* **179**, 921–931.

Emshwiller MG, Gleeson TT (1997) Temperature effects on aerobic metabolism and terrestrial locomotion in American alligators. *Journal of Herpetology* **31**, 142–147.

Farmer CG (2010) The provenance of alveolar and parabronchial lungs: insights from paleoecology and the discovery of cardiogenic, unidirectional airflow in the American alligator (*Alligator mississippiensis*). *Physiological and Biochemical Zoology* **83**, 561–575.

Farmer CG, Carrier DR (2000a) Ventilation and gas exchange during treadmill locomotion in the American alligator (Alligator mississippiensis). *Journal of Experimental Biology* **203**, 1671–1678.

Farmer CG, Carrier DR (2000b) Pelvic aspiration in the American alligator (*Alligator mississippiensis*). *Journal of Experimental Biology* **203**, 1671–1678.

Farmer CG, Sanders K (2010) Unidirectional airflow in the lungs of alligators. *Science* **327**, 338–340.

Farmer CG, Uriona TJ, Steenblik M, Olsen D, Sanders K (2008) The right-to-left shunt of crocodilians serves digestion. *Physiological and Biochemical Zoology* **81**(2), 125–137.

Franklin CE, Grigg GC (1993) Increased vascularity of the lingual salt glands of the Estuarine Crocodile, *Crocodylus porosus* kept in hyperosmotic salinity. *Journal of Morphology* **218**, 143–151.

Gans C, Clark B (1976) Studies on ventilation of *Caiman crocodilus* (Crocodilia: Reptilia). *Respiration Physiology* **26**, 285–301.

Gatten RE Jr (1980) Metabolic rates of fasting and recently fed spectacled caimans (*Caiman crocodilus*). *Herpetologica* **36**, 361–364.

Gienger CM, Tracy CR, Brien ML, Manolis SC, Webb GJW, Seymour RS, *et al.* (2011) 376 Energetic costs of digestion in Australian crocodiles. *Australian Journal of Zoology* **59**, 416–421.

Glazier DS (2010) Activity affects intraspecific body-size scaling of metabolic rate in ectothermic animals. *Journal of Comparative Physiology. B, Biochemical, Systemic, and Environmental Physiology* **179**, 821–828.

Gonzalez R, Mcdonald D (1994) The relationship between oxygen uptake and ion loss in fish from diverse habitats. *Journal of Experimental Biology* **190**, 95–108.

Grigg GC (1978) Metabolic rate, RQ and Q_{10} in *Crocodylus porosus* and some generalisations about low RQ in reptiles. *Physiological Zoology* **51**, 354–360.

Grigg GC, Cairncross M (1980) Respiratory properties of the blood of *Crocodylus porosus*. *Respiration Physiology* **41**, 367–380.

Grigg GC, Gruca M (1979) Possible adaptive significance of low red cell organic phosphates in crocodiles. *Journal of Experimental Zoology* **209**, 161–167.

Grigg GC, Wells RMG, Beard LA (1993) Allosteric control of oxygen binding by haemoglobin during embryonic development in the crocodile *Crocodylus porosus:* The role of red cell organic phosphates and carbon dioxide. *Journal of Experimental Biology* **175**, 15–32.

Grubb BR, Jorgensen DD, Conner M (1983) Cardiovascular changes in the exercising emu. *Journal of Experimental Biology* **104**, 193–201.

Gruca M, Grigg GC (1980) Methemoglobin reduction in crocodile blood: are high levels of MetHb typical of healthy reptiles? *Journal of Experimental Zoology* **213**, 305–308.

Hancock TV, Gleeson TT (2008) Contributions to elevated metabolism during recovery:

Dissecting the excess postexercise oxygen consumption (EPOC) in the Desert Iguana (*Dipsosaurus dorsalis*). *Physiological and Biochemical Zoology* **81**, 1–13.

Hartzler LK, Munns SL, Bennett AF, Hicks JW (2006) Recovery from an activity-induced metabolic acidosis in the American alligator, *Alligator mississippiensis. Comparative Biochemistry and Physiology. A. Comparative Physiology* **143**, 368–374.

Hicks JW (2002) The physiological and evolutionary significance of cardiovascular shunting patterns in reptiles. *News in Physiological Sciences* **17**, 241–245.

Hicks JW, White FN (1992) Pulmonary gas exchange during intermittent ventilation in the American alligator. *Respiration Physiology* **88**, 23–36.

Hochachka PW, Mommsen TP (1983) Protons and anaerobiosis. *Science* **219**(4591), 1391–1397.

Huxley TH (1882) On the respiratory organs of *Apteryx. Proceedings of the Zoological Society of London* **1882**, 560–569.

Jackson DC, Andrade DV, Abe AS (2003) Lactate sequestration by osteoderms of the broad-nose caiman, *Caiman latirostris*, following capture and forced submergence. *Journal of Experimental Biology* **206**, 3601–3606.

Jelkmann W, Bauer C (1980) Oxygen binding properties of caiman blood in the absence and presence of carbon dioxide. *Comparative Biochemistry and Physiology A* **65**, 331–336.

Jensen FB, Wang T, Jones DR, Brahm J (1998) Carbon dioxide transport in alligator blood and its erythrocyte permeability to anions and water. *The American Journal of Physiology* **274**, R661–R671.

Kay FR (1977) 2,3-Diphosphoglycerate, blood oxygen dissociation and the biology of mammals. *Comparative Biochemistry and Physiology A* **57**, 309–316.

Killen SS, Costa I, Brown JA, Gamperl AK (2007) Little left in the tank: metabolic scaling in marine teleosts and its implications for aerobic scope. *Proceedings. Biological Sciences* **274**, 431–438.

Kirshner D (1985) Buoyancy control in the estuarine crocodile, *Crocodylus porosus* Schnieder. PhD thesis. University of Sydney.

Kleiber M (1932) Body size and metabolism. *Hilgardia* **6**, 315–353.

Lance VA, Elsey RM, Coulson RA (1993) Biological activity of alligator, avian, and mammalian insulin in juvenile alligators: plasma glucose and amino acids. *General and Comparative Endocrinology* **89**(2), 267–275.

Maina JN (2000) What it takes to fly: the structural and functional respiratory refinements in birds and bats. *Journal of Experimental Biology* **203**, 3045–3064.

McCue MD, Bennett AF, Hicks JW (2005) The effect of meal composition on specific dynamic action in Burmese pythons (*Python molurus*). *Physiological and Biochemical Zoology* **78**, 182–192.

Millard RW, Johansen K, Milsom WK (1973) Radio-telemetry of cardiovascular responses to exercise and diving in penguins. *Comparative Biochemistry and Physiology A* **46-A**, 227–240.

Munns SL, Frappell PB, Evans BK (1998) The effects of environmental temperature, hypoxia, and hypercapnia on the breathing pattern of saltwater crocodiles (*Crocodylus porosus*). *Physiological Zoology* **71**, 267–273.

Munns SL, Hartzler LK, Bennett AF, Hicks JW (2005) Terrestrial locomotion does not constrain venous return in the American alligator, *Alligator mississippiensis. Journal of Experimental Biology* **208**, 3331–3339.

Naifeh KH, Huggins SE, Hoff HE, Hugg TW, Norton RE (1970) Respiratory patterns in crocodilian reptiles. *Respiration Physiology* **9**, 21–42.

Ottaviani G, Tazzi A (1977) The lymphatic system. In *Biology of the Reptilia Volume 6.* (Eds C Gans and TS Parsons) pp. 315–462. Academic Press, New York.

Owerkowicz T, Baudinette RV (2008) Exercise training enhances aerobic capacity in juvenile estuarine crocodiles (*Crocodylus porosus*). *Comparative Biochemistry and Physiology. Part A, Molecular & Integrative Physiology* **150**, 211–216.

Perry SF (1988) Functional morphology of the lungs of the Nile crocodile *Crocodylus niloticus*: non-respiratory parameters. *Journal of Experimental Biology* **134**, 99–117.

Perry SF (1989) Morphometry of crocodilian lungs. *Fortschritte der Zoologie* **35**, 546–549.

Perry SF (1990) Gas exchange strategy in the Nile crocodile: a morphometric study. *Journal of Comparative Physiology. B, Biochemical, Systemic, and Environmental Physiology* **159**, 761–769.

Perry SF (1998) Lungs: comparative anatomy, functional morphology and evolution. In *Biology of the Reptilia* (Eds C Gans and AS Gaunt) pp. 1–92. Society for the Study of Amphibians and Reptilians, New York.

Powell FL, Gray AT (1989) Ventilation–perfusion relationships in alligators. *Respiration Physiology* **78**, 83–94.

Randall DJ, Baumgarten D, Malyusz M (1972) The relationship between gas and ion transport across the gills of fishes. *Comparative Biochemistry and Physiology. A. Comparative Physiology* **41**, 629–637.

Rathke H (1866) *Untersuchungen über die Entwickelung und den Körperbau der Krokodile.* Wilhelm von Wittich, Braunschweig, Germany.

Read MA, Grigg GC, Irwin SR, Shanahan D, Franklin CE (2007) Satellite tracking reveals long distance coastal travel and homing by translocated Estuarine Crocodiles, Crocodylus porosus. *PLoS ONE* **9**, 1–5.

Reese AM (1914) The vascular system of the Florida alligator. Proceedings of the Academy of Natural Sciences of Philadelphia 1914, 413–425.

Reese AM (1915) *The Alligator and Its Allies.* GP Putnam's Sons, New York.

Robergs RA, Ghiasvand F, Parker D (2004) Biochemistry of exercise-induced metabolic acidosis. *American Journal of Physiology. Regulatory, Integrative and Comparative Physiology* **287**, R502–R516.

Robertson SL, Smith EN (1979) Thermal indications of cutaneous blood flow in the American alligator. *Comparative Biochemistry and Physiology A* **62**, 569–572.

Romão MF, Santos ALQ, Lima FC, De Simone SS, Silva JMM, Hirano LQ, *et al.* (2011) Anatomical and topographical description of the digestive system of *Caiman crocodilus* (Linnaeus 1758), *Melanosuchus niger* (Spix 1825) and *Paleosuchus palpebrosus* (Cuvier 1807). *International Journal of Morphology* **29**(1), 94–99.

Sanders RK, Farmer CG (2012) The pulmonary anatomy of *Alligator mississippiensis* and its similarity to the avian respiratory system. *Anatomical Record (Hoboken, N.J.)* **295**(4), 699–714.

Santos CM, Abidu-Figueiredo M, Teixeira MJ, Nascimento AA, Sales A (2011) Light microscopic and immunohistochemical study of the trachea of the broad-snouted caiman. *Veterinarni Medicina* **56**(1), 48–54.

Schachner ER, Hutchinson JR, Farmer C (2013) Pulmonary anatomy in the Nile crocodile and the evolution of unidirectional airflow in Archosauria. *PeerJ* **1**, e60.

Scheid P (1979) Mechanisms of gas exchange in bird lungs. *Reviews of Physiology, Biochemistry and Pharmacology* **86**, 137–186.

Schmidt-Nielsen K (1972) *How Animals Work.* Cambridge University Press, Cambridge, UK.

Secor SM (2009) Specific dynamic action: a review of the postprandial metabolic response. *Journal of Comparative Physiology. B, Biochemical, Systemic, and Environmental Physiology* **179**, 1–56.

Seymour RS, Bennett AF, Bradford DF (1985) Blood gas tensions and acid-base regulation in the salt-water crocodile, *Crocodylus porosus*, at rest and after exhaustive exercise. *Journal of Experimental Biology* **118**, 143–159.

Seymour RS, Bennett-Stamper CL, Johnston S, Carrier DR, Grigg GC (2004) Evidence of endothermic ancestors of crocodiles at the stem of Archosaur evolution. *Physiological and Biochemical Zoology* **77**(6), 1051–1067.

Seymour RS, Gienger CM, Brien ML, Tracy CR, Manolis SC, Webb GJ, *et al.* (2013) Scaling of standard metabolic rate in estuarine crocodiles *Crocodylus porosus*. *Journal of Comparative*

Physiology. B, Biochemical, Systemic, and Environmental Physiology **183**(4), 491–500.

Stacy BA, Whitaker N (2000) Hematology and blood biochemistry of captive mugger crocodiles (*Crocodylus palustris*). *Journal of Zoo and Wildlife Medicine* **31**(3), 339–347.

Sullivan B (1974) Reptilian hemoglobins. In *Chemical Zoology IX, Amphibia and Reptilia.* (Eds BT Scheer and M Florkin). pp. 377–398. Academic Press, New York.

Uriona TJ, Farmer CG (2006) Contribution of the diaphragmaticus muscle to vital capacity in postprandial American alligators (*Alligator mississippiensis*). *Journal of Experimental Biology* **209**(21), 4313–4318.

Uriona TJ, Farmer CG (2008) Recruitment of the *diaphragmaticus, ischipubis*, and other respiratory muscles to control pitch and roll in the American alligator (*Alligator mississippiensis*). *Journal of Experimental Biology* **211**(7), 1141–1147.

Warburton SJ, Hastings D, Wang T (1995) Responses to chronic hypoxia in embryonic alligators. *Journal of Experimental Zoology* **273**, 44–50.

Weber RE, Jensen FB (1988) Functional adaptations in hemoglobins from ectothermic vertebrates. *Annual Review of Physiology* **50**, 161–179.

Weber RE, White FN (1986) Oxygen binding in alligator blood related to temperature, diving and 'alkaline tide'. *The American Journal of Physiology* **251**, R901–R908.

Weibel ER, Bacigalupe LD, Schmitt B, Hoppeler H (2004) Allometric scaling of maximal metabolic rate in mammals: Muscle aerobic capacity as determinant factor. *Respiratory Physiology & Neurobiology* **140**, 115–132.

White CR (2011) Allometric estimation of metabolic rates in animals. *Comparative Biochemistry and Physiology. A. Comparative Physiology* **158**, 346–357.

White FN, Hicks JW, Ishimatsu A (1989) Relationship between respiratory states and intracardiac shunts in turtles. *The American Journal of Physiology* **256**, R240–R247.

White CR, Phillips NF, Seymour RS (2006) The scaling and temperature dependence of vertebrate metabolism. *Biology Letters* **2**, 125–127.

Wright JC (1985) Diving and exercise physiology in the estuarine crocodile, *Crocodylus porosus*. PhD thesis. University of Sydney.

Wright JC (1986) Effects of body mass, temperature and activity on aerobic and anaerobic metabolism in the juvenile crocodile, *Crocodylus porosus*. *Physiological Zoology* **59**, 505–513.

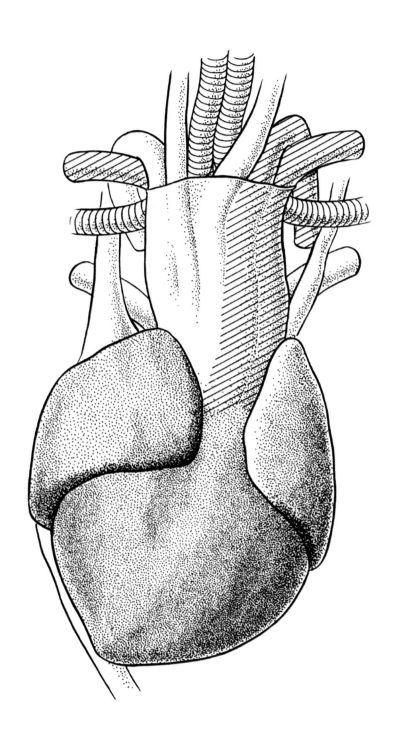

THE WORLD'S MOST EXTRAORDINARY HEART

I first learned of the puzzling complexity of the crocodylian heart as a third year student at the University of Queensland in 1962 when Dr Maurice Bleakly brought a crude plaster model into a vertebrate zoology lecture. He told us that it was a very peculiar heart and nobody knew how it worked. Jabbing a stubby finger at the model, he explained its unique features and why it was puzzling. I don't recall the details, but I must have tucked away the fact that the structure was mysterious because years later, when an opportunity came up to do research on crocodiles, looking at the heart was certainly on my list. When the renowned comparative cardiovascular physiologist, Kjell Johansen took a sabbatical from Denmark's University of Aarhus, he visited my laboratory at The University of Sydney in 1981. Kjell was a Norwegian I had known since 1965 when he came to Queensland to work on Queensland lungfish, which I was studying for my Honours degree. Kjell's considerable surgical skills made exploring the crocodile heart a logical project for his Sydney visit. We implanted fine cannulae into major blood vessels and monitored blood pressure and oxygen content and, with considerable technical help from Lyn Beard, were able to unravel many of its puzzles. Just as importantly, we were able to develop hypotheses that became the basis of further work, some now confirmed and one in particular, provocatively, still remaining to be explored.

Because 'the crocodile heart' always gets attention in good vertebrate zoology courses and textbooks, and because all of the text book descriptions I know of have shortcomings, it is worth describing the anatomy in detail. The chapter will also review the heart's functional flexibility, as well as speculating about its functional significance. Significant differences have not yet been found between the hearts of crocodiles and alligators, so we can discuss them together.

Although both points are controversial, one way to understand the heart is to think of it as a four-chambered heart evolved to support endothermy, which subsequently evolved the capacity for pulmonary by-pass shunting, quite possibly to support long submergences by this now aquatic ectotherm. This and the next two chapters ('Diving' and 'Thermal relations') will expand those concepts in detail.

A TOUR OF THE CROCODYLIAN HEART

GROSS ANATOMY

The heart lies very close to the ventral surface, in the midline and half way between the insertions of the front and rear legs. When a crocodylian can be persuaded to lie quietly on its back, the beating of the heart can usually be seen. Apart from the heart, the rest of the blood system of crocodylians, including its renal portal system (Chapter 7), is broadly similar to that in other reptiles. However, the

---

An evolutionary wonder, the crocodylian heart is the most complex of any vertebrate. (Illustration by June Jeffery)

heart and major aortas of crocodylians are unique in anatomy and flexibility of operation. I often assert that crocodylians have the most functionally sophisticated and elegant of all vertebrates' hearts and by the end of this and the next chapter readers may agree. Looking at the heart *in situ* or at a specimen in hand it is sometimes difficult to identify what is what, so it is usually represented in a simplified diagrammatic way (Fig. 8.1).

Crocodylians are the only living reptiles with a four-chambered heart – two atria and two ventricles (Figs 8.1, 8.2) – a feature shared with birds and mammals. Thus, while oxygenated blood can be supplied to the body at hydrostatic pressures between 85/65 and 60/40 mmHg (systolic/diastolic) (11.3/8.7 and 8.8/5.3 kPa) in air-breathing, resting *C. porosus*, the lungs receive 'venous' blood at lower hydrostatic pressures (systolic 15–20 mmHg, 2–2.7 kPa). Efficient lungs depend upon a supply of blood at low hydrostatic pressure, or else the delicate lung tissues may be damaged. In humans, high pulmonary arterial pressure (pulmonary hypertension) leads to a range of serious disorders requiring treatment.

So, like mammals and birds, crocodylians show a steep pressure gradient between 'systemic' and pulmonary circuits, aided by the heart being four chambered. High systemic pressures support higher metabolism and, although crocodylians are usually thought of as having only moderate aerobic capacity (see Chapter 7), satellite tracking data show that they are well capable of sustained swimming (see Chapter 4). Their phylogenetic position as archosaurs, along with birds, suggests that dinosaurs too may have had four-chambered hearts, with possible implications for ideas about likely dinosaur metabolism and body temperatures (see Chapter 10).

Apart from their four-chambered heart, the rest of the anatomy is very different from avian and mammalian hearts because crocodylians have a string of unique anatomical features which, collectively, allow them to be selective about how much 'venous' blood flows to the lungs. The rest

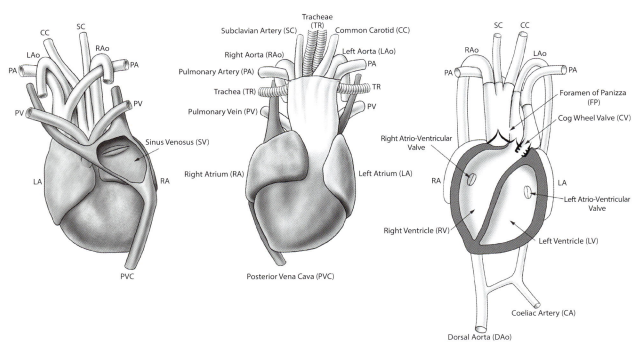

Fig. 8.1. *(Left) Dorsal and (centre) ventral views of the heart of* C. porosus. *(Right) The cut-away schematic ventral view shows the left ventricle exiting to the right aorta, and the right ventricle with exits to both the pulmonary artery and left aorta. Note that the foramen of Panizza will be covered by the medial leaf of the semilunar valve on the right aorta side during systole, but cannot be reached by the medial valve on the left aorta side. Note too the extraordinary 'cog-wheel' valve at the base of the pulmonary arch. (Left and centre schematics adapted from artwork by June Jeffery in Grigg 1989)*

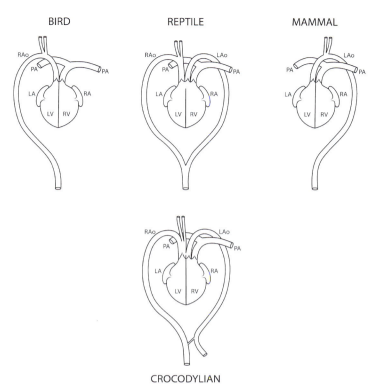

Fig. 8.2. *Schematic representation of the heart and major outflow vessels in typical reptiles, mammals, birds and crocodylians. Lettering as in Fig. 8.1.*

'overflows' from the right ventricle (RV) and recirculates to the systemic circuit. That is, they have a 'pulmonary by-pass shunt' (PBS), also called 'pulmonary-systemic shunt', which is both fascinating and enigmatic. It is fascinating because, although such shunting is typical of most reptiles, crocodylians do it in a unique way; it is enigmatic because, although we now know a lot about the directions in which blood can flow, what flexibilities exist and even about their control, we still know too little about when the shunt is activated, and why. Most people think it relates to diving, but there are other hypotheses and more research is called for, particularly on large unrestrained animals (see later).

Anatomists have been puzzled by the crocodylian heart and associated structures and speculated about function for more than 180 years. The first paper on the topic was published in 1833 by an Italian Count and Professor at the University of Pavia, Bartolomeo Panizza (Panizza 1833). He and I were both born on 15 August, which might appeal to astrologers. Dissecting an American alligator, Panizza described and speculated about a foramen joining two of the outflow tracts and it now bears his name (Figs 8.1, 8.3). As we shall find out, the foramen of Panizza is very significant functionally, but in a preserved specimen it is small and not easy to see and I have often wondered how Panizza even noticed it. In the last 50 years, with new experimental approaches and new and better technologies available, there has been a gradual increase in understanding, particularly in the last 25 years.

To understand the crocodylian heart and the remaining puzzles about its functioning, detailed knowledge of its anatomy is essential.

SEVEN UNIQUE OR STRIKING FEATURES OF THE CROCODYLIAN HEART AND OUTFLOW CHANNELS

1. Origin of the left aorta

In reptiles there are typically two aortas: the left and right (Fig. 8.2). They give off branches anteriorly to supply arterial blood to the head and front end

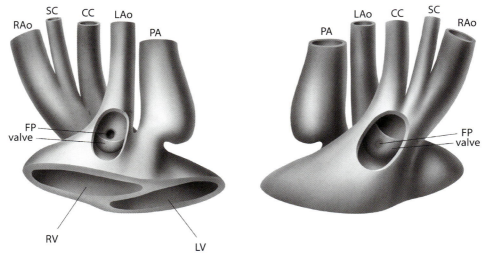

Fig. 8.3. *Schematic, cut-away views of the foramen of Panizza within the left (LAo) and right (RAo) aortas. The medial leaf of the semilunar valve cannot obstruct flow through the foramen from the LAo at any stage of the cardiac cycle (left) but, on the RAo side, the foramen sits deep in the pocket of the medial leaf of the semilunar valve (right), so that the foramen can allow flow from the RAo side occur only during late systole/early diastole when it is exposed as the bicuspid semilunar valves snap closed.*

of the animal and then turn posteriorly above the lungs and join to form the dorsal aorta: the main aortic trunk supplying blood to most of the viscera and the hind end. In birds, only the right aorta (RAo) has survived evolutionary selection pressures, and in mammals only the left survives (a reminder of the separate evolution of birds and mammals from reptiles). In crocodylians, however, both are still present (Figs 8.2, 7.18, 7.19) but are of different sizes. The RAo supplies the head and anterior body and sweeps dorsally and caudally to supply the posterior body (Figs 7.18, 7.19, 8.1, 8.2), as in birds: a reminder of their close relationship. The left aorta (LAo) is a smaller vessel, which also sweeps dorsally and caudally to join the RAo asymmetrically, dorsal to the lungs, at the 'anastomosis' (see 4 below) and appearing to provide the main blood supply to the gut via the coeliac artery (Figs 7.18, 7.19, 8.2).

It is worth noting that the heart can be thought of as being twisted 180° about the long axis, so that the RAo (on the right side of the body) comes from the LV, on the left side of the body and the LAo connects with the RV, with the main aortic trunks crossing each other. Close to where they cross we find the foramen of Panizza (Fig. 8.3).

Although the anatomy of the aortic outflows is broadly similar to most reptiles, except for the difference in size and the asymmetry at the anastomosis, there is a very striking difference. In most reptiles, both aortas are positioned to receive blood from the left ventricle (LV), but in Crocodylia the LAo arises discretely from the right ventricle (RV), alongside the pulmonary aorta (PA)! The anatomy suggests that the LAo should receive 'venous' blood, which would then be sent to the posterior body, particularly to the gut. Quite reasonably, this perplexed the early anatomists. Why would the posterior body and gut receive venous blood or, at best, a mixture of oxygenated and deoxygenated blood?

2. Foramen of Panizza

This is an opening in the common wall between the left and right aortas close to where they leave the heart (Figs 8.1, 8.3). Looking from inside the RAo, the foramen is deep within the pocket of the medial flap of the semilunar aortic valve and covered by it during systole, so a flow through the foramen from that side can occur only at the end of systole when the LV aortic valves snap shut (Fig. 8.3). On the other hand, the medial flap of the semilunar valve at the base of the LAo cannot reach high enough to occlude the foramen. The positions of these structures in relation to the foramen has considerable functional significance: flow from RAo to LAo can occur only during diastole, but 'reverse' flow, from LAo to RAo can occur whenever there is a pressure gradient in that direction. Additionally, Grigg and Johansen (1987) proposed that the diameter of the foramen is variable, and that has since been confirmed: it has associated muscles and its diameter is under the control of the autonomic nervous system (Axelsson and Franklin 2001). The foramen thus provides 'conditional' communication between LAo and RAo and the functional significance of this will be discussed below.

Christina Bennett-Stamper conducted a very detailed study of the embryological development of the foramen of Panizza in American alligators. She sought particularly to determine whether it is a remnant of incomplete development of the septum between the LAo and RAo during embryogenesis (Goodrich 1958) or, as suggested by Greil (1903) as a secondary perforation formed late in development. She showed clearly that Greil's interpretation is correct. The first sign of the foramen during embryonic development is at Ferguson stage 21 (Fig. 12.55) and it opens at stage 24–25, ~50 days into a 65-day incubation (Bennett-Stamper 2003). The resolution of that uncertainty was significant because it implies that the foramen is a novel structure. This provided one of many pieces of evidence assembled by Seymour *et al.* (2004) in support of their conclusion that today's crocodylians have endothermic ancestry. This is discussed in Chapter 10 but, briefly, Seymour and colleagues proposed that, in re-evolving ectothermy, the 'endothermic' four-chambered heart has been maintained while a suite of specialisations have evolved to provide a PBS, which is advantageous to their semi-aquatic predatory lifestyle (see Chapter 10).

Related specialisations are cartilaginous struts supporting the region of the aortic outflow tracts and an extraordinary structure at the base of the PA, the cog-wheel valve.

3. Cartilaginous struts

Located in the vicinity of the outflow vessels is a complex arrangement of cartilaginous struts (Fig. 8.4). Fred White, whose name crops up in numerous places in reptilian cardiovascular physiology, provided the first substantial description (White 1956), but they were reported

first in 1866 by the eminent German comparative embryologist Martin Rathke. White interpreted them as structures that provide rigidity to the heart in the vicinity of the outflow tracts, and sites of attachment for valves, notably the right atrio-ventricular and left aortic semilunar valves. Of particular interest, he noted that the process of the central cartilage that sweeps over the foramen of Panizza is embedded in the wall between LAo and RAo and, due to its rigidity, holds the foramen open. Additionally, he suggested that 'the cartilages act in such a manner during ventricular systole as to stretch taut the semilunar valves of the left aortic arch'. This is easy to accept; note their apparent tautness in Fig. 8.16 (left).

Bennett-Stamper (2003) found them useful landmarks as she interpreted her many thin sections of embryonic hearts and she agreed with White's interpretation that they provide support for the outflow tracts. White thought a description of

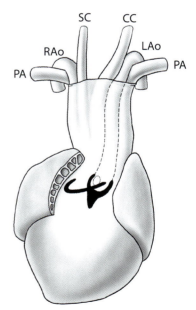

Fig. 8.4. *The region of the heart in the vicinity of the outflow tracts is stiffened by struts of cartilage that also serve as attachment points for semilunar valves. They may stretch taut the semilunar valves in the LAo and the process of one of them, embedded in the wall between the left and right aorta and arching over it, appears to hold the foramen open. Because the calibre of the foramen is known to be under active control, it could be that its diameter increases with relaxation of smooth muscle anchored to this cartilaginous process. (Adapted from Grigg 1989 and White 1956, with permission)*

the struts was warranted because of an 'apparent importance to the circulatory dynamics of the crocodilian heart' but, so far, their possible function has attracted less attention than the other unusual features.

4. 'Cog-wheel valve'

Guarding the exit from the RV to the PA is an extraordinary ring of 'cogs': connective tissue protuberances into the vessel lumen that mesh together and whose disposition suggests that their function is to regulate blood flow into the PA. These 'cogs' were described first by Grahame Webb in 1979 (Fig. 8.5). They project from within the muscular sub-pulmonary conus, just upstream of the leaf-like semilunar valves, and are visible from within the RV (Fig. 8.6) (Axelsson *et al*. 1996, 1997). The structure is perhaps most descriptively referred to as a 'cog-wheel valve'. The 'cog wheel' has a sphincter-like arrangement of smooth muscle and is amply innervated (Karila *et al*. 1995). It is well placed to

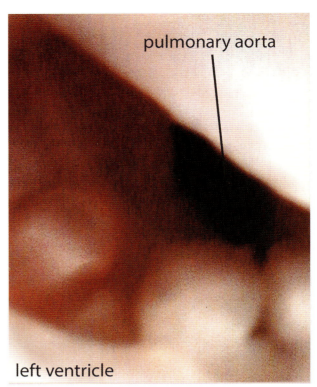

Fig. 8.6. *This is a view using a fibre optic angioscope to look towards the pulmonary aorta from within the right ventricle of a still-beating crocodylian heart, the blood having been replaced with physiological saline (Axelsson* et al. *1996). Note the 'cogs' of the cog-wheel valve surrounding the lumen of the pulmonary aorta at its entrance. Muscular constriction of the base of the pulmonary aorta brings the teeth closer together, restricting and thus regulating blood flow to the lungs. This is an 'active' heart 'valve', which is unique to crocodylians. (Photo Michael Axelsson)*

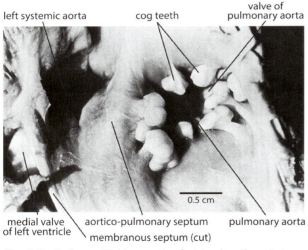

Fig. 8.5. *Grahame Webb was the first to describe what are now known as the cog teeth, comprising the cog-wheel valve. This picture is from his 1979 paper and was made from a fairly large, preserved specimen (so the 'cogs' are a bit shrunken). The annular disposition of these 'scrotum-like' nodules growing from the tissue at the base of the pulmonary aorta is strongly suggestive of a capacity to effect its partial or complete closure. This causes pressure in the right ventricle to rise and, if the cog-wheel valve constricts sufficiently, RV blood will be pushed through the semilunar valves into the LAo, providing the pulmonary by-pass shunt (see Figs 8.12, 8.13 and relevant text). (Photo from Webb 1979, © 1979 Wiley-Liss, Inc.)*

control blood flow into the PA (Axelsson *et al*. 1997) and – because the teeth fit together so snugly when the conus is contracted – can even block the flow completely. Partial constriction of the cog-wheel valve initiates the secondary pressure rise seen in the RV (e.g. Fig. 8.8) and, when more fully developed, leads to the shunting of 'venous' blood away from the lungs and into the systemic circuit: a pulmonary-systemic shunt (Figs 8.12, 8.13, see below).

The cog-wheel valve is the only known 'active' heart valve within the vertebrates and its constriction is responsive to the β-adrenergic tone of the heart (Franklin and Axelsson 2000; Axelsson 2001) and must be under tight control. Syme *et al*. (2002) explored the basis for the beat-to-beat, phasic constriction of the cog-wheel valve, developing as it does part way through systole in each cardiac cycle and producing the biphasic pressure trace in the RV (Fig. 8.8). They measured electrical and cardiovascular pressure events in anaesthetised American alligators and found a distinct and variable delay between the depolarisation of the RV (which triggers RV contraction) and the electrical stimulus for constriction of the cog-wheel valve. Furthermore, the extent of the delay was influenced by vagal and cholinergic stimulation and they implied that the length of the delay could determine the extent of the secondary rise in RV pressure during systole. That is, a shorter delay between the RV and cog-wheel ECGs may lead to higher RV pressures, which, when RV pressure exceeds LAo pressure, spill blood through to the LAo, thus initiating the

pulmonary-systemic shunt. It is worth noting, too, that the PA is enlarged at the base (Figs 8.1, 8.3) and somewhat compliant, so that the flow through the lungs is smoothed (Fig. 8.9 and see Chapter 7).

Like the foramen of Panizza, the cog-wheel valve develops late; indeed, it is still undeveloped in hatchling *C. porosus* (Webb 1979), but was present in a 1.1 m juvenile (Seymour *et al.* 2004). There can be little doubt that the cog-wheel valve is a secondary, crocodylian invention.

5. The 'dorsal anastomosis'

Where the LAo joins the RAo, dorsal to the lungs, there is a short, muscular section of vessel: the dorsal anastomosis (Fig. 8.1). It completes what Jones and Gardner (2010) called 'the ring around the

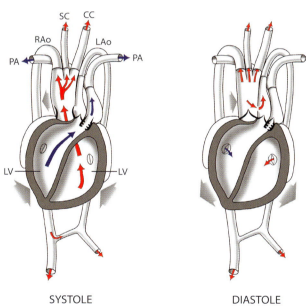

SYSTOLE   DIASTOLE

Fig. 8.7. *Pattern of blood flow during (left) systole and (right) diastole in an alert, resting crocodylian breathing air. The heart is functioning as a fully divided four-chambered heart, with no pulmonary-systemic shunting; the cog-wheel valves are open and there is no mixing of oxygenated and 'venous' blood between the systemic and pulmonary circuits. The left aorta is pressurised from the right aorta via the dorsal anastomosis ('around the loop') and from the surges of blood coming from the right aorta through the Foramen of Panizza during diastole (right). Because blood pressure in the right ventricle remains lower than in the left aorta throughout the cycle (Fig. 8.8), the semilunar valves between the right ventricle and the left aorta remain closed and the 'venous' blood (lower in oxygen) exits to pulmonary aorta with each systole.*

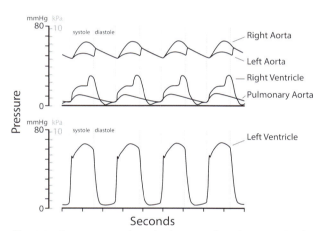

Fig. 8.8. *Pressure records in the heart and outflow vessels of a 7.0 kg* C. porosus *in a large tank, free to submerge or surface at will. Pressures in the pulmonary circuit are lower than in the systemic circuit. Heart rate 27 bpm. The peaks of the left ventricular (LV) pressure pulses map onto the pulses in the right aorta for the periods when the LV semilunar valves are open. High pressure in the left aorta prevents the valves at its base opening and spilling 'venous' blood from the right ventricle. The early pressure rise in the LAo during systole is not yet fully explained: it coincides with a small anterograde flow too (Axelsson* et al. *1989; Petersson* et al. *1992) so it is unlikely to come 'around the loop' via the anastomosis. It may derive from the base of the LAo being 'squeezed' by expansion of the RAo and PA during systole. Note the 'foramen spikes' in the left aorta trace: that is, the sharp pressure increases as oxygenated blood surges through from the right aorta via the foramen of Panizza at the end of systole. The pressure rise in each right ventricular pulse is biphasic, responding to the phasic control of blood flow into the pulmonary aorta (PA) by the cog-wheel valve. (Adapted from Grigg and Johansen 1987)*

heart'. It is sphincter-like and under active control, responding to a variety of regulatory substances, including adrenaline and nitrous oxide (Karila *et al.* 1995; Axelsson *et al.* 2001). It has been shown experimentally that its diameter has an effect on blood flow and pressure in the LAo and on the balance of flows between the gut (coeliac artery) and the rest of the organs and posterior body (Axelsson *et al.* 1997). More needs to be done to establish fully the role of this unusual structure.

6. A common carotid artery

Unusually among reptiles, the brain receives blood only from a branch of the RAo, the common carotid, and not from the LAo (Fig. 8.1). This provides a direct source of well-oxygenated blood, and also opens opportunities for preferential distribution of blood to the brain during submergence (see below).

7. Size and muscularity of the right ventricle

Typically the RV among amniotes is conspicuously less muscular than the left, driving only the lower pressure, pulmonary circuit. In crocodylians, the RV is nearly as thick as the left and the pressure pulse in the RV, which is typically biphasic even at rest (Fig. 8.8), shows that it has surplus 'capacity'. Indeed, working on an isolated, perfused heart, Franklin and Axelsson (1994) showed that it is capable of generating hydrostatic pressures in the same range as the LV. As we shall see, it is the RV that drives the PBS, pushing open the semilunar valves and delivering blood to the LAo and, in a more fully developed shunt, driving the systemic circuit via the foramen of Panizza.

NON-SHUNTING AND SHUNTING PATTERNS OF BLOOD FLOW

Having introduced the anatomical attributes of the heart, and some of their capacity for flexibility, it is now appropriate to review what is known about patterns of blood flow under different situations, and the way in which these patterns depend upon the heart's unusual features, before discussing ideas about functional significance.

BLOOD FLOW DURING NORMAL AIR BREATHING OR SHORT SUBMERGENCES

Patterns of flow during normal air breathing were worked out initially from observations of pressures in the major blood vessels, but it is convenient to show the patterns first (Fig. 8.7) and the pressures later (Fig. 8.8). During normal air breathing, the systemic and pulmonary blood supply circuits are in series, as in birds and mammals, with no cross-flow between them (Fig. 8.7).

Venous blood flows from the anterior and posterior vena cavae (Chapter 7, Figs 7.18, 7.19, 8.1) into the right atrium and RV to be pumped to the lungs via the PA (Fig. 8.9). Because of the expanded and compliant nature of the PA close to the heart, flow to the lungs is continuous throughout the cardiac cycle, with little variation between diastole and systole (Petersson *et al.* 1992). Oxygenated blood flows from the pulmonary veins (PV) to the left atrium, through atrio-ventricular valves to the LV and out through semilunar valves to the RAo. Flows in the RAo and its three outflow tracts (Fig. 8.1) are more pulsatile than in the pulmonary circuit. Most of that blood, freshly oxygenated in the lungs, supplies the anterior and posterior systemic circuits, perfusing the whole body, then returning via the major veins to the right atrium and thus to the RV whence it is pumped to the lungs. Some of the oxygenated blood, however, flows through the foramen and into the LAo at the end of each systolic beat of the heart, during diastole (systole) (Figs 8.7, 8.8, 8.10, and 8.11). That is, blood surges into the RAo during systole and then into the LAo via the foramen during diastole once the RAo semilunar valves retreat and expose the foramen. That is, surges into RAo and LAo alternate (this was controversial for many years; see Box 8.1). The sudden pressure surge in the LAo at the end of systole is referred to as the 'foramen spike' (Grigg and Johansen 1987), from its origin and shape on a LAo pressure trace (Fig. 8.8). The pressure spike results because one half of the semilunar valve, which opens from the LV, covers the foramen on its RAo side throughout systole, so blood cannot flow through the foramen until the very end of the power stroke when the

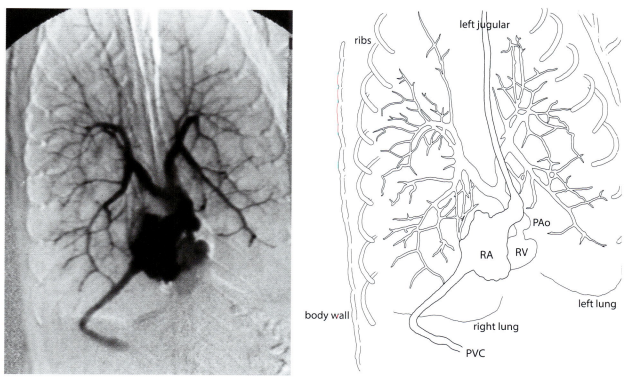

Fig. 8.9. *Digital subtraction angiogram, with interpretive sketch, made a few seconds after injecting contrast medium into the left jugular vein in an alert, resting* Crocodylus johnstoni. *The contrast medium has entered the right atrium, with some overflow into the posterior vena cava, and travelled into the right ventricle from where it has been pumped into both pulmonary aortas, illuminating the arborescent arterial system in the lungs. Also visible is a constriction at the base of the pulmonary aorta where the cog-wheel valve is located (open in this case). The ribs and the body wall are visible as a slight shadow. (Angiography performed in 1985 at the Flinders Medical Centre, Adelaide, whose staff are acknowledged gratefully, as well as the assistance of colleagues Russell Baudinette, Peter Frappell and Bren Gannon)*

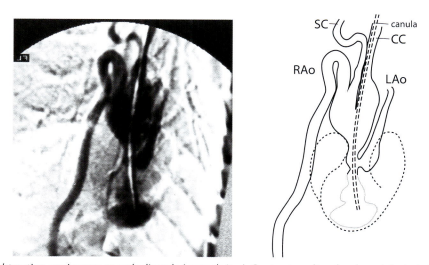

Fig. 8.10. *Digital subtraction angiogram at early diastole (ventral view). Contrast medium has been injected via a cannula inserted into the left ventricle via the common carotid, from whence it has exited into the right aorta and its branches the subclavian and the common carotid and, early in diastole, through the foramen of Panizza into the left aorta. (Angiography was performed in 1985 at the Flinders Medical Centre, Adelaide, whose staff are acknowledged gratefully, as well as the assistance of colleagues Russell Baudinette, Peter Frappell and Bren Gannon)*

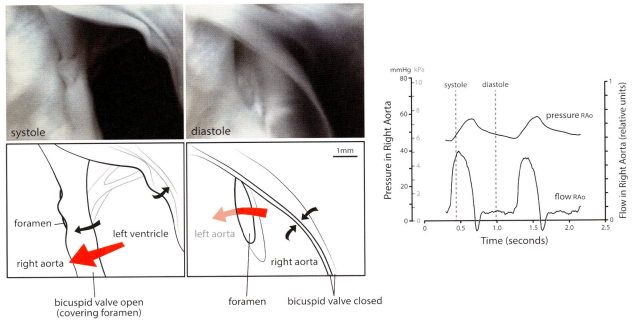

Fig. 8.11. *Alternate covering and uncovering of the foramen of Panizza during the cardiac cycle. These spectacular views were obtained using a fibre optic angioscope inserted into the common carotid of a large, freshly killed, artificially ventilated Cuban crocodile (Crocodylus rhombifer), with its heart* in situ *but isolated, continuing to beat normally and perfused with an oxygenated saline solution (Axelsson et al. 1996). The scope was threaded towards the heart, almost to the base of the right aorta, to a point where the foramen of Panizza was visible. During the ventricular contraction phase of the cardiac cycle (systole, left image), the semilunar valves are pushed open by the exiting blood (red arrow) and we see down into the bowels of the left ventricle. Note that during systole the medial semilunar valve closes across the foramen (a tiny bit of it is showing), preventing flow into the left aorta. At the end of systole, the two semilunar valves snap closed (right image, black arrows), exposing the foramen through which (oxygenated) 'blood' is now free to flow (red arrow) into the left aorta during diastole. (Right) Blood pressure and flow in the right aorta through one cardiac cycle are shown. Compare these photos with those taken of the foramen from its other (left aortic) side (Fig. 8.16). (Graph from Axelsson et al. 1996, photos Michael Axelsson)*

valve flap snaps away. Although the diameter of the foramen is under active control (see review by Axelsson 2001), there is no evidence to suggest that any active dilation of the foramen accompanies the cardiac cycle in crocodylians under these conditions.

In this pattern, typical of normal air breathing, the semilunar valves between the RV and the LAo remain closed throughout the cardiac cycle, blood does not 'leak' from the RV into the LAo and low oxygen blood in the RV all flows to the PA (Fig. 8.7). That is, the heart is operating as a 'normal' four-chambered heart, oxygenated arterial and 'venous' streams of blood are kept separate, there is no 'venous' blood bypassing the lungs and flowing to the LAo (Fig. 8.7) and hydrostatic pressure in the 'systemic' circuit is much higher than in the pulmonary circuit (Fig. 8.8). A good supply of well-oxygenated blood is delivered throughout the body

at a high hydrostatic pressure, while low hydrostatic pressure in the pulmonary circuit protects the fragile lung tissues but generates sufficient pulmonary blood flow for effective gas exchange (Fig. 8.9).

Under these circumstances, flow in the LAo is low and somewhat tidal. Shelton and Jones (1991) estimated net flow in resting American alligators to be 2–6% of the flow in the RAo; the flow also alternates in direction, away from the heart in diastole as blood flows in from the RAo via the foramen and back during systole, so that net flow may even be zero (Axelsson *et al.* 1991; Shelton and Jones 1991). The latter authors concluded from their pressure and flow measurements of small alligators at rest and breathing air that the LAo 'has very little significance in the overall cardiac cycle, except as a mechanism to prevent total stagnation of blood in the left aorta at times when the left aortic valves

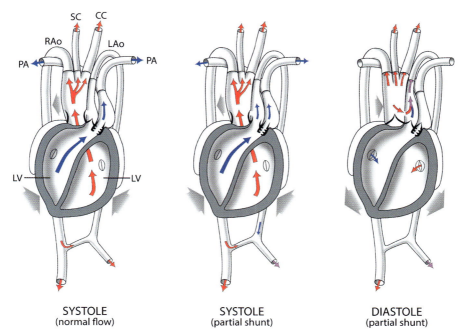

Fig. 8.12. *(Centre and right) Partial (small) pulmonary by-pass shunt in operation compared with (left) non shunt. (Centre) Partial constriction at the base of the pulmonary aorta (i.e. partial closure of the cog-wheel valve) is restricting flow, leading to a pressure build up in the right ventricle sufficient to spill some of the 'venous' blood into the left aorta with each systolic beat of the heart, mixing with blood coming through the foramen of Panizza from the right aorta during diastole. In C. porosus we measured a decrease in blood oxygen saturation in the LAo whenever blood pressure events implied mixing, providing confirmation of this.*

are closed. The slow flow of blood prevents clot formation and damage to the artery'.

In this 'normal', non-shunting pattern, there is flow through the anastomosis from the RAo to the gut via the coeliac artery during systole, but little or no flow during diastole (Axelsson *et al*. 1997). Presumably this is because of resistance under the influence of the flow into the LAo through the foramen of Panizza during diastole (Fig. 8.7).

SHUNTING BLOOD AWAY FROM THE LUNGS (PULMONARY BY-PASS SHUNTING)

The separation between pulmonary and systemic circuits in crocodylians depends upon pressure in the LAo remaining high enough to keep the LAo semilunar valves closed. Should RV pressure rise, or LAo pressure fall, some of the 'venous' blood will exit to the LAo with each beat of the heart, re-entering the systemic circuit instead of flowing to the lungs. A convincing proof is that when the anastomosis is occluded experimentally, it causes a fall of blood pressure in the LAo and a shunt results (Axelsson *et al*. 1997). The pattern of blood flow in a partially

developed PBS is illustrated in Fig. 8.12 and, in Fig. 8.13, the pressure patterns of two partial shunts, each developed to a different extent. By controlling the cog-wheel valve and, thus, RV pressure, crocodylians thus have the capacity to be selective about how much blood goes to the lungs. Other reptiles have that capacity too, but in those the shunt is within the heart, between incompletely divided ventricles (see review by Hicks and Krosniunas 1996). In crocodylians, extraordinarily, it is outside the heart (extra-cardiac), contrived by a suite of crocodylian innovations, including the thick-walled ventricle, the unusual exit of the LAo from the RV next to the PA, the cog-wheel valves, the foramen of Panizza and, probably, the vasoactive anastomosis, as well as vagal nerve control mechanisms. The obvious questions are how did such an extraordinary combination of elements evolve, and what for? Before tackling these questions, we will review the nature of the shunt and its occurrence.

Fred White, by then a Professor of Physiology at UCLA, was able to induce PBSs in American alligators either by diving them forcibly or injecting acetylcholine (White 1969). He noted

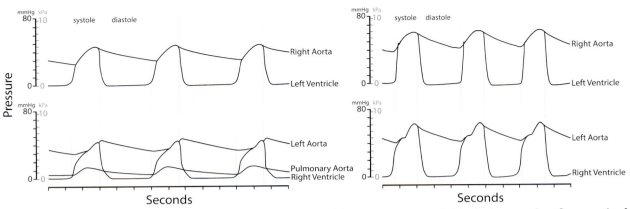

Fig. 8.13. *Two examples of pulmonary by-pass shunting, diagnosed from pressures recorded in the heart and outflow vessels of a 7.0 kg* C. porosus *resting on the bottom of a shallow aquarium (heart rates 13 and 16 bpm, respectively). The animal was free to surface or submerge at will. In both examples, each recorded several minutes into a dive, right ventricular (RV) pressures are high enough to push open the semilunar valves at the base of the left aorta (LAo) allowing some 'venous' blood to flow out. The shunt is not 'all or none': (left panel) the LAo pressure trace is still dominated by inflow from the RAo through the foramen, with just a small 'nose' showing from the RV; (right panel) with the shunt more fully developed, the LAo trace is dominated by the RV pressure. (Adapted from Grigg and Johansen 1987)*

BOX 8.1   HISTORICAL NOTE: RESOLUTION OF TWO LONG RUNNING CONTROVERSIES

This beat-by-beat surge of oxygenated blood through the foramen during diastole, so well recognised now, was controversial for some time, along with a controversy about whether the LAo would hold well-oxygenated 'arterial' or low oxygen 'venous' blood. The French anatomist Sabatier (1873) realised that the exit of the LAo from the RV would provide an opportunity for what we now call a pulmonary by-pass shunt (PBS). However, he also thought that the LAo would normally carry oxygenated blood, which would come from the RAo via the foramen of Panizza. This was insightful, but the Englishman Edwin Goodrich, a giant among early comparative anatomists, disagreed. He saw the exit of the LAo from the RV as counterproductive, even deleterious, as it would mix 'venous' blood in with oxygenated blood (Goodrich 1919). Working on *Caiman crocodilus*, Fred White put that matter to rest when he measured oxygen content of blood in both vessels and found they matched each other, as well as blood in the left atrium (White 1956). Further work by Greenfield and Morrow (1961) on *A. mississippiensis*, Khalil and Zhaki (1964) on *C. niloticus* and Grigg and Johansen (1987) on *C. porosus* produced similar results: the LAo carries oxygenated blood during normal air breathing. As for the source of that oxygenated blood, White envisaged that it surged through the foramen into the LAo at the same time as into the RAo. However, the puzzle was resolved by pressure recordings in the RAo and LAo by Greenfield and Morrow (1961) working on supine American alligators with the thorax opened, and Grigg and Johansen (1987) working on alert *C. porosus* in an aquarium but otherwise unrestrained. Both of these studies showed a surge of blood through the foramen from the RAo to the LAo during diastole, when the semilunar valve on the RAo side retreated, as described above and in Figs 8.7, 8.8. This was confirmed by measurements of blood flow too, in studies of *Caiman crocodilus* (Axelsson *et al.* 1989) and American alligators (Shelton and Jones 1991). Finally, this pattern of flow and the blocking of the foramen by semilunar valves on the RAo side during systole were visualised beautifully by high-resolution angioscopy (Axelsson *et al.* 1996) (Fig. 8.11). There is no longer any doubt: oxygenated blood gets into the LAo through Count Panizza's famous foramen, just as had been deduced by Sabatier about 100 years earlier.

that development of the PBS was associated with bradycardia and an increase in RV pressure, which he attributed to contraction of the pulmonary outflow tract under vagal control. He noted that in turtles, too, bradycardia and a PBS are associated with diving and interpreted the shunt as a mechanism to redistribute cardiac output during dives.

In the mid 1980s, Kjell Johansen, Lyn Beard and I were able to observe shunting in juvenile *C. porosus* (1.7–7.6 kg) diving voluntarily in an experimental tank. The animals spent most of their time under water, rising briefly to the surface every few minutes to take a breath before sinking again to the bottom. They had blood vessels cannulated so that we could monitor blood pressure and take blood samples periodically to measure oxygen levels (Grigg and Johansen 1987).

Shunting was diagnosed by systolic RV pressures rising progressively until their peaks mapped onto LAo pressures, with 'venous' blood exiting to the LAo during systole (Fig. 8.13). This was confirmed by the LAo having oxygen partial pressures higher than the PA but lower than the RAo. Most, but not all, such events were accompanied by substantial bradycardia. As discussed in Chapter 9, there has been an erroneous assumption that diving in crocodylians depends upon anaerobic metabolism, and that shunting is associated with that, but these shunts developed in aerobic dives (Chapter 9). The shunting and bradycardia ceased when the animal surfaced for the next breath (Fig. 9.16).

Shunts can be implemented to different degrees and Fig. 8.13 shows pressure events associated with small (left) and moderate (right) shunts. In the same 7 kg *C. porosus* that generated the data in Fig. 8.13, we obtained initially enigmatic blood pressure traces in which pressures in left and right aortas were coincident throughout the cardiac cycle (Fig. 8.14). This pattern developed ~20 min into a voluntary 30 min dive during which significant bradycardia developed (from 20 to 7–12 bpm) and blood pressure fell (from ~80/60 in RAo pre-dive to 60/40, 25°C) (Grigg and Johansen 1987). We saw the same pattern in several animals, associated with voluntarily generated severe bradycardia, and low central blood pressures or low water temperature. We could prompt similar responses if we extended submergences artificially by holding animals underwater to prevent them surfacing. It is the same pattern described by White (1969) in force-dived alligators. To account for the flexibility of patterns seen, Grigg and Johansen (1987) suggested that the calibre of the foramen of Panizza must be variable, able to be dilated and contracted in a controlled way, and this has now been demonstrated experimentally (Axelsson and Franklin 2001).

With the onset of shunting, less blood returns to the left atrium, so RAo output will be less and, in more fully developed shunts, with much of the pulmonary circuit by-passed, the dominant cardiac output will be that from the RV to LAo. The diameter of the LAo is smaller than the right, but larger at its base (Webb 1979) as far as the foramen of Panizza. Accordingly, the flow pattern in a more fully developed shunt must be as depicted in Fig. 8.15, with the circulation being driven mostly by the unusually large, muscular RV and through a dilated foramen into the RAo, in the 'reverse'

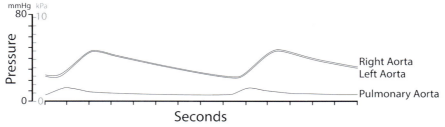

Fig. 8.14. *Pressure records in the three outflow vessels of a 7.0 kg* C. porosus *recorded during a prolonged voluntary dive (30 min) in which a severe bradycardia developed (7 bpm in this example). Pressure traces in LAo and RAo coincided in both magnitude and timing, as if they were in open communication through a wide open foramen of Panizza. The pressure trace implies that there is still some flow in the pulmonary aorta. Observations such as these prompted us to propose that the foramen could be dilated and contracted (the 'variable calibre' hypothesis), which was subsequently confirmed experimentally (Axelsson and Franklin 2001). (Adapted from Grigg and Johansen 1987)*

direction (Grigg 1989, 1991). This was at first seen as a radical idea; indeed one of the referees to the paper in which the hypothesis was (ultimately) published said that it was preposterous. However, that the 'plumbing' will support such a flow was demonstrated by Pettersson *et al.* (1992). They interrupted the pulmonary flow experimentally in an 'open chest', anaesthetised caiman by placing temporary ligatures on both pulmonary arteries and found that flow increased in the LAo and decreased in the RAo, yet systemic blood pressure was maintained. A 'reverse' flow was subsequently demonstrated *in vivo* by Axelsson *et al.* (1996) in *C. porosus* (Fig. 8.16).

Several anatomical features are well suited to the foramen permitting flow in the so-called 'reverse' direction (Fig. 8.15). The controlled, variable calibre of the foramen, its 'open access' from the LAo throughout the cardiac cycle and the enlarged base of the LAo (in large *C. porosus*, Webb 1979) as though it only ever needs to accommodate a large flow as far as the foramen, all seem to be 'designed' to provide a continuing perfusion of the head in particular during well-developed pulmonary by-pass shunting, with blood sourced from the RV via the foramen and the common carotid. Additionally, the coronary artery exits from near the semilunar valve of the RAo, well placed to maintain perfusion of the heart during such times. Whereas it is thought that the heart muscle derives blood supply primarily by direct perfusion from the ventricle in most reptiles,

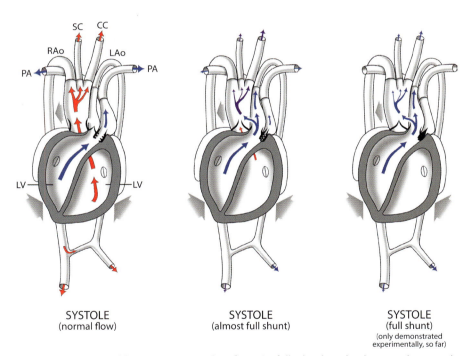

Fig. 8.15. *(Centre and right) Patterns of flow in two examples of a more fully developed pulmonary by-pass shunt, with a 'reverse' flow of blood through the foramen of Panizza, compared with (left) normal, non-shunted flow. Only systolic patterns are shown. A pressure pattern such as that seen in Fig. 8.14 could be expected from that shown in the centre panel, with some pulmonary flow maintained; with significantly reduced pulmonary flow, LV output can be only very small and most of the circulation will be driven by the RV. (Right) The pattern of flow resulting from experimental ligation of both pulmonary arteries (Pettersen et al. 1992) is also shown. In both these patterns, the circulation is being driven primarily or exclusively by the RV, blood reaching the systemic circuit via the left aorta and into the RAo via the foramen of Panizza in the 'reverse' direction to that shown in Figs 8.7 and 8.12. Whether full closure of the cog-wheel valve (right) is ever realised in natural situations is so far unknown; the best evidence is the cessation of weight loss by submerged animals in response to a sudden 'fright' (see text). In these patterns, redistribution of blood away from the skeletal muscles (50% of body weight in crocodylians) and other organs through peripheral vasoconstriction leaves the preferential distribution of blood and its stored oxygen to the head and brain. Such patterns have been demonstrated experimentally: the left pattern in voluntary dives (Grigg and Johansen 1987). Further elucidation is likely to depend upon radio-telemetric or ultrasonic monitoring of blood flow in large animals submerging voluntarily in natural or naturalistic situations.*

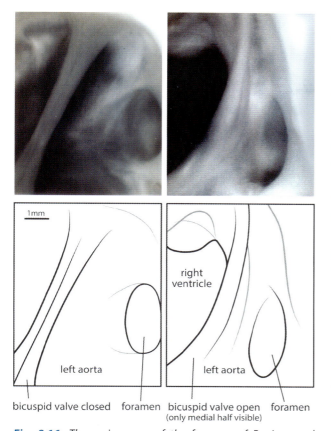

bicuspid valve closed   foramen   bicuspid valve open   foramen
                                    (only medial half visible)

Fig. 8.16. *These views are of the foramen of Panizza and semilunar valves from within the left aorta during a full pulmonary by-pass shunt initiated experimentally by occluding both pulmonary arteries, generating the flow pattern shown in the right-hand diagram of Fig. 8.15. The images were obtained in the same manner as those in Fig. 8.11, but threading the angioscope into the left aorta almost to its base. In diastole, blood pressure in the left aorta exceeds that in the right ventricle, so the semilunar valves remain closed, as is the case with no shunt operating. (Right panel) During systole, resistance in the PA leads to a rise in right ventricular pressure, exceeding that in the left aorta, and the valves open and blood spills into the left aorta with each beat of the heart. In this example, with complete occlusion of the pulmonary circuit, all of the right ventricular blood by-passes the lungs and re-enters the systemic circuit via the dilated foramen of Panizza. Such a flow is regarded as a 'reverse' flow through the foramen. In striking contrast to the anatomy on the right aortic side (Figs 8.11, 8.3), the foramen cannot be obscured by the medial semilunar valve at any stage of the cardiac cycle and there is no anatomical impediment to the flow of blood through the foramen from left aorta to right aorta if that is what the pressure gradient dictates. (Photos Michael Axelsson)*

in American alligators this is supplemented by substantial perfusion via the coronary arteries (Kohmoto *et al.* 1997).

It seems that the extent of shunting can be controlled over a continuum from none (Fig. 8.7), though partial shunts (Fig. 8.12) to those seen in Fig. 8.15 (centre) and, possibly, even 8.15 (right). Shunting leads to a lowering of systemic blood oxygen levels, with a concomitant reduction in metabolic rate (Chapter 9). The longest voluntary dives made by crocodylians, such as the resting dives described by Campbell *et al.* (2010; Fig. 9.17), are likely to be accompanied by shunting of the type shown in Fig. 8.15 (centre), but that remains to be determined by field studies. During normal air breathing, the foramen seems to have little or no function (indeed several studies have confirmed this). It seems likely that its main significance, in conjunction with the other specialisations of the crocodylian heart, is to allow this shunting of RV blood to the RAo via the base of the LAo: 'reverse' flow.

There is no doubt that crocodylians do show pulmonary by-pass shunting and that it occurs during long dives. The question is why?

WHAT IS THE SIGNIFICANCE OF PULMONARY BY-PASS SHUNTING IN CROCODYLIANS?

Crocodylians are not unique among reptiles in having a PBS. Indeed, snakes, lizards and turtles all have undivided ventricles and, although they manage to keep arterialised and venous blood from mixing as it traverses the heart, all have the capacity for shunting blood between the two circuits. The functional significance of the shunting capacity is for the most part enigmatic: satisfying hypotheses are not yet available except for those in crocodylians and turtles. But there has been no shortage of suggestions. Farmer and Hicks (2002) tabulated 12 postulated functions of which eight were possibly relevant to crocodylians. These were: thermoregulation (to facilitate warming); digestion (to facilitate HCl secretion); protecting the lungs against fluid filtration; reducing CO_2 flux into the lungs; saving cardiac energy; facilitating recovery from extracellular acidosis; triggering hypometabolism (for diving or aestivation); and for metering the oxygen store (to extend dive time).

These last two are linked and are supported by experimental evidence, as will be discussed, and Farmer *et al.* (2008) have reported experimental data that suggest a possible benefit to digestion through facilitating acid secretion in the stomach.

The lack of convincing explanations and data prompted Jim Hicks at the University of California, Irvine, to question whether pulmonary-systemic shunting has any adaptive significance at all in modern reptiles (Hicks 2002), implying it may be an evolutionary hangover like the human appendix, which has survived because there have been no selection pressures to remove it. He has reaffirmed that position in a more recent review (Hicks and Wang 2012); 'Experimental evidence supporting the notion that cardiac shunts are a derived state, or provide an adaptive advantage does not exist. More likely, the unique cardiovascular system of reptiles simply represents an ancestral condition and/or an embryonic character trait that combined with low metabolic demands persists because there has been no selection pressure against the presence of cardiac shunts'. Unfortunately, they did not speculate about what circumstances might have provided the selection pressures leading to the evolution of the PBS in reptiles in the first place. But it is there.

Scepticism in science is always praiseworthy and, as Hicks has pointed out, 'story telling' is all too frequent in comparative physiology, with patterns that seem to make good functional sense being interpreted as adaptive, without good evidence. Nevertheless, 'story telling' or 'just so' stories in biology often provide coherent hypotheses that are useful in stimulating and guiding further research.

Two separate, and not mutually exclusive, functions have been given consideration to explain the PBS in crocodylians: that it provides a mechanism to extend the time for which they can remain submerged; and, more recently, that it serves digestion. These two proposals will now be reviewed.

DOES THE SHUNT FACILITATE DIGESTION?

As Webb (1979) pointed out, the vascular anatomy suggests that, whereas the head and most of the body are supplied by the RAo, the gut seems to be served particularly from the LAo (Fig. 8.1). This anatomy is somewhat misleading, however, because flow in the LAo is normally quite small (Axelsson *et al.* 1991) and the coeliac artery must receive most of its flow in crocodylians at rest via the anastomosis. If this is so, flow can be expected to increase in the LAo during digestion and this has been shown in *C. porosus* (Axelsson *et al.* 1991) and American alligators (Farmer *et al.* 2008). Jones and Shelton (1993), discussing the possible benefit/s of shunting, suggested that the shunted blood, being more acidic, may facilitate the secretion of HCl into the stomach and this hypothesis was tested in *A. mississippiensis* by Farmer *et al.* (2008) in a series of elegant and technically challenging experiments. She and her colleagues measured rates of gastric acid secretion and digestion of bone in animals with the shunt disabled, by blocking the exit from the LAo upstream of the foramen of Panizza, and compared them with intact animals. The animals with the shunt disabled showed lower rates of acid secretion and slower digestion of bone, which they suggested could result from a compromised supply of CO_2 to the gut via the LAo. Carbon dioxide is one of the providers of protons for HCl generation by combining with water to form carbonic acid, which dissociates immediately. Farmer and colleagues suggested that the pulmonary-systemic shunt enhances HCl formation in the stomach and that the shunt, therefore, serves digestion. Further, they suggested that input of oxygenated blood from the foramen of Panizza ensured sufficient oxygen. This could be the result of dilation of the foramen under low adrenergic stimulation, leading to increased blood flow from the RAo (Axelsson and Franklin 2001). The source of the increased flow in the LAo needs to be clarified. Axelsson *et al.* (1991) also described an increase in blood flow in the LAo following feeding, but assumed the source was the RAo, via the foramen of Panizza. If blood to the gut is supplied by shunting, its oxygen content would be lower and, given the increased oxygen consumption that accompanies feeding (Chapter 7), a well-oxygenated source from the RAo seems more likely. Farmer *et al.* (2008) envisaged constructive mixing from both sources.

At first glance, this is a compelling proposal. However, Eme *et al.* (2010) and Jones and Gardner (2010) found that growth by *A. mississippiensis* and *Caiman crocodilus* was not compromised by long-term experimental occlusion of the LAo. Eme *et al.* (2010) occluded the LAo both upstream and downstream of the foramen of Panizza in hatchling (~70 g) alligators and measured growth rates over 2 years, compared with a control group. Any possible effect was exacerbated by exercising them regularly on a treadmill or swimming in a flume. Both groups grew to 1.3–1.4 kg, with little difference between: presumably the lack of a LAo was not a hindrance to digestion, or anything else. Taking a similar approach, Jones and Gardner (2010) measured growth rates and food conversion rates in hatchling alligators growing to ~6 kg, with and without the LAo sectioned and found no effect on either growth or food conversion rate. Additionally, they sectioned either the LAo or RAo (downstream of the common carotid and right subclavian branches) in six juvenile (≈2.5 kg) *Caiman crocodilus* and compared growth and food conversion rates over 20 weeks. They found no statistically significant differences, but a strong trend to lower performance in those with the RAo cut (!). Over a longer time period, 4 years, two survivors with the RAo cut were more compromised (average 15 kg) than two survivors with the LAo cut (average 24 kg). The implication is that, deprived of flow via either the LAo or the RAo, flow to the gut and the rest of the body could be maintained via the dorsal anastomosis, part of what they termed the 'ring around the heart'. It would be fascinating to know the circulatory patterns in each of these two groups of caiman. Their survival and growth after such major intervention makes a very striking statement about the adaptability of the crocodylian circulatory system!

Despite these growth studies, the idea that the PBS may facilitate digestion is interesting and, as noted by Farmer *et al.* (2008), further work is warranted. It might even have implications for explaining the enigmatic occurrence of a PBS in reptiles that do not dive. Could this be a more generic, ancestral role for the ubiquitous shunting capacity seen in reptiles?

DOES THE SHUNT EXTEND AEROBIC DIVE LIMIT?

Crocodylian diving physiology is the subject of the next chapter, but it is convenient to review here the evidence for (and against) the PBS having a role in diving, as background for Chapter 9.

There is an array of experimental and behavioural evidence that implementation of the crocodylian PBS during a dive leads to a reduction in metabolic rate, spinning out the oxygen stored in the blood and lungs and extending maximum aerobic dive time.

1. The most direct evidence comes from two separate sets of experimental observations on juvenile *C. porosus*, in combination.

   The first came serendipitously and was overlooked until recently, when David Kirshner reminded me of it from his PhD research (Kirshner 1985). As part of his work on buoyancy (Chapter 4), David was measuring the increase in density that accompanies submergence as oxygen is used and replaced with a lesser volume of CO_2 (RQ < 1, Chapter 7). He was weighing juvenile *C. porosus* fully submerged and resting on a mesh screen slung below an electronic balance. Their mass increased steadily as the minutes ticked by. Occasionally the study animals would be disturbed accidentally by a sudden loud noise, such as someone entering the room loudly, and whenever that occurred the rate of change in lung volume 'decreased appreciably, becoming almost negligible in some cases'. It has been known for a long time that sudden disturbance to a submerged crocodylian typically prompts a sudden fall in heart rate – that is, a 'fright bradycardia' (e.g. Gaunt and Gans 1969; Smith *et al.* 1974; Wright *et al.* 1992) and this is always interpreted as the initiation of the PBS. These chance observations, made in the pursuit of answers to quite different questions, provide independent evidence that the response to 'fright' is associated with reduced removal of oxygen from the lungs, likely associated with a PBS.

   The second set of observations was provided by Wright *et al.* (1992) who reported a 6-fold increase in dive duration following deliberately induced 'fright' bradycardia in submerged juvenile *C. porosus*. Without any disturbance, the animals spent most of the time submerged, lying quietly on the bottom of the tank and surfacing every few minutes for a breath. Dive lengths were 3.08 ± 1.87 min (mean ± s.d.) and heart rates during these short dives fell by an average of ~15% compared with the rates during the breath. After several hours of this, the crocodiles were disturbed abruptly by the experimenters entering the room and banging loudly on the side of their tank As expected, this caused a rapid onset of 'fright bradycardia', with heart rate dropping precipitously to be, on average, 35% lower than surface rates and dive durations were prolonged to 19.6 ± 1.8 min.

Twenty minute dives would still be within the time frame of dives that could be undertaken aerobically without any special physiological response (Fig. 9.17). However, put together with David Kirshner's observations, these two studies carry the strong implication that the extended dive times reported by Wright *et al.* (1992) were the result of operation of the PBS.

Incidentally, Wright *et al.* (1992) noted that 'C. porosus did not develop a bradycardia in response to disturbance on land, suggesting that submergence is a necessary pre-condition for the development of 'fright' bradycardia'. Because crocodylians routinely use submergence as a refuge, this can be taken as further circumstantial evidence that the 'fright' response and its accompanying PBS facilitate extension of aerobic dive times.

2. A second line of reasoning follows from the very long voluntary dives reported by recent field studies. Field data from *C. johnstoni*, reviewed in Chapter 9, show that, although most of their dives are quite short (a couple of minutes), many are very long. The longest voluntary dives recorded from 11 individuals (5– 0 kg), ranged from ~50 to 150 min (Campbell *et al.* 2010). These animals spent 28% of their time submerged and 95% of that was spent at rest (!).

Quite clearly, such long breath hold dives are much longer than could be supported aerobically at resting rates of oxygen consumption (Fig. 9.17)

and some physiological mechanism/s must be in play to support them. The traditional explanation has been that they rely on anaerobic support, and some current literature still makes that assumption. However, there are compelling arguments that those long resting dives are undertaken aerobically, as are most dives by diving birds and mammals (see Chapter 9 and Box 9.1). It is much more likely that the long dives involve a reduction in metabolic rate, and there is experimental evidence to show that. Wright (1987) showed that, in contrast to short dives by juvenile *C. porosus*, which involved no reduction in oxygen consumption, prolonged dives were accompanied by a 60–90% reduction.

Diving birds and mammals lack a PBS, but, in addition to reducing metabolic rate, they show bradycardia and a redistribution of blood, including peripheral vasoconstriction and reduced perfusion of muscle and many internal organs (Box 9.1). The association of bradycardia with diving in crocodylians is well known and Weinheimer *et al.* (1982) reported a marked reduction in blood flow to the skin and tail musculature of American alligators force dived or in response to disturbance. Vertebrate physiology is generally conservative and the parallels between physiological correlates of diving seen in crocodylians, birds and mammals are striking, but not surprising.

3. Informative research about the probable role of the PBS in extending dive duration has come from studies of freshwater turtles. The North American painted turtle, *Trachemys* (= *Pseudemys* = *Chrysemys*) *scripta* has been the main study animal, with much of the work undertaken by Don Jackson from Brown University and Warren Burggren at the University of North Texas and their colleagues. Most of the dives in *T. scripta* are 'short' (< 30 min) and are accompanied by a modest slowing of the heart, a decline in oxygen in the lung and systemic blood, and reduced pulmonary perfusion during dives, with the shunting of blood within the heart away from the lungs. Jackson (1968) showed that metabolic rate falls during dives and attributed it to a reduction in tissue oxygen supply. Most

dives are supported aerobically and, in more prolonged dives, 2–3 h (!), oxygen stored in the blood declines more rapidly than lung oxygen: lung oxygen is conserved, with considerable control exercised over its release (Burggren and Shelton 1979; Burggren 1988). These authors interpreted their data on the patterns of decline of oxygen in the lungs and arterial blood as the result of 'intermittent perfusion of the lungs, 'metering' oxygen to the systemic circulation during the dive'. The taxonomically distant Australian freshwater turtle, *Chelodina longicollis*, also yielded compelling data, showing increased arterial oxygenation correlating with increased pulmonary perfusion (Burggren *et al.* 1989). The implication is that control can be exercised over the perfusion of the lung by the extent to which the (intracardiac) PBS is implemented. They suggested that this 'may be a general phenomenon among diving reptiles', which is in harmony with the suggestion by Grigg and Johansen (1987) the PBS enables crocodylians to match lung perfusion to oxygen requirements as the dive progresses.

4. The crux of extending breath hold dive time without engaging anaerobic support seems to be a down-regulation of oxygen requirements during the dive. In diving mammals and birds, the explanatory model for long dives relies upon the down-regulation of metabolism as a consequence of hypoperfusion of peripheral tissue and some organs and skeletal muscle (Box 9.1). Mammals and birds do this without a PBS. Hicks and Wang (1999) found in anaesthetised freshwater turtles that lowered levels of arterial oxygen resulted from the intracardiac PBS. They concluded that the PBS, in recirculating low oxygen blood to the systemic circuit instead of sending it to the lungs 'may trigger the development of a hypometabolic state and therefore contribute to the prolongation of aerobic dive times'. Their review in 2004 extended this concept (Hicks and Wang 2004), although they have since backed away from it after finding that partial occlusion of the PA of non-anaesthetised turtles did not lead to reduced oxygen consumption (Wang and Hicks 2008).

The putative model that emerges from these studies of freshwater turtles is that their prolonged aerobic dives involve an increase in pulmonary resistance driving an intracardiac (R-L) shunt, bradycardia, reduced cardiac output, a selective redistribution of blood away from the skeletal muscle and many of the dorsal aorta-serviced organs, which, through hypoperfusion, suffer substantially reduced oxygen consumption. Blood supply is maintained to the anterior part of the systemic circuit (carotids), supplying the brain and also to the heart (coronaries). Reduced pulmonary perfusion conserves oxygen stored in the lung, from which it can be drawn periodically.

Freshwater turtles also show outstanding capacities for very extended submergences, many hours and even many months, particularly in cold climates where they may overwinter under ice, torpid and tolerating complete anoxia, with limited ATP production coming from anaerobic metabolism. Interested readers can follow this up in a substantial recent review by Jackson and Ultsch (2010).

Hicks' and Wang's scepticism relied partly upon a study by Eme *et al.* (2009), which compared oxygen consumption and apnoeic periods between juvenile alligators with a disabled shunting capacity and a sham operated group. Measurements were made over a couple of days with the animals in a dive chamber similar to that used by Wright (1987) (Chapter 9) and no differences were found between the two groups. The authors concluded that occlusion of the LAo and, thus, removal of the pulmonary-systemic shunt 'does not alter normal diving behaviour or reduce the voluntary apnoeic periods of alligators (Eme *et al.* 2009). Putting these results with their growth study discussed above, Eme *et al.* (2010) concluded that the 'crocodilian R-L cardiac shunt does not provide an adaptive advantage for juvenile alligator growth and supports the logic that cardiac shunts persist in crocodilians because they have not been selected against'. The implication in the Eme *et al.* (2009) study is that the surgically compromised animals would have shown shorter 'dives' than the sham operated animals if the latter extended their dives by employing their PBS and reducing aerobic metabolism consumption accordingly. This did not occur. However, the comparative brevity of the apnoeic periods (dives?) in both groups suggests that neither had any incentive to extend the dives. Assuming American alligators are generally similar to crocodiles, almost all of the dives by both groups were shorter than 1.7 kg animals could make without either employing the shunt or reducing metabolic rate (Chapter 9, Fig. 9.17). Modal apnoeic period in both groups was 2 min. Some animals in each group showed longer apnoeic periods: 27 min in one; 25 min in the other. The Eme *et al.* (2009) study highlights one of the difficulties of studying diving physiology: there is no way to ask a crocodylian to 'do its best' and 'please stay down for as long as you can'.

Resolving whether or not the PBS allows crocodylians to extend their submergence time is not straightforward. Captive animals cannot be relied upon to behave normally and small ones are not necessarily representative. The question will probably be resolved in the field, either by a telemetric study of blood flow and other variables in large crocodiles or alligators free ranging in natural or naturalistic surroundings (Grigg 1989; Axelsson and Franklin 2011) or by a comparison between maximum dive times observed in surgically compromised and intact animals. Neither of these will be simple, or cheap. The technical capability already exists for the former (Axelsson *et al.* 2007) and the latter (Campbell *et al.* 2010), and if the former could be combined with pulmonary and blood gas monitoring, so much the better. Until then we have a very good working hypothesis: that the PBS plays a significant role in extending crocodylian dive time. Note: a telemetric study such as that referred to here is in progress (Craig Franklin *pers. comm.* June 2014).

A WORKING HYPOTHESIS

The following working hypothesis seems reasonable. Most foraging dives are short – a minute or two –Chapter 9) and pulmonary by-pass shunting

is unlikely to be necessary. In short resting dives (the length being size dependent, Fig. 9.17) the pattern of blood flow would be as represented in Fig. 8.7, with no shunting and metabolism continuing at pre-dive rates. A modest bradycardia can be expected, compared with the tachycardia that accompanies the respiratory event in these intermittent breathers (Chapter 7). The longest voluntary dives described under field conditions have been interpreted as resting dives (Chapter 9). In such dives, tightening of the cog-wheel valve initiates an early expression of the shunt (Fig. 8.12 centre and right), with some blood by-passing the lungs and spilling from the RV through into the LAo. Blood flow through the foramen of Panizza from RAo to LAo continues and the blood in the LAo is a mixture of well-oxygenated and less-well-oxygenated blood, leading to a lower level of oxygen in the systemic outflows and the dorsal aorta. A smaller flow of oxygenated blood is now entering the LV. In response to lower systemic oxygen levels, selective vasoconstriction leads to the beginning of cardiovascular redistribution away from the periphery, skeletal muscles and some internal organs. Oxygen consumption is down-regulated by the lower systemic oxygen levels. There is a reduced rate of oxygen uptake from the lungs.

In a more prolonged resting dive, tighter constriction of the cog wheel reduces pulmonary blood flow even more, and systemic vasoconstriction is more intense. Reduction of oxygen consumption is greater and, with much less blood in the pulmonary circuit, blood flow and pressure in the RAo falls to a point where the direction of blood flow through the foramen is reversed (Fig. 8.15 centre). The cardiovascular system, much of it constricted or with a considerably reduced flow, becomes primarily 'right-heart driven' by the extraordinarily muscular RV, while oxygen remaining in the blood plus some from the lungs is delivered selectively to the heart and brain. Conveniently, the coronary artery exits from the RAo near the semilunar valves and the brain is supplied by the common carotid, also via the RAo.

These changes can be implemented progressively or very rapidly as would be the case in 'fright' dives (Chapter 9) in which severe bradycardia occurs as an immediate response (Fig. 9.16). We know that banging on the experimental tank of a crocodile submerged in a resting dive stimulates an immediate and severe bradycardia, indicative of a substantial PBS.

If there is a prey capture that involves a struggle, the prey item is usually brought to the surface. It would be very interesting to monitor heart rate of an animal in the field when a short, aerobic foraging dive leads to prey capture and subsequent underwater struggling to the point of needing anaerobic support.

Complete closure of the cog-wheel valve would result in the flow pattern shown in Fig. 8.15 (right), which can be induced experimentally (see above) but whether or not the readjustment ever reaches that stage under natural conditions is so far unknown. If it did, the RV-driven circuit could maintain perfusion at low rates through much of the systemic circuit, relying on periodic opening of pulmonary flow to re-oxygenate the brain and heart, as in turtles (Burggren 1988) and as speculated for crocodylians (Grigg 1989, 1991). The ability of both the foramen and the dorsal anastomosis to be opened and closed raises all sorts of possibilities that can only be elucidated with further work.

Almost all of the elements of this interpretive hypothesis have been observed and described, and it is a hypothesis that can be tested and which I hope will provoke further research. Crocodylian diving behaviour and physiology and further discussion about the possible significance of the unusual anatomy of the crocodylian heart will be explored in the next chapter.

References

Axelsson M (2001) The crocodilian heart: more controlled than we thought? *Experimental Physiology* **86**, 785–789.

Axelsson M, Franklin CE (2001) The calibre of the foramen of Panizza in *Crocodylus porosus* is variable and under adrenergic control. *Journal of Comparative Physiology. B, Biochemical, Systemic, and Environmental Physiology* **171**, 341–346.

Axelsson M, Franklin CE (2011) Elucidating the responses and role of the cardiovascular system in crocodilians during diving: fifty years on from the work of CG Wilber. *Comparative Biochemistry and Physiology. Part A, Molecular & Integrative Physiology* **160**, 1–8.

Axelsson M, Holm S, Nilsson S (1989) Flow dynamics of the crocodile heart. *American Journal of Physiology* **256**, R875–R879.

Axelsson M, Fritsche R, Holmgren S, Grove DJ, Nilsson S (1991) Gut blood flow in the estuarine crocodile, *Crocodylus porosus. Acta Physiologica Scandinavica* **142**, 509–516.

Axelsson M, Franklin C, Loefman C, Nilsson S, Grigg G (1996) Dynamic anatomical study of cardiac shunting in crocodiles using high-resolution angioscopy. *Journal of Experimental Biology* **199**, 359–365.

Axelsson M, Franklin CE, Fritsche R, Grigg G, Nilsson S (1997) The subpulmonary conus and the arterial anastomosis as important sites of cardiovascular regulation in the crocodile, *Crocodylus porosus. Journal of Experimental Biology* **200**, 804–814.

Axelsson M, Olsson C, Gibbins I, Holmgren S, Franklin CE (2001) Nitric oxide, a potent vasodilator of the aortic anastomosis in the estuarine crocodile, *Crocodylus porosus. General and Comparative Endocrinology* **122**, 198–204.

Axelsson M, Dang Q, Pitsillides K, Munns S, Hicks J, Kassab GS (2007) A novel fully implantable multi-channel biotelemetry system for measurement of blood flow, pressure, ECG and temperature. *Journal of Applied Physiology (Bethesda, Md.)* **102**, 1220–1228.

Bennett-Stamper C (2003) Structural development of the Foramen of Panizza in embryonic alligators. MSc thesis. New Mexico State University, Las Cruces.

Burggren W (1988) Cardiovascular responses to diving and their relation to lung and blood oxygen stores in vertebrates. *Canadian Journal of Zoology* **66**, 20–28.

Burggren W, Shelton G (1979) Gas exchange and transport during intermittent breathing in chelonian reptiles. *Journal of Experimental Biology* **82**, 75–92.

Burggren W, Smits A, Evans B (1989) Arterial O$_2$ homeostasis during diving in the turtle *Chelodina longicollis. Physiological Zoology* **62**(3), 668–686.

Campbell HA, Sullivan S, Read MA, Gordos MA, Franklin CE (2010) Ecological and physiological determinants of dive duration in the freshwater crocodile. *Functional Ecology* **24**, 103–111.

Eme J, Gwalthney J, Blank JM, Owerkowicz T, Barron G, Hicks JW (2009) Surgical removal of right-to-left cardiac shunt in the American alligator (*Alligator mississippiensis*) causes ventricular enlargement but does not alter apnoea or metabolism during diving. *Journal of Experimental Biology* **212**, 3553–3563.

Eme J, Gwalthney J, Owerkowicz T, Blank JM, Hicks JW (2010) Turning crocodilian hearts into bird hearts: growth rates are similar for alligators with and without right-to-left cardiac shunt. *Journal of Experimental Biology* **213**, 2673–2680.

Farmer CG, Hicks JW (2002) The intracardiac shunt as a source of myocardial oxygen in a turtle, *Trachemys scripta. Integrative and Comparative Biology* **42**, 208–215.

Farmer CG, Uriona TJ, Steenblik M, Olsen D, Sanders K (2008) The right-to-left shunt of crocodylians serves digestion. *Physiological and Biochemical Zoology* **81**, 125–137.

Franklin CE, Axelsson M (1994) The intrinsic properties of an in situ perfused crocodile heart. *Journal of Experimental Biology* **186**, 269–288.

Franklin C, Axelsson M (2000) An actively controlled heart valve. *Nature* **406**, 847–848.

Gaunt AS, Gans C (1969) Diving bradycardia and withdrawal bradycardia in Caiman crocodiles. *Nature* **223**, 207–208.

Goodrich ES (1919) Note on the reptilian heart. *Journal of Anatomy* **53**, 298–304.

Goodrich ES (1958) *Studies on the Structure and Development of Vertebrates*. Dover, New York.

Greenfield LJ, Morrow AG (1961) The cardiovascular dynamics of Crocodilia. *Journal of Surgical Research* **1**, 97–103.

Greil A (1903) Beiträge zur vergleichenden anatomie und entwicklungsgeschichte des herzens und des truncus arteriosus der wirbelthiere. *Gegenbaurs Morphologisches Jahrbuch* **31**, 123–310.

Grigg GC (1989) The heart and patterns of cardiac outflow in Crocodilia. *Proceedings of the Australian Physiological and Pharmacological Society* **20**, 43–57.

Grigg G (1991) Central cardiovascular anatomy and function in Crocodilia. In *Physiological Adaptations in Vertebrates: Respiration, Circulation, and Metabolism* (Eds SC Wood, RE Weber, AR Hargens and RW Millard) pp. 339–353. Dekker, New York.

Grigg G, Johansen K (1987) Cardiovascular dynamics in *Crocodylus porosus* breathing air and during voluntary aerobic dives. *Journal of Comparative Physiology. B, Biochemical, Systemic, and Environmental Physiology* **157**, 381–392.

Hicks J (2002) The physiological and evolutionary significance of cardiovascular shunting patterns in reptiles. *Physiology (Bethesda, MD)* **17**, 241–245.

Hicks JW, Krosniunas E (1996) Physiological states and intracardiac shunting in non-crocodilian reptiles. *Experimental Biology Online* **1**, 1–19.

Hicks JW, Wang T (1999) Hypoxic hypometabolism in the anesthetized turtle, *Trachemys scripta. The American Journal of Physiology* **277**, R18–R23.

Hicks JW, Wang T (2004) Hypometabolism in reptiles: behavioural and physiological mechanisms that reduce aerobic demands. *Respiratory Physiology & Neurobiology* **141**, 261–271.

Hicks JW, Wang T (2012) The functional significance of the reptilian heart: new insights into an old question. In *Ontogeny and Phylogeny of the*

Vertebrate Heart. (Eds D Sedmera and T Wang) pp. 207–227. Springer Science and Business Media, Philadelphia.

Jackson DC (1968) Metabolic depression and oxygen depletion in the diving turtle. *Journal of Applied Physiology* **24**(4), 503–509.

Jackson DC, Ultsch GR (2010) Physiology of hibernation under the ice by turtles and frogs. *Journal of Experimental Zoology A Ecology, Genetics and Physiology* **313**, 311–327.

Jones DR, Gardner M (2010) Ring around the heart: an unusual feature of the crocodilian central circulatory system. Australian Zoologist 35(2), 146–153.

Jones DR, Shelton G (1993) The physiology of the alligator heart: left aortic flow patterns and right-to-left shunts. *Journal of Experimental Biology* **176**, 247–269.

Karila P, Axelsson M, Franklin C, Fritsche R, Gibbins I, Grigg G, *et al.* (1995) Neuropeptide immunoreactivity and co-existence in cardiovascular nerves and autonomic ganglia of the estuarine crocodile, *Crocodylus porosus*, and cardiovascular effects of neuropeptides. *Regulatory Peptides* **58**, 25–39.

Khalil K, Zhaki K (1964) Distribution of blood in the ventricle and aortic arches in Reptilia. *Zeitschrift fur Vergleichende Physiologie* **48**, 663–689.

Kirshner D (1985) Buoyancy control in the estuarine crocodile, *Crocodylus porosus* Schnieder. PhD thesis. University of Sydney.

Kohmoto T, Argenziano M, Yamamoto N, Vliet KA, Gu A, De Rosa CM, *et al.* (1997) Assessment of transmyocardial perfusion in alligator hearts. *Circulation* **95**, 1585–1591.

Panizza B (1833) Sulla struttura del cuore e sulla cicolazione del sangue del *Crocodilus lucius. Bibllioteca Italiana* **70**, 87–91.

Pettersson K, Axelsson M, Nilsson S (1992) Shunting of blood flow in the Caiman: blood flow patterns in the right and left aortas and pulmonary arteries. In *Physiological Adaptations in Vertebrates: Respiration, Circulation, and Metabolism*, (Eds SC Wood, RE Weber, AR Hargens and RW Millard) pp. 355–362. Dekker, New York.

Sabatier A (1873) Etudie sur le coeur et la circulation centrale dans la série des vertebrés. *Annales Des Sciences Naturelles (Zoologie)* **18**, 1–89.

Seymour RS, Bennett-Stamper CL, Johnston S, Carrier DR, Grigg GC (2004) Evidence of endothermic ancestors of crocodiles at the stem of Archosaur evolution. *Physiological and Biochemical Zoology* **77**, 1051–1067.

Shelton G, Jones DR (1991) The physiology of the Alligator heart. The cardiac cycle. *Journal of Experimental Biology* **158**, 539–564.

Smith ED, Allison RD, Crowder WJ (1974) Bradycardia in a free ranging alligator. *Copeia* **1974**, 770–772.

Syme DA, Gamperl AK, Jones DR (2002) Delayed depolarization of the cog-wheel valve and pulmonary-to-systemic shunting in alligators. *Journal of Experimental Biology* **205**, 1843–1851.

Wang T, Hicks JW (2008) Changes in pulmonary blood flow do not affect gas exchange during intermittent ventilation in resting turtles. *Journal of Experimental Biology* **211**, 3759–3763.

Webb GJW (1979) Comparative cardiac anatomy of the Reptilia. III. The heart of crocodilians and a hypothesis on the completion of the interventricular septum of crocodilians and birds. *Journal of Morphology* **161**, 221–240.

Weinheimer CJ, Pendergast DR, Spotila JR, Wilson DR, Standora EA (1982) Peripheral circulation in *Alligator mississippiensis*: effects of diving, fear, movement, investigator activities and temperature. *Journal of Comparative Physiology* **148**, 57–63.

White FN (1956) Circulation in the reptilian heart *(Caiman sclerops). The Anatomical Record* **125**, 417–431.

White F (1969) Redistribution of cardiac output in the diving alligator. *Copeia* **1969**(3), 567–570.

Wright JC (1987) Energy-metabolism during unrestrained submergence in the saltwater crocodile *Crocodylus porosus*. *Physiological Zoology* **60**(5), 515–523.

Wright JC, Grigg GC, Franklin CE (1992) Redistribution of air within the lungs may potentiate "fright" bradycardia in submerged crocodiles *(Crocodylus porosus)*. *Comparative Biochemistry and Physiology A* **102**, 33–36.

9

DIVING AND SUBMERGENCE BEHAVIOUR AND PHYSIOLOGY

Although crocodylians have long had a strong reputation for being able to remain submerged for long periods, there has been a dearth of detailed information. In 1981, Peter Harlow, Bill Farlow, Kjell Johansen and I tried to get some information about natural diving patterns in free ranging, undisturbed Crocodylus porosus. We caught 12 crocodiles, 6–51 kg, in the downstream tidal Liverpool River, Northern Territory, Australia and attached what would now be seen as very clumsy devices, designed and built by Bill to monitor whether the head was above or below water. It is a story to weep over. We fitted devices to all of the animals, held in place with snazzy neoprene 'jackets' with radio transmitters attached, and released them at their capture sites. We trialled the jacket on the smaller crocs in captivity before release and they seemed unaffected, but when we

recaptured seven of them, 5–16 days later, only one croc was still wearing one!

We got excellent data from that one, a 9.5 kg female, which made the loss of what we could have had even more acute. Fortunately, really sophisticated depth loggers are now easily available and, nearly 30 years later, Frank Seebacher, Craig Franklin and Hamish Campbell got some great data from both 'freshies', C. johnstoni and 'salties', C. porosus (see below). But a lot more data are needed, and from more species and in different habitats, before we have a good understanding of those parts of crocodylian lives that are spent under water.

WHEN? WHY? HOW DEEP? HOW, AND HOW FOR SO LONG?

These are questions about crocodylian diving and submergence that it would be good to have answers to. Crocs have long been famous for the ability to stay under water for long periods (Figs 9.1–9.6), for 'a couple of hours even'. This has always attracted interest, probably because humans cannot, yet there are surprisingly few satisfactory answers to any of these questions. Until recently, most reports about natural diving behaviour have been anecdotal, but happily there is a wealth of new information from field studies on Australia's freshwater crocodile, *Crocodylus johnstoni*. One significant finding was very enlightening, and it confirmed my long held

Reproduced from Grigg et al. 1985, with permission. (Illustration GCG)

Large crocodylians, like this adult male Crocodylus porosus, *are capable of resting on the bottom for extraordinarily long periods. (Photo DSK, modified from image taken at WILD LIFE Sydney Zoo)*

309

suspicion: in undisturbed animals, the longest submergences were associated with periods of rest. Crocs always see the water as a place of safety, so where better to 'sleep' than under water? Active dives, such as might be expected in foraging, were very short. But this is information from one species at one location. There is a bit known about field submergences in *C. porosus* too, but with the suitable technology now available, it is hoped that a lot more will follow.

As well as reviewing what is known of their diving behaviour, this chapter will focus particularly on the longer dives and suggesting an explanation about how long dives can be supported physiologically. This discussion is based on physiological data gained from studies on juvenile *C. porosus* and American alligators, interpreted against the behavioural data gained in the field. Particularly, the focus will be on estimating the likely maximum duration of oxygen-supported submergences. A major conclusion will be that, as in other diving vertebrates, reliance on anaerobic metabolism to extend submergence is likely to be rare, with surprisingly long dives being well able to be supported aerobically.

The duration of time under water is only one aspect of diving capacity: depth is another. Even less is known about how deep crocodylians can and do dive. This has received little attention, presumably because most crocodylians live in relatively shallow habitats, so a capacity for deep diving has not been presumed. But *C. porosus* have been recorded at depths of 15 m (Craig Franklin *pers. comm.* 2012) and American alligators at 18 m (Lazaro Ruda; http://www.thelivingsea.com/), and they can probably go much deeper. Again, it would be wonderful to have more data.

DIVING BEHAVIOUR OF CROCODYLIANS IN THE WILD

Hugh Cott (1961), referring to Nile crocodiles, *C. niloticus*, observed that it is extremely difficult

Fig. 9.1. *An adult Cuvier's dwarf caiman,* Paleosuchus palpebrosus *at rest on the bottom of a freshwater spring in Mato Grosso do Sul, Brazil. (Photo Michael Patrick O'Neill)*

Fig. 9.2. *This American alligator was spotted by Lazaro Ruda in 18 m of water off the Florida Keys. Note how it is partly buoyed by the water; buoyancy issues are discussed in Chapter 4. Little or nothing is known about how deep crocodylians can dive and future studies are awaited with interest. Alligators are not equipped physiologically to live in sea water, but large animals can tolerate it for some hours and many alligators feed routinely in brackish or salty habitats (but they need periodic access to fresh water, as discussed in Chapter 11). (Photo Lazaro Ruda, TheLivingSea.com)*

to get reliable field data on submergence times and that little was known about diving endurance. He was clearly impressed by anecdotal accounts of large individuals remaining submerged and motionless on the bottom in clear water for an hour or so, and he expressed surprise at how long even small individuals could remain submerged. He made some observations in a pond on four small individuals (290–992 mm total length) and noted that the longest dives (to 44 min), which he thought were likely to be the maximum, were made by the largest individual. He graphed a trend line, which implied that adults could submerge 'far in excess' of the hour-long dives reported to him anecdotally. A crude extrapolation of his graph indicates that a 4-m individual (~300 kg) would have a maximum submergence time of ~2.25 h. As we shall see, this is probably an underestimate.

The extent to which the patterns described below are typical of crocodylians in general is unknown. There are few studies so far, but what there are show that patterns will be different in different habitats and that probable underlying circadian rhythms are modified by current flow and the state of the tide. Hopefully new technology will soon reveal a lot more information about crocodylian diving behaviour in a wider diversity of both habitats and species.

AMERICAN ALLIGATORS, *ALLIGATOR MISSISSIPPIENSIS*

There is little information about the diving behaviour of American alligators, *A. mississippiensis*, in the wild: Norbert Smith and his co-authors acquired some incidentally in the early 1970s. At that time, it was thought that any dive would be accompanied by severe bradycardia (see Box 9.1, p. 328). After

Fig. 9.3. *A 6-foot (2 m) American crocodile, C. acutus, photographed at night off the Turneffe Atoll, Belize. It shows the palatal valve closed and excluding water entry, which is always assumed but seldom observed. (Photo Louis Guillette)*

attaching a home-made heart rate transmitter to a 45 kg alligator, they found that, free-ranging and undisturbed, the animal surfaced every 5–7 min to breathe, showing no bradycardia during the dives (Smith *et al*. 1974). If spooked, however, it remained under water for 20–30 min with the heart rate falling as low as 2 beats per minute (Fig. B9.1). This is a tiny piece of information but it was very significant when it was published and it was the first study to report that most of the dives made voluntarily are for only short periods and without bradycardia.

ESTUARINE CROCODILES, *CROCODYLUS POROSUS*

There were no studies aimed at reporting the duration, frequency and daily (or tidal) and seasonal rhythms of natural dives in any crocodylians until our own (frustratingly disappointing) 1981 study on *C. porosus* (mentioned above) (Grigg *et al.* 1985) and the next came 20 years after that. The single

C. porosus (9.75 kg) from which we obtained data in the Liverpool River, Northern Territory, Australia was in a mangrove-lined and conspicuously tidal section of the river. The state of the tide had a marked influence on its behaviour. It was released and recaptured twice, each over a 4-day period. From the first period, we learned that it emerged from the water for several hours on each of the afternoon low tides, presumably to bask (Chapter 10). In the second release, diving was recorded and most of them were very short, 1–3 min (Fig. 9.7), in daylight hours and on the upper half of the tide (Fig. 9.8). We speculated that these short dives were foraging dives and Campbell *et al.* (2010a, 2010b) thought similarly about short dives made by *C. johnstoni* (see below). The daily patterns progressed with the tidal cycle. It is unlikely that a crocodile living in a tidal river with its comparatively rapid flows would rest under water as Campbell and colleagues described for *C. johnstoni* in a billabong but, when frightened,

Fig. 9.4. *An adult American alligator,* Alligator mississippiensis, *at rest under water in Everglades National Park, Florida, USA.* (Photo DSK)

C. porosus certainly use the water as a refuge, submerging for long periods. What prompted the 22 min and 30 min dives by this animal is unknown.

One other study of diving behaviour in free-ranging *C. porosus* has been reported. In addition to studies of movement by *C. porosus* (Chapters 4 and 12), Campbell *et al.* (2010c) recorded diving as well. They used acoustic tags to monitor the movements of 27 large (2–5 m) *C. porosus* over ~60 km in the North Kennedy River, a tidal river on the west side of Cape York, North Queensland. Some animals were also fitted with dive loggers. The crocodiles made good use of tidal flows to facilitate their travel. When current direction was unfavourable, they would either haul out onto the bank, or submerge 3–5 m to the bottom and await a change of tide. Periods between breaths averaged 18 min and some lasted an hour. Maximum submergences recorded by Campbell *et al.* in *C. porosus* are discussed below (Hamish Campbell *pers. comm.* and see Fig. 9.17).

FRESHWATER CROCODILES, *CROCODYLUS JOHNSTONI*

Studies of the diving physiology of crocodylians have mostly been of *C. porosus*, but most of the information about natural diving behaviour has come from *C. johnstoni* (Fig. 9.9). How typical the submergence patterns seen in *C. johnstoni* are of crocodylians in general is so far unknown, but diving behaviour is likely to be much more diverse across the group than the physiology that supports it, and data from crocodylians living in different habitats is sorely needed.

Field study of dive times and frequencies in C. johnstoni

The first systematic and quantitative study of crocodylian diving was undertaken by Seebacher *et al.* (2005) on *C. johnstoni* and it was followed by more work on the same species at the same location (Campbell *et al.* 2010a,b). All were conducted in a permanent freshwater billabong, ~350 m × 20 m

Fig. 9.5. *A large male* Crocodylus porosus, *'Smaug', submerged in clear water. His nostrils are closed and nictitating membranes cover his eyes (see Chapters 3, 5). (Crocodile courtesy of Adam Britton, photo DSK)*

Fig. 9.6. *Another view of Smaug, with a better view of the nictitating membrane and ISOs conspicuous. (Photo Adam Britton)*

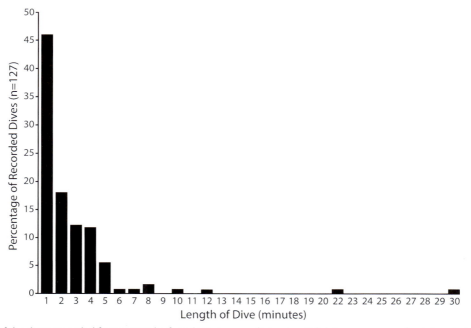

Fig. 9.7. *Most of the dives recorded from a 9.75 kg female* C. porosus *living in a tidal river were very short. In a 4-day period, only a few dives were for longer than 5 min and dives of 1–2 min were in the majority by far. (Adapted from Grigg et al. 1985, with permission)*

and up to 3.5 m deep, in Lakefield National Park in North Queensland.

The studies employed miniature time–depth, temperature and heart rate recorders/transmitters, either attached or implanted, that allowed remote collection of diving behaviour and other relevant information. It is difficult to study the behaviour of wild crocodylians because they are so cryptic and so easily disturbed by the presence of humans (Read *et al.* 2007), so these tiny microprocessor-driven recorders open vast new opportunities. However, even though they allow large quantities of

valuable information to be collected, interpretation without simultaneous visual observations is sometimes problematic. Nevertheless, the recent data on *C. johnstoni* and *C. porosus* provided significant information about their natural patterns of submergence, as well as added insights about daily and seasonal patterns and the effects of temperature.

As in *C. porosus*, most dives by *C. johnstoni* are quite short, lasting only a few minutes (Figs 9.10, 9.11). The proportion of time spent submerged by the five animals in the study by Seebacher *et al.*

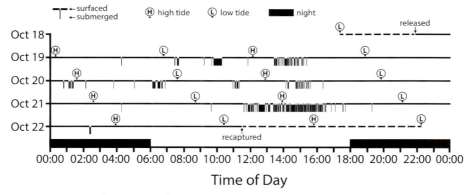

Fig. 9.8. *Dives recorded over 4 days from a 9.5 kg female* C. porosus *living in a tidal river. Most of her diving activity was in the form of short dives concentrated during the afternoon high tides. A separate dataset, recorded 4 days earlier, showed her out of water during the afternoon low tides, perhaps associated with basking (Chapter 10). (Redrawn from Grigg et al. 1985, with permission)*

Fig. 9.9. *Ruchira Somaweera snapped this 2–3 foot (60–90 cm) 'freshie' at about midday in 80 cm of water in a creek running into Lake Argyle in the north of Western Australia. It was showing no interest in fish that were swimming around it. Anecdotal reports of crocodylians remaining under water for very long periods have established their reputation as prodigious divers, but quantitative data from free-ranging crocodylians in the wild have become available only recently. Recent studies of* C. johnstoni *('freshies') have reported voluntary dives of 60– >120 min spent without activity, apparently resting (Fig. 9.17). (Photo Ruchira Somaweera)*

(2005) ranged from 18 to 41% and, overall, 55% of the dives lasted less than 15 min. Nearly 70% were in water less than 40 cm deep, from which it is easy

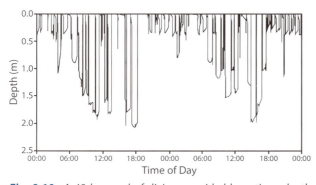

Fig. 9.10. *A 48-h record of diving, provided by a time–depth recorder attached to a* C. johnstoni. *Much can be learned, but without contemporaneous visual observations interpretation can be problematic. (From Seebacher et al. 2005, with permission. © 2005 University of Chicago)*

to raise the snout to the surface for a breath. The longest voluntary dives recorded from their five *C. johnstoni* were 68, 75, 95, 118 and 120 min, the last by a 14.8 kg individual. Seebacher and colleagues also reported that average dive durations were less at warmer temperatures, as could be expected at a warmer body temperature with higher metabolic rate, and that diving was accompanied by slight bradycardia, around a 12% reduction of the pre-dive rate. It is a moot point whether these small declines in heart rate are 'bradycardias' or whether the heart rates during the surface ventilation are 'tachycardias', facilitating rapid replenishment of the oxygen stores. It depends what is taken as the reference point, and the cycling in heart rates seen in repetitive surfacing events is paralleled by a similar cycling during intermittent breathing events on land (see Chapter 7).

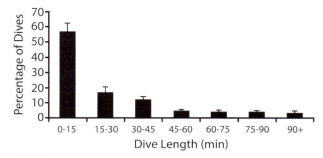

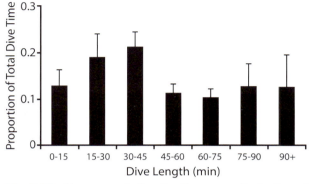

Fig. 9.11. *(Upper panel) Duration/frequency distribution of voluntary dives made by* C. johnstoni *in a fresh water billabong in Lakefield National Park, North Queensland during the dry season (mean ± s.e.). (Lower panel) The same data plotted a different way: proportion of time spent under water categorised by dive length. Much of the time under water is the result of the longer dives; in this example, 70% of the time submerged was the result of only 30% of the number of individual dives. (From* Seebacher *et al. 2005, with permission. © 2005 University of Chicago)*

In a more comprehensive study, Campbell *et al.* (2010a) recorded more than 13 000 dives by 17 *C. johnstoni* over a size range from 5 to 42 kg in the tropical winter dry season on Cape York, northern Australia (Fig. 9.12). They followed it up with further study in the summer at the same location (Campbell *et al.* 2010b). They were able to distinguish two different types of dives: 'active' dives, which were mostly less than 1 min and probably related to foraging, and 'resting' dives (Campbell *et al.* 2010a). 'Resting' dives were much longer than active dives and were conducted mostly in deeper water, with the animal presumably at the bottom of the lagoon until surfacing briefly for a breath (Figs 9.13, 9.14). Diagnosis of the short dives as 'active' and the longer dives as 'resting' was possible by noting whether or not there were changing depths during the dive and because some of the study animals also

had an attached telemetric device to record tail beat, an indication of swimming.

Resting dives averaged between 11 and 13 min in duration and showed no significant correlation with body mass. These accounted for 63% of the dives logged, but resulted in 97% of the time spent submerged! After each 'resting' dive, surface time was quite brief, usually ~50 seconds, and in animals larger than 15 kg even less, often 20 seconds or so. Some of the apparently resting dives were very long. The longest in each of eleven individuals ranged from 60 to 149 min (Hamish Campbell *pers. comm.* 2011). These are plotted in Fig. 9.17.

Campbell *et al.* (2010a) noted that the short, 'active' dives coincided with times likely to be suitable for prey capture, and they were followed by longer periods at the surface than 'resting' dives. The reason/s for the longer periods at the surface between short foraging dives is unknown. Campbell and colleagues reasoned that active dives were terminated before anaerobic metabolism was required and, considering the calculations plotted in Fig. 9.17, it seems unlikely that the longer surface times could be due to any need to 'pay back an oxygen debt' (EPOC, see Box 9.1, p. 328). Presumably most foraging dives would be short, speculative, unsuccessful, and end well before plasma lactate levels rise to a compromising extent. More likely time at the surface is part of their 'sit-and-wait' foraging strategy. They forage both under water and at the surface, monitoring above and below the water surface visually and with their integumentary sense organs (ISOs) (see Chapter 5). A prey capture that involves a struggle could certainly lead to a build up in lactate, which might dictate the need for a subsequent recovery period (Chapter 7).

Disturbance by the investigators prompted the very longest dive times. The most spectacular were a 344 min dive by a 5 kg animal and a 402 min dive by a 42 kg animal in response to recapture attempts (Fig. 9.14 lower panel). During these dives, the time–depth recorder showed that the animals remained sedentary at the bottom. Such long dives in crocodiles of this size must surely be beyond any aerobic capacity, and this was confirmed by a sample taken at the end of the 344 min dive, which gave a plasma lactate reading of 24.3 mmol/L

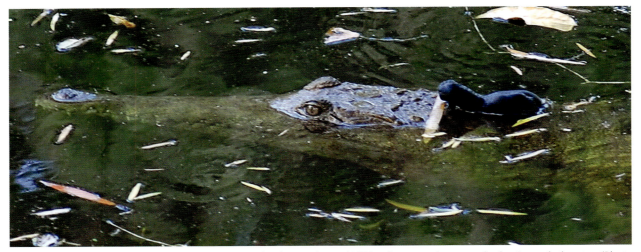

Fig. 9.12. *A large freshwater crocodile with a time-depth recorder and VHF radio-transmitter attached to its nuchal scutes. (Photo Hamish Campbell)*

(compared with less than 1 mmol/L in a resting *C. porosus*, Wright 1987).

Daily rhythm in 'resting' and 'active' submergences by C. johnstoni

In both winter and summer, the animals showed circadian rhythms in their daily patterns of behaviour (Fig. 9.15). It was possible to construct a daily activity pattern of the *C. johnstoni* in these studies, and to quantify the extent to which time spent underwater was part of it. The data in Campbell *et al.* (2010a)

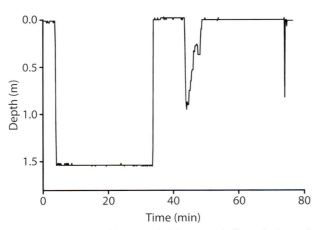

Fig. 9.13. *Dive profiles recorded by Campbell et al. (2010a) from* C. johnstoni *living in a permanent waterhole in north Queensland. All profiles fitted one of these three patterns: a lengthy dive at a single depth (judged to be a resting dive); or 'active' dives either with or without changing depth. The short, active dives may be associated with foraging behaviour. (Modified from Campbell et al. 2010a, © British Ecological Society)*

show distinct daily rhythms in the occurrences of 'resting' and 'active' dives and the depths to which the animals submerged (Fig. 9.15). The early daylight and mid-morning hours (0400–0900 h) for *C. johnstoni* in that lagoonal habitat were characterised by 'resting', with fewer than 13% of dives logged in those hours classified as 'active'. The animals spent 50–60% of the morning submerged, in deeper water than at other times of the day. As the day wore on, the proportion of 'active' dives increased, along with proportionally more time spent at the surface. The latter was interpreted as foraging time. In the early evening (1700–1900 h), more than 65% of the logged dives were 'active' and during the evening and night more than 80% of the time was spent at the surface. These are averaged patterns, and individuals may behave differently and vary from time to time. One example of a different day's activity is shown in Fig. 9.14 (upper panel); for whatever reason, this croc spent most of the day submerged. What these records do not record, of course, is any time spent out of the water; presumably some of the 'surface' time is spent on land, basking or foraging.

In a different habitat, behaviour patterns for this species could be quite different and there is likely to be a wide variety of submergence patterns across the different species. Capacity for diving can be expected to vary with body size (see later) but this trend was not seen in *C. johnstoni* during winter (Campbell 2010a), presumably their behavioural requirements were not constrained by aerobic

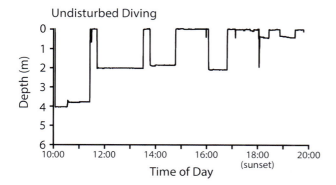

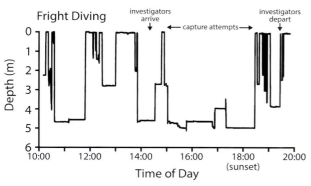

Fig. 9.14. *(Upper panel) This 42 kg* C. johnstoni *must have been having a lazy day, spending most of the time resting on the bottom. Crocodylians are not robotic and there is variability between and within individual behaviour patterns. (Lower panel) The arrival of the investigators setting their nets to capture him prompted the animal to remain submerged for one of the longest dives recorded, 344 min. (Graphs provided by Hamish Campbell)*

limitations. In summer, however, with higher oxygen consumption at warmer temperatures, smaller animals dived for less time than large ones (Campbell *et al.* 2010b). Perhaps they were more constrained than larger ones because of their relatively smaller lungs (Chapter 4 and below).

SELECTION PRESSURES FAVOURING CAPACITY FOR MAKING VERY LONG DIVES

The water is a refuge for crocodylians. Campbell *et al.* (2010a) found that *C. johnstoni* spend daily rest periods on the bottom, surfacing to breathe at intervals, and sometimes these periods of submergence are very long. In today's world, crocodylians need have little fear of predation once they reach a reasonable size (Chapter 13), but they evolved in times when there were many large reptiles about, including larger crocodylians. Apart from

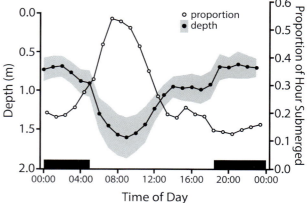

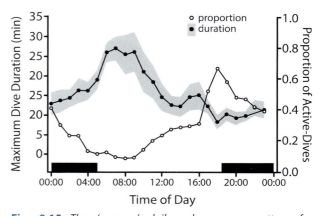

Fig. 9.15. *The 'average' daily submergence pattern for* C. johnstoni *in a Queensland tropical lagoon, derived from more that 13 000 dives by 17 individuals (5–42 kg) recorded automatically by Campbell et al. (2010a). For each animal, and for each hour of the daily cycle, the longest dive, the proportion of the hour spent submerged, the average depth of submergence and the proportion of dives which were active (as opposed to resting) were noted. The analysis revealed that, in this season and in this habitat, the majority of the animals spent the mornings resting, submerged on the bottom and, in the afternoon and evenings, made short active dives and/or pursued other activities, presumably including foraging. Grey shading shows ± SE. (Adapted from Campbell et al. 2010a, with permission)*

resting and taking refuge on the bottom, perhaps crocodylians also use their ability to submerge for long periods as a tactic for ambush, lying on the bottom with eyes, ears and integumentary sense organs (ISOs) all alert (Chapter 5).

It is easy to imagine selection pressures favouring the evolution of attributes that enable them to extend their time under water. It is less easy to see any benefit if prolonged dives were to require significant anaerobic metabolism because of the long recovery time this implies (Chapter 7 and below). Anaerobic support for long dives undertaken voluntarily would seem to be counter-productive. As in birds and mammals, it is much more likely that almost all crocodylian diving is undertaken aerobically, and that evolutionary selection pressures have favoured extending submergence time aerobically rather than anaerobically; hence, probably, the evolution of their pulmonary by-pass shunt (below and Chapter 8). If their terrestrial ancestors were endothermic, as argued by Seymour *et al.* (2004) (Chapter 10), an additional benefit from re-evolving ectothermy would have been a capacity for longer submergences as they made the transition to an aquatic lifestyle (Chapter 10).

The next section explores the physiology of crocodylian diving, leading to projections about the likely maximum capacities for aerobic dives. It will develop calculated aerobic dive limits (cADL) over a wide range of body mass, based on data from *C. porosus*, and compare these with recorded field observations.

Physiological support for crocodylian diving

As intermittent breathers (Chapter 7), crocodylians can be thought 'pre-adapted' to diving. Many, if not most, of their dives occur within or close to normal non-ventilatory periods. However, many of the dives last for much longer, and these attract particular attention and demand explanation, which is the subject of this section.

Knowledge about the diving physiology of crocodylians has grown alongside that for birds and mammals and this forms a useful reference (see Box 9.1, p. 328). Many of the conclusions about diving mammals are applicable to crocodylians as well: most of their dives are short, with little physiological adjustment, and in long dives they exhibit bradycardia and an increase in peripheral vascular resistance with a reduction and redistribution of cardiac output. During long dives, they also show a reduction in metabolic rate. However, apart from being intermittent breathers, they also show some other significant differences from mammals, as follows:

Differences between crocodylians and diving mammals

1. Crocodylians have a lower metabolic rate (Chapter 7)

As ectotherms, crocodylians have much lower oxygen requirements than birds and mammals, (~80% less), so longer aerobic dives can be expected. This is a quantitative, not a qualitative difference.

2. Crocodylians dive with a substantial lung oxygen store

Long dives by crocodylians tend to be spent at the bottom, for resting or refuge, and slightly negatively buoyant. They often make a partial exhalation leaving the surface or on the way down, presumably to adjust buoyancy. Buoyancy adjustment must be very important for them: measurements on juvenile *C. porosus* showed that voluntary submergences occur with a specific gravity of 1.028 with almost no variation (Kirshner 1985, see Chapter 4). This means that, unlike diving mammals, crocodylians dive with a substantial pulmonary oxygen store. A 500 kg *C. porosus* diving in shallow water could be expected to have a submergence lung volume of ~50 L compared with a maximum inflation volume (such as in HOTA posture) of ~75 L (Donald Henderson *pers. comm.* 2011). This is a stark contrast to diving mammals, in which long dives tend to be deep dives, and before which they exhale most of the air from their lungs. This protects them against 'the bends', caused by nitrogen dissolving in the blood under pressure at depth and then bubbling out of solution at re-surfacing, causing tissue damage. In Weddell seals, elephant seals and sperm whales, all deep divers, the lungs contribute very little to oxygen store (only 4–6%), the blood about two-thirds and the rest in the muscle as myoglobin (Ponganis *et al.* 2011). In *C. porosus*, the lungs hold at least half of the total oxygen store at the start of a dive, almost all the rest being in the blood (see below). All the research so far has been on shallow diving individuals. It would be fascinating to look at the physiology of deep diving in a crocodylian, such as the alligator at 18 m in Fig. 9.2. This animal shows no sign of collapse in the thoracic region under an additional 1.8 atmospheres of hydrostatic pressure, so it must have left the surface quite positively buoyant, as if anticipating a deep dive, and swum down quite actively. Do crocodylians get 'the bends'? Sea snakes also dive with a lot of air in the lungs, and they avoid the bends by losing nitrogen to the water across the skin (Seymour 1974). Oxygen does not diffuse readily through crocodylian skin (Wright (1986a) and nitrogen is less diffusible than oxygen, so this 'escape' is unavailable to them.

It is worth noting that hatchlings are poor divers, tending to be very positively buoyant and apparently needing to exhale significantly to allow submergence; studies of diving physiology in hatchling crocodylians are unlikely to be representative.

3. Crocodylians have much less myoglobin

Rich red muscles are a feature of diving mammals and may contribute 30–50% of the oxygen store (Ponganis *et al.* 2011). Most crocodylian muscles, however, are creamy white and a few are pale pink at best. Their oxygen store, therefore, comprises what is bound to haemoglobin in the blood plus whatever is taken down in the lungs, with a little dissolved in the tissues.

4. Crocodylians have a pulmonary by-pass shunt

This was explored in Chapter 8. Briefly, diving in crocodylians is often associated with the implementation of a pulmonary by-pass shunt in which blood low in oxygen that would normally be directed to the lungs is recirculated to the systemic arterial circulation, lowering arterial oxygenation. A likely interpretation is that this leads to a reduction in aerobic metabolic rate (hypometabolism), thus eking out the oxygen store and extending their aerobic dive limit. This interpretation is not without controversy (Chapter 8), but finds support from work on freshwater turtles, which also have a pulmonary by-pass shunt and show reduced metabolism during long dives.

THE DIVING PHYSIOLOGY OF CROCODYLIANS

A question to settle early is whether or not long dives of crocodylians could be explained by significant oxygen uptake through the skin, as in some other diving reptiles.

Is there cutaneous uptake of oxygen?

No. Although sea snakes can take up oxygen through the skin (Seymour 1982), Wright (1986a) measured gas exchange across the skin in *C. porosus* and found it to be negligible. Because *C. porosus* is a lightly armoured crocodylian, with comparatively little dermal ossification, it seems safe to assume that cutaneous gas exchange in all crocodylians is so small that it can be ignored.

Physiological correlates of inactive dives, voluntary and 'disturbed'

Wright (1985a, 1985b, 1987) measured a range of physiological variables in juvenile *C. porosus* (1.81–.2 kg) either diving voluntarily or 'persuaded' to extend their period of submergence. Measurements were made in a large tank, shallow enough for the animals to reach the surface without swimming. The water surface was excluded to them, except into a respiratory hood at one corner. The crocodiles learned to surface and breathe in this hood, through which a known flow of air was drawn and oxygen content recorded, enabling oxygen consumption to be calculated. The animals soon habituated to the situation and would lie quietly at the bottom of the tank for many hours, raising themselves periodically to breathe. Wright measured pre- and post-dive oxygen consumption and, in some individuals, heart rate, blood and lung gases and plasma lactate sampled through indwelling cannulae. Most measurements were made during quiet, undisturbed submergences made voluntarily, but dives could be extended artificially by denying animals access to the hood. In some cases, a 'fright' was delivered when the subject was either at the surface or on the bottom.

Undisturbed crocodiles in these experiments spent most of their time on the bottom, surfacing to breathe every few minutes. Average submergences

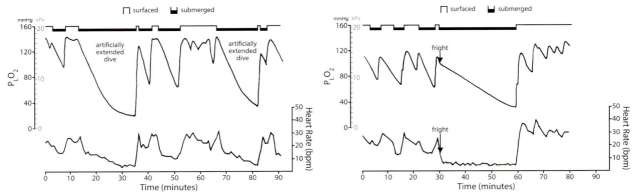

Fig. 9.16. *Changes in lung gas oxygen level and heart rate during three types of aerobic dive shown experimentally by a juvenile* C. porosus. *The slope of the fall in lung oxygen is an indicator of oxygen consumption. Several voluntary undisturbed dives, two 'artificially extended' (by denying access to air) and a single 'fright' dive (by banging on the tank just after a submergence) are shown. Plasma lactate was in the normal range after each of these dives. It was elevated if dives were prolonged artificially for 50–60 min, which was stressful. Even these small crocodiles clearly have a great capacity to conserve oxygen and extend their periods of submergence without needing to rely on anaerobic metabolism, which appears to be recruited only as a last resort. (Adapted from Wright 1987, with permission. © University of Chicago)*

in different individuals ranged from 1.3 to 5.7 min, 90% of dives were less than 5 min and some lasted 9–10 min. These short dives used 6–12% of the calculated total oxygen store, several 20% or so and the maximum was 30.2%. Surface times between dives were mostly brief, and the proportion of time submerged averaged 39.4% (12–74% in different individuals). These dives were similar in length to most of the dives recorded in the field by Grigg *et al.* (1985) for *C. porosus* (Fig. 9.7) and Seebacher *et al.* (2005) for *C. johnstoni* (Fig. 9.10). A small, but statistically significant, bradycardia developed during these 'relaxed', undisturbed, voluntary dives. Heart rate fell by ~14% from the 20 bpm averaged at the surface ($t = 25°C$). This is very similar to the way heart rate cycles during the typical intermittent ventilatory pattern of a crocodylian on land (Chapter 7) and similar to the 12% drop reported in *C. johnstoni* diving voluntarily in the field (Seebacher *et al.* 2005). Lung oxygen increased with each surfacing (Fig. 9.16).

A crucial question, of course, is about oxygen consumption during the dive. Lewis and Gatten (1985) found that there was no reduction in oxygen consumption during short voluntary dives by juvenile *A. mississippiensis*. Wright (1987) found the same in juvenile *C. porosus*. He assessed the rate of oxygen consumption during the dive by comparing the post dive oxygen uptake with pre-dive levels,

finding the increase matched the replenishment of the depleted oxygen store. A shortfall would have indicated a reduction in metabolic rate, and an increase would suggest an 'oxygen debt' to 'pay back'. Oxygen consumption remained at pre-dive levels during undisturbed, voluntary dives (range 2–13 min, average 5 min) and, because plasma lactate did not increase and there was no EPOC, he concluded that the dives had been conducted aerobically. Indeed, the animals surfaced from the majority of these dives with most of the oxygen store still intact, lung P_{O_2} typically at 70 mmHg, compared with 123 mmHg at submergence (see Chapter 7), so there was considerable room to extend the dives further aerobically. At the end of the longest (13 min) dive, lung P_{O_2} was 40 mmHg and the arterial blood ~75% saturated (Wright 1985a,b).

To encourage longer dives, Wright denied the animals access to the hood and, rather than immediately employing anaerobic metabolism, crocodiles reduced oxygen consumption. Two such dives are shown in Fig. 9.16 (left). Continuously monitored heart rate and pulmonary oxygen levels are revealing (Wright 1987) (Fig. 9.16). Lung P_{O_2} values fell and so did heart rate, the slope of the decline in lung oxygen values being a good indicator of oxygen uptake. Oxygen consumption declined slowly in the latter stages, accompanied by a further slowing of heart rate (Fig. 9.16 left). In such

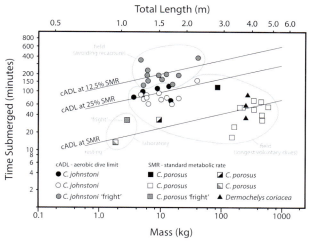

Fig. 9.17. *Longest submergences observed in the field (except as marked) in the two species of Australian crocodylians compared with predicted maximum aerobic submergence limits (cADL) calculated conservatively for* C. porosus *across a size range. Predicted limits were calculated at 25°C, assuming use of 68% of the oxygen store (Wright 1985a) and that oxygen consumption is maintained at pre-dive (resting) rates (lower line), at 25% (middle line) and 12.5% of pre-dive rates.* C. johnstoni *diving voluntarily (black circles) (Seebacher et al. 2005);* C. johnstoni *diving voluntarily (open circles) and with disturbance as the animals evaded recapture (shaded circles) (Campbell et al. 2010a);* C. porosus *diving voluntarily (open squares) (Hamish Campbell and Craig Franklin pers. comm. 2010);* C. porosus *(black square) (Craig Franklin pers. comm. 2009);* C. porosus *diving voluntarily (black and white square) (Grigg et al. 1985);* C. porosus *laboratory, resting (grey and white square) (Wright 1985b);* C. porosus *laboratory 'fright' dive generated by banging on the tank (grey square) (Wright 1987). Data from leatherback turtles,* Dermochelys coriacea, *are shown for comparison (black triangles) (López-Mendilaharsu et al. 2008 (upper point) and Southwood et al. 1999). Broad similarity across crocodylian species in metabolic rate, lung volumes and general physiology suggest that data gained on other species could usefully be compared with this graph.*

artificially prolonged dives, oxygen consumption did not decline until lung oxygen reached partial pressures of 40–50 mmHg, as if the fall in internal oxygen levels had provoked metabolic depression. The implication is that whereas crocodylians mostly dive with little or no bradycardia and little or no decrease in oxygen consumption, they can also extend their aerobic dive by progressively reducing heart rate and oxygen consumption during the dive.

If a 'fright' was administered, a severe bradycardia developed immediately, with a drop in heart rate to

3–4 beats per minute and a coincident fall in the rate of oxygen removal from the lungs (Fig. 9.16 right) suggesting operation of the PBS and a slowing in aerobic metabolism. This occurred whether the animal was already submerged or at the surface, in which case it would dive immediately after the fright was delivered. Dive times were prolonged to an average of ~20 min (Wright *et al.* 1992). Coincident with the onset of the 'fright' bradycardia there was a redistribution of air in the lungs anteriorly, without any exhalation of air, which may have influenced the initiation of bradycardia (Wright *et al.* 1992). The dramatic bradycardia in response to a 'fright' was similar to that reported by Smith *et al.* (1974) from a free-ranging alligator (Fig. B9.1). At the end of the fright-induced prolonged submergence, lasting 30 min in the example in Fig. 9.16 (right), plasma lactate levels were in the normal range (less than 1 mmol/L), indicating that these dives, too, were conducted aerobically. Upon surfacing, heart rate and ventilatory rate were increased, reflecting replenishment of the oxygen stores to pre-dive levels. Post-dive oxygen consumption was not elevated beyond that required to replenish oxygen stores.

These results, from animals at rest or nearly so, are in stark contrast to the long prevailing view that diving in crocodylians needs to be supported anaerobically: a view that, for some reason, persisted longer in crocodylians than in mammals and birds.

Challenging the animals further, Wright (1987) found that plasma lactate levels were elevated after only the longest artificially extended dives. After denying them access to the hood for 50–60 min, with the animals becoming quite restless, he found post-dive lactate levels of up to 3 mmol/L. This was a several-fold increase over resting levels, but small in comparison to values recorded after exhaustive exercise in *C. porosus* (up to 15 mmol/L, Wright 1985b, and 20–30 mmol/L, Seymour *et al.* 1985) and 22–24 mmol/L after 3–5 h submergence (!) by *C. johnstoni* seeking refuge from the investigators (Campbell *et al.* 2010a).

Putting this series of experiments together, it seems reasonable to conclude that quite long resting dives can be supported aerobically and strategies to extend dive time aerobically, including metabolic depression, are employed before an animal resorts to anaerobic metabolism with its subsequent costs.

By how much can oxygen consumption be reduced? Wright's data imply reductions in oxygen consumption over a wide range: by ~60–90% of pre-dive levels. There are few data to compare this with directly, but Jackson (1968) measured metabolic reduction calorimetrically in painted turtles, *Chrysemys picta*, and recorded comparable drops of 60–85%, although some of these were during forced dives. These reductions are also consistent with the conclusion of Guppy and Withers (1999) who, in an extensive review of metabolic reductions in aestivating or hibernating animals, concluded that 'metabolic depression to approx. 0.05–0.4 of rest is a common and remarkably consistent pattern for various non-cryptobiotic animals (e.g. molluscs, earthworms, crustaceans, fishes, amphibians, reptiles)'. Such reductions parallel the reductions in heart rate seen commonly in diving reptiles quite well, so it seems reasonable to conclude, for calculations of cADL (see Box 9.1, p. 328), that aerobic metabolic rate in crocodylians during prolonged dives may fall by 60–90%, depending upon the nature of the dive. The reduction in metabolism during the dive is almost certainly a consequence of reduced perfusion with only partly oxygenated blood of many of the organs and tissues. The role of reduced blood flow in depressing metabolic rate has been proposed in diving mammals (Hochachka 1992, 2000 and see Box 9.1, p. 328), the result of a redistribution of blood from the skin, muscle and many of the organs, along with reduced cardiac output. Redistribution of blood during prolonged dives has been observed in crocodylians too. Weinheimer *et al.* (1982) measured blood flow in the muscle and skin of the lateral tail region of juvenile alligators and found little or no evidence of peripheral vasoconstriction during 2–15 min dives undertaken voluntarily, but a marked reduction or cessation of blood flow during dives provoked by disturbance, or forced under for up to 45 min. These authors reported that the alligators appeared stressed at the end of forced dives, which is not surprising.

In summary, although much more data are needed the following interpretation seems reasonable.

For submerged, inactive animals there seem to be four 'modalities' of submergence, three of which are aerobic and, probably employed only in unusual situations, a fourth invoking anaerobic metabolism. All are depicted in Fig. 9.16.

- Most inactive dives are short (up to 10–15 min in 2–3 kg juveniles, longer in larger animals), involve no reduction in oxygen consumption and little alteration in heart rate or blood flow. These are similar in pattern to the intermittent respiration seen in a croc at rest on land.

- Longer inactive aerobic dives can be made, to 30–40 min (in 2–3 kg juveniles), longer in larger animals. They are accompanied by a slowing of the heart rate, reduced cardiac output and reduced oxygen consumption, typically once lung P_{O_2} values fall below ~40 mmHg. Reduced oxygen consumption is initiated by reduced flow of less-well-oxygenated blood, the result of a pulmonary by-pass shunt triggered by tightening of the cog-wheel valve at the base of the pulmonary aorta (as discussed in Chapter 8). This is accompanied by peripheral vasoconstriction and redistribution of blood flow away from the skeletal musculature (which comprises about half of the body weight) and elsewhere within the systemic circulation, as described in force-dived or disturbance-dived alligators (Weinheimer *et al.* 1982).

- So called 'fright' dives, in response to a sudden disturbance, are similar to prolonged aerobic dives except that the precipitous bradycardia and associated events, including metabolic depression, occurs immediately, extending the life of the oxygen store. Such dives can be undertaken aerobically, but, as shown in the very long dives by *C. johnstoni* avoiding recapture (Fig. 9.14 lower panel) can progress to having an anaerobic contribution too.

- The very longest inactive dives are those that extend beyond the extended aerobic limit; that is, beyond the 'dive lactate threshold', DLT, and in which the support of anaerobic metabolism is recruited and lactate accumulates. These are dives in which an animal is taking severe steps to avoid capture or predation, as in Fig. 9.14 (lower panel), after which a high plasma lactate value was reported. Such long dives, it seems, could not be associated with significant exertion as well, as will be discussed in the next section.

These patterns can grade into each other and it is probable that a pattern of progressive recruitment of dive-extending strategies is typical of crocodylians in general.

Physiological support of 'active' dives

Whereas most foraging dives will be short and unsuccessful, even the successful ones are likely to be short too, with the prey item brought to the surface for swallowing. Depending upon the extent of the struggle associated with capture, there may be some lactate accumulation. The limit of anaerobically supported activity is set by the glycogen reserve of skeletal muscle (Chapter 7) and exhaustive activity can be sustained for a period that depends upon body size: maybe 10 min by a 20 kg animal or 25 min by a 100 kg animal (Fig. 7.4). It seems likely that the time limits of active dives are set more by the glycogen reserve and its rate of consumption than the oxygen store, and very energetic dives will inevitably be short in comparison to resting dives supported aerobically. This is confirmed subjectively by the ease of tiring a harpooned croc.

The brevity of the 'active' dives described in *C. johnstoni* by Campbell *et al.* (2010a) is easily explicable as a consequence of a preference (like other diving vertebrates, Box 9.1, p. 328) for remaining aerobic: typical foraging dives are short and repetitive, avoiding a significant build up of lactate and the inevitable need to spend time recovering, which may constrain further activity. After exhaustive exercise, lactate levels in the plasma may reach 15–30 mL/L. It is worth noting that Wright (1987) showed that *C. porosus* can pay back an oxygen debt from a forced, anaerobic dive by resting on the bottom and surfacing briefly from time to time. During this, however, further activity (pursuit or escape) will be compromised. Emperor penguins (*Aptenodytes forsteri*), too, need

not remain surfaced to recover from dives lasting beyond their aerobic dive limit (Sato *et al.* 2011) and this is likely to be a common practice among diving vertebrates.

When more data become available, they will probably show that significant lactate build up in the everyday lives of crocodylians is uncommon.

WHAT ARE THE LONGEST SUBMERGENCES BY CROCODYLIANS THAT CAN BE SUPPORTED AEROBICALLY?

Aerobic dives are probably taken to the absolute maximum only rarely, but it is interesting to speculate about what a maximum may be, especially as it has been assumed erroneously for a long time that long dives depend on anaerobic metabolism. Our focus will be on dives (submergences) made voluntarily. Could the long, undisturbed dives reported above for *C. johnstoni* and *C. porosus* have been undertaken aerobically?

PRACTICAL DIFFICULTIES OF MEASURING MAXIMUM DIVE CAPABILITY

We may never be able to measure a maximum duration directly with any confidence: it is just not possible to ask a crocodylian to stay under until it can remain there no longer! Furthermore, crocodylians may never actually stay submerged voluntarily to their maximum – even direct observations can provide only minimum estimates of any maximum.

One measure of aerobic dive limit would be a determination of the longest dive possible under natural conditions without any build up of plasma lactate. This would be the dive lactate threshold, DLT (Butler 2006) (see Box 9.1, p. 328). Such data are not yet available in any crocodylian and, anyway, it would be very difficult to collect in the field because of the difficulty of avoiding a capture (and activity) artefact. One approach might be to monitor plasma pH telemetrically. Nor is it known whether the transition from aerobic to anaerobic metabolism would be as marked, as in seals. It could be a gradual transition: a naturally occurring prolonged dive might typically be accompanied by underwater muscular activity (e.g. holding onto a struggling wildebeest!). A further drawback is that dives taken naturally to reach the limit are probably quite rare in crocodylians, adding an additional practical difficulty.

In the absence of information about a DLT, we can only make speculative calculations based on reasonable assumptions. Because there is more information available for *C. porosus* than other species, the modelling will be based on it. However, crocodylians have broadly similar physiology, so conclusions drawn for *C. porosus* will be indicative of what is probable in other species too.

CALCULATIONS OF AEROBIC DIVE LIMIT, cADL

Calculated aerobic dive limit (cADL) is typically determined by dividing the useable oxygen available at the start of the dive by an estimate of the rate of oxygen consumption. But aerobic dives do not usually use all of the 'useable' oxygen: Wright (1987) showed that juvenile crocodiles routinely surface with substantial oxygen 'still in their tank'. Our estimates will be therefore be conservative because they are based on examples of actual oxygen useage, measured on voluntary dives. Whether cADL is a suitable term for this is debatable; maybe 'calculated **voluntary** aerobic dive limit', cvADL, would be more correct. Body size needs to be taken into account too, because metabolic rate and lung volume scale differently with mass and, because crocodylians are ectotherms whose metabolic rate varies with temperature (Chapter 10), dive time depends also on body temperature. Indeed, the effect of temperature has been demonstrated – Seebacher *et al.* (2005) measured shorter dive times in *C. johnstoni* at warmer body temperatures, over a range of body temperature from 22 to 32°C.

Without substantial myoglobin, the oxygen store at submergence is made up of what is held in the lungs, blood, and dissolved in the tissues (including the blood plasma). Some of the calculations depend upon assumptions about certain values that are difficult or impossible to measure, so the results presented should be regarded as indicative rather than definitive.

Oxygen in the lungs at submergence

Lung volume in a swimming or diving crocodylian is determined primarily by buoyancy requirements and not respiratory requirements (Kirshner 1985 and Chapter 4). Wright and Kirshner (1987) measured the volume of air in the lungs (V_L) at submergence by plethysmography in 24 juvenile *C. porosus* over a size range from 271 to 3762 g at 23–25°C. From their work, a 1 kg *C. porosus* could be expected to submerge with a lung volume of 46.8 mL, at which it would be slightly negatively buoyant (See Chapter 4). The density of bony tissue can be expected to increase with growth, as in most other vertebrates, so the volume of the lungs at submergence can be expected to scale with body mass with an exponent greater than 1.0 to maintain the same, slightly negative buoyancy (as specific gravity of 1.028 in fresh water). Lacking a crocodile-specific exponent, 1.09 was chosen, the same as for mammals and birds (Calder 1984). Considering the bone density of crocodylians and their osteoderm development with growth, this is likely to be an underestimate. Whereas juvenile crocodiles are less likely to have them, adults almost always have gastroliths, acquiring them progressively towards adulthood (Fig. 4.30). The acquisition of 1% body weight of gastroliths implies an increase in lung volume of 20% to maintain the same buoyancy state (Kirshner 1985, Chapter 4). Taking scaling and the acquisition of gastroliths into account, a 500 kg *C. porosus* can be expected to submerge in fresh water with a lung volume of 49 L and a 1000 kg individual with 105 L. Confirmation that these values are sensible comes from an entirely different approach. Following on from his exploration of a possible role for gastroliths as stabilisers in the HOTA posture (see Chapter 4), Donald Henderson at the Royal Tyrrell Museum of Palaeontology, Alberta, calculated that a 462 kg alligator would be floating at the surface with a lung volume of 48 L, but would sink if that were reduced to 45.6 L (Henderson *pers. comm.* 2012). It is satisfying that two completely independent approaches, each involving several uncertain assumptions, produced reasonably similar outcomes.

Lung gas, of course, differs from air. An average partial pressure of oxygen in the lungs of juvenile *C. porosus* of 124 mmHg after a ventilation event (see Chapter 7) equates to a fractional oxygen concentration of 0.168 (Wright 1985b). If this is typical of large crocodiles as well, it enables V_L to be converted to lung oxygen content at submergence for a crocodile of any body mass. In these three examples above, the 1 kg crocodile would dive with 7.86 mL oxygen in its lung, a 500 kg croc with ~8.2 L and a 1000 kg croc with ~17.6 L of oxygen.

Oxygen in the blood and other tissues at submergence

Estimates of the amount of oxygen carried bound to haemoglobin in the blood at submergence can be made from knowledge of the blood volume, the haemoglobin concentration and the percent saturation of arterial and venous blood. In the following calculations, 25% of the blood is assumed to be arterial and the measurement of blood volume (7.27%) made by Huggins (1961) on *A. mississippiensis* is used and assumed to scale directly with body mass, as in other vertebrates. Field-measured haematocrits averaged 24.8% from 96 *C. porosus*, which translates to 7.73 g/100 mL haemoglobin and thus 11 vol % O_2 (4.92 mmol/L) (Grigg and Gruca 1979; Grigg and Cairncross 1980). Percentage saturations of arterial and venous blood were assessed from resting values for oxygen and CO_2 partial pressures measured in arterial and venous blood (90 and 30 mmHg, respectively, for O_2 and 19 mmHg for CO_2, Wright 1985a,b), using the oxygen equilibrium curves measured by Grigg and Cairncross (1980, see Chapter 7).

This leads to a blood oxygen store of 5.71 mL/kg of body weight at submergence. Because blood volume, packed cell volume (PCV, or haematocrit) and, therefore, haemoglobin concentration are all thought to scale isometrically with body mass, this can be taken as appropriate over a range of crocodylian sizes. Wright calculated that ~3.8% of the oxygen store at submergence is dissolved in tissues and blood plasma, and it is reasonable to assume that this also scales directly with body mass. The relative contribution of the lungs as oxygen store, compared with the blood, increases with body size, from 56%, with 40% in the blood in a 1 kg crocodile to 73% with 24% in the blood in a 1000 kg animal.

Projected lengths of inactive aerobic dives: cADL

Aerobic dive limits may be calculated by dividing the useable oxygen store at submergence by an appropriate rate of oxygen uptake. Wright (1985a) found that the longest voluntary dive he observed used 68% of the oxygen store and that was without any reduction in metabolic rate. The longest voluntary dives made in the field by *C. johnstoni* (Fig. 9.14 upper panel) were interpreted by Campbell *et al.* (2010a) as resting dives, so SMR measured as oxygen consumption by Seymour *et al.* (2013) (Chapter 7) should be appropriate, converted to 25°C, which is more appropriate than 30°C for northern Australian waters, using a Q_{10} of 1.8 (Wright 1986b). Calculated aerobic dive lengths (cADL) over a wide range of body size are shown in Fig. 9.17. These are conservative estimates because it is unlikely that 68% usage of the oxygen store is a maximum. Longer aerobic dives are achieved by metabolic depression, so ADLs were calculated for 75% and 87.5% reductions in SMR as well (Fig. 9.17).

The increase in cADL with mass is explained by the difference between scaling exponents for lung volume at submergence (1.09) and SMR (0.83). That is, buoyancy requirements dictate that larger crocodylians dive with more oxygen than they would have if the size of the lungs were related only to metabolic requirements.

It is reassuring that the 13 min dive by a 1.81 kg animal, during which oxygen consumption was not reduced (above), falls close to the value for cADL. Most of the longest voluntary dives that have been recorded so far from crocodylians in the field are within the limit suggested by a 75% metabolic depression (Fig. 9.17). Additionally, about half of the longest *C. johnstoni* dives known to be associated with some disturbance could be accommodated by a reduction to 12.5% of SMR. They were known to be evading capture, so were probably supported

BOX 9.1   THE GROWTH OF UNDERSTANDING ABOUT VERTEBRATE DIVING PHYSIOLOGY

An understanding of diving in reptiles grew alongside that in mammals and birds, and each informed the other, so a brief summary (covering diving mammals in particular) is appropriate as background to a review of the diving behaviour and physiology of crocodylians. I have drawn from excellent reviews by Butler and Jones (1997), Kooyman and Ponganis (1998) and Hochachka (2000) and have taken a historical approach because I feel it to be particularly informative.

Early days

About 75 years ago, the famous Scandinavian physiologist Per (Pete) Scholander and the American physiologist Laurence Irving and colleagues began experiments to learn how some birds and mammals cope with breath-hold dives for such surprisingly long periods: much longer than humans can. Initial experiments were conducted by forcing animals to dive, sometimes strapped to a board. Often only the head was submerged if the animal was large. The researchers observed that diving showed three 'reflex' responses: 'apnoea', in which ventilation was inhibited when the nose was submerged; 'bradycardia' in which the heart rate slowed markedly; and associated with that, there was 'peripheral vasoconstriction' – a reduction or even cessation of blood flow in the skin and skeletal muscles. Scholander referred to these collectively as the 'diving response', and it could be induced even in non-diving birds and mammals, including humans, though to a lesser extent. They also discovered that plasma lactate was elevated after a dive, suggesting that some of the energy used during the dive was supported by anaerobic metabolism.

So the Irving-Scholander model arose to explain how air-breathing vertebrates are adapted to remaining under water without ventilating their lungs. Their model proposed that there is a redistribution of blood away from the muscles and the periphery as a result of selective constriction of blood vessels in these tissues so that the available oxygen is preferentially delivered to the heart and brain. In order to prevent a large increase in blood pressure arising from the selective vasoconstriction, cardiac output falls, mainly as a result of a reduction in heart rate, known as bradycardia. The bradycardia is usually taken as an indicator of all the other adjustments taking place during submersion. Anaerobic metabolism occurs in the poorly perfused muscles, with lactate accumulating there until surfacing, and a period of recovery at the surface during which the excess lactate is metabolised aerobically to 'pay back' the oxygen debt (i.e. the 'excess post exercise oxygen consumption, EPOC; see Chapter 7) (Irving 1939; Scholander 1940). The first studies on crocodylians were of forced dives. Wilbur (1960) submerged juvenile alligators for short dives, apparently several minutes until they struggled, and monitored heart rate. He reported a 'profound' bradycardia – as low as 1 bpm – with rapid return to normal when the animal was returned to the surface. He concluded that alligators showed the same response as had been found in other diving vertebrates up to that time. In a PhD study at Scripps Institute of Oceanography in California, with Scholander as supervisor, Harald Andersen conducted a much more substantial study, also on juvenile alligators (Andersen 1961). Again, the alligators were strapped to a board, but these forced dives were very long: 35–120 min. Measurements included heart rate, femoral arterial blood pressure, oxygen content and plasma lactate, with oxygen consumption assessed from lung oxygen content. The paper makes interesting reading. His alligators may have become habituated to some extent, and one wonders what might have been if he had given some short dives; he did note much less bradycardia during some dives made by an animal swimming freely in a tank. He concluded that alligators fitted the pattern described for other vertebrates, including circulatory adjustments that 'enable the alligator to make the limited oxygen stores last throughout prolonged periods of submersion' and a flood of lactate into the blood after a dive. He also noted that in some dives there must have been a reduction in oxygen consumption and marvelled that they could remain 'in a state of useful consciousness until the oxygen stores are almost exhausted'. All indications were that the

BOX 9.1 CONTINUED

Irving–Scholander model applied to crocodylians too. Note that the model envisaged significant dependence on anaerobic metabolism to support the dive. However, Scholander and colleagues, including Andersen, also found that the oxygen debt was often smaller than expected, implying some reduction of aerobic metabolism during the dive. They also realised that many natural dives were too short and too repetitive to be accompanied by lactate build up, so they must have realised that their model was not relevant for short dives. Nevertheless, the model became so well accepted that in his review of the physiology of diving in vertebrates, Andersen (1966) observed that 'the physiological modifications that enable the divers to remain under water for long periods of time are largely understood'. This was a very premature conclusion.

Studies of unrestrained animals – it turned out that dives are mostly aerobic

Soon after Anderson's review, doubts began to accumulate about some attributes of the Irving–Scholander model and reptile studies were at the forefront. It was realised that forcing an animal to dive adds a 'fright'-induced bradycardia, which is typically much larger than in dives made voluntarily. Belkin (1968) found this in turtles. Also, Gaunt and Gans (1969) monitored heart rates in *Caiman crocodilus* left undisturbed in a quiet room and found that their short voluntary dives, and even some of the longer dives, showed insignificant bradycardia. However, if someone entered the room the heart slowed significantly. It was becoming clear that profound bradycardias could be more an expression of the animal's psychological state than just a strategy for diving, and the implication was that responses monitored in forced dives were not representative of what a vertebrate animal does when it dives voluntarily. Many of the reported 'severe' bradycardias were actually artefacts. Norbert Smith explored this in the wild. A biologist *cum* electronics engineer, he built a heart rate transmitter and attached it to a 2.44 m, 45.4 kg male *A. mississippiensis* in a lake on a Texan wildlife refuge (Smith *et al.* 1974). The alligator showed little or no bradycardia during quiet, undisturbed dives, which typically lasted 5–7 min, but when a boat approached or in response to some other disturbance a very pronounced bradycardia resulted (Fig. B9.1). He confirmed in the field what Gaunt and Gans (1969) had reported in the laboratory. He also noted that 'attempts to induce bradycardia by human approach near a captive alligator removed from the water were unsuccessful'. The 'fright' bradycardia in crocodylians was apparently associated with being submerged as well as 'frightened'.

The realisation that studies of forced dives could be very misleading prompted more emphasis on studying unrestrained animals in the field and, with new techniques emerging, biologists were able to get information

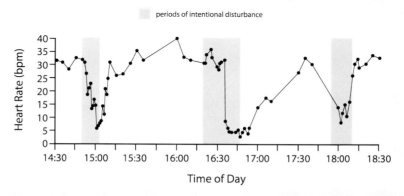

Fig. B9.1. *In this seminal field study by Norbert Smith, a free-ranging American alligator with a heart rate transmitter attached made short voluntary dives with little or no slowing of its heart, but responded to disturbance with a severe bradycardia, with the heart rate falling precipitously. This confirmed in the field what was being found in the laboratory: studying forced dives was not yielding results that were representative of natural diving physiology. (Adapted from Smith et al. 1974, with permission)*

(Box continued)

BOX 9.1 CONTINUED

about the natural diving behaviour of animals, from cormorants and ducks to penguins and seals. They also frequently monitored physiological correlates such as heart rate and blood chemistry. It gradually became clear that many dives were actually quite short and, often, repeated in rapid succession with insufficient time at the surface to pay back any putative oxygen debt. Further, associated bradycardia was slight or moderate rather than severe, and sometimes absent. Researchers studying reptiles were at the forefront again; Seymour (1979) bled sea snakes captured at sea and found low levels of lactate. In what Ponganis *et al.* (2011) described as a landmark paper, Kooyman *et al.* (1980) showed that most dives in Weddell seals are conducted aerobically, and do not result in elevated levels of lactate. Evidence was accumulating that most dives are conducted aerobically.

The next model to emerge accepted that most dives are supported aerobically and, for a time, the anaerobiosis and lactate components, as well as the severe bradycardia described by the Irving–Scholander model looked almost irrelevant. Researchers drew attention to many diving vertebrates having higher blood oxygen capacity, and therefore enhanced oxygen stores compared with non-divers, plus an ability to recruit additional red blood cells quickly from the spleen, and higher concentrations of myoglobin in muscle. However, seeking explanations for some of the longest dives, researchers compared observed dive durations with maximum aerobic dive times calculated from the size of oxygen stores and the rate of oxygen consumption during the dive, and found shortfalls.

Learning from Weddell Seals: aerobic dive limit (ADL), diving lactate threshold (DLT) and calculated ADL (cADL)

The growing understanding of diving physiology was helped enormously by studies of Weddell seals (*Leptonychotes weddellii*) in Antarctica in the 1970s and since (Fig. B9.2). The data could be collected because seals at a breathing hole in the sea ice could be fitted with recording packages which were then retrieved, and blood samples taken, at the end of the dive when the seal re-surfaced at the same hole (see reviews by Kooyman *et al.* 1981; Kooyman and Ponganis 1998).

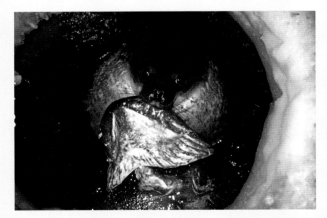

*Fig. B9.2. This Weddell seal (*Leptonychotes weddellii*) has just surfaced to breathe in a hole bored through summer sea ice near Scott Base, Antarctica, to allow scientists access to the ocean (Ross Sea). It has brought from the bottom, 300 m below, an Antarctic cod (*Dissostichus mawsoni*). Studies of these seals have added a lot to the understanding of diving physiology. They can forage at great depth, remaining underwater for up to 25 min without any build up of plasma lactate, or longer by recruiting anaerobic metabolism. Because seals will echo-locate such holes and use them repeatedly, scientists have been able to attach recording devices and recover them on a subsequent surfacing. Crocodylian biologists are not so well favoured and recovering attached recording packages can be a significant challenge. (Photo GCG)*

BOX 9.1 CONTINUED

In their seminal work, Kooyman *et al.* (1980) found that plasma lactate levels were still at resting levels at the end of dives lasting up to ~25 min (Fig. B9.3). Weddell seals could dive aerobically for up to 25 min! What was more, 97% of their dives were less than 25 min – and they could make another dive immediately, without any need to rest at the surface. It became clear that almost all of the dives made by Weddell seals are supported aerobically, using the oxygen store they have at submergence.

On the other hand, dives longer than 25 min showed an accumulation of lactate (Fig. B9.3). Of the nearly 3% of dives by Weddell seals in the Kooyman *et al.* (1980) study that were longer than 25 min, some were as long as an hour and, after prolonged dives, plasma lactate was elevated considerably. Furthermore, after making these very long dives, the seals would haul out onto the ice and lie there recovering for longer than the length of the dive while plasma lactate slowly returned to normal: 'paying back the oxygen debt'. The same pattern has been found in other long divers too, making occasional use of a capacity to go anaerobic and, when they do, needing longer at the surface to recover. This is reminiscent of the mechanisms described by the Irving–Scholander model and, as Butler and Jones (1997) wrote, the story had come 'full circle' and the evolving model included a role for anaerobic metabolism in supplementary support of the longest dives.

The seal studies led to a useful way of assessing maximum aerobic dive time in terms of the maximum breath hold that is possible without an increase in plasma lactate, during or after the dive, referred to by Kooyman *et al.* (1980) as the aerobic breath-hold limit. This soon became known as the 'aerobic dive limit' (ADL) (Fig. B9.3). Although this is a useful concept, such data would be very difficult or impossible to collect in many species. Accordingly, researchers must rely on calculating ADL from the total useable oxygen store and the rate of oxygen uptake during the dive, both of which are hard to assess with any accuracy, so the result has to be regarded as a 'best estimate'. Butler (2006) suggested that, to distinguish the difference between these two ways to measure aerobic limit, the ADL should be renamed 'diving lactate threshold' (DLT) and that calculated

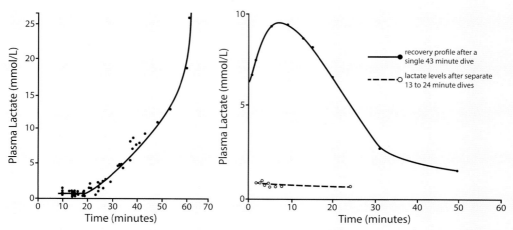

Fig. B9.3. *Plasma lactate levels after voluntary dives by large male Weddell seals show (left) that dives up to ~25 min can be undertaken aerobically. Longer dives show accumulation of lactate in the plasma. The dive length beyond which accumulation occurs defines the aerobic dive limit (ADL), which Butler (2006) suggested would be better called the 'dive lactate threshold', DLT. Direct measurement of plasma lactate immediately following dives is difficult and aerobic dive limits are usually calculated from knowledge of, or assumptions about, the oxygen store and the rate of oxygen consumption. Butler (2006) suggested that calculated limits are better designated as cADL to indicate they have been calculated, not measured directly. (Right panel) Dives longer than the ADL require metabolism of the excess lactate (see Chapter 7) and this may dictate a long period on the surface. Hence, there is a penalty associated with dives longer than the ADL and, in nature, such dives are uncommon. (Adapted from Kooyman et al. 1981, with permission. © 1981 Annual Reviews Inc.)*

(Box continued)

BOX 9.1 CONTINUED

values for ADL should be designated as cADL. Ponganis *et al.* (2011) have provided a useful review of what constitutes the ADL (or DLT), focussing on seals and penguins.

Diving aerobically rather than anaerobically is energetically advantageous and, for many species, foraging time can be maximised by making short dives repeatedly, with little or no recovery periods in between. Following longer dives, supported anaerobically, activity could be compromised during the subsequent recovery period. In birds and mammals after an experimentally forced submergence, it may take even six times as long as the dive before plasma lactate returns to the pre-dive level (Butler and Jones 1982). Kooyman *et al.* (1981) pointed out that a series of short, repetitive aerobic dives yields much more underwater foraging time than fewer, more prolonged dives with long recovery periods. They concluded that this is why 'prolonged dives are rare in nature'. Interestingly, in forced dives, Weddell seals showed lactate accumulation even in short dives, perhaps because of the 'fright' component referred to by Belkin (1968).

Additionally, a reduction in oxygen consumption

The most recent addition to the progressively evolving model to explain vertebrate diving has been the acceptance that diving can be accompanied by a reduction in oxygen consumption. Again, this can be illustrated by data from Weddell seals. A 20–25 min aerobic dive is very impressive for a mammal, particularly because its oxygen store could be expected to last only about one-third to half that long at resting rates of oxygen consumption. A substantial reduction in oxygen consumption during the dive would provide an explanation, and this is actually a very old idea. The Nobel Prize winning French physiologist Charles Richet suggested in 1899 that aerobic metabolism may be reduced during diving (Richet 1899), but it has only comparatively recently become widely accepted. One of its strong proponents was Peter Hochachka at the University of British Columbia who saw the reduced perfusion of skin, muscle and many organs as a result of the redistribution of peripheral blood centrally as 'the proximate physiological cause of suppressed metabolism of peripheral tissues and organs' (Hochachka 1992; Hochachka 2000).

The current model

In summary, the current model for diving in birds and mammals recognises that most dives are aerobic and that many divers have enhanced oxygen storage in blood and myoglobin, which may be managed 'judiciously' (quoting Butler and Jones 1997) to extend the dive by cardiovascular adjustments. These adjustments redistribute blood away from some organs (signalled by a modest bradycardia) and reduce oxygen consumption in the hypoperfused tissues. In diving marine mammals, variable exertion during the dive is accommodated by reducing blood flow to the muscles to match their variable oxygen requirements, so maximising the ADL (e.g. Davis and Kanatous 1999, working on Weddell seals). The reduced blood flow is accompanied by a moderate bradycardia. Hence, ADL is not a fixed value, but varies depending on the pattern of exertion undertaken during the dive. Dives by some (probably all) species can be further extended by anaerobic metabolism, leading to a release of lactate into the blood after the dive. This is then metabolised by a post-dive increase in oxygen consumption. Anaerobic support is uncommon in nature because it is energetically disadvantageous and because of the need for a longish recovery phase. With vertebrate physiological evolution having been so conservative, it is reasonable to expect crocodylians to fit the same general pattern as mammals and birds.

partly anaerobically. The lengths of the longest voluntary, undisturbed dives by *C. porosus* are in striking contrast. All but one could be accomplished with little or no reduction in SMR, although several are close to the projected limit. It is probable that this is a reflection of different behaviour rather than because *C. porosus* have less capacity than *C. johnstoni*. These were travelling animals waiting for the tidal current to change. Admittedly, the cADL lines on Fig. 9.17 are derived from *C. porosus* data, but physiological differences between them are unlikely to explain the difference.

To answer the question posed at the beginning of this section, all of the longest voluntary dives recorded from these two species, ranging freely in their natural habitat, are likely to have been supported aerobically. The longest dives of all, made by *C. johnstoni* evading capture, probably relied on some anaerobic support (Fig. 9.17), as suggested by limited sampling of their plasma lactates.

And how long can the largest crocs remain submerged for? The calculations suggest that a 1 tonne, 6 m animal could submerge at rest for almost an hour without reducing oxygen consumption, 3.5 h with a 75% reduction and nearly 9 h if oxygen consumption were reduced to one-eighth of resting.

Body temperature will have a big influence on maximum submergence time, and the thermal relations of crocodylians are taken up in the next chapter.

References

Andersen HT (1966) Physiological adaptations in diving vertebrates. *Physiological Reviews* **46**, 212–243.

Andersen HT (1961) Physiological adjustments to prolong diving in the American alligator, *Alligator mississippiensis*. *Acta Physiologica Scandinavica* **53**, 23–45.

Belkin DA (1968) Bradycardia in response to threat. *American Zoologist* **8**, 775

Butler PJ (2006) Aerobic dive limit. What is it and is it always used appropriately? *Comparative Biochemistry and Physiology. Part A* **145**, 1–6.

Butler PJ, Jones DR (1982) The comparative physiology of diving in vertebrates. *Advances in Comparative Physiology and Biochemistry* **8**, 179–364.

Butler PJ, Jones DR (1997) Physiology of diving in birds and mammals. *Physiological Reviews* **77**, 837–899.

Calder WA III (1984) *Size, Function and Life History*. Harvard University Press, Cambridge, Massachusetts.

Campbell HA, Sullivan S, Read MA, Gordos MA, Franklin CE (2010a) Ecological and physiological determinants of dive duration in the freshwater crocodile. *Functional Ecology* **24**, 103–111.

Campbell HA, Dwyer RG, Gordos M, Franklin CE (2010b) Diving through the thermal window: implications for a warming world. *Proceedings. Biological Sciences* **277**, 3837–3844.

Campbell HA, Watts ME, Sullivan S, Read MA, Choukroun S, Irwin SR, *et al.* (2010c) Estuarine crocodiles ride surface currents to facilitate long-distance travel. *Journal of Animal Ecology* **79**, 955–964.

Cott HB (1961) Scientific results of an enquiry into the ecology and economic status of the Nile Crocodile (*Crocodylus niloticus*) in Uganda and Northern Rhodesia. *Transactions of the Zoological Society of London* **29**, 211–356.

Davis RW, Kanatous SB (1999) Convective oxygen transport and tissue oxygen consumption in Weddell seals during aerobic dives. *Journal of Experimental Biology* **202**, 1091–1113.

Gaunt AS, Gans C (1969) Diving bradycardia and withdrawal bradycardia in *Caiman crocodilus*. *Nature* **223**, 207–208.

Grigg GC, Cairncross M (1980) Respiratory properties of the blood of *Crocodylus porosus*. *Respiration Physiology* **41**, 367–380.

Grigg GC, Gruca M (1979) Possible adaptive significance of low red cell organic phosphates in crocodiles. *Journal of Experimental Zoology* **209**, 161–167.

Grigg GC, Farwell WD, Kinney JL, Taplin LE, Johansen K, Johansen K (1985) Diving behaviour in a free ranging *Crocodylus porosus*. *Australian Zoologist* **21**, 599–606.

Guppy M, Withers P (1999) Metabolic depression in animals: physiological perspectives and biochemical generalizations. *Biological Reviews of the Cambridge Philosophical Society* **74**, 1–40.

Hochachka PW (1992) 1992 Metabolic biochemistry and the making of a mesopelagic mammal. *Experientia* **48**, 570–575.

Hochachka PW (2000) Pinniped diving response mechanism and evolution: a window on the paradigm of comparative biochemistry and physiology. *Comparative Biochemistry and Physiology. A. Comparative Physiology* **126**, 435–458.

Huggins SE (1961) Blood volume parameters of a poikilothermic animal in hypo- and hyperthermia. *Proceedings of the Society for Experimental Biology and Medicine* **108**, 231–234.

Irving L (1939) Respiration in diving mammals. *Physiological Reviews* **19**, 112–134.

Jackson DC (1968) Metabolic depression and oxygen depletion in the diving turtle. *Journal of Applied Physiology* **24**(4), 503–509.

Kirshner D (1985) Buoyancy control in the estuarine crocodile, *Crocodylus porosus* Schnieder. PhD thesis. University of Sydney.

Kooyman GL, Ponganis PJ (1998) The physiological basis of diving to depth: birds and mammals. *Annual Review of Physiology* **60**, 19–32.

Kooyman GL, Wahrenbrock EA, Castellini MA, Davis RW, Sinnett EE (1980) Aerobic and anaerobic metabolism during voluntary diving in Weddell seals: evidence of preferred pathways from blood chemistry to behaviour. *Journal of Comparative Physiology* **138**, 335–346.

Kooyman GL, Castellini MA, Davis RW (1981) Physiology of diving in marine mammals. *Annual Review of Physiology* **43**, 343–356.

Lewis LY, Gatten RE (1985). Aerobic metabolism of American alligators, Alligator mississippiensis, under standard conditions and during voluntary activity. *Comparative Biochemistry and Physiology a-Physiology* **80**(3), 441–447.

López-Mendilaharsu M, Rocha CFD, Domingo A, Wallace BP, Miller P (2008) Prolonged deep dives by the leatherback turtle, *Dermochelys coriacea*, pushing their aerobic dive limits. *Marine Biodiversity Records* **6274**, 1–3.

Ponganis PJ, Meir JU, Williams CL (2011) In pursuit of Irving and Scholander: a review of oxygen store management in seals and penguins. *Journal of Experimental Biology* **214**, 3325–3339.

Read MA, Grigg GC, Irwin SR, Shanahan D, Franklin CE (2007) Satellite tracking reveals long distance coastal travel and homing by translocated Estuarine Crocodiles, *Crocodylus porosus. PLoS ONE* **2**, 1–5.

Richet C (1899) De la resistance des canards a l'asphyxie**.** *Journal de Physiologie et de Pathologie Générale* **1**, 641–650.

Sato K, Shiomi K, Marshall G, Kooyman GL, Ponganis PJ (2011) Stroke rates and diving air volumes of emperor penguins: implications for dive performance. *Journal of Experimental Biology* **214**, 2854–2863.

Scholander PF (1940) Experimental investigations on the respiratory function in diving mammals and birds. *Hvalrads Skrifta Norske VidenskapS-Akud. Oslo* **22**, 1–131.

Seebacher F, Franklin CE, Read M (2005) Diving behaviour of a reptile *(Crocodylus johnstoni)* in the wild: Interactions with Heart rate and Body temperature. *Physiological and Biochemical Zoology* **78**(1), 1–8.

Seymour RS (1974) How sea snakes may avoid the bends. *Nature* **250**, 489–490.

Seymour RS (1979) Blood lactate in free-diving sea snakes. *Copeia* **1979**, 494–497.

Seymour RS (1982) Physiological adaptations to aquatic life. In *Biology of the Reptilia. Volume 13.* (Eds C Gans and FH Pough) pp. 1–51. Academic Press, New York.

Seymour RS, Bennett AF, Bradford DF (1985) Blood gas tensions and acid-base regulation in the saltwater crocodile, *Crocodylus porosus*, at rest and after exhaustive exercise. *Journal of Experimental Biology* **118**, 143–159.

Seymour RS, Bennett-Stamper CL, Johnston S, Carrier DR, Grigg GC (2004) Evidence of endothermic ancestors of crocodiles at the stem of Archosaur evolution. *Physiological and Biochemical Zoology* **77**(6), 1051–1067.

Seymour RS, Gienger CM, Brien ML, Tracy CR, Manolis SC, Webb GJ, *et al.* (2013) Scaling of standard metabolic rate in estuarine crocodiles *Crocodylus porosus Journal of Comparative Physiology. B, Biochemical, Systemic, and Environmental Physiology* **183**, 491–500.

Smith ED, Allison RD, Crowder WJ (1974) Bradycardia in a free ranging alligator. *Copeia* **1974**, 770–772.

Southwood AL, Andrews RD, Lutcavage ME, Paladino FV, West NH, George RH, *et al.* (1999) Heart rates and diving behavior of leatherback sea turtles in the Eastern Pacific Ocean. *Journal of Experimental Biology* **202**, 1115–1125.

Weinheimer CJ, Pendergast DR, Spotila JR, Wilson DR, Standora EA (1982) Peripheral circulation in *Alligator mississippiensis*: effects of diving, fear, movement, investigator activities and temperature. *Journal of Comparative Physiology* **148**, 57–63.

Wilbur CG (1960) Cardiac response of *Alligator mississippiensis* to diving. *Comparative Biochemistry and Physiology* **1**, I64–167.

Wright JC (1985a) Oxygen consumption during voluntary undisturbed diving in the salt water crocodile, *Crocodylus porosus*. In *Biology of Australian Frogs and Reptiles* (Eds GC Grigg GC, R Shine and H Ehmann) pp. 423–429. Surrey Beatty & Sons, Sydney.

Wright JC (1985b) Diving and exercise physiology in the estuarine crocodile, *Crocodylus porosus*. PhD thesis. University of Sydney.

Wright JC (1986a) Low to negligible cutaneous oxygen uptake in juvenile *Crocodylus porosus*. *Comparative Biochemistry and Physiology. A. Comparative Physiology* **84**, 479–481.

Wright JC (1986b) Effects of body mass, temperature and activity on aerobic and anaerobic metabolism in the juvenile crocodile, *Crocodylus porosus*. *Physiological Zoology* **59**, 505–513.

Wright JC (1987) Energy-metabolism during unrestrained submergence in the saltwater crocodile *Crocodylus porosus*. *Physiological Zoology* **60**(5), 515–523.

Wright JC, Kirshner DS (1987) Allometry of lung volume during voluntary submergence in the saltwater crocodile *Crocodylus porosus*. *Journal of Experimental Biology* **130**, 433–436.

Wright JC, Grigg GC, Franklin CE (1992) Redistribution of air within the lungs may potentiate "fright" bradycardia in submerged crocodiles *(Crocodylus porosus)*. *Comparative Biochemistry and Physiology A* **102**, 33–36.

THERMAL RELATIONS

When in 1971 the opportunity came to work on crocodiles, I was particularly interested in doing a field study of their body temperatures and thermal relations. In my mind's eye I saw myself on a rocky escarpment overlooking a Northern Territory billabong watching C. porosus *cavort while I monitored body temperatures using implanted radio-transmitters. It was nearly 20 years before I got that opportunity. Ironically, the enjoyment was vicarious, because the opportunity came in 1989 while helping Frank Seebacher, then a post-graduate student, get set up for just such a study on 'freshies',* C. johnstoni, *at 'The Croc Hole' in the Lynd River in North Queensland. Subsequently, Frank, Lyn Beard and I finally got to look at* C. porosus *living in very naturalistic surroundings in the large breeding lagoon at the Edward River Crocodile Farm on Cape York, Queensland.*

The questions about crocodylian body temperatures and thermal relations that sprang to my mind in 1971 were to do with describing daily and seasonal patterns of body temperature and how much and how well crocs regulate that. I was also particularly interested in the effect of body size on body temperature. Large crocs, because of their greater thermal inertia, would obviously take a longer time to heat and cool than small crocs. Would this cause the core body temperature (Tb) of very large crocs to be high and stable like typical mammals and birds, though for a different reason? The question was

particularly provocative because there has always been speculation about the body temperature of dinosaurs. Would crocs be a good model for thinking about dinosaur Tb? Were dinosaurs 'warm blooded'? If they were, could that pattern have accrued without the metabolic heat production typical of birds and mammals: that is, with only reptilian physiology and behaviour, simply as a consequence of being very large? Crocodylians grow to only 1–2 tonnes at most, much less than many of the dinosaurs, but they are the best living model we have. I thought it would be wonderfully interesting to get some data on daily and seasonal patterns of Tb in very large crocs and, in due course, we did.

INTRODUCTION: CROCODYLIANS ARE NOT LIKE OTHER REPTILES

'During the four winter months they eat nothing; they are four footed, and live indifferently on land or in the water. The female lays and hatches her eggs ashore, passing the greater portion of the day on dry land, but at night retiring to the river, the water of which is warmer than the night-air and the dew. Of all known animals this is the one which from the smallest size grows to be the greatest, for the egg of the crocodile is but little bigger than that of a goose, and the young crocodile is in proportion to the

A group of American alligators, Alligator mississippiensis, *basking in the winter sun in Everglades National Park, Florida, USA. (Photo DSK)*

egg; yet when it is full grown the animal measures frequently seventeen cubits and even more'.

Herodotus BC 484–425, Book II, 68.

Herodotus wrote the above about crocodylians nearly 2500 years ago, and included some of the features particularly relevant to thermal relations: amphibious behaviour, thermally related movements between land and water and the enormous increase in size they make between hatching and maturity. Each of these characteristics contributes to crocodylians having different daily, seasonal and developmental patterns of changing body temperatures (Tb) from those seen in 'typical' reptiles (if such an entity exists).

Temperature is important because most metabolic processes are temperature sensitive (see Chapter 7) and this usually has an impact on performance. A warm crocodylian will fare better than a cold one during predation or in a fight. Because the existing crocodylians are all ectotherms, their body temperature (Tb) depends on the environment they are in (ecto = outside) and not on heat produced metabolically as in mammals and birds, which are endotherms (endo = inside). Therefore we can expect crocodylians to show seasonal variations in their Tb and to grow faster in summer, which they do, and in most species the benefits of being warm lead to them mating and nesting in the warmer season.

However, crocodylians are not slaves to daily and seasonal changes in ambient temperature. Through behavioural, physiological and biochemical adjustments, they modify Tb to avoid being too warm or too cool and what they do and how they do it will be the subject of this chapter. The field of reptilian thermal relations has a lot of terminology which is often used differently by different authors, so readers should familiarise themselves with the terminology followed in this chapter (Box. 10.1).

As mentioned above, crocodylians do not fit the 'typical' reptilian pattern of body temperature regulation. Although there are many exceptions, the 'typical' pattern in a heliothermic reptile is usually thought of as having body temperature regulated to a reasonably stable level during the day by shuttling between sun and shade, making postural adjustments, changing the angle of presentation to the sun and augmenting these behaviours by altering blood flow through the skin (see reviews by Huey 1982; Bartholomew 1982; Avery 1982; Seebacher and Franklin 2005). This pattern, with a daily period of 'behavioural homeothermy' is much easier for small terrestrial reptiles to achieve than for large aquatic ones, so the patterns we see in crocodylians are mostly very different, without daily plateaus. Only *C. johnstoni*, so far, has been reported to show behaviour approaching this pattern (see below), but it is very likely that it will also be found in other smallish species, or large species when small, in due course.

Most crocodylians show a quite different pattern, with daily sinusoidal or saw-tooth cycles in Tb, often after a night spent in the water. Like other reptiles, however, they do show behaviour that is convincingly thermoregulatory (but see Box 10.2), and they achieve thereby a degree of 'uncoupling' Tb from the ambient thermal conditions by employing similar mechanisms: basking, shade seeking, moving into or out of water, as well as a limited range of physiological mechanisms such as changes in peripheral blood flow and evaporative cooling.

Their growth to large body size and aquatic habits have a big influence on their thermal relations. Water and air have very different thermal properties: the high specific heat and high thermal conductivity of water mean that Tb responds to a thermal gradient much more quickly than in air. Hence, it is not surprising that daily and seasonal movements between air and water have been assumed to have thermal significance, and they often do. Large body size leads to increased thermal inertia and slower rates of temperature change and, as we shall see in *C. porosus*, daily and even weekly thermostability.

Grigg and Seebacher (2001) and Seebacher and Grigg (2001) reviewed crocodylian thermal relations a decade ago, building on a review by Lang (1987). New data gathered in the last decade make another review timely. Implanted data loggers have become more widely available and more information about Tb patterns is now available. More is known about some of the underlying thermoregulatory mechanisms, new analytical approaches suitable for

BOX 10.1   TERMINOLOGY

The gold standard for terminology in thermal biology should be the most recent edition of the *Glossary of Terms for Thermal Physiology* (IUPS Thermal Commission 2003). Unfortunately, the breadth of its applicability is constrained by having come primarily out of mammalian and human physiology, so some of the definitions do not accommodate the diversity of patterns and mechanisms that biologists need (and love) to deal with. Accordingly, modern zoologists and comparative physiologists tend to use some terms in a different way. But terminology in thermal biology is often confusing, and sometimes even used in contradictory ways, so this section describes the meanings of thermally related terminology used in this book.

A big step forward in terminology occurred more than 50 years ago, when two terms descriptive of a **pattern** of body temperature – **'homeothermy'** (= same temperatures) and **'poikilothermy'** (= many temperatures) – were replaced by terms relating to the **mechanism** by which a particular pattern is displayed. These 'new' terms are **ectothermy** (source of heat primarily from the environment) and **endothermy** (source of heat primarily from the animal's metabolism) (Cowles 1962). This allowed the old term 'homeothermy', previously, and still occasionally used as a synonym for the sort of temperature regulation you and I enjoy, to be used in a different sense. For example, you and I are **homeothermic endotherms**: the main source of our heat being internal (metabolic) and the pattern of our Tb being regulated to be homeothermic: that is, regulated at the same, constant temperature, (usually taken to imply stability within 2°C). In contrast, hibernating and torpidating endotherms are **heterothermic endotherms**: that is, displaying different temperatures at different times. They may be homeothermic in one season, heterothermic in another. We and other endotherms can also say that we are **tachymetabolic** (fast metabolisers) because our metabolism is high enough to produce heat that affects Tb. The antonym is **bradymetabolic** (slow metabolism), meaning slower than mammals and birds and more like reptiles, amphibians, fishes and invertebrates.

To reiterate, homeothermy is not used here as synonymous with homeothermic endothermy, which was its old fashioned usage, but as a term descriptive of a stable pattern of Tb. Thus, the gaining of a stable Tb through being very large, because of thermal inertia, is called **passive homeothermy**, **inertial homeothermy** or, sometimes, **gigantothermy**. It should be noted that the first two terms are unfortunate because, although they do imply stability, they do not necessarily imply 'warmth' (to drive this point home, Antarctic fish too are passive homotherms) and, perhaps worse, they tend to perpetuate the misuse of 'homeothermy' as being equivalent to 'homeothermic endothermy'. The term 'gigantothermy' carries least baggage (forgive the choice of phrase) and is therefore preferable.

Other important words in thermal biology are acclimatisation and acclimation. These refer to the slow (days or weeks) biochemical and/or physiological responses that many animals make to changed conditions, and which ameliorate the effects of the change. If the response occurs during exposure to naturally changing circumstances, for example, to a seasonal change in temperature, it is an **acclimatisation**. Responses are often studied under experimental conditions, in which case the response is an **acclimation**. The nature of the response may be the same, or it may not and the difference/s may be important; hence the value of a slightly different term.

One word that is used frequently in studies of thermal relations is **thermoregulation**, and defining it satisfactorily is not straightforward. In homeothermic endotherms, such as humans, its meaning is clear: it embraces the processes by which our stable Tb is maintained. Unfortunately, the term is usually imbued with an implication that evidence of thermoregulation can be judged by presence or absence of a stable (and usually warm) Tb. This is quite satisfactory for humans and many other endotherms that maintain a warm and stable Tb throughout their lives. However, hibernating or torpidating mammals and birds may show large daily and/or seasonal changes in Tb, all under close thermoregulation, but certainly not presenting a stable

(Box continued)

large ectotherms have been developed (Seebacher *et al.* 2003, see Box 10.2) and there has been increased awareness of seasonal thermal acclimatisation (Seebacher and James 2008; Guderley and Seebacher 2011). Also, the provocative proposal has been made that ancestral crocodylians were endothermic and have re-evolved ectothermy secondarily along with their evolution, to be able to function as sit-and-wait predators in aquatic habitats (Seymour *et al.* 2004, discussed below and in Chapter 8).

In this chapter, daily and seasonal patterns of Tb revealed by field studies of two crocodylids (*C. johnstoni* and *C. porosus*) and three alligatorids (*A. mississippiensis*, *Caiman yacare* and *Paleosuchus palpebrosus*) will be reviewed first. Then will follow a discussion of the behavioural and physiological mechanisms by which crocodylians may actively thermoregulate by modifying rates of heat exchange with their surroundings. The focus will then shift to crocodylians as models for dinosaur thermal relations and gigantothermy and, finally, the provocative question of whether or not ancestral crocs may have been endothermic.

TEMPERATURE AND THE WORLD DISTRIBUTION OF CROCODYLIANS

The majority of crocodylians occur in the tropics (most) and subtropics where they experience daily maximum ambient temperatures of 25°C–35°C. Several species extend into temperate climates but, for those that do, populations are usually denser

and have longer breeding seasons in the warmer parts of their distribution. Alligatorids have a wider latitudinal range than crocodylids, but only just, from ~36°N (*Alligator* in the USA) to 36°S (*Caiman yacare* in Argentina), compared with 32°N (*C. palustrus* in northern India) to 34°S (*C. niloticus* in South Africa). *Gavialis gangeticus* are found as far north as 34°N. The few remaining wild Chinese alligators, *A. sinensis*, occur in the Yangtze River at ~31°N latitude but had a wider distribution in the comparatively recent past (1000–7000 years), to as far as 35°N (Thorbjarnarson and Wang 2010).

Despite such similar latitudinal ranges, alligators encounter cooler habitats and tolerate cold better than crocodylids. In parts of its range, *Alligator mississippiensis* experiences winters with below freezing air temperatures and body temperatures as low as 5°C have been recorded. They have even been recorded in pools that freeze over, with the animals maintaining a breathing hole through the layer of ice (Fig. 10.27), although prolonged exposures to such conditions may cause death (Brisbin *et al.* 1982). This has been observed also in captive Chinese alligators in China's breeding centres (Thorbjarnarson and Wang 2010).

Nevertheless, crocodylians are typically found in warm climates, so the presence of crocodylian or crocodylomorph fossils, often at higher latitudes than today's crocodylians, has been used to indicate a warm climate at higher latitudes in the Late Cretaceous (Markwick 1998, Tarduno *et al.* 1998).

Inevitably there has been speculation about what effect global climate change may have on

BOX 10.2   THERMAL RELATIONS AND THE DIAGNOSIS OF THERMOREGULATION

The daily and seasonal patterns that an animal's body temperature (Tb) shows in relation to its thermal environment are often referred to as its 'thermal relations'. Although escaping a place too hot or too cold might be judged an example of thermoregulation, the term is usually taken to mean that Tb is regulated to a suitable, often reasonably stable, level through active behavioural and physiological processes. Applied to ectotherms, equating thermoregulation with thermostability is a reflection of its meaning for endotherms, and is usually assumed to enhance fitness. However, as discussed in the main text, crocodylians do not generally show thermostability, at least not on land. Indeed, the shuttling between sun and shade seen in many lizards, thereby achieving a stable Tb during the day, has been described only in *C. johnstoni* so far (see main text). More typically, because even the juveniles are much larger than most reptiles (Fraser and Grigg 1984), crocodylians show daily oscillations in temperature that are roughly sinusoidal in shape as they respond slowly to the daily cycling thermal environments they expose themselves to. Despite their size, crocodylians do have a great capacity to modify Tb by behaviour and by controlling their rates of heating and cooling. **Perhaps thermoregulation as applied to reptiles should be redefined as the active application of physiological and/or behavioural mechanisms to modify Tb to be different from what it would be if they did nothing about it, rather than defining it in terms of the extent of stability achieved.**

Demonstrating that thermoregulation is operating (in either the conventional or this 'redefined' sense) is not simple. Many early studies of reptile thermal relations assumed that thermoregulation was occurring when measurements of Tb in lizards seen out and about were above air temperature ('warm') and the figures clustered together in what seemed to be a reasonable range. However, Heath (1964) found, in an elegantly simple way, that the temperature of 13 beer cans filled with water and placed in the sun showed a similar distribution of temperatures. Such data obtained from a reptile could be, and have been, misconstrued as being the result of thermoregulation. Heath pointed out the need for making a comparison between the measured Tbs of some 'control': for example, a non-thermoregulating ectotherm behaving randomly, before the occurrence of active thermoregulation could be concluded. That is to say, thermoregulation can be inferred only when thermoregulatory behaviour has been observed explicitly or when it has been demonstrated that patterns of body temperature are non-random (Seebacher *et al.* 2003). Heath's paper made a very significant methodological step forward in reptilian thermoregulatory studies.

The other difficulty that attended early studies was that of measuring a relevant environmental temperature. Tbs were sometimes compared with 'black bulb' temperatures, which give an indication of solar radiation. In modern studies, 'operative temperature' (Te) is regarded as the most relevant measure of the ambient thermal environment and Tb values are compared with 'null distributions representing body temperature distributions of non-thermoregulating, randomly moving hypothetical animals' (Seebacher *et al.* 2003).

Operative environmental temperature, Te, is 'the temperature of an inanimate object of zero heat capacity with the same size, shape and radiative properties as an animal and exposed to the same microclimate' (Bakken and Gates 1975). Seebacher *et al.* (2003) pointed out that this equates to the surface temperature of an animal, taking into account all heat transfer mechanisms acting at the animal surface. Te provides an estimate of the Tb that the reptile would achieve if it came instantly to thermal equilibrium with the prevailing thermal conditions. It is therefore a good measure of the heating or cooling opportunity being offered by the environment and it is, of course, an ever changing value. For a submerged animal, the relevant Te is the same as water temperature. But in air, because Te is shape dependent and increases with body size, a size range of animals under identical ambient conditions have different Te values. Sometimes an approximation can be gained by measuring the temperature inside a very thin-walled copper model of the animal (to simulate

(Box continued)

BOX 10.2   CONTINUED

'zero heat capacity'). More accurately, as described by Seebacher et al. (1999), Te can be calculated from heat exchange equations using knowledge of air temperature, solar radiation and surface area. Calculations can take into account the amount of body surface in contact with the ground or in the water. For an example of how Te may change for a particular animal during the day, see Fig. 10.6, which shows winter and summer patterns of Te to which a C. johnstoni could be exposed in a daily behavioural cycle of night spent in the water and periods of basking during the day.

To tackle the shortcoming of a lack of comparison with a control (Heath 1964), Hertz et al. (1993) proposed that the effectiveness of thermoregulation could be assessed by comparing field-measured core Tbs in small ectotherms with what would be expected from a 'null' model; that is, temperatures that would result if there were no thermoregulation. The null model would be derived from values of Te determined at a variety of sites selected randomly within the habitat and this became a standard approach. However, Seebacher et al. (2003) and Seebacher and Shine (2004) pointed out that, although the standard Hertz et al. (1993) approach may be adequate for small skinks, it has a significant drawback when applied to larger animals. This is because Te is, by definition, surface temperature and, because of thermal lags and the ever-changing thermal environment and movement by the animal, core body temperature will rarely equal surface temperature. A range of randomly measured operative temperatures can therefore substantially over-estimate the range of Tbs that large reptiles could achieve, meaning that measured Tbs are always likely to be different from the null model, as a methodological artefact. The Hertz et al. (1993) approach is therefore problematic for reptiles too large to come very rapidly into equilibrium with their surroundings and certainly problematic for any crocodylian, even hatchlings.

In a study of seasonal thermal relations of American alligators (1.59–53.64 kg), Seebacher et al. (2003) developed and applied a method to deal with this difficulty posed by large reptiles, taking into account rates of heat transfer within the animal. They calculated core Tb null distributions for alligators moving randomly and adopting random postures in relation to sun and shade exposure, as well as to water and air, and compared them with measured Tbs. To determine core Tb null distributions, they first calculated Te (equating to surface temperature in animals this size) for each of the animals and in each of the random postures and then calculated core Tb by transient heat transfer analysis, taking into account conduction through the body wall and convection due to blood flow. For details of the method and a review of some of the issues surrounding thermoregulatory questions, see Seebacher et al. (2003).

Using their new approach, they were able to show that alligators were cooler in summer and warmer in winter than a comparable alligator moving randomly (Fig. B10.1). This could be taken to imply that the alligators were thermoregulating (in the 'redefined' sense, see above), even though Tb followed a nearly sinusoidal daily oscillation with no daily 'plateau' such as is seen in many lizards. However, there is no evidence that the differences between measured and random values were the result of any thermoregulatory motivation by the alligator, so demonstrating the difference is really only the first step. In this context, it is notable that summer Tbs were warmer than winter Tbs, so regulation between summer and winter was not 'perfect', and the lower winter Tbs were not because of a shortage of warmth: the alligators could have chosen more sun exposure in winter and achieved a warmer Tb (Seebacher et al. 2003). Could the difference between the measured values and the null model be related to some other, so far unknown factor? The most parsimonious interpretation is that the differences are thermoregulatory, but it is worth bearing in mind that there could be other explanations. Demonstrating that there is a difference between measured values of Tb and the 'control' null model is one thing, but proving that the difference is a consequence of some thermoregulatory motivation by the animal is another matter and a conclusion usually relies on a combination of judgment and 'Occam's Razor'.

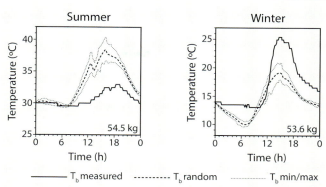

Fig. B10.1. *(Left) Summer and (right) winter comparisons between body temperatures (Tb) measured in American alligators (A. mississippiensis) and values calculated for a hypothetical alligator moving randomly on the same day (Seebacher et al. 2003). Body masses were as shown. The animals were cooler in summer and warmer in winter than temperatures calculated for an alligator moving randomly. This could reasonably be taken to imply active thermoregulation. Note also that ambient conditions would have allowed winter animals to achieve warmer Tb, so the large differences between winter and summer Tb did not result from environmental constraints. It is as if lower Tb in winter was a choice, for whatever reason/s, among which may be a capacity of alligators to undertake seasonal acclimatisation (see text). C. johnstoni showed a similar pattern, with lower Tbs in winter than in summer when more warmth was available (see below) (From Seebacher et al. 2003, with permission. © 2003 by The University of Chicago.)*

These results are discussed further in the main text. Here, they illustrate the power of this analytical approach when applied to an exploration of thermal relations in large reptiles and they also serve as a reminder that the study of thermal relations in ectotherms is far from straightforward!

crocodylians, and any *C. porosus* seen in Australia further south than usual prompts a newspaper article implicating global warming. Range extensions are likely in a warmer world and, unless there is compensatory behavioural or evolutionary adaptation, their temperature-dependent sex determination could lead to gender-skewed populations. However, crocodylians currently do well in the world's warmest climates and, since they evolved during the much warmer Cretaceous, they should cope well in a warmer world.

DAILY AND SEASONAL PATTERNS OF BODY TEMPERATURE

RECORDING BODY TEMPERATURE AND BEHAVIOURAL DATA IN THE FIELD

Much of the early research on crocodylians was on captive animals and focussed on physiological mechanisms of (presumed) thermoregulation

rather than on describing thermoregulatory patterns. Field studies of body temperature came later, initially from measurements on freshly shot animals, later using implanted radio-transmitters and, most recently, with newer technology, the surgical insertion of miniature data loggers. The latter technique is not a panacea, however, because even though a continuous record of Tb can be obtained, for months or even years at a time, it often cannot be associated with information about coincident behaviour. In addition, no data are collected from many types of loggers until the logger is retrieved and this is not always easy from free-ranging crocs. A further drawback is that in animals as large as crocodylians, more than one logger would be desirable to record internal thermal gradients, which may be substantial. Abercrombie *et al.* (2002) have written charmingly about the frustrations and limitations of using data loggers to study body temperature.

As with so much crocodylian biology, Hugh Cott (1961) was also a pioneer in field studies of body temperature. It was before radio-telemetry of course, but he took an opportunity to measure the cloacal temperatures of seven fairly large *Crocodylus niloticus* (both sexes) that had been freshly shot. The size range was 2.15–4.22 m and Tb ranged from 23.0–29.0°C. The warmest was shot while it was (apparently) basking. Most were shot in the water. Cott interpreted daily movements between water and land in a thermal context (as did Herodotus), discussed the implications of a large body size on rates of heat exchange and recorded observations on the thermal significance of gaping (see below). His conclusion, which must have been based more on behavioural cycles and intuition than on his actual Tb data, was that *C. niloticus* are able to 'exercise a remarkable degree of control over the thermal level of the body'. As much as I admire Hugh Cott's work, he would have trouble floating that past a referee these days!

Some of the earliest research on crocodylian thermal relations was done on American alligators, the pioneers being Jim Spotila, Jeff Lang and Norbert Smith, and various aspects of their studies are mentioned in this chapter. Before getting to the main focus in this section, a seminal study by Norbert Smith is worth noting. Norbert, whose name appears from time to time in this book,

undertook a research master's project in the early 1970s on the thermal relations of American alligators (Smith 1975a). He used a multichannel radio-transmitter of his own design and construction (Smith 1974) to monitor heart rate, dorsal and ventral subcutaneous temperatures and 'core' Tb from a probe inserted deep into the muscle adjacent to the cloaca. The study was conducted on free-living animals at a wildlife refuge in southern Texas, where, because they were so wary, he encountered the difficulties of making simultaneous observations of behaviour. He worked also on habituated animals at a zoo in Waco, Texas, where he was able to observe responses of dorsal, ventral and core temperatures to behavioural events. Fig. 10.1, reproduced from his 1975 paper, plots a dataset recorded from a free ranging alligator for just over a day and shows a nice saw-tooth cycle in core Tb. The difference between 'dorsal' and 'core' temperatures is a reminder that body temperatures are not homogeneous throughout. His core temperatures were measured with a thermistor placed surgically beside the cloacal chamber, 10 cm from the surface, and the ventral probe was just below the skin of the thorax. Most modern studies insert telemeters or data loggers into the body cavity, where they

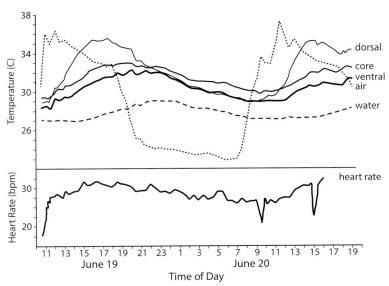

Fig. 10.1. *A 30-hour run of temperature and heart rate data telemetered in July 1971 from a mature adult male* A. mississippiensis *free-ranging in a wildlife refuge in Texas. The three pronounced bradycardias were 'fear evoked' (see Chapter 9), the first at the initial release. This illustration is from one of the earliest field studies of thermal relations in any crocodylian. (Adapted from Smith 1975a, with permission. © 2003 by The University of Chicago)*

may sink ventrally. Likewise, telemeters fed as pseudogastroliths would be close to the ventral surface. The approximate match between Smith's 'ventral' and 'core' temperatures is reassuring. Not all of Smith's conclusions have stood the test of time, but he drew attention to the importance of both behaviour and physiological mechanisms in alligator thermal relations.

Ideally, thermal relations are studied by monitoring body temperatures and behaviour as continuously as possible in free-living individuals over a good size range, in both winter and summer, along with a diverse and appropriate range of environmental data. This is a lot harder than it sounds. Tb is usually sampled using implanted temperature sensitive radio-transmitters or data loggers. Radio signals travel well in fresh water, but are absorbed in brackish and salt water. The availability of long-lived loggers seems the perfect solution, but they need to be retrieved before any data are collected, so some losses can be expected and this needs to be anticipated in the sampling design. By contrast, radio-telemetry can provide a continuous stream of data if the observer is there to collect it (unless transmitters fail). Sometimes an automated receiving system can be used (see Grigg and Beard 2000). Regardless of the sampling technique, Tb data collected without simultaneous behavioural observations can be difficult to interpret, and this is a particular problem with crocodylians because they are so wary in the wild. If you are there to observe the croc, the behaviour you observe will probably be 'waiting patiently for you to leave' behaviour! It is not surprising that many studies have focussed on captive animals in naturalistic situations such as at large crocodile/alligator farms where the inmates may be at least partly habituated to humans. For large valuable animals in a farm, retrieval difficulties often prevent the use of data loggers and Lyn Beard, Frank Seebacher and I resorted to feeding encapsulated transmitters to large *Crocodylus porosus* in a 60 ha enclosure and lagoon at the Edward River Crocodile farm at Pormpuraaw, Cape York, Australia. The crocs had opportunities for normal behaviour (see below) but the range of transmitters was reduced inconveniently by the water being just slightly brackish.

The most informative data on natural patterns have come from field studies and the easiest way to come to terms with crocodylian body temperatures and thermal relations is probably to consider some case studies, then draw out some generalisations.

Two species of Crocodylidae have had close study: the comparatively small *C. johnstoni* (15 individuals, 3–20 kg) and the much larger *C. porosus* (11 individuals, 32–1010 kg), representing a wide size range. For Alligatoridae, studies on *A. mississippiensis* and *Caiman yacare* will be summarised.

BODY TEMPERATURE PATTERNS IN AUSTRALIA'S FRESHWATER CROCODILE

The first substantial long-term field study was of the freshwater crocodile, *Crocodylus johnstoni*, in a PhD programme by Frank Seebacher. Frank tracked changes in Tb using radio-transmitters implanted into crocs living in The Croc Hole (17° 07'S, 144° 03'E): an anabranch off the Lynd River in northern Queensland. He could watch marked individuals from a perch on a rocky outcrop high above the billabong (Fig. 10.2) and record behavioural and Tb data simultaneously. All the crocs spent most of their time in the water, but many of them emerged during the day and basked in the sunshine (Fig. 10.3), raising their body temperature. He also measured environmental data: water temperature at several depths, air and ground temperatures and solar radiation. These allowed calculation of operative temperatures (see Box 10.2), which give a measure of the warmth or 'coolth' offered by the environmental conditions at a particular time for that particular crocodile (i.e. the Tb it would reach if everything stayed the same and the croc came into thermal equilibrium with its surroundings). The larger the croc, the greater the thermal inertia and the slower the rate of change so, in practice, reaching that equilibrium was rare on land, but common when they spent overnight in the water, which they usually did.

Daily patterns of body temperature (Tb)

Frank studied a range of sizes from ~3–20 kg (large adult *C. johnstoni* are small compared with other crocs). None of these crocs was large enough to

Fig. 10.2. *To understand an animal's thermal relations, it is of great value to have direct behavioural observations to go with the body temperature (Tb) data gained from implanted radio-transmitters or data loggers. This is often difficult with crocodylians in the wild, because they are so wary, but at the Croc Hole on the Lynd River in northern Queensland, Frank Seebacher was able to observe marked* C. johnstoni *and record their behaviour while logging Tb from implanted temperature-sensitive transmitters. (Photo GCG)*

Fig. 10.3. *Photographed from the rocky viewing perch, a male* Crocodylus johnstoni *basks on a rocky island in a billabong connected to the Lynd River in northern Queensland. The crocs here apparently prefer to bask on small rocky islands. These are in short supply so there is competition for them. Crocs higher in the peck order chase subordinates off and so gain the benefit of higher temperatures and, presumably, higher growth rates and increased reproductive potential. (Photo GCG)*

store heat overnight and all the crocs in The Croc Hole study started each day at water temperature.

Some of the crocs showed patterns similar to what is often regarded as a typical 'terrestrial lizard' pattern: that is, morning basking in the sun until some warm point is reached, then 'shuttling' between sun and shade (or in this case water) so that a regulated 'plateau Tb' is achieved (Fig. 10.4, left panel), much warmer than the water. This temperature was in the low 30s but varied seasonally (see next section).

Others, however, showed a very different pattern, remaining in the water all day so their Tb conformed to water temperature (Fig. 10.4 right panel). Most interestingly, at least some of these crocs were apparently kept in the water by social pressure! Frank saw basking crocs chased back into the water by apparently higher status animals, curtailing their attempt to get warm. Some of the crocs stayed in the water and did not even attempt to emerge. Presumably these lower status animals grow more slowly and their maturation and breeding opportunities are postponed (see Chapter 13).

A finding with implications for crocodylians that grow larger than *C. johnstoni* is that larger individuals have fewer basking events each day (Fig. 10.5). Larger crocs take longer to heat but they also retain the heat longer once it is gained so they can spend more time doing other things before choosing to bask again to restore their heat load. The implications of this are that crocs larger than ~30 kg might be expected to show only a single peak in daily Tb, and this is what we found subsequently in *C. porosus* from 32 kg and upwards (see below).

The other finding of particular interest was that there was abundant solar radiation throughout the year (Fig. 10.6) and crocodiles that basked routinely were not limited by lack of solar radiation. Operative temperatures were sufficiently high that, had they remained basking indefinitely, even on cold winter days, they would have reached lethal levels (in the vicinity of 35°C).

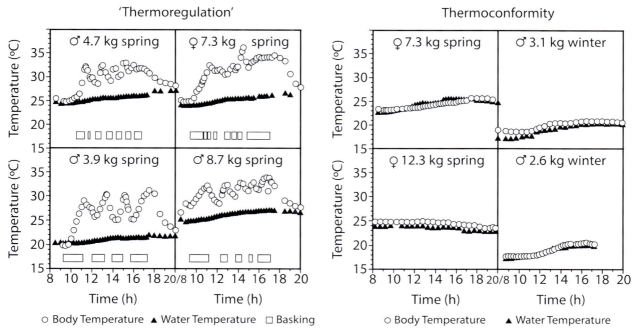

Fig. 10.4. Crocodylus johnstoni. *All crocs in The Croc Hole became equilibrated with water temperature overnight. Some emerged to bask in the mornings and alternated between land and water during the day (left half), so Tb was warmer than the water during the day. Basking periods are indicated by the boxes. The difference between the temperature peaks, when they sought the water, and the troughs, when they emerged again to bask, defines the preferred body temperature range (PBTR). Some crocs remained in the water all the time so Tb equilibrated with environmental temperature (i.e. a pattern of thermoconformity) (right half). These showed much lower Tb. Competition for basking sites resulted in crocs higher up the pecking order gaining the benefit of higher Tb. (Adapted from Seebacher and Grigg 1997, with permission)*

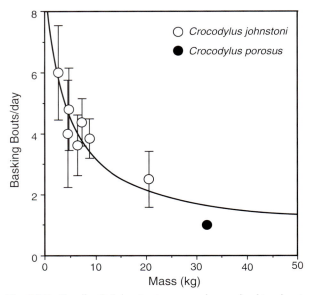

Fig. 10.5. *Smaller* C. johnstoni *averaged more basking bouts and, correspondingly, more Tb peaks per day than larger ones (open circles; average values and 95% confidence limits). Interestingly, the curve fitted to the* C. johnstoni *data (Seebacher and Grigg 1997) predicts that animals of 30–40 kg are likely to have only a single bout daily, which fits well with the daily patterns observed in* C. porosus *(closed circle, from Grigg* et al. *1998) all of which (from 32–1010 kg) showed a single daily peak. (Adapted from Grigg* et al. *1998)*

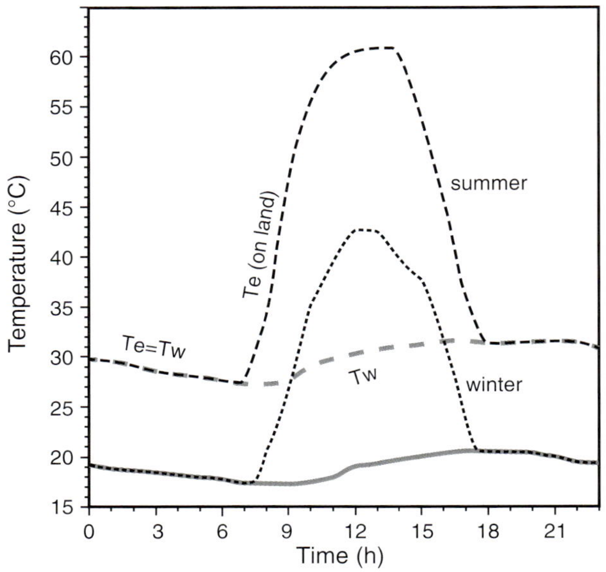

Fig. 10.6. *Operative temperatures available to* C. johnstoni *undergoing their normal behavioural cycle of spending the night in the water and the day alternating between a basking site on land and the water. Note that in both winter (lower line) and summer (upper line) the daytime operative temperatures (Te) are well above their preferred range of Tb (Fig. 10.7) so they had abundant access to warmth. Despite this, they chose to begin and cease basking at cooler body temperatures in winter than in summer! (Redrawn from Seebacher and Grigg 1997, with permission)*

Seasonal pattern

Although most *C. johnstoni* populations live at lower elevations and lower latitudes, at 17°S and 400 m AMSL the Croc Hole is far enough south and elevated enough to have a strong seasonal temperature cycle, with water temperatures almost always below optimum. Hence, basking patterns similar to those shown in Fig. 10.4 were seen in both winter and summer.

However, it became clear that the crocs that showed the daily 'shuttling', thermoregulatory pattern were choosing to cease basking at a lower body temperature in winter than in summer. In the daily patterns seen in Fig. 10.4 (left panel), basking crocodiles were making sequential decisions about when to retreat to the cooler water and when to re-emerge and bask. Averaging these values across all crocodiles and comparing winter and summer yielded the seasonal pattern seen in Fig. 10.7. The difference between the two lines represents a range of Tb at which, apparently, they felt 'comfortable'.

This defined a preferred body temperature range (PBTR) and it showed a 4°C change between summer and winter, the midpoint of the range changing from ~33°C to 29°C, while water temperature underwent a 7–9°C seasonal cycle.

This seasonal cycle of preferred Tb was unexpected and could not be explained by any shortage of available heat in the winter because there were long periods every sunny day for which operative temperatures were well above the Tb at which crocs ceased basking (Fig. 10.6). Clearly the crocs were not limited by lack of available heating opportunities in winter and were changing their behaviour by choice to operate at lower temperatures. This could be an example of seasonal acclimatisation. *C. johnstoni* spend most of each 24 h day in the water and, with water temperature undergoing such substantial change seasonally, by nearly 10°C, some seasonal acclimatisation of their metabolism can be expected, with concomitant

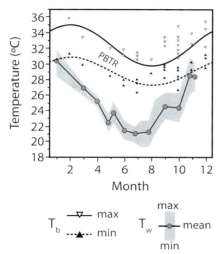

T_b —▽— max ···▲··· min

T_w max mean min

Fig. 10.7. *Seasonal changes in the average Tb at which basking C. johnstoni chose to return to the water (upper solid line) or to re-emerge to bask (dotted line). The difference between the two lines represents a range of Tb at which, apparently, they felt 'comfortable'. This can be referred to as a preferred body temperature range (PBTR). The seasonal change in water temperature (Tw) at this elevated location is conspicuous. (Modified from Seebacher and Grigg 1997, with permission)*

changes in the thermal optimum. Thermal metabolic acclimation has been demonstrated experimentally in *C. porosus* (Glanville and Seebacher 2006) and American alligators (Guderley and Seebacher 2011) and more will be said about seasonal metabolic acclimatisation later.

BODY TEMPERATURE PATTERNS IN THE ESTUARINE CROCODILE

A study of Tb in large, free-ranging estuarine crocodiles, *C. porosus,* was impractical with the resources we could muster, so Lyn Beard, Frank Seebacher and I opted for a study in the very large breeding lagoon at the Edward River Crocodile Farm at Pormpuraaw, (14°54'S, 141°37'E) North Queensland: a wonderful fenced off wetland in which very large animals go about their daily business in a very naturalistic setting (Figs 10.8, 10.9). Surgical implantation was impractical in these very large and also very valuable beasts and,

Fig. 10.8. *Aerial view of the Edward River Crocodile Farm showing the natural wetland fenced off to form extensive breeding lagoons. (Photo GCG)*

Fig. 10.9. *Hymn, α male at the Edward River Crocodile Farm. This was the venue for our study of thermal relations of large* Crocodylus porosus *and, at a calculated body mass of ~1 tonne, Hymn was the largest subject. (Photo GCG)*

even if data loggers had been invented by then, getting them back would have been a daunting prospect. So, with farm manager Don Morris' very considerable help, we fed temperature sensitive radio-transmitters encased in dental acrylic to 11 *C. porosus* over a size range from 2.1 to 5.5 m (estimated to weigh 32–1010 kg), a logical upward extension from the *C. johnstoni* study. It was impractical to measure or weigh any of the animals, so lengths were estimated, sometimes by measuring an imprint on a sandy bank and converting that to mass from a mass–length relationship. The one tonner, Hymn (Fig. 1.37), was a particular target of the study. He was the largest at the farm and had been boss cocky among the 200+ adult animals in the lagoon since his capture in 1978 in the Malaman River (Robbie Bredl *pers. comm.* 2012).

Measurements of environmental variables were made as in the study of *C. johnstoni,* and operative temperatures calculated. A slight brackishness in the water reduced the range of the transmitters

to 50–75 m so we could not use an automated recording system in the large lagoon and had to collect data manually. Because the crocs were somewhat habituated to humans, this did not disturb them. Some of the observations were made on animals in a smaller lagoon, using the automated system. We collected data in both summer and winter.

Daily and seasonal patterns of Tb and correlations with body size

There were pronounced daily cycles in body temperature in all animals, showing a saw-toothed shape, with the heating phase steeper than the cooling phase (Figs 10.10, 10.11). We saw no sign of the multiple (shuttling) basking events that typified thermoregulating *C. johnstoni*; indeed, the single daily cycle was pretty much as predicted by the trend towards fewer basking bouts with increasing size (Fig. 10.5), reaching a single bout above ~30–40 kg. The habitats in these two studies

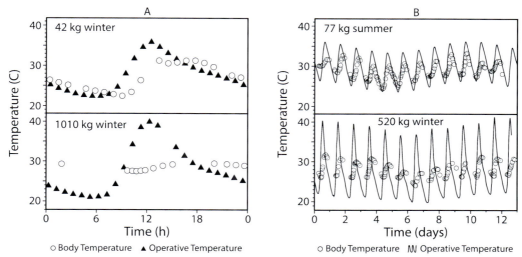

Fig. 10.10. (Left) Typical data collected in winter at the Edward River Crocodile farm from a 'small' (42 kg) and a large (1010 kg) C. porosus *over a single day. Note that the larger animal stores heat overnight and is still 'warm' in the morning. (Right) Typical data collected over several days from a not so large (77 kg) animal in summer and a much larger (520 kg) individual in winter. Body temperatures (open circles) show a typical saw-toothed daily cycle, responding to the daily cycling of operative temperature (Te, solid triangles or line) and the larger animals are less variable due to their larger thermal inertia. We were perplexed to find a 10–13 day cycle in Te (and, consequently, in Tb as well). The study lagoon is close to the sea, behind a frontal dune system (Fig. 10.8) and there are extensive exposed mudflats at low tide. Perhaps this unanticipated cycle is related to tidal and wind patterns. Whatever the reasons, it does show that crocodile Tb is very responsive to prevailing environmental conditions. (Adapted from Grigg* et al. *1998)*

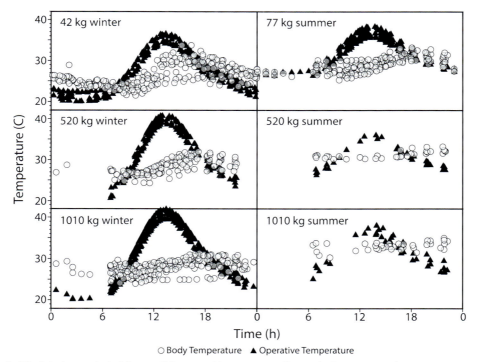

Fig. 10.11. *Pooled Tb data (open circles) for small, large and largest crocodiles and relevant operative temperature cycles (Te, closed triangles) in winter and summer at the Edward River Crocodile farm. Note that the smallest animal shows the largest daily range in Tb and that all showed a noticeable seasonal cycle, although the largest animal was essentially thermostable over a single day (as in Fig. 10.10, left). All animals were warmer in summer than in winter; that is, the large animals could store heat from day to day but not from season to season. The single day's run of cool Tb for the 1010 kg animal in winter was the result of a cold snap. (Modified from Grigg* et al. *1998)*

are, of course, quite different, and they are different species, but this 'fit' is quite satisfying and invites further exploration in other species.

The smallest animal in our study was 32 kg and it showed a daily cycle in Tb of ~7°C in both summer and winter. The largest individuals reached or approached a state of daily homeothermy. The magnitude of the daily cycles in Tb decreased progressively with increasing body mass from ~7.0°C in 30–40 kg animals to ~2°C in summer and 3°C in winter in animals larger than 600 kg or so (Fig. 10.12). That is, in summer, crocs larger than 600 kg in this habitat met the criterion sometimes applied to define homeothermy: that is, thermostability within 2°C. In winter they fell just outside that. The greater daily range shown by large animals in winter compared with summer (Fig. 10.12) is a reflection of longer periods spent in the sun, hence higher Te values (e.g. Figs 10.10, 10.11). These results confirmed the prediction from a mathematical model by Spotila *et al.* (1973) that large reptiles would have a relatively stable body temperature because of a combination of thermal lag and heat storage.

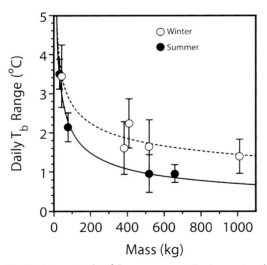

Fig. 10.12. *In our study of* C. porosus *captive in a naturalistic situation on Cape York, Queensland, at 15°S latitude, the larger the croc, the smaller the daily range in body temperature. In winter they experienced a larger range than in summer. In both summer and winter, individuals larger than ~300 kg were 'homeothermic' (in the sense defined in Box 10.1), but even the largest individuals showed a seasonal cycle in Tb outside the 2°C criterion for homeothermy (Fig. 10.11 and Table 10.1). (Adapted*

If large crocs can be homeothermic (or nearly so) on a daily basis, and if stability within 2°C is used as a criterion for homeothermy (see Box 10.1), the notion of large crocs as 'passive homeotherms' fails (just) when seasonal changes are taken into account. On average, both small and large crocs were 3–4°C warmer in summer than in winter. In smaller animals, the seasonal difference between modal Tb in winter and summer was 25.1 to 28.4°C (3.3°C) and in larger animals 28.7 to 33.6°C (3.9°C). They could store heat from day to day, but at the latitude of the study (15°S), did not maintain a constant temperature across the seasons, and this has implications for extrapolating from crocodiles to dinosaurs (see below). Closer to the equator, in a less seasonal climate, large *C. porosus* could very well be homeothermic throughout all seasons.

One idea about how homeothermic endothermy evolved in mammals and birds is that it arose via 'passive' or 'inertial homeothermy' (McNab 1978), for which the data collected in this study of a large crocodylian present a good model. McNab envisaged that reptiles ancestral to mammals showed passive homeothermy (not a preferred term, see Box 9.1) at body masses of 30–100 kg, evolved external insulation and, under the combined influence of selection pressures that favoured smaller size and the maintenance of metabolic rate, made the conversion from passive homeothermy to endothermy. The data from *C. porosus* suggest that passive homeothermy, even on a daily basis, will not eventuate except at much larger body size: larger than any of the therapsid creatures on the cynodont line along which mammals, and endothermy, are thought to have arisen (Fig. 2.3).

Not only were the larger *C. porosus* more thermostable, they were also warmer. The one tonner was ~4°C warmer than the 30–40 kg animals, in both summer and winter. I had predicted long ago, before we had any data, that larger crocodiles would turn out to have higher body temperatures than smaller ones (Grigg 1977), which is what we found. However, my reasoning turned out to be wrong. Starting from first principles, Frank Seebacher was able to predict the same increase that we measured empirically (Seebacher *et al.* 1999). The accumulation of warmth as a consequence of

TABLE 10.1   SUMMARY OF THE PATTERNS OF BODY TEMPERATURE AND THERMALLY SIGNIFICANT BEHAVIOUR SEEN IN OUR STUDY OF *C. POROSUS* AT THE EDWARD RIVER CROCODILE FARM

Winter	Summer
Basking common, on land/exposed to air and sun during day, in water at night	Little or no basking; in water/shade by day, often on land at night
Saw-tooth daily cycles in Tb of approx. 7°C (40 kg croc) declining with body size to 3°C (1000 kg)	Saw-tooth daily cycles in Tb approx. 7°C (40 kg croc) declining with body size to 2°C (1000 kg)
Large crocs warmer than small crocs; modal Tb increased from 25.1°C (40 kg) to 28.7°C (1000 kg)	Large crocs warmer than small crocs; modal Tb increased from 28.4°C (40 kg) to 33.6°C (1000 kg)

large body size is a consequence of the increasing importance of radiative heat gain over convective heat gain as body size increases. This finding too has implications for speculations about the Tb of dinosaurs (see below).

There are about nine species of crocodylian that can grow larger than 500 kg (Table 1.1) – large enough to gain a measure of day to day thermostability. However, it is worth remembering that they are not large all their lives: they start off very small and the physics of heat exchange dictate changing patterns of thermal relations as they grow. A hatchling crocodylian may weigh less than 100 g, yet grow to 1000 kg or more: a mass increase of four orders of magnitude. In contrast, most mammals increase during growth by only one or two orders of magnitude. As they grow, their heat capacity (i.e.

thermal inertia) increases, and the surface area to mass ratio (i.e. the proportional surface area over which heat exchange with the environment can occur) decreases. The thickness of the boundary layer across which heat exchange between the animal surface and the surrounding air and/or water that can occur also increases with length, so that the rate of heat exchange by convection decreases as the animal grows and radiant heat exchange comes to dominate.

Seasonal changes in behaviour

All of the individuals in the study at Pormpuraaw showed seasonal changes in daily patterns of thermally significant behaviour, reversing their daily cycles of exposure to air (sun) and water. In winter, they typically basked during the day

Fig. 10.13. *(Left)* C. porosus *basking on a winter morning at the Edward River Crocodile Farm. (Right) Big crocs seem to be particularly reluctant to leave the water and often bask only partly exposed. Seebacher (1999) showed that* C. johnstoni *4 kg and larger needed to be more than 60% exposed above water to gain heat, and the required percentage decreases as the animal gets larger because radiant heat gain gradually comes to dominate (see below). The large croc in this photo is ~50% exposed, but, because of its large size, it can be expected to be gaining heat. This animal, Hymn, was likely to spend all day basking like this in winter. (Photos GCG)*

Fig. 10.14. *(Left) In summer, the water holds the heat while the air cools down and the crocs often move onto land at night or into very shallow water. Large crocs often seek shade during the day without leaving the water. (Right) The large male Hymn has sought shallow water in the shade, minimising insolation and exposed to any cooling breezes. (Photos GCG)*

and many spent the whole day out of the water, not necessarily in the sun all the time (Fig. 10.13). The proportion of the croc exposed to air varied quite a lot. Basking can be effective if much of the animal is exposed: a consequence of its well-vascularised scutes, which take up heat readily and convect it internally, transported by the blood circulation. Floating animals, however, cannot float high enough to expose sufficient of themselves to the air so, despite some claims to the contrary, 'aquatic basking' is ineffective (there is a more comprehensive discussion of this topic below).

Fig. 10.15. *This croc has found a comfortable, shady spot beside the lagoon at the Edward River Croc Farm in which to spend much of the summer. Water in this so-called 'stinkhole' is substantially shaded and will not get too hot. (Photo GCG)*

The animals usually spent the winter nights in the water, which was warmer than air. In the summer, the crocs avoided exposure to the sun, spending much of their time in the shade, and/or in the water, and they often emerged from the water to spend the night on land because nocturnal air temperatures in summer were often cooler than the water (Figs 10.14, 10.15).

BODY TEMPERATURE PATTERNS IN THE AMERICAN ALLIGATOR

At the Rockefeller Wildlife Reserve in Louisiana, which has been the focus of much research into *A. mississippiensis* for nearly 50 years, Frank Seebacher teamed up with Ruth Elsey and Philip Trosclair to look at alligator thermal relations (Seebacher *et al.* 2003). They retrieved data from seven individuals in summer (1.93–54.54 kg) and seven more (1.59–53.64 kg) in winter, using data loggers implanted surgically into the body cavity.

Apart from the description of thermal patterns in summer and winter, the study became a vehicle for the development of the method described in Box 10.2 that extends the capacity for estimating body temperature null distributions to large reptiles, for comparison with measured values. This advance in methodology provides a way to answer critical questions about the extent to which an observed pattern of Tb is random, or the result of active thermoregulation.

Based on the way that the pattern of daily cycles across the two species of *Crocodylus* changes with size (Fig. 10.5), it might have been predicted that the small alligators would show shuttling behaviour, with several basking bouts during a day, and single bouts providing a saw-tooth daily pattern in the larger alligators (Fig. 10.16), such as that seen in *C. porosus* at 32 kg and larger. This was not the case. Representative examples of the patterns they recorded, in both seasons, are shown in Fig. 10.17

Fig. 10.16. *Like large* C. porosus *(Fig. 10.13) and other crocodylians, large American alligators spend a lot of time in winter basking. These were photographed in Everglades National Park in winter. Alligators can be seen basking at the same location in summer, too, but fewer individuals (three or four, compared with 30 or so) and for shorter periods. (Photo DSK)*

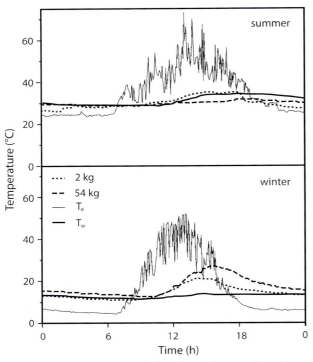

Fig. 10.17. Representative daily cycles in Tb in small and large American alligators in summer and winter and a comparison between Tb of small and large alligators on the same day in each of the two seasons in relation to water temperature (Tw) and maximum operative temperature (Te). Maximum Te is calculated for an animal fully exposed in sun and on land; note that it exceeded 30°C for 12 h on 30 June and 6–7 h on 14 February. (From Seebacher et al. *2003, with permission. © The University of Chicago)*

and all animals, small and large, showed only a single daily oscillation peaking in the afternoon. Furthermore, none showed a daily plateau in Tb, which might suggest thermoregulation by shuttling as was seen in *C. johnstoni* (Seebacher and Grigg 1997, Fig. 10.4). Seebacher *et al.* (2003) observed that this is not to say that shuttling never occurs in alligators and it could be that the habitat in the Rockefeller Reserve (extensive, continuous marshes) constrains access to land for basking. Alternatively, it could imply a difference between Alligatoridae and Crocodylidae. Grigg and Seebacher (2001) referred to suggestions of fundamental differences in thermal behaviour between alligators and crocodiles, and this could be another example, but too few species have been studied so far to draw any conclusions.

In summer, *A. mississippiensis* Tbs differed little from water temperature, averaging ~30°C, whereas in winter, with much cooler water temperature (12–16°C), Tb values were elevated (Fig. 10.17) by spending less time in the water. For all alligators, daily Tbs averaged ~30°C in the summer sampling and 16°C in winter.

Whereas *C. porosus* showed an increase in modal Tb with increasing body size in both summer and winter, alligators showed this only in winter: 54 kg alligators were ~2°C warmer than 2–3 kg ones. Seebacher *et al.* (2003) explained this to be a consequence of the alligators spending so much of the time in the water, whereas in winter animals were more exposed to solar radiation. The accumulation of warmth at larger body sizes results from the increasing importance of radiative heat gain over convective heat gain as body size increases.

Were the alligators thermoregulating? It would be easy to conclude from these data that the alligators are not showing much evidence of thermoregulation, but simply passively following the daily cycles. If thermoregulation is defined in terms of stability of Tb, they would fail (see Box 10.2). However, using the methodology developed for large crocodylians and applied by Seebacher *et al.* (2003) (Box 10.2), it became clear that alligators were thermoregulating actively during both sampling periods; that is, their Tbs were cooler than the calculated random null distributions in summer and warmer than the equivalent nulls in winter.

Furthermore, even the small alligators seem not to be taking full advantage of the heat available in the environment in winter (Seebacher *et al.* 2003), an observation that parallels the observation that *C. johnstoni* exhibits a lower 'preferred body temperature range' (PBTR) in winter (Fig. 10.7) (Seebacher and Grigg 1997). One explanation might be that the alligators in this part of their range were simply inactive, yet they were seen feeding, moving onto land when available and they were in good condition. The authors suggested, as for the seasonal fall in PBTR in *C. johnstoni*, that in winter the alligators might have undergone a thermal metabolic acclimatisation, with the thermal sensitivity of metabolic enzymes changing so that metabolic rate can be maintained at a lower Tb. More about this later.

Fig. 10.18. Caiman yacare *basking in the Brazilian Pantanal. Two of them are facing the sun and are gaping (see below) so they may be minimising further heat uptake. The third is better positioned to absorb heat. (Photo GCG)*

BODY TEMPERATURE PATTERNS IN THE PARAGUAYAN CAIMAN

A study of *Caiman yacare* in the enormous tropical wetland of the Brazilian Pantanal by Zilca Campos and colleagues measured Tb in 51 individuals (3.3–42 kg) in both warm and cool seasons using implanted transmitters and data loggers (Campos *et al.* 2005). They also measured cloacal temperatures in several hundred caiman (40 g–43 kg) captured in the evenings (1830–2200) as an adjunct to a population study (Coutinho 2001) (Fig. 10.18).

Like the three species discussed above, there were marked seasonal differences of nearly 5°C in body temperature between winter and summer (mean Tb 25.7–30.1°C). The lowest temperature measured was 16.9°C, the highest 37.4°C and no gender related differences were apparent. The Pantanal is subject to the periodic arrival of cold fronts in the cool season (May–September), which drop ambient temperatures dramatically, and in such an event

the mean Tb of telemetered caiman fell to ~22°C. Body temperatures were more variable in smaller individuals, as would be expected.

Caiman are typically aquatic at night, and water temperature usually determines their nocturnal Tb in both winter and summer. Cloacal temperatures measured on caiman caught in the water during the evening hours were surprisingly variable, but generally lower in the cool season. In winter, Tbs measured by telemetry were frequently above both water and air temperatures, up to 15°C warmer than the water, indicative of basking. Indeed, large numbers of caiman lying around the shorelines of water bodies, apparently basking, are a major attraction for cool season visitors to the Pantanal (Fig. 4.2). In summer, daytime Tbs were lower than water temperature in many individuals, suggesting shade seeking, but with water temperatures generally above 28°C, management of Tb was apparently not seen by the caiman to be a priority, with the data suggesting thermoconformity at these times. Campos and colleagues concluded that thermoregulation in the warmer months has a lower priority than foraging and social interactions. A follow up publication from the study (Campos and Magnusson 2011) emphasised that, although thermoregulatory basking was implied by daytime emergences from the water in winter, much of the movement by caiman between air and water at other times apparently had little to do with thermoregulation.

The caiman were in a size range at which, if they behaved like *C. johnstoni*, might be expected to shuttle between land and water in the cool season. In actuality, their seasonal changes in emergence behaviour are more like those reported for both *C. porosus* and small *A. mississippiensis*, discussed above. The study exemplifies the conundrum posed by the question 'are they thermoregulating?' (see Box 10.2 and below). Their winter behaviour carries that implication strongly, but not in the 'lizard' sense of shuttling to maintain a 'Tb plateau', and at warmer times of the year the caiman make little or no effort at modifying Tb. They do not need to. They are free to move between air and

Fig. 10.19. *Cuvier's dwarf caiman,* Paleosuchus palpebrosus. *(Photo John White)*

water without any constraint imposed by thermal considerations. It seems to be a common pattern for crocodylians.

BODY TEMPERATURE PATTERNS IN CUVIER'S DWARF CAIMAN

Paleosuchus palpebrosus (Fig. 10.19) at the southern end of their range show little evidence of thermally related behaviour. Body temperatures were monitored by Campos and Magnusson (2013) in

Fig. 10.20. *(Left) A 1.5 m* Caiman yacare *and (right) a 5.5 m* C. porosus *relaxing in winter sunshine, part in and part out of the water. Crocodylians will often spend time this way, with much of the upper surface exposed to the sun and it is more common among larger animals, which often seem reluctant to leave the water. Are they basking? Without making measurements, it is not straightforward to predict whether this is leading to a rise or fall in Tb. Heat transfers are occurring by convection in water and air, radiation in air and by conduction with the substrate, so an appropriate overall operative temperature is not straightforward to calculate. Seebacher (1999) calculated that, under typical sunny winter conditions,* C. johnstoni *lying in water with a 60% exposure to air in sunshine would be gaining heat, but losing heat at 40% exposure. This suggests that the* Caiman *is unlikely to be gaining heat. The larger animal is ~50% exposed, but radiation has a more dominant role in heat exchange in larger animals. This is a typical posture adopted by this particular individual in winter and, although we know that its Tb does not alter much on a daily basis, we do know empirically that its Tb rises during the day (Fig. 10.10). This behaviour was not seen in summer. In general, a crocodylian can achieve a balance between heat gain and heat loss by choosing a particular balance between the proportion above and below water. (Photos GCG)*

a population of *P. palpebrosus* living in a hillside stream under dense vegetation and flowing into the Brazilian Pantanal. The study was at 19°S, in habitat subject to low winter temperatures, which are exacerbated by the arrival of periodic cold fronts. Seven male and six females were captured (2.5–20 kg) and fitted with implanted temperature-sensitive radio-transmitters. Mean monthly body temperatures ranged seasonally between 20.1 and 25.6°C, close to stream temperatures and ~5°C lower than in the Pantanal's caimans, *Caiman yacare* (above), reflecting cooler ambient temperatures. This is very cool for crocodylians, but not as cold as experienced by American alligators at their northern extent, ~33°N latitude. The lowest Tb measured in *P. palpebrosus* was 13.7°C, the highest 31.1°C. Summer Tbs were often lower than both air and water, as would be expected if they were in their burrow. Winter Tbs were often a little above water temperatures, again implying use of a burrow and, unlike in *C. yacare*, there was no indication of the sort of increase in Tb that would result from basking. Some animals were seen in the sun in late winter months but no correlated Tb data were available. The density of the vegetation made direct observation of telemetered animals difficult. The authors concluded that, in this habitat, *P. palpebrosus* remain in their burrows during cold periods and do not seek to warm themselves by basking. This is similar to behaviour by both species of alligator in cold conditions. Noting that this species has been reported in altitude-generated cool habitats in Columbia, the researchers suggested that their capacity to tolerate low temperatures enables them to extend their range into habitats inhospitable to other crocodylians.

What would be interesting would be to compare these results with similar data from *A. mississippiensis* and *A. sinensis* at their northern limits and with *P. palpebrosus* in the more tropical parts of its range. The pattern seen in *P. palpebrosus* may be more a result of having behavioural and physiological flexibility than any intrinsic differences from other alligatorids. The data that would allow these comparisons are, unfortunately, unavailable so far. No self respecting crocodile, of course, could cope so well!

Synthesis: are crocodylians thermoregulators?

Crocodylians are conspicuously capable of activity over a wide range of Tb and, in the field, seem to give a low priority to seeking any 'preferred' Tb, even though they may show a preferred temperature on an artificial thermal gradient. The highest temperatures observed are in the mid 30s, the lowest in the mid teens, except in the two species of *Alligator*, which have the highest latitudinal distributions and have been observed to survive water temperatures well below 5°C (more below).

Their thermal relations are different from what is usually regarded as being typical for reptiles. Convincing thermoregulatory shuttling between sun and shade (or water in this case) in free-ranging individuals is described so far in only one species, *C. johnstoni*, studied in a cool part of its range and, even in that species, the resultant Tb pattern is rather loose compared with that of many squamate reptiles. Typically, crocodylians show a conspicuous daily Tb cycle over several degrees, with a single peak towards mid afternoon. Thermostability is conferred only by very large size, a very stable habitat or, perhaps, a voluntary or forced choice to remain in the water.

So can we say that crocodylians are thermoregulators? Well, it depends on how that is defined. Compared with the way the term is applied in the mammal and bird literature, with the implied maintenance of a high and stable Tb, the answer has to be no. But this 'traditional' definition has shortcomings when applied to ectotherms, as discussed in Box 10.2. Even though Tb patterns showing daily stable plateaus are not typical of crocodylians, there is a wealth of observations of behaviour that is convincingly 'thermoregulatory'. The conspicuous and 'thermally sensible' seasonal changes in behaviour shown by *C. porosus*, *A. mississippiensis* and *Caiman yacare* provide examples. The shuttling behaviour shown by *C. johnstoni* comes closest to being an example of thermoregulation in the typical 'reptilian' sense, but even that fails the usual mammalian and avian definition. For ectotherms, Heath (1964) pointed out that the relevant criterion of thermoregulation should, more properly, be that measured Tbs are different from what an identical animal moving randomly would show. That is easier to explore in small reptiles, which have little thermal inertia, but Seebacher *et al.* (2003) developed a method that allows it to be explored in larger species. Using that method, they showed convincingly that daily Tb patterns in the American alligators in their study were different from an alligator behaving randomly, so, by that criterion, they were thermoregulating. A pedant might argue that even this criterion is insufficient unless it can be shown that the real motivation for the observed behaviour is thermoregulatory, but in the long run perhaps that is a matter of judgment.

To me, it seems quite clear that many of the observed patterns of body temperature observed in crocodylians are the result of three processes: behavioural and physiological mechanisms by which they influence the uptake, or loss, of heat from their bodies and biochemical adjustments that minimise the effect of Tb on performance. These mechanisms form the subject of the next section.

Behaviours that modify body temperature

Heat will be flowing either into or out of any body, any time, any where, unless it is already in equilibrium with its thermal surrounds. In animals, the rates of heat flow can be modified to some extent physiologically, but mainly by behaviour, particularly by adjusting the amount of exposure to the sun. In most crocodylians, the sun is their major source of warmth and they use it or avoid it in diverse ways, some quite subtle.

Movement between land and water; daily and seasonal basking

Movements between water and land are part of many crocodylians' daily behavioural repertoire and, as the quote from Herodotus shows, recognition that they are probably choosing between available thermal environments goes back more than 2000 years. This is now well confirmed, as shown by the case studies described above. In the cool, upland habitat of The Croc Hole *C. johnstoni* spend the nights in the water and, year round, rely on basking to elevate Tb above water temperature. Further north, in a coastal lagoon, *C. porosus* seek shade in summer and are more likely to be on land at night, to take advantage of the cooler night air (Fig. 10.14). In winter, however, they seek the sun's warmth, either leaving the water completely or exposing a large part of themselves to it (Fig. 10.13). Although there has been a tendency for some reports to refer to daily patterns as though they are fixed, the reality is that they are quite flexible, with the patterns varying to suit local conditions. *C. niloticus* in Uganda were reported to haul out onto land to bask during the morning, return to the water during the heat of the day, bask again in the late afternoon and return to the water in the evening (season not specified) (Cott 1961). Working on Central Island in Lake Rudolf (now Lake Turkana), with ~500 *C. niloticus* to count in the focal 0.65 km^2 lake and just 1.2 km of shoreline, Modha (1968) described an opposite pattern of movement between water and land. With soil temperatures above 40°C for most of the daylight hours, black bulb temperatures above 50°C and a lack of shade surrounding the lake in that volcanic bowl, being in the water, with a surface temperature of around 30°C, was obviously a more attractive proposition.

This is not to say, of course, that all movements to and from land are related to choosing comfortable thermal conditions and just because a crocodylian is on land and in the sun, it may not necessarily be 'basking'. Not all crocs rely on basking. Smooth-fronted caiman, *Paleosuchus trigonatus*, live in streams in the Amazonian rainforest where opportunities for basking are few and, according to Magnusson and Lima (1991), they bask rarely (see more about this fascinating species in Chapter 13). *Osteolaemus* also live in many places with a closed forest canopy.

Knowing when basking is most likely to occur is important for planning population surveys. Downs *et al.* (2008) wanted to know the optimum times to find *C. niloticus* basking on the shorelines of Lake St Lucia in KwaZulu-Natal, South Africa, so they could time their winter aerial surveys when the animals were most easily seen. They recorded winter Tb cycles in four *C. niloticus* (1.1–1.9 m) at the nearby St. Lucia Crocodile Centre and found that all four animals showed a single daily oscillation in Tb from the low 20s to high 20s (°C), very similar to the daily cycles measured in *C. porosus* (Grigg *et al*. 1998). (As an aside, the St. Lucia Crocodile Centre was an initiative of Tony Pooley, whose seminal work on Nile crocodiles is referred to elsewhere.)

Basking partly exposed

Crocodylians are far more subtle in their selection of a suitable microclimate than just being either in or out of the water and/or the sun; they often choose to be in the shade, if available, or partly exposed above water (Fig. 10.20, p. 358). Large individuals may be particularly reluctant to leave the water. In these situations, heat is being exchanged simultaneously with the air and water, each with very different thermal properties and with conduction, convection and radiation all operating in an ever changing thermal environment. So, what implication does this have for Tb? How much of a croc needs to be exposed for it to be gaining heat from the sun? Spotila *et al*. (1972) made the first serious attempt at quantifying heat budgets of a crocodylian and Seebacher (1999) took it much further by modelling heat exchange in a series of scenarios, comparing the predictions of the model with field measurements of Tb in *C. johnstoni*. Crucial to a quantitative understanding of heat exchange in such complex situations is calculation of the relevant operative environmental temperature, Te (see Box 10.2). Te varies with the proportion of the body exposed to air, and the rate of change of temperature depends on the gradient between Te and Tb. If the air is warmer than the water, there will be a particular proportion of emergence from the water at which the animal starts to gain heat and, of course, a point at which heat is being neither gained nor lost. In round figures, Seebacher (1999) found that a moderate sized *C. johnstoni* (above ~4 kg), emerging from cool water under sunny skies, warmed if 60% or more of its body was exposed. Conversely, a warm animal lost heat if 60% or more of its body was submerged. Neutrality was at ~50% exposure. Obviously these values will vary a bit with conditions and body size.

Note that heat gained by basking can be retained for a period after a return to the water, longer in larger crocs. Small individuals, 1–2 kg have very little capacity to retain heat and will essentially be thermoconformers (Seebacher 1999). If a partial exposure to air is maintained, this will slow the rate of heat loss.

Surface basking?

These findings imply the need for caution about reports of 'surface basking' by aquatic reptiles. It has been claimed, not only for crocodiles and alligators (Smith 1979), but for marine and freshwater turtles and various other aquatic vertebrates, even sea snakes, that they can raise their body temperature by swimming with the back exposed. This is sometimes referred to as 'surface basking', but the analysis referred to above shows that raising Tb in

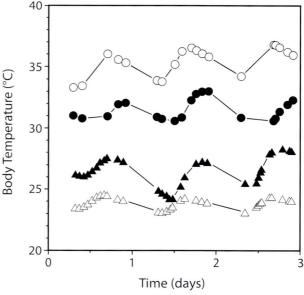

Fig. 10.21. *Body temperatures measured in a 520 kg* C. porosus *undergoing normal summer and winter behaviour (closed symbols) compared with temperatures that would result if winter behaviour was employed in summer (open circles) and* vice versa *(open triangles). Clearly the seasonal behavioural cycle they practice minimises the seasonal temperature range, whereas a winter behaviour in summer would lead to lethal levels of Tb. (Adapted with permission from Seebacher* et al. *1999)*

that way is extremely unlikely. About 50% of the body would probably need to be exposed above water and none of these animals can float anywhere nearly high enough in the water to make a useful difference. This is confirmed by Fish and Cosgrove (1987) who found that, by adopting a high float posture under an infrared lamp, small American alligators (315–990 g) in 15°C water could raise their Tb by only ~0.4°C.

However, if a heat load has been gained from a basking event, surface basking may slow the rate of return to water temperature. Seebacher (1999) showed, for example, that a 20 kg *C. johnstoni* can extend the time it takes to cool to the lower limit of its PBTR from about 1 h if fully immersed to 2 h if it retains an exposure of 20%. By fully inflating its lungs, a crocodylian could probably float with ~15% of its surface area exposed (Henderson *pers. comm.* and Chapter 4).

Seasonal changes to basking patterns may be crucial

There is no doubting the importance of the seasonal changes in basking behaviour and the use of shade and water described above for *C. porosus*. In summer, they spent the day in water or sought shade and emerged onto land at night, while in winter the behaviour was reversed: they basked on land during the day and remained in the water at night. The implications for their body temperature are far more than just 'comfort': they are necessary to survive. The heat budget analysis by Seebacher (1999) allowed an exploration of what would happen to them if their behaviour cycle was reversed: that is, if they basked in summer and sought cooler conditions in the winter. The results were striking (Fig. 10.21). In 520 kg individuals, choosing summer behaviour patterns in winter would lower Tb by up to 4°C. Choosing winter behaviour patterns in summer would raise temperatures by at least 6°C, with much of the time above 35°C, which would be lethal (Seebacher *et al.* 1999).

This brings up an interesting consideration. Because of the way Tb lags behind changes in Te, and because of the dynamics of heat exchange, much thermally related behaviour has to be anticipatory. A large croc cannot wait until its Tb is almost at a preferred level before moving into the shade, or a lethal level, because core temperature will continue to rise for some time.

If a croc gets caught in a situation where options for cooling are unavailable, it can be lethal. I have seen a dead 2 m *C. porosus* up against the fence outside the Edward River Farm. It had escaped somehow (they sometimes did) and was apparently trying to get home before feeding time, but, in searching for a hole in the fence it had apparently heated up and died in the afternoon sun. This also happens in the wild (Fig. 13.25).

OSTEODERMS: CROCODYLIAN 'SOLAR PANELS'

This topic is, perhaps, more physiological than behavioural, but its relevance to basking makes its inclusion here relevant.

In addition to their roles as body armour, as a source of calcium for egg shells and for the attachment of tendons as part of the dorsal bracing system (Chapter 3) and as a buffer for lactate (Chapter 7), the osteoderms have caught the public imagination as crocodylian 'solar panels'. Seidel (1979) noted that they are very well vascularised and well placed to absorb radiant heat, which can then be transported around the body by the circulatory system. The evidence for osteoderms acting as solar panels is mainly circumstantial (Farlow *et al.* 2010), but compelling. It is well known that exposure to heat in reptiles can lead to peripheral vasodilation. In *C. johnstoni*, Grigg and Alchin (1976) monitored washout curves for Xe[133] dissolved in saline and injected under the skin. These showed that cutaneous perfusion increased under exposure to a heat lamp. That observation was confirmed and extended substantially in *C. porosus* by Seebacher and Franklin (2007). As will be discussed below, there is abundant evidence that reptiles, including crocodylians, are able to heat more rapidly than they cool (thermal hysteresis) and that the heating phase is accompanied by increased peripheral blood flow, the cooling phase by a decrease. Most reptiles, of course, lack osteoderms and whether or not the osteoderms in crocodylians facilitate heat exchange beyond what would occur across the skin alone is still unknown.

The osteoderms are certainly well vascularised (Seidel 1979), so it seems very likely that control of blood flow through the osteoderms has a significant influence on the heat flow between crocodylians and their environment. Farlow *et al.* (2010) speculated about a supposed thermoregulatory role for the spectacular dermal plates that occur in two dorsolateral rows on *Stegosaurus* by using alligator osteoderms as a model. Apart from documenting structural similarities, including good blood supply, Farlow and colleagues used infrared thermographic imaging to measure the distribution of dorsal surface temperatures of basking *Caiman latirostris*. Skin overlying the osteoderms appeared cooler than the skin between them, suggesting that cool blood from the core may have been flowing preferentially to the osteoderms, which might be expected if osteoderms play a role as heat collectors. Farlow *et al.* (2010) acknowledged that this evidence is indirect. However, as osteoderms underlie such a large proportion of the dorsal surface area of most crocodylians, it would be hard to argue that they are not important conduits for heat exchange.

SPECULATION … SUBMERGENCE TO COOLER WATER?

The results from both *C. johnstoni* and *C. porosus* show that crocodylians probably need to take steps in summer to avoid overheating. Many of the swamps and lakes that make suitable habitat are devoid of fringing tree cover. In numerous lakes and lagoons, a thermocline develops in summer, and the surface waters in many places may become too hot for comfort. With their prodigious diving capabilities (Chapter 9), it would be surprising if they do not take advantage of the cooler conditions provided by water at the bottom, under the thermocline. Submergence to cooler water below a thermocline could have more benefit than just avoiding overheating, because it could also save energy. Grigg (1978) and Smith (1975b) have reported quite high Q_{10} values in crocodylians (2.7–3.1) (Chapter 7), suggesting markedly enhanced energy conservation at low body temperatures. I am unaware of any data that show this, so far.

MUD AND WATERWEED PARASOLS

Shade-seeking is common in summer, but we frequently saw *C. porosus* at the Edward River Crocodile Farm swimming in the lagoon in summer in full sun and carrying a thick coating of mud (Fig. 10.22). Presumably they collect it as a result of burrowing into the soft muddy substrate. We do not know if this mud sun shield is collected intentionally, but it must give some protection from incoming radiant energy: a mud parasol. The breeding lagoon at the Edward River Crocodile Farm is relatively shallow and the water can get quite warm right to the bottom. An alternative explanation for pushing under mud on the bottom may be that they are

Fig. 10.22. C. porosus *at the Edward River Crocodile Farm coated with mud.* John Loveridge noted that Nile crocodiles, C. niloticus, *basking on the shore of Lake Kariba in Zimbabwe, often had a load of the floating waterweed,* Salvinia molesta, *collected when they left the water. Measurements confirmed that this slowed the rate of heat uptake (Loveridge 1984), so a load of mud, however gained, is likely to have a similar effect. (Photos GCG)*

Fig. 10.23. *These* C. porosus *were caught on a drying floodplain and seem to be using mud for protection from the sun. Death through overheating in such situations is discussed in Chapter 13. (Photo Garry Lindner)*

seeking a cooler microenvironment. Under the mud, however, it can be expected to be cooler. Perhaps both explanations are correct, one leading to the other. *C. porosus* apparently caught unaware on a drying flood plain showed the same behaviour, in the last of drying pools (Fig. 10.23), and there can be no doubt that these animals covered themselves deliberately.

Another example of apparent 'parasol use' has been explored by John Loveridge in Zimbabwe (Loveridge 1984). He noted that Nile crocodiles often came ashore from Lake Kariba 'festooned' with the floating aquatic weed, *Salvinia molesta*. *C. porosus* in northern Australia show the same behaviour (Fig. 10.24) and in both cases the waterweed is a non-native species. John noticed that the crocs made no effort to remove the weed and wondered what thermal implications there might be, so he measured rates of heating with and without the weed. As expected, the presence of the weed slowed the heating rate considerably. He made no suggestion that the collection of *Salvinia* was contrived, interpreting it

as a consequence of emerging from under a tightly packed floating mat. But the result implies that the mud packs we saw on *C. porosus*, whether contrived or an accidental by-product of some other behaviour, would also be effective in reducing insolation.

It is worth noting that if it were to be shown convincingly that the mud pack or a deck of *Salvinia* or some other parasol were acquired deliberately, such behaviour might qualify as tool use, adding another example to that described by Dinets *et al.* (2013) (Fig. 5.31), discussed in Chapter 5.

SOCIAL EFFECTS

Crocodylians are often aggressive to each other so, not surprisingly, interactions between individuals can disrupt patterns of behaviour in a way that leads to a different pattern of Tb. Seebacher and Grigg (2001) reviewed examples from both *C. johnstoni* and *C. porosus*. In the study of *C. johnstoni* at the Croc Hole (see above), individuals higher up the peck order attained access to preferred basking

Fig. 10.24. C. porosus *carrying a load of waterweed,* Salvinia molesta. *Whether gained accidentally or by design is unknown, but the weed is native to Brazil. It is easy to imagine crocodiles learning to load a coating of damp, cool* Salvinia *in the summer heat.* (Photo Garry Lindner)

sites by chasing subservient animals into the water. These lower status individuals' peak Tb may be up to 6°C lower as a result. Fig. 10.25 shows a series of modelled scenarios representing various Tb trajectories that a *C. johnstoni* might display after being displaced from a basking site.

In *C. porosus,* we saw aggression by two large males (1010 and 408 kg), in different lagoons, targeting two particular smaller individuals (820 and 233 kg, respectively). Body temperatures in the harassed individuals were lower by 3–4°C.

Subordinate animals may therefore experience much lower average Tb and, presumably, grow more slowly and have a delayed entry to adulthood. A dominant individual can thus suppress the development of potential rivals. The implications of this at a population level are discussed in Chapter 13. It is easy to imagine a constraint such as this being a stimulus for dispersal by sub-adults.

OPPORTUNISTIC USE OF WARMTH

It is well known that crocodylians take advantage of unexpected additional warmth if it becomes available. Well-celebrated examples include the use of warmed water, by *A. mississippiensis,* in Par Pond at the US Department of Energy's Savannah River nuclear reactor plant (Murphy and Brisbin 1974) and by *C. acutus* at the Turkey Point Nuclear Generating Station in southern Florida, near Homestead. In the latter instance, this protected population, close to the northernmost limit of its range, is credited with seeding the recovery of this previously endangered species in the USA, now de-listed to vulnerable (Chapter 14).

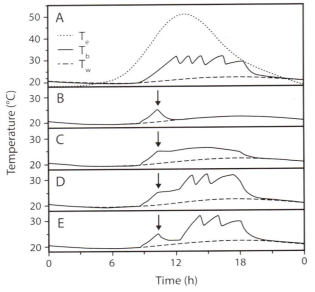

Fig. 10.25. *Projected trajectories of Tb in a subordinate C. johnstoni basking at a preferred site and then chased (arrow) into the water by a dominant individual. The projections include different amounts of submergence following retreat. (A) Set point regulation of Tb, undisturbed (set-point thermoregulation by shuttling between sun and water has so far not been described in any other species of crocodylian). Chased off, an animal may (B) remain in the water for the rest of the day, (C) stay in the water but remain 50% exposed for the rest of the day or (D) resume basking later, or (E) return to the water with 20% exposed until resuming basking later on. (Adapted from Seebacher and Grigg 2001)*

A particular example familiar to me was the largest (1 tonne) male *C. porosus* at the Edward River Crocodile Farm, who would, in winter, periodically take command of a channel through which slightly warmer groundwater was pumped into the main lagoon. We soon realised why: when he used that channel his Tb was warmer than his normal daily pattern (Fig. 10.26).

COPING WITH COLD, 'HIBERNATION'

Many reptiles survive long periods of winter cold in a refuge of some sort. Such behaviour in reptiles is often called hibernation, which is unfortunate in some ways because it is very different from the hibernation that occurs in many mammals. In mammalian hibernation, there is an active down-regulation of metabolism and heat production and a consequent reduction in Tb. Metabolism slows also in reptiles taking refuge in a burrow or cave during winter, but this is by the Q_{10} effect (Chapter 7) as they cool to ambient temperature, not from a cessation of thermoregulatory heat production. Nevertheless, in reptiles 'hibernation' is defined by their behaviour, not by physiological events. In reviewing reptilian hibernation, Gregory (1982) noted that the term is somewhat ill defined, particularly in relation to aestivation. With low

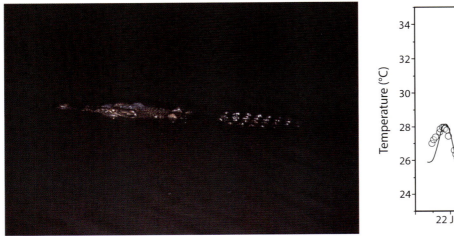

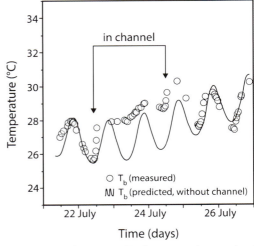

Fig. 10.26. *In winter, the 1 tonne C. porosus 'Hymn' at the Edward River Crocodile Farm chose periodically to spend a couple of nights in a channel through which slightly warmer water was being pumped into the lagoon, raising his body temperature above the usual winter pattern. (From Grigg and Seebacher 2001, photo GCG)*

metabolism and low evaporative water loss in the high ambient humidity of a cold burrow, reptiles can sit out a long cold period. Reptiles often take shelter for long periods not because of low temperature but because of aridity and Gregory referred to that as aestivation. There is not always a clear distinction between the two and, anyway, the term aestivation often implies physiological correlates (think of aestivating African Lungfish, *Protopterus*). In short, reptiles taking refuge to avoid long periods of cold can be said to be hibernating or, if avoiding aridity, aestivating. But the terms are behavioural descriptors, applied to reptiles usefully for want of better ones, but used outside their strict meaning. A pedant would probably prefer to avoid using them for reptiles and might prefer to use brumation instead, in recognition that the lower metabolic rate arises differently from the way it does in 'true' hibernation.

Taplin (1988) reviewed the prevalence of burrowing, hibernation and aestivation in crocodylians. Burrowing was listed for 16 species and aestivation for nine of them, across both alligatorids and crocodylids. The list is probably an underestimate. How crocodylians cope with aridity, will be discussed in Chapter 11. Both species of *Alligator* are more cold tolerant than crocodiles, extending as they do to higher latitudes. They cope with long periods of cold mainly by taking refuge in dens or burrows. The champion crocodylian burrowers seem to be Chinese alligators, with good descriptions provided by Thorbjarnarson and Wang (2010). Their burrows run more or less horizontally, with an entrance either above or below water level,

Fig. 10.27. *A cold snap in the marshlands in southern Louisiana leads to shallow ponds and gator holes freezing. The American alligator in this shallow pond is taking advantage of the water being warmer than the air and keeping a breathing hole open. Under thicker ice, a small alligator could be too weak to maintain the opening and may asphyxiate. Even at very low temperatures, alligators are able to respond, slowly, to disturbance but temperatures lower than ~4°C pose a serious risk to their survival. (Photos supplied by Ruth Elsey)*

sometimes both, and often screened with vegetation. The burrows may be short, but some are very long, many metres, with side branches and terminating in expansions that allow room to turn. Older animals have longer and more elaborate burrows. American alligators also make burrows, often associated with a hole dug to keep an area of open water in the marsh. McIlhenny (1935) described one digging by grasping and tearing at soil and roots in its jaws, backing out to deposit the spoil, and sweeping material away with its tail. Deeper in it raked soil out with the hind feet.

During short cold events, it often makes sense for alligators to remain in the water because it is usually warmer than the air. American alligators can survive in very cold water, with their temperatures dropping to 5°C and below. Brisbin *et al.* (1982) reported some mortalities and concluded that the lowest temperature from which they can recover is ~4°C. That would kill most other crocodylians. American alligators have been reported by several authors to maintain a breathing hole through the ice, but small ones may lack the strength to do so (Brisbin *et al.* 1982; Hagan *et al.* 1983; Brandt and Mazotti 1990; Lee *et al.* 1997) (Fig. 10.27). Chinese alligators show similar behaviour (Thorbjarnarson and Wang 2010).

Nile crocodiles, *C. niloticus*, also become dormant in the cooler, dry season in southern Africa (Pooley 1982), retreating into caverns excavated or enlarged in river banks. They show the same behaviour in warmer parts of their range too and, in this and other species, whether it is stimulated by 'coolth' or aridity may be a moot point. With its distribution well south in Argentina, *Caiman yacare* too is likely to show both 'hibernation' and 'aestivation'.

THERMOPHILY AFTER FEEDING AND TO DEAL WITH INFECTION

By placing crocodylians in an artificial thermal gradient, it has been demonstrated that they can select specific body temperatures with a degree of precision. Thus larger hatchlings of both Alligatoridae and Crocodylidae in a thermal gradient select higher temperatures (+1.5–4.0°C) after feeding (Lang 1979; Lang, 1987). Warmer Tb assists in digestion and assimilation of food and

fighting infections. This is a common phenomenon in reptiles. For example Slip and Shine (1988) described Diamond Pythons (*Morelia spilota spilota*) on a thermal gradient increasing Tb by 2–5°C after feeding, and there are many other examples. Associated with digestion, there is an increase in oxygen consumption that produces heat (the so called specific dynamic action, SDA) and Stuginski *et al.* (2011) found by infrared imaging that the surface temperature of a Brazilian viper, *Bothrops moojeni*, increased ~1–1.5°C, depending on meal size, within a day or so after experimental feeds. In contrast, Slip and Shine noted that, without access to an external heat source, there was no increase in Tb. Snakes eat very large meals proportional to their body size. It seems quite unlikely that body temperature in crocodylians may be increased significantly by the SDA following a meal, whereas choosing longer exposure to a heat source following a feed is quite likely. Field data seem to be lacking so far.

Another example of selection for higher temperatures on a thermal gradient is what Lang (1987) has described as 'behavioural fever', in which juvenile American alligators selected temperatures 2–5°C higher following experimental infection with *Hydrophila*, a pathogenic bacterium. That this occurs in more natural situations was shown by Merchant *et al.* (2007) who experimentally infected juvenile alligators living outdoors with *Aeromonas* and observed a febrile response of 2–4°C. Indoors, however, with alligators kept at a stable environmental temperature, there was no response, providing strong evidence that the febrile response to infection is induced behaviourally. The increase in Tb has been interpreted as a mechanism for resisting infection, in a manner analogous to the endothermic fever developed in infected birds and mammals.

PHYSIOLOGICAL MECHANISMS OF THERMOREGULATION

Although behaviour is the dominant way by which body temperature is modified in reptiles, two physiological mechanisms may operate in concert. These are control of heat flow through the skin by varying the amount of blood flow, and opening the mouth to promote cooling by evaporation from the moist surfaces.

Both of these occur in crocodylians, but neither is particularly well understood quantitatively, particularly in relation to body size. Additionally, crocodylians are tolerant of a wide range of body temperatures (that is, they are eurythermal), and there are no data to show that metabolic heat production is influential. They are solely ectothermal.

CONTROL OVER PERIPHERAL BLOOD FLOW

The analogy between osteoderms and solar panels was discussed above. It is obvious that a basking crocodylian could increase the rate of heat uptake by flowing more blood through the dorsal surface and transporting it internally, warming the whole animal more quickly. Likewise, a heat load could be retained longer in a cooler environment if there were a peripheral vasoconstriction. This capacity to adjust peripheral blood flow is a common feature of both birds and mammals and our own changes in peripheral blood flow in relation to temperature are familiar. We evolved from a common stem with reptiles (see Fig. 2.3) and this is one capacity that we still share with them.

The phenomenon was first described in reptiles by Bartholomew and Tucker (1963) in the Australian Agamid lizard, *Pogona barbata* (previously *Amphibolorus barbatus*). They noted that these lizards heated faster than they cooled when exposed to a warm and then a cool environment. Smith (1976) described the same pattern in American alligators. This difference between heating and cooling rates is usually referred to as 'heating and cooling hysteresis'. The functional interpretation of it is that the reptile benefits because it needs to spend less time basking and can spend longer foraging, or other activities, until it needs to bask again. That is, it is normally interpreted as significant to reptiles that are shuttling set-point regulators.

Heating and cooling hysteresis is usually demonstrated in the laboratory by monitoring Tb during exposure to a step-function increase and then a decrease in ambient temperature. The rate of change of temperature to reach equilibrium with the new environment is exponential and is described by the thermal time constant, tau (τ), which is usually expressed in minutes and is an expression of the animal's thermal inertia. Large beasts have larger values of τ. By monitoring heart rates and Tb in *P. barbata* under a range of conditions in the field, Grigg and Seebacher (1999) concluded that these animals routinely employ their capacity for adjusting rates of heating and cooling.

The situation in crocodylians is not so clear cut. It is certainly important in small individuals, but its value to large animals remains uncertain. In a laboratory study, Grigg and Alchin (1976) reported hysteresis between heating and cooling rates in *C. johnstoni* (2–8 kg) in both air and water, with higher heart rates during heating. Monitoring the washout of saline loaded with Xe^{133}, we noted that localised heating of the body surface with a heat lamp increased the washout, indicating local vasodilation. Turning off the lamp slowed the washout. The increased washout was accompanied by increased heart rate. The experiments gave direct evidence that crocodylians can show heat-related changes in peripheral blood flow. Seebacher and Franklin (2004, 2007) explored the control over blood flow and heart rate pharmacologically in yearling *C. porosus*. One of their findings was that heart rate does not necessarily correlate with increased peripheral flow: changes in blood flow during heating and cooling can occur independently from changes in heart rate.

The size-dependence of heating and cooling rates has particular significance for crocodylians. Norbert Smith was the first to explore this, on American alligators up to 10 kg, and in water as well as air. He found a log linear relationship between thermal time constant and body mass (Smith 1976). Grigg *et al.* (1979) combined Smith's data with data from *C. johnstoni* and *C. porosus* and nine other 'lizard-shaped' reptiles and formulated generalised equations describing thermal time constants over a size range from 0.02–50 kg. Knowledge of thermal time constants is required for modelling heat budgets.

The significance of having control over peripheral blood flow in large crocodylians is uncertain. The Grigg *et al.* (1979) equations imply that it would be insignificant in reptiles smaller than ~20 g and, although this was confirmed by direct measurements (Fraser and Grigg 1984), a theoretical explanation for it is not easy to find (Seebacher 2000). The same equations imply that even very large crocs would show heating and cooling hysteresis, but Turner and Tracy (1985) considered that it would be unimportant in very large animals, maximising at ~5 kg and tapering down from there in larger animals. Contrary to those findings, Seebacher (2000) used heat transfer theory to assess the importance of heat transfer by blood flow relative to conduction in two lizards: *Pogona barbata* (< 1 kg) and *Varanus varius* (4–6 kg). He found that the relative importance of convection by blood flow in total animal heat transfer increased with mass, which is consistent with the generalised equations in Grigg *et al.* (1979). Whether that trend continues as reptiles get larger and larger remains unknown, and it

is worth noting that Seebacher *et al.* (1999) were able to predict their measured Tbs in very large *C. porosus* very satisfactorily using only the physical attributes of the animals, their behaviour and environmental data, without allowing for any physiological adjustments.

In short, the importance of heating and cooling hysteresis for large crocodylians under field conditions remains to be determined. There seems to be little doubt that *C. johnstoni* and other small crocodylians are able to accelerate heat uptake and retard heat loss by regulating peripheral blood flow, but the importance of this capacity in very large individuals remains an open question. Resolving it under natural conditions poses an interesting and challenging research project, made more challenging with the observation by Seebacher and Franklin (2007) that changes in heart rate are not a reliable indication of changes in peripheral blood flow. Given those well-vascularised osteoderms,

Fig. 10.28. *A large* C. porosus *gapes on a sandy beach. Limited observational and experimental evidence imply that gaping has thermoregulatory significance and a threat function. It is also known to occur without any obvious context so may serve other functions as well. It remains one of those interesting, but so far unresolved, puzzles. (Photo Garry Lindner)*

Fig. 10.29. *A juvenile American alligator gaping on a warm spring afternoon in Big Cypress National Preserve, southern Florida. (Photo DSK)*

my bet is that even the largest animals employ it to their benefit.

MOUTH GAPING

Although this is behavioural, evaporative cooling of moist buccal membranes by reptiles and birds (e.g. gular flutter) is usually categorised as a physiological mechanism of thermoregulation, so it is discussed here.

Many crocodylians gape: that is, hold their mouth open for a long time when on land or partly submerged (Figs 10.18, 10.28, 10.29). It seems to be more prevalent after a long period of basking. Some species do it more often than others and some seem not to do it at all. Because gaping exposes the moist buccal membranes to evaporation by any drying breeze, it has inevitably been suggested to be a cooling mechanism. Analogy has been drawn with the well-known fluttering of the gular region seen in many birds at the hottest time of day. Nile crocodiles often come ashore to bask in large numbers (or they

used to, when they were more common) and Cott (1961) recorded the prevalence of gaping behaviour at different times of day and whether they were in the sun or the shade. He found that gaping increased as the day warmed up, decreased as the day cooled and was more common in crocs lying in the sun than the shade. He concluded that it was thermoregulatory. Diefenbach (1975) considered its thermoregulatory significance was negligible. Spotila *et al*. (1977) did some simple experiments on American alligators, measuring head, body and palate temperatures in three alligators, 2.2, 4.7 and 14.5 kg, which were heated by both radiation and conduction in a controlled environment chamber, with mouths closed and mouths propped open. They found that the heads of the alligators with the mouths propped open heated more slowly than those with the mouth closed, and concluded that 'gaping was important in reducing the rate of heat gain to the head region'. Cooling by evaporation is likely to be important in crocodylians, whether or not the mouth gapes open. For example, Grigg and Alchin (1976) heated *C. johnstoni* in a water bath and noted periodic decreases in head temperature as core Tb rose, which they attributed to an 'active cooling which could result from transient increase in respiratory rate leading to evaporative cooling in the nasal passages'. Loveridge (1984) found that *C. niloticus* did not gape spontaneously when put in experimental situations, but he did measure lower surface temperatures on the tongue and oral surfaces in warmed crocs with the mouth propped open.

Kofron (1993) drew attention to a different type of role for gaping: as a threat display. He noted 54 gaping occurrences of which the overwhelming majority were clearly threat displays directed at other crocodiles, hippopotamuses or humans. However, it is very clear that gaping occurs without any threat implication and it could, of course, have more than one purpose.

So gaping remains somewhat enigmatic and the need for more work is clearly indicated.

Most charming for me is the fanciful explanation that dates back to Herodotus, that (Nile) crocodiles are in a neat symbiosis with the 'trochilus bird' (probably Egyptian Plover), as follows: 'Since he has his living in the water he keeps his mouth all full within of leeches; and whereas all other birds and beasts fly from him, the trochilus is a creature which is at peace with him, seeing that from her he receives benefit; for the crocodile having come out of the water to the land and then having opened his mouth (this he is wont to do generally towards the West Wind), the trochilus upon that enters into his mouth and swallows down the leeches, and he being benefited is pleased and does no harm to the trochilus'. (from George Campbell Macaulay's 1904 translation of Herodotus' The Histories Book 2, Euterpe, digitised by Project Gutenberg).

DO LARGE CROCODYLIANS GAIN WARMTH FROM METABOLIC HEAT PRODUCTION?

Metabolism in crocs (and other reptiles) is typically only ~20% of what is seen in endotherms (Chapter 7), which seems too small to make an impact on Tb. Large crocodylians, however, have much smaller surface area/mass ratios than small ones and various authors have speculated that this leads to metabolic heat being retained, so that large crocs will be warmer. Norbert Smith (1975a) claimed that large American alligators could maintain Tb several degrees above water temperature by endogenous heat production. Terpin *et al.* (1979) disagreed, but Bartholomew (1982) concluded that in very large crocodiles 'one might expect to find core body temperature elevated at least a few degrees above ambient by endogenous heat production'. Large crocs are, indeed, warmer than small ones. Grigg *et al.* (1998, and above) found that a very large, 1 tonne *C. porosus* was ~4°C warmer than 30–40 kg crocs in the same habitat in both summer and winter. Could this be the result of metabolic heat, retained because of their large surface area/mass ratio? Or is there an entirely physical explanation? Starting from first principles, Frank Seebacher was able to predict the same increase that we measured empirically (Seebacher *et al.* 1999). Frank's modelling predicted our empirically measured values of Tb very accurately and over a wide range of conditions and body size, without the need for the inclusion of any term to represent heat production. This confirms that even very large crocodylians are simple ectotherms, without any significant endothermic contribution to their thermal budget. The explanation is that in modelling their heat exchange, higher Tb values are predicted as a consequence of the way in which routes of heat exchange scale with increasing body mass. The accumulation of warmth is a consequence of the increasing importance of radiative heat gain over convective heat gain as body size increases.

This has implications for speculations about the Tb of dinosaurs as well; see below.

One situation where endogenous heat production may be important, however, is in nests. Ewert and Nelson (2003) reported eggs incubated in vermiculite in a cluster to be 2–3°C warmer than eggs incubated in a more open arrangement. They attributed this to the metabolism of the eggs and noted the implication for the sex ratio of hatchlings because of temperature dependent sex determination (TSD) (see Chapter 12). In the wild, Magnusson *et al.* (1990) implicated embryonic heat production as significant during the later stages of incubation of *P. trigonatus* nests that were not associated with termite mounds (Chapter 12).

TOLERANCE FOR A WIDE RANGE OF BODY TEMPERATURES

Despite the various mechanisms by which crocodylians can modify their Tb, most routinely show a substantial daily range. A 40 kg *C. porosus* may show a 7°C range in both winter and summer. *C. johnstoni* in The Croc Hole commonly showed a 10°C range and a similar range was shown in summer by small American alligators in Louisiana. Even casual observations show convincingly that crocodylians are fit and feisty in the mornings as well as in the afternoons or evenings. They must be capable of operating over a wide range of body temperatures and, indeed, several physiological studies have confirmed this. For example, maximum performance of hatchling *A. mississippiensis* on a treadmill was essentially independent of temperature over a temperature range from 25°C to 35°C (Emshwiller and Gleeson 1997). In *C. porosus*, yearlings showed a plateau in maximum swimming performance between 23°C and 33°C (Elsworth *et al.* 2003) and this was confirmed by Campbell *et al.* (2013) who also found that maintaining performance at the higher temperature incurred an anaerobic debt.

Understandably, these studies have been made on small individuals, but it seems reasonable to assume that many crocodylian processes are conspicuously less sensitive to temperature in the range of Tb they experience.

THERMAL ACCLIMATISATION (SEASONAL BIOCHEMICAL ADJUSTMENTS)

The seasonal change in preferred body temperature range reported in *C. johnstoni* by Seebacher and Grigg (1997) prompts thinking about seasonal changes in their metabolic machinery. Thermal acclimatisation of metabolism is well known among aquatic ectotherms. These have little capacity to regulate body temperature to counter seasonal changes in water temperature, yet measurements of summer and winter metabolic rates are often

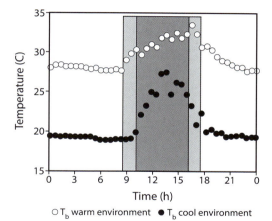

Fig. 10.30. *Daily body temperature profiles for yearling* C. porosus *acclimated for 33 days at either 20°C or 30°C (water and air temperatures), with five individuals in each group. The warm-acclimated group had radiant heat available for 9 h each day (light shading), the cool group 6 h (dark shading). Note that the cool-acclimated group ceased basking while radiant heat was still available, indicating their choice of a lower Tb after cool acclimation. (Adapted from Seebacher* et al. *2009, with permission. © The University of Chicago)*

similar. The phenomenon has been explored most thoroughly in fishes. The explanation is that winter and summer acclimatised individuals make complementary adjustments to mitochondrial oxidative capacities and membrane properties so that the effects of seasonal differences in water temperature are minimised. This can be seen as a biochemical alternative to 'thermoregulation', enabling an animal whose Tb is obliged to follow water temperature to escape from the impact that those seasonal changes might otherwise dictate. That is, they adjust the internal machinery to suit the prevailing ambient temperatures.

Thermal acclimatisation in reptiles has been much less studied, presumably because most are terrestrial and can uncouple their Tb from environmental temperature. Hence, the focus has been mainly on their thermoregulatory skills. However, many of them spend much of their lives in the water. *C. johnstoni* at The Croc Hole commonly spent 16 h or so in the water, and many also spent much of the day at water temperature. It seemed quite reasonable, therefore, to propose that the seasonal cycles in preferred body temperature range (PBTR) were a sign of thermal acclimatisation (Seebacher and Grigg 1997).

Frank Seebacher has taken this idea much further, showing that yearling *C. porosus* acclimated to water at ~20°C for a month and given a daily basking opportunity, chose lower daily maximum temperatures (~28°C) than a control group acclimated at ~30°C. The latter basked until Tb was ~34°C (Fig. 10.30). The response must have been a consequence of the majority of their time being spent at the cooler temperature. This is very reminiscent of the seasonal change in PBTR seen in *C. johnstoni*. Further, the cold-acclimated animals showed the same capacity for prolonged swimming as well as an increased protein-specific oxidative capacity of muscle and liver mitochondria (Glanville and Seebacher 2006).

There is also good evidence for seasonal thermal acclimatisation in American alligators. Guderley and Seebacher (2011) kept hatchlings at water temperatures of 20°C and 30°C with daily basking opportunities and found that they grew at similar rates and, from a series of biochemical comparisons, that they 'optimised metabolic processes for the seasonally altered, preferred body temperature'.

Very interestingly, as Guderley and Seebacher (2011) pointed out, the results from these studies show that the combination of behavioural (plus physiological) thermoregulation and the biochemical adjustments leading to seasonal acclimatisation result in crocodylians compensating more completely for seasonal temperature change than is typical of fishes and other solely aquatic ectotherms. Despite this, alligators in North Carolina, 35°N, at the northern limit of their range, have short active seasons and take ~18 years to reach maturity, compared with 10 years in Louisiana (25°N) (Lance 2003). Obviously the availability of sufficient solar radiation becomes a limiting factor.

Studies on thermal acclimation in crocodylians have all been undertaken on small individuals, but there is no reason to expect they would not apply also to large ones. We described a seasonal cycle in Tb in very large *C. porosus*, with modal Tbs lower in winter than in summer by 3–4°C (Grigg *et al.* 1998 and see Table 10.1). We interpreted this at the time in terms of a response to winter temperatures, even though Te values frequently reached higher peak values in winter (Fig. 10.10 right, 10.11). This begs

the question: could the lower winter temperatures actually be an expression of choosing a lower PBTR? However, as the figures show, there is no indication of the animals seeking shade or the water while Te values are still high, unlike the animals in Fig. 10.30, so our original interpretation is probably correct.

CROCODYLIANS, GIGANTOTHERMY, AND DINOSAUR ENDOTHERMY

Among many puzzles about dinosaur biology, none is as persistent or as much in the public curiosity as their body temperature. Were they 'cold-blooded' ectotherms (and bradymetabolic) or 'warm-blooded' endotherms (tachymetabolic)? The early days of this debate can be learned from Thomas and Olson (1980) and Bakker (1986). The former is a multi-authored conference publication; the latter a passionately written treatise proselytising the wonder of dinosaurs and in support of their endothermy. Since these seminal publications, there has been a huge volume of literature on the topic and a comprehensive review is well beyond our scope. The case for the negative has been put frequently and strongly by Willem Hillenius and John Ruben (2004a) and Chinsamy and Hillenius (2004) while Padian and Horner (2004) presented a strong case in favour. There seems now to be acceptance that the highly active bipedal theropods (in the same clade as birds) were endothermic. That acceptance grew even before the recognition that so many of them had feathers, which are widespread within the coelurosaurs (Currie *et al.* 2004; Nespolo *et al.* 2011; Godefroit *et al.* 2013) – the clade holding most of the theropods, including *Compsognathus*, tyrannosauroids, the 'raptors' (Dromaeosauridae) and the birds (and many others). As evidence accumulates, bit by bit, it seems possible that endothermy will become accepted as typical of dinosaurs as a group. There is good evidence too that early crocodylomorphs were endothermic (Seymour *et al.* 2004, reviewed below), with the implication that endothermy may have been a characteristic of the basal archosaurs: a suggestion made by Robert Bakker in 1986 which caused many palaeontologists to look askance at him.

Reviewing the huge literature on this topic is beyond our scope, but information from crocodylians has been informative about some of the puzzles surrounding the dinosaur endothermy puzzle and this will be our focus.

CROCODYLIAN DATA VALIDATE THE CONCEPT OF 'GIGANTOTHERMY' ('INERTIAL HOMEOTHERMY')

Although the weight of opinion (and data) now favours at least some, and maybe all, of the dinosaurs having had endothermic physiology, in the early 1970s that was seen as radical. Also during the 1970s grew the realisation that very large dinosaurs did not necessarily have to be endothermic to be warm: they could be warm and stable simply by being very large, without any assistance from metabolically produced heat. This is intuitively straightforward and attention was drawn to it in a quantitative way by Spotila *et al.* (1973) who showed by biophysical modelling that the Tb of a very large reptile such as a large dinosaur, with only reptilian physiology, would be higher than ambient temperature and stable through time. In other words, these authors showed that very large reptiles, with no more metabolic frills than other reptiles have, could be warm homeothermic ectotherms (using those terms in the modern sense, see Box 10.1). In due course, McNab (1978) dubbed the concept 'inertial homeothermy' and it is sometimes (and better) referred to as 'gigantothermy' (Paladino *et al.* 1990).

Gigantothermy would not, however, allow the many small dinosaurs to be warm, or the juveniles of large ones. In my lectures to biology students at the University of Sydney in the 1970s, I used to say that sticking a thermometer up the backside of a sauropod and finding it warm would not tell us anything about its metabolism. But sticking it up a very small dinosaur early on a cold morning and finding it warm would imply strongly that its physiology was endothermic.

How large would a dinosaur with only ectothermic physiology have to be before gigantothermy could convey stable 'warm bloodedness' throughout the daily and seasonal cycles? An empirical study of the thermal relations of the largest living reptiles, *C. porosus*, might answer that question and that was

one of the questions that prompted our study of the thermal relations of *C. porosus* discussed earlier in this chapter. We expected to see at least the signs of gigantothermy in the largest *C. porosus*, and we did.

As presented earlier in this chapter, data from 32–1010 kg *C. porosus* in the tropical climate of Cape York, Australia (Grigg *et al.* 1998; Seebacher *et al.* 1999) showed that a 1 tonne crocodile in that habitat is homeothermic on a daily and weekly basis but, at that latitude (15°S), even that animal was not homeothermic across all seasons. Extrapolating the size relationships to the very large, late Cretaceous crocodylian *Deinosuchus* (Fig. 2.37), Frank Seebacher's biophysical model predicted that, at 4000 kg, Tb would have been stable within 1°C in any season throughout its geographic range from 30 to 55°N palaeolatitude (Grigg and Seebacher 2001). It would still have shown a seasonal cycle in Tb and, at the most southerly part of its range,

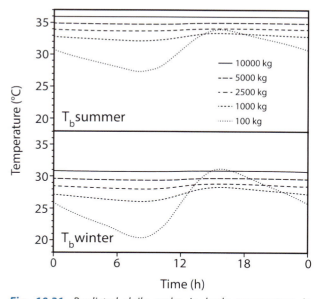

Fig. 10.31. *Predicted daily cycles in body temperature in summer and winter for very large reptiles with crocodylian characteristics, based on models developed from first principles and tested against* C. porosus *living in naturalistic conditions at 17°S on Cape York, Queensland. Note that with increasing body size, body temperatures are predicted to rise. Note also that, whereas daily thermostability is shown by animals from ~600 kg upwards, even 10 000 kg animals can be expected to show a conspicuous seasonal cycle. Extrapolation of the empirical data from crocodiles to larger reptiles demonstrates the reality of the gigantothermy concept. (Adapted from Seebacher* et al. *1999, with permission)*

would have had to eschew sun for part of each day to avoid Tb rising above 36°C, even in winter. This is a consequence of its larger body size dictating higher daily maxima for Te. At 55°N, however, the calculations predicted winter Tbs of ~23°C when they, like the modern American alligators at their northern extent (Lance 2003), might have stopped feeding for several months each year.

Extrapolating to the larger dinosaurs, even a 10 000 kg individual at the latitude of the *C. porosus* study and behaving in the same way would still show a seasonal cycle of ~5°C, from 31 to 36°C. However, because increasing size also brings increased warmth, crocodile-like reptiles heavier than ~5000 kg, avoiding heat in the summer and seeking sun in winter, would have a 'comfortable' body temperature ranging from 30 to 35°C year round (Fig. 10.31). Many dinosaurs, of course, weighed more than that: some sauropods were well above 50 000 kg.

The analysis showed also that thermoregulatory behaviour must have been very important to the very large dinosaurs, as it is in *C. porosus*. In the tropics, they would have especially needed ways to escape direct sunshine to avoid overheating and their behaviour was probably very different at different latitudes. They may have been crepuscular or even nocturnal in the tropics in summer (Seebacher *et al.* 1999) or migrated to higher latitudes, where different behaviour would have been appropriate (Spotila *et al.* 1991). This would be so regardless of whether dinosaurs were ectothermic or endothermic.

Our crocodylian conclusions about giganto-thermy were backed up by Gillooly *et al.* (2006), seeking evidence for it among dinosaurs. They used a generalisation that relates Tb to growth rates, deriving these from bones from seven species of dinosaurs over a size range from 12 to nearly 13 000 kg. Their calculated Tbs increased with mass, from ~25°C in the ceratopsian *Psittacosaurus mongoliensis* to 41°C in the sauropod *Apatosaurus excelsus*. Interestingly, they validated the relevance of their model by comparing its performance with our data on *C. porosus*, finding that it matched quite well. They concluded 'these results provide direct evidence that dinosaurs were reptiles that exhibited inertial homeothermy'.

The implication is that, via gigantothermy and appropriate behaviour, the large dinosaurs could have been ectothermic homeotherms over much of their range. However, although our results and also Gillooly and colleagues' showed that gigantothermy **could** lead to large dinosaurs having a warm and stable Tb, they do not get to the nub of the question; were the large dinosaurs warm, gigantothermic ectotherms, or were they warm tachymetabolic endotherms? As discussed below, the answer is that gigantothermy by itself cannot confer the energetic characteristics of a tachymetabolic endotherm (Seymour 2013). If the dinosaurs had such energetics, they could not have got there via gigantothermy.

Meanwhile, several researchers were exploring ingenious ways to assess the temperatures at which dinosaur bones had actually been formed when the animal was alive, and inventing a new field, the intriguingly termed palaeothermometry.

MEASUREMENTS OF DINOSAUR BODY TEMPERATURE: PALAEOTHERMOMETRY

There are convincing data from palaeothermometry to suggest that dinosaurs did have warm (high 30s) body temperatures. Several researchers have found ways to measure dinosaur Tb indirectly, using stable isotopes. All have concluded that 'dinosaur' Tb was warm and reasonably stable. However, most of the bone samples that have been analysed came from large dinosaurs. Barrick and Showers (1994) took fossilised bone from a 6–7 tonne *Tyrannosaurus rex* and measured its oxygen isotopic composition as an indicator of the temperature at which the bone was formed. They concluded that the animal's Tb would have been stable within ~4°C and that it must have been not only homeothermic but also an endotherm. This quickly drew negative responses pointing out that a reptile that large could be both warm and have a stable Tb simply by gigantothermy, so the endothermic conclusion was problematic at best. Using further developed methodology, Eagle *et al.* (2011) measured the isotopes carbon-13 and oxygen-18 in well preserved teeth of two sauropods, *Camarosaurus* and *Brachiosaurus*, ~12.5 and 50 tonnes, respectively. The way these isotopes bond or 'clump' depends on temperature at the time of tooth or bone formation, thereby preserving a record of Tb. They dubbed this method 'clumped isotope thermometry' and graphed average temperatures of 35°C and 38°C, respectively, for *Camarosaurus* and *Brachiosaurus*.

Because of gigantothermy, measuring high Tb in large dinosaurs does not resolve their metabolic status as either tachymetabolic (endothermic) or bradymetabolic (ectothermic). We need to know what the Tb of small ones would have been and Amiot *et al.* (2006) provided some particularly relevant data. They analysed samples, particularly teeth, from theropod, sauropod, ornithopod and ceratopsian dinosaurs from both high and low latitudes. Most of their samples came from large critters, but they also included three small (≈2 m) theropods, the 'raptors' made famous by the film Jurassic Park, a small ceratopsid (3 m) and a small (2.5 m) ornithopod. The oxygen isotope signatures implied stable Tb in all groups, including the small ones, at a level typical of endotherms and over a wide range of latitudes and ambient temperatures. They interpreted this as an indication that high metabolic rates were widespread among Cretaceous dinosaurs and that endothermy may have been an ancestral characteristic of dinosaurs or, alternatively, acquired independently 'in the studied taxa' by convergent evolution. The former seems much more likely.

Going outside the Archosauria, some very intriguing results have recently intimated that ichthyosaurs, plesiosaurs and even mosasaurs may have been homeothermic endotherms. Whereas inertial homeothermy without any tachymetabolism can easily explain palaeothermometric data showing very large terrestrial reptiles having a high and stable Tb, similar results from marine reptiles would be much harder to explain. However, Bernard *et al.* (2010) measured oxygen isotope compositions of tooth phosphates in ichthyosaurs, plesiosaurs and mosasaurs and compared them with those of fish living in the same seas. They concluded that 'these large marine reptiles were able to maintain a constant and high body temperature in oceanic environments ranging from tropical to cold temperate'. They estimated stable body temperatures of 35°C ± 2°C in ichthyosaurs

and plesiosaurs even down to Te = 12°C. Mosasaur temperatures were estimated at up to 39°C ± 2°C, but more affected by ambient water temperature. Reference to Fig. 2.3 emphasises the cladistic distances between these three groups and distance from the Archosauromorpha (mosasaurs were squamates). All were large, ocean-going predators. Could they have been 'warm blooded' in the way that today's tunas and lamnid sharks are? Or could they have been endothermic like today's dolphins? Has endothermy arisen multiple times? Or is there some so far unrecognised flaw in the assumptions underlying the isotope methodology?

So, palaeothermometry has provided a lot of provocative data, reporting warm, endotherm-like body temperatures but without convincingly resolving the conundrum about whether those temperatures were an expression of endothermy or gigantothermy. The question is not just about temperature, it is also (and more importantly) about metabolic rate. A new paper on *C. porosus* by Roger Seymour at the University of Adelaide and colleagues may help to resolve this issue…

GIGANTOTHERMY DOES NOT CONFER THE ENERGETICS OF ENDOTHERMY

Much of the impetus to thinking that dinosaurs must have been endothermic has come from reconstructions of their probable lifestyles. From their body forms, many were clearly active, energetic predators and had lifestyles that would be more easily explicable if they were 'warm blooded'. An implication runs through some of the discussions about gigantothermy, and also the reports of large dinosaurs with warm body temperatures, that this would be enough to explain their apparently energetic lifestyles and that there is consequently no need to invoke any hypotheses about endothermy.

With this in mind, Seymour (2013) wondered about the energetic capabilities of an ectothermic large dinosaur compared with an endothermic one, using crocodiles as a model. He estimated maximal power output from measurements of aerobic and anaerobic metabolism in burst-exercising estuarine crocodiles, *Crocodylus porosus*, up to 200 kg and compared the results with similar data from endothermic mammals. Given the significant way in which gigantothermy depends upon large body size, he was particularly interested to take into account of the way that these two components of metabolism scale with mass and this turned out to be crucial. At 30°C, a 1 kg crocodile produces ~16 Watts aerobically plus anaerobically during the first 10% of exhaustive activity, which is 57% of that expected for a similarly sized mammal. However, a 200 kg crocodile produces only 14% of that for a mammal, ~400 Watts. Also, crocodiles tire very much more quickly than mammals. He concluded that 'ectothermic crocodiles lack not only the endurance, but also the absolute power for exercise, that are evident in endothermic mammals' and that 'if dinosaurs had similar exercise physiology as crocodiles, then it is unlikely that they would have been as successful as predators and prey for 185 million years, while coexisting with mammals in completely terrestrial environments'.

His analyses imply strongly that gigantothermy cannot have imbued large ectothermic dinosaurs with the energetic capabilities typical of endothermy.

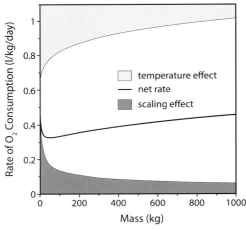

Fig. 10.32. *Due to scaling effects (Chapters 1 and 7), mass-specific oxygen consumption at rest can be expected to decrease with increasing mass. However, empirical data from* C. porosus *showed that Tb increases with mass, and that could be expected to lead to an increase in mass-specific oxygen consumption (Q_{10} effect, Chapter 7). Combining the two, the temperature effect wins: with increasing body size, very large ectotherms can be expected to have an increase in resting mass-specific metabolic rates. The increase is, however, small, and would still be more typical of an ectotherm than an endotherm. As Seymour (2013) has shown, gigantothermy does not lead to endothermy. (Adapted from Seebacher* et al. *1999, with permission)*

The parsimonious interpretation is that the large dinosaurs, as well as the small ones, were probably tachymetabolic endotherms.

The implication that gigantothermy (passive 'homeothermy') could equate to endothermy has been driven by confusion arising from the previous use of the term 'homeothermy' to mean homeothermic endothermy (see Box 10.1),

Seymour's conclusion that gigantothermy does not lead to endothermy is supported by speculative calculations reported by Seebacher et al. (1999), arising from their empirical observation that the Tb in C. porosus increases with body mass. They noted that this would raise resting metabolic rates, which would be offset to some extent by the coincident lowering of mass specific oxygen consumption (Chapter 7). Would these competing effects cancel each other out? Calculations showed that mass-specific oxygen consumption could be expected to rise with increasing mass, but only slightly (Fig. 10.32). In other words, the resting aerobic metabolic rate of an ectothermic reptile, scaled up to many tonnes, would be warm and homeothermic (by gigantothermy), but its metabolic rate would remain more typical of an ectotherm than an endotherm. Admittedly, Seebacher and colleagues took only aerobic metabolism into account, but their approach led independently to exactly the same conclusion as Seymour (2013): gigantothermy does not lead to endothermy.

COULD ALL OF THE DINOSAURS HAVE BEEN ENDOTHERMIC?

I do not warm to the idea that endothermy arose several times within the Ornithodira (Avemetatarsalia). It is more parsimonious to think that it arose further back and data in support of that view are accumulating. In a biomechanical study, Pontzer et al. (2009) presented results which 'support the hypothesis that endothermy was widespread in at least larger non-avian dinosaurs', and quite possibly the smaller ones as well. They applied two different methods based on data linking hindlimb structure and muscle mass to energetic costs of walking and running in order to calculate metabolic rates of 12 species of bipedal dinosaur over a size range from 1 to

6000 kg. Calculations for the five species above 200 kg (*Gorgosaurus*, *Dilophosaurus*, *Plateosaurus*, *Allosaurus* and *Tyrannosaurus*), all saurischian dinosaurs, showed metabolic rates consistent with tachymetabolism and, thus, endothermy. Results for the smaller dinosaurs were more ambiguous: one method supported endothermy, the other did not. Among the smaller dinosaurs (1–20 kg) were three species of ornithischians and this, coupled with data on high growth rates, was put forward as evidence that endothermy was 'plausibly ancestral for all dinosauriforms'. That hypothesis has been strengthened by recent interpretations of lung structure and function in both dinosaurs and crocodylians (Farmer and Sanders 2010 and Chapter 7).

Other evidence is accumulating. Also in support of dinosaurs having had a high metabolic rate, Seymour *et al.* (2012) noted that the foramina through which blood vessels enter the long bones are larger in mammals than in most reptiles. The size of the foramina correlates well with exercise metabolic rate. They found that mammals and varanid lizards had larger foramina than non-varanid reptiles. They looked at bones from 10 species of dinosaur, representing Theropoda, Ornithopoda, Sauropodomorpha, Ceratopsidae and Stegosauridae and found long bone foramina indicative of a highly active and aerobic lifestyle. Evidence long thought to show that dinosaurs must have been ectothermic can now be discounted or, at least, seriously questioned. The occurrence of growth rings in dinosaur bones has long been interpreted to indicate ectothermy, with marked, temperature-related seasonal growth spurts. Köhler *et al.* (2012) examined bone structure in a wide diversity of mammalian ruminants and concluded that 'cyclical growth is a universal trait of homoeothermic endotherms'. Although this might seem a big extrapolation from observations on a single clade of mammals, no matter how convincing the data, but the finding certainly undermines the argument that dinosaurs could not have been endothermic because their bone structure records cyclical growth rates.

Elements of the respiratory system can sometimes be judged from fossil material and results gleaned so far seem to support tachymetabolism being widespread among dinosaurs. Sereno *et al.* (2008) described a late Cretaceous theropod from Argentina with pneumatised axial bones post-cranially, thought to be a reliable correlate of respiratory air sacs. They concluded that this new genus, *Aerosteon*, may have had bird-like respiration. Butler *et al.* (2012) looked for evidence of pneumaticity in a wide range of dinosaurs and a pterosaur using micro-computed tomography. They found 'unambiguous' evidence of internal pneumatic cavities linked to the exterior by foramina in bones of sauropods, theropods and pterosaurs, suggesting that air sacs were present ancestrally in the Ornithodira (Avemetatarsalia), possibly implying that endothermy was ancestral to both Dinosauria and Pterosaurs. I have little trouble imagining pterosaurs as endotherms and, indeed, Claessens *et al.* (2009) presented arguments in favour of them having flow-through ventilation of their respiratory surfaces.

Crocodylians too have lungs like birds in many ways (Chapter 7) and if crocodylomorphs were endotherms, as suggested by Seymour *et al.* (2004), it would support the idea that endothermy is plesiomorphic in archosaurs, as suggested by Bakker (1986). We may never know for sure, but there is quite good evidence that endothermy was a feature of some of the early Crocodylomorpha: today's crocodylians may have endotherms in their ancestry, as will now be discussed.

CROCODYLIANS MAY HAVE ENDOTHERMIC ANCESTRY

Some of the anatomical and physiological attributes in modern crocodylians stimulated Roger Seymour to wonder whether they may have had endothermy in their ancestry. He pursued the idea in consultation with several colleagues, leading to the conclusion that 'physiological, anatomical, and developmental features of the crocodylian heart support the palaeontological evidence that the ancestors of living crocodylians were active and endothermic, but the lineage reverted to ectothermy when it invaded the aquatic, ambush predator niche' (Seymour *et al.* 2004).

EVIDENCE FOR CROCODYLOMORPH ENDOTHERMY

The evidence that at least some of the early crocodylomorphs may have been endothermic can be summarised as follows.

Palaeontological evidence

Just as the reconstructions of many of the dinosaurs suggested active and endothermic lifestyles, so it is with some of the archosauromorphs. Early Triassic stem archosaurs such as *Proterosuchus* (quadrupedal, 1.5 m), which resembled modern crocs and was an amphibious ambush predator, the mid-Triassic bipedal *Euparkeria* (1 m) (Fig. 2.4) and the quadrupedal erythrosuchids (2.5–5 m) (Fig. 2.5) were all agile terrestrial predators. In the mid-Triassic, the archosaur lineage split into the Pseudosuchia (= Crurotarsi) (Crocodyliformes and relatives) and the Ornithodira (= Avemetatarsalia) (pterosaurs, dinosaurs and eventually birds) as discussed in Chapter 2 (Fig. 2.3). Both of these lines had small and large terrestrial predators, some obviously very agile. Among early Pseudosuchia were several discussed in Chapter 2: *Gracilisuchus* (Fig. 2.8), *Terrestrisuchus* (Fig. 2.14) and, in the late Triassic, very early Crocodylomorphs such as *Protosuchus* (Fig. 2.15). Most of these were small, long-legged, and presumably cursorial, terrestrial predators. The lineage from early crocodylomorphs to the Eusuchia, which contains the extant crocodylians, is reviewed in Chapter 2 and Fig. 2.3, and two overall trends are apparent: an increase in the range of body sizes and a shift from terrestrial to aquatic lifestyles.

The structure and assumed lifestyle of these early crocodylomorphs suggest that at least some of them could have been endothermic. The structure, upright stance, bone histology and presumed lifestyle of the basal archosaurs have been argued by Bakker (1986) to imply endothermy, but that conclusion has always been controversial. Seymour *et al.* (2004) pointed out that, although the idea of endothermy among basal archosaurs has not gained general acceptance, neither has it been disproven. There is nothing obvious in the early evolutionary history of crocodylians that is inconsistent with the idea of ancient, endothermic terrestrial ancestry giving way to ectothermic aquatic descendants.

Heart anatomy and a pulmonary by-pass shunt

Substantial evidence comes from the heart anatomy and its embryological development, and bone histology. The anatomy and functioning of the heart is discussed in Chapter 8 and will not be reviewed here. Suffice it to say that their four-chambered heart, unusual for a reptile, allows crocodylians to have high pressures in the systemic circuit and low pressures in the pulmonary circuit. This is characteristic of mammals and birds, which are, of course, endothermic. Lungs operate best at low blood pressure, so a four-chambered heart allows high systemic pressures and resultant higher cardiac output without compromising low pressures in the pulmonary circuit. Endothermic ancestry would provide an easy explanation for crocodylians having a four-chambered heart.

But there is a lot more. Uniquely among reptiles, crocodylians have a suite of anatomical attributes that allow shunting between these two circuits, particularly an extra-cardiac pulmonary by-pass shunt, which provides an opportunity to divert blood away from the lungs and into the systemic circuit via the foramen of Panizza (Fig. 8.12). These attributes are the left aorta arising from the right ventricle instead of the left (Fig. 8.1), the variable calibre foramen of Panizza providing conditional connectivity between the left and right aortas where they cross over each other as they exit the heart, and the active (Fig. 8.5), controllable cog-wheel valve at the base of the pulmonary aorta (Figs 8.1, 8.6), which regulates blood flow to the lungs. These and other features are discussed in Chapter 8. The capacity to shunt blood away from the lungs is usually interpreted as an adaptation to diving. Resultant lowering of the oxygen levels in the systemic circuit would induce hypometabolism, thus spinning out the oxygen store in the lungs and blood (Chapter 9). Turtles and other diving reptiles can shunt blood between the two circuits within the heart itself, because their ventricle is incompletely divided, whereas in crocodylians the shunt pathway is between the outflow tracts, that is, outside the heart itself.

If crocodylians are secondarily aquatic, as is clear from the fossil record, and if they are derived from endothermic ancestry, which is the hypothesis under discussion, it could be argued they would benefit from having a pulmonary by-pass shunt. One way to 'invent' one would be to open the septum between the two ventricles, creating (or re-creating) the situation more typical of a diving reptile. But that could lead to less pressure differential between the systemic and pulmonary circuits, perhaps compromising the functioning of the lungs because of congestion from fluid accumulation. The complex anatomy by which crocodylians achieve this shunt capability solves the problem. How did it evolve?

Shunting capacity is a secondary development

Evidence from the embryological development of the heart suggests that the shunting capacity has evolved secondarily. The early development of the crocodylian heart from a simple tube is typical of vertebrate hearts in general and an unpublished MS thesis by Christine Bennett-Stamper at the University of New Mexico showed that the foramen in embryonic American alligators develops as a secondary perforation in the septum dividing the left and right aortas (Bennett-Stamper 2003). That is, it is a novel structure. Further, the cog-wheel valve is not present in hatchling *C. porosus*, but it is present in 1.1 m individuals (Seymour *et al.* 2004; Webb 1979). That fits with my own observations, because I have known for a long time that hatchling *C. porosus* are very poor at diving. They take a while to learn to dive, which means that studies on small crocodylians that are aimed at elucidating the significance of shunting are likely to be uninformative; see Chapter 8. Taking these observations together, it seems clear that the capacity to shunt is derived secondarily. It has been 'added on' to a pre-existing four-chambered heart.

Good lung ventilation during locomotion

Whereas respiration in lizards is compromised during activity, making it hard for them to run and breathe at the same time, Farmer and Carrier (2000) have shown that American alligators and, by implication, other crocodylians too, are not so constrained. Seymour *et al.* (2004) noted that Carrier (1987) described seven characteristics associated with the ability to enhance breathing and stamina during locomotion: (1) diaphragmatic muscles; (2) enlarged transverse processes and epaxial muscles; (3) bipedal locomotion; (4) upright posture; (5) bounding gait; (6) lateral stability of the vertebral column; and (7) endothermy. This suite evolved in the synapsids, leading to mammals and in the diapsids, leading to birds. Carrier (1987) pointed out that at least elements 1, 2, 4, and 5 are evident in living crocodylians and that, along with a four-chambered heart and a reduced fifth digit on the hind foot (reduced digits in non-fossorial vertebrates usually being associated with an active gait), these characteristics are inconsistent with low stamina and ectothermy. He proposed that a shift in lifestyle, from terrestrial endothermy to the amphibious ectothermy, could explain these features of modern crocodylians, as reinforced by Farmer and Carrier (2000).

Fibrolamellar bone

Embryonic and juvenile American alligators have fibrolamellar bone, which is characteristic of birds and mammals and whose presence in some dinosaurs is often cited as an indicator of endothermy. One interpretation of this could be that its early presence and subsequent loss in alligators during development could be a vestige from endothermic ancestry.

Unidirectional airflow in alligator lungs

Since the publication by Seymour *et al.* (2004) has come the very striking discovery that air in the lungs of American alligators flows through parabronchi in the same direction during both inspiration and expiration (Farmer and Sanders 2010). The parabronchi are presumably homologues of those seen in birds and unidirectional airflow through them implies a capacity for very efficient gas exchange, as discussed in Chapter 7. Farmer and Sanders suggested that unidirectional airflow may have evolved in the basal archosaurs and was present in their descendants, including Crocodylomorpha (Fig. 2.3). The lungs of the bipedal Crurotarsi probably had lungs well able to support endothermic metabolism, providing strong support for the Seymour *et al.* (2004) hypothesis that today's crocodylians have endotherms in their ancestry.

This finding provides a probable explanation for a conundrum noted in crocodylian lungs by Perry (1990). He observed that the lungs are complex and appear very well designed, reminiscent of bird lungs, and he wondered why they seemed to be 'over-engineered'. The notion that crocodylians are over-engineered was taken up by Colleen Farmer and David Carrier in relation to other attributes. Before the discovery of unidirectional airflow, they noted that the apparently competent respiratory system in alligators, with such a complex musculo-skeletal system devoted to maintaining ventilation, seemed out of place in a sit-and-wait predator with a fairly low metabolic rate and 'a poor capacity for sustained vigorous terrestrial locomotion' (Farmer and Carrier 2000). They wondered if there had been strong selection early in the crocodylian lineage for a high aerobic metabolism (!) and speculated that these highly evolved systems were 'a legacy from cursorial ancestors'.

The discovery that the lungs of alligators have unidirectional airflow that are so much like birds is a stunning addition in support of the notion that the extant crocodylians seem over-engineered for their present lifestyle and may have inherited it from a more tachymetabolic ancestry.

EVIDENCE AGAINST CROCODYLOMORPH ENDOTHERMY

Seymour *et al.* (2004) noted two significant items of contrary evidence: the supposedly endothermic prehistoric crocodylomorphs lacked nasal turbinate bones and, as far as is known, lacked filamental or feather-like structures that imply insulation. The presence of nasal turbinates is often promoted as a solid requirement for endothermy, and their lack in Dinosauria is often used as evidence against any having been endothermic (e.g. Hillenius and Ruben 2004a,b). As for external insulation, there are many living endotherms without external insulation,

including humans, and Grigg *et al.* (2004) pointed out that insulation can be internal, such as a fatty layer under the skin.

The Seymour and colleagues' proposal that the early crocodylomorphs may have been endothermic provoked a spirited response from Hillenius and Ruben (2004b) and, in turn, an even more spirited defence by Seymour (2004), both conveniently published in the same issue of the journal and both well worth reading.

Clearly the notion that modern crocodylians had endothermic ancestry in their dim background is provocative but, interestingly, the research on lungs, published since the Seymour and colleagues proposal has strengthened the argument even further.

IF CROCODYLIANS' ANCESTORS HAD ENDOTHERMY, WHY LOSE IT?

There is an anthropocentric tendency to regard homeothermic endothermy as a 'pinnacle' evolutionary achievement, so the idea that endothermy would ever be lost through evolutionary processes seems counterintuitive. Why might crocodyliforms lose it? Seymour *et al.* (2004) drew attention to palaeontological evidence of crocodyliforms evolving from terrestrial to aquatic lifestyles and from small sizes to a larger range of body sizes, both exemplified within the Eusuchia.

One explanation for their loss of endothermy may be that, energetically, it would be very challenging for a 'sit-and-wait' aquatic predator to be endothermic. The high specific heat and high conductivity of water mitigate against that and, indeed, there seem to be no examples of endothermic aquatic sit-and-wait predators.

Although this is probably the most compelling explanation, there are contributory advantages. It could be that part of the tool kit of an aquatic ambush predator is the ability to remain under water for a long time (Chapter 9) and if so, a lower metabolic rate would be an advantage. And ectothermy should not, anyway, be thought of as such a poor way to make a living. Pough (1980) wrote a paper entitled 'The advantages of ectothermy for tetrapods', making the point that ectothermy is very efficient and should be thought of as an alternative, low-energy lifestyle, not a less successful one, and he gave several reasons. One is that ectotherms do not need to provide heat for endothermic thermoregulation, so their food requirements are lower and, consequently, they have a much higher conversion of food into biomass than birds and mammals. This also means that they are able to survive for long periods in places where food is available only periodically and their capacity to tolerate low body temperatures reduces food requirements further, providing another advantage. The only crocodyliform survivors of the Cretaceous–Paleogene extinction event were the semi-aquatic ones, which were presumably all ectothermic.

A modern example of endotherms returning, temporarily, to their ancestral ectothermic physiology may be the occurrences of torpor and hibernation in so many birds and mammals. Based particularly on studies of the thermal relations of echidnas and their hibernation biology, and the occurrence of hibernation in all three groups of modern mammals, Grigg *et al.* (2004) suggested that torpor and hibernation are plesiomorphic. These 'survival skills' that enable them to live in places where food supply is not continuous may not be the 'advanced' adaptations to cope with cold, as is usually assumed, but revisited echoes of their evolutionary history. When mammals and birds abandon homeothermic endothermy and become torpid or hibernate, they may be simply re-visiting the ectothermic physiology of their ancestors.

Could even the basal archosaurs have been endothermic?

The suggestion that basal archosaurs were endothermic was made by Bakker (1986) based on interpretations of likely lifestyles from body shape and bone histology. The idea has been controversial ever since, sometimes discussed very passionately. In the previous section, evidence was presented that endothermy is likely to be ancestral to Ornithodira (Dinosauria plus Pterosauria), relying particularly on the work by Sereno *et al.* (2008), Pontzer *et al.* (2009) and Butler *et al.* (2012).

But Butler and colleagues went further, suggesting that the basal archosaurs may have had pulmonary air sacs. Although their micro-computed tomography revealed no unambiguous pneumaticity in Crurotarsi (Pseudosuchia) they noted that many of them 'possess the complex array of vertebral laminae and fossae that always accompany the presence of air sacs in ornithodirans' and that 'a reduced, or non-invasive, system of pulmonary air sacs may be have been present … (and secondarily lost in extant crocodylians) and was potentially primitive for Archosauria as a whole'.

The presence of air sacs is probably a good indicator of a functional, unidirectional flow-through gas exchange system, so unidirectional air flow across the respiratory surfaces was probably typical of pterosaurs and dinosaurs, as well as of the modern crocodylians and, in all probability, their (possibly endothermic) ancestors. Convergent evolution? More likely is that the discrete air sacs that characterise modern birds and many of the dinosaurs and pterosaurs, evolved by elaboration of the sort of lung system seen in modern crocodylia. That is, of evolution from a system lacking air sacs but having lungs with sac-like chambers that are unlikely to be of significance for gas exchange and that may be functionally equivalent to air sacs.

Of course, the presence of unidirectional airflow past the respiratory surfaces during both inspiration and expiration does not prove endothermic metabolism, but the newest information strengthens, rather than detracts from, the proposal put forward by Seymour *et al.* (2004) that modern crocodylians have endothermic ancestry.

If someone turns up a fossil crocodylomorph with feathers … well that will be interesting!

References

Abercrombie CL, Howarter S, Morea C, Rice K, Percival HF (2002) Everglades alligator thermoregulations: unanswered questions. In *Proceedings of the 16th Working Meeting of the Crocodile Specialist Group*. 7–10 October 2002 pp. 131–140. Gainesville, Florida. IUCN–The World Conservation Union, Gland, Switzerland.

Amiot R, L'Ecuyer C, Buffetaut E, Escarguel G, Fluteau F, Martineau F (2006) Oxygen isotopes from biogenic apatites suggest widespread endothermy in Cretaceous dinosaurs. *Earth and Planetary Science Letters* **246**, 41–54. doi:10.1016/j.epsl.2006.04.018

Avery RA (1982) Field studies of body temperatures and thermoregulation. In: *Biology of the Reptilia. Volume 12* (Eds C Gans and FH Pough) pp. 93–166. Academic Press, New York.

Bakken GS, Gates DM (1975) Heat transfer analysis of animals: some implications for field ecology, physiology, and evolution. In *Perspectives of Biophysical Ecology.* (Eds DM Gates and RB Schmerl) pp. 255–290. Springer, New York.

Bakker RT (1986) *The Dinosaur Heresies.* Bath Press, Bath, UK.

Barrick RE, Showers WJ (1994) Thermophysiology of *Tyrannosaurus rex*: evidence from oxygen isotopes. *Science* **265**, 222–224.

Bartholomew GA (1982) Physiological control of body temperature. In *Biology of the Reptilia. Volume 12* (Eds C Gans and FH Pough) pp. 167–213. Academic Press, New York.

Bartholomew GA, Tucker VA (1963) Control of changes in body temperature, metabolism, and circulation in the agamid lizard, *Amphibolorus barbatus. Physiological Zoology* **36**, 199–218.

Bennett-Stamper C (2003) Structural development of the foramen of Panizza in embryonic alligators. MSc thesis. New Mexico State University, La Cruces.

Bernard A, Lécuyer C, Vincent P, Amiot R, Bardet N, Buffetaut E, *et al.* (2010) Regulation of body temperature by some Mesozoic marine reptiles. *Science* **328**(5984), 1379–1382.

Brandt LA, Mazotti ANDFJ (1990) The behavior of juvenile *Alligator mississippiensis* and *Caiman crocodilus* exposed to low temperature. *Copeia* **1990**, 867–871.

Brisbin IL Jr, Standora EA, Vargo MJ (1982) Body temperatures and behavior of American alligators during cold winter weather. *American Midland Naturalist* **107**, 209–218.

Butler RJ, Barrett PM, Gower DJ (2012) Reassessment of the evidence for postcranial skeletal pneumaticity in Triassic archosaurs, and the early evolution of the avian respiratory system. *PLoS ONE* **7**(3), e34094.

Campbell HA, Sissa O, Dwyer RG, Franklin CE (2013) Hatchling crocodiles maintain a plateau of thermal independence for activity, but at what cost? *Journal of Herpetology* **47**(1), 11–14.

Campos Z, Magnusson WE (2011) Emergence behaviour of yacare caimans (*Caiman crocodilus yacare*) in the Brazilian Pantanal. *The Herpetological Journal* **21**, 91–94.

Campos Z, Magnusson WE (2013) Thermal relations of dwarf caiman, *Paleosuchus palpebrosus*, in a hillside stream: evidence for an unusual thermal niche among crocodilians. *Journal of Thermal Biology* **38**, 20–23.

Campos Z, Coutinho M, Magnusson W (2005) Field body temperatures of caimans in the Pantanal, Brazil. *The Herpetological Journal* **15**, 97–106.

Carrier DR (1987) The evolution of locomotor stamina in tetrapods: circumventing a mechanical constraint. *Paleobiology* **13**, 326–341.

Chinsamy A, Hillenius WJ (2004) Physiology of non-avian dinosaurs. In *The Dinosauria* (Eds DB Weishampel, P Dodson and H Osmolska) pp. 643–659. University of California Press, Berkeley.

Claessens LPAM, O'Connor PM, Unwin DM (2009) Respiratory evolution facilitated the origin of pterosaur flight and aerial gigantism. *PLoS ONE* **4**(2), e4497.

Cott HB (1961) Scientific results of an enquiry into the ecology and economic status of the Nile crocodile (*Crocodilus niloticus*) in Uganda and Northern Rhodesia. *Transactions of the Zoological Society of London* **29**, 211–356.

Coutinho ME (2001) Reproductive biology and its implications for management of caiman, *Caiman yacare*, in the Pantanal wetland, Brazil. In *Crocodilian Biology and Evolution.* (Eds GC Grigg, F Seebacher and CE Franklin) pp. 229–243. Surrey Beatty & Sons, Sydney.

Cowles RB (1962) Semantics in biothermal studies. *Science* **135**, 670.

Currie PJ, Koppelhus EB, Shugar MA, Wright JL (2004) *Feathered Dragons: Studies on the Transition from Dinosaurs to Birds.* Indiana University Press. Bloomington, Indiana.

Diefenbach COda C (1975) Thermal preferences and thermoregulation in *Caiman crocodilus. Copeia* **1975**, 530–540.

Dinets VJC, Brueggen JC, Brueggen JD (2013) Crocodilians use tools for hunting. *Ethology Ecology and Evolution*(in press).

Downs CT, Greaver C, Taylor R (2008) Body temperature and basking behaviour of Nile crocodiles (*Crocodylus niloticus*) during winter. *Journal of Thermal Biology* **33**, 185–192.

Eagle RA, Tütken T, Martin TS, Aradhna K, Tripati AK, Fricke HC, *et al.* (2011) Dinosaur body temperatures determined from isotopic (13C–18O) ordering in fossil biominerals. *Science* **333**, 443–445.

Elsworth PG, Seebacher F, Franklin CE (2003) Sustained swimming performance in crocodiles (*Crocodylus porosus*): effects of body size and temperature. *Journal of Herpetology* **37**(2), 363–368.

Emshwiller MG, Gleeson TT (1997) Temperature effects on aerobic metabolism and terrestrial locomotion in American alligators. *Journal of Herpetology* **31**, 142–147.

Ewert MA, Nelson CE (2003) Metabolic heating of embryos and sex determination in the American alligator, *Alligator mississippiensis. Journal of Thermal Biology* **28**, 159–165.

Farlow JO, Hayashi S, Tattersall GJ (2010) Internal vascularity of the dermal plates of *Stegosaurus* (Ornithischia, Thyreophora). *Swiss Journal of Geosciences* **103**, 173–185.

Farmer CG, Carrier DR (2000) Ventilation and gas exchange during treadmill locomotion in the American alligator (*Alligator mississippiensis*). *Journal of Experimental Biology* **203**, 1671–1678.

Farmer CG, Sanders K (2010) Unidirectional airflow in the lungs of alligators. *Science* **327**, 338–340.

Fish FE, Cosgrove LA (1987) Behavioral thermoregulation of small American alligators in water postural changes in relation to the thermal environment. *Copeia* **1987**(3), 804–807.

Fraser S, Grigg GC (1984) Control of thermal conductance is not significant to thermoregulation in most reptiles. *Physiological Zoology* **57**(4), 392–400.

Gillooly JF, Allen AP, Charnov EL (2006) Dinosaur fossils predict body temperatures. Public Library of Science: Biology *l* **4**, 1467–1469.

Glanville EJ, Seebacher F (2006) Compensation for environmental change by complementary shifts of thermal sensitivity and thermoregulatory behaviour in an ectotherm. *Journal of Experimental Biology* **209**, 4869–4877.

Godefroit P, Demuynck H, Dyke G, Hu D, Escuillie F, Claeys P (2013) Reduced plumage and flight ability of a new Jurassic paravian theropod from China. *Nature Communications* **4**, 1394.

Gregory PT (1982) Reptilian hibernation. In *Biology of the Reptilia. Volume 13.* (Eds C Gans and FH Pough) pp. 53–154. Academic Press, New York.

Grigg GC (1977) The body temperature of dinosaurs and crocodiles. In *Australian Animals and their Environment.* (Eds H Messel and ST Butler) pp. 355–367. Shakespeare Head Press, Sydney.

Grigg GC (1978) Metabolic rate, RQ and Q_{10} in *Crocodylus porosus* and some generalisations about low RQ in reptiles. *Physiological Zoology* **51**, 354–360.

Grigg GC, Alchin J (1976) The role of the cardiovascular system in thermoregulation of *Crocodylus johnstoni. Physiological Zoology* **49**, 24–36.

Grigg GC, Beard LA (2000) Application of radiotelemetry to studies of the physiological ecology of vertebrates. in *Biotelemetry 15: Proceedings of 15th International Conference on Biotelemetry.* 9–14 May 1999, Juneau, Alaska, USA. (Eds JH Eiler, DJ Alcorn and MR Neuman) pp. 535–551. International Society on Biotelemetry, Wageningen, The Netherlands.

Grigg GC, Seebacher F (1999) Field test of a paradigm: hysteresis of heart rate in thermoregulation by a free-ranging lizard, *Pogona barbata*. *Proceedings. Biological Sciences* **266**, 1291–1297.

Grigg GC, Seebacher F (2001) Crocodilian thermal relations. In *Crocodile Biology and Evolution*.(Eds GC Grigg, F Seebacher and CE Franklin). pp. 297–309 Surrey Beatty & Sons, Sydney.

Grigg GC, Drane CR, Courtice GP (1979) Time constants of heating and cooling in the Eastern Water Dragon, *Physignathus lesueruii*, and some generalizations about heating and cooling in reptiles. *Journal of Thermal Biology* **4**, 95–103.

Grigg GC, Seebacher F, Beard LA, Morris D (1998) Thermal relations of large crocodiles, *Crocodylus porosus*, free-ranging in a naturalistic situation. *Proceedings. Biological Sciences* **265**, 1793–1799.

Grigg GC, Beard LA, Augee M (2004) The evolution of endothermy and its diversity in mammals and birds. *Physiological and Biochemical Zoology* **77**(6), 982–997.

Guderley H, Seebacher F (2011) Thermal acclimation, mitochondrial capacities and organ metabolic profiles in a reptile (*Alligator mississippiensis*). *Journal of Comparative Physiology. B, Biochemical, Systemic, and Environmental Physiology* **181**, 53–64.

Hagan JM, Smithson PC, Doerr PD (1983) Behavioral response of the American alligator to freezing weather. *Journal of Herpetology* **17**, 402–404.

Heath JE (1964) Reptilian thermoregulation: evaluation of field studies. *Science* **146**, 784–785.

Hertz PE, Huey RB, Stevenson RD (1993) Evaluating temperature regulation by field-active ectotherms: the fallacy of the inappropriate question. *American Naturalist* **142**, 796–818.

Hillenius WJ, Ruben JA (2004a) The evolution of endothermy in terrestrial vertebrates: who? when? why? *Physiological and Biochemical Zoology* **77**, 1019–1042.

Hillenius WJ, Ruben JA (2004b) Getting warmer, getting colder: reconstructing crocodylomorph physiology. *Physiological and Biochemical Zoology* **77**, 1068–1072.

Huey RB (1982) Temperature, physiology, and the ecology of reptiles. In *Biology of the Reptilia, Volume 12*. (Eds C Gans and FH Pough) pp. 25–91. Academic Press, New York.

IUPS Thermal Commission (2003) Glossary of terms for thermal physiology. 3rd edn. *Journal of Thermal Biology* **28**, 75–106.

Kofron CP (1993) Behavior of Nile crocodiles in a seasonal river in Zimbabwe. *Copeia* **1993**(2), 463–469.

Köhler M, Marin-Moratalla N, Jordana X, Aanes R (2012) Seasonal bone growth and physiology in endotherms shed light on dinosaur physiology. *Nature* **487**, 358–361.

Lance VA (2003) Alligator physiology and life history: the importance of temperature. *Experimental Gerontology* **38**, 801–805.

Lang JW (1979) Thermophilic response of the American alligator and the American crocodile to feeding. *Copeia* **1979**, 48–59.

Lang JW 1987. Crocodilian thermal selection. In *Wildlife Management. Crocodiles and Alligators*. (Eds GJW Webb, SC Manolis and PJ Whitehead) pp. 301–317. Surrey Beatty & Sons, Sydney.

Lee JR, Burke VJ, Gibbons JW (1997) Behavior of hatchling *Alligator mississippiensis* exposed to ice. *Copeia* **1997**, 224–226.

Loveridge JP 1984 Thermoregulation in the Nile crocodile, *Crocodylus niloticus*. In *The Structure, Development and Evolution of Reptiles*. (Ed. MWJ Ferguson.) pp. 443–468. Academic Press, London.

Magnusson WE, Lima AP (1991) The ecology of a cryptic predator, *Paleosuchus trigonatus*, in a tropical rainforest. *Journal of Herpetology* **25**, 41–48.

Magnusson WE, Lima AP, Hero JM, Sanaiotti TM, Yamakoshi M (1990) *Paleosuchus trigonatus* nests: sources of heat and embryo sex ratios. *Journal of Herpetology* **24**, 397–400.

Markwick PJ (1998) Fossil crocodilians as indicators of Late Cretaceous and Cenozoic climates:

implications for using palaeontological data in reconstructing palaeoclimate. *Palaeogeography-Palaeoclimatology-Palaeoecology* **137**, 205–271.

McIlhenny EA (1935) *The Alligator's Life History.* Christopher Publishing House, Boston, Massachusetts.

McNab BK (1978) The evolution of endothermy in the phylogeny of mammals. *American Naturalist* **112**, 1–21.

Merchant M, Williams S, Trosclair PL III, Elsey RM, Mills K (2007) Febrile response to infection in the American alligator (*Alligator mississippiensis*). *Comparative Biochemistry and Physiology. A. Comparative Physiology* **148**, 921–925.

Modha ML (1968) Basking of the Nile crocodile on Central Island, Lake Rudolph. East African Wildlife Journal **6**, 81–88.

Murphy TM, Brisbin IL Jr (1974) Distribution of alligators in response to thermal gradients in a reactor cooling reservoir. In *Thermal Ecology.* (Eds JW Gibbons and RR Sharitz) pp. 313–321. Technical Information Center, US Atomic Energy Commission, Oak Ridge, Tennessee.

Nespolo RF, Bacigalupe LD, Figueroa CC, Koteja P, Opazo JC (2011) Using new tools to solve an old problem: the evolution of endothermy in vertebrates. *Trends in Ecology & Evolution* **26**, 414–423.

Padian K, Horner JR (2004) Dinosaur physiology. In *The Dinosauria:* (Eds DB Weishampel, P Dodson and H Osmolska) pp. 660–671. University of California Press, Berkeley.

Paladino FV, O'Connor MP, Spotila JR (1990) Metabolism of leatherback turtles, gigantothermy, and thermoregulation of dinosaurs. *Nature* **344**, 858–860.

Perry SF (1990) Gas exchange strategy in the Nile crocodile: a morphometric study. *Journal of Comparative Physiology. B, Biochemical, Systemic, and Environmental Physiology* **159**, 761–769.

Pontzer H, Allen V, Hutchinson JR (2009) Biomechanics of running indicates endothermy in bipedal dinosaurs. *PLoS ONE* **4**(11), e7783.

Pooley AC (1982) *Discoveries of a Crocodile Man.* William Collins, London.

Pough FH (1980) The advantages of ectothermy for tetrapods. *American Naturalist* **115**, 92–112.

Seebacher F (1999) Behavioural postures and the rate of body temperature change in wild freshwater crocodiles, *Crocodylus johnstoni. Physiological and Biochemical Zoology* **72**, 57–63.

Seebacher F (2000) Heat transfer in a microvascular network: the effect of heart rate on heating and cooling in reptiles (*Pogona barbata* and *Varanus varius*). *Journal of Theoretical Biology* **203**, 97–109.

Seebacher F, Franklin CE (2004) Integration of autonomic and local mechanisms in regulating cardiovascular responses to heating and cooling in a reptile (*Crocodylus porosus*). *Journal of Comparative Physiology. B, Biochemical, Systemic, and Environmental Physiology* **174**, 205–210.

Seebacher F, Franklin CE (2005) Physiological mechanisms of thermoregulation in reptiles: a review. *Journal of Comparative Physiology. B, Biochemical, Systemic, and Environmental Physiology* **175**, 533–541.

Seebacher F, Franklin CE (2007) Redistribution of blood within the body is important for thermoregulation in an ectothermic vertebrate (*Crocodylus porosus*). *Journal of Comparative Physiology. B, Biochemical, Systemic, and Environmental Physiology* **177**, 841–848.

Seebacher F, Grigg GC (1997) Patterns of body temperature in wild freshwater crocodiles, *Crocodylus johnstoni*: thermoregulation vs. concormity, seasonal acclimatisation and the effect of social interactions. *Copeia* **1997**, 549–557.

Seebacher F, Grigg GC (2001) Social interactions compromise thermoregulation in crocodiles, *Crocodylus johnstoni* and *Crocodylus porosus*. In *Crocodilian Biology and Evolution.* (Eds GC Grigg, F Seebacher and CE Franklin. Surrey Beatty & Sons, Sydney.

Seebacher F, James RS (2008) Plasticity of muscle function in a thermoregulating ectotherm (*Crocodylus porosus*): biomechanics and metabolism. *American Journal of Physiology.*

Regulatory, Integrative and Comparative Physiology **294**, R1024–R1032.

Seebacher F, Shine R (2004) Evaluating thermoregulation in reptiles: the fallacy of an inappropriately applied method. *Physiological and Biochemical Zoology* **77**, 688–695.

Seebacher F, Grigg GC, Beard LA (1999) Crocodiles as dinosaurs: behavioural thermoregulation in very large ectotherms leads to high and stable body temperatures. *Journal of Experimental Biology* **202**, 77–86.

Seebacher F, Elsey RM, Trosclair PLIII (2003) Body temperature null-distributions in large reptiles: seasonal thermoregulation in the American alligator (*Alligator mississippiensis*). *Physiological and Biochemical Zoology* **76**, 348–359.

Seebacher F, Murray SA, Else PL (2009) Thermal acclimation and regulation of metabolism in a reptile (Crocodylus porosus): the importance of transcriptional mechanisms and membrane composition. *Physiological and Biochemical Zoology* **82**, 766–775.

Seidel MR (1979) The osteoderms of the American alligator and their functional significance. *Herpetologica* **35**, 375–380.

Sereno PC, Martinez RN, Wilson JA, Varricchio DJ, Alcober OA, Larsson HCE (2008) Evidence for avian intrathoracic air sacs in a new predatory dinosaur from Argentina. *PLoS ONE* **3**(9), e3303.

Seymour RS (2004) Reply to Hillenius and Ruben. *Physiological and Biochemical Zoology* **77**, 1073–1075.

Seymour RS (2013) Maximal aerobic and anaerobic power generation in large crocodiles *versus* mammals: implications for dinosaur gigantothermy. *PLoS ONE* **8**(7), e69361.

Seymour RS, Bennett-Stamper CL, Johnston S, Carrier DR, Grigg GC (2004) Evidence of endothermic ancestors of crocodiles at the stem of Archosaur evolution. *Physiological and Biochemical Zoology* **77**(6), 1051–1067.

Seymour RS, Smith SL, White CR, Henderson DM, Schwarz-Wings D (2012) Blood flow to long bones indicates activity metabolism in mammals, reptiles and dinosaurs. *Proceedings. Biological Sciences* **279**, 451–456.

Slip DJ, Shine R (1988) Thermophilic response to feeding of the diamond python, *Morelia s. spilota* (Serpentes, Boidae). *Comparative Biochemistry and Physiology A* **89**, 645–650.

Smith EN (1974) Multichannel temperature and heart rate radio telemetry transmitter. *Journal of Applied Physiology* **36**(2), 252–255.

Smith EN (1975a) Thermoregulation of the American alligator, *Alligator mississippiensis*. *Physiological Zoology* **48**, 177–179.

Smith EN (1975b) Oxygen consumption ventilation and oxygen pulse of the American alligator during heating and cooling. *Physiological Zoology* **48**, 326–337.

Smith EN (1976) Heating and cooling rates of the American alligator, *Alligator mississippiensis*. *Physiological Zoology* **49**, 37–48.

Smith EN (1979) Behavioural and physiological thermoregulation of crocodilians. *American Zoologist* **19**, 239–247.

Spotila JR, Soule OH, Gates DM (1972) Biophysical ecology of the alligator- heat energy budgets and climate spaces. *Ecology* **53**, 1094–1102.

Spotila JR, Lommen PW, Bakken GS, Gates DM (1973) A mathematical model for body temperatures of large reptiles: implications for dinosaur ecology. *American Naturalist* **107**, 391–404.

Spotila JR, Terpin KM, Dodson P (1977) Mouth gaping is an effective thermoregulatory device in alligators. *Nature* **265**, 235–236.

Spotila JR, O'Connor MP, Dodson P, Paladino FV (1991) Hot and cold running dinosaurs: body size, metabolism and migration. *Modern Geology* 16, 203–227.

Stuginski DR, Fernandes W, Tattersall GJ, Abe AS (2011) Postprandial thermogenesis in Bothrops moojeni (Serpentes: Viperidae). *Journal of Venomous Animal Toxins including Tropical Diseases* **17**, 287–292.

Taplin LE (1988) Osmoregulation in crocodilians. *Biological Reviews of the Cambridge Philosophical Society* **63**, 333–377.

Tarduno JA, Brinkman DB, Renne PR, Cottrell RD, Scher H, Castillo P (1998) Evidence for extreme climatic warmth from Late Cretaceous artic vertebrates. *Science* **282**, 2241–2243.

Terpin KM, Spotila JR, Foley RE (1979) Thermoregulatory adaptations and heat energy budget analysis of the American alligator, *Alligator mississippiensis. Physiological Zoology* **52**, 296–312.

Thomas RDK, Olson EC (1980) *A Cold Look At The Warm-Blooded Dinosaurs. AAS Selected Symposium 28.* Westview Press, Colorado.

Thorbjarnarson J, Wang X (2010) *The Chinese Alligator; Ecology, Behavior, Conservation, and Culture.* The Johns Hopkins University press, Baltimore, Maryland.

Turner JS, Tracy CR (1985) Body size and the control of heat exchange in alligators. *Journal of Thermal Biology* **10**, 9–11.

Webb GJW (1979) Comparative cardiac anatomy of the Reptilia. III. The heart of crocodilians and an hypothesis on the completion of the interventricular septum of crocodilians and birds. *Journal of Morphology* **161**, 221–240.

11

SALT AND WATER BALANCE

It was during the dry season of 1973 that I first became intrigued about how crocs manage their body water and electrolytes. This was after seeing very small estuarine crocodiles, Crocodylus porosus, *in the lower reaches of the Liverpool River, in water as salty as the Arafura Sea into which it flows. There they were, looking sassy, splashing in the shallows against the muddy exposed bank at low tide and snapping at prawns. There was no fresh water nearby and it was hard to imagine these small crocs swimming far enough to find any. Their common name in Australia is 'salties', and everything I was seeing suggested a physiological capacity for living full time in salt water, not just visiting. But the research to that time reported that crocodiles lacked salt glands: the few species that could be found in salt water were assumed to need fresh water nearby if they were to survive in the sea. A very big crocodile (or alligator) could presumably go for a long time and, perhaps, long distances between drinks, but not a 200 g hatchling. That chance observation of baby crocs swimming around in salt water diverted me for a few years from what I initially wanted to study – their thermal relations – and a fascinating story of discovery, suspenseful at times, gradually unfolded (see Grigg 1993, http://espace.library.uq.edu.au/view/ UQ:9724).*

SALT AND WATER BALANCE IN *CROCODYLUS POROSUS:* AN INTRODUCTION TO CROCODYLIAN OSMOREGULATION

Estuarine or saltwater crocodiles, 'salties' in the Australian vernacular, *Crocodylus porosus*, thrive across a very wide range of salinities, from fresh to hypersaline waters. A description of the patterns and mechanisms of their osmoregulation provides a convenient way to introduce salt and water balance (osmoregulation) in crocodylians in general and discuss some of the procedures by which it can be explored. With that as background ,we will then look at three species which live mostly in fresh water but are sometimes found in salt or brackish water: *C. johnstoni, Alligator mississippiensis* and *Caiman latirostris* (the latter two being alligatorids). Other species will be mentioned briefly. Relevant terminology is explained in Box 11.1. The most recent review of crocodylian osmoregulation is that by Leslie and Taplin (2001).

Events leading to the discovery of salt glands provide classic examples of the value of studying

---

An estuarine or saltwater crocodile, Crocodylus porosus, *proudly displaying its lingual salt glands. (Photo DSK)*

BOX 11.1   EXPLANATION OF TERMS

Osmoregulation: Literally, regulation of water, but in practice an umbrella term referring to the electrolyte composition and concentrations of body fluids and the processes by which they are regulated; 'osmoregulation', as a field of study within comparative physiology, is also known as 'salt and water balance'.

Homeostasis: If the concentrations of water and electrolytes in the body fluids are regulated to be constant, or nearly so, this is an example of homeostasis (literally 'same state').

Osmotic concentration: Water moves by osmosis across semi-permeable membranes from the more dilute side to the more concentrated. Biologists often want to know about gradients in osmotic concentration across membranes and are typically interested in the gradient between blood plasma and the water outside.

One easy way to express this gradient is with reference to the freezing point of a solution, because the more concentrated a solution, the lower its freezing point. A 1 molar solution of a simple solution freezes at $-1.86°C$, compared with $0°C$ for pure water. (A 1 molar solution contains the gram molecular weight of the solute made up to 1 L with water.) Biological solutions are mixtures of many solutes, both electrolytes and non-electrolytes, with a wide range of molecular weights, so biologists talk instead about Osmolar concentrations or, more usually, milliOsmolar (mOsm) concentrations. A 1000 mOsm (1 Osmolar) solution has the same osmotic concentration as a 1 molar solution, and so also freezes at $-1.86°C$. Because the relationship between freezing point and osmotic concentration is linear, by measuring the freezing point of a solution of unknown concentration one can know its osmotic pressure. Machines called osmometers do this easily enough on small samples. Solutions also have a higher vapour pressure than pure water, and vapour pressure osmometers are also available, which take advantage of that effect.

Convenient comparisons: By a convenient coincidence, sea water (SW) has a concentration very close to 1000 mOsm, so it is easy to go between different expressions of concentration. Crocodylian blood plasma is ~300 mOsm, which is very close to 30% SW. Salinity of water is often expressed in parts per thousand, ppt, and SW is 35 ppt, sometimes expressed as $°/_{oo}$. So the osmotic concentration of crocodylian plasma could also be said to be equivalent to 10.5 $°/_{oo}$ (or ppt).

Hyperosmotic, hypoosmotic, isosmotic: refer to a solution, such as plasma, that has a higher, lower, or the same osmotic concentration as water'.

Hypersaline and **hyposaline:** refer to the concentration of water in comparison to SW (i.e. more or less concentrated than SW, respectively).

Euryhaline, stenohaline: describe an animal's tolerance to changes in salinity (eury = wide, steno = narrow).

Brackish water: This usually means water that is salty, but hypoosmotic to body fluids.

---

physiological ecology in the field, of how 'natural experiments' can be very useful, and how laboratory results can be very misleading, so the story will be recounted chronologically.

The logical explanation for that 1973 observation of very small crocodiles living in salty water (Fig. 11.1) would be that they have salt glands, as described by Schmidt-Nielsen and Fänge (1958) in marine turtles and Galapagos iguanas. These authors speculated that estuarine crocodiles would

be found to have salt glands and implied that would provide an explanation for the literary references to 'crocodile tears'. But this had been explored by 1973 and the answer had come up negative. Bill Dunson from Pennsylvania State University, who published much of the original research on reptilian salt glands, had kept hatchling American crocodiles, *Crocodylus acutus,* and estuarine crocodiles, *Crocodylus porosus,* experimentally in salt water, but found that they quickly lost condition (Dunson 1970). He also salt loaded them and monitored sodium chloride in tears, because marine turtles excrete excess salt that way, and he tried injecting methacholine – a drug likely to stimulate secretion – but none of his results suggested the operation of salt glands. He must have been reluctant, though, to give up the idea, because he noted that his hatchlings may have been unrepresentative through being so young and never before having experienced salt water. There was one very intriguing result: Bill had kept a 3.4 kg *C. acutus* in full strength sea water and reported that after 5 months 'the animal was still vigorous and was feeding well' (!). He attributed this to its larger size, not to specialised excretory organs. Peaker and Linzell (1975) in their comprehensive monograph on salt glands in birds and reptiles were also obviously hopeful that salt glands would be found in crocodylians, observing that 'no really large salt-adapted specimens have been studied and it

is hoped that some brave person will try to collect tears from a twenty foot Estuarine Crocodile'. In an almost contemporaneous review, Dunson (1976) reported that, although there were salt glands in all other reptiles groups, they were lacking in crocodylians. However, he also added 'studies of adult crocodiles freshly captured from water of high salinity would be most interesting …' Unknown to all these authors, that is exactly what we were doing in the northern Australian dry season in 1975.

There was abundant anecdotal evidence that large *C. porosus* are at home in the sea, just as they are in estuaries. They are not uncommon visitors to Barrier Reef islands and there are anecdotal reports of them being seen well out to sea. We came across a large one carrying barnacles, which is a good sign of a long time spent at sea (Fig. 11.2) and there was plenty of information about them being on beaches (Fig. 11.3). But a large crocodile could surely spend

Fig. 11.1. *A yearling* Crocodylus porosus *among mangrove pneumatophores in the estuary of the Liverpool River, Northern Territory, Australia. The salinity of the water here, near where it enters the Arafura Sea, was ~30 ppt, nearly three times the concentration of crocodylian body fluids. The 1973 sighting of such small crocs in such salty water, with no fresh water nearby, prompted a study of their salt and water balance physiology. (Photo GCG)*

Fig. 11.2. *Barnacles on a* Crocodylus porosus. *Barnacles cannot live long in fresh water, so crocodiles with barnacles must have been in sea water for substantial periods of time. (Photos DSK)*

a long time in sea water even without any special physiology, so these observations by themselves did not prove much. A hatchling or a yearling is an entirely different matter: their larger surface area to mass ratio makes them more vulnerable to osmotic stress and dehydration. So the focus in our 1975 field trip was on small crocs, partly because of that and partly for practical reasons: they are much easier to catch!

THE PATTERN OF REGULATION BY *C. POROSUS* IN AN ESTUARY

In that 1975 dry season field trip, we were able to take advantage of a superb set-up for a natural experiment: the salinity gradient of the Liverpool River from pure fresh water down to full strength sea water (Figs 11.4, 11.5). We sampled blood and urine from 110 juvenile *C. porosus* (900 g–46 kg) captured across the salinity gradient and found that they maintained plasma osmotic and electrolyte

homeostasis from one end to the other. Plasma osmotic concentration averaged 304 mOsm (close to 30% the concentration of sea water) and was obviously regulated very tightly. *C. porosus* emerged as a very competent osmotic and ionic regulator (Grigg 1981) (Fig. 11.5, Table 11.1).

Upstream, where the water is substantially more dilute than body fluids, *C. porosus* deal with osmotic flood – dilution of body fluids by water absorbed or ingested. Downstream they face dehydration and excessive salt intake. In the middle reaches (e.g. between 22 and 32 km upstream in Fig. 11.5 at the time of our study), they have to manage both situations alternately as the tide ebbs and flows, so that a crocodile living at its chosen spot will cope alternately with hypo- and hyperosmotic conditions every few hours. It is useful therefore to identify four zones in the river, based on the nature of the salinity cycle experienced by a crocodile living there (Fig. 11.5). The plasma homeostasis data show *C. porosus* in fresh water can excrete excess water and retain electrolytes, as we might expect. Those nearer the river mouth must be excreting excess salt and conserving water that would otherwise be lost by osmosis. How, though, were they doing it?

Urine electrolytes offered no immediate explanation – not as we understood them then anyway. Their concentration increased in hyperosmotic water but, as in other reptiles, remained hypoosmotic to the blood. A conspicuous change in urine between fresh and salt water was from clear and copious in SAL 1 to milky and less copious in SAL 4. In salt water, the urine was often more of a paste with little or no free liquid (Fig. 11.12). This change reflected a shift in nitrogen excretion, from ammonotely (excretion as ammonium bicarbonate) to uricotely (excretion as insoluble uric acid): a common strategy in reptiles lacking ready access to fresh water. Realising that sodium and potassium urates are insoluble, we wondered if crocodiles had invented a new way to excrete sodium in these solids. Analysis showed more potassium in SAL 4, but virtually no sodium. Urine looked unlikely to be the prime route for sodium excretion. These results were quite clear cut and we were sure they could be explained by salt glands, but where were they?

Fig. 11.3. *An estuarine, saltwater or Indo–Pacific crocodile,* Crocodylus porosus, *relaxes on a northern Australian beach. (Photo Garry Lindner)*

Our next field trip was in July 1979. We worked in the salty end of the Liverpool River's main tributary, the Tomkinson (Fig. 11.4) and this time we focussed on hatchlings. The association between nest sites and fresh water was well known, and Bill Magnusson (1978) had found that hatchlings in captivity died more rapidly in sea water than in fresh water. It was logical, therefore, to suggest that *C. porosus* select nest sites that provide a source of freshwater for hatchlings during the 6 months of dry season that follows the summer monsoonal breeding season. We caught, measured, marked and released 15 hatchlings, 160–270 g, in salinities ranging tidally from 25 to 34 ppt (2–3 times plasma concentration). Eleven of them were recaptured 2 weeks later. Nine had grown noticeably in length and weight despite having had no evident access to fresh water during the period (Grigg *et al*. 1980). Their growth matched what had been measured for hatchling *C. porosus* elsewhere. Six of the animals were captured again in November, towards the end

of the dry season, and their growth had continued. There had been rain on only 3 days, in October. We were even more convinced, as speculated by Schmidt-Nielsen and Fänge (1958), and Dunson (1970), that *C. porosus* had special mechanisms, probably salt glands, but where?

DISCOVERY OF THE LINGUAL SALT GLANDS

In June 1980, nearly a year after the Tomkinson River study, Laurence Taplin, who had been part of that work, discovered the salt glands. They are on the tongue (Taplin and Grigg 1981) and the story of their discovery is told in a semi-popular review (Grigg 1994 at http://espace.library.uq.edu.au/view/UQ:9724). With the wisdom of hindsight, the presence of the glands on the surface of the tongue now seems obvious and we, or others, should have found them earlier. They probably escaped notice for so long because researchers usually tape crocodylians' mouths closed to avoid being bitten!

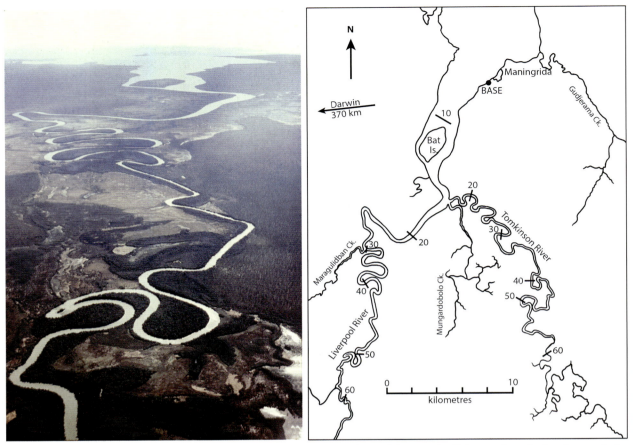

Fig. 11.4. *(Left) Aerial view of the Liverpool River, Northern Territory: a beautiful, natural salinity gradient. In the distance is the Arafura Sea. (Right) This river system, including its tributary the Tomkinson River, was a perfect location for a field study of salt and water balance in* C. porosus. *From our base in Maningrida, we could access the whole of it by boat, from the sea right up into fresh water (river distances marked in kilometres). In 1975, we captured more than 100* Crocodylus porosus, *over quite a size range, and found they maintained their osmotic and electrolyte homeostasis over the whole salinity range (Grigg 1981). The same rivers were the venue for many studies of* C. porosus *through the University of Sydney's crocodile research programme instigated by Professor of Physics, Harry Messel (see also Chapter 13). (Photo Laurence Taplin)*

By 1980, we had explored and rejected ideas of orbital and nasal glands, rectal glands like those in sharks, and generalised excretion across the integument (which would have been both curious and novel!). On the day the glands were discovered, however, Laurence had turned attention to the mouth – a seemingly unlikely place for a crocodile to have excretory organs – and was working with a crocodile whose mouth was taped but propped open. When he injected methacholine chloride, a drug likely to stimulate glandular secretion (see later), he was rewarded with a flow of clear, salty fluid from glands on the surface of the tongue (Fig. 11.16). It was a clever and very well executed piece of detective work.

Finding the salt glands provided an immediate explanation for the ability of even very small *C. porosus* to live in full strength sea water. Subsequent work by Taplin showed that they can even tolerate hypersaline situations. He studied hatchlings and juveniles (140–500 g) in Mungardobolo Creek: a long, sinuous and very salty tidal tributary of the Liverpool River, Northern Territory (Fig. 11.4). This creek has little or no freshwater inflow throughout the long dry season and becomes severely hypersaline through evaporation. The animals maintained their osmotic and ionic homeostasis even at salinities of 64 ppt – almost twice that of sea water (Taplin 1984a). This is sufficiently salty to cure hides! Obviously *C. porosus* are very euryhaline.

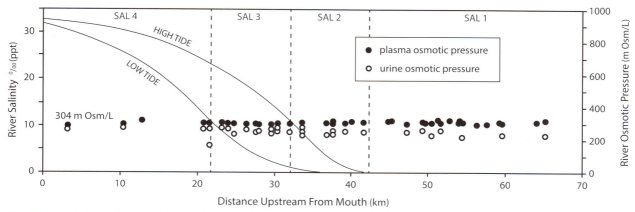

Fig. 11.5. *Salinity profile in the Liverpool River at the time of the study (July 1975) with plasma and cloacal urine osmotic pressures from juvenile* Crocodylus porosus *sampled at their capture sites. Crocodiles in the SAL 1 section of the river experience only fresh water and in the SAL 4 section only hyperosmotic conditions. In SAL 3, they experience hyperosmotic and hypoosmotic conditions alternately with each tidal cycle and in SAL 2 an increase in salinity as the tide floods, but not to hyperosmoticity. We found that plasma osmotic concentration is well regulated across this wide range of salinity and, as in other reptiles (but unlike mammals), that urine osmotic pressure is always less concentrated than the blood. (Adapted from Grigg 1981)*

Indeed, they may be the most euryhaline of any reptile because the marine turtles, those champion reptiles of saltwater living, do not do very well in fresh water while salties will live and breed in fresh water at huge distances from the sea.

Armed with an explanation for the saltie's salinity tolerance, our attention turned quite quickly to other crocodylians. We rather expected to find the saltie exceptional in the salt gland department, given it is one of the most sea-going species and most other species are found principally or solely in fresh water. Instead, we found salt glands in all of the crocodylids we examined, yet none in the spectacled caiman or the American alligator (or later in the broad-snouted caiman). This seemingly deep-seated distinction between crocodylids and alligatorids, of which more later, led us to speculate about its possible evolutionary and zoogeographical implications and to follow-up work on other parts of the osmoregulatory system, especially the renal-cloacal complex, that reinforces the physiological differences between these taxa.

OSMOREGULATORY ORGANS AND PROCESSES

The active processes most relevant to osmoregulation in all crocodylians are feeding, drinking (or not), and excretion. There is limited evidence that active uptake of electrolytes may play some role in conserving salts in fresh water, the site being unknown (Taplin 1988). Passive diffusion of water and electrolytes across the skin is a significant factor. The main osmoregulatory organs are the skin (as a low permeability barrier), kidneys, cloaca and (in Crocodylidae) the salt glands. Figure 11.6 provides a schematic summary.

Skin

Compared with fishes and amphibians, crocodylians have skin that is much less permeable to water and salts. The low permeability results from a layer of keratin: a protein secreted by the epidermis (Fig. 3.6). Osmotic and diffusional fluxes of water and salts across the skin are, nevertheless, greater than in terrestrial reptiles. Measurements of evaporative water loss in air, and water flux in both fresh and salt water, suggest that alligatorid skin may be more permeable to water than crocodylid skin (Taplin 1988). The buccal epidermis of crocodylids is more keratinised than in alligatorids (Mazotti and Dunson 1989). Permeability of the skin to sodium ions is very low and apparently similar in both groups, although better data are needed (Taplin 1988).

Drinking (or not)

All crocodylians probably drink fresh water, supplementing the small osmotic influx across the skin, but whether or not they drink salt water has

TABLE 11.1 COMPARISON OF ELECTROLYTES AND OSMOTIC CONCENTRATIONS IN PLASMA AND LIQUID AND URINE FROM *CROCODYLUS POROSUS* CAPTURED IN FRESH (SAL 1) AND SALT (SAL 4) WATER IN THE LIVERPOOL RIVER, NORTHERN TERRITORY, AUSTRALIA

An animal living in a SAL 4 section of the river experiences hyperosmotic conditions throughout the whole tidal cycle. In a SAL 1 section it is exposed only to fresh water (Fig. 11.5) (data from Grigg 1981).

	Plasma		Liquid urine		Urine solids	
	Fresh water (SAL 1)	Salt water (SAL 4)	Fresh water (SAL 1)	Salt water (SAL 4)	Fresh water (SAL 1)	Salt water (SAL 4)
Osmotic pressure (mOsm/kg)	307 (305–309)	302 (298–305)	252 (240–265)	277 (259–296)	Not applicable	Not applicable
Sodium (mmol/L)	136 (133–139)	130 (126–135)	7.9 (0–17.5)	9.8 (2.4–17.2)	0.0	0.3 mg/g (0.1–0.5)
Potassium (mmol/L)	4.1 (3.7–4.4)	3.8 (3.6–4.0)	2.2 (1.6–2.9)	8.8 (5.5–12.0)	0.2 mg/g (0–0.04)	12.7 mg/g (4.1–21.3)
Calcium (mmol/L)	2.8 (2.7–3.0)	2.8 (2.6–3.0)	1.0 (0.6–1.4)	4.4 (0–9.0)	21.5 mg/g (11.3–31.7)	1.3 mg/g (0.6–2.0)
Magnesium (mmol/L)	1.4 (1.3–1.5)	1.5 (1.2–1.8)	0.06 (0.03–0.08)	11.7 (0–28.2)	34.0 mg/g (23.8–44.1)	5.0 mg/g (1.2–8.8)
Ammonia (mmol/L)	0.8 (0.5–1.1)	0.8 (0.4–1.1)	98 (85–111)	43 (31–56)	Not applicable	Not applicable
Chloride (mmol/L)	121 (117–125)	114 (111–117)	9.1 (4.4–13.9)	9.7 (5.0–14.3)	–	–
Bicarbonate (mmol/L)	14.9 (13.7–16.1)	17.8 (16.1–19.5)	41.8 (33.7–50.0)	0.8 (0–1.84)	–	–
% solid in urine			2.7 (clear)	30.5 (milky)		

important implications for osmoregulation. Here again we see differences between alligators and crocodiles.

In hyperosmotic water, *Crocodylus porosus* avoid drinking (Taplin 1984b). Unfed, they dehydrate rather than drink, even to the extent that plasma sodium levels rise (Taplin 1985). Left unfed they will die. When fed, they can maintain homeostasis in sea water, with the food providing sufficient water as well as nutrients. Under natural conditions, their water turnover is much less in salt water than in fresh. We measured water turnover isotopically in free-ranging hatchlings, juveniles and sub-adults in brackish hypoosmotic (2–7 ppt) and hyperosmotic (25–35 ppt) conditions in the Tomkinson River in

1979 and 1981. Water turnover in the brackish part of the river was twice that further downstream (~160 mL/kg$^{0.63}$, compared with ~80 mL/kg$^{0.63}$) (Grigg *et al.* 1986). Presumably, drinking salt water would incur too much of an electrolyte penalty. This makes an interesting contrast with the loggerhead turtle, which drinks sea water routinely (Bennett *et al.* 1986) and whose salt glands can excrete sodium chloride at 4–6 times the rate of those in *C. porosus* (see below).

C. porosus, and all the crocodylids tested so far, apparently know instinctively not to drink hyperosmotic water (Taplin 1984b). In contrast, naïve alligators and caimans do not discriminate between fresh and salt, but they can apparently

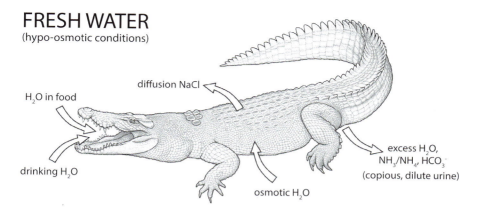

FRESH WATER
(hypo-osmotic conditions)

H_2O in food

diffusion NaCl

drinking H_2O

osmotic H_2O

excess H_2O,
NH_3/NH_4, HCO_3^-
(copious, dilute urine)

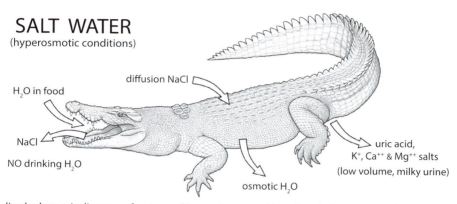

SALT WATER
(hyperosmotic conditions)

H_2O in food

diffusion NaCl

NaCl

NO drinking H_2O

osmotic H_2O

uric acid,
K^+, Ca^{++} & Mg^{++} salts
(low volume, milky urine)

Fig. 11.6. *Generalised schematic diagram of water and ion exchanges taking place in* C. porosus *in fresh water (hypoosmotic) and salt water (hyperosmotic). The reasonably impermeable crocodylian skin buffers them against short-term exposure to changes in salinity, but maintenance of body fluid homeostasis in the longer term can be a challenge because of water loss or gain by osmosis. A hatchling crocodile dies of dehydration in about 3 weeks in sea water if unfed. In fresh water they gain water by osmosis, by drinking and in food. Excess is excreted via the kidneys, losing some electrolytes in the process, which are made up by feeding. In salt water, they lose water by osmosis and face an excess of electrolytes, particularly NaCl, by diffusion and through feeding. They rely on their food for water and need to excrete excess electrolytes without losing much water. Unlike mammals, crocodylian kidneys cannot excrete concentrated urine so they face dehydration in salt water. Salt glands in crocodylids, however, work in concert with the kidneys and cloaca, excreting excess sodium chloride and minimising water loss. Alligatorids lack salt glands and in hyperosmotic conditions rely on drinking fresh or mildly brackish water to rehydrate and flush out excess electrolytes.*

learn to do so. Jackson *et al.* (1996) tested how captive-raised, dehydrated juveniles of several species responded to the choice to re-hydrate by drinking water at a range of salinities. Drinking was indicated by a significant weight gain after 15 min. *Caiman crocodilus* and *Alligator mississippiensis* captive-raised in fresh water all drank hyperosmotic water. However, individuals captured from a naturally occurring estuarine population, where they had a range of salinities to choose from, avoided drinking the hyperosmotic water. We found similar results in an estuarine population of *Caiman latirostris* (Grigg *et al.* 1998, see below). It is likely that alligatorids in general lack

the instinct to avoid drinking hyperosmotic water, but learn fairly easily. Jackson *et al.* (1996) also tested three species of Crocodylidae – *Crocodylus porosus, Crocodylus siamensis* and *Osteolaemus tetraspis.* All three were captive-raised in fresh water and all avoided drinking hyperosmotic water. These three species all have salt glands and *C. siamensis* is the only one among them which is not known to frequent saline habitats in the wild (Taplin 1988). *Crocodylus acutus*, too, is known to avoid drinking hyperosmotic water (Mazotti and Dunson 1984). The instinct to avoid drinking salty water seems to be deeply embedded in Crocodylidae, but not Alligatoridae.

How crocodylians sense and judge salinity is not yet clear. There are taste buds on the tongue, palate and on the walls of the buccal cavity and pharynx (Weldon and Ferguson 1993). Jackson and Brooks (2007) suggested that thicker keratinisation in the buccal cavities of crocodylids compared with alligatorids would compromise their buccal sense of taste and they might rely on their integumentary sense organs (ISOs) instead (Chapters 3, 5). This was, however, refuted by Leitch and Catania (2012). The structure and function of the ISOs are discussed in Chapter 5.

The differences between the Alligatoridae and Crocodylidae raise interesting evolutionary questions that will be discussed later about whether the three 'marine' attributes of crocodylids – salt glands, strongly keratinised buccal cavities, avoidance of drinking hyperosmotic water – are ancient (plesiomorphic) or have evolved secondarily and independently from freshwater ancestors.

Feeding

Most food that crocodylians eat has a lower electrolyte content than sea water. Vertebrate prey, which most adult crocs rely on (Chapter 6), ranges in concentration from 20–30% as concentrated as the sea, broadly similar to a crocodylian's own body concentrations. This means that food represents a water gain as well as a nutritional gain, as long as the animal can deal with any extra electrolytes and excrete the nitrogen without undue water loss. Insects and other invertebrates often have higher levels of electrolytes than vertebrates, but still proportionally less than sea water. The value of crustaceans (crabs and shrimps) as a source of water for crocodylians in marine situations has sometimes been underestimated to the point of assuming that a diet of crabs and prawns would give a marine croc no water benefit. However, many crustaceans remain hypoosmotic to their aquatic medium, commonly by 20% or so (Potts and Parry 1964) and, furthermore, that is based on measurements on the blood, ignoring the intracellular fluid that comprises the bulk of the body's water. Taplin (1985) tabulated whole body water, sodium and potassium data for a range of crabs, prawns and fish known to be preyed on by *C. porosus* in northern Australia.

The data confirmed that fish are a better choice than crustaceans, but the latter also can provide a net water benefit. Prawns are a better choice than crabs, and fiddler crabs (*Uca* spp.) were the most 'expensive' of the species analysed. Interestingly, Taylor (1979) noted that fiddler crabs seemed to be under-represented in the diet of juvenile crocs, despite their abundance, suggesting the possibility of selectivity on this basis. The message is that crustaceans cannot be dismissed automatically as unable to provide a water benefit in addition to nutrition. The strong aversion of crocodylids to drinking salt water implies that, in hyperosmotic conditions, they must have a heavy reliance on water they gain from their food. In fresh water too, of course, food will be a significant source of water gain, additional to what enters by osmosis. Any excess is excreted via the kidneys.

The analysis of sodium and potassium in a range of food items (Taplin 1985) shows that they contain nearly as much potassium as sodium. This is not an issue in fresh water, where electrolyte retention is necessary. But excess potassium, like excess sodium, must be excreted in hyperosmotic situations, which explains the much higher levels of potassium in both liquid and solid fractions of the urine in saline conditions (Table 11.1).

Kidneys

As in all vertebrates, the kidneys are the amazing organs that carry out most of the active regulation of electrolytes and water, as well as the excretion of excess nitrogen. Paired lobular kidneys are situated dorsally and posteriorly in the body cavity, one on each side of the midline (Figs 11.7, 11.8). Each drains to the cloaca via its ureter.

The functional units of the kidneys are the nephrons (Fig. 11.9): thousands of fine tubules blind at one end and opening into the collecting ducts, which, in turn, discharge urine into the ureters. The processes occurring in the nephrons are filtration of fluid from the blood plasma by the glomerulus, selective reabsorption of ions and metabolites across the walls of nephron tubules and secretion of some excretory products into the tubules. The resultant fluid is urine, which, in crocodylids in particular, is subject to further processing in the cloaca before

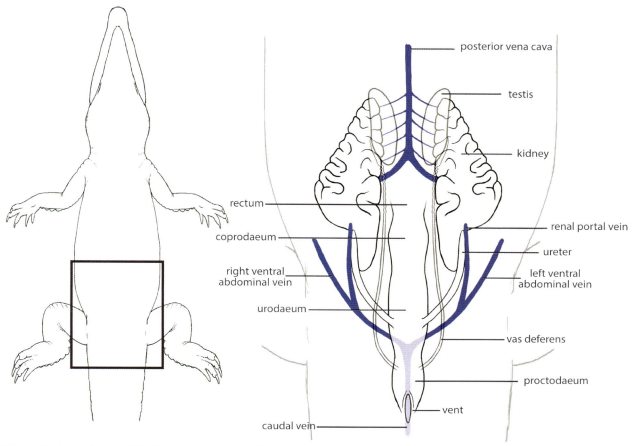

Fig. 11.7. *Ventral view of the kidneys and associated ducts and their entry points to the cloaca in* Crocodylus porosus. *Unlike the situation in mammals, some of the blood processed in the kidneys is supplied by the renal portal vein, whose capillaries bathe the kidney tubules. Arterial blood is supplied to the glomeruli by branches from the dorsal aorta. (From various sources including Kuchel and Franklin 2001 and Chiasson 1962)*

release (see below). The 'filtration kidney' structure is fairly typical of freshwater vertebrates and similarities between A. *mississippiensis* and *C. acutus* suggest that anatomical details of the kidneys are likely to be similar across all crocodylians. Vertebrate nephrons typically have histologically and functionally distinct regions and are comparatively short: 4–5 mm (Huber 1917). Their structure in crocodylians has been reviewed by Davis and Schmidt-Nielsen (1967) working on *C. acutus* and Ventura *et al.* (1989) and Moore *et al.* (2009) working on American alligators, *A. mississippiensis*. These last authors provided a comprehensive histological and histochemical description of each of the tubule segments. At the blind end is the renal corpuscle, with Bowman's capsule containing a knot of capillaries, the glomerulus. As in other reptiles and in birds, the renal corpuscles are smaller in relation to tubule diameter than in mammals. In the glomerulus, arterial blood supply comes into contact with a thin membrane through which an ultrafiltrate of water, electrolytes, ammonia and other small molecules flows into the lumen of the nephron. Unlike mammalian kidneys, which have only arterial blood supply, the nephron tubules in reptiles are bathed also by blood from the renal portal veins (Chapter 7 and Fig. 7.19). Active exchanges of electrolytes occur across the nephron walls by selective secretion and reabsorption into this venous flow. Of the sodium and chloride that enter through the glomerulus, most is reabsorbed in the proximal tubule (Fig. 11.9) further down the nephron, so the kidneys play a major role in retaining electrolytes when crocodiles are in fresh water and electrolyte retention is required. Work by Pidcock *et al.* (1997) and Kuchel and Franklin (1998)

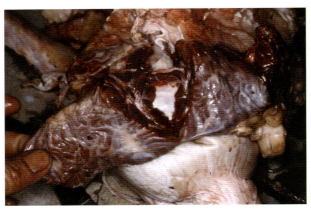

Fig. 11.8. *A partly transected kidney of* C. porosus *showing its lobulated nature. (Photo GCG)*

on *C. porosus* shows that the probable mechanism in fresh water is similar to that worked out by Coulson and Hernandez (1983) for *Alligator mississippiensis*: that is, rapid excretion of nitrogen is achieved by exchanging NH_4^+ for Na^+, and HCO_3^- for Cl^-. Thus, NaCl is reclaimed from the urine and much of the ammonia is excreted as ammonium bicarbonate, NH_4HCO_3, which is more benign than ammonia. The kidneys of crocodylians (and all other reptiles) lack the functionally spectacular counter-current multiplier loops of Henle that feature so prominently in any biology textbook, and that enable mammals to produce urine more concentrated than the blood.

The kidneys of reptiles in general and crocodylians in particular cannot therefore produce liquid urine that is hyperosmotic to blood, and our data on *C. porosus* in the Liverpool River confirmed this expectation (Fig. 11.5, Table 11.1), even after secondary processing in the cloaca (see below).

Sodium is the dominant cation in the body (and in salt water), but excess calcium, magnesium and potassium are also acquired in the diet, or from the water itself by diffusion. Because they mostly form soluble salts, their excretion could pose a problem. However, they all form insoluble salts with uric acid, so the formation of insoluble calcium, magnesium and potassium urates provides a route for their excretion, and one that aids nitrogen excretion as well. These excess cations are excreted either in solution in the urine or as precipitated urates. Even in fresh water, the urine is likely to have a small amount of precipitated urates (see later). In salt water, the urine leaving the kidneys is a slurry of insoluble urate spherules (see Fig. 11.13).

The ureters discharge a flow of ureteral urine into the urodaeum, one of the three chambers of the cloaca, through two dorsolateral papillae (Fig. 11.10) where it can be further modified by selective exchange of ions and secondary reabsorption of water across the cloacal epithelium. In alligatorids,

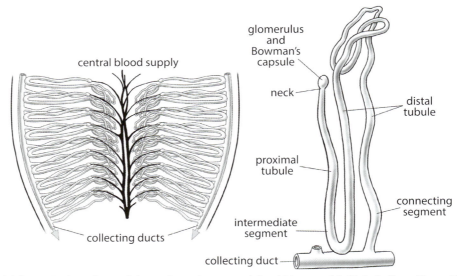

Fig. 11.9. *(Right) Schematic view of one of the nephrons in a crocodylian kidney, and (left) their disposition within a lobule of the kidney. Numerous collecting ducts from the lobules join to form the paired ureters that convey urine to the cloaca. (Schematic based on diagrams and descriptions of* Crocodylus acutus *by Davis and Schmidt-Nielsen 1967, and* Alligator mississippiensis *by Huber 1917, Ventura et al. 1989 and Moore et al. 2009)*

there is little or no reprocessing of urine, but in crocodylids the cloaca plays a much more active role.

Cloaca

It is fair to say that placental mammals, including humans, are unusual among vertebrates in not having a cloaca. Cloacas (from Latin, meaning sewer or canal) are typical of elasmobranchs, lungfish, amphibians, reptiles, birds, monotremes and marsupials. In crocodylians, the cloaca is a medial tubular chamber subdivided by muscular sphincters into three distinct chambers, the coprodaeum, urodaeum and proctodaeum. It has a single opening to the exterior (Fig. 3.39) through which faeces, urine and reproductive products all pass. Its anatomy and histology have been described for *C. porosus* by Kuchel and Franklin (2000, 2001). Like birds, crocodylians have no bladder, but they store quantities of urine in the urodaeum (Figs 11.7, 11.10). There is little or no mixing between the three compartments. Faecal pellets are formed in the

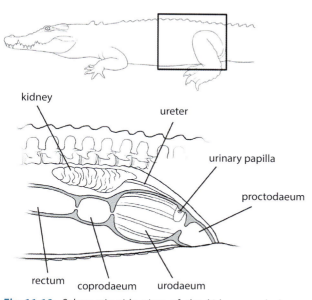

Fig. 11.10. *Schematic side view of the kidneys and cloaca and relevant ducts in a generalised crocodylian (see also Fig. 12.15). The ureters convey urine from the kidneys and discharge it via the urinary papillae into the middle cloacal chamber, the urodaeum. In a crocodylid living in salt water, excess sodium in the kidney ultrafiltrate will be pumped actively across the wall of the urodaeum into the blood, causing water to follow. The excess sodium chloride is then excreted in a concentrated solution via the lingual salt glands, thus retaining water. (From various sources including Kuchel and Franklin 2001)*

coprodaeum, water being removed in the process. Faeces and urine are voided separately, and semen from the vas deferens enters the base of the penis, which everts from the proctodaeal wall (see Chapter 12) without the possibility of it mixing with urine, which would compromise its viability. The penis has no role in urination.

The cloaca in crocodylids is not simply a passive tube down which urine and faeces pass. Rather, the urodaeum is an important and very active regulatory organ: a site for the secondary processing of urine that flows in from the kidneys (Pidcock *et al.* 1997; Kuchel and Franklin 1998). Most particularly, it is here that water is reclaimed by crocodylids with active salt glands, and by other reptiles living in salt water, or terrestrially where water retention is critical. The nature of cloacal modification of urine has been explored in *Crocodylus porosus* and *Alligator mississippiensis* (Pidcock *et al.* 1997).

In *C. porosus*, and probably in other crocodylids, water is extracted from the stored urine by active pumping of sodium ions across the cloacal membranes in exchange for potassium, creating locally higher osmotic pressure in the tissues. This draws water out of the cloaca and into the perfusing blood by osmosis. The result is a removal of sodium and reclamation of water, with a reduction in urine volume, an increase in potassium concentration and an increase in osmotic concentration that approaches, but does not exceed, that of the plasma. Because of this secondary processing, it is necessary to distinguish between ureteral and cloacal urine: that is, the refined filtrate excreted from the kidneys and the urine that is finally excreted from the body. Pidcock *et al.* (1997) collected samples of ureteral urine by inserting a specially crafted tube into the cloaca to gather it directly from the ureters. Cloacal urine was collected by inserting a fire-polished glass tube carefully through the cloacal sphincter into the urodaeum. Unfortunately, there are very few, if any, data on the composition of urine voided naturally by *C. porosus* or, indeed, by any of the crocodylians. Unless noted otherwise, data reported about crocodylian urine typically refer to samples of cloacal urine taken from the cloaca at a time chosen by the investigator.

The excess sodium chloride that has been pumped across the wall of the urodaeum is excreted via the salt glands. So, in crocodylids in hyperosmotic water, the cloaca works in tandem with the lingual salt glands to reclaim water that has been excreted by the kidneys (Taplin 1988; Kuchel and Franklin 1998). Similar mechanisms occur in some birds, but their cloacal physiology is different. In birds, cloacal urine is refluxed into either the coprodaeum or rectum, and it is here that water is withdrawn. There is no evidence so far that reflux from the urodaeum to the coprodaeum occurs in crocodiles (Kuchel and Franklin 1998, 2000, 2001).

These authors have also shown that, when *C. porosus* are initially raised in fresh water, they can adapt to salt water by increasing the surface area of the internal walls of the urodaeum, which presumably increases their capacity to retain water. Salt glands too, are larger in animals kept in salt water (Franklin and Grigg 1993), so it can be assumed that the physiological capacities of crocodylids in general acclimatise with exposure to life at a different salinity. We saw examples of this in field studies of *C. johnstoni* living in hyperosmotic water (see below).

Urine characteristics

There was a time when physicians diagnosed diseases by way of inspecting a patient's urine: the most famous in history is probably that of King George III of England whose urine was port wine colour, a consequence of the porphyria he inherited from Mary Queen of Scots. Diagnosing illness by urine colour has largely gone by the wayside, although a more intense yellow colour is a useful warning of dehydration. In *C. porosus*, the regulatory and excretory processes lead to conspicuously different looking (and tasting!) urine between fresh water and salt water. The differences are worth noting because, by extrapolation, simple observations of the urine can be a useful indication of the physiological processes underway in crocodylians whose physiology is less well known. Urine characteristics of *Crocodylus porosus* in both fresh water and salt (hyperosmotic) water are presented in Table 11.1. In freshwater, the cloacal urine is fairly clear and copious, as it is

in alligatorids (see Fig.11.39) and, if left to stand, only a very small quantity of white precipitate will settle. The precipitate is mostly small quantities of insoluble calcium and magnesium urates. In *C. porosus* caught from salt water, however, the urine is an opaque white/cream liquid, much smaller in volume and with a significant amount of insoluble, crystalline suspended solids comprising the nitrogenous excretory product uric acid and its urate salts (Figs 11.11, 11.12, 11.13), . Urine collected from crocodiles living in salt water is often a semi-solid cream-yellowish pellet rather than a liquid (Fig. 11.12), most of the water having been removed in the kidneys and cloaca. It is significant that, due to the operation of the lingual salt glands, the urine of *Crocodylus porosus* captured from salt water does not taste salty. In alligatorids captured in salt water, however, the urine is conspicuously salty to the taste (see below, and Fig. 11.40).

Nitrogen excretion is interlinked closely with water and electrolyte regulation. Crocodylians, being carnivores, have a high protein diet and metabolism of the component amino acids leads to surplus nitrogen, much in the form of ammonium ions, which are toxic and must be excreted. Carbon dioxide is a by-product of respiratory metabolism and it too needs to be excreted. In fresh water, crocodylians excrete much of their excess ammonium ions bound to carbon dioxide as ammonium bicarbonate. This is formed in the kidney tubules in exchange for sodium and chloride ions which, in fresh water, need to be retained (Coulson and Hernandez 1964, 1983). The excretion of CO_2 this way leads to a lower whole body respiratory exchange ratio (RER, Chapter 7) than would otherwise be expected (Grigg 1978). Ammonium bicarbonate is also very soluble in water, but in fresh water there is no shortage of dilute urine to carry it. Crocodylians are therefore said to be ammonotelic in fresh water (Table 11.1). Even in fresh water, however, some nitrogen is excreted as urates, synthesised in the liver and excreted via the kidneys.

In salt water, where water retention is the problem, most of the nitrogen is excreted as low solubility uric acid and urate salts. These are formed within the kidneys and travel down the ureters as a slurry

Fig. 11.12. A pellet of urine comprising mostly uric acid and its salts, voided by a juvenile Crocodylus porosus *captured in salt water. (Photo GCG)*

Fig. 11.11. Cloacal urine being sampled from a large Crocodylus porosus *captured in salt water. Note the milky suspension of uric acid and urate salts. Because of the salt glands, this urine will be very low in sodium chloride. In contrast, urine from an alligatorid captured in salt water is clear and salty to the taste (compare with Figs 11.39, 11.40, showing clear, salty urine from a* Caiman latirostris *captured in salt water). (Photo GCG)*

Nielsen *et al.* 1957; Schmidt-Nielsen *et al.* 1958). They were trying to understand how marine birds can survive such long periods at sea without fresh water to drink. Knut speculated subsequently that crocodiles, too, would have salt glands near their

of spherules (Fig. 11.13). This is known as uricotelic excretion. Although the actual concentration of the liquid cloacal urine remains hypoosmotic to the blood, water is reclaimed from this slurry across the cloacal wall by sodium pumping to such an extent that if all the precipitates were redissolved the resultant solution would be hyperosmotic. The combined kidney–cloaca–salt gland axis is quite effective at excreting excess nitrogen and salts with a minimum of water expended in the process.

Salt glands

The great comparative physiologist Knut Schmidt-Nielsen and his colleagues discovered the first known salt glands – in cormorants (Schmidt-

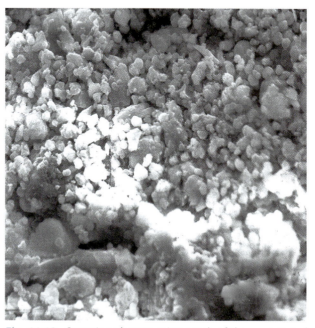

Fig. 11.13. Scanning electron micrograph of the precipitated uric acid and urate spherules in a sample of urine from a Crocodylus porosus *captured in hyperosmotic salt water. (Image made in Electron Microscope Unit, University of Sydney)*

eyes (Schmidt-Nielsen 1959). In due course, it was our great pleasure to host Knut and his wife Margareta in our laboratory at the University of Sydney where Laurence Taplin was able to show him where they really are (Fig. 11.14).

Many birds and reptiles that live in or are associated with the sea have salt glands. They allow the excretion of excess sodium chloride with minimal water loss: personal desalination plants. They have evolved many times and different animals have co-opted different glands for the purpose. Albatrosses, petrels and seagulls have nasal salt glands that excrete a clear concentrated fluid that dries to a white residue of sodium chloride. Marine turtles have orbital salt glands that discharge near their eyes, explaining the streams of tears when a female turtle makes her way up the beach to lay eggs. Galapagos marine iguanas have nasal salt glands and sneeze a fine salty spray. In sea-snakes, the salt glands are under the tongue (sub-lingual) and crocodylids have them on the surface of the tongue (lingual) (Figs 11.15, 11.16, 11.17, 11.18). There is a vast literature on reptilian and avian salt glands. Useful reviews are by Peaker and Linzell (1975), Dunson (1976) and Shuttleworth and Hildebrandt (1999).

It is the presence of salt glands, working in concert with active sodium pumping in the cloaca, that explains how it is that *C. porosus*, *C. niloticus* and *C. acutus* (almost certainly – see below) and quite probably other crocodylids can live in water more concentrated than the body fluids without

Fig. 11.15. Pores on the surface of the tongue are the exits through which a salty secretion is excreted. (Photo GCG)

needing to drink fresh water. Almost all of the solute in the secretion is sodium chloride, at about the same concentration as sea water (Table 11.2). There is a very small amount of potassium chloride. Taplin *et al*. (1999) calculated that, with salt gland secretions being three times more concentrated than the plasma, each mL of secretion represents a gain of 2 mL of free water. This represents a water gain of ~48 mL/day to a 1 kg croc – more than enough to compensate for the water loss in sea water.

In *C. porosus*, the salt glands are 20–40 discrete, lobulated salt-excreting glands lying in the tongue mucosa, each opening onto the surface through a conspicuous pore (Figs 11.15, 11.16) (Taplin and Grigg 1981). Glands have 10–14 lobules, each surrounded by collagenous connective tissue (Franklin *et al*. 2005). Each lobule of the gland is densely packed with branching secretory tubules whose ultrastructure shows that they are essentially indistinguishable from salt glands in marine turtles and are very similar to avian salt glands. The tubules comprise columnar epithelial cells with interdigitating membranes surrounding minute ducts lined with microvilli. The cells have abundant mitochondria and the interstitial tissue surrounding the tubules is richly supplied with blood vessels (Figs 11.19, 11.20) and unmyelinated nerve fibres (Taplin and Grigg 1981; Franklin and Grigg 1993; Franklin *et al*. 2005). The ducts from each lobule empty into a short, but wide, excretory duct (Franklin *et al*. 2005), which opens to the surface of the tongue. The pores are so conspicuous

Fig. 11.14. Laurence Taplin shows Knut and Margareta Schmidt-Nielsen lingual salt glands on the tongue of a hatchling Crocodylus porosus (c. 1981). (Photo GCG)

Fig. 11.16. *Beads of salty secretion accumulate over the pores. Note the keratinised nature of the surface of the tongue. By slightly opening and closing the mouth a few times, the secretions can be flushed away. (Photo GCG)*

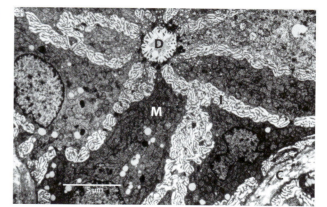

Fig. 11.18. *Transmission electron micrograph showing a secretory tubule of a lingual salt gland from Crocodylus porosus in cross section. Note the narrow central duct (D), with microvilli, interdigitating lateral cell membranes (I), abundant mitochondria (M), and surrounding connective tissue (C). Scale bar = 5 μm. (From Grigg and Gans 1993 with permission from the Australian Biological Resources Study, Canberra. Image made in Electron Microscope Unit, University of Sydney)*

that, given the crocodylian habit of lying around with their mouths wide open (see Chapter 10 and Figs 10.28, 10.29), it is easy to wonder in retrospect why they were not suspected earlier. Certainly they were described earlier, but as salivary glands (quite reasonably) (Owen 1866; Von Bayern 1884). Franklin *et al.* (2005) and Cramp *et al.* 2007) explored the machinery present for controlling salt glands in *Crocodylus porosus* and found it broadly similar to birds, with both cholinergic and adrenergic innervation.

It may seem paradoxical that the salt glands discharge onto the surface of the tongue, but the functional mouth of crocodiles is, after all, made by closure of the palatal and gular flaps at the back of the jaws (Figs 6.21, 9.3). The inside of the mouth is really part of a crocodylian's external surface. A salt-loaded *C. porosus* in water will periodically open and close the jaws very slightly several times

Fig. 11.17. *Scanning electron micrograph showing a vertical section through a single lingual salt gland of Crocodylus porosus. Note the individual lobules separated by connective tissue, with their ducts opening to a pore on the surface of the tongue via what Franklin et al. (2005) called an excretory duct. (From Taplin and Grigg 1981 and Grigg and Gans 1993 the latter with permission from the Australian Biological Resources Study, Canberra. Image made in Electron Microscope Unit, University of Sydney)*

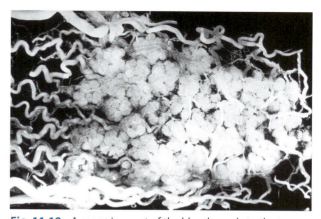

Fig. 11.19. *A corrosion cast of the blood supply to the tongue of a small Crocodylus porosus (dorsal view). A casting resin has been injected into the arterial blood supply and, after it is set, the flesh has been dissolved away in acid, leaving a beautifully detailed cast of the blood supply. The blood supply to the salt glands is so extensive that the individual lobules are outlined. (From Franklin and Grigg 1993, © 1993 Wiley-Liss, Inc.)*

TABLE 11.2   PLASMA SODIUM AND LINGUAL SALT GLAND CONCENTRATION AND SECRETORY RATE IN THREE CROCODYLIDS KNOWN TO LIVE IN HYPEROSMOTIC ENVIRONMENTS

(Means ± s.e.). (From Taplin *et al.* 1982, 1985)

Species	Plasma sodium concentration (mmol/L)	Average maximum Na+ concentrationof salt gland secretions (mmol/L)		Average maximum salt gland Na+ secretory rate under methacholine chloride stimulation (μmol/100 g$^{0.7}$.h)	
		In hypoosmotic water	In hyperosmotic water	In hypoosmotic water	In hyperosmotic water
C. porosus 3.6–32 kg	145 ± 4.8	510 ± 100	592 ± 40	45 ± 19	84 ± 23
C. johnstoni 3.6–8.1 kg	158 ± 5.9	386 ± 65	580 ± 33	6 ± 5[1] 35 ± 2.4 [2]	47 ± 2.3
C. acutus 3.6–6.2 kg	155 ± 7.3	454 ± 90	unreported	47 ± 18	Unreported

[1]*In fresh water distant from the river.*
[2]*In brackish water in the estuary.*

in quick succession, flushing excess salt from the vicinity of the tongue (Taylor *et al.* 1995).

The proof that a gland is a functional salt gland is usually its secretion of a fluid saltier than the blood plasma, often stimulated by loading the animal with salt. In *C. porosus* this turned out to be problematic because injecting a salt load was not effective in stimulating their secretion. However, Taplin found

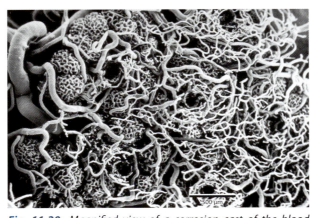

Fig. 11.20. *Magnified view of a corrosion cast of the blood supply to the tongue of a small* Crocodylus porosus (*dorsal view*). *The blood supply to individual salt glands shows clearly, as do the positions of excretory pores. (From Franklin and Grigg 1993, © 1993 Wiley-Liss, Inc.)*

that by injecting methacholine chloride – a drug that is an analogue of acetylcholine – the salt glands on the tongue produce a secretion with about the same strength as sea water (Taplin and Grigg 1981), and this has become an effective technique for diagnosing salt glands in other crocodylians. The lack of response to a salt load was puzzling. Perhaps not surprisingly, crocodile salt glands seemed reluctant to secrete in an animal removed from the water and tied down with its mouth propped open. Taplin thought that the handling might have provoked an inhibition of spontaneous secretion, which was being overridden by the methacholine. Subsequently, Greg Taylor set up a situation where an unrestrained crocodile in a tank could be relaxed and have a salt load delivered remotely via a cannula implanted in a blood vessel. In this way, he was able to demonstrate a flow of salty fluid from the tongue immediately after a salt load by measuring increased electrical conductance at the tongue surface using a custom-made sensor (Taylor *et al.* (1995).

From field data, we can be confident that the salt glands are handling the excess salt for *C. porosus* living in salt water. Using radioisotopes, we measured sodium effluxes from free-ranging crocs in brackish and hyperosmotic conditions in the Tomkinson River. Maximum rates of sodium efflux in hatchlings over a couple of weeks were 7.5 mmol/kg$^{0.63}$.day (Grigg *et al.* 1986), compared with salt gland secretory rates of 7.9 mmol.kg$^{-0.63}$/day measured in the laboratory. Salt gland capacity is more than enough to account for the field fluxes. Juveniles and sub-adults, on the other hand, had lower sodium efflux rates, 5.0 mmol.kg$^{-0.63}$/day but, unlike the hatchlings, which maintained their condition and actually grew over the 2 weeks, they lost weight. This was quite in contrast to the freshly captured ones in the same habitat. It is notoriously difficult to maintain captive *C. porosus* in salt water because they are so reluctant to feed, and these free-ranging animals lost condition in the same way. We concluded that the juveniles and sub-adults were showing effects from their capture and handling. Hatchlings are much more naïve and were apparently unaffected.

Reliance on the salt glands is indicated, too, by hypertrophy of the glands following salt water exposure (Franklin and Grigg 1993: a well-known phenomenon in ducks too when reared on brackish drinking water. Cramp *et al.* (2007, 2008) showed that secretory rate increased in saltwater-acclimated crocs and that the increase in gland size is the result of an increase in size of the cells in the secretory tubules. The number of pores on the tongue and the number of secretory tubules remained the same after acclimation to sea water.

Summary and implications for crocodylian dispersal

Estuarine crocodiles can live and maintain their ionic and osmotic homeostasis in fresh water, sea water and in hypersaline situations without access to fresh water. The following summarises what makes it possible. They have low skin permeability to water and sodium, as do all extant crocodylians. In fresh water, they rely on the kidneys for electrolyte retention and the excretion of excess water, producing copious dilute urine. They are ammonotelic in fresh water, excreting nitrogen as ammonia, mostly bound with CO_2 to form ammonium bicarbonate, resulting in slightly alkaline urine. Some nitrogen is excreted also in the form of insoluble calcium, magnesium and potassium urates. *Crocodylus porosus* drinks fresh water if available but will not drink salt water. In hyperosmotic situations they face water loss and the need to excrete excess sodium chloride and other ions. They achieve this by supplementing kidney functions with an osmotically active cloaca and lingual salt glands. Excess sodium chloride leaves the kidneys in the ureteral urine. Sodium ions are pumped across the cloacal wall, with chloride and water following, so water is retained in the body. The resultant cloacal urine has a much reduced volume and its liquid fraction is more concentrated, although never more so than the blood. In salt water, *Crocodylus porosus* is uricotelic and the cloacal urine has a high proportion of solids, uric acid and its insoluble urate salts of calcium, magnesium and potassium. The excess sodium chloride is excreted by the lingual salt glands, which produce a surprisingly pure solution of sodium chloride at a concentration typically 3–4 times its concentration in the blood. However, their salt glands are less competent than those of marine turtles and, unlike turtles, they depend on gaining water from their food.

Interestingly, although the finding of salt glands might seem to explain their geographic distribution from India and throughout South-East Asia to the Solomon Islands (Fig. 4.36), we now know that it does not necessarily depend on that physiology. Measurements of water and sodium fluxes (Taplin 1985) have shown that a 10 kg *C. porosus* could survive in sea water without feeding or drinking for up to 4 months. It would by then have lost one-third of its body water, but the implication is clear: adult *C. porosus* could go a long time between drinks, and even without salt glands they could have easily island-hopped their way to their present distribution, finding fresh water on the way. Indeed, large *A. mississippiensis* are seen out to sea off North America's Caribbean coast (see below) and off the Florida Keys (Fig. 9.2). Likewise, the long coastal journeys revealed by recent satellite telemetry (Read *et al.* 2007 and Chapter 4) do not by themselves prove a marine physiological capacity; only large animals so far have been subjected to satellite tracking. This is not to say that seafaring crocodiles do not use their salt glands – undoubtedly they do and thereby maintain their homeostasis throughout their journeys. Do smaller crocodiles make oceanic journeys? There seems to be almost no data, but Fig. 11.21 provides one example. The observations of estuarine crocodiles visiting Adele and other remote, waterless islands (Chapter 4) show that even quite modestly sized individuals can make journeys that must involve many days spent in the ocean.

This section on *C. porosus* has showcased many of the elements involved in its salt and water balance and, by implication, in other crocodylids. The complementary functions of the kidneys, cloaca and salt glands in Crocodylidae have been

Fig. 11.21. *Monash University scientists Shana Nerenberg, Rohan Clarke and Ashley Herrod watch a small sub-adult* C. porosus *at Adele Island: a small cay ~80 km off the Kimberley Coast of Western Australia. Crocodiles are seen commonly by visitors to this and other remote offshore islands, presumably to take advantage of seabird colonies and other feeding opportunities such as hatchling turtles (see also Chapter 4 and Fig. 4.37). (Photo George Swann)*

explored, and reference made to the lack of salt glands in Alligatoridae. In the following sections, we will look at two other crocodylids, *C. acutus* and *C. johnstoni*, and in the gavialids *Gavialis gangeticus* and *Tomistoma schlegelii* (briefly) and two alligatorids, *Alligator mississippiensis* and *Caiman latirostris*. As mentioned above, alligatorids lack salt glands and are therefore equipped only for fresh water. Nevertheless, they do have some capacity for temporary excursions into salt or brackish water, even the small ones, and the way they do this will be described.

SALT AND WATER BALANCE IN OTHER CROCODYLIANS

OTHER CROCODYLIDAE

Crocodylus acutus

The American crocodile (Fig. 11.22) occurs in rivers and estuaries and is seen routinely in the near-shore ocean. It has a wide distribution around the Florida Keys and many places in the Caribbean, including the less saline reaches of the hypersaline Lake Enriquillo in the Dominican Republic. The females nest on sandy beaches (Fig. 12.42) so the hatchlings must frequently cope with hyperosmotic water very early in their lives, perhaps right from the outset. In other words, *C. acutus* are found in similar habitats to those occupied by *C. porosus*. However, laboratory experiments had found that *C. acutus* did poorly in hyperosmotic water, suffering dehydration (Dunson 1970, 1982; Evans and Ellis 1977) so the significance of their salt glands, when described by Taplin *et al.* (1982), was downplayed by some authors on the basis that their secretion had been observed and measured only under methacholine stimulation. Accordingly, explanations for their survival in salt water continued to rely on them finding lenses of fresh water floating on sea water after rainstorms (Mazotti and Dunson 1984, 1989; Mazotti *et al.* (1986). However, as discussed above, our experience with *C. porosus* was that they too do poorly in laboratory exposure to salt water, mainly because of a reluctance to feed. The same is true of *C. niloticus* (Leslie and Spotila 2000), but both species can be coaxed gradually into feeding well and then maintaining their homeostasis. In short, poor performance in the laboratory is, in these crocodylians, a very poor guide to field performance. The hesitation to accept the salt glands of *C. acutus* as functionally significant continues (see below), yet measured concentrations and secretory rates in *Crocodylus acutus* under methacholine chloride stimulation are comparable with those in *C. porosus* (Taplin *et al.* 1982) (Table 11.2) and, although no histological studies have been reported, even a casual inspection of the tongue reveals salt glands every bit as conspicuous as in *C. porosus* (Fig. 11.23).

The likelihood of marine capability in *C. acutus* has been reinforced recently by a study of the diet of *C. acutus* captured on atolls and cays off the coast of Belize (Platt *et al.* 2013), where the animals have no access to fresh water during the dry season. Platt and colleagues found that hatchlings and juveniles fed mainly on insects and crustaceans, particularly crabs and some prawns. They commented that, because of the sodium load it poses, their diet is unlikely to be a source of water for rehydration and

that 'there is no *a priori* reason for dismissing an excretory role for the lingual salt glands of *C. acutus* dwelling in hyperosmotic environments of coastal Belize'. The ecological situation Platt and colleagues described in Belize and also their results are very similar to the situation reported by Taylor (1979) for hatchling and juvenile *C. porosus* at the mouth of the Liverpool River in the Northern Territory, and explored physiologically as described above. These animals were growing happily in hyperosmotic conditions on a diet of insects, crabs and prawns, with no access to fresh water in the dry season. As discussed above, such a diet can provide a sufficient water source for *C. porosus* (Taplin 1985) and, with similarly competent salt glands (Taplin *et al.* 1982), there is every reason to think that *C. acutus* can be as competent in sea water as *C. porosus*.

Crocodylus niloticus

Nile crocodiles grow nearly as large as *Crocodylus porosus* and they too occur frequently in saline waters. In St Lucia Estuary in northern Natal,

Fig. 11.22. *An American crocodile,* Crocodylus acutus, *Wasa lagoon, Ambergris Caye, Belize. (Photo Brandon Sideleau)*

Fig. 11.23. *Salt glands on the tongue of the American crocodile,* Crocodylus acutus. *Under methacholine chloride stimulation, the salty secretions issuing from the pores of* C. acutus *salt glands are comparable in maximum sodium secretory rate and average maximum osmotic concentration with those of* C. porosus. *(Photos GCG)*

where Tony Pooley did a lot of observational work (Chapter 12 and elsewhere), they are seen even in seasonally hypersaline situations. The observations made on *C. niloticus* emphasise similarity to *C. porosus* and *C. acutus* so there is no need for a comprehensive review here. The secretory rate of the salt glands is about the same as in *C. porosus* and there are typically more pores on the tongue (Taplin *et al.* 1985; Taplin 1988). Like *C. porosus*, they do not do well immediately on being taken into hyperosmotic water in captivity. However, after slow acclimation to hyperosmotic conditions over 3 months, and if they continue to feed well, both hatchling and juvenile *C. niloticus* osmoregulate effectively in both 17 and 35 ppt sea water (Leslie and Spotila 2000). Like *C. porosus*, these crocodiles avoided drinking salt water and gained their water from food. Lovely *et al.* (2007) have reported haematological and biochemical characteristics of *C. niloticus* captured in the Okavango Delta, Botswana, including some electrolyte data. This provides a set of baseline 'normal' data, useful in diagnosing pathologies.

Crocodylus johnstoni

The common names of Australia's two crocodiles; the freshwater crocodile and the estuarine (or saltwater) crocodile (or just 'freshies' and 'salties')

reflect different typical habitats and distributions and imply differences in physiology as well. However, the names are a bit misleading. Salties occur in both fresh and salt waters and, although much less commonly, freshies can also be found in estuaries, even in hyperosmotic conditions. It also turns out that their physiologies are surprisingly similar: like all Crocodylidae, *Crocodylus johnstoni* has salt glands (Taplin *et al.* 1982) and almost certainly a cloaca that can actively reabsorb sodium and water. However, unlike the three species discussed above, it is not known to occur routinely in full strength sea water or hypersaline conditions and is therefore of particular interest.

The presence of salt glands in a freshwater reptile poses a bit of a puzzle. Why have them? *Crocodylus johnstoni* occur almost entirely in freshwater habitats, typically in the more upstream rocky stretches of rivers and in billabongs, where they nest on sand banks. However, Neill (1971), in describing their range, implied that they may be found in brackish water and, indeed, surveys of the distribution of *C. porosus* across all the river systems of northern Australia by Harry Messel and his team (Messel *et al.* 1979, 1980, 1981, and see Chapter 13) reported *C. johnstoni* in brackish water in four rivers very distant from each other. These were the Victoria River in Western Australia, the Adelaide River east of

Darwin, Duck Creek on the west side of Cape York, and the Limmen Bight River in the south-west corner of the Gulf of Carpentaria, south of Roper River in the Northern Territory (Fig. 11.24). There are sure to be more occurrences. It is unlikely that this explains why *C. johnstoni* have salt glands. Much more likely, it reflects their evolutionary history (Taplin and Grigg 1989), a topic that will be taken up later.

Messel's surveys included measurements of salinity in the rivers. In the Limmen Bight River, three juvenile *C. johnstoni* (600–900 mm total length) were seen at salinities hyperosmotic to their blood: 12, 20 and 22 ppt. The survey was done in the dry season and these animals were unlikely to have had fresh water nearby. In Duck Creek, three larger individuals were seen (900–1500 mm total length) at 17, 19.5 and 26.5 ppt, also in the dry season.

These observations provoked our interest. Laurence Taplin, Lyn Beard and I went to the tidal

section of the Limmen Bight River in 1982 and again in 1984 to have a look (Taplin *et al.* 1985, 1993, 1999) (Fig. 11.25). We found freshies living in the tidal part of the estuary and the first one we saw in hyperosmotic water was at 22 ppt (630 mOsm): twice the osmotic concentration of crocodile blood. Out of 36 caught in the area subject to tidal influence, 20 were in hypoosmotic water (< 10.5 ppt) and 16 were further downstream in hyperosmotic conditions, up to ~24 ppt (680 mosm/L), which is 60% the concentration of sea water and twice the concentration of their body fluids. We sampled a similar number of *C. johnstoni* from fresh water also, well above the possible influence of salt water. We took blood and urine samples from freshly captured animals and measured salt gland secretions (Fig. 11.26).

With daytime ambient air temperatures up to the mid 40s (Celsius), on our second trip we could

Fig. 11.24. *The mouth of the Limmen Bight River, Northern Territory, Australia, where it flows into the Gulf of Carpentaria. In the dry season, the tidal influence reaches ~60 km upstream and well up side creeks too. We caught healthy, normally conditioned* Crocodylus johnstoni *living at salinities as high as 24 ppt, which is more than twice the concentration of their blood and a long way from any fresh water. (Photo GCG)*

Fig. 11.25. *While some locals look on, Laurence Taplin and Lyn Beard at our field camp on the Limmen Bight River sample blood and urine from* C. johnstoni *captured in saline conditions. With Australia's* C. porosus *being nicknamed 'salties' and* C. johnstoni *known commonly as 'freshies', we christened these crocs 'brackies'. (Photo GCG)*

Fig. 11.26. *Collecting secretions from the tongue of a* Crocodylus johnstoni *captured in the estuary of the Limmen Bight River, Northern Territory, Australia. Secretions absorbed onto filter paper for a known time were removed with forceps, stored, and analysed later. This allowed secretory rates to be calculated. For measuring their concentration, secretions were collected directly into fine glass tubes. (Photo Laurence Taplin)*

sample tongue secretions only in the early hours of the morning, when air temperatures fell low enough to have the animals out of the plastic lined holding tanks we dug into the sandy ground (Fig. 11.26). What we found left us in no doubt that *C. johnstoni* have functional salt glands and are able to regulate their body fluids very well in this potentially hostile environment: two-thirds the concentration of sea water. They maintained essentially the same water and electrolyte concentrations as their relatives in completely fresh waters in the same locale (Fig. 11.27). Their stomachs were full of crabs (Fig. 6.5). Presumably this was a good place to live,

with plenty of food. Like *C. porosus*, *C. johnstoni* would not drink 18 ppt salt water (see above and Chapter 5) but we obtained some equivocal data that suggested they would drink water at 8 ppt.

Two pieces of evidence showed they were using their salt glands. Salt gland secretory rates were much higher in the animals caught in saline water (40–50 μmol/100 $g^{0.7}$.h) than in *C. johnstoni* from fresh water (5–7 μmol/100 $g^{0.7}$.h), and the cloacal urine had low values of sodium and chloride, consistent with what would be expected if salt is being pumped across the cloacal walls and exiting

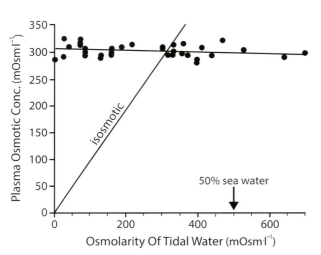

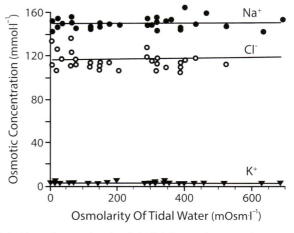

Fig. 11.27. *(Left) Plasma osmotic concentrations and (right) sodium, chloride and potassium levels in* C. johnstoni *captured across a salinity gradient in the tidal section of the Limmen Bight River in 1982 and 1984. (Adapted from Taplin* et al. *1993)*

via the salt glands. Low sodium chloride in the urine of a crocodile caught in salt water is a dead give away. However, although the concentration of the secretions from the salt glands was not much lower than in *C. porosus*, and *C. johnstoni* has a similar number of glands, the volume produced was less, so the total secretory rates were lower (Table 11.2). The salt glands in these *C. johnstoni* had less capacity than those in *C. porosus*. This suggests that these *C. johnstoni* are not well equipped for living in full strength sea water. All the animals in our study that were caught in hyperosmotic water were within reasonable distance of the hypoosmotic parts of the waterway. We do not know if they depended on getting a drink of less salty water from time to time to supplement the 'desalination' provided by the salt glands, or above what level of salinity this would become essential.

It is worth noting, though, that *C. johnstoni*, like *C. porosus* (and probably all crocodylids), have the ability to acclimatise to hyperosmotic conditions. Salt gland secretory rates averaged 37 µmol/100 g$^{0.7}$.h in the tidal fresh water, hypoosmotic section of the river and 55 µmol/100 g$^{0.7}$.h further downstream at 70% SW (24 ppt). In fresh water billabongs away from the river, the secretory rate was much lower, lower even than upstream in the fully fresh water (SAL 1) part of the river (Taplin *et al.* 1999).

Despite this, no populations of *C. johnstoni* are so far known to live in continually hyperosmotic conditions where they could not access fresher water, but the possibility cannot be discounted. There is one tantalising record: a 5 kg *C. johnstoni* was captured by Laurence Taplin in hypersaline conditions in the Albert River in north Queensland at a salinity of 43 ppt. The animal appeared healthy and was shipped to Sydney where salt gland performance was measured at higher concentrations and higher rates than we had measured in any other *C. johnstoni*. Over several months this performance declined as it adapted to freshwater conditions. Never underestimate a crocodile!

In short, physiological differences provide a partial explanation for the observed differences in the distribution of *C. johnstoni* and *C. porosus*. With less capable salt glands, *C. johnstoni* are apparently constrained from colonising river mouths and the coastline. But their salt glands do enable a few isolated populations to adopt an estuarine lifestyle, allowing them to capitalise on the food resources there. Social interactions are also important in explaining the different distributions of the two species, because *C. porosus* are typically aggressive to *C. johnstoni* (Fig. 13.7) and there may also be feeding preferences (Chapter 6). As will be discussed at the end of this chapter, the presence of salt glands in *C. johnstoni* is more likely a consequence of their ancestry than a response to current needs.

Other crocodylids

Salt glands and their secretory characteristics have been examined in small numbers of other crocodylids, including *Osteolaemus tetraspis*, *Crocodylus palustris* and *Mecistops cataphractus*, all of them kept in captivity in fresh water (Taplin *et al.* 1985). All showed secretory rates and concentrations that suggest they have, or can have, an osmoregulatory function. None of the measures is necessarily representative of field values. *Osteolaemus* is of particular interest because it is a survivor from a much earlier group of Crocodylidae, before the evolution of the modern genus *Crocodylus* (Chapter 2).

Further, even though it is common to read that most crocodiles occur only in fresh water, Taplin (1988) reviewed natural history information across the group and found that, although most are found mainly in fresh water habitats, 10 out of the 12 species then recognised are known to visit estuaries or salty situations, sometimes even the sea.

Tomistoma schlegelii, Gavialis gangeticus

The lack of information about these two species is perhaps the biggest gap in crocodylian osmoregulatory biology. Convincing data are lacking for both of them. Plasma osmotic pressure in three *Tomistoma* held captive in fresh water ranged from 291 to 300 mOsm/kg with sodium, potassium and chloride levels of 143–154, 3.2–4.6 and 64–113 mmol/L respectively: all similar to other crocodylians (Leslie and Taplin 2001). Although *Tomistoma* has convincing-looking salt glands on the tongue and a keratinised buccal cavity, the three captive specimens we viewed in Sarawak produced

Fig. 11.28. *Laurence Taplin collecting secretions from the tongue of a* Tomistoma schlegelii, *its mouth propped open and filter paper on the tongue to collect secretions. (Photo GCG)*

Fig. 11.29. *The tongue of* Tomistoma schlegelii, *with the secretory pores of lingual glands visible. The (long captive) specimens we examined showed little secretory capability. (Photo GCG)*

negligible secretions (Taplin *et al*. 1985) (Figs 11.28, 11.29). *T.* schlegelii is considered to be strictly confined to fresh water, but extinct tomistomines are known from coastal and estuarine deposits (Piras *et al.* 2007). The tongue and buccal cavity in *Gavialis* look more like a crocodile's than an alligator's and are well keratinised. However, it has only a few, minute glands along the anterior margin of the tongue adjacent to the mandibular symphysis and they showed little secretory capacity (Leslie and Taplin 2001).

ALLIGATORIDAE

Alligator mississippiensis

American alligators cannot survive indefinitely in hyperosmotic conditions (Mazotti and Dunson 1984; Lauren 1985; Lance *et al*. 2001). Like all Alligatoridae examined (e.g. Figure 11.30), they lack salt glands (Taplin *et al*. 1982). Nevertheless, alligators may be found in brackish situations bordering freshwater marshes in many places (Fig. 11.31). For example, Birkhead and Bennett (1981) reported juveniles in saline canals in southern Louisiana, apparently taking advantage of fish and shellfish at low tides when the salinity decreased. Adult alligators seen occasionally at sea (Lauren 1985, quoting *pers comm.* from Ted Joanen). Their smaller surface area/mass ratio would confer on large alligatorids a capacity to survive at sea longer than small ones. The skins of large individuals are thicker, too, which probably helps.

The American alligator in Fig. 9.2 is off the coast of Florida and Elsey (2005) reported a sighting, also with a photograph (Fig. 11.32), of a large alligator that spent several hours swimming at and around an offshore oil rig in the Gulf of Mexico, 63 km from the Louisiana coast and 56 km south of the nearest land, Marsh Island. It appeared healthy. Workers at the rig regarded this as highly unusual: one worker had previously seen an alligator just once, 16 km offshore, another had never seen one in 29 years of working on the rigs. These animals are probably strays, unlikely to survive unless they return to the coast and find fresh (or brackish) water (see also the section on *Caiman latirostris*, below). In

Fig. 11.30. *It is thought that no alligatorid (alligators and caimans) has salt glands. All of those examined so far have tongues similar to this* Caiman crocodilus, *with a smooth surface having a glutinous sheen (see also Fig. 11.38,* Caiman latirostris*). (Photo GCG)*

Fig. 11.31. *American alligator in a saltwater marsh in St Marks National Wildlife Refuge, Florida. Alligators here have access to both saline and freshwater marshes and cross back and forth readily between them. (Photo DSK)*

hurricanes such as Rita in November 2005, many alligators along the Gulf Coast of Louisiana died from incursions of salt water into the normally freshwater marshes and survivors in the months following were severely stressed with very high levels of plasma osmotic pressure, sodium and corticosterone (Lance *et al.* 2010).

Experimentally, Lauren (1985) found that mortalities began after 3 weeks when captive 380 g juveniles of ~500mm length were exposed to salinities of 15 ppt or greater. Normal plasma osmotic pressure in alligators is ~280 mOsm.kg^{-1}, equivalent to ~10 ppt. It reached nearly 400 after 4 weeks exposure to 20 ppt (Fig. 11.33).

Whereas crocodylids apparently know instinctively not to drink hyperosmotic salt water, alligatorids have to learn, and naïve alligators will drink both fresh water and hyperosmotic water indiscriminately (Jackson *et al.* 1996). There is uncertainty about the sense organs involved (see Chapter 5).

For quantitative data on the osmoregulatory biology of alligators, we have to rely mainly on studies of captive alligators because there has been little field work. In fresh water, they regulate well, with plasma osmotic pressure and electrolyte levels similar to those in crocodiles and, indeed, to other fresh water or amphibious reptiles (Mazotti and Dunson 1989).

The kidneys of alligatorids and crocodiles appear to operate quite similarly, retaining sodium, chloride and other electrolytes quite well and, when in fresh water, producing copious dilute urine low in sodium chloride and containing a substantial amount of ammonium bicarbonate. However, alligators differ greatly in their response to salt water: urinary sodium chloride increases as the kidneys strive to remove the excess salt from the blood (Lauren 1985; Pidcock *et al.* 1997). Unlike in crocodylids (above), the urine is not modified substantially in the cloaca. Experimentally, Lauren (1985) exposed juvenile

Fig. 11.32. *An American alligator swimming around an oil rig 63 km off the Louisiana coastline in ~23 m of water, far out into the Gulf of Mexico. (Photo Johnny Migues* per *Ruth Elsey)*

alligators to a range of salinities. At 5 ppt, plasma osmotic pressure and sodium chloride were the same as in fresh water. Above a salinity of 10 ppt, both rose dramatically, showing that alligators have difficulty osmoregulating even at modest salinities (Fig. 11.33). Their urine osmotic pressure increased also, but, as in all crocodylians, remained hypoosmotic to the blood. At higher salinities, urine volume was reduced and some uric acid was produced. At and above 10 ppt, they refused to feed and became progressively dehydrated.

Caiman latirostris

A letter in 1991 from Bill Magnusson in Manaus, Brazil, told of a population of broad-snouted caiman, *Caiman latirostris*, living in a very salty estuary on Ihla do Cardoso, off the coast to the south of Sao Paulo (Fig. 11.34). Bill knew that alligatorids lack salt glands and recognised that this population would be interesting so he suggested it was worth a look. By extraordinarily good fortune, Tim

Moulton, another biologist friend from University of Sydney days, was living on the island with his family. This solved language and logistic problems simultaneously.

Lyn Beard and I left Brisbane in May 1992 for Ihla do Cardoso, laden with everything we needed for the work except a boat, a 12V battery, an outboard motor and fuel. In the complex of mangrove-lined estuaries at the western end of the island, we caught 12 *Caiman latirostris*, all > 2 kg, eight of which were in salinities of 15–24 ppt, which would be hyperosmotic to the blood throughout the tidal cycle (Figs 11.35, 11.36) Grigg *et al*. 1998. They were all found, however, in the upstream reaches of the estuaries and fresh water streams were reasonably close by. We also caught several much smaller individuals in a fresh water river at the eastern end of the island. The situation was reminiscent of *Crocodylus johnstoni* in the Limmen Bight River (see above). As in *Crocodylus johnstoni*, their stomachs were full of crabs, which were numerous.

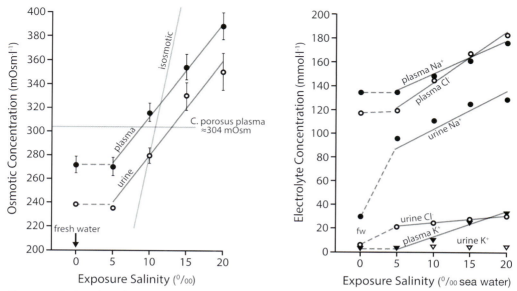

Fig. 11.33. *Plasma and urine electrolytes and osmotic pressure in juvenile* Alligator mississippiensis, *~400 g, held experimentally for 4 weeks at a range of salinities. Unlike the Crocodylidae studied so far, alligators in salt water cannot maintain plasma homeostasis. (Adapted from Lauren 1985)*

Fig. 11.34. *The estuary of the Jacareú River, Ihla do Cardoso, southern Brazil. The estuary winds sinuously through the mangroves in the foreground. Across the strait are other islands and in the hazy distance the Brazilian mainland. (Photo GCG)*

Fig. 11.35. *In the mangrove-lined estuary of the Jacareú River, Ihla do Cardoso, southern Brazil, Tim Moulton tags a capture site so we can return the animal to the same spot later. (Photo GCG)*

Fig. 11.37. *Drawing a blood sample from the cervical sinus of the largest* Caiman latirostris *we captured, 15.2 kg. It was harpooned by Tereca Queirol Melo well down in the estuary. The plasma osmotic pressure of 18 animals captured across the salinity spectrum from fresh water to 24 ppt (hyperosmotic to blood) averaged 293 mOsm, similar to other crocodylians, and there was no correlation with salinity. (Photo Tereca Queirol Melo)*

Blood samples were taken from the cervical sinus (Fig. 11.37; Olson *et al.* 1975), and we quickly found typical crocodylian plasma osmotic pressures, tightly regulated around 293 mOsm/kg and not at all correlated with salinity at the capture sites. This too was reminiscent of the data from *Crocodylus johnstoni* in hyperosmotic sites in the Limmen Bight River. But here the similarity ended, because it was obvious that *Caiman latirostris* have the less keratinised tongue of an alligatorid, no salt glands (Fig. 11.38) and, most tellingly, the urine was both copious and dilute (Fig. 11.39). Very interestingly, and unlike naïve alligatorids, they would not drink sea water when dehydrated, but drank fresh water readily when it was offered.

If there were any operational salt glands, we would expect the urine from a crocodylian caught in salty water to be less copious, rich in suspended solids, and very low in sodium, as it would be in a *C. porosus*. We had no flame photometer to measure sodium chloride until we got the samples back to the laboratory in Brisbane, but the human tongue is a good field substitute and, without doubt, the urine samples we were collecting from *Caiman latirostris* caught in the estuary were salty (Fig. 11.40). This confirmed our conclusion from visual observations of the tongue: like other

Fig. 11.36. *A broad-snouted caiman,* Caiman latirostris, *in hypersaline water in the estuary of the Jacareú River, Ihla do Cardoso, southern Brazil. (Photo GCG)*

Fig. 11.38. *The tongue of a broad-snouted caiman,* Caiman latirostris, *showing its typical alligatorid features. Compare with Fig. 11.30. (Photo GCG)*

Fig. 11.39. *Lyn Beard taking the clear, copious urine from a broad-snouted caiman captured in salty water in the estuary of the Rio Jacareú, Ilha do Cardoso, Brazil. (Photo Tereca Queirol Melo)*

Alligatoridae, *C. latirostris* has no salt glands. *Caiman latirostris* was behaving in salt water like an alligator.

How are these caiman maintaining their homeostasis at plasma concentrations well below that of the outside water? The explanation seems to be that they excrete the excess salts in the urine, using water gained by visits to a source of fresh (or hypoosmotic) drinking water. As it happened, we sighted two of our study animals again, just by chance, 2–3 days after their release. Each had travelled at least 600 m, sufficient to put the fresh water streams at the heads of the estuaries within easy reach. Periodic visits by these two 15 kg caiman were apparently sufficient for them to maintain their plasma homeostasis. It would be very interesting to look at movement patterns of *Caiman latirostris* (or *A. mississippiensis*, see above) in an estuarine situation such as this. My prediction is that they would visit fresh or dilute water periodically, the larger ones less frequently than the smaller ones.

What are the *Caiman latirostris* doing in the estuary anyway, with all those fresh water streams on the island? Demographically, it is the juveniles and sub-adults that are in the estuary, with the adults being found in pools up in the fresh water streams, which they may be dominating. Also in the streams are the hatchlings and very small juveniles. We postulated that, if adults are dominating the small streams because they provide the suitable breeding habitat, the estuary may provide good foraging habitat for the displaced and/or sexually immature individuals and, when they are ready, they could challenge for a breeding spot upstream.

LIVING OUT OF WATER, 'AESTIVATION'

We are used to thinking about crocodylians living around water, but many of them spend quite a lot of time away from it, and some have to cope without it for considerable periods in habitats subject to long dry seasons. Two primarily rainforest genera, *Osteolaemus* and *Paleosuchus*, often spend the day time in burrows (Fig. 11.41) and forage considerable distances over land at night.

In the dry season, the western species of *Osteolaemus* copes without water for long periods, during which they continue to feed, becoming terrestrial predators (Matt Shirley *pers. comm.*). This will provide a limited source of water. Almost certainly, they will also have the capability to remain inactive in a burrow for long periods if necessary. These are obviously very interesting and unusual crocodylians and we can look forward to learning more about their lifestyle.

Dry times can occur in cool or warm seasons, with water loss by evaporation exacerbated in the warm. The pros and cons of using the terms 'hibernation' and 'aestivation' was discussed in Chapter 10. Regardless of their limitations, these behavioural terms are widespread in reptilian literature and spending periods of time in a refuge of some sort while coping with aridity is often called aestivation. This is in tune with that term's application to desert

Fig. 11.40. *In the absence of a flame photometer to measure sodium chloride, the human tongue is a good field substitute. The urine of* Caiman latirostris *captured from the hyperosmotic environment of the estuary was salty, as would be expected in a crocodylian lacking salt glands. (Photos Lyn Beard)*

frogs and African lungfish, although in these it is accomplished with significant physiological change including down-regulating metabolism. In reviewing the occurrence of hibernation and aestivation in crocodylians, Taplin (1988) tabulated occurrences in nine species. Because of their size and reptilian physiology, most crocodylians would seem to be capable of coping with harsh conditions for long periods by remaining inactive in a burrow, cavern or similar shelter. One obvious question is to wonder whether or not this is accompanied by

a reduction in metabolism, the other is about the water economy.

The physiology of coping through long dry seasons without access to water has been studied in 'freshies': Australia's freshwater crocodile, *C. johnstoni*. It followed observations by Bryan Walsh – an officer with the (then) Conservation Commission of the Northern Territory – of several *C. johnstoni* in a drying pool of Saunders Creek, 200 km south of Darwin. Once the pool had dried up, they were found in three low-ceilinged caverns

Fig. 11.41. *Two burrows of* Osteolaemus *sp. cf.* tetraspis *in Guinea. These animals become terrestrial predators when there is no water nearby in the dry season. Note the snare set around the mouth of the burrow on the right. (Photo and information Matthew H. Shirley)*

undercut along the bank of the creek (Walsh 1989). There were ~17 individuals, both males and females, over a size range from 2 to 30 kg. The nearest permanent water was 15 km downstream. His agency had received reports previously of 'freshies' in drying waterholes and had sometimes relocated them to water. Walsh visited the caverns periodically. On one visit he noted tracks made by one of them to and from a small pool that had resulted from a rainstorm. As the weeks passed he came to realise that these animals were not at risk, but were merely waiting out the dry months. When the rains came, the caverns were quickly vacated. He recommended against any future relocations. Keith Christian and Rod Kennett from Charles Darwin University followed up Walsh's observations with a physiological study (Fig. 11.42) (Kennett and Christian 1993; Christian *et al.* 1996). They found that animals spent 3–4 months of the year in these dry season refuges without access to water. The animals lost ~13% of their body weight over 3 months, while body water remained about the same proportion of body mass ($\approx 75\%$). Water fluxes in late aestivation were ~25% of those for active animals in water, with effluxes slightly higher than influx throughout, reflecting the decrease in body water. After 3 months, plasma osmotic pressure was only slightly higher (a mean 310, cf. 297 normally), as were both plasma sodium and chloride. Cloacal urine osmotic pressure increased with dehydration, due particularly to a 10-fold increase in potassium. Dissolved nitrogenous wastes were not more concentrated, but uric acid solids increased, so that by November the cloaca had mostly white paste and little liquid. These changes are similar to what is seen in crocodylids living in hyperosmotic water (see above). Christian and colleagues noted that *C. johnstoni* have lingual salt glands, but noted that it is unlikely they could have played any role without water being available to rinse away the excreted salt. Although I agree with this, urinary sodium and chloride values were comparatively low and this, combined with the large increase in potassium is strikingly similar to freshies living in brackish water (Taplin *et al.* 1999) or naïve *C. niloticus* freshly exposed to brackish water (Taplin and Loveridge 1988). Taplin

Fig. 11.42. *Taking full advantage of the local vegetation, Rod Kennett and Agio Pereira weigh a freshwater or Johnston River crocodile,* C. johnstoni, *that has been removed from the undercut cavern behind them. It and others were spending the months of the dry season holed up in low-ceilinged caverns along the dry creek bed, with the nearest permanent water 15 km away. On average, the animals lost ~13% of their mass over 3 months, well within their capacity to tolerate. There was no evidence of metabolic rate being reduced below normal resting levels and no ill effects were reported. (Photo Keith Christian)*

and colleagues suggested a reciprocal relationship between potassium and sodium, with salt gland involvement implied, so some role for salt glands in coping with months spent without water cannot be dismissed completely.

Unlike an alligator spending the winter in a burrow or cave to 'hibernate' in temperate latitudes, and depending upon when they enter the retreat, *C. johnstoni* may not get a significant Q_{10}-driven decrease in metabolic rate (Chapter 10). During the Christian and colleagues' study, in that tropical climate, mean body temperatures increased from 22°C in August to 28°C in November with the gradual seasonal change. Christian *et al.* (1996) also assessed CO_2 production isotopically, as a measure of field metabolic rate (FMR) and calculated a rate of 26 kJ/kg/day. Comparing this with resting metabolic rates for other crocodylians, they concluded that the animals relaxing in caverns at Saunders Creek had not down-regulated their metabolic rate. They concluded that holing up in a cave or burrow 'is a viable response to the extreme conditions of the dry season in tropical Australia, and given an adequate refuge, crocodiles can survive many months in aestivation'.

SALT GLANDS AND IMPLICATIONS FOR CROCODYLIAN EVOLUTIONARY HISTORY

The extant Crocodylidae are all of comparatively recent origin (Miocene) and closely related (Chapter 2) and, because they all have salt glands, it has been suggested that they may be derived from marine-capable ancestor/s akin to *Crocodylus porosus* (Taplin and Grigg 1989). There need be little doubt that salt glands are plesiomorphic in Crocodylidae, but too little is known about the osmoregulatory biology of *Tomistoma* and *Gavialis* to be bold about how far back that evidence of marine ancestry can be taken within eusuchians. The apparent presence of salt glands in *Tomistoma* and the buccal morphology of it and *Gavialis* led to Taplin and Grigg (1989) postulating that they shared the same marine ancestry as crocodylids. The overall thrust of that paper was that the physiology of the Crocodylidae was amenable to explanations of biogeography involving transoceanic dispersal. That biogeographical proposition holds regardless of the phylogenetic position of *Tomistoma* in relation to *Gavialis*.

The problem of how far back the marine ancestry is shared among extant crocodylians is clouded by several uncertainties. One derives from uncertainty about the affinity of *Tomistoma* (see Chapter 2). If it is a crocodylid, a marine ancestry would be easier to accept because it comes from a diverse lineage of extinct tomistomines, many of which were in coastal and estuarine deposits (Piras *et al.* 2007). If *Tomistoma* is a gavialid, on the other hand, and if *Gavialis* has a marine ancestry, that would imply that alligatorids too have a marine ancestry and have become secondarily restricted to fresh water (Fig. 2.3).

An intriguing approach to the question was taken recently by Patrick Wheatley at the University of California Santa Cruz. He was able to calibrate carbon- and oxygen-stable isotope ratios in the carbonate portion of tooth bioapatite of extant crocodylians against their marine resource use (prey and water) and then apply that to teeth of fossil crocodylomorphs (Wheatley 2010; Wheatley *et al.* 2012). The results implied marine capability in fossil Crocodyloidea, Tomistominae, Gavialoidea and the dyrosaurid *Hypsosaurus*, and ambiguous results in *Deinosuchus*, a fossil alligatoroid. He concluded that the origin of a salt water tolerance was 'at least as old as the common ancestor of Crocodylia + Dyrosauridae (Cretaceous) and perhaps as old as the common ancestor of Crocodylia + Metriorhynchidae (Jurassic)' depending upon which phylogeny is accepted. (Dyrosaurs were large marine neosuchian crocodylomorphs that appeared in the Cretaceous, survived the K–Pg extinction event and went extinct in the Eocene.) That is, he concluded that marine capability is plesiomorphic to Eusuchia, including Alligatoridae which, he suggested, lost it as long ago as the Mesozoic. His argument would be stronger if we knew that the Dyrosauria had lingual salt glands and is weaker because we know that the Metriorhynchidae (Thallatosuchia) did not. The (presumed) salt glands in the metriorhynchids *Metriorhynchus superciliosus* were in front of their eyes: antorbital salt glands (Gandola *et al.* 2006) (Figs 2.16, 2.17). Fernández and Gasparini (2008) and Fernández and Herrera (2009) described similar traces of salt glands in an Argentinian metriorhynchid, *Geosaurus araucanensis*, and concluded from their size that, like marine turtles, they would have had sufficient excretory capacity to maintain homeostasis by drinking sea water.

Salt glands have arisen many times across the reptiles and birds, and various glands in the head have been co-opted as salt glands from time to time: nasal, lachrymal, antorbital, sub-lingual and lingual. It is too early to conclude that the Alligatoridae too had marine ancestry, based on the oxygen isotope evidence. The Crocodylia have substantial terrestrial ancestry and kidneys to match, so, until there is more compelling information, it seems more parsimonious to note that, at least, we can be sure that lingual salt glands are plesiomorphic to Crocodylidae. To resolve the question, we probably need not only more data from *Tomistoma* and *Gavialis*, but more species of tomistomines and gavialids as well. Bring it on!

The situation we described in *Caiman latirostris*, with breeding focussed in fresh water and a reliance on adjacent estuaries as feeding grounds, could provide hints about how salt glands evolved. It is easy to see that selection pressures could favour increased capacity for salt excretion in such situations.

Finally, it is worth recalling that Peaker and Linzell (1975), in their very substantial review of salt glands, lamenting how little was then known about crocs, observed that 'no really large salt-adapted specimens have been studied and it is hoped that some brave person will try to collect tears from a twenty foot Estuarine Crocodile'. Ironically, but for very good reasons, it was the small ones that provided the answers.

REFERENCES

Bennett JM, Taplin LE, Grigg GC (1986) Sea water drinking as a homeostatic response to dehydration in hatchling loggerhead turtles, *Caretta caretta*. *Comparative Biochemistry and Physiology* **83A**(3), 507–513.

Birkhead WS, Bennett CR (1981) Observations of a small population of estuarine-inhabiting alligators near Southport, North Carolina. *Brimleyana* **6**, 111–117.

Chiasson RB (1962) *Laboratory anatomy of the alligator.* WC Brown, Dubuque, Iowa.

Christian KA, Green B, Kennett R (1996) Some physiological consequences of aestivation by freshwater crocodiles, *Crocodylus johnstoni*. *Journal of Herpetology* **30**, 1–9.

Coulson RA, Hernandez T (1964) *Biochemistry of the Alligator: A Study of Metabolism in Slow Motion.* Louisiana State University Press, Baton Rouge.

Coulson RA, Hernandez T (1983) *Alligator Metabolism: Studies on Chemical Reactions in vivo.* Pergamon Press, London.

Cramp RL, Hudson NJ, Holmberg A, Holmgren S, Franklin CE (2007) The effects of saltwater acclimation on neurotransmitters in the lingual salt glands of the estuarine crocodile, *Crocodylus porosus. Regulatory Peptides* **140**, 55–64.

Cramp RL, Meyer EA, Sparks N, Franklin CE (2008) Functional and morphological plasticity of crocodile (*Crocodylus porosus*) salt glands. *Journal of Experimental Biology* **211**, 1482–1489.

Davis LE, Schmidt-Nielsen B (1967) Ultrastructure of crocodile kidney (*Crocodylus acutus*) with special reference to electrolyte and fluid transport. *Journal of Morphology* **121**, 255–276.

Dunson WA (1970) Some aspects of electrolyte and water balance in three estuarine reptiles, the Diamondback Terrapin, American and 'saltwater' crocodiles. *Comparative Biochemistry and Physiology* **32**, 161–174.

Dunson WA (1982) Salinity relations of crocodiles in Florida Bay. *Copeia* **1982**, 374–385.

Dunson WA (1976) Salt glands in reptiles. In *Biology of the Reptilia. Volume 5.* (Eds C Gans and WR Dawson) pp. 413–445. Academic Press, London.

Elsey RM (2005) Unusual offshore occurrence of an American Alligator. *Southeastern Naturalist (Steuben, ME)* **4**(3), 533–536.

Evans DH, Ellis TM (1977) Sodium balance in the hatchling American crocodile, *Crocodylus acutus. Comparative Biochemistry and Physiology. A. Comparative Physiology* **58**, 159–162.

Fernández M, Gasparini Z (2008) Salt glands in the Jurassic metriorhynchid *Geosaurus*: implications for the evolution of osmoregulation in Mesozoic marine crocodyliforms. *Naturwissenschaften* **95**, 79–84.

Fernández MS, Herrera Y (2009) Paranasal sinus system of *Geosaurus araucanensis* and the homology of the antorbital fenestra of metriorhynchids (Thalattosuchia: Crocodylomorpha) *Journal of Vertebrate Paleontology* **29**, 702–714.

Franklin CE, Grigg GC (1993) Increased vascularity of the lingual salt glands of the estuarine crocodile, *Crocodylus porosus* kept in hyperosmotic salinity. *Journal of Morphology* **218**, 143–151.

Franklin CE, Taylor G, Cramp RL (2005) Cholinergic and adrenergic innervation of lingual salt glands of the estuarine crocodile, *Crocodylus porosus. Australian Journal of Zoology* **53**, 345–351.

Gandola R, Buffetaut E, Monaghan N, Dyke G (2006) Salt glands in the fossil crocodile *Metriorhynchus. Journal of Vertebrate Paleontology* **26**, 1009–1010.

Grigg GC (1978) Metabolic rate, RQ and Q_{10} in *Crocodylus porosus* and some generalisations about low RQ in reptiles. *Physiological Zoology* **51**, 354–360.

Grigg GC (1981) Plasma homeostasis and cloacal urine composition in *Crocodylus porosus* caught along a salinity gradient. *Journal of Comparative Physiology* **144**, 261–270.

Grigg GC (1993) Twenty years of wondering and worrying about how crocodiles live in salt water. In *Herpetology in Australia: A Diverse Discipline*. (Eds D Lunney and D Ayers) pp. 265–277. Royal Zoological Society of New South Wales and Surrey Beatty & Sons, Sydney. <http://espace.library.uq.edu.au/view/UQ:9724>

Grigg GC, Gans C (1993) Crocodilia: morphology and physiology. In *Fauna of Australia. Volume 2A. Amphibia and Reptilia.* (Eds Glasby CJ, Ross GJB and PL Beesley) pp. 326–336. Australian Government Publishing Service, Canberra.

Grigg GC, Taplin LE, Harlow P, Wright J (1980) Survival and growth of hatchling *Crocodylus porosus* in salt water without access to fresh drinking water. *Oecologia* **47**, 264–266.

Grigg GC, Taplin LE, Green B, Harlow P (1986) Sodium and water fluxes in free-living *Crocodylus porosus* in marine and brackish conditions. *Physiological Zoology* **59**(2), 240–253.

Grigg GC, Beard LA, Moulton T, Queirol Melo MT, Taplin LE (1998) Osmoregulation by the broad-snouted caiman, *Caiman latirostris*, in estuarine habitat in southern Brazil. *Journal of Comparative Physiology. B, Biochemical, Systemic, and Environmental Physiology* **168**, 445–452.

Huber C (1917) On the morphology of the renal tubules of vertebrates. *The Anatomical Record* **13**, 305–339.

Jackson K, Brooks DR (2007) Do crocodiles co-opt their sense of "touch" to "taste"? A possible new type of vertebrate sense organ. *Amphibia-Reptilia* **28**, 277–285.

Jackson K, Butler DG, Brooks DR (1996) Habitat and phylogeny influence salinity discrimination in crocodilians: implications for osmoregulatory physiology and historical biogeography. *Biological Journal of the Linnean Society. Linnean Society of London* **58**, 371–383.

Kennett R, Christian K (1993). Aestivation by freshwater crocodiles (*Crocodylus johnstoni*) occupying a seasonally ephemeral creek in tropical Australia. In *Herpetology in Australia: A Diverse Discipline*. (Eds D Lunney and D Ayers) pp. 315–319. Royal Zoological Society of New South Wales and Surrey Beatty & Sons, Sydney.

Kuchel LJ, Franklin CE (1998) Kidney and cloaca function in the estuarine crocodile (*Crocodylus porosus*) at different salinities: evidence for solute-linked water uptake. *Comparative Biochemistry and Physiology A* **119**, 825–831.

Kuchel LJ, Franklin CE (2000) Morphology of the cloaca in the estuarine crocodile, *Crocodylus porosus*, and its plastic response to salinity. *Journal of Morphology* **245**, 168–176.

Kuchel LJ, Franklin CE (2001) Osmoregulatory plasticity in a crocodile: the physiological and morphological responses of the estuarine crocodile *Crocodylus porosus* acclimated to fresh water and salt water. In *Crocodilian Biology and Evolution*. (Eds GC Grigg, F Seebacher and CE Franklin) pp. 280–285. Surrey Beatty & Sons, Sydney.

Lance VA, Morici LA, Elsey RM (2001) Physiology and endocrinology of stress in alligators. In *Crocodilian Biology and Evolution*. (Eds GC Grigg, F Seebacher and CE Franklin) pp. 327–340. Surrey Beatty & Sons, Sydney.

Lance VA, Elsey RM, Butterstein G, Trosclair PL III, Merchant M (2010) The effects of Hurricane Rita and subsequent drought on alligators in southwest Louisiana. *Journal of Experimental Zoology* **313A**, 106–113.

Lauren DJ (1985) The effect of chronic saline exposure on the electrolyte balance, nitrogen metabolism, and corticosterone titer in the American alligator, *Alligator mississipiensis*. *Comparative Biochemistry and Physiology* **81A**(2), 217–223.

Leitch DB, Catania KC (2012) Structure, innervation and response properties of integumentary sensory organs in crocodilians. *The Journal of Experimental Biology* **215**, 4217–4230.

Leslie AJ, Spotila JR (2000) Osmoregulation of the Nile crocodile, *Crocodylus niloticus*, in Lake St. Lucia, Kwazulu/Natal, South Africa. *Comparative Biochemistry and Physiology A – Molecular & Integrative Physiology* **126**(3), 351–365.

Leslie AJ, Taplin LE (2001) Recent developments in osmoregulation in crocodilians. In *Crocodilian Biology and Evolution*. (Eds GC Grigg, F Seebacher and CE Franklin) pp. 265–279. Surrey Beatty & Sons, Sydney.

Lovely CJ, Pittman JM, Leslie AJ (2007) Normal haematology and blood biochemistry of wild Nile crocodiles (*Crocodylus niloticus*) in the Okavango Delta, Botswana. *Journal of the South African Veterinary Association* **78**(3), 137–144.

Magnusson WE (1978) Nesting ecology of *Crocodylus porosus*, Schneider, in Arnhem Land, Australia. PhD thesis, University of Sydney.

Mazotti FJ, Dunson WA (1984) Adaptations of *Crocodylus acutus* and *Alligator* for life in saline water. *Comparative Biochemistry and Physiology A* 79, 641–646.

Mazotti FJ, Dunson WA (1989) Osmoregulation in crocodilians. *American Zoologist* **29**(3), 903–920.

Mazotti FJ, Bohnsack B, McMahon MP, Wilcox R (1986) Field and laboratory observations on the effects of high temperature and salinity on hatchling *Crocodylus acutus*. *Herpetologica* **42**(2), 191–196.

Messel H, Gans C, Wells AG, Green WJ, Vorlicek GC, Brennan KG (1979) *The Victoria and Fitzmaurice river Systems. Monograph 2. Surveys of Tidal Waterways in the Northern Territory of Australia and their Crocodile Populations*. Pergamon Press, Sydney.

Messel H, Vorlicek GC, Wells AG, Green WJ, Johnson A (1980) *Tidal Waterways on the South-Western Coast of the Gulf of Carpentaria. Monograph 12. Surveys of Tidal Waterways in the Northern Territory of Australia and their Crocodile Populations*. Pergamon Press, Sydney.

Messel H, Vorlicek GC, Wells AG, Green WJ, Curtis HS, Roff CRR, *et al.* (1981) *Southwestern Cape York Peninsula. Northern Cape York Peninsula. Monograph 16. Surveys of Tidal Waterways on Cape York Peninsula, Queensland, Australia and their Crocodile Populations*. Pergamon Press, Sydney)

Moore BC, Hyndman KA, Cox A, Lawler A, Mathavan K, Guillette LJ (2009) Morphology and histochemistry of juvenile American alligator (*Alligator mississippiensis*) nephrons. *The Anatomical Record: Advances in Integrative Anatomy and Evolutionary Biology* **292**(10), 1670–1676.

Neill WT (1971). *The Last of the Ruling Reptiles: Alligators, Crocodiles and their Kin*. Columbia University Press, New York.

Olson GA, Hessler JR, Faith RT (1975) Techniques for blood collection and intravascular infusion of reptiles. *Laboratory Animal Science* **25**, 783–786.

Owen R (1866) *On the Anatomy of Vertebrates. Volume I. Fishes and Reptiles*. Longman, London.

Peaker M, Linzell JL (1975) *Salt Glands in Birds and Reptiles. Monographs of the Physiological Society 32*. Cambridge University Press, Cambridge, UK.

Pidcock S, Taplin LE, Grigg GC (1997) Differences in renal-cloacal function between *Crocodylus porosus* and *Alligator mississippiensis* have implications for crocodilian evolution. *Journal of Comparative Physiology. B, Biochemical, Systemic, and Environmental Physiology* **167**(2), 153–158.

Piras P, Delfino M, Del Favero L, Kotsakis T (2007) Phylogenetic position of the crocodylian *Megadontosuchus arduini* (de Zigno, 1880) and tomistomine palaeobiogeography. *Acta Palaeontologica Polonica* **52**, 315–328.

Platt SG, Thorbjarnarson JB, Rainwater TR, Martin DR (2013) Diet of the American crocodile (*Crocodylus acutus*) in marine environments of coastal Belize. *Journal of Herpetology* **47**, 1–10.

Potts WTW, Parry G (1964) *Osmotic and Ionic Regulation in Animals*. Pergamon Press, Oxford, UK.

Read MA, Grigg GC, Irwin SR, Shanahan D, Franklin CE (2007) Satellite tracking reveals long distance coastal travel and homing by translocated estuarine crocodiles, *Crocodylus porosus*. *PLoS ONE* **9**, 1–5.

Schmidt-Nielsen K (1959) Salt glands. *Scientific American* **200**, 109–116.

Schmidt-Nielsen K, Fänge R (1958) Salt glands in marine reptiles. *Nature* **182**, 783–785.

Schmidt-Nielsen K, Jörgensen CB, Osaki H (1957) Secretion of hypertonic solutions in marine birds. *Federation Proceedings* **16**, 113–114.

Schmidt-Nielsen K, Jörgensen CB, Osaki H (1958) Extrarenal salt excretion in birds. *The American Journal of Physiology* **193**, 101–107.

Shuttleworth TJ, Hildebrandt JP (1999) Vertebrate salt glands: short- and long-term regulation of function. *Journal of Experimental Zoology* **283**, 689–701.

Taplin LE (1984a) Homeostasis of plasma electrolytes, sodium and water pools in the estuarine crocodile, *Crocodylus porosus*, from fresh, saline and hypersaline waters. *Oecologia* **63**, 63–70.

Taplin LE (1984b) Drinking of fresh water but not seawater by the estuarine crocodile, *Crocodylus porosus*. *Comparative Biochemistry and Physiology* **77A**, 763–767.

Taplin LE (1985) Sodium and water budgets of the fasted estuarine crocodile, *Crocodylus porosus*, in sea water. *Journal of Comparative Physiology* **155B**, 501–513.

Taplin LE (1988) Osmoregulation in crocodilians. *Biological Reviews of the Cambridge Philosophical Society* **63**, 333–377.

Taplin LE, Grigg GC (1981) Salt glands in the tongue of the estuarine crocodile. *Science* **212**, 1045–1047.

Taplin LE, Grigg GC (1989) Historical zoogeography of the eusuchian crocodilians: a physiological perspective. *American Zoologist* **29**, 885–901.

Taplin LE, Loveridge JP (1988) Nile crocodiles, *Crocodylus niloticus*, and estuarine crocodiles, *Crocodylus porosus*, show similar osmoregulatory responses on exposure to seawater. *Comparative Biochemistry and Physiology A.* **89**, 443–448.

Taplin LE, Grigg GC, Harlow P, Ellis TM, Dunson WA (1982) Lingual salt glands in *Crocodylus acutus* and *C. johnstoni* and their absence from *Alligator mississipiensis* and *Caiman crocodilus*. *Journal of Comparative Physiology* **149**, 43–47.

Taplin LE, Grigg GC, Beard LA (1985) Salt gland function in fresh water crocodiles: evidence for a marine phase in eusuchian evolution? In *Biology of Australasian Frogs and Reptiles* (Eds GC Grigg, R Shine and H Ehmann) pp. 403–410. Surrey Beatty & Sons, Sydney.

Taplin LE, Grigg GC, Beard LA (1993) Osmoregulation of the Australian freshwater crocodile, *Crocodylus johnstoni*, in fresh and saline waters. *Journal of Comparative Physiology. B, Biochemical, Systemic, and Environmental Physiology* **163**(1), 70–77.

Taplin LE, Grigg GC, Beard LA, Pulsford T (1999) Osmoregulatory mechanisms of the Australian freshwater crocodile, *Crocodylus johnstoni*, in freshwater and estuarine habitats. *Journal of Comparative Physiology. B, Biochemical, Systemic, and Environmental Physiology* **169**, 215–223.

Taylor JA (1979) The foods and feeding habits of sub-adult *Crocodylus porosus* Scheider in northern Australia. *Australian Wildlife Research* **6**, 347–359.

Taylor GC, Franklin CE, Grigg GC (1995) Salt loading stimulates secretion by the lingual salt glands in *Crocodylus porosus*. *Journal of Experimental Zoology* **272**, 490–495.

Ventura SC, Northrup TE, Schneider G, Cohen JJ, Garella S (1989) Transport and histochemical-studies of bicarbonate handling by the alligator kidney. *The American Journal of Physiology* **256**, F239–F245.

Von Bayern LF (1884) *Zur Anatomie der Zunge.* Reidel, Munich, Germany

Walsh B (1989) Aestivation in the freshwater crocodile? *Australian Zoologist* **25**, 68–70.

Weldon PJ, Ferguson MWJ (1993) Chemoreception in crocodilians: anatomy, natural history and empirical results. *Brain, Behavior and Evolution* **41**, 239–245.

Wheatley PV 2010 Understanding salt water tolerance and marine resource use in the

Crocodylia: a stable isotope approach. PhD thesis, University of California, Santa Cruz.

Wheatley PV, Peckham H, Newsome SD, Koch PL (2012) Marine resource use in the American crocodile (*Crocodylus acutus*) in Southern Florida. *Marine Ecology Progress Series* **447**, 211–229.

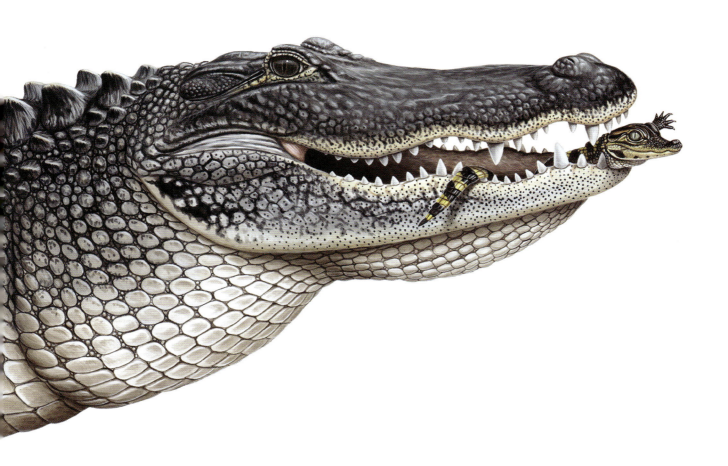

12

REPRODUCTION

For a number of years Lyn Beard and I and other colleagues visited the Edward River Crocodile Farm at Pormpuraaw on the west side of Cape York, working particularly on thermal relations of C. porosus *(Chapter 10), changes in the embryo's blood respiratory properties during development (Chapter 7) and the environment within nests (this chapter). Throughout almost all of that time, the naturalistic 20 ha breeding lagoon was presided over by the dominant male: a conspicuously large 5 m male named Hymn (Fig. 1.37). Early one breeding season, we watched him once again see off the next biggest male, Toothy, in a late afternoon aggressive encounter, responding to Toothy's open mouthed advances with menacing-looking blowing of bubbles from his nostrils and, when really provoked, by impressive and explosive sideways chops of his head (Fig. 12.2 and http:// crocodilian.com/books/grigg-kirshner). A couple of days later, and from a distance, we watched Hymn mate with a female much smaller than he was, and in quite shallow water. The female would swim away from him repeatedly, then stop and wait for him to catch her until, in the fading twilight, they mated (video at http://crocodilian.com/books/grigg-kirshner). Subsequently, Nancy Fitzsimmons and I had a student, Mona Lisa Jamerlan, who analysed genetic material from every male she could get samples from (not a simple matter), including Hymn, and compared it with that from eggs that had been removed from nests and taken to the farm's incubator. She found little sign of Hymn's genetic signature. Either he* was too old or, as in many birds and mammals, the dominant male does not get most of the matings. Mona Lisa found also that each clutch of eggs often involved more than one father: that is, multiple paternity. Crocodiles enjoy interesting reproductive lives, and that will be the subject of this chapter.*

INTRODUCTION: THEY'RE MORE LIKE BIRDS …

All crocodylians have internal fertilisation, lay eggs and show complex reproductive behaviour, including maternal care. Their reproductive anatomy and physiology are much more like birds than other reptiles and so is much of their behaviour. The females make nests, lay eggs and often defend them (Fig. 12.1). The hatchlings call from within the egg and the mother helps them hatch, takes the hatchlings to the water where they form a crèche. She continues to show maternal care, usually for a few weeks but sometimes much longer. Nowhere else among reptiles is parental care so elaborate (Shine 1988). Some reptiles show very limited care, but crocodylians are the only group in which parental care is characteristic. Elaborate parental care is, of course, a characteristic of birds too. Male crocodylians always grow much larger than females, and gender is determined not by chromosomes but by the temperatures they experience during embryonic

An American alligator, Alligator mississippiensis, *with gular pouch distended, carries one of her hatchlings. (Illustration DSK, courtesy Weldon Owen Publishing)*

development in the nest (a big difference from birds, in which gender is determined genotypically). Males are often aggressive towards each other and there is a complex social hierarchy, which is established by aggression and by elaborate social signalling: visually, vocally and with pheromones (Garrick and Lang 1977; Garrick *et al.* 1978; Vliet 1987, 1989, 2001 and Chapter 3). Large males are dominant and territorial and, although large males have been thought to fertilise most of the females in a particular area, multiple paternity in clutches of eggs is common. Most observations have been made on a relatively small number of species and mostly in captivity, so the relevance of observations to other species and behaviour in the wild can often only be inferred. Most work has been done on *A. mississippiensis*, *Caiman crocodilus*, *Crocodylus niloticus* and *C. porosus* but enough elements of similar behaviour have been recorded in other species to suggest that all crocodylians are broadly similar in their reproductive biology. Because alligatorids and crocodylids have been apart since the Mesozoic, this similarity implies that

reproductive biology in crocodylians has been very conservative and is very old. Its similarity to birds has implications for interpreting dinosaur biology: a recurrent theme in this book.

The large size and difficulty of handling sexually mature individuals, particularly males, combined with the fact that most animals slaughtered at crocodile farms are juveniles, assure that both heroic and Herculean efforts will be needed to answer many of the remaining questions about the physiology of crocodylian reproduction.

CROCODYLIAN REPRODUCTIVE LIVES

SEX, MATURITY AND REPRODUCTIVE LONGEVITY

Crocodylians may be sexed by parting the cloaca enough to expose either a penis or a clitoris anteriorly. This is more difficult in hatchlings than in older animals, but Webb *et al.* (1984) were able to

"The chooks have layed a lot of eggs by the river Gran."

Fig. 12.1. *A cartoon by Malcolm Evans, part of his 'Edna' series, which ran last century in* The Land: *an Australian rural newspaper. (Reproduced with permission from the artist and the newspaper)*

Watching large males competing for superiority is memorable. In November 1992 Lyn Beard and I were at the Edward River Crocodile Farm in North Queensland, early in the breeding season, when a 4.5 m *C. porosus* ('Toothy') challenged the 5 m alpha male Hymn (the 'Boss Croc', see text). We videotaped the contest until it was too dark to see. The story is told below in stills and the video can be viewed at http://crocodilian.com/books/grigg-kirshner. The contest began in late afternoon when the challenger approached Hymn, to the right of our field of view. Hymn's response was a swift chopping side swipe directed at Toothy's head, causing a mighty splash.

1. After his initial approach was rebuffed with a head swipe from Hymn, Toothy (left) backed off, with Hymn advancing slowly but emphatically towards him. Although backing away, Toothy kept his aggressive posture, with his head raised, mouth held wide open and his raised tail 'wagging' slowly.

2. Closing in, Hymn (right) quickly delivered another head swipe, sending a spray of water high into the air but not connecting. As far as we could tell, none of the subsequent 'head chops' struck its target, either. Toothy's retreats brought them closer to us and they eventually passed within 20 m.

3. Hymn's advance continued, punctuated several times by head swipes towards Toothy. Toothy (left) struck at Hymn only once, but his slow retreat continued and there was never any doubt about who was The Boss. Nevertheless Toothy was not giving up easily and he maintained his head high, mouth open aggressive posture throughout.

4. During Hymn's slow, almost menacing advances he often emitted barely audible, low pitched but energetic sounds, accompanied by flurries of disturbed water dancing momentarily above each flank and presumably sending out sub-audible vibrations (SAVs, Chapter 5). We noticed Toothy do this only once.

5. There were also episodes of prolonged 'narial geysering' by Hymn, from close in, making a 20 cm bubbling waterspout by exhaling forcibly through his nares. The rush of air was quite audible and he sank lower as his lungs deflated. On one occasion Toothy, too, tried geysering, but mostly he retreated, keeping a safe distance.

6. Later, Toothy looked less aggressive during some of Hymn's geysering, as if conceding, but he soon provoked yet another, final head chop as the light faded. Apart from the very impressive headstrikes, all of this behaviour appeared measured and even low key, but we were never in any doubt that we were watching a significant contest.

After an hour or so of this encounter it was getting too dark for us to see much more. Hymn must have been an old hand at these contests. He was fully grown and already looking old and toothless even in 1980 when I first met him. He was boss of the lagoon then, lost that crown (to Toothy!) for a couple of years in the mid 1990s but regained it and remained the Boss Croc of the lagoon for most of the rest of his life. He died in December 2012. He was the star of our study on thermal relations discussed in Chapter 10 (Fig. 10.9). Whenever he was out and about in this large, naturalistic area, all of the other crocs were well aware of him and kept out of his way. So did we.

Fig. 12.2. *An alpha male 'boss croc' rebuffs a challenge by a subordinate.*

diagnose the sex of *C. porosus* hatchlings with a very high success rate. Ziegler and Olbort (2007) have provided a guide to sexing crocodylians, with an extended, illustrated version at http://www.wmi.com.au/csgarticles/Genital_Structure_Sex_Identification_L.pdf.

Determining maturity is less straightforward than one might think, even if carcasses are available for examination. Alistair Graham had an abundance of carcasses to examine after shooting 500 mature *C. niloticus* in Lake Turkana for a population study (Chapter 13). In males, he distinguished between sexual maturity, based on testis weight, and readiness for sexual activity as indicated by the presence of spermatozoa in the vas deferens (Graham 1968). Mostly a researcher wishes to know about the maturity or seasonal readiness of a living individual and typically this is judged by everting the penis carefully with a finger, taking a smear from the penile sulcus and examining it under a microscope for sperm (Fig. 13.29). However, Coutinho *et al.* (2001) pointed out that, because of social factors, this may lead to an underestimation of the age at which males may be functionally active in the population. Although many males may be capable of mating, competition between males has a big influence on who actually gets to breed.

VA Lance *et al.* (2009) successfully used ultrasonography to distinguish previtellogenic follicles, vitellogenic follicles, recently shelled eggs, fully developed well-calcified eggs, and atretic follicles in mature female alligators, so that technique has potential for assessing state of maturity, as well as reproductive stage in females.

In the hope of using hormone cycles as an indication of sexual maturity, Lance and Elsey (2002) tracked testosterone levels in sub-adult male alligators. They were surprised that even immature individuals showed a seasonal hormonal cycle similar to fully mature breeding males, initially at lower levels and increasing gradually with body size, therefore giving no information about the onset of maturity. Immature female alligators show no seasonal cycle in oestradiol levels (Lance 2003).

Most crocodylians live long lives with reproductive activity spanning many seasons. Typically they take more than 10 years to reach sexual maturity,

except some caimans which mature in 5–6 years or even less (Chapter 13). There may be large differences in different parts of a species' range because of the effects of temperature and habitat quality. For example, alligators reach sexual maturity at ~10 years in Louisiana, but not until 18 years in the much cooler environment of North Carolina (Lance 2003). On that basis, maturity age in Florida might be expected to be similar to Louisiana, but food is much scarcer in the Everglades and maturity there takes 12–14 years. Sometimes differences are not easy to explain. For example, *Caiman crocodilus* females mature in 5–6 years at Alter do Chao, near where the Tapajos River enters the Amazon in Brazil (Magnusson and Sanaiotti 1995), whereas females of the very closely related female *Caiman latirostris* on Ilha do Cardoso, take ~13 years to mature (Moulton *et al.* 1999). As a result, *Caiman crocodilus* has been regarded as a 'weedy' species (see discussion Chapter 13).

There may also be large differences in age at maturity between individuals in the same population. In the Lynd River in North Queensland, youngest ages of *C. johnstoni* at maturity ranged from 12 to 15 years in females and 12 to 16 years in males, yet some individuals were found to be still immature at 24 years (Tucker *et al.* 2006). The presence of a dominant male in The Croc Hole in the Lynd River inhibited growth of younger males (Chapters 10 and 13) (Tucker *et al.* 1998). The 10 oldest females sampled by Tucker and colleagues in the Lynd River population averaged 40 years (range 35–54), males 52 years (range 48–64). For comparison, female *C. johnstoni* in the McKinlay River in Australia's Northern Territory, more than 200 km via the coast to the west, mature at 11–14 years, males at 16–17 years and are thought to remain reproductively active until 40–45 years (Webb *et al.* 1987a). These statistics are surprisingly similar in the two areas. *C. porosus* females typically mature at 12 years, males at 16 years, but females can mature as young as 8 years. For example, it is thought that females matured younger and at a smaller size after *C. porosus* populations had been severely reduced by hunting in the post Second World War years (Chapter 13).

The possible maximum longevity of crocodylians is more in the realm of opinion than based on data, as discussed in Chapter 1, but many are certainly long lived. How long they remain reproductively active is also unknown. Clearly many females breed over a period of many years, but they do become senescent. Graham (1968) found that six out of 33 female *C. niloticus* longer than 290 cm from Lake Turkana had ovaries that he judged to be inactive and therefore senescent. There was no way to determine age accurately, but his 'eye lens weight' versus age graph implies that these females were aged ~40 years. Despite the behavioural dominance of the large male, Hymn, at the Edward River Crocodile Farm (Fig. 12.2), Lewis *et al.* (2013) could find little or no sign of a genetic contribution to 10 clutches in his lagoon. Hymn had been in captivity since 1978, when he was already a very large, conspicuously mature individual. Nancy Fitzsimmons, team leader on that study, commented that rather than being a dominant 'dad', he was a dominant 'dud'. He died in December 2012, having had a period of some months without eating, during which he became thin and lethargic. He then rallied for a few weeks and seemed to 'come good', but his body was then spotted floating in the lagoon in water too shallow for a boat (and wading through the mud was out of the question!), so it could not be recovered (*pers. comm.* via Adam Britton, May 2013). The cause of death is unknown; perhaps he just died of old age.

COMMUNICATION AND SOCIAL STRUCTURE

The communications that dictate social structure, negotiate courtship and facilitate hatchling and crèche formation are all ultimately related to reproduction, so this is the appropriate chapter in which to discuss crocodylian visual, acoustic and olfactory language.

As anyone knows who has picked one up, crocodylians are quite good at expressing themselves. From hatching onwards, they can indicate their displeasure by emitting distress calls or angry sounding squawks and similar acoustic protests. Left to themselves, they use a range of sounds (Chapter 5), visual signals and release of pheromones (Chapter 3) to structure their sociality and negotiate their

courtship and mating. A lot of their signalling is universal. Roaring at an adversary, tilting back the head to expose the throat and show submission, displaying teeth, frolicking with, and stroking the face of, a potential sexual partner are all gestures humans can identify with. Scary eh! We do not make sub-audible vibrations or head slaps, but we could probably read those messages if we did.

Papers by Garrick and Lang (1977) and Garrick *et al.* (1978) provided comprehensive descriptions, reviewed earlier literature and introduced a nomenclature for different signalling behaviours. They compared American alligators with *C. acutus* and *C. niloticus* and concluded that there was great similarity across all three species, with alligators standing just a little apart, differing in some details. More recently a PhD thesis and publications by Vladimir Dinets (2011a, 2011b, 2013) have broadened our knowledge across more species and work on the Chinese alligator (Wang *et al.* 2007) showed their similarity to the American alligator.

All the information confirms the similarity of social signals across the extant crocodylians, which suggests that these are extremely old. An analysis by Senter (2008), who mapped 14 communication signals seen in seven species, concluded that 11 of them are ancestral in both Alligatoridae and Crocodylidae, implying they were already present by the late Cretaceous. Different species would be highly likely to read each other's signals easily enough: a *C. porosus* translocated to a Louisiana marsh would likely communicate with the locals quite well.

Acoustic signalling is the most effective way to communicate over long distances, whereas visual signals become more applicable at short range.

SOUND PRODUCTION: THE CROCODYLIAN LARYNX

Most reptiles are not vocal, but there are two striking exceptions: the two extant archosaur groups, crocodylians and those modern theropod dinosaurs, the birds. Vergne *et al.* (2009), Brazaitis and Watanabe (2011) and Riede *et al.* (2011) all drew attention to this and its implication that dinosaurs could have been vocal. Birds make sound through their syrinx, which is unique to them, but crocodylians use their larynx to make a range of squawks, grunts, roars and bellows. They are often said to lack 'vocal cords' but Müller (1839, quoted in Riede *et al.* 2011) showed that laterally infolded 'vocal folds', below the glottis, vibrate when air is blown through an alligator larynx. Riede *et al.* (2011) reviewed early investigations, which showed that these folds can be drawn together (adducted) or apart (abducted) by laryngeal musculature, but their length could not be modified, which is the basis for frequency modulation in mammals. These authors showed experimentally, however, that frequency modulation is brought about by interplay between sub-glottal pressure (force generated by exhalation) and abduction and adduction of the vocal folds. It seems quite likely that dinosaurs could have produced sounds with a similar, comparatively simple larynx.

LONG RANGE SIGNALLING: MOSTLY ACOUSTIC

Most crocodylians can make loud vocal sounds (bellows or roars) or, non-vocally, slap the head down on the water surface, snap the jaws together underwater or vibrate their flanks to produce infrasound, sub-audible to humans. These signals all travel some distance, beyond line of sight. Dinets (2011a, 2011b) has referred to these long-distance signals as advertisement calls, by analogy with the terminology used for frogs and, as in frogs, they are heard most frequently in the mating season.

Bellowing and roaring

All crocodylians are able to make loud vocalisations, although many do so quite rarely. They are described as bellows when they are made by alligatorids and roars when they are made by crocodylids, even though both sounds seem to be made similarly and probably have a similar function. I will maintain the same convention. Bellowing is very common in both species of *Alligator*, at dawn and, most actively, during courtship (*A. mississippiensis* http://crocodilian.com/cnhc/images/!amis10a. wav; *A. sinensis* http://crocodilian.com/cnhc/images/!asin2a.wav). Female alligators also bellow, but not as loudly as males, and apparently do not produce infrasound (SAVs, Chapter 5 and below) as a prelude. In courtship, males and females bellow together, one call triggering others

until choruses may form. In Crocodylidae, roaring, probably the equivalent vocalisation, is much less common. I have never heard it in either *C. porosus* or *C. johnstoni*. When I was in primary school, I lived in North Queensland and there were always tales about 'crocs roaring up the river', which used to get my heart thumping because I too was often 'up the river', in a homemade canoe. More likely there was confusion with the calls of Great Billed Herons that did, indeed, live up the river and that make a throaty prolonged roar-like call. Modha (1967) reported Nile crocodile alpha males roaring as they chased a rival out of their territory. It has been described in both *C. acutus*, and *C. intermedius*, and different authors have reported various occurrences, so it is quite possible that roaring occurs in all species of crocodylian, though less conspicuously and much less frequently than in alligators.

Uniquely, male gharials have a bulbous hollow outgrowth at the tip of the snout, overlying the nares (Fig. 12.3). According to Whitaker *et al.* (2007) it amplifies 'hisses' snorted through the compressed underlying nostrils and 'can be heard for nearly a kilometre on a still day'.

Bellows and roars are almost always made in the water, usually from a 'head oblique, tail arched' (HOTA) posture (Fig. 12.4) and in some species a roar or bellow follows a head slap (see below). Without the SAV component, bellows are quite ineffective at transmitting sound underwater, but they may have a transmission distance in air of several

Fig. 12.3. *The ghara, a distension surrounding the nares at the tip of the snout of male gharials,* Gavialis gangeticus, *which functions as a resonating chamber. Its name comes from the Nepali word for 'pot'. (Photo Ruchira Somaweera)*

Fig. 12.4. *The bellowing cycle of an American alligator. Starting with head and tail lifted in the so-called 'head oblique tail arched' (HOTA) position, and riding high in the water on inflated lungs, the animal sinks into the water as air is expelled through the larynx to create the bellow. Resonance driven by the sound produced makes the water dance from the flanks and then from the dorsum when it submerges towards the end of the event. Photographed in Everglades National Park, Florida. (Photo Vladimir Dinets)*

hundred metres and so are particularly useful in fragmented habitats (Dinets 2011a, b). In alligators, Todd (2007) measured dominant frequencies in air between 20 and 250 Hz at a sound pressure level at 1 m of 91–94 dB from which he calculated an 'active space' (crudely, the detection distance) of ~160 m in the 125–200 Hz band. In water, he measured dominant frequency between 20 and 100 Hz with a source sound pressure level (SPL) of 121–125 dB at 1 m, from which he calculated an active space of up to 1.5 km in the 63–100 Hz band and up to 80 m at 16–50 Hz. Waveform and frequency (sonographic) analyses of a typical alligator bellow are shown in Fig. 5.14.

Vergne *et al.* (2009) reviewed and described categories of calls made by crocodylians and drew attention to a dearth of information, particularly about their function. Typical speculation is that bellows and roars function as assertive expressions of dominance, aggression or territoriality, perhaps leading to the spacing out of individuals in the non-breeding season. In addition, despite the lack of ex-

perimental studies, observations on both species of *Alligator* suggest that one important function may be to attract breeding individuals to a breeding area.

Comprehensive descriptions of bellowing behaviour in male and female *A. mississippiensis* have been provided by Garrick and Lang (1977), Garrick *et al.* (1978) and Vliet (1987, 1989). Todd (2007) has provided an analysis of both the sound and its transmission characteristics. Alligators often bellow in a chorus, which may include females. A chorus may be started by one individual, often by a female, and others join in. A bellow cycle begins with the alligator adopting an elevated posture and inhaling, raising the head, and then adopting the HOTA position, which may be held for a few seconds until SAVs are produced, with water dancing off the back (Figs 12.4, 12.5). Then the bellow is emitted, during which the animal sinks deeper into the water with the exhalation of air and the cycle may then be repeated. Bellowing by females does not include emission of SAVs. The dominant frequency of the bellows is very low, ~200 Hz, which is suitable for

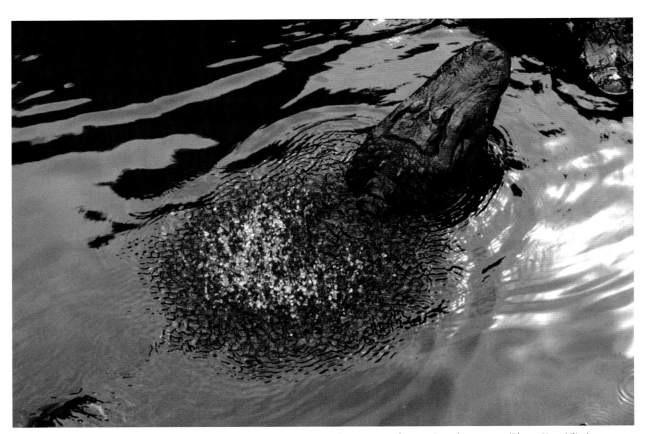

Fig. 12.5. *American alligator producing infrasound, showing up at the water surface as Faraday waves. (Photo Kent Vliet)*

long distance signalling. A musky odour is often detected from a bellowing chorus (Vliet 1989, Chapter 3). Most observations have been made on captive animals, but nocturnal courtship aggregations in the wild have also been described (Dinets (2010), followed by bellowing choruses in the morning.

Broadly similar behaviour has been described in Chinese alligators. Wang *et al.* (2006, 2007, 2009a, b) reported research conducted at the Anhui Research Center for Chinese Alligator Reproduction in Southern Anhui Province, China. These alligators are very vocal and have quite an acoustic repertoire, said to comprise 'tooting, bubble blowing, hissing, mooing, head slapping and whining' (!) and, for long distance transmission, they bellow during their active season in the HOTA manner of American alligators. Wang and colleagues analysed sonograms of bellows, using terminology adapted from the frog literature and recorded dominant frequencies below 500 Hz, suitable for radial transmission over a long distance. They discussed how a call with these characteristics could be useful for bringing together normally dispersed individuals in the mating season. Because bellowing occurs in the water, it also indicates the availability of a site suitable to congregate and mate. Both males and females bellow and Wang *et al.* (2009b) described morning chorusing by animals in a large naturalistic pond, usually initiated by a single individual. Sometimes even thunder would get a chorus going. By using playback experiments, they showed that males and females responded equally to recorded bellows and, subsequently, that they responded to a wide range of artificially varied bellows. The authors pointed out that Chinese alligators have no sympatric relatives. It would be very interesting to compare sonograms of bellows or roars from sympatric crocodylian species: are there interspecific differences and can they be differentiated, as they are in frogs?

There appear to be no reports of bellowing choruses in caimans or crocodylids, which seem to be much less vocal in their reproductive behaviour.

Infrasound, sub-audible vibrations (SAVs)

Bellowing by males is usually preceded by the production of infrasound, sub-audible vibrations (SAVs) generated non-vocally (Fig. 5.14). Details of the production mechanism appear to be lacking. Todd (2007) suggested it may be analogous to sound production by fish, which use rapid muscle contractions to vibrate their swim bladder. There is no more spectacular crocodylian display than the way in which sub-audible vibrations (SAV) make water dance above the submerged back of *A. mississippiensis*, often preceding a bellow (Fig. 5.14). The dancing water droplets are caused by Faraday waves and I am grateful to Andrew Taylor, University of New South Wales, for pointing this out.

Faraday waves are standing waves that can result where a liquid overlies a vibrating surface. Spectacular though it is to a human observer, we miss most of it because we cannot detect the high-energy infrasound being transmitted both underwater and above, at around 20–50 Hz (Todd 2007, Chapter 5). The SAVs can travel a long way under water. In alligators, Todd (2007) estimated an 'active space' (space within which another alligator may hear) of 1.5 km. What function this has at long range seems to be unknown, but it is used at close range such as in male–male contests, displayed by the dominant male Hymn in the contest described in Fig. 12.2. At close quarters too, the reception of SAVs by the sacculus may be sexually stimulating during courtship, if Todd's (2007) interpretation is correct (Chapter 5 and in the section on Courtship, below).

According to Dinets (2011a), all species except *Gavialis gangeticus* produce infrasound (although information is lacking for *C. mindorensis* and *O. tetraspis*). In non-*Alligator* species, however, it is not so spectacular.

Many questions remain about infrasound production and the contexts and formats in which it is used. Apart from the familiar example when it precedes bellowing by alligators (Fig. 5.14), short bursts of it are produced as signals to other males and females (Fig. 12.6). Toothy and Hymn (Fig. 12.2) each sent out short bursts periodically during their contest, and Hymn produced apparently similar pulses as he chased and courted a female (see below and at http://crocodilian.com/books/grigg-kirshner).

Head slaps, jaw slaps and jaw claps

In a head slap as defined by Garrick *et al.* (1978) from observations of alligators, the head is raised,

Fig. 12.6. *The role of infrasound (SAVs) is multifaceted and not yet completely understood. Here, a female with her snout lifted in apparent submission is accompanied by two males. A third male approaches and produces a short burst of infrasound. Although it is easy to speculate about the meaning in the burst of infrasound and conjure up a scenario, we do not know what is really going on here. SAVs travel well in water and are useful in long distance signalling. Their near field use is less well explored. (Photo Adam Britton)*

leaving the lower jaw submerged, then slapped down onto the water surface with a mighty splash and a 'pop' as the jaws come together. Other signals are often combined. They described jaw claps as rapid closing of the open jaws to produce a loud deep 'pop', observing that they are also an element of head slaps. Dinets (2011b) noted that jaw slaps are made in some species by slapping the jaws together at or below the water surface, but categorised them with head slaps. Detailed comparisons may lead to them being recognised as variants of the same behaviour. Head slaps produce a sound that travels well under water (Dinets 2011a,b). They seem to be typical of all crocodylians and may have similar functions. Vliet (1989) described head slaps in alligators as an assertion display comprising several elements: HOTA, production of SAVs, the head slap, a jaw clap, a growl, inflated posture and growl, any of which may vary in intensity. These are combined acoustic and visual signals. As with

bellowing, there is sometimes a musky odour and sometimes there is also an oily sheen on the water that seems to be associated with the cloacal region. Presumably pheromones are being released (Vliet 1989 and see Chapter 3).

Sometimes head slaps are repeated several times in quick succession and in some species (e.g. *C. intermedius*, Thorbjarnarson 1993) a roar and head slap are combined. Head slaps are effective at transmitting sound both above and below water. Dinets (2011a) measured transmission by alligator head slaps up to 200 m in air and, depending on the conformation of the water body, up to 500 m under water. It is easy to see that clapping the jaws together beneath the surface would produce a sound that travels well under water.

In reviewing these relatively long-distance acoustic signals, Dinets (2011a,b) tabulated and published his observations about which species employ which signal/s. Most behaviours have been seen in most species, conspicuous exceptions being a lack of either vocalisation or infrasound by *Gavialis gangeticus,* and infrasounds but no reported vocalisation in *Tomistoma schlegelii.* In some species, some roars and bellows are not at all loud and there are large differences between species in the frequency of use. Dinets concluded that slaps were more suited to continuous habitats and vocalisations to fragmented habitats. The function of these long-range aerial and sub-surface signals, unless employed close to a target individual/s, is not easy to determine. In naming them 'advertisement calls', Dinets is probably near the mark.

SIGNALLING AT CLOSE RANGE: VISUAL, ACOUSTIC, OLFACTORY

Visual signals dominate at close range, but acoustic and olfactory signals are also employed. The two significant contexts in which social signalling is important are maintenance of the social structure (a 'pecking order') and during courtship and mating,

which is, of course, the ultimate significance of all of these social interactions. Courtship and its behaviour will be discussed in the next section. Here we will talk about the hierarchy and the role of males in particular.

Long-distance signalling is mostly acoustic through air or under water, but visual signals dominate when crocs are within sight of each other. The American alligator also makes a range of coughs, grunts, growls and hisses (Garrick *et al.* 1978), while the Chinese alligator adds tooting, mooing and whining (Wang *et al.* 2007). These last few I have difficulty imagining, but crocodiles that I have met certainly hiss and grunt on occasion, particularly to indicate displeasure at being interfered with (grunt) or in threat (hiss).

Snout-lifting

Snout-lifting with the jaws closed is a visual signal that observations suggest signifies submission (Fig. 12.7). It is displayed by both males and females, apparently to acknowledge an animal close

Fig. 12.7. *A (presumed) subordinate male snout-lifts in submission in the presence of the alpha male (a 'boss croc'). (Photo GCG)*

by is higher up in the hierarchy. It is also shown commonly by females in courtship and copulation.

Contests between males: high float, wide gape, narial geysering, tail wagging, mock fighting

The 'fight' between Toothy and Hymn (Fig. 12.2) provides a good example of how expressive crocodylians can be in their visual social signalling, augmented by the non-vocal sounds made in narial geysering and infrasound production. An animal that thinks itself superior may show it by inflating its lungs to carry itself higher in the water. This may be accompanied by a HOTA posture and, in *C. porosus* and probably others, a wide open mouth, showing the teeth (Fig. 12.8).

Narial geysering (Fig. 12.2) is seen in alligators and caimans, *Gavialis*, *Tomistoma* and in at least several species of *Crocodylus* when males face off. This is both visual and acoustic (non-vocal), the

exhalation through the external nares producing a conspicuous sound. The sounds produced by narial geysering, although not apparently yet measured, may travel some distance under water. David Lindner (*pers. comm.* 2012) blew bubbles into the water through a PVC pipe at Goose Camp and was rewarded a few minutes later by the arrival of the 'Boss Croc' (see below) as if to see what was going on. Males may also chase each other, or lunge, and engage in mock and sometimes serious fighting (Fig. 12.10).

Gharials employ a particularly spectacular combat method; they use their extraordinarily long snouts (Fig. 12.9). Simon Maddock described this, from observations made at the Madras Crocodile Bank Trust south of Chennai, India, as follows 'The males moved towards each other (head on), propelled themselves out of the water, and using their jaws like swords, tried to push each other to the water's edge. No biting or mouth gaping was

Fig. 12.8. *Two large male* C. porosus *facing off in shallow water. Open mouth and a larger body size, height out of the shallow water, signify the likely winner in this case. This open mouth behaviour in* C. porosus *seems not to have been described in alligators and its prevalence in other Crocodylidae is uncertain. (Photo GCG)*

Fig. 12.9. *Male combat between male gharials,* Gavialis gangeticus *described appropriately as 'fencing behaviour' by Simon Maddock (2010) from observations made at the Madras Crocodile Bank Trust in India. One male eventually retreated and the winner 'moved closer towards the females'. Note the bulbous gharas at the tips of their snouts. (Photo Simon Maddock)*

observed, only strong swipes to the jaw of the opponent. The fight continued for approximately 30 minutes before one male retreated to the far side of the enclosure whilst the winner (in this case, also the larger) male moved closer toward the females'. He and others at the Trust called it, appropriately, 'fencing behaviour' (Maddock 2010). It occurs also in alligators, the same name being used.

Mostly, however, crocodylians convey information about themselves by behaviour that is almost too subtle for the human observer to notice or categorise: height and angle of the head, height of the float in water, position and movement of the tail (whether arched or wagging). Less subtle is the combination of a raised head and lifted tail, which a male might use to show assertiveness, and still less subtle is a swift chop by a dominant male at the head of a rival, unlikely to escape the atten-

tion of even a disinterested human observer. Table 12.1 lists some of the visual and acoustic signals by which crocodylians signal their status or intentions.

In most social animals signalling avoids injurious physical conflict and this is usually the case in crocodylians. Despite all the signals, however, many crocodylians do carry scars from fighting and often suffer serious wounds (Fig. 13.24). They have a great capacity to recover, but the damage is sometimes too great and they die (Chapter 13, Fig. 12.11). Not long after Hymn was captured in the Malaman River on Cape York in 1978, a big croc was caught in the nearby Edward River, christened Psalm and released into the large lagoon. A couple of years later Psalm died from wounds inflicted by Hymn. The contest would have been terrible to behold.

TABLE 12.1   CROC TALK: VISUAL AND ACOUSTIC (VOCAL AND NON-VOCAL)

The following signals are seen throughout the Crocodylia, albeit with different frequencies of occurrence in different species, or in different situations.

Aggressive behaviour by males

- alertness shown by the 'head oblique tail arched' (HOTA) position
- inflated posture (i.e. floating high in the water)
- combining HOTA with wide open jaws (*C. porosus*, otherwise uncommon?)
- tail wag (commonly with HOTA)
- chasing, lunging, mock or actual fighting, head swipes
- sub-audible vibrations (SAVs)
- narial geysering
- head slaps (which generate a sound that travels well)
- jaw claps
- patrolling an area (alpha males)

Appeasement, submission

- low profile in the water
- lifting the head with the mouth closed (sign of submission)
- disinterest, departure

SOCIAL STRUCTURE: 'BOSS CROCS' AND TERRITORIALITY

The reader might conclude from the preceding sections that a well-stocked river is active with crocodylian sounds and semaphore day in and day out, but an observer can spend a long time in the field and see very little. Most of the observations have been made in captivity, where higher densities of animals increase the levels of interaction. Most of the day-to-day life of a croc population involves individuals that already know their position in the hierarchy and, consequently, generate and respond to more subtle messages that humans do not notice. Tracking studies on *C. porosus* (Kay 2004; Brien *et al.* 2008; Read *et al.* 2007; Campbell *et al.* 2013) all show that large males and many females inhabit the same stretches of water, presumably displaying their own body language, as well as noting that of conspecifics, without any conspicuous action being necessary. In the breeding season, the equation must change and, driven by hormones, alpha males are likely to assert

their position in the hierarchy more emphatically. It is interesting, however, that despite the lengths alpha males go to in maintaining their position as top croc, multiple paternity in clutches of eggs is reasonably common (see below). Genetic studies have shown that, just as in many mammalian and avian species where dominant males might have been thought to dominate the mating as well, the 'sneaker male' phenomenon is alive and well. Body size is obviously very important and larger animals usually win, so it is perhaps not surprising that crocodylians in general show conspicuous sexual dimorphism. Position in the hierarchy is very important and can even affect their body temperature and, consequently, growth, as shown in *C. johnstoni* (Chapter 10). A low status 'freshie' that is chased frequently from a basking site will have less time at an optimal Tb, a slower growth and thus a slower entry to maturity.

With this diversity of signals, crocodylians surely have a social structure that is well understood by

Fig. 12.10. *Two male* C. porosus *fighting. (Photos Adam Britton)*

Fig. 12.11. *Actual physical fights are usually avoided but this animal has lost its left hind leg, probably torn off by another male. Note the distended cloaca as the carcass starts to decompose. (Photo Bill Green)*

them, even if not yet fully described. Observations provide substantial clues. Dave Lindner and his son Garry, both very astute observers, have had many opportunities to observe *C. porosus* in the rivers and wetlands of Kakadu National Park and have got to know some of the big crocs individually. One animal in particular, 'Roughnut' – at least 5 m (~16 feet) in length (Fig. 12.12) – is known to have been the alpha male since 1975 in a floodplain and billabong area known as Gingela (Fig. 12.13). Aboriginal people first became acquainted with him while spending time at Goose Camp to hunt and gather magpie geese, file snakes, feral pig, water pythons, saratoga, barramundi, catfish, freshwater turtles and aquatic plants (Lindner 2004). Subsequently, Aboriginal people were employed to irrigate areas being replanted and rehabilitated following damage by feral buffalo (*Bubalis bubalis*) and had further opportunities to observe Roughnut, the only large crocodile to survive locally from the days of extensive hunting of crocodiles for hides.

The billabong is ~3 km long and connects to the South Alligator River in the wet season. In the late dry season, ~100 *C. porosus* congregate there as the wetlands dry out. Dave has been keeping an eye on Roughnut since 1993.

Observations on Roughnut, as well as other alpha males showing similar behaviour, have led to these very large, dominant males being given the term 'boss croc'. Long-term observations at Gingela have been described by Garry Lindner (2004) and elaborated by Dave (*pers. comm.* 2010 and 2012) and they provide fascinating insights into crocodylian social structure that are worth recording here, and which may in future be extended by replicate observations and possibly some experiments.

By late dry season, daytime counts in the east end of Gingela billabong yield up to 30 males (3.5–4 m). To quote from Lindner (2004):

'Roughnut's territory includes a couple of 4 m plus crocodiles. This is followed by a number

Fig. 12.12. *Roughnut, the 5 m (plus) boss croc C. porosus in the billabong at Goose Camp, Kakadu National Park, Northern Territory, Australia. He has been well known to the locals since 1975 and under closer observation since 1993. (Photo Garry Lindner)*

of crocodiles about the 3 m size class. The most abundant size range in this group of 100 crocodiles is the 1 to 2.5 m size class. Typically, as with other 'boss crocodile' situations the majority of these crocodiles (about 60% of the population) occupy about 25% of the space available, at one end of the billabong system. Roughnut and the remaining 40% of the population occupy the other 75% of the billabong. The crocodiles 'accommodated' by Roughnut are usually female animals up to 2.7–3 m in length and are fairly evenly dispersed throughout the billabong. The remaining 60% of Roughnut's peer group exist in the shallow end, which is ~2 m deep in the middle sections. These crocodiles co-exist, albeit the 40% of smaller ones with a nervous disposition to the larger size classes, in fairly high densities in ~750 m of waterway that is ~150 to 200 m in width at its widest point ….

Roughnut's wives use his presence when feeding, other crocodiles in the 3 m plus size class

leave when Roughnut approaches, his harem of females remain in his presence when feeding albeit at a close distance. Roughnut's presence affects peer crocodiles when he approaches. These smaller crocodiles are often engaged in predatory observation and stalking of magpie geese flocks, but their attention is immediately diverted to Roughnut when he approaches the area. What is not understood is how crocodiles sense each other, particularly when Roughnut is approaching from 500 m to a kilometre away. The smaller crocodiles sense his approach and seem to divert their behaviour from predatory observation to wariness and avoidance of the approaching boss crocodile'.

Dave Linder describes large males being attracted to feeding opportunities, such as gunshots or activity by other crocodiles. This leads to a chain reaction. They repeatedly try to snatch food from each other (Fig. 6.17) and vigorous chases occur. Raids by 3.5 m to 4 m males result in Roughnut

Fig. 12.13. *Roughnut's domain: a permanent billabong in an area known as Gingela in Kakadu National Park. (Photo GCG)*

giving chase and then patrolling and circling an area, with feeding abandoned for the time being.

The implication is clear. 'Roughnut' exerts a strong influence over the crocodylian social structure in 'his' billabong. He dominates a large portion of it and tolerates females. Animals not tolerated do enter his area to forage but, sensing his presence or impending arrival, become wary after being chased and soon depart. Dave has written that Roughnut's females wait until he has had a feed before they will come in. When feeding, he deters close feeding interest from females with snap and lunge rebukes. They recognise when he is full and when it is safe to move in and feed.

This brief summary provides a compelling example of the influence that a large dominant male crocodile can have in a seasonally dense population. An animal such as this exercises a stabilising influence over the local crocodylian social structure and the detailed observations made and recorded at the Goose Camp billabong over more than 20 years deserve a full description elsewhere.

The situation with Roughnut at Goose Camp is not very different from the social system described by Modha (1967) from his observations of ~500 *C. niloticus* living in a lake of ~65 ha on Central Island in Lake Turkana, Kenya. Here, a dozen dominant males patrolled and defended contiguous sections of shoreline of 60–230 m and adjacent water to ~50 m offshore, right around the lake. Entry by other large, presumed males resulted in them being chased out. They often fled onto the shore, hastened on their way by the half-emerged dominant male snapping and roaring. Actual fights were rare. Modha observed that smaller adults in a territory showed submission by snout lifting and either fled, often chased by the alpha male (Modha thought these fugitives were sexually mature males), or did not flee (these were probably females; they were sometimes courted). If other males entered his territory and attempted to court/mount (presumed) females, they were chased out, but the females were not resident in any particular territory and Modha saw no sign of a dominant male trying to herd or 'collect' them. Kofron (1990, 1991) also reported large bull Nile crocodiles chasing subordinates away from a mating 'arena', but tolerating adult and sub-adult males and sub-adult females that showed submissive behaviour by snout lifting.

In different situations different 'rules' will certainly apply: crocodylians are not robots. Roughnut certainly defends a territory, and aggressive territorial defence is well embedded in the public perception of alligators and crocodiles and consequently often blamed for attacks on humans. However, many crocodylians live in rivers or similarly 'linear' aquatic systems along which they make extensive movements (e.g. Read *et al*. 2007; Campbell *et al*. 2010, 2013 and Chapter 4) to acquire food, mate and nest. Marking off and defending a section of these 'highways' as a territory in the conventional sense would be impractical. Indeed, radio-telemetry data on *C. porosus* in rivers flowing into Cambridge Gulf, Western Australia (Kay 2004), and in a long waterhole in North Queensland (Brien *et al*. 2008) showed that home ranges of large males frequently overlapped and that they commonly passed through each other's 'territories'. Likewise, Campbell *et al*. (2013) tracked eight male *C. porosus* in the Wenlock River on Cape York, also in North Queensland, and found five with core areas, some overlapping, while the other three travelled throughout the river system, passing through all the others' space. Perhaps travelling individuals communicate their status and intentions and so gain safe passage.

SPERM AND EGG PRODUCTION
MALE ANATOMY AND PHYSIOLOGY

The testes are paired, situated between the kidneys while overlapping them anteriorly and are attached by the mesorchium to the dorsal wall of the body cavity (Figs 12.14, 12.15). They are elongated, flattened and more rounded anteriorly than posteriorly, their size dependent upon the state of maturity and their seasonal cycle. They can be asymmetrical; the left testis is always shorter than the right in *Caiman yacare* (Coutinho *et al*. 2001). The testes in *Caiman yacare* are creamy yellow, becoming reddish when suffused with blood during full

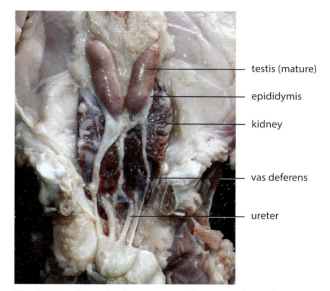

Fig. 12.14. *Reproductive system of a reproductively active 2.7 m male* C. porosus. *Note the paired testes, suffused with blood at this time of year, each with a vas deferens containing spermatozoa and seminal fluid. (Photo Steve Johnston)*

testis (mature)
epididymis
kidney
vas deferens
ureter

spermatogenesis. Internally, they comprise many fine, coiled seminiferous tubules within which sperm are formed, surrounded by layers of connective tissue and Leydig cells, which secrete testosterone.

Sperm exits the testes via numerous coiled efferent ducts, which discharge into a coiled epididymis on the dorsal surface of each testis. In the breeding season, each epididymis becomes distended with many millions of stored sperm (Lance 1989). The duct system by which sperm travel from the seminiferous tubules to the vas deferens secretes seminal fluid and is differentiated histologically into several regions, described in *Caiman crocodilus* by Guerrero *et al*. (2004). The epididymis tapers to the vas deferens, which conveys sperm to the penis via the cloaca at the time of mating. Left and right vasa deferentia open independently and dorsolaterally into the urodaeum just posterior to the ureters (Figs 12.14, 12.15). The testes undergo a conspicuous annual cycle in size and activity under hormonal influence, which peaks during courtship and mating (Fig. 12.30) (Lance 1987, 1989; Kofron 1990; Hamlin *et al*. 2011).

The penis is single, with a curved, semi-rigid shaft and a somewhat tulip shaped glans penis (Fig. 12.16). It originates from the ventral wall of

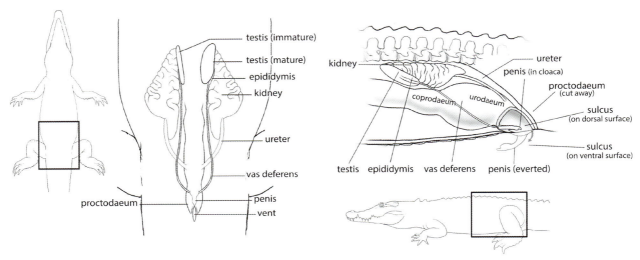

Fig. 12.15. *Schematic ventral and lateral views of crocodylian male anatomy. The penis, normally retracted into the ventral wall of the proctodaeum, is shown here partially everted.*

the proctodaeum and, in operation, is rotated to the exterior by muscles that pull it forward and out. Its anatomy and presumed functioning have been described in *A. mississippiensis* by Moore *et al.* (2012) and Kelly (2013) and in *C. porosus* by Johnston *et al.* (2014a). Superficially, the structure in both species seems similar so a single description will probably suffice. The shaft is densely fibrous with collagen and retains its shape and size, even when the penis is erect and functional. The shaft has an open groove, the sulcus, in the dorsal midline, which follows the outer margin of the curved penis so that it is functionally ventral when the penis is everted. The vasa deferentia open via papillae into the base of the sulcus and, in operation, deliver sperm into the sulcus. During penile erection and intromission, engorgement of a longitudinal blood-filled sinus is thought to close the sulcus to form a tube that delivers sperm to the tip of the tulip-shaped glans (Figs 12.16, 12.18, 12.19, 12.37).

Spermatogenesis

There are good reasons to assume that hormonal regulation of reproductive events in crocodylians is similar to that in other amniotes, perhaps closest to birds, even though their hormones are not structurally identical. For example, Lance *et al.* (1985) found that injection of synthetic mammalian gonadotropin-releasing hormone (GnRH) led to an increase in testosterone in alligators, as it does

in birds, but not turtles or snakes. Further, in hypothalamic tissue they identified a GnRH-like peptide similar to that in chickens. Mammalian follicle stimulating hormone (FSH) also stimulated an increase in testosterone, but not mammalian luteinising hormone (LH) (Lance and Vliet 1987). Additionally, Leydig and Sertoli cells are identifiable within testicular tissue and testosterone levels surge during courtship and mating (Fig. 12.17).

The hormonal regulation of reproduction in male crocodylians is likely to be as follows. In response to photoperiod, temperature and other environmental influences, the hypothalamus secretes a crocodylian equivalent gonadotropin-releasing hormone (GnRH), a peptide that stimulates the pituitary to produce glycoprotein gonadotropins equivalent to FSH and LH. The LH equivalent stimulates Leydig cells (in among the seminiferous tubules, Fig. 12.17) to produce the androgenic steroid hormones testosterone and dehydroepiandrosterone (DHEA) (although most of the DHEA is probably produced in the adrenal cortex). Spermatogenesis occurs in the presence of high concentrations of testosterone and Lance (1989) and Hamlin *et al.* (2011) have shown it peaking during courtship and mating (Fig. 12.30). Presumably testosterone is also responsible for the secondary sexual characteristics of crocodylians, as in other vertebrates. FSH stimulates activity by Sertoli cells (Fig. 12.17), which have a diversity of 'supporting' roles for spermatogenesis in mammals

Fig. 12.16. *The penis of* C. porosus, *everted manually (curved anteriorly). The sulcus (visible ventrally, left) is glandular and may secrete seminal fluids and lubricatory mucins, so contributing to semen quality. The penis is located in a very microbe-rich environment and, by analogy with birds, the mucins may form an anti-microbial barrier (Moore* et al. *2012). (Photos GCG)*

and, probably, crocodylians too. In alligators, DHEA also peaks with testosterone, but, unlike testosterone, it remains present in substantial levels throughout the year (Hamlin *et al.* 2011). These authors pointed out that DHEA is linked to non-breeding aggression and territoriality in birds. In mammals, it is a major circulating steroid with diverse functions so far not fully resolved. Nothing is known with certainty about the function of DHEA in crocodylians. The annual cycles of these two important androgenic steroids are shown in Fig. 12.30.

With the release of these steroidal hormones, sperm production is stimulated as well as the onset of sexually related aggressive behaviour. In the testes, spermatogonia in the walls of the seminiferous tubules divide repeatedly by mitosis to produce spermatocytes, which then undergo two stages of meiosis to produce haploid spermatids (Fig. 12.17). These in turn differentiate into spermatozoa, which are stored in the epididymis – a wider section

of the duct dorsal to the testis. The epididymis tapers to the vas deferens, which conveys sperm to the penis via the cloaca at the time of mating. As mentioned above, the left and right vasa deferentia open independently and dorsolaterally into the urodaeum just posterior to the ureters (Fig. 12.15).

Ultrastructural details of spermatogenesis in *A. mississippiensis* have been described by Gribbins *et al.* (2010). Ultrastructure of the tubules and duct system in which sperm are formed, stored and delivered has been described in *Caiman crocodilus* by Guerrero *et al.* (2004). They drew attention to structural similarities to birds and dissimilarities from all other extant Reptilia and also from Mammalia, emphasising again the sister group relationship between crocodylians and birds. It is highly probable that the dinosaurs and pterosaurs had structurally and functionally similar male reproductive ducts, following the logic expressed by Brazaitis and Watanabe (2011).

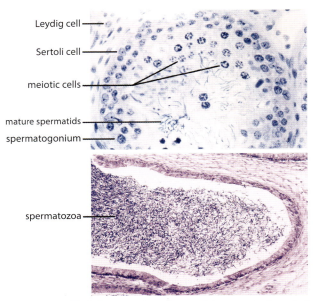

Fig. 12.18. *Steve Johnston, University of Queensland, massages sperm from a C.* porosus *for research into sperm storage and artificial insemination. (Photo supplied by Steve Johnston)*

Fig. 12.17. *(Top) Cross section of one of the repeatedly coiled seminiferous tubules that make up the testes of crocodylians and in which spermatogenesis proceeds, with meiosis resulting in haploid spermatids, the precursors of mature spermatozoa. (Bottom) Cross section of the epididymis of a mature crocodylian showing masses of stored spermatozoa. Both of these photographs are of histological sections taken from* Caiman yacare. *(Photos Marcos Coutinho)*

Spermatozoa

Like those of all vertebrates, crocodylian spermatozoa have a head containing a haploid nucleus and a flagellum whose sinuous beating makes the sperm motile (Fig. 12.20). The spermatozoa of *C. johnstoni*, *A. mississippiensis* and *Caiman crocodilus* have been examined so far and Jamieson (1999) noted that they are very similar to what is envisaged as the basic amniote sperm. Further, they are almost identical to spermatozoa from Chelonia (turtles and relatives) and *Sphenodon* (tuataras) and also share similarities with ratite and lower passerine birds (Jamieson 2007). Considering the similarity of crocodylian sperm to those of basal birds, the implication is that the sperm of dinosaurs must have been similar.

Gribbins *et al.* (2011) described the ultrastructure of *A. mississippiensis* sperm and reviewed descriptions of sperm from *Caiman crocodilus* (Saita *et al.* 1987) and *C. johnstoni* (Jamieson *et al.* 1997). In all three, the sperm are elongated and filiform (thread-like) with a pointed acrosome anteriorly, a cylindrical

Fig. 12.19. *A trickle of viscous sperm emerges from the tip of the glans penis and drips into the collection tube. Note how the inflated glans forms a hollow cuff which, if engorged after insertion during copulation through the sphincter at the upper end of the proctodaeum would be held in place within the urodaeum. (Photo supplied by Steve Johnston)*

nucleus, a 'midpiece' enclosing the two centrioles and mitochondria, and the flagellum (tail). Most of the length of the flagellum (and of the sperm) forms the 'principal piece', followed by a short, terminal 'endpiece'. A schematic representation of a spermatozoon from *C. johnstoni* is shown in Fig. 12.21. The following functional interpretations are inferred from what is known, or assumed, about other vertebrate sperm. The pointed acrosome contains digestive enzymes and facilitates entry of the sperm to the egg. The acrosome fits over the elongated cylinder of the nucleus. Penetration of the egg envelope to allow fertilisation may be assisted by rod-like 'perforatoria' lodged in canals within the nucleus. The centrioles consist of triplets of microtubules. The anterior (proximal) centriole is short and sits at the base of the nucleus whereas the posterior (distal) centriole, from which the axoneme arises, extends throughout the midpiece. The midpiece contains tiered rings of mitochondria, which are the 'energy batteries' of the flagellum. The axoneme consists of nine pairs of microtubules (called 'doublets'), surrounding a pair of microtubules (called 'singlets'). Coordinated movement of proteinaceous 'arms' (dynein) extending between adjacent doublets causes the axoneme filaments to slide against each other, so driving the flagellar beating. This is called a 9 × 2 + 2 axoneme, and it is characteristic of cilia and flagella in general. It has a fibrous sheath for most of its length, defining the 'principal piece' beyond which is the unsheathed 'endpiece'

Gribbins *et al.* (2011) noted that whereas *C. johnstoni* has one to three endonuclear canals, *A. mississippiensis* and *Caiman crocodilus* each have only one. Also, the two alligatorids show slightly fewer rings of mitochondria than *C. johnstoni*. These differences suggest possible familial differences and highlight the need for more species to be examined.

Cryopreservation of spermatozoa, combined with a capacity for artificial insemination, would be a great boon to crocodile and alligator farmers. Big males are more difficult to keep and much more expensive to feed than mature females, so if sperm could be taken from a recognised stud, diluted to a satisfactory concentration, stored frozen until needed and then inseminated artificially to a substantial number of females, the savings could be very worthwhile. Stephen Johnston at the University of Queensland is working on that in association with John Lever at the Koorana Crocodile Farm near Rockhampton in Central Queensland. So far, they have developed a reliable protocol for acquiring (Figs 12.18, 12.19) and assessing semen quality (Johnston *et al.* 2014a), found a suitable diluent and made a good start on cryopreservation (Johnston *et al.* 2014b). Preliminary attempts at artificial insemination in the saltwater crocodile have thus far resulted in only six fertile banded eggs and one hatchling (Johnston *pers. comm.* 2013). Further characterisation of reproductive seasonality of both male and female crocodiles, along with assessment of the importance of sperm storage in this species, will no doubt lead to better knowledge about the best timing for insemination. Apart from improvements to farming, these techniques have potential benefits for conservation efforts involving captive breeding.

FEMALE ANATOMY AND PHYSIOLOGY

Ovaries and oogenesis

The Aboriginal people who helped our work at Maningrida in the 1970s and 1980s had a good understanding of female crocodylian anatomy (Figs 12.22, 12.23). Females have paired ovaries posteriorly within the body cavity, ventral and medial to the kidneys at their anterior end (Figs 12.24, 12.25). They

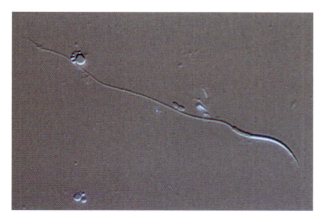

Fig. 12.20. *A filiform spermatozoon of* C. porosus. *The acrosome and haploid nucleus make up the head (bright blue), in front of the mid-piece and flagellum (tail). (Photo supplied by Steve Johnston)*

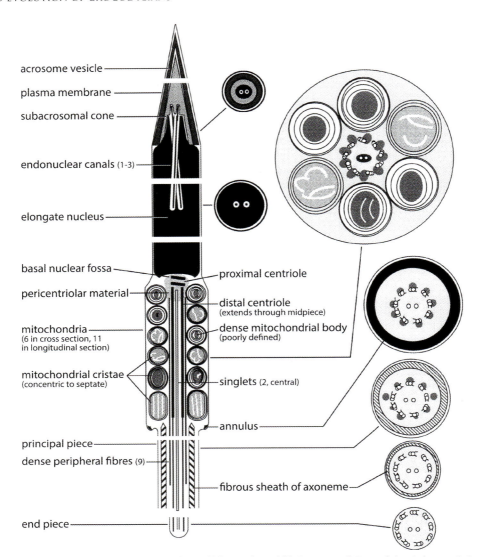

Fig. 12.21. *Schematic diagram of the structure of the filiform (thread-like) sperm of* Crocodylus johnstoni. *Its 'war head', the haploid nucleus, sits behind the pointed acrosome, with the midpiece (its 'battery pack') and the flagellum (propeller) a bit further back. The flagellum occupies most of the length of the sperm. It is made up of the 'principal piece' and a much shorter 'endpiece', both abbreviated here because the interesting bits are at the anterior end. In crocodylians, sperm have to swim or be conveyed by muscle contractions 1–1.5 m up the oviduct to be able to fertilise eggs before they become invested with membranes and the calcareous shell. Arriving at the still naked ovum, the digestive enzymes contained in the acrosome are thought to facilitate entry, perhaps assisted in some way by perforatoria housed in the endonuclear canals. (Figure provided by Barrie Jamieson from Jamieson* et al. *1997)*

are creamy yellow in young and non-reproductive females, reddish when suffused with blood during the reproductive season. Histologically, there is a cortical zone, in which ovarian follicles develop and from which, ultimately, yolky ova (singular: ovum) are shed, surrounding a medullary zone of spongy well-vascularised tissue and fluid-filled spaces. As with males, most of our knowledge about reproductive biology of female crocodylians comes from alligators (reviews by Lance 1989; Guillette

et al. 1997; Guillette and Milnes 2001; Uribe and Guillette 2000; Moore et al. 2008) but similar data on C. niloticus (Kofron 1990) provide reassurance that generalisations can be made across crocodylians.

The early stages of ovarian development have been described in alligators by Moore et al. (2008). At hatching, the cortex contains germinal epithelium containing oogonia (diploid primordial germ cells) and scattered oocytes (paused at the first stage of meiosis). Soon after hatching, follicles begin forming in the thickening cortex and by 3 months the ovary has an expanded cortex containing many oogonia (all she will ever have) and oocytes, some enclosed within tiny previtellogenic follicles: spherical cellular sacs that can be seen through the thin transparent covering of the peritoneum enveloping the ovaries, giving it a granular appearance. The process by which oocytes undergo meiosis and develop into mature eggs (ova) is known as oogenesis. Once a female alligator is sexually mature, her ovaries can be expected to have follicles at three size classes: 1 mm, 1–4 mm and a crop of vitellogenic pre-ovulatory follicles 5–45 mm in diameter (Figs 12.26, 12.29), suggesting a follicle life of at least 3 years (Lance 1989).

As in the males, environmental cues initiate reproduction, stimulating the hypothalamus to produce a crocodylian-equivalent GnRH, which stimulates the pituitary to produce glycoprotein gonadotropins equivalent to FSH (Follicle Stimulating Hormone) and LH (Luteinising Hormone). Whereas in males LH stimulates the production of testosterone by the Leydig cells, in females it stimulates the production of oestradiol, a derivative of testosterone, by the ovaries, which results in extensive growth of the oviducts and hypertrophy of the oviducal secretory glands in anticipation of ovulation. It also has a role in precipitating ovulation. Some testosterone is also produced by the ovaries, peaking at the time of

Fig. 12.23. *Kelly Bardgudubu with one of his bark paintings, c. 1975. This is in the X-ray style and depicts a gravid female saltwater crocodile,* C. porosus, *common in the Liverpool and Tomkinson Rivers, which enter the Arafura Sea at Maningrida, in Arnhem Land, Northern Territory, where this photograph was taken. Maningrida is one of several major centres for Australian Indigenous art. (Photo GCG)*

Fig. 12.22. *Kelly Bardgudubu explains how eggs are formed in saltwater crocodiles,* C. porosus. *Maningrida, Northern Territory, Australia, 1975.*

mating (Fig. 12.30). Growth of the season's ovarian follicles is stimulated by FSH and these also secrete oestradiol. Circulating oestradiol stimulates the production in the liver of the protein vitellogenin, which is transported to the ovary where it is taken up by the ovum and converted to the egg yolk proteins and lipoproteins. In due course, these provide nutrient for the developing embryo. The amount of vitellogenin is massive: Lance (1987) described plasma being cloudy because of the amount of vitellogenin being transported. When fully yolked up, the ova in alligators and similarly large crocodylians are 40–50 mm in diameter, which is huge! (Fig. 12.24 right, 12.25). Along with the growth of the follicles, primary oocytes divide

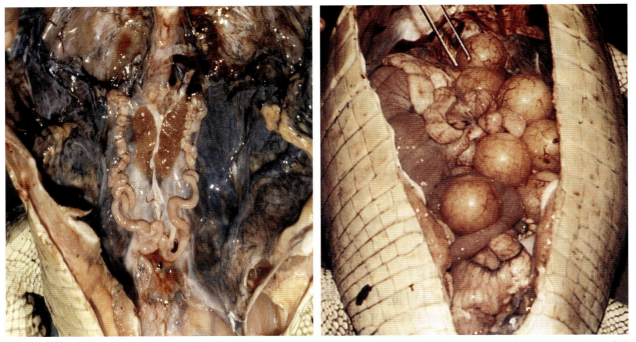

Fig. 12.24. *Reproductive anatomy of non-breeding 1.5 m (left) and breeding 1.6 m (right)* Caiman yacare. *(Photos Marcos Coutinho)*

(presumably still within the ovary) to become haploid, mature ova and, with their accumulated yolk are shed from the ovarian follicles into the top of the oviducts (ovulation) where fertilisation occurs.

In the ovaries, what remains of each follicle after its ovum is shed becomes a corpus luteum, which is similar to birds rather than to other reptiles, and which has a temporary endocrine function, secreting mainly progesterone, which, by analogy

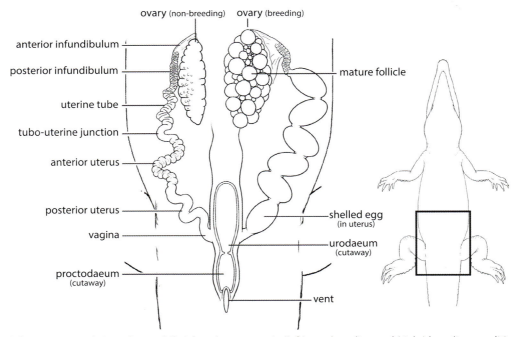

Fig. 12.25. *Schematic ventral view of crocodylian female anatomy in (left) non-breeding and (right) breeding condition.*

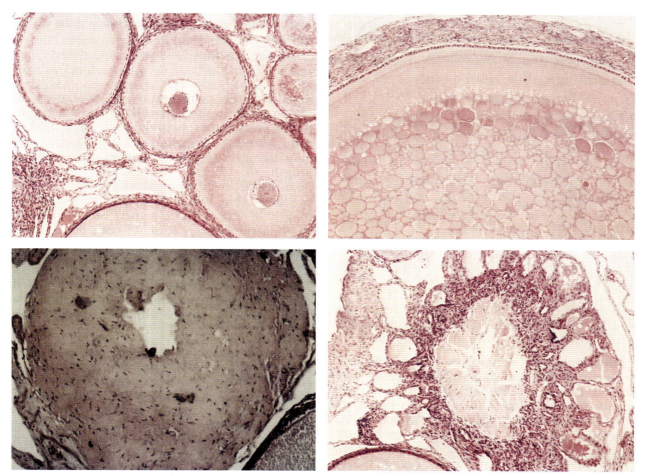

Fig. 12.26. *Elements to be found within a crocodylian ovary, in this case a 1.6 m* Caiman yacare *captured in May at the end of the summer breeding season. (Top left) Pre-vitellogenic follicles, (top right) vitellogenic follicle with yolk globules, (bottom left) corpus luteum and (bottom right) atretic follicle. (Photos supplied by Marcos Coutinho)*

with birds, supports the maintenance of gravidity. The volume of the corpus luteum and the levels of plasma progesterone both decline rapidly after the eggs are laid (Fig. 12.30) (Lance 1989; Guillette *et al.* 1995), but its remains are visible for some months, allowing determination of whether or not a female laid eggs during the previous nesting season.

Most research describing the histology of ovaries and the development of oocytes has been done on *A. mississippiensis*, particularly by Lou Guillette and colleagues (Guillette *et al.* 1995; Guillette *et al.* 1997; Uribe and Guillette 2000; Moore *et al.* 2008) but results from work on *Caiman yacare* (Coutinho *et al.* 2001) and *Caiman crocodilus* (Calderón *et al.* 2004) have emphasised similarity within alligatorids at least. Similarities in reproductive biology across

the Crocodylia, from gross anatomy to behaviour suggest that it is highly likely that ovarian development, vitellogenesis and oogenesis are similar across the extant crocodylians.

Oviducts, ovulation, fertilisation and the gravid stage

At ovulation, which may be stimulated by mating, the large yolky eggs are shed by the ovaries and wafted into the oviducts, probably by the funnel-shaped, thin-walled, cilia-lined infundibula, which sweep over the surface of the ovaries. As in other reptiles (Girling 2002) and in birds, the oviducts are differentiated longitudinally into specialised regions that have different functions (Palmer and Guillette 1992; Uribe and Guillette 2000; Guillette and Milnes 2001). This similarity to birds endorses the archosaurian connection yet again, although birds have only a single oviduct, presumably to lighten their bodies for flight.

Papers by Palmer and Guillette (1992), Buhi et al. (1999), Guillette and Milnes (2001) and Bagwill et al. (2009) have described many aspects of the morphology, histology and functioning of alligator oviducts. Some differences in terminology and interpretation need to be accommodated. Palmer and Guillette (1992) recognised seven regions along the length of the oviduct: anterior infundibulum, posterior infundibulum, tube (or 'uterine tube'), uterotubular junction (called isthmus by Bagwill et al.), anterior uterus, posterior uterus and vagina (Fig. 12.25). Most of what follows is drawn from their paper and one by Buhi et al. (1999). The oviduct is lined throughout with an epithelium of ciliated cells and non-ciliated secretory cells. Ciliated cells presumably assist in moving the eggs along. The posterior infundibulum is a muscular tube with a folded mucosal lining that lacks true glands. It also has secretory cells that secrete proteins that contribute to the egg's albumen: an important source of water for the developing embryo. The next section is the so-called tube (uterine tube), the transition marked histologically by the presence of branched acinar glands within a longitudinally folded mucosal lining. The tube is long and convoluted and secretes more of the proteins that make up the albumen and also mucopolysaccharides (glycosaminoglycans or GAGs), which are probably important in the elaboration of the egg shell. They may also be important for retention of water within the albumen. The uterotubular junction is a short translucent tube that lacks glands and whose main function seems to be to connect the tube to the anterior uterus, which is flattish, pale cream or grey, and has a glandular endometrial lining. The function of the anterior uterus in crocodylians seems to be the production of the proteinaceous fibres that make up the egg membranes, whereas the shell is calcified in the posterior uterus. Wink and Elsey (1986) and Wink et al. (1987) showed that egg-laying alligators lose calcium from the femurs, presumably as a source for shelling the eggs. Calcium is sourced from other skeletal elements as well, including the

osteoderms, which accounts for the formation of growth 'rings' used for skeletochronology (Chapter 13 and Klein et al. 2009).

Each oviduct opens separately and dorsolaterally into the posterior of the cloacal urodaeum via a vagina-like structure with a short narrow lumen, which spirals through the muscular cloacal walls (Palmer and Guillette, 1992) (Fig. 12.25). How these relate functionally to the traditional role of a vagina, as a receptacle for a penis, is unclear and the appropriateness of the term is uncertain. Females have a clitoris, which is much smaller than the penis but lies in the same position. Its function is unresolved (to us that is, but maybe not to a female crocodylian).

Gist et al. (2008) observed aggregations of sperm in the folds of the vaginal walls of recently mated alligators. Fertilisation must occur before the ovum becomes enveloped by the egg membranes and calcareous shell, which are secreted in the anterior and posterior uteri respectively. This is quite a long swim, but it will be assisted by muscular contractions of the oviduct, probably mediated by high levels of oestradiol (Steve Johnston pers. comm.). With large yolky eggs descending the oviducts sequentially, presumably distending them as they go and threatening to displace sperm, how could sperm remain so high in the oviduct? How would the last eggs in the clutch become fertilised? The answer seems to be that sperm is not only present there, but is stored there too, in crypts in the walls. Gist et al. (2008) and Bagwill et al. (2009) found sperm in the vaginal folds and also as far up the oviducts as the 'downstream' end of the tube, apparently stored in the lumina of oviducal glands (not the same ones that secrete the shell and shell membranes) (Figs 12.27, 12.28). Presumably this ensures an adequate supply of sperm throughout the period of ovulation. The authors did not imply that sperm can be stored from season to season, even though Davenport (1995) reported a P. palpebrosus that laid a fertile egg among a clutch of 16, 488 days after it was last with a male: an observation that remains unexplained (see also 'Multiple paternity' below).

At fertilisation, a haploid sperm unites with a haploid ovum to form a diploid zygote.

The fertilised egg now receives its quota of albumen from the walls of the posterior infundibulum and the tube of the oviduct. The proteinaceous fibres of the egg membranes are deposited as the egg travels through the anterior uterus (Fig. 12.25), and the calcareous shell is secreted around the egg within the posterior uterus. Thus, the whole clutch is treated sequentially, egg by egg, in different parts of the oviduct, until the female is said to be gravid. Eggs accumulate in the posterior uterus (Fig. 12.29) while she is constructing her nest. Towards the end of the period of gravidity, the female can look quite distended with eggs, and laying will soon follow.

The oviducts remain regressed in non-breeding adult female alligators (Fig. 12.25 left) but they have a great capacity to distend, growing 3–5 times longer when gravid, with thickened walls (Figs 12.25 right, 12.29).

Palmer and Guillette (1992) saw a similarity between the longitudinal separation of egg membrane fibre production and calcium deposition into different regions of the alligator oviduct and the differentiation of these functions into the isthmus and shell gland in birds. In other reptiles, the uterus is homogeneous. They described egg production as an 'assembly line' in both birds and alligators (and presumably all crocodylians). There is, however, one very great difference between them: in birds,

eggs are laid at a rate of one daily, whereas in crocodylians a large clutch of fertilised and shelled eggs are stored in the uterus – up to 60 or so in some species (see later) – and, as in turtles, are laid in one 'sitting'.

Seasonality and the hormonal control of reproduction

All crocodylians (except perhaps *P. palpebrosus*) are strongly seasonal breeders. Joanen and McNease (1989) published a comprehensive review of the seasonal reproductive events in alligators. Figure 12.30 summarises schematically the accompanying seasonal changes in the levels of major reproductive hormones in relation to the behavioural and other physiological events of the reproductive cycle. It is worth noting that even immature male alligators show a seasonal hormonal cycle similar to fully mature breeding adults, but at lower hormone concentrations, increasing gradually with body size. In contrast, female alligators show no seasonal cycle in oestradiol until they are mature (Lance 2003).

Although the hormonal cycle of the American alligator has come under the closest study (Lance 1987, 1989), similar cycles have been recorded in *Caiman yacare* (Coutinho *et al.* 2001) and *C. niloticus* (Kofron 1990) and, more recently, a new study of hormonal cycles in male alligators by Hamlin *et al.* (2011) has drawn attention to the role of DHEA. This last study was based on monthly samples taken

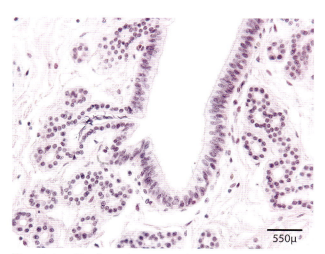

Fig. 12.27. *Histological section in the oviduct at the junction between the vagina and the uterus of a recently mated alligator. Sperm are visible in the lumen of one of the uterine glands. (Photo Dan Gist)*

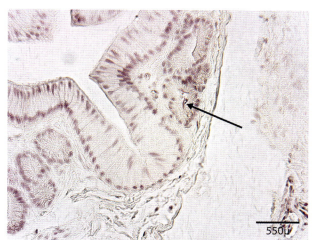

Fig. 12.28. *Histological section high in the oviduct at the tubal-isthmus junction showing sperm (arrow) in one of the glands in the wall of the tube. (Photo Dan Gist)*

Fig. 12.29. *Shelled eggs in the oviducts of a gravid alligator, shortly before laying. Note also the ovaries, with follicles of different sizes. (Photo Ruth Elsey)*

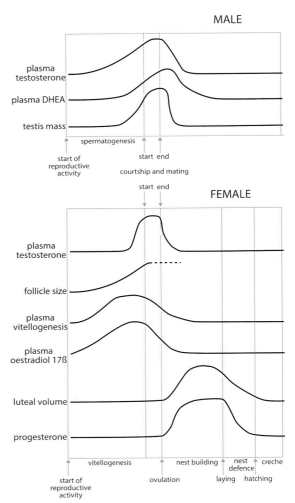

Fig. 12.30. *The anatomical, physiological and behavioural events that lead to the union of haploid gametes to form a diploid zygote and the subsequent deposition and care of the developing embryo are controlled by endocrine regulation, under the influence of environmental factors. Courtship leads to copulation, which may stimulate ovulation, and, while the fertilised eggs are passing down the oviducts and acquiring albumen and a shell, the female constructs a mound or hole nest. After laying, there is an incubation period of 2–3 months before the young hatch. Post hatching, most are 'looked after' in a crèche for a few weeks or much longer in some species. The endocrine systems that drive these events are homologues of those found in other tetrapods. This schematic synthesis is derived primarily from data on alligators published by Lance (1987, 1989), Guillette et al. (1997) and Hamlin et al. (2011), but evidence suggests broadly similar pattern across the crocodylians. In most species, including A. mississippiensis and C. porosus, the cycle begins towards the end of winter, with mating and nest building in spring, and hatchlings produced towards the end of summer. In strong contrast, C. johnstoni, a hole nester, lays eggs in late winter, which is the northern Australian dry season. This timing mostly avoids flooding of its*

from alligators captured throughout 2008–2009 in waterways surrounding the Kennedy Space Center in Florida. That population has been protected for many years and is disturbed only by the occasional blast of a departing rocket. The overall similarity between all these studies suggests that a synthesis of them, summarised in Fig. 12.30, is likely to be representative of crocodylians in general.

Photoperiod, temperature, rainfall, nutritional status and social factors are presumed to influence the level of circulating reproductive hormones and, thus, the initiation and timing of reproductive cycles. Most crocodylians live in habitats subject to large seasonal changes in temperature and/or water level, so it is not surprising that reproductive cycles are different between species according to habitat characteristics, and also within the same species in different parts of their range. As a general rule, the timing seems to fit production of hatchlings at a time of year that maximises their chances of survival, but there are plenty of exceptions. Indeed, egg loss from the flooding of *C. porosus* nests is so common (Chapter 13), one might think a different strategy should have evolved long ago through natural selection (perhaps Fig. 13.18). Temperature can be a major factor. Both American and Chinese alligators cope with colder winters better than any other crocodylians and the start of their reproduction depends upon seasonal temperatures. *A. mississippiensis* occurs as far as 35°N latitude, but they range to Florida at 25°N. The current, very restricted (see Chapter 14) distribution of *A. sinensis* is around 31°N. Both species may spend up to 6 months or so dormant (Chapter 10) and the onset of egg laying varies by 2–4 weeks depending on when the warm spring weather arrives (Lance 2003; Thorbjarnarson and Wang 2010). In these cooler climates, the whole reproductive cycle from vitellogenesis and spermatogenesis through to hatching has to be accomplished within the non-dormant months.

The production of sperm and eggs, mating and nesting all take time, so the cycle must begin long before the outward signs of the reproductive season and often in response to environmental cues most likely quite different from the propitious conditions anticipated for hatching. Day length is the most likely environmental cue in crocodylians, but how they measure that it is so far unknown. Measurement of day length in mammals involves the circadian cycle of melatonin produced by the pineal gland, but pineal melatonin seems not to be involved in birds (Meddle *et al.* 2002). With so many parallels between them, the way birds judge photoperiod may be relevant to crocodylians, but this is not yet fully understood either. Although it would seem logical that photoperiod reception would rely on vision, in birds the most likely source of the most relevant photoreception for determining day length is extra-retinal, in the hypothalamus, where there is also some type of circadian clock. Sharp (2005) has proposed that interaction between hypothalamic photoreceptors and the clock leads to the conversion of thyroxine to triiodothyronine (T3) and neurosecretion by the hypothalamus of gonadotropin-releasing hormone (GnRH) which, in turn, stimulates the production of hormones that initiate reproductive events. Although birds do have a pineal gland, and also show circadian cycles

in melatonin, it is not essential for the reproductive photoperiodic response. Like birds, crocodylians also have extra-retinal photoreception, in the brain and possibly in the hypothalamus (Kavaliers and Ralph 1981). Unlike birds though, crocodylians lack a pineal gland and, although they do have melatonin in the plasma, source unknown, it has long been thought not to show a circadian cycle (Roth *et al.* 1980). Recent work has shown, however, that there is a daily cycle in plasma melatonin in *C. johnstoni* kept in a natural photoperiod (Firth *et al.* 2010). It could be that further work will reveal that birds and crocodylians monitor seasonal changes in photoperiod by a similar mechanism.

This is not to imply that changes in photoperiod alone initiate reproductive events. In alligators, it has been proposed that temperature is influential (Lance 1989) and it is probably involved in all species, as well as other factors such as nutritional status. Reproduction incurs an enormous physiological drain on a female, who produces a large clutch of eggs followed by nest guarding, crèche management, and so forth, so all females are unlikely to breed every year. In alligators, about half of the adult females can be expected to breed in any single year (Lance 1989; Guillette *et al.* 1997, 2003). Thorbjarnarson and Wang (2010) tabulated the percentages of individuals breeding per year in three species of alligatorid and four species of crocodylids. The overall range was from <10% (alligators in North Carolina) to 21–88% (*C. niloticus* in different studies). Overall, *Crocodylus* species seem to have a higher proportion of individual females breeding in any single year. Maybe this is because in general they occur in warmer climates. In captivity, with plenty of food, it is common for female crocodylians to breed every year.

BEHAVIOUR IN THE BREEDING SEASON: MATING AND NESTING
COURTSHIP AND COPULATION

In his usual charming style, Tony Pooley described his excitement the first time he saw Nile crocodiles mating (Pooley 1982). This species is quite social and for some weeks Tony had been watching a group on a sandbank in the Pongola River, Ndumu Game

Reserve, in what was then Tongaland, in the north of the South African province of (now) KwaZulu-Natal. The crocs, mostly females, had become partially habituated to his quiet, seated presence and hauled out of the water each day to bask and doze, ignoring him. They were often joined by a large male (nicknamed Hans) from whom the other crocs kept 'a respectful distance'. Then suddenly one day one of the females, 'Skewjaw', approached the big male in the water and, submerging head and tail, displayed her rump. He responded by circling her, she 'gave a toad-like warbling growl' and Hans 'began a rough caress, rubbing the underside of his jaw and throat backwards, forwards and sidewards across Skewjaw's neck'. Tony recognised the similarity to what he had read about alligators and had no doubt about what he was watching. After 'many minutes', Skewjaw became rigid and raised her head almost vertically. Hans 'approached her from behind, brought a huge webbed hind foot over her rump, then a forefoot across the back of her neck and mounted her'. He 'rode' her for a couple of minutes, holding tightly with his claws with their tails intertwined, his bent downward and rotated to bring their cloacas together. Tony reported going home elated! The next afternoon a different female approached Hans, also displaying her rump, then raised her snout, jaws open, prompting Hans to give 'three mock bites at her bared throat in quick succession, creating a spectacular splashing as he half-circled her'. The mounting pattern was the same as the previous day, lasting just a couple of minutes. While this was going on, two other females, plus Skewjaw again, presented themselves to Hans and he was able to oblige, with about an hour's gap between each, before the end of the afternoon (!). He broke off his foreplay once to chase away a couple of males that Tony assumed were sub-dominant.

This description agreed pretty much with the report by Modha (1967) who watched eleven copulations lasting from 30–100 (average 58) seconds. He noted that copulation usually occurred in shallow water, the male sometimes lying on his side, but they also mate in deeper water.

I have seen mating in *C. porosus* only once, at a distance. The 5 m alpha male, Hymn, courted and (apparently) mated in quite shallow water in the very large lagoon at the Edward River Crocodile Farm. A short video sequence, filmed from a distance, is lodged at http://crocodilian.com/books/grigg-kirshner. The female drew herself to his attention by swimming close and lifting her head, then fleeing, but not too far. His interest aroused, he followed in the shallow water with speed and purpose, using his limbs and tail on the muddy bottom and making several low coughing sounds accompanied by a dance of water from his flanks as he sent out some SAVs while travelling. After some distance, she slowed as if allowing him to catch up, circled a little with head lifted and, after a little more circling and to-ing and fro-ing she appeared to allow herself to be caught and he swam over the top of her in the shallow water. His body was tilted towards the right side and dorsoventrally flexed. His mouth was slightly agape, at least part of the time. Water vibrated periodically from his flanks and his tail lifted intermittently, swishing from side to side. She must have been under water for much of the time. Glimpses of her head showed it was about level with his front legs and tilted upwards as if seeking a breath. Occasional spouts of water rose from the vicinity of her nostrils. They were still together when the twilight faded …

This serendipitous observation of mating by *C. porosus* fitted the general pattern except there was little if any extensive head rubbing and snout touching: Hymn may have been one of the more perfunctory lovers. Behaviour probably varies between individuals and in different circumstances.

As in other crocodylian social behaviours, there are differences between species but also strong general similarities overall. The earliest substantial description of courtship and mating in a crocodylian seems to have been by Modha (1967) after 8 months observing a large population of *C. niloticus* in a lake on Central Island, Lake Turkana. A more detailed and comparative study was undertaken by Leslie Garrick and Jeff Lang on captive populations of *A. mississippiensis* and *C. acutus* at Gatorama, near Palmdale, Florida (Garrick and Lang 1977). They included in their analysis data on *C. niloticus* from Modha, and described a sequence of stages applicable to all three species. There have been further descriptions: in the wild, more on *C. niloticus* (Pooley 1982; Kofron 1991) and of *C. palustris*, the mugger (Whitaker and Whitaker 1984); and, in captivity, *Caiman crocodilis* (Staton and Dixon 1977), *C. johnstoni* (Compton 1981), *C. intermedius* (Thorbjarnarson and Hernández 1993a,b) and a very detailed study by Kent Vliet at the St Augustine Alligator Farm (Vliet 1987, 2001). Brief descriptions of bellowing and courtship by Wang *et al.* (2009a,b) show that the Chinese alligator courts very similarly to their American relatives.

The pattern of courtship and copulation across all the species so far described is broadly similar to that described by Garrick and Lang (1977), with the two species of *Alligator* a bit different from the *Crocodylus* species. A full, comparative review of the behaviour of all the species is outside our scope.

Courtship and mating occur in the water and auditory, visual, tactile and olfactory stimuli are all involved. The sequence comprises phases of attraction (advertisement behaviour), pair formation, precopulatory behaviours and copulation. These patterns are not stereotypic, showing a degree of flexibility and there are differences between individuals and between occasions in the same individual. Perhaps they like variety.

Among crocodylids, depending on which species and, probably, the local population/situation, a male's interest may be signified by roars, head slaps and/or release of SAVs. The male may approach a female or she may approach him, and frequently it is the female that approaches first. Chasing may ensue and the female is likely to snout-lift in submission (Fig. 12.31). Unless one of them breaks off, the pair is formed and courtship ensues, sometimes for a prolonged period (Fig. 12.33). Typically, individuals swim together, often in circles, with a lot of body contact and rubbing of heads over each other's head and neck, presumably distributing pheromones secreted by the chin glands (see comments on mandibular and paracloacal glands, Chapter 3). There may be loud exhalations, bubbling, submerging and re-surfacing and observers report an impression of gentleness, even enjoyment of each other's company. Sometimes, the female flees suddenly, but not too far, drawing the male into a chase before circling is resumed. The male may rest

his head on her back or shoulders (called 'riding'), periodically sending out SAVs. Unless one of them breaks it off, and if there is no interruption from a more dominant male, the courtship may lead to copulation (Fig. 12.36).

In alligators, which have been studied in much more detail, the pattern is similar, the main difference being the bellowing choruses which, at least in captive alligators, seem to be a prelude to courtship. Kent Vliet at the University of Florida, Gainesville conducted a very detailed study of a captive population of alligators at the St Augustine Alligator Farm in northern Florida, sometimes entering the water to make observations at alligator eye line (Vliet 2001) (Fig. 12.32).

Courtship aggregations occur in captive *A. mississippiensis*, with bellowing choruses made up of males and females together, usually early in the morning, with the males showing the famous dancing of water off their backs as they exhale, generating SAVs through the water as well as the vocal sounds through the air. *A. sinensis* show similar behaviour (Wang *et al.* 2009a,b).

Choruses of bellowing alligators, so well known in captive alligators, are less well known in the wild. Indeed, nocturnal courtship and breeding aggregations have been reported only recently (Dinets 2010), from several sites in Florida, Texas, South Carolina, Mississippi and Louisiana (Fig. 12.34). He reported bellowing, never during the courtship gatherings and more likely on the morning after them than the morning before. Courtship behaviour in gharials involves similar head pressing and touching behaviour as in other crocodylians (Fig. 12.35).

Bellowing in chorus may be an alligatorid feature: it occurs in *Caiman* and *Melanosuchus* but not, apparently, in *Paleosuchus*. It has not been reported among crocodylids, but may occur in muggers (*Crocodylus palustris*). If so, that would be interesting because, as implied by their specific name, muggers live in marshland and one theory about bellow chorusing is that it is an adaptation to living in the marshes, to summon adults to suitable free water sites.

It is easy to accept that bellowing could act as a 'rallying cry' to bring in dispersed individuals

Fig. 12.31. *Courtship in* C. porosus. *This male (left) is rubbing his head back and forth along the female's neck and flanks and slowly circling, while she assumes a submissive, head-raised posture. (Photo Adam Britton)*

Fig. 12.32. *Kent Vliet, University of Florida, during his PhD study of alligator courtship behaviour at the St Augustine Alligator Farm in northern Florida. In the foreground is a pair of courting alligators. (Photo supplied by Kent Vliet)*

acknowledgement (male on left)

'facing off' (female on left)

tactile behaviour on head and neck (male above)

mutual touching, bumping, rubbing (female on left)

female mounting male (female on left)

pressing down on snout (female on left)

copulation (male on left)

pair *in copulo* (male above)

Fig. 12.33. *Elements of courtship behaviour in the American alligator. (Photos Kent Vliet)*

Fig. 12.34. *A courtship gathering of American alligators in Big Cypress National Preserve, Florida. (Photo Vladimir Dinets)*

from surrounding marshlands. It is more difficult to imagine what role this plays once the congregation is assembled. Neil Todd, from his study of alligators at the Australian Reptile Park near Gosford, NSW, has a very intriguing idea about this (see also Chapter 5) (Todd 2007). He proposed that the SAVs are detected in the ear not by the basilar papilla in the cochlear, which detects sounds in the human audible range, but by the sacculus (Fig. 5.11). This is the main sound detection organ in fishes, sensitive at low frequencies, and in fishes and amphibians it is thought to be related to reproductive activity. Todd has suggested (*pers. comm.* 2007) that the sacculus may have an auditory function in all vertebrates and he has gone as far as suggesting that the appeal of thumping bass percussion in rock music may lie in its stimulus to the sacculus, with ancient sexual overtones. Could a second function of the bellowing in alligators, in addition to a role in attraction to an aggregation site, be sexual stimulation?

The actual mechanics of copulation are improperly understood but we can make a darn good guess. Most (squamate) reptiles have two hemipenes, which they use alternately, whereas crocodylians have only one penis. The connections between the uteri and the urodaeum are identified as vaginas, but the term may be inappropriate because it is unlikely that the penis enters either of them. More likely, the sperm is deposited into the urodaeum, as in other reptiles, from whence it travels high up into both oviducts (Fig. 12.25). The anatomy of the penis and its functional aspects have been described above. Presumably during copulation the glans penis is inserted through the sphincter at the upper end of the proctodaeum and into the urodaeum where it becomes engorged with blood and acts as a plug (shaped a bit like a 'plumber's friend'), which holds the penis in place and prevents contamination by water while semen is ejaculated from the tip (Fig. 12.37) adjacent to the ostia of the female's vagina. Perhaps the urodaeum undergoes a muscular contraction during copulation, which squeezes the semen into the lower ends of the uteri, from where they swim

Fig. 12.35. *Courting gharials. Note the male's ghara and eyes protruding above the water. They show similar behaviour to other crocodylians. (Photo Vladimir Dinets)*

and are 'squished' by muscular contraction up the oviducts to the tubo-uterine junction (Figs 12.25, 12.27, 12.28) where many are stored in crypts of glands lining the walls (Gist *et al.* (2008); Bagwill *et al.* (2009).

MATING SYSTEMS

Determination of paternity in a range of wildlife populations has been made possible by techniques in molecular genetics and, as in human populations, there have been some surprises. For example, dominant male ungulates guarding harems have been revealed to be cuckolded quite often, leading to the invention of the term 'sneaky fucker' (or, for polite company, 'sneaker male' or kleptogyny). Even more interestingly, the same techniques have revealed contributions from more than one father to the litters or clutches of a diversity of mammals (rodents, bears, shrews, deer, antelopes and even a primate), birds (a long list) reptiles (skinks, turtles and now crocodylians – see below) and some frogs.

With external fertilisation, multiple paternity in a frog comes as no surprise, but it has become clear that multiple paternity can be a trait selected for, conveying a fitness benefit. If a female is going to put all that energy and time into raising a litter or clutch of young, perhaps it is better to have a few fathers in the hope that at least one will be worth all that effort!

Multiple paternity

Genetic analysis of the clutches from several crocodylians has shown that it is common for more than one father to have been involved. This was first shown in alligators by Davis *et al.* (2001). Using five polymorphic microsatellite loci, they genotyped hatchlings from 22 wild clutches in the Rockefeller Wildlife Reserve in Louisiana, as well as the attending females so that the maternal genetic profile was available. Multiple paternity was diagnosed in seven of the clutches, with one showing evidence of three fathers, the others, two. They also

Fig. 12.36. *Mating in Cuban Crocs,* Crocodylus rhombifer. *The female has her head lifted. (Photo John White)*

found evidence of a male contributing to more than one clutch. A much longer and more comprehensive study, also conducted on alligators in the wild at the Rockefeller Wildlife Reserve, examined eggs from 114 nests over eight breeding seasons (SL Lance *et al*. 2009). Genotypes at five microsatellite loci provided evidence of multiple paternity in 51% of clutches, variable from year to year (40–67%) but always high. Additionally, clutches produced by 10 females over 2 or 3 years were monitored to see if any females mated exclusively with the same male for more than one season. One female was sampled in 3 years and all her offspring were sired by the same male. Another had clutches in two sequential seasons with the same father. The other eight all had at least one multiple-fathered clutch.

Hu and Wu (2010), also studying alligators, used microsatellite analysis and found multiple paternity in at least three out of ten clutches in Chinese alligators (*Alligator sinensis*) at the Xuanzhou Nature Reserve. In another study of alligatorids, Amavet *et al*. (2008) used microsatellite markers to sample

genotypes in four nest-guarding broad-snouted caiman (*Caiman latirostris*) and their hatchlings in Santa Fe province, Argentina. They concluded that at least two of the four showed evidence of more than one father. Multiple paternity has also been

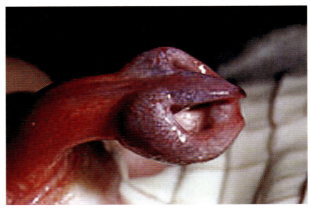

Fig. 12.37. *Penis of the spectacled caiman,* Caiman crocodilus. *The speculation is that the tulip-shaped glans is inserted as far as the female's urodaeum and held in place when it becomes engorged with blood. (Photo Adam Britton)*

shown to occur in black caiman (*Melanosuchus niger*) (Muniz *et al.* 2011).

Multiple paternity occurs in crocodylids too. McVay *et al.* (2008), using microsatellite data, inferred contributions from at least two fathers in hatchlings incubated from five out of 10 nests of Morelet's crocodile (*C. moreletii*) in Belize. In *C. porosus*, Lewis *et al.* (2013) also used microsatellite markers to examine 386 hatchlings from 13 wild clutches (Melacca Swamp, Adelaide River, Northern Territory) and 364 hatchlings from 21 clutches laid in the Edward River Crocodile Farm (North Queensland). Multiple paternity showed up in 69% of the wild and 38% of the captivity-produced hatchlings. Up to three fathers were indicated in some clutches. There was nothing to suggest a higher hatching success in multiple-fathered clutches.

Promiscuity

Crocodylians have been said to be polygynous (Lang 1987): that is, a male will mate with several females, and we know from Pooley (1982) that this occurs in the wild, even in a single afternoon (!). Indeed, such high proportions of multiple paternity might not be expected in a species whose mating system seems to be so dominated by alpha males ('boss crocs', see above). What is unclear, however, is the extent to which alpha males maintain a 'harem' by their territorial defence and Modha (1967) said quite specifically that females moved through and were not resident in a male's territory. In a smaller area, however, such as in the billabong dominated by Roughnut (see above), a boss croc may be able to control access more effectively. The high frequency of multiple paternity revealed by all the studies so far suggests that if the males are polygynous the females are certainly polyandrous and Lewis *et al.* (2013) have concluded that the mating system might be better defined as promiscuous.

'Site-fidelic' and roving males

On a larger geographic scale than that reported by Brien *et al.* (2008 above), an example of males moving through other males' territories has recently been reported from a study of movement by *C. porosus*

in the Wenlock River on the west side of Cape York, Queensland. Eight males were followed. Five of them remained in comparatively small areas, some overlapping, but the other three, dubbed 'nomadic' by Campbell *et al.* (2013), travelled the rivers extensively. The 'site-fidelic' males travelled just as far as the nomads, but almost all of it was within their core area, as if patrolling. Their 'home' areas were much larger than the females' areas. Three of them overlapped extensively with females, the other two were adjacent. No observational behaviour is available, but the movements of these five males are reminiscent of boss crocs, discussed earlier in this chapter, and it is easy to postulate that patrolling, territoriality and dominant status in their patch leads to a higher chance of mating success. Consistent with this, four of the five 'sedentary' males were larger than 4 m, but only one of the three nomads. The nomadic males passed through the 'territories' of all the others, perhaps showing appropriate obeisance. The high proportion of multiple paternity seen among clutches of crocodylians in the wild (above) implies that passing males are unlikely to pass up any opportunity to mate, should it arise.

Mate fidelity: female choice

In contrast to rampant promiscuity, SL Lance *et al.* (2009) found examples of mate fidelity within alligators in a 10-year study in the freshwater marshes of the Rockefeller Wildlife Reserve in Louisiana. Although they reported multiple paternity in about half of the clutches they examined, they also found evidence of mate fidelity, by females; the males were not so faithful, siring offspring with different females. Lance and colleagues. found matings by female alligators with the same male in different seasons in seven out of the 10 females they monitored over several seasons and concluded that these were examples of mate fidelity. They argued against the alternative explanation (that matings were random): the result of proximity alone. First, there are data that male alligators in this marsh habitat have larger home ranges and travel more than females. Also, taking into account the relative population densities of reproductively active males and females in the study area, they calculated that there is only a very small chance of repeated mating between the same pair occurring randomly. There is therefore the intriguing possibility of females showing active mate choice, which is well known in birds but not in reptiles, except if monogamous. The idea is given strength by records of captive females rejecting a prospective partner repeatedly, but mating successfully after a different male is provided. Lance and colleagues wondered if females may be choosing dominant males and identified the need for further work to include genetic typing of males as well as females, and behavioural data, presumably including telemetric or recording GPS data, to resolve the issue.

Sperm storage

If crocodylians could store sperm from season to season, that could lead to false conclusions about a female mating with more than one male in a season. Gist *et al.* (2008) and Bagwill *et al.* (2009) found evidence of sperm storage in alligators, in the folds of the vaginal walls and far up in the oviducts (Fig. 12.28). Both groups concluded, however, that sperm could not be stored from one year to the next because there was so little sperm remaining in the storage areas at the end of the season. Sperm storage could hardly explain the results from SL Lance *et al.* (2009) anyway, with some contributions from the same father 3–5 years after the first one was identified.

The apparently widespread occurrence and high incidence of multiple paternity in crocodylians, with no negative studies reported so far, and examples from both alligatorids and crocodylids, suggests that it conveys benefits that have been selected for. These are likely to be the maintenance of genetic variability and ensuring fertilisation with at least some of the most viable sperm. There may be conservation implications too, because resultant increased genetic diversity may increase a population's recovery rate after a bottleneck or large loss of genetic diversity (McVay *et al.* (2008). Finally, it is worth noting that its widespread occurrence – across alligatorids and crocodylids – suggests that multiple paternity is probably very old. With birds showing a similar trait, that of extra pair mating, this may be a mating system that is

very old indeed. Maybe it was even characteristic of the basal archosaurs and dinosaurs …

NESTING AND NESTS

Holes and mounds

About one-third of crocodylians nest in a hole they dig in a suitable substrate; the rest build a mound (see below). All Alligatoridae and about half of the Crocodylidae (plus *Tomistoma*) nest in mounds; the remaining species use holes (Table 12.2).

Mound nesters lay their eggs in an organic nest mound constructed by the female with great care and over a period of a few days or weeks, following ovulation and while the eggs are being invested with albumen and a shell. She uses grass or other vegetation, litter, and sand, shell grit or soil in a wide range of proportions, depending upon what is available at the site (Figs 12.38, 12.39). I have seen nests of *C. porosus* that are almost completely sand and shell grit, soil and leaf litter or grass, or mixtures of all of these. The mound is constructed using the hind legs, with the vegetation being torn off using the teeth or dug up with the hind feet. The female frequently leaves a tail drag over the top of the nest after laying (Fig. 12.39). Sometimes a female will discontinue construction before a nest is finished and move elsewhere to start and then use another, presumably because something about the first mound did not suit. Mound nests are usually built by species that nest in a monsoon or rainy season and they often have one or more wallows adjacent, in which the female will lie up.

In mound nesters, the eggs are usually laid at night into a nesting cavity in the top of the mound, formed by the female with her hind legs and dug immediately under her cloaca as she lies over the nest, using each leg alternately. The eggs, wet and shiny with mucus, drop from the cloaca singly or several at a time and in quick succession. No nest material or soil is placed between them, so the eggs are surrounded by air spaces (Fig. 12.40). They are then covered and any sign of the nest cavity obscured by the female's finishing touches. When eggs are laid in a hole, the nest chamber is dug and formed with the hind legs, like the familiar and ritualised-looking nest-digging by marine turtles

TABLE 12.2 NESTING HABITS OF EXTANT CROCODYLIANS (FROM A RANGE OF SOURCES, INCLUDING BRAZAITIS AND WATANABE 2011)

	Hole nesters	Mound nesters
Gavialidae	*Gavialis gangeticus*	*Tomistoma schlegelii*
Alligatoridae		*A. mississippiensis* *A. sinensis* *Caiman crocodilus* *Caiman yacare* *Caiman latirostris* *Melanosuchus niger* *Paleosuchus trigonatus* *P. palpebrosus*
Crocodylidae	*C. acutus* (primarily) *C. intermedius* *C. rhombifer* (primarily) *C. niloticus* (communal) *C. suchus* *C. johnstoni* *C. palustris*	*C. acutus* (less commonly) *C. moreletii* *C. rhombifer* (less commonly) *C. siamensis* *C. porosus* *C. mindorensis* *C. novaeguineae* *Mecistops* (both species) *Osteolaemus* (3 species)

(Fig. 12.41). *C. acutus* sometimes nests in a hole dug into a mound of sand (Fig. 12.42). Once egg-laying has started the females go into a trance-like state and are not easily interrupted (Fig. 12.43). After laying, the nest is filled in to the level of the surroundings and tamped down so that it is no longer noticeable.

The eggs seem to have inherent bactericide and fungicide properties, because in natural nests they usually stay relatively pristine and free from infection throughout the considerable incubation period, despite what must be legions of potential infectants in the surrounding nest material. This apparent natural bio-protection must be compromised when eggs are removed from a nest and taken to an incubator, which is the normal practice on crocodile farms, because infections, including fungal penetration, are then more common unless preventative steps are taken.

Bayani *et al*. (2011) reported the interesting occurrence and diversity of burrowing and hole nesting by muggers, *C. palustris*, (Fig. 12.44) in the north-western Indian state of Gujarat. They noted that the Vishwamitri River supports a healthy breeding population of muggers where it passes through densely populated areas of Vadodora, a city of about 2 million people. The authors distinguished different types of burrows, each several metres long, which are apparently used as retreats throughout the year. In one type, designated a chamber nest, egg fragments suggested its use for nesting. They reported the presence of a mud sill in front of some nests and, because its height increased from ~30 to 60 cm in the monsoon season, suggested its use for excluding water.

The explanation for this strong dichotomy in behaviour between hole nesting and mound nesting within a group that is otherwise so conservative in its reproductive behaviour remains unresolved. Allen Greer (1970) noted a match between crocodylian phylogeny as it was understood at that time and suggested that hole nesting is primitive. However, it has since become accepted that the extant members of *Crocodylus* have radiated more recently than the Alligatoridae, all of which make mounds. Further, seven out of 12 (currently recognised) *Crocodylus* species nest in holes, while *Mecistops* and

Fig. 12.38. *A* C. porosus *nest on the South Alligator River. Last year's nest is nearby. Note wallows and track between nest and water. This is the nest we spotted from the air and that I was later looking for on the ground when the female 'galloped' past me on her way to the water (Chapter 4). (Photo GCG)*

Osteolaemus (Crocodylidae) use a mound, as does *Tomistoma* (Gavialidae), while *Gavialis gangeticus* (Gavialidae) nests in a hole. Greer's hypothesis that hole nesting is primitive has therefore not stood the test of time. Grigg *et al.* (2010) observed that any explanation for the adoption of one or other nesting habits is more likely to be found by looking among ecological correlates. Various authors have put ideas forward. Brazaitis and Watanabe (2011) noted that mound nesters tend to nest in the rainy season, to raise the eggs above the damp, with hole nesters nesting in the dry season, with *C. siamensis* being one exception (Platt *et al.* 2011). Thorbjarnarson (1996) suggested that rising water levels might isolate mound nests from predation, and that mound nests may be less prone to flooding (Fig. 12.50). Another difference between them is that some hole

nesters nest communally (*C. niloticus*), which could provide advantage against predation. *C. johnstoni* may also nest close to each other but that may be a consequence of availability of suitable sandbanks because they do not defend their nests (Somaweera and Shine 2012). In contrast, Woodward *et al.* (1984) reported clumping of the mound nests of alligators in Orange Lake, Alachua County, Florida and noted that this could not be explained easily by focussing on a particular habitat. They suggested that clumping could result from social factors. Mound nesting seems to provide a more favourable gaseous environment (see below), but if that were an explanation, why is it that hole nesting persists at all?

The conclusion has to be that no explanation has yet emerged that accounts for this striking dichotomy in behaviour in a group that, otherwise, shows strikingly similar patterns of reproductive behaviour, physiology and ecology. And, as with most 'rules' in biology, there are exceptions: *C. acutus* is primarily a hole nester but sometimes constructs a mound, as do captive *C. rhombifer* and *C. palustris*. Medem (1981) observed that *C. intermedius* will build a mound if suitable substrate is unavailable. So the behaviour seems not to be completely hard-wired – absence of nesting material may provoke a captive mound-builder to dig a hole, or an unsuitable substrate may cause a hole nester to scrape up whatever can be found to cover her eggs.

Choice of nest site

Crocodylians are fussy about where they choose to nest. Brazaitis and Watanabe (2011), having tabulated the preferred nesting habitat of all extant species, pointed out that nearly all prefer freshwater wetlands or grasslands. They noted that those American alligators that live, unusually, in saline and brackish habitats in Louisiana (Fig. 11.31), move to freshwater-fed marshes to nest and rear young. This is not surprising, considering that alligators lack salt glands and have little capacity for life in saline situations (see Chapter 11). However, *C. porosus*, competent as they are in sea water, also seek freshwater sites for nesting. Brazaitis and Watanabe (2011) referred to them leaving saline mangrove and forest habitats on Palau to nest in

Fig. 12.39. *Some riverside mound nests of C. porosus in the Northern Territory. (From top left) Two views of a grassy nest and wallow on the East Alligator River; the female, tail just visible in the first photo, chose to leave hurriedly. Another well made grassy nest on the East Alligator River, then three on the Liverpool River, one with its tail drag, another with egg cavity exposed and (bottom right) a very shaded nest made mostly of soil and sticks. (Photos: top two Buck Salau, remainder supplied by GCG)*

Fig. 12.40. *Egg cavity opened out in a soil and grass mound nest of C. porosus. The eggs are in an air space. (Photo Garry Lindner)*

freshwater grasslands, and throughout their range they seek nesting habitat associated with fresh water. They can, however, still nest where the hatchlings' first taste of water will be salty, because *C. porosus* hatchlings survived from a nest beside the Northern Territory hypersaline Mungardobolo Creek (Fig. 11.4) where salinity varied from 36 to 64 ppt during the tidal cycle (Taplin 1984).

Additional attributes of good nesting sites for mound-builders are suitable vegetation or other material for nest construction, whereas hole nesters need a substrate in which a suitable hole can be dug.

Suitable sites may be close by. American alligators living in marshes, for example, mate in open water and may build their mound nests nearby. They can, however, be flexible or, perhaps, determined. Just as captive crocodylians may 'make do' if suitable material for a nest is not available, wild ones, too, will use poor habitat. The hole-nesting Orinoco crocodile, *C. intermedius*, for example, will dig a hole in rocky soil if a sandy beach is not available (Thorbjarnarson and Hernández 1993a) and, according to Medem (1981), will construct a mound if necessary. *C. johnstoni*, hole nesting in Australia's northern dry season, prefer sandbanks or shaded moist sandy soil and can usually find that without having to undertake a long journey. If suitable sandbanks are not available, they make the best of what there is. There are very few, if any, suitable sites on Lake Argyle: a huge artificial lake formed in the early 1970s for irrigation, by damming the upstream reaches of the Ord River in northern Western Australia. Figure 12.45 shows a typical water's edge and Fig. 12.72 the type of substrate in which *C. johnstoni* may dig their nest holes. Somaweera and Shine (2012) described nest site selection by females in the very large population that has grown since the lake was created. They found that the distribution of nests was far from random and that females often abandoned digging a hole, apparently having decided its moisture and thermal characteristics were unsuitable. Comparing these abandoned 'test'

Fig. 12.41. *Nest of a Nile crocodile dug into sand beside the Luangwa River in Zambia. Nile crocs often nest communally, but there was no sign of this here. Hole nesters dig with the hind feet, alternately, and usually deeply enough to reach moist soil or sand. They fill it in to the level of the surrounding substrate and pack it down, obscuring it. (Photos GCG)*

Fig. 12.42. *Frank Mazotti examines a mound nest of* C. acutus *on one of the Florida keys. This species is primarily a hole nester but both holes and mounds such as this are used in Florida and some parts of Mexico. The mounds made by this species are typically made of sand rather than organic matter. (Photo GCG)*

holes with nests with eggs, the authors found that the latter had lower and more stable temperatures, lower moisture and were situated in more friable, gravelly, sandy and loamy soil, rather than clay or rocky ground. Despite the lack of 'first choice' nesting sites, the lake now holds an enormous population of *C. johnstoni*.

Crocodylians do not always nest close to where they normally live: tracking studies have reported some substantial journeys. Kay (2004) radio-tracked *C. porosus* in the Ord River and other rivers flowing into Cambridge Gulf in northern Western Australia. Females occupied small core areas in the dry season and moved to nesting habitat in the wet season, making round trip journeys to sites 15–62 km away (Figs 12.46, 12.47). One of the females visited the same site in sequential nesting seasons, more than 20 km from her dry season core area. Across the top of Australia and far to the east in the Wenlock River and its tributaries on Cape York, Campbell

et al. (2013) used GPS systems that relayed data via satellite to monitor movements by male and female *C. porosus* throughout September–February 2010, in the southern summer monsoon: their breeding season. As in the Ord River, females left their core areas and travelled 23–55 km, presumably to nesting sites; they spent a long time there and showed very little movement, consistent with guarding a nest. Intriguingly, three of the four females made a preliminary journey to their supposed nest sites early in the season, remaining up to 48 h only, before returning 'home', and then 1–2 weeks later they again travelled to these presumed nest sites. They were still there at the end of February when the study ended, as would be expected if they were guarding a nest. The reason for these preliminary journeys is unexplained. It is very tempting to postulate that they choose a suitable site and then return to their core area or other suitable place, court and mate and then travel again to the nesting site. As pointed out

Fig. 12.43. *A 1.87 m female* C. johnstoni *in the process of covering her eggs after depositing them in a hole dug into the sandy bank of the Ord River, Western Australia. She had spent 23 min laying 14 eggs. Nests are dug with the hind feet, each one alternately to scoop out the soil. (Photo Ruchira Somaweera)*

by Campbell and colleagues, there is no obvious reason for a female to pre-select a nest site, or check out one already known, particularly when a round trip journey of more than 100 km may be involved.

Also unexplained so far is the precise benefit of these long journeys to a nesting area. This species nests solitarily and, although *C. porosus* seems to prefer broad-bladed grasses for nesting (Webb *et al.* 1977; Magnusson 1980a; Harvey and Hill 2003), they manage quite well with a wide range of material (e.g. Figure 12.39). It is not straightforward to explain why they expend such energy when, from our limited human perspective, one would think there are suitable sites much closer to home. Nest sites are often high up in side creeks (Figs 12.46, 12.47), associated with fresh water. But, because they nest in the wet season, all the rivers are likely to be running fresh right to the mouth at this time of year, and hatchlings that emerge late in the season into even hypersaline waterways still survive and

grow (see Chapter 11). Clearly there is a lot more to learn about the journeys undertaken by this species and the reasons for them.

Egg mass, clutch size and frequency, double clutching

Although crocodylians are conservative in many ways, adult female sizes range from less than 10 kg (*P. trigonatus*) to ~160 kg (*C. porosus*). Despite this size disparity hatchling size ranges only from ~20 to 90 g. Since hatchling size relates to egg size, it might then be expected that smaller species would produce fewer eggs than larger species and, within a species, younger, smaller females would produce fewer eggs. Hatchling production has big implications for recruitment to the population, so constraints imposed by body size could be significant. John Thorbjarnarson (1996) reviewed available data in a comprehensive review, considering possible correlations between clutch characteristics and

Fig. 12.44. Crocodylus palustris, *photographed in the Katarniaghat Wildlife Sanctuary on the Indo-Nepalese border. (Photo Vladimir Dinets)*

type of nest and phylogeny. He found no nest type or phylogenetic correlations, but concluded that larger females do produce larger eggs, as well as more of them, and thus drop heavier clutches. Also, egg mass declines as a proportion of body mass, explaining the lower proportional range in hatchling size than in female body mass. The stand out species are *Mecistops cataphractus*, laying small numbers of very large eggs and *C. porosus* and *C. niloticus*, which lay very large clutches of large eggs. In a study of the correlations between female body size and clutch and hatchling characteristics in a single species, Verdade (2001) found that larger female *C. latirostris* produced larger eggs and larger hatchlings, but not larger clutches.

I have been unable to find any confirmed example of a female in the wild producing two clutches in a single season, but there are instances of its occurrence in captivity, which could be a consequence of captivity providing a more reliable and sub-

stantial supply of food. Platt *et al.* (2011) reported a double-clutching by captive *C. siamensis* on a crocodile farm in Cambodia and referred to anecdotal reports of occasional occurrences on nearby farms. Double-clutching has also been reported in captive *C. palustris* by Whitaker and Whitaker (1984), who noted that the second clutch was on average 41 days later, always smaller than the first and had a lower hatching success. Second clutches have been reported from captive *C. mindorensis* too, 4–6 months later and of similar size. In our early days working with *C. porosus* in northern Australia, we found some nests late in the season and wondered whether they could be second efforts by the same female, but none has been confirmed. There have been hints of the possibility in *C. niloticus*. In Uganda, Cott (1961) reported fresh eggs in August and early September and again in December and January. Graham's (1968) findings in Lake Turkana (previously L. Rudolf) make this feasible: dissect-

Fig. 12.45. *Ruchira Somaweera checking out a nest of* C. johnstoni *beside Lake Argyle: an artificial lake behind a dam on the Ord River, Kimberley region, northern Western Australia. (Photo supplied by Ruchira Somaweera)*

ing carcasses, he noted two sets of differently sized ovarian follicles and that there was viable sperm over a sufficiently long period.

When a negative assertion is made, it is commonly followed by someone popping up with an exception. However, if double-clutching ever does occur in the wild it must be uncommon. Certainly the production of large clutches of eggs makes a considerable energetic impost on a female and double-clutching could occur only in a habitat with abundant food. The most likely candidates for two clutches in a season would be large members of the larger species, whose clutch mass is a lower proportion of body mass, and any reports of it would be more in the category of curiosity than evidence of any significant recruitment boost to populations.

DEFENCE OF THE NEST

Nest attendance by the female (Figs 12.48, 12.49, 12.50) is likely to be universal but the proportion of

time in residence is variable, as is the likelihood of her defending it against predators (Table 12.3). In many species, including *A. mississippiensis*, *C. porosus* and *C. siamensis*, the female may dig a wallow close by which becomes filled with water and in which she lies, often submerged and hidden (Fig. 12.39). Assistance in defence from the male seems to have been reported from only *Caiman crocodilus*.

A female may not remain right at the nest, but lurk in an adjacent lagoon or billabong, handy in the event of a predator showing up. However, just because nest attendance and defence are reported for a species, it does not follow that all individuals will show it. Crocodylian eggs are targeted by a wide range of invertebrates and vertebrates (reviewed by Somaweera *et al.* 2013, see Chapter 13) and it is clear that nest defence is far from effective. There are few quantitative data, but in one recent study Charruau and Hénaut (2012) used camera traps to record attendance by *C. acutus* from laying to hatching at four nests on two Mexican islands in the Caribbean. One female visited often, two less often and one not at all. Visitation was more common early and late in incubation and, despite visits from nine known egg predators, no defence of the nest was recorded. With the current explosion in the use of camera trapping right across wildlife biology, we can expect to see a lot more data of this type. Until then, we have mainly anecdotal information and most relates to failure of defence against humans. In Australia, Aboriginal people have been robbing nests for millennia and the frequent practice of removing freshly laid eggs from nests in crocodile farms and in the wild provides an opportunity to observe nest defence behaviour. Often there is no sign of the female, but sometimes she emerges suddenly and in spectacular fashion from her hiding place and rushes with great speed towards the nest (Fig. 12.51). Frequently she will scamper right up onto it and adopt a threatening stance with mouth open. People collecting eggs soon learn to expect this and act accordingly. The attack is seldom pressed home (although I would not count on it!) and she can usually be chased off with a broom or light pole, often only after she has made a few angry snaps at it. Females do not defend the nest after the hatchlings have left: they're not silly. They will, however, continue to defend a

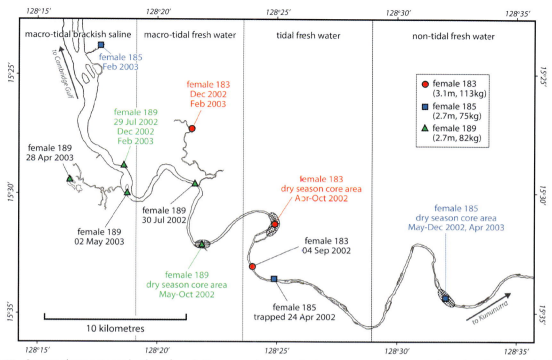

Fig. 12.46. *Seasonal movements by three female* C. porosus *between their core areas and presumed nesting sites. Females 183 (red) travelled 20 km and 189 (green) 15 km and did so in two successive years. Female 185 (blue) travelled 62 km to a presumed nest site and is known to have returned to the same core area at the end of the breeding season. (Adapted from Kay 2004, with permission)*

nest from which all the eggs have been removed for incubation in a commercial farm.

Kushlan and Kushlan (1980) took an experimental approach, presenting models of a raccoon and a human to alligators in Everglades National Park and noting the responses. They distinguished 10 separate component behaviours among the responses they observed from visits to more than 100 nests. These were divided into four phases: 'attentiveness' (approach, submerged, head emergent); 'threat' (position on or near nest, open mouth with or without hissing, open mouth lunge); 'aggressive' (mock bite, hard bite); and 'withdraw'. In the 'threat' and 'aggressive' phases, 'open mouth lunge' may be interspersed, providing a graded response depending on the stimulus and its persistence. Presented experimentally with the rag doll raccoon (investigators distant) 'The alligator initially responded … with open-mouthed posture and hissing. After these responses the alligator gave two hard bites, the second of which tore the raccoon model from its control line. The alligator then ate the model'. Presented with a polystyrene block the same size as the raccoon, the alligator behaved similarly except did

not eat it. They concluded that 'nest attendance may be effective in guarding against the alligator's most important contemporary nest predator'. Presented with a rag doll human on the end of a 3 m pole, the alligator showed a similar series of responses, culminating in a hard bite and shaking, and then withdrew. The authors concluded this would have been an effective deterrent. In a trial with a human model the day after hatching, an alligator that had previously attacked showed only an attentive response.

Nest defence is an enigma because, despite what seems to be an elaborate and rapid response, coupled with undoubted capacity to deal with almost any intruder, predation on nests is substantial. Adding to the enigma, it is characteristic of crocodylids, alligatorids and gavialids so it seems quite safe to assume that the crocodylian behaviours of nest attendance and defence are extremely old.

BIOLOGY OF THE EGGS AND EMBRYONIC DEVELOPMENT

To this point, our story has concerned the adults and their visible reproductive and nesting behaviour, as

Fig. 12.47. *This is the wetland, probably spring fed, to which Kay's* C. porosus *#183 travelled in the 2002 and 2003 nesting seasons (Fig. 12.46), 20 km from her core area on the Ord River. The stream exiting at lower right joins a creek that flows into the Ord River. (Photo Winston Kay)*

well as the internal physiological events that lead to the union of haploid eggs and sperm high in the oviducts. The tiny diploid zygote (embryo) so formed becomes enclosed with yolk (nutrient) and a water source (albumen) within a hard, calcareous shelled egg and is then deposited in the nest. The story now shifts to the embryos and their support systems, as well as the environment in which they grow, develop and, usually with parental assistance and care, emerge as hatchlings to start their independent lives.

EGGS AND DEVELOPMENT

Through eating hen's eggs (most boringly infertile), we are so familiar with the idea of egg-laying that we probably never stop to think about an egg as a biological entity. Not much seems to be happening when you look at a fertilised crocodile (or other) egg, yet within its own pool of fluid (the amnion)

the embryo, with all of its genetic blueprints dictating a capacity to grow to adulthood, is taking up oxygen (O_2) and excreting carbon dioxide (CO_2), gaining water from the albumen and nourishment from the yolk, excreting nitrogenous waste (mainly ammonia when the embryo is very small, then urea) into its allantoic sac … and growing. It was the evolution of internal fertilisation and the so-called 'cleidoic' (meaning 'box-like') egg, with its shell and packaged water and nutrients, that allowed the reptiles, unlike their immediate ancestors the amphibia (although there are exceptions), to avoid the necessity of returning to water to breed. When the embryo is tiny, these exchanges with its surroundings occur by diffusion but, with growth, the heart and other organs develop additional capacity and an extensive blood system develops **outside** the embryo. This provides a transport system that conveys deoxygenated blood from the

Fig. 12.48. *A female American alligator emerges from the water in response to a human visitor arriving at her nest. (Photo Ruth Elsey)*

embryo to a vascular network in the membranes (mainly the chorion) just under the shell, where O_2 that has diffused in through the pores in the shell can be loaded onto the embryonic haemoglobin (see Chapter 7) and conveyed back to the embryo. Likewise, an arterial supply sends out a network of blood vessels over the yolk, which absorb nutrient from the yolk, while the embryonic kidney excretes excess nitrogenous waste (not much, because there is little protein catabolism) into a sac attached near the cloaca, the allantois.

This is reminiscent of the connection of a mammalian embryo to the maternal blood system via a placenta, except that the extra-embryonic vessels and structures are connected directly to the nutrients and gases within, and immediately outside, the egg's shell. At hatching, the last of the yolk is taken up into the stomach of the hatchling and, sure enough, just as mammals retain a remnant of their lifeline 'stalk' to their mother via

the placenta, so baby crocs end up with an umbilical scar: a remnant of their lifeline to the nutrients within the egg. Sometimes little crocs hatch with a sac of yolk still not internalised completely.

With their hard shells and large yolks, crocodylian eggs are very much like bird eggs, and the similarities extend to their embryological development (Packard *et al*. 1977). Even the circadian pattern of yolk deposition during vitellogenesis emphasises their affinity with birds and their distance from other reptiles (Astheimer *et al*. 1989).

The literature on crocodylian eggs and embryonic development is extensive, reflecting the explosion of information in the area in the 1980s particularly, when the farming and ranching of crocodylians was becoming commonplace. Farmers were choosing to incubate eggs artificially in controlled conditions and the need for basic research was recognised, to maximise the advantages of these new facilities. There was an early review of reptile eggs and

Fig. 12.49. *An adult* Crocodylus siamensis *basking on her nest in Mesangat Lake, East Kalimantan, Indonesia.* Crocodylus siamensis *is one of the rarest crocodylians and also very shy. This photograph is also rare; Agata Staniewicz, who studied them in East Kalimantan, says she knows of no other photos showing* C. siamensis *nesting in the wild, at least not in Indonesia. This one was taken with a camera trap. (Photo Agata Staniewicz and Yayasan Ulin)*

embryos in general by Packard *et al.* (1977) and a more recent review by Deeming (2004a). A cluster of nine papers (Webb *et al.* 1987b) provides a very substantial review focussing on crocodylians.

Anatomy of the egg and its contents

Crocodylian eggs are about the size and shape of a goose egg (or a very large hen's egg in smaller species) and, unlike the soft shells of most other reptiles, are hard like birds' eggs.

An egg's internal components are easiest to view about half way through incubation. For our study of the O_2 transport properties of embryonic crocodylian blood (Grigg *et al.* 1993), we opened eggs and sampled blood from the chorio-allantoic vein using a fine needle. I always had a feeling of awe as we lifted off the top third of the shell to see this beautiful little creature on its pillow of yolk (Fig. 12.52). The components are shown

schematically in Fig. 12.53 and it is convenient to introduce these now to provide a vocabulary for what follows. Conspicuous and mostly at each end of the egg by mid-incubation, albumen provides a source of water, which becomes reduced to rubbery pads as development proceeds. Also conspicuous is the yolk, contained within the yolk sac, and the embryo developing within its amniotic sac. Less noticeably, the whole contents of the egg are surrounded by the membranous chorion. The embryo is nourished by the yolk via the vitelline artery and vein, which exit and enter the embryo via the umbilical stalk. This arrangement is reminiscent of the way in which developing placental mammals acquire nourishment from the maternal blood supply across the placenta. The allantois is a balloon-like out-pocketing from the gut, which stores waste products. Early in development, gaseous ammonia is excreted but

Fig. 12.50. A C. porosus *at her nest. This was taken in captivity in PNG. (Bill Green)*

this is gradually replaced by urea, which is stored in the allantoic sac.

The allantois, small at first, gradually expands until it envelops most of the egg contents. Fusing with the chorion, where it is pressed against the external shell, it forms the chorioallantois, providing for O_2 and CO_2 exchange between the embryo and the atmosphere of the clutch cavity, via the shell and shell membranes. As the embryo's O_2 requirements increase, the chorioallantois progressively increases in size, forming an expanding opaque band around the equator of the egg (Fig. 12.54). The gases are carried by the chorioallantoic artery and vein, which exit and enter the embryo via the umbilical stalk.

At the time of laying, the embryo, atop the yolk, is only a few millimetres in length (Fig. 12.55) and the egg contents are almost entirely yolk and albumen. As the embryo grows, these are consumed progressively, taking up less space, as the volumes of both embryo and allantoic sac increase. Webb

et al. (1987c,d) have described these changes quantitatively.

Opaque banding

The events shortly after laying are of particular interest to crocodile and alligator farmers because they dictate the way new eggs must be handled and allow fertile and infertile eggs to be distinguished. The events are described by Webb *et al.* (1987c,d), based mainly on *C. porosus* and *C. johnstoni*, with reference to *A. mississippiensis*. At the time of laying, the embryo's amnion and yolk sac are already formed and they attached to the vitelline membrane, which surrounds the yolk. Soon after laying, with the egg coming to rest in the nest's egg cavity, the yolk with the small embryo attached rotates within the surrounding albumen until it comes to rest at the top. This occurs in birds too and the rotation is presumably driven by gravity acting on a density gradient within the yolk. Within ~24 h, the embryonic membranes become attached to the

TABLE 12.3   SUMMARY OF NESTING CHARACTERISTICS AMONG CROCODYLIANS (F = FEMALE, M = MALE)

	Hole/Mound?	Mean egg mass (g)	Number of eggs	Incubation period (days)	Attend and defend nest?	Open nest and carry to water?	Creche?
Alligator mississippiensis	Mound	77	30–50		F	F	F
A. sinensis	Mound	48	20–30	65–82	F	F	Unknown
Caiman crocodilus	Mound	63	15–30		F, M assists defence	F, M opens, M? carries	F and M
Caiman yacare	Mound	61	30–60		F	F opens, but carriage not reported	F
Caiman latirostris	Mound	76	20–70		F	Opens, carriage not reported	Yes
Melanosuchus niger	Mound	144	30–65		F	F opens, carriage not reported	F
P. trigonatus	Mound	67	10–20		Unknown	Opens, carriage not reported	Unknown
P. palpebrosus	Mound	69	10–15		F	Unknown	Unknown
C. rhombifer	Mound/ Hole	104	30–40	60–75	unknown	Unknown	Unknown
C. moreletii	Mound	79	20–40	80	F	F	F and M
C. acutus	Mound/ Hole	113	30–60	90–100	F	F	F
*C. intermedius**	Hole	110	40–70	80–90	F, defence not reported	Unknown	F
C. niloticus	Hole	107	25–80	90–100	F	F,M open and carry	F and M
C. suchus	Hole	Unknown	Unknown	Unknown	Unknown		Unknown
C. siamensis	Mound	91	6–43		F?, defence not reported	Unknown	Uncertain
C. palustris	Mound/ Hole	100	25–30		F	F and M open and carry	F and M
C. porosus	Mound	110	40–60	80–90	F	F	F and M
C. mindorensis	Mound	73.6	10–20	85	F defends (in captivity)	Unknown	Unknown
C. novaeguineae	Mound	88.5	22–45	80	F	F opens, F and M carry	M
C. johnstoni	Hole	69.7	5–21	63–98	F	F	F
Mecistops cataphractus	Mound	146.0	13–27	90–100	F, defence not reported	F opens, F? carries	F?
O. tetraspis	Mound	55.0	10–20	85–105	F	F open, F and M carry	F
O. osborni	Mound				F	F open, F and M carry (presumably)	F (presumably)
T. schlegelii	M	139.9	20–60	90	F?, defence not reported	F opens, carriage uncertain	Unknown
G. gangeticus	H	161.4	30–50	83–94	F	F opens, but not transport	F and M

* *C. intermedius* can switch to mound if suitable substrate not available (Medem 1981).
From various sources, including: Brazaitis and Watanabe (2011); Thorbjarnarson (1996); Thorbjarnarson et al. (2001); Thorbjarnarson and Wang (2010); Platt et al. (2011); Webb et al. (1983a); Somaweera and Shine (2012); Waitkuwait (1985); Whitaker and Whitaker (1984); Whitaker (2007); Riley and Huchzermeyer (1999); Platt et al. (2008).

Fig. 12.51. *An airborne female* C. porosus *leaving the water speedily to defend her nest (right foreground). Note the broom, at right, being carried by the (rapidly departing) intending nest robber. Commonly, however, humans can approach crocodylian nests and provoke little or no response and, with care and sometimes after dissuasion, eggs can be collected for ranching (see Chapter 14). (Photo GCG)*

shell membrane, establishing the 'top' of the egg and once this has happened the orientation must be maintained for at least the first couple of weeks or the embryo will likely suffocate. On crocodile

and alligator farms, eggs are usually removed from the nests and transferred to an incubator, safe from predators and at controlled temperature and humidity. To avoid killing the embryos, farmers routinely mark the top of the egg at collection and maintain that orientation (Fig. 12.54). At this early stage, the egg has a certain translucency, but soon

Fig. 12.52. *An embryonic* C. porosus *about half way through incubation lies within its amniotic sac, surrounded by fluid in the allantois and lying on a pillow of yolk. Pads of albumen not used so far are visible at the ends of the egg. (Photo GCG)*

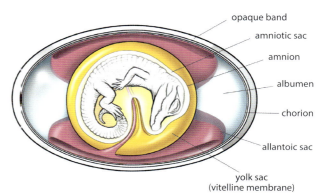

opaque band
amniotic sac
amnion
albumen
chorion
allantoic sac
yolk sac
(vitelline membrane)

Fig. 12.53. *Schematic representation of a crocodylian egg.*

Fig. 12.54. *Partly banded eggs of* C. porosus. *(Photo GCG)*

after the attachment of the embryo at the top of the yolk a white (opaque) patch develops above the developing embryo. This is probably the result of water being transported from albumen above the embryo to the sub-embryonic space below the embryo, drying out the shell in that region and thus opening pores in the shell and the shell membranes, so facilitating gas exchange. Within a few days the patch expands to encircle the egg around its equator (Fig. 12.54). At this stage, the respiratory and excretory functions of the chorio-allantois are adequately developed and embryos are more able to cope if the egg orientation should alter.

The development of this 'opaque band' allows farmers to sort the eggs, rejecting any that are infertile or in which the embryo has died. As development proceeds, the width of the opaque band expands as water is removed progressively from pores over more of the shell, opening up for gas exchange by diffusion between the atmosphere and the embryo. The chorio-allantois expands below the opaque band. By keeping an eye on the progress of the white band, until it encompasses the whole of the surface area, a farmer can remove any eggs whose development stops. The formation of the band and its progressive increase in size thus provides a croc farmer with a visible sign of satisfactory progress in development.

Embryonic development

By the time eggs are laid, embryos have undergone considerable development: The zygote will have divided repeatedly to form a 2–3 mm diameter circle of cells on top of the yolk (the blastodisc), divided further into upper and lower layers of cells, developed the 'primitive' streak and undergone gastrulation to form into a tiny bilaterally symmetrical embryo with a distinct head, a notochord (precursor of the vertebral column) and segmental muscle blocks (somites) (Fig. 12.55, e.g. D2). The external embryonic membranes, which form the embryonic 'support system' (yolk sac, amnion, allantois and chorion), are also in place by the time the egg is laid. For a reader curious about the detail of the very early development of the embryo, it is essentially identical to that of chickens, of which descriptions are readily available.

The earliest substantial descriptions of the embryological development of a crocodylian were by Clarke (1891) and Reese (1915), describing it in *A. mississippiensis* and Voeltzkow (1899) in *C. niloticus*. These are all available digitally and worth chasing up, if only to salute these pioneers and the wonderful artwork, even though Ferguson (1985, 1987) has drawn attention to significant shortcomings. These early descriptions defined numbered developmental stages based on morphological stages defined for the embryology of chickens. A comprehensive study by Ferguson (1985, 1987) of the embryological development of alligators incubated at 30°C led to a detailed staging system based on their progressive acquisition of visible morphological attributes. Comparisons with other species, also incubated at 30°C, led him to conclude that the system is also applicable to *C. porosus*, *C. johnstoni* and *C. niloticus* and is probably useful for all crocodylians. These papers and others by Mark Ferguson provide the 'gold standard' for crocodylian embryology. In a beautifully illustrated paper, Iungman *et al.* (2008) followed *C. latirostris* embryos incubated at 31°C and found a close correlation with *A. mississippiensis*, but also some differences. Standardised stages are necessary for comparative embryological studies. However, field biologists sometimes want to be able to sample an egg, assess the age of the embryo and judge when the clutch was laid. Because the rate of development depends on temperature, the relationship between morphological stages and time since laying is not constant and the stage does not necessarily provide a good estimate of age. The effects of temperature are far from trivial. Webb *et al.* (1983b) pointed out

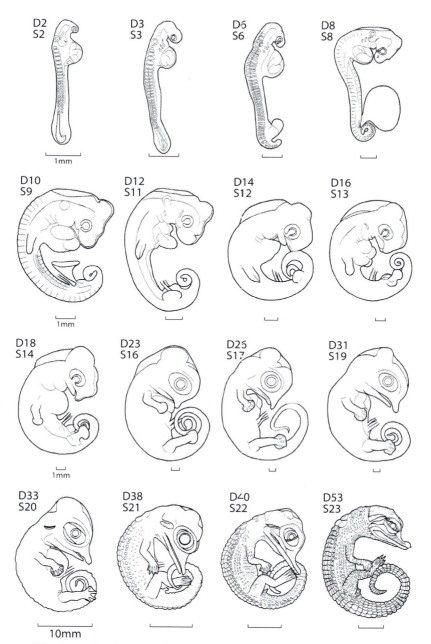

Fig. 12.55. *Development of* C. johnstoni *embryos incubated at 30°C. D = age in days since laying, S = morphological stage, deduced from Ferguson (1985). At 30°C,* C. johnstoni *can be expected to hatch at 85–90 days. By 50–55 days, the embryo looks like a miniature version of a hatchling, with the yolk sac still large and three more Ferguson stages to go. In these latter stages, the main visible change is increase in size and the gradual, partial or complete, incorporation of the yolk sac into the hatchling's body, leaving a mid-ventral 'umbilical scar'. Note: this figure is indicative only. Ferguson (1985) provided detailed descriptions of* A. mississippiensis *embryos at each stage, with accompanying photographs. (Adapted from Webb* et al. *1983a, with permission)*

that the rate of development in *C. johnstoni* at 26°C is half the rate at 30°C and that field incubation times in *C. porosus* range from at least 71 to 114 days, almost certainly the consequence of variable nest temperatures. Magnusson and Taylor (1980) drew embryonic development and recorded nest temperature in *C. porosus* and, subsequently, Webb *et al.* (1983 a, b) documented a developmental series of known-age embryos in *C. porosus* and *C. johnstoni* (e.g. Figure 12.55) incubated at 30°C and offered a way to take temperature into account and calculate real age. Further research on both these species provided a more robust approach. In a wide-ranging paper on the effects of temperature on embryonic development, Webb *et al.* (1987e) provided information that allows the age equivalent of an embryo at a particular morphological Ferguson stage to be translated to actual age. That is, a field biologist could judge the morphological stage of an embryo and translate that to an age at the relevant nest temperature.

THE ENVIRONMENT OF THE NEST

Eggs cannot walk, so their survival depends upon them being placed somewhere suitable for the 2–3 months of their incubation. Not all choices of nest site are wise: many eggs die through predation or flooding (Chapter 13) and some from overheating or drying out. Here we will discuss the environmental conditions in which embryonic crocodylians develop: the temperature, humidity and levels of O_2 and CO_2 that characterise hole and mound nests. Humidity should be high and fairly stable in both types of nest, and temperature reasonably well buffered, but one might think that O_2 levels would get very low and CO_2 very high, particularly around eggs buried in a mound of rotting vegetation. However, the successful production of hatchling crocodylians after 3 or so months in a hole or a mound of rotting vegetation suggests that these situations provide conditions that are mostly suitable. The next section compares and contrasts the attributes of hole and mound nests as providers of a suitable environment for developing embryos.

Oxygen and carbon dioxide and their exchanges

The persistence of both hole nesters and mound nesters for millions of years proves that both practices provide gaseous environments satisfactory for embryonic development. Levels of O_2 and CO_2 have been measured in both hole nesters (*C. acutus*, Lutz and Dunbar-Cooper 1984; *C. johnstoni*, Whitehead 1987) and mound nesters (*C. porosus* and *A. mississippiensis*, Grigg *et al.* 2010).

Hole nests of two *C. acutus* dug into sand/shell substrates showed lowest O_2 partial pressures of ~130 mmHg (compared with ~150 mmHg in saturated air) and CO_2 pressures of ~16 mmHg

(cf. 0.3 mmHg in air). (Note: 1 mmHg = 0.133 kPa.) A third nest in similar substrate showed lowest O_2 of 116 mmHg, and some nests in marl (mudstone rich in calcium carbonate) ranged from 120 to 140 mmHg O_2 and 13 to 20 mmHg CO_2. Lutz and Dunbar-Cooper concluded that such gas levels in successful nests must be favourable. Whitehead (1987) measured changes in gas partial pressures during incubation in natural hole nests of *C. johnstoni*. In coarse/medium grain sand O_2 levels fell to 130–135 mmHg and CO_2 levels rose to 12–15 mmHg as the embryos grew and required more O_2. Less favourable gas levels were found in fine-grained substrates: 105–125 mmHg O_2 and 20–40 mmHg CO_2, respectively. Hole nesting in a sandy or friable clay substratum apparently provides an adequate gaseous environment. Oxygen levels fell lower than usual following heavy rain.

What about eggs in mound nests of rotting vegetation? Surely we could expect to find challenging levels of O_2 and CO_2 here. But, again, the data show otherwise. Measurements of O_2 and CO_2 in gas samples from mound nests of *Alligator mississippiensis* averaged ~140 mmHg and 16 mmHg, respectively (Booth and Thompson 1991) and, in another study, the lowest and highest levels were 133 mmHg and 23 mmHg, respectively, and changed little during the incubation period (Grigg *et al.* 2010). This latter study also reported data from nine 'wild' mounds of *C. porosus* on the Tomkinson River, Northern Territory. The lowest P_{O_2} measured was 128 mmHg and the highest P_{CO_2} 22 mmHg. Two nearly full-term nests had gas partial pressures, respectively, of 146 and 5 mmHg (a grass and soil nest) and 130 and 22 mmHg (a soil and sticks nest).

In contrast to hole nests, the gaseous environment of mound nests is determined mainly by the metabolism of the mound itself. Gases in 12 mound nests with the eggs removed to incubators were measured at the Edward River Crocodile Farm, Pormpuraaw, Queensland (Grigg *et al.* (2010). The nests were constructed of soil, grass, shell-grit and vines in various combinations and the ranges of O_2 and CO_2 partial pressures were the same as in wild nests with eggs. The O_2 demand of the decomposing nest was 4–7 times that of the clutch at the end of incubation, with a respiratory exchange ratio (RE) of ~0.9, so it is not surprising that the metabolism of mound material has a bigger influence than the eggs on internal gas tensions, as in the mound nests of megapode birds (Seymour *et al.*, 1986).

The two types of nest emerge as having different gas exchange attributes. In mounds, but not holes, the nest material has a major influence on the gaseous environment, and in holes, but not mounds, O_2 levels fall progressively as the embryos grow. What is the explanation for this? Is the mechanism of exchange different between them? Diffusion explains gas exchange in the hole nests of sea turtles (Ackerman 1977) and must surely apply to crocodylian hole nesters too. Does gas exchange in mounds occur by convection rather than diffusion?

It seems not. Intuitively, convective gas exchange might be expected in mound nests, with metabolism warming the eggs slightly (see below) and generating a gentle upcurrent of air. However, it turns out that in mounds, as well as in holes, gas exchange between the clutch and the outside air occurs by diffusion. Grigg *et al.* (2010) found that their O_2 and CO_2 data fitted a steady-state diffusion model, but not a convective model. To put this another way, taking the respiratory exchange ratio (RER, Chapter 7) into account, the modelling showed that more CO_2 accumulates in relation to O_2 consumed than would be expected if gas exchange were by convection. The difference between them must be, therefore, that mounds have a larger surface area for gas exchange, in proximity to the clutch, than a hole nest, and that the diffusion

resistance of mound material is likely to be less than that of sand.

These considerations suggest that mounds may provide a more beneficial gaseous environment than hole nests, but leaves unexplained the dichotomy in nesting habits seen within Crocodylia.

The gaseous environment within the mound, however, is only part of the story. What about within the developing embryo itself? To be useful to the embryo developing within its amniotic pool, O_2 molecules must traverse several diffusion barriers: the shell and shell membranes being the largest. The diffusion barrier must prevent dehydration by retarding loss of water vapour while allowing the exchange of O_2 and carbon dioxide.

These are contradictory requirements, and a balance is struck between them, a balance that changes during incubation as embryonic O_2 requirements increase.

Increasing O_2 requirements during embryonic growth and development have been measured in *C. johnstoni* and *C. porosus* by Whitehead and Seymour (1990) and in *A. mississippiensis* and *C. porosus* by Grigg *et al.* (2010). A typical pattern is shown in Fig. 12.56. In these species and probably all crocodylians, O_2 requirements peak 80% of the way through incubation, coinciding with the maximum rate of embryonic growth, and then decrease towards hatching (Whitehead *et al.* 1990). This decrease is typical of reptiles with hard-shelled eggs and it may allow slower developers to catch up, thus facilitating synchrony of hatching.

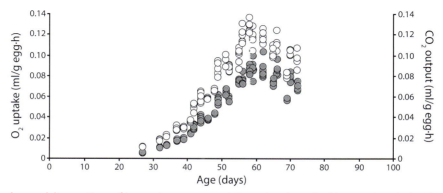

Fig. 12.56. *A typical crocodylian pattern of increasing oxygen uptake and carbon dioxide excretion during development, peaking at ~80% of the way through incubation. The decline as hatching approaches may allow slower developing hatchlings to catch up so all can emerge together. Respiratory exchange ratio (RER) equals 0.75. Measurement temperature 30°C. (Adapted from Grigg et al. 2010, with permission)*

Exchange of O_2, CO_2 and water vapour occur across tiny pores in the rigid calcareous eggshell and the underlying two layered, leathery shell membrane (Fig. 12.57). The structure of the shell and shell membrane have been described in detail by Ferguson (1981, 1982, 1985) and Kern and Ferguson (1997). There has been interest in finding the layer that forms the main barrier to diffusion and how its permeability changes as development proceeds. Counter-intuitively, the thinnest component of the shell and shell membranes provides the most significant barrier. The calcareous shell is pierced by numerous pores and has an outer layer of calcite crystals arranged vertically and an inner, thicker 'honeycomb' layer with the crystals arranged horizontally (Fig. 12.57) (Ferguson 1982). It could easily be assumed that the calcareous shell with its visible pores provides the diffusion barrier. However, Whitehead (1987) deduced that the calcareous shell contributes little to diffusion resistance, and that most of it resides in the underlying shell membrane. Kern and Ferguson (1997) came to the same conclusion. They described the ultrastructure of the shell membrane, resolving it into two layers: a 'fibrous membrane'; and, below that, a very thin, filmy 'limiting membrane' (Fig. 12.57). This has 'an immense population of tiny pores', ~340,000/cm² of them, 0.51μm in diameter. There are also slightly larger pores (diameter ~35 μm and 190/cm²). The evidence points to the limiting membrane providing the main barrier

to gaseous diffusion. At laying, only ~6% of these are open, letting in sufficient O_2 for the very small embryo but retarding water loss. At hatching time, removal of water from the shell membrane leads to 22–4% of the pores being open, allowing more O_2 in to accommodate it greater need by the embryo.

Even towards the end of incubation, however, the shell and attached membranes still form a significant barrier. In late term eggs of *C. johnstoni* surrounded by air, Whitehead (1987) measured P_{O_2} values of ~80 mmHg in air spaces between the shell membrane and the chorio-allantois. This seems surprisingly low – a drop of 70 or so mmHg across the shell and shell membranes – and it prompts questions about the 'oxygen cascade' between the clutch environment and the embryo, which will now be explored.

Developing embryos exchange gases via an external vascularised structure, the chorio-allantois, which is pressed against the inside of the shell membrane and increases in surface area with development, eventually lining almost all of the inside of the egg (Fig. 12.58).

The chorio-allantois is the embryo's respiratory organ and receives O_2-poor, CO_2-rich blood via the chorio-allantoic artery which, therefore, has 'venous' type blood, with the 'arterialised' blood returning to the embryo via the chorio-allantoic vein. Grigg *et al.* (1993) measured O_2 and CO_2 partial pressures in these vessels in full-term embryos of *C. porosus*. Blood samples were taken through a window cut in

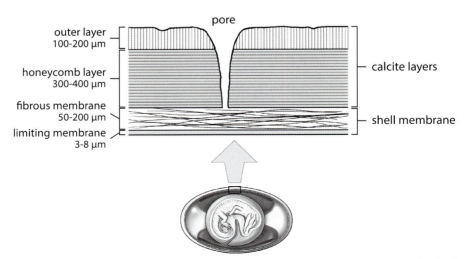

Fig. 12.57. *Schematic diagram of the structure of the shell and shell membranes. Counter-intuitively, the limiting membrane provides the main barrier to gaseous diffusion. (Based on Kern and Ferguson 1997)*

Fig. 12.58. *The well vascularised chorio-allantois within the egg of a nearly full term* C. porosus *embryo. (Photo GCG)*

the shell, with respiration occurring via the chorio-allantois. Oxygen partial pressure in the chorio-allantoic artery averaged 22 mmHg (3.0 kPa) and CO_2 32 mmHg (4.2 kPa), providing steep partial pressure gradients for exchange. After exchanges across the chorio-allantois, O_2 in the arterialised blood in the chorio-allantoic vein averaged 52 mmHg (6.9 kPa) and CO_2 averaged 12 mmHg (1.6 kPa), implying a blood oxygenation of only ~85%. These measurements were made with eggs in air, biasing O_2 levels higher and CO_2 levels lower than would be the case in a mound (or hole). It seems unlikely that embryos receive fully oxygenated arterial blood towards the end of incubation when O_2 demand peaks (Fig. 12.56). David Booth found that in *C. porosus* eggs holding late-term embryos, exposure to hypoxia of half atmospheric O_2 led to a halving of O_2 consumption, from ~10 to 5 mL O_2/h. He noted also that four late term eggs exposed to an O_2 partial pressure of 75 mmHg, half atmospheric, responded by pipping the shell and, presumably, breathing air (Booth 2000).

Humidity

The main source of water for the developing embryo is the albumen at each end of the egg and, if the hole or mound is too dry, the egg and its contents can dehydrate. On the other hand, in a very moist atmosphere, the egg can take up water by osmosis. Both holes and mounds usually provide a humid environment and the eggs are tolerant of a wide range. If it gets very dry, large air spaces can develop between the calcareous shell

and the leathery shell membrane. If very moist, eggs may swell and develop longitudinal cracks (Fig. 12.64), yet eggs with either of these symptoms commonly hatch successfully. However, if the pores in the shell become blocked with free surface water, diffusion is drastically reduced (Grigg 1987) and the embryos can asphyxiate. Presumably this is the cause of embryo mortality when nests are flooded (Chapter 13).

Crocodylian eggs are more permeable than bird eggs to water vapour exchange. Packard *et al.* (1979) found that eggs of *A. mississippiensis* eggs lose water 5 times faster than bird eggs of similar size and Grigg and Beard (1985) measured a 4–5 fold rate of loss in *C. porosus* eggs. Presumably the difference is that birds' eggs are frequently exposed to air in an open nest. Grigg and Beard (1985) also found that full-term eggs after exposure for a time to dry air can rehydrate in a humid environment, gaining water about 8 times faster than the rate at which water could be produced metabolically (Chapter 7) (Grigg 1987). The permeability of *C. porosus* eggs to water did not increase during incubation until cracks developed in the calcareous shell very late in incubation. Grigg and Beard (1985) measured high internal hydrostatic pressures in eggs towards the end of incubation: 105–122 mmHg above ambient. We attributed this to release by the developing embryo of low molecular weight metabolites into the allantois causing a higher osmotic pressure within the egg, leading to an increase in turgor. High pressures have been measured also in the western rat snake, *Pantherophis obsoletus* (Lillywhite and Ackerman 1984) and the veiled chameleon, *Chamaeleo calyptratus* (Adams *et al.* 2010). The shell is a source of calcium as the crocodylian embryo grows (Packard and Packard 1989), so the shell becomes weakened and more fragile towards the end of incubation. The degradation and cracking of the shell are further discussed below, along with the events related to hatching.

There have been few measurements of humidity within nests. Waitkuwait (1985) measured water content of the mound material of wild nests of *Mecistops cataphractus* in Côte d'Ivoire. He found them almost constant at 37%, with relative humidities of the outside air mostly from 85 to 95%

saturated and only rarely below 70%. He concluded that the relative humidity in the air space of the egg chamber should have been almost always at the saturation point.

Temperature

To develop successfully, crocodylian embryos need relatively stable and warm temperatures. Incubation temperature is important and affects the rate of development and, thus, the length of incubation. For example, Lang *et al.* (1989) found experimentally that, at 28°C, *C. palustris* embryos took 100 days from laying to pipping and 65 days at 33°C. The effect of temperature on the growth and development of *C. johnstoni* has been reported by Whitehead *et al.* (1990). Because of this, there is no 'fixed' time for incubation, and only a likely range of times can be given for wild nests. The rate of development has implications for hatchling fitness (Joanen *et al.* 1987). However, the most striking influence of temperature is that the temperature the embryos experience during a particular period

of incubation determines their sex (see temperature dependent sex determination, TSD below). This means that a single clutch may produce all females, all males or, due to temperature gradients within the clutch, a mixture of both. Nests laid early or late in a nesting season may experience a different thermal regimen, so that the sex ratio of recruits to a population can be dependent on when the eggs were laid. Interest in these variables has led to a very large literature on nest temperatures and the factors that influence them, but only a summary will be given here.

Temperatures recorded from successful nests across the incubation period range from ~28 to 34°C, with no apparent difference between hole and mound nests (Magnusson *et al.* 1985). Both types of nest insulate the clutch well from the daily fluctuations in ambient temperature, sometimes so well that the clutch temperature may be quite stable over a period of days (Figs 12.59, 12.60, 12.61). Overall trends in ambient temperature, however, eventually do affect clutch temperature. Webb *et al.* (1977) showed in *C. porosus* nests that late in the season, with ambient temperatures falling towards autumn, nest cavity temperatures also declined. Lang *et al.* (1989), compared temperature records from *C. palustris* nests built early and late in the northern spring–summer. They concluded that early nests were cooler, took longer to hatch and produced females, whereas later nests were warmer, would hatch in less time and were more likely to produce males. Rainfall lowers clutch temperature, if there is enough of it (Webb *et al.* 1977; Magnusson 1979).

In general, clutch temperatures are warmer than average daily temperature by 1–3°C, so there is heat coming from somewhere. Its source has been a topic for speculation and, sometimes, experimentation. In mound nests, sunshine plus the decomposition of vegetable material are the probable sources of heat, whereas in hole nests, lacking vegetable material, insolation alone is usually assumed. However, the effect of heat produced by metabolism of late stage embryos within such a well-insulated space may have been underestimated. Without experiments, it is difficult to resolve the separate effects of decomposition, embryo metabolism and trends in ambient temperature, because all will be operating

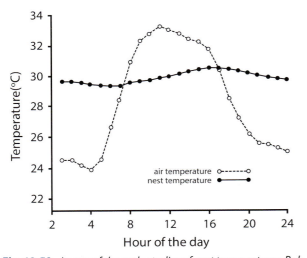

Fig. 12.59. *In one of the early studies of nest temperatures, Bob Chabreck monitored hourly nest and air temperatures in three grass and soil mound nests of* A. mississippiensis *over 20 days in each of three seasons at the Rockefeller Wildlife Refuge in southwestern Louisiana. This graph presents the results, averaged. The temperature of all nests averaged higher than average air temperature (1.4°C), which he attributed to the heat produced by vegetable matter decomposition, with the insulating properties of the nest material accounting for its stability (1.2°C daily). (From Chabreck 1973 with permission, Allen Press Publishing Services)*

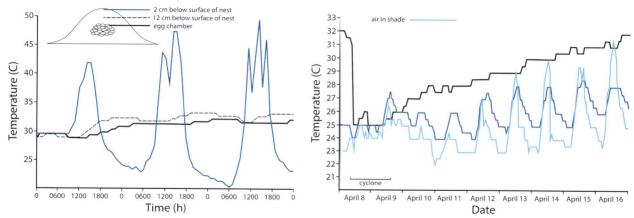

Fig. 12.60. *Temperatures of grass* C. porosus *nests in two different situations: (left) one experiencing no rainfall; and (right) one before, during and following cyclonic rainfall. The slight daily increases in egg cavity temperature (left) show that despite the nest being well insulated, external temperature does have an influence. Cavity temperatures in the 7 days preceding the cyclone had ranged from 32.5 to 34°C and fell to 25°C in the cyclone (right). Over the following days, cavity temperature returned to the pre-cyclone level in a pattern that implies a contribution from endogenous heat production. This could be from decomposition of vegetation, embryonic metabolism, or a combination of the two. (Redrawn from Magnusson 1979, copyrighted material included courtesy of SSAR)*

simultaneously. Vegetable decomposition may be a factor early in incubation, but may decline in the later stages, as embryonic metabolism increases. That the latter can be significant has been demonstrated by Ewert and Nelson (2003), who set up an experiment comparing groups of alligator eggs either dispersed or clustered in vermiculite. The eggs in the cluster generated a warmer environment, produced more males and hatched earlier. It is quite likely that, in both hole and mound nests, heat produced as a by-product of embryo metabolism has an influence on the clutch temperature, as suggested by Modha (1967) to explain increasing temperatures with depth in the nest holes of C. niloticus.

Noting that nobody had studied crocodylian nests in closed canopy forests, Magnusson *et al.* (1985) located nests of *Paleosuchus trigonatus* in Reserva Florestal Ducke and elsewhere near Manaus, Amazonas, Brazil. They located 13 nests, which comprised mounds of loosely compacted leaf litter, of which nine had been built against or on top of large termite mounds (Fig. 12.62). Sometimes construction by the termites envelopes some of the eggs. Of the other four nests, three had diffuse termite workings within and one was built against the base of a palm tree. Exploring a nest against a termite mound and measuring temperature profiles, they found the eggs were being warmed

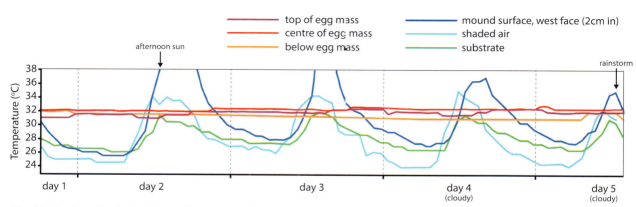

Fig. 12.61. *Four days' data from a large, natural sand, soil, shell and grass mound built by a* C. porosus *at the western edge of riparian vegetation fringing the breeding lagoon at the Edward River Crocodile Farm, Pormpuraaw, Queensland. Data collection started on 2 March 1983. (Gordon Grigg and Lyn Beard, unpublished)*

Fig. 12.62. *Bill Magnusson crouches at a leaf-litter mound nest of* Paleosuchus trigonatus *built in deep shade, under a forest canopy and against a termite mound, from which it captures some warmth. The increased temperature accelerates the development of the eggs (see also Chapter 10). However, although most nests seem to be built against termite nests, that does not seem to be a requirement. (Photo GCG)*

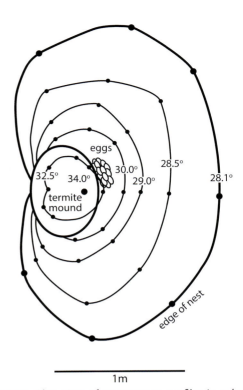

Fig. 12.63. *Plan view of temperature profiles in a leaf litter mound of* Paleosuchus trigonatus *and the termite mound that it has been built against. (From Magnusson* et al. *1985, copyrighted material included courtesy of SSAR)*

by heat from the termite mound (Fig. 12.63). To confirm this, they moved the clutch to the outside of the nest mound and found that clutch temperature fell to equilibrate with the leaf litter. A nest they constructed themselves, distant from any termite mound, showed no sign of warmth, but a recently abandoned nest against a decomposing tree trunk was nearly 3°C warmer than the surrounds. In a subsequent study they found a further 12 nests, four reusing old nest sites, but only five were associated with termite mounds (Magnusson *et al.* 1990). Nests not associated with termites were also successful, so the use of termite mounds is obviously only part of the story. Magnusson and colleagues suggested for one nest that heat may have been captured from a decomposing tree stump. They also wondered whether *Osteolaemus* would be similar to *P. palpebrosus* in its nesting habits. Since then, Riley and Huchzermeyer (1999) have studied *O. osborni* in the Likouala Swamp Forests of the Congo Basin. They found seven nest mounds, all built against tree trunks and tree roots and 'made from a slurry of peat, twigs, and debris swept up from the bed of an adjoining pool'. The mounds were 'interwoven by plant roots mostly from the ubiquitous ginger plant (*Aframomum angustifolium*)'. The authors did not discuss nest temperature but implied that the mounding of material allowed the raised eggs to remain relatively dry within the mound, aided by

'the desiccating effect of the roots'. With average temperatures in the dry season of 26°C at Impfondo (climatezone.com), the nearest population centre, it could be that any additional warmth required is supplied by the metabolism of the developing embryos.

Puzzles remain concerning the assumed preeminent role of nest material decomposition. Eggs in hole nests manage without it. Decomposition would seem to be so variable between nests of different material, in the time course of its heat production and in its lack of any regulatory mechanism, that more work probably needs to done in this area.

TEMPERATURE-DEPENDENT SEX DETERMINATION

Patterns

Apart from the temperature of a nest influencing the rate of development, the length of incubation and the fitness of the hatchlings, it also determines their sex. Among the amniotes, birds and mammals

Fig. 12.64. *A late-term egg of C. porosus. Towards the end of incubation, the hard shell is weakened by the slightly acidic nest environment and the embryo's removal of calcium. The egg takes up water from the humid environment, swells a little and longitudinal cracks develop. The calcareous shell fragments flake away from the leathery shell membrane, often assisted by movement of the embryo within. (Photo GCG)*

all have genotypic sex determination (GSD), but most turtles, some lizards, tuataras and probably all crocodylians have temperature-dependent sex determination (TSD), which is a particular type of environmentally determined sex determination (ESD). Given the numerous anatomical, physiological and behavioural similarities between crocodylians and birds, it is very interesting that they have a different mechanism of sex determination.

TSD was described first in an agamid lizard, *Agama agama*, the red headed rock lizard, by Madeleine Charnier in Dakar, Senegal (Charnier 1966). Peter Harlow (2004) told of her chance discovery and related the subsequent early history of the recognition that sex is determined in many reptiles by their thermal history during embryonic incubation. Charnier noticed that eggs of *Agama agama* left to incubate in the cool sand of her lizard cages produced only one male among 46 hatchlings, while all 30 eggs in a warmed incubator produced males. She suggested that the sex of the hatchlings had somehow been influenced by their thermal experience. But her publication was overlooked. Perhaps that is not surprising: it was a short paper in the *Comptes Rendus des Séances de la Société de Biologie de l'ouest Africain*. Five years later Pieau (1971) found that different incubation temperatures

produced different sex ratios in two species of chelonian, *Testudo graeca* and *Emys orbicularis*, and researchers began to look at other species. TSD was soon found in some other lizards and turtles and the reality of its occurrence in some reptiles, but not others, was recognised and explored, providing a really fascinating and enduring puzzle. Despite a lot of enquiry and speculation, there has still not been any 'eureka' moment, and there is no satisfying explanation about the adaptive significance of sex being determined by thermal history rather than genotypically. For me, the importance of TSD first struck when long ago I read a paper by Mrosovsky and Yntema (1980), which pointed out that 'head-starting' sea turtles by collecting eggs and incubating them in safety but at male-producing temperatures was absolutely counter productive as a conservation strategy!

Ferguson and Joanen (1982, 1983) provided the first description of TSD in a crocodylian, *A. mississippiensis*, in a comprehensive study in marshland on the Rockefeller Wildlife Refuge, Louisiana. They incubated eggs in the laboratory at a series of temperatures, monitored temperatures in natural nests at locations chosen to span a range of likely incubation temperatures, mapped the positions of eggs within these nests and sexed the hatchlings produced. They also performed switching experiments to reveal when an embryo's sex became fixed. The data showed unequivocally that alligators have TSD. Low incubation temperatures produce females, higher temperatures produce males, and sex is determined by the time an embryo is about half way through incubation. However, this first crocodylian study of TSD missed some important elements: notably that the thermal 'window' for producing males is very narrow and that, above it, females are produced, as recounted below.

It is worth remembering that there is only a reasonably narrow range of temperatures over which successful development can occur at all. For example, in a finer resolution study of alligators, Lang and Andrews (1994) found that incubation at constant temperatures below 29°C and above 34°C produced few, if any, viable hatchlings, although brief exposures to lower or higher temperatures could be tolerated. Within this narrow thermal window, it turned out that at 31.5°C 100% females were produced, but just 0.3°C higher at 31.8°C 50% males resulted. This defined a female–male (FM) 'pivotal temperature' of 31.8°C, tightened considerably from the Ferguson and Joanen study. At 32°C and 33°C males were produced, but they also found a male–female transition with a MF pivotal temperature of 33.8°C. There were also some differences between clutches. After a complicated set of experiments shifting eggs between male-producing and female-producing temperatures at different times Lang and Andrews established that the temperature sensitive period (TSP) for alligators is over Ferguson stages 21–24, which, at an incubation temperature of 31°C, is 30–45 days after laying and, not surprisingly, this coincides with the time when the gonads are differentiating.

As well as reporting more detailed results from alligators, Lang and Andrews (1994) reviewed what was then known about patterns of TSD in 10 other species of crocodylian and there seems to be little to add since that time. They added to, and interpreted, studies on *Caiman crocodilus* and *Crocodylus palustris* (Lang *et al.* 1989), *C. johnstoni* and *C. porosus* (Webb *et al.* 1987e; Webb and Cooper-Preston 1989) (Fig. 13.30), as well as *C. niloticus* (Hutton 1987), and presented new data on *C. moreletii*, *C. siamensis* and *Gavialis gangeticus*. All of these species have an FMF pattern, with most known about the first five. The FM pivotal temperatures show less variability than the MF transitions and all show some variation between individual clutches. Where resolved, the TSP is at a similar period of incubation, when the gonads are differentiating. Only in *A. mississippiensis* is there a temperature at which only males are produced; in the other species, at least some females are produced at all temperatures. Both *Paleosuchus trigonatus* and *Alligator sinensis* also have TSD, with females at low temperatures and males at higher, but data are insufficient to be sure about the presence of an FMF pattern and, in the case of *A. sinensis*, the matter remains unresolved (Thorbjarnarson and Wang 2010). Deeming (2004b) noted that crocodylids have a narrower range of male-producing temperatures than alligatorids. From his graph, 50% or more males can be expected

in alligatorids from 31.5 to 34.2°C, compared with 32.1 to 32.9°C in crocodylids.

Not all crocodylians have yet been studied, but with data from about half the species, and alligators, caimans, crocodiles and gharials all being similar, it can be reckoned that TSD with an FMF pattern is a crocodylian characteristic. Concerning the so-far-unstudied species, we do know that at least 21 species lack visibly identifiable sex chromosomes. Cohen and Gans (1970) examined chromosome complements in 21 species of crocodylian – both males and females in 13 of them – and found no sign of 'consistent heteromorphism of any given pair which might have indicated the existence of sex chromosomes'.

With TSD, it stands to reason that some nests produce only a single sex. Webb and Smith (1984) observed male only hatchlings from 37% of 19 clutches of *C. johnstoni*, and Lang *et al.* (1989) noted all-male and all-female outputs from *C. palustris* nests at the Madras Crocodile Bank. But, because the nest cavities of crocodylians commonly show temperatures in the low 30s, with little diurnal variation and close to the pivotal temperatures, it is not uncommon for natural nests to produce a mixture of males and females. Concern has been expressed that climate change has major implications for species that have TSD. Somaweera and Shine (2012) showed elegantly how well a crocodylian (*C. johnstoni*) living in a harsh environment that provided few prime nesting sites, was able to nest successfully by making careful choices among what was on offer. It seems likely that crocodylians will do quite well in their present distributions and, after all, they all have higher latitudes, which will become available for them to colonise.

Mechanism

The occurrence of TSD is distributed haphazardly and enigmatically within reptile groups, offering few clues about its evolution or its adaptive significance. The FMF pattern is referred to as TSD Pattern 2 (Ewert *et al.* 2004) and it is common in some chelonians as well as crocodylians. Among other chelonians a MF pattern (Pattern 1a) is common, but New Zealand's tuataras, *Sphenodon punctata* and *S. guntheri*, are apparently unique in showing Pattern 1b: that is, FM (females at low temperature and males above), in this case with a transition temperature of 21.6°C (*S. guntheri*) or 22°C (*S. punctata*) (Mitchell *et al.* 2006; Nelson *et al.* 2010). In squamates, the dominant form of sex determination is genotypic, but TSD is known in the gecko families Eublepharidae and Gekkonidae, as well as in dragons (Agamidae) and skinks (Scincidae), all of which are the Pattern 2 type, which is the FMF variety observed in crocodylians. Each of these lizard families has species with GSD as well (reviewed by Harlow 2004). There are anecdotal accounts of TSD occurring in other lizard families, but these are so far unverified. Not only is both GSD and TSD known to occur within the same family of lizards, but among the Agamidae there are at least two genera in which examples of both GSD and TSD occur within a single genus. Within the Scincidae, there is a particularly interesting example that takes this one step further, showing both means of sex determination within a single species. The sex of populations of the viviparous Tasmanian snow skink, *Niveoscincus ocellatus,* is determined genotypically in the highlands but shows TSD in the lowlands (Pen *et al.* 2010). TSD has not been found in snakes (Teller 2010). Patterns of sex determination in chelonians were reviewed by Ewert *et al.* (2004). They reported TSD in 64 species from 10 families, including sea turtles. Both Pattern TSD 1a and TSD 2 are well represented. All six species of sea turtle show Pattern 1a (MF) TSD (Wibbels 2003).

The dominant occurrence of TSD in Sphenodontia, Testudines and Crocodylia has suggested to some authors that it may be the primitive reptilian (sauropsidan) system. However, modern Amphibia have GSD, as do many invertebrates, and the apparent haphazard distribution of TSD suggests that it may have evolved or been adopted many times. This is reinforced by its occurrence in some agamid lizards and not others, and examples of both GSD and TSD in the same genus (*Amphibolorus*, Harlow 2004) suggest that the evolution of TSD from GSD, or vice versa, may be a reasonably short step. The work on *Niveoscincus* (above) – with either GSD or TSD depending upon whether it is the highland or lowland population (above) – provides further evidence, as does the recognition since the late 1980s

that treatment with oestradiol at male-producing temperatures during the thermosensitive period (TSP) can override the effect of temperature and lead to females. Sarre *et al.* (2004) noted that the process of differentiation of the gonads into either male or female, involving the hormones androgen and oestrogen and the enzyme aromatase, is highly conserved across vertebrates and they showed ways by which GSD could be modified to TSD by comparatively minor changes, even as small as the expression of a single gene. Noting that female is the default sex, Deeming and Ferguson (1989) speculated that maleness may eventuate from the temperature-dependent production of some 'male-determining factor'. Deeming (2004b) noted that the *Dmrt1* gene was a possible candidate: it has temperature-dependent differential expression in alligator embryogenesis (Smith *et al.* 1999). That idea is still in play, with Matsumoto and Crews (2012) putting forward a model based on red-eared slider turtles (*Trachemys scripta*) in which, at male-producing temperatures, *Dmrt1* suppresses the *FoxL2* gene, which would otherwise stimulate aromatase production and female gonads. In their model, this leaves *Dmrt1* to act on the *Sox9* gene whose expression leads to the differentiation of testes.

THE LEAD-UP TO HATCHING

With its sex determined and looking already like a fully formed hatchling, the visible changes in the last third or so of embryonic life are mainly its growth (Fig. 12.55). The yolk sac and albumen gradually get smaller and the yolk sac and any remaining yolk are usually drawn fully into the body, leaving a mid-ventral umbilical scar with the closure. If the incubation has been at a lower than optimal temperature, remaining yolk can give a new hatchling quite an inflated belly.

Shell degradation, egg tooth and pipping

During embryonic growth, the shell is an important source of calcium, magnesium and phosphorus for the developing embryo (Jenkins 1975; Packard and Packard 1989). This includes sequestration of calcium in the yolk late in development for use after hatching. Additionally, decomposition of the nest materials, by-products of microorganisms

Fig. 12.65. *A hatchling pips from the egg. The egg 'tooth' is visible between the narial button and the tip of the snout. A wriggling hatchling can open a split in the shell membrane by raking the egg tooth against it from the inside. (Photos DSK, GCG)*

and the excretion of CO_2 from the egg to the moist environment, all contribute to an acidic environment, which further degrades the shell (Ferguson 1981, 1982). All these processes weaken the shell and, late in incubation, longitudinal cracks often develop (Fig. 12.64). Grigg and Beard (1985) described the development of cracking in *C. porosus* eggs in moist vermiculite from ~70% of the way through incubation. Longitudinal, and then transverse, cracks segment the shell into calcareous plates. High internal hydrostatic pressure causes the egg to become turgid and swell (see above) and this may help loosen the calcareous plates from the underlying leathery shell membrane so that they flake away easily, assisted by any wriggling by the embryo. In hole nests, such as in *C. niloticus*, which lack decomposing vegetable material, external degradation of the shell is much less marked and fine cracks develop later and may be widened by the wriggling of the embryo (Pooley 1977). Incubated in a drier environment, the development of cracks occurs later, egg swelling may not occur and the shell may remain robust. In some cases hatching

from dry nests cannot occur without parental assistance.

Just before hatching, movements of the hatchling inside the leathery shell can sometimes be seen, even the raking of the embryo's nose against the inside wall, armed with its 'egg tooth'. This is an elaboration of the epidermis into a 1–2 mm thorn-like structure at the tip of the snout (Figs 12.65, 12.68) with which they 'pip' the shell if it is still intact and/

Fig. 12.66. *Half hatched, a hatchling* C. porosus *shows how it fits, curled up within the egg. (Photo GCG)*

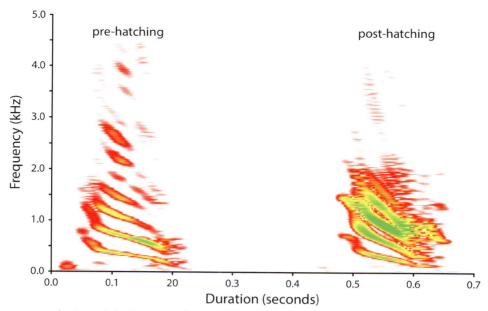

Fig. 12.67. *Sonograms of calls made by C. porosus from (left) within the egg and (right) after hatching. Calling from inside the egg is possible only once the animal is using its lungs to breathe, either after pipping, with the snout out or, possibly, from an airspace. Calls made from within the egg have slightly different characteristics, including less sound energy, probably from the sound being modified by the egg. Calling from within the egg stimulates others within the clutch to call and often attracts the female to excavate the nest and carefully carry the hatchlings to the water. After hatching, these calls assist the crèche to stay together. (Graphic supplied by Adam Britton)*

or tear through the shell membrane. It is resorbed or lost within a few weeks after hatching. Birds have an egg tooth on the beak, which is functionally, but not developmentally, similar. Many snakes have an egg tooth, which is a modified bony tooth.

Helping an embryo to hatch provides an opportunity to see how the fully developed embryo, now a hatchling, fits curled within the egg. There is little room for anything else (Fig. 12.66). Respiration via the lungs, however, is usually established before hatching and we can be sure of that because it is common for hatchlings to call from within the egg. This requires an air space within the egg, much like that of birds, as discussed below.

Fig. 12.68. *Newly emergent C. porosus hatchlings. (Left) Photographed in a wild nest in the Northern Territory, this hatchling and its clutchmates pipped and started hatching when eggs (removed earlier from the same clutch) began grunting on being returned to the nest after use in a gas exchange study. (Right) A hatchling peers cautiously at the photographer, its tail still partly within the membranous sac and a bit of remnant albumen caught in its teeth. Note the prominent egg tooth. (Photos DSK, GCG)*

Fig. 12.69. *A freshly hatched Philippine crocodile, C. mindorensis. Critically endangered, there is a big conservation programme underway and hatchlings of this species are particularly valuable (Chapter 14). (Photo Merlijn van Weerd)*

Sounds from the egg

An early report of hatchlings calling from within the egg was made by Grabham (1909), writing from Khartoum and describing Nile crocodiles on a tributary of the Blue Nile, south-west of Gallabat near the border between Sudan and Ethiopia (then Abyssinia). He wrote 'About a yard away, again, the presence of another nest was made evident by the croaking of young crocodiles beneath the sand, and it would appear that this enables the parent to know when to release its young by excavating a hollow to such a depth that only a thin cover of sand is left over the eggs'. This was in a letter to the journal *Nature*, placed immediately below a note that 'The editor does not hold himself responsible for opinions expressed by his correspondents' (perhaps the editor doubted the veracity of the report).

Pooley (1962) described Nile crocodiles making 'plaintive 'eeauw' grunts' just before hatching, 'audible for 6 feet even when the egg is buried a foot under the sand' and that others of the same clutch

Fig. 12.70. *A hatchling gharial,* Gavialis gangeticus: *a critically endangered species whose survival depends on active conservation measures. (Photo Simon Maddock)*

would join in, and the calling of hatchlings from within the egg is well substantiated (Pooley and Gans 1976; Magnusson 1980b Vergne and Mathevon 2008; reviewed by Britton 2001; Vergne *et al.* 2009). Britton (2001) noted that there are reports of calls from inside eggs that appear intact (Fig. 12.67). For this to occur, there must be an airspace, perhaps formed by the shell membranes being torn away, maybe associated with some dehydration shrinking the membrane away from the shell late in incubation. Calling can be heard from several metres away and usually starts a day or so before hatching, but can be initiated earlier by disturbance. Somaweera and Shine (2012) described calling by *C. johnstoni* 1–5 days before hatching. Taking an experimental approach, playing recorded sounds to eggs nearly ready to hatch, Vergne and Mathevon (2008) established that pre-hatch groups responded to the recordings by calling, and often moving in response, and that the playback stimulated hatching. It seems certain that the calls assist with synchronising hatching. Furthermore, there are numerous reports of females responding to those same calls and excavating the nest. Grabham went very close to implying this in 1909. Pooley (1977) demonstrated females responding to hatchling calls when he found he could take a hatchling to the safe side of a wire mesh fence, play recorded hatchling calls and having a captive female *C. niloticus* leave the water, take the hatchling carefully from his hand and release it into the water. Subsequently, Vergne and Mathevon (2008) showed experimentally in captive *C. niloticus* that the sound of pre-hatching calls deep within a nest hole could stimulate the mother to open the nest by digging.

By the time a female crocodylian responds to calls coming from deep within the nest, she will have been occupied for months with the behavioural phases of reproduction: courting; copulating; nest building; egg laying; and maintaining and guarding the nest. Calls made from within the nest trigger the final stages: releasing the hatchlings from the nest; transporting them to the water; and, in some species at least, maintaining some protection over them in the first few weeks or, in others, months of their lives. It is a wonderful experience to see a hatchling suddenly emerge from its egg, glistening wet, with every detail perfect (Figs 12.69, 12.70). I once heard Grahame Webb, enthusing about this on a radio programme, say 'They look like something that'd be made by a jeweller'.

PARENTAL CARE

Some of the most interesting behaviour shown by crocodylians was for a long time disregarded because it sounded too fanciful. Parental care, which is mostly but not exclusively maternal care, is very uncharacteristic of reptiles (see review by Shine 1988) and zoologists initially had trouble believing that a 'mere reptile' could guard its nest for months, respond to squawking hatchlings, open the nest, perhaps even help them from the egg and then carry them to the water and stay with them for weeks after that, apparently guarding against predation. Perhaps it was just too difficult to accept that a female weighing several hundred kilograms, with jaws that could subdue a wildebeest or a buffalo, was also able to delicately collect and transport 100 g hatchlings in those same jaws. However, with more and better reports, acceptance was growing slowly and it was a frequent topic of discussion in the very early 1970s. 'Belief' was helped along by Howard Hunt at Atlanta Zoo who provided a careful description of the behaviour shown by a captive *C. moreletii* presented with eggs from a different female and buried in her nest mound. She responded to a playback at the mound of recorded hatchling grunts by leaving the water and carrying the young to the water (Hunt 1975).

The person most responsible for the observations that convinced everyone that crocodylians actually do these things was probably Tony Pooley (1974, 1977), employed by the Natal Parks Board and making observations and conducting experiments in the far north-east corner of South Africa from the late 1950s. His story is told in an engaging book *Discoveries of a Crocodile Man* (Pooley 1982) and his scientific publications dating from 1962 convinced the world about the occurrence of substantial maternal care by Nile crocodiles and, by implication (proven by subsequent studies) crocodylians in general. His first paper on crocodiles observed that it was 'likely' that 'the parent crocodile may help its young to escape from the nest'. He would almost

Fig. 12.71. *(Left panel, top) A* C. niloticus *female gently picks up a hatchling that has been offered to her through the fence and (centre and bottom) carries it to the water in her gular pouch to release it into the water. (Right panel, top) A female* C. niloticus *responds to calls from experimentally buried full term eggs and takes one into her mouth. (Centre) She manipulates an egg between palate and tongue, releasing the hatchings. (Bottom) The male also responded to calls and manipulated an egg until the hatchling was released, after 14 min, and took it to the water. (From Pooley 1977, © The Zoological Society of London)*

certainly have been unaware then that he shared this opinion with Cott (1961), who noted that the eggs were often trapped under hard packed soil, so that release without help would be impossible. Pooley's opinion that a parent must help the young escape was confirmed subsequently (Pooley 1969) and then expanded (Pooley 1974). His review paper (Pooley 1977), which contained photographs and more detailed description (Fig. 12.71), referred to the disbelief that had prevailed. Its opening paragraph reads 'The apocryphal and observational literature on the Nile crocodile has long contained scattered reports that this species and some other crocodilians dig up their eggs at the time of hatching and carry

Fig. 12.72. *A female* C. johnstoni *takes a hatchling from her nest hole on the shore of Lake Argyle, an artificial lake in Western Australia's Kimberley region, before carrying it carefully to the water to release it. (Photo Ruchira Somaweera)*

the young to the water (Bartram 1791 in Bartram and Harper 1943; Goldsmith 1805 in Cott 1961)'. It went on 'In recent summaries of the crocodilians, such reports have generally been ignored or set aside as untrustworthy (Neill 1971; Guggisberg 1972)'. He then explained that the paper would show that 'some of the original observations were correct and the skepticism of later authors unfounded'.

Part of the doubt probably arose because many nests hatch successfully without parental assistance, as pointed out by Neill (1971), referring to American alligators when he expressed doubt about reports of any maternal involvement. But maternal care for Nile crocodiles, as described by Pooley, is now well accepted to be characteristic of crocodylians in general and it appears to be remarkably similar throughout the group. What is more, the realisation has grown that, although most care is performed by females, there are some examples of males, or both parents, performing some of the 'care' functions, so

a name change to 'parental care' has been deemed appropriate.

It is worth noting that, although parental care is very rare in other reptiles, but *de rigueur* in crocodylians and birds, it is likely to have occurred in at least some of the dinosaurs as well (Reisz *et al.* 2005; Varricchio *et al.* 2008). Before getting excited that parental care might have been ancestral for archosaurs, the latter paper concludes that paternal, not maternal care, was ancestral in birds, in contrast to the situation in crocodylians, which mostly show maternal care. More data are awaited with interest.

Somaweera and Shine (2012) have provided a comprehensive review from which it is clear that parental care has been described across all crocodylian groups (Table 12.3). Citing research and observations reported in nearly 80 published reports, they tabulated the known occurrences of the five parental behaviours: attending the nest during incubation; defending the nest; opening the nest; carrying the young to the water; and defending the young after hatching. With only a handful of exceptions, in every genus, each of the five parental behavioural categories is represented and there are few gaps: mostly as 'unknowns' rather than 'known not'. Even *Gavialis gangeticus* is reported to transport hatchlings to the water, a feat thought unlikely because of the jaw structure and dentition. Likewise, *C. johnstoni* (Fig. 12.72) and *Mecistops cataphractus* transport their young, despite their slender snouts, although *Tomistoma schlegelii* may not (Table 12.3).

Whereas most of the early observations were made in zoo settings, zoologists now enjoy access to new technologies such as remotely triggered video cameras and similar devices that enable direct observations of crocodylians in the field. A couple of recent studies employing such devices have greatly added to the detail of parental behaviour. Somaweera and Shine (2012) used motion-sensitive infrared triggered digital still/video cameras to record eight incidences of nest excavation and hatchling transport to the water by *C. johnstoni* from nest holes on the shore of the artificial Lake Argyle in Western Australia's Kimberley region. Unlike a typical zoo situation, the authors did not know the gender of the adults that released and transported the hatchlings and, indeed, could not confirm that they were the relevant parent. However, the use of the remote surveillance technology added detail that would have been very difficult to get any

Fig. 12.73. A dingo takes advantage of a freshly excavated nest of C. johnstoni, feeding on hatchlings and unhatched eggs while the mother (probably) transports hatchlings to the water one at a time. (Photo Ruchira Somaweera)

other way in the wild. Vocalisations in response to an investigator walking nearby were recorded from the nest holes 1–5 days before nests were opened. Adults dug with a forelimb and then raked away the soil with a hindlimb, using the front and hindlimb from the same side while lying with the head and body on the ground, before switching to the other side. They commenced digging mostly in daylight hours and stopped frequently, often changing position. They took 0.4–2.2 (average 1.4) h to excavate the nests and then 1.1–6.2 (average 2.0) h to transport the hatchlings to the water. Some of them returned up to about a day later to dig again, even though there were no longer any eggs. The authors concluded that, in this difficult substrate (see earlier), successful escape of hatchlings from their nests depended upon excavation by the adults. On emerging from the nest, hatchlings adopted a 'head up' posture until the adult picked them up with the tip of the snout. They were then repositioned and transported singly, usually hidden within the gular pouch.

Somaweera and Shine also reported several observations of the adult chewing and manipulating an egg in its mouth, releasing the hatchling, as was described in *C. niloticus* by Pooley (1977), who noted also that a rotten egg would be swallowed. Helping hatchlings to escape from the egg by oral manipulation has been reported also in *Caiman crocodilus* (Alvarez del Toro 1969), *Alligator mississippiensis* (Garrick and Lang 1977; Hunt 1987; Kushlan and Simon 1981), *A. sinensis* (Huang and Watanabe 1986), *Caiman yacare,* (Cintra 1989) and *C. novaeguineae* (Lang 1987). Somaweera and Shine (2012) noted that *C. johnstoni* adults tended to ignore infertile eggs, extending other reports of an apparent ability to differentiate. None of the hatchlings was seen to walk by itself to the water. During the transport phase, hatchlings still at the nest suffered predation by crows and a dingo (Fig. 12.73 and see Somaweera *et al.* 2011). Whereas the presence of the female may deter some predators, others will be attracted to a freshly opened nest with squirming hatchlings. The risk may be reduced if multiple hatchlings are carried, and *C. moreletii* (Hunt 1977) and *C. niloticus* (Pooley 1977) have been reported to carry several hatchlings to the water at once. Somaweera *et al.* (2013) have provided a comprehensive review of predation of all crocodylian life stages, including a discussion about anti-predator strategies around nests and hatchlings, as well as the length of parental care.

Somaweera and Shine (2012) also pointed out that the level of parental care varies between individual parents and that nests commonly fail because an adult does not show up. Modha (1967) observed the fate of 150 *C. niloticus* nests and reported that no hatchlings managed to crawl out of the nest without maternal help. Joanen (1969) recorded that more than half the nests of alligators failed for the same reason, so it is not a problem restricted to hole nests. Villamarín-Jurado and Suárez (2007) found that failure of the female to release hatchling *Melanosuchus niger* was a significant cause of mortality. It depends on how hard packed the nest or substrate is; for example, Webb *et al.* (1983a) noted *C. johnstoni* emerging unassisted from nests dug in friable sand.

Once in the water, hatchlings face a different set of problems.

THE FRAGILE HATCHLINGS' NEW WORLD

Having survived to exit the nest, hatchlings are a tasty morsel for a wide diversity of predators (Fig. 13.19). Somaweera *et al.* (2013) tabulated a long list of known predators; ants, crabs, any local predatory fish with a large enough gullet,

large frogs, other crocodylians and just about any sympatric predatory lizard, snake, turtle, bird or mammal for which a hatchling croc is a suitable size for a meal. Not surprisingly, hatchling mortality in their first year is often very high, as discussed in Chapter 13.

Survival from predation depends upon camouflage (Fig. 3.8) and cryptic behaviour and, often, some protection from a parent. When hatchlings enter the water, in most species they are known to form a crèche (Fig. 12.74), or pod, near the bank or among vegetation, watched over by an adult, usually the mother (Table 12.3). This has been recognised for a long time. In his field study on Central Island in Lake Rudolf (now Lake Turkana, Kenya), Modha (1967) noted that hatchlings of the Nile crocodile stay together after hatching. Hunt (1975) reported that the seven survivors of a clutch of *C. moreletii* hatched in a multi-species crocodylian exhibit at Atlanta Zoo stayed together for a month, feeding on cockroaches and crickets, and that the foster mother chased other crocs away from them except the dominant male, who allowed the hatchlings to bask on his back. After the small crèche dispersed, the hatchlings gradually disappeared, presumably eaten by other crocodylians in the exhibit. *C. porosus* hatchlings also form crèches. Webb *et al.* (1977) maintained a watch on several crèches in the Liverpool and Tomkinson Rivers, Northern Territory, with adults in attendance for up to 2.5 months after hatching. Alligatorid hatchlings

Fig. 12.74. *Crèches of* Mecistops *(left) and* C. suchus *hatchlings in Gabon. In both cases, the mother (presumably) was in attendance. (Photos Matthew H. Shirley)*

Fig. 12.75. *Hatchling of the West African species of* Mecistops, *photographed in Côte d'Ivoire. (Photo Matthew H. Shirley)*

in the wild also form crèches (usually called pods), with periods of attendance by the parent very variable, from as little as a few days to 21 months or more (Deitz and Hines 1980; Ouboter and Nanhoe 1987; Campos *et al*. 2012; DSK unpublished). Long-term pods in some alligatorids may persist beyond the attendance of the parent.

Not surprisingly, alligator pods are usually made up of siblings (Deitz 1979) and that must be true of other species too. However, it seems they cannot recognise each other as such. Pasek and Gillingham 1999) noted that some animals benefit from being able to recognise kin, but experiments with alligator hatchlings revealed no evidence of that. They did show preference for grouping with other hatchlings, but without any distinction. Likewise, a mother alligator will respond to calls made by any hatchling alligator and chase other alligators away (Dietz 1979). A crèche can therefore be made up of hatchlings from more than one nest. When Bartram (1791 in Bartram and Harper 1943) reported seeing 'a train of young' alligators, up to 100, following

an adult, and wrote about maternal care, he was ridiculed, but Wayne King at the Florida Museum of Natural History has defended several of Bartram's observations that attracted scorn but have since been confirmed, including that one (King 2008). Not only may crèches include hatchlings from more than one nest, in at least some alligatorids, they may

Fig. 12.76. *A yearling gharial,* Gavialis gangeticus, *and some of a crocodylian ecologist's basic tools of trade: measuring scales, a Vernier caliper and a ruler or tape. (Photo Simon Maddock)*

also include yearlings from the previous cohort (Ouboter and Nanhoe 1987).

How do they stay together? Obviously, vision may play a role, but sound is also important. Hunt and Watanabe (1982) described vocal exchanges between hatchling alligators and their mother, even using it to lead them from a nursery near the hatching site to another location. Hatchlings rely on acoustic signalling to remain in the crèche. Magnusson (1980b) released hatchlings of *C. porosus* and played recordings of hatchling calls on one side of a river and lured hatchlings from the other side. They did not form a crèche, and Magnusson suggested that, without an adult being present, dispersing may be the best strategy. Britton (2001) has referred to such calls as 'contact' calls: a term adopted by Vergne *et al.* (2009) for calls that are thought to maintain cohesion among juveniles. They are similar to the post-hatching call shown in Fig. 12.67. Vergne *et al.* (2012) analysed contact calls of juvenile Nile crocodiles (*C. niloticus*), black caiman (*Melanosuchus niger*) and spectacled caiman (*Caiman crocodilus*) and conducted playback experiments to determine responses to juvenile contact calls by juvenile and adult Nile crocodiles and juvenile spectacled caiman. Despite overall similarity in their acoustic spectra, there were significant differences between species. However, the test animals did not discriminate. They did, however, recognise the calls as 'crocodylian', because they did not respond similarly to a 'white noise' control. Further trials using synthetic calls revealed that the extent of frequency modulation of the call was the key to recognition (in both juveniles and adults), particularly its slope. The results confirm that contact calls carry information that is intelligible to other crocodylians and have acoustic characteristics likely to allow effective acoustic communication.

Is olfaction involved in maintaining the crèche? Heaphy (1985) and Richardson and Park (2001) noted that the dorsal integumentary glands (Chapter 3), whose function is so far unknown, may secrete a pheromone that helps keep hatchlings together. The hypothesis was prompted by the finding that the glands appear to be functional mainly in the first few months of life. Passek and Gillingham (1999) found experimentally that hatchling alligators showed no preference for water scented by siblings over water scented by non-siblings. Maybe the question could be resolved by similar experiments, to determine whether water scented by hatchlings is preferred over unscented, control water.

Reports vary about how well a parent defends a crèche and there seem to be few reports from the wild. One example is that by Hunt and Watanabe (1982) who reported that a mother alligator twice drove away blue herons from a crèche they were observing.

The benefit of being in a group is explicable only if the putative anti-predator watch by an adult is effective to some extent. Without that, presumably, hatchlings would be more conspicuous in a group and a potential predator could scoff the lot. Without benefit from the attending adult, one would expect dispersal to be a better strategy. Crèche formation is found in both crocodylids and alligatorids, and is very ancient. Some benefit from it seems logical, but the effectiveness of crèche behaviour and its defence by an adult seems to have not yet been quantified and would make an interesting study.

The 'period of grace' in the crèche comes to an end after a few weeks or months. At some point, the hatchlings become wary of their mother and the parent starts to show signs of aggression (Hunt and Watanabe 1982). The hatchlings disperse and have to fend for themselves (Figs 12.75, 12.76). Their new world includes many threats, including larger crocodylians (see cannibalism, Chapter 13).

REFERENCES

Ackerman RA (1977) The respiratory gas exchange of sea turtle nests (Chelonia, Caretta). *Respiration Physiology* **31**, 19–38.

Adams GK, Andrews RM, Noble LM (2010) Eggs under pressure: components of water potential of chameleon eggs during incubation. *Physiological and Biochemical Zoology* **83**, 207–214.

Alvarez del Toro M (1969) Breeding the spectacled caiman, *Caiman crocodilus*, at Tuxtla Gutierrez Zoo. *International Zoo Yearbook* **9**, 35–36.

Amavet P, Rosso E, Markariani R, Pina CI (2008) Microsatellite DNA markers applied to detection of multiple paternity in *Caiman latirostris* in Santa Fe, Argentina. *Journal of Experimental Zoology A. Ecological Genetics and Physiology* **309A**, 637–642.

Astheimer LB, Manolis SC, Grau CR (1989) Egg formation in crocodiles: avian affinities in yolk deposition. *Copeia* **1989**, 221–224.

Bagwill A, Sever DM, Elsey RM (2009) Seasonal variation of the oviduct of the American alligator, *Alligator mississippiensis.* (Reptilia: Crocodylia). *Journal of Morphology* **270**, 702–713.

Bartram W, Harper F (1943) Travels in Georgia and Florida, 1773–1774. A report to Dr. John Fothergill. *Transactions of the American Philosophical Society. New Series* **33**(2), 121–242, <http://www.jstor.org/stable/1005614>

Bayani AS, Trivedi JN, Suresh B (2011) Nesting behavior of *Crocodylus palustris* (Lesson) and probable survival benefits due to varied nest structures. *Electronic Journal of Environmental Sciences* **4**, 85–90.

Booth DT (2000) The effect of hypoxia on oxygen consumption of embryonic estuarine crocodiles (*Crocodylus porosus*) *Journal of Herpetology* **34**(3), 478–481.

Booth DT, Thompson MB (1991) A comparison of reptilian eggs with those of megapode birds. In *Egg Incubation. Its Effects on Embryonic Development in Birds and Reptiles.* (Eds DC Deeming and MWJ Ferguson) pp. 325–344. Cambridge University Press, Cambridge.

Brazaitis P, Watanabe ME (2011) Crocodilian behaviour: a window to dinosaur behaviour? *Historical Biology* **23**(1), 73–90.

Brien ML, Read MA, McCallum HI, Grigg GC (2008) Home range and movements of radio-tracked Estuarine Crocodiles (*Crocodylus porosus*) within a non-tidal waterhole. *Wildlife Research* **35**, 140–149.

Britton ARC (2001) Review and classification of call types of juvenile crocodilians and factors affecting distress calls. In *Crocodilian Biology and Evolution* (Eds GC Grigg, F Seebacher and CE Franklin), pp. 364–377. Surrey Beatty & Sons, Sydney.

Buhi WC, Alvarez IM, Binelli M, Walworth ES, Guillete LJ Jr (1999) Identification and characterization of proteins synthesized de novo and secreted by the reproductive tract of the American alligator, *Alligator mississippiensis. Journal of Reproduction and Fertility* **115**, 201–213.

Calderón ML, De Pérez GR, Ramírez-Pinilla MP (2004) Morphology of the ovary of *Caiman crocodilus* (Crocodylia: Alligatoridae). *Annals of Anatomy* **186**, 13–24.

Campbell HA, Watts ME, Sullivan S, Read MA, Choukroun S (2010) Estuarine crocodiles ride surface currents to facilitate long-distance travel. *Journal of Animal Ecology* **79**(5), 955–964.

Campbell HA, Dwyer RG, Irwin TR, Franklin CE (2013) Home range utilisation and long-range movement of estuarine crocodiles during the breeding and nesting season. *PLoS ONE* **8**(5), e62127.

Campos Z, Sanaiotti T, Muniz F, Farias I, Magnusson WE (2012) Parental care in the dwarf caiman, *Paleosuchus palpebrosus* Cuvier, 1807 (Reptilia: Crocodilia: Alligatoridae). *Journal of Natural History* **46**(47–48), 2979–2984.

Chabreck RH (1973) Temperature variation in nests of the American alligator. *Herpetologica* **29**, 48–51.

Charnier M (1966) Action de la température sur la sex ratio chez l'embryon d'Agama agama (Agamidae, Lacertilien). *Société de Biologie de L'ouest Africain* **160**, 620–622.

Charruau P, Hénaut Y (2012) Nest attendance and hatchling care in wild American crocodiles (*Crocodylus acutus*) in Quintana Roo, Mexico. *Animal Biology* **62**(1), 29–51.

Cintra R (1989) Maternal care and daily pattern of behavior in a family of caimans, *Caiman yacare*, in the Brazilian Pantanal. *Journal of Herpetology* **23**, 320–322.

Clarke SF (1891) The habits and embryology of the American alligator. *Journal of Morpholgy* **5**, 182–214.

Cohen MM, Gans C (1970) The chromosomes of the order Crocodilia. *Cytogenetics* **9**, 81–105.

Compton A (1981) Courtship and nesting behaviour of the freshwater crocodile, *Crocodylus johnstoni*, under controlled conditions. *Australian Wildlife Research* **8**, 445–450.

Cott HB (1961) Scientific results of an inquiry into the ecology and economic status of the Nile crocodile (*Crocodilus niloticus*) in Uganda and Northern Rhodesia. *Transactions of the Zoological Society of London* **29**, 211–356.

Coutinho M, Campos Z, Cardoso F, Massara P, Castro A (2001) Reproductive biology and its implications for management of caiman (*Caimans yacare*), in the Pantanal wetland, Brazil. In *Crocodilian Biology and Evolution* (Eds G Grigg, F Seebacher and CE Franklin) pp. 229–243. Surrey Beatty & Sons, Sydney.

Davenport M (1995) Evidence of possible sperm storage in the caiman, *Paleosuchus palpebrosus*. *Herpetological Review* **26**, 14–15.

Davis LM, Glenn TC, Elsey RM, Dessauer HC, Sawyer RH (2001) Multiple paternity and mating patterns in the American alligator, *Alligator mississippiensis*. *Molecular Ecology* **10**, 1011–1024.

Deeming C (2004a) *Reptilian Incubation: Behaviour and Environment*. Nottingham University Press, Nottingham, UK.

Deeming C (2004b) Prevalence of TSD in crocodilians. In *Temperature-Dependent Sex Determination in Vertebrates* (N Valenzuela and VA Lance, eds) pp. 33 –41 Smithsonian Books, Washington DC.

Deeming DC, Ferguson MWJ (1989) The mechanism of temperature dependent sex determination in crocodilians: a hypothesis. *American Zoologist* **29**, 973–985.

Deitz DC (1979) Behavioral ecology of young American alligators. PhD thesis University of Florida, Gainesville.

Deitz DC, Hines TC (1980) Alligator nesting in North-central Florida. *Copeia* **1980**, 249–258.

Dinets V (2010) Nocturnal behavior of the American alligator (*Alligator mississippiensis*) in the wild during the mating season. *Herpetological Bulletin* **111**, 4–11.

Dinets V (2011a) The role of habitat in crocodilian communication. *Open Access Dissertations*. Paper 570. University of Miami, Florida, <http://scholarlyrepository.miami.edu/oa_dissertations/570>

Dinets V (2011b) Effects of aquatic habitat continuity on signal composition in crocodilians. *Animal Behaviour* **82**(2), 191–201.

Dinets V (2013) Do individual crocodilians adjust their signaling to habitat structure? *Ethology Ecology and Evolution* **25**(2), 174–184.

Ewert MA, Nelson CE (2003) Metabolic heating of embryos and sex determination in the American alligator, *Alligator mississippiensis*. *Journal of Thermal Biology* **28**, 159–165.

Ewert MA, Etchberger CR, Nelson CE (2004) Turtle sex-determining modes and TSD patterns, and some TSD pattern correlates. In *Temperature-Dependent Sex Determination in Vertebrates* (Eds N Valenzuela and VA Lance) pp. 21–32. Smithsonian Books, Washington DC.

Ferguson MWJ (1981) Extrinsic microbial degradation of the alligator eggshell. *Science* **214**, 1135–1137.

Ferguson MWJ (1982) The structure and composition of the eggshell and embryonic membranes of *Alligator mississippiensis*. *Transactions of the Zoological Society of London* **36**, 99–152.

Ferguson MWJ, Joanen T (1982) Temperature of egg incubation determines sex in *Alligator mississippiensis*. *Nature* **296**, 850–853.

Ferguson MWJ (1985) The reproductive biology and embryology of the crocodilians. In *Biology of the Reptilia. Volume 14*. (Eds C Gans, FS Billett and PFA Mader-son. John Wiley and Sons, New York.

Ferguson MWJ (1987) Post-laying stages of embryonic development for crocodilians. In *Wildlife management: Crocodiles and Alligators*. (Eds GJW Webb, SC Manolis and PJ Whitehead) pp. 427–444. Surrey Beatty & Sons, Sydney.

Ferguson MWJ, Joanen T (1983) Temperature-dependent sex determination in *Alligator mississippiensis*. *Journal of Zoology* **200**, 143–177.

Firth BT, Christian KA, Belan I, Kennaway DJ (2010) Melatonin rhythms in the Australian freshwater crocodile (*Crocodylus johnstoni*): a reptile lacking a pineal complex? *Journal of Comparative Physiology. B, Biochemical, Systemic, and Environmental Physiology* **180**, 67–72.

Garrick L, Lang J (1977) Social signals and behavior of adult alligators and crocodiles. *American Zoologist* **17**, 225–239.

Garrick L, Lang J, Herzog HA (1978) Social signals of adult American alligators. *Bulletin of the American Museum of Natural History* **160**, 155–192.

Girling JE (2002) The reptilian oviduct: a review of structure and function and directions for future research. *Journal of Experimental Zoology* **293**, 141–170.

Gist DH, Bagwill A, Lance V, Sever DM, Elsey RM (2008) Sperm storage in the oviduct of the American alligator. *Journal of Experimental Zoology* **309A**, 581–587.

Grabham GW (1909) A crocodile's nest. *Nature* **80**, 96.

Graham A (1968) 'The Lake Rudolf crocodile (*Crocodylus niloticus* Laurenti) population'. Report to the Kenya Game Department by Wildlife Services Limited. Nairobi, Kenya, <http://ufdcimages.uflib.ufl.edu/aa/00/00/75/90/00001/lakerudolfcrocodilesalistairgraham.pdf>

Greer AE (1970) Evolutionary and systematic significance of crocodilian nesting habits. *Nature* **227**, 523–524.

Gribbins KM, Siegel DS, Anzalone ML, Jackson DP, Venable KJ, Rheubert JL, *et al.* (2010) Ultrastructure of spermiogenesis in the American Alligator, *Alligator mississippiensis* (Reptilia, Crocodylia, Alligatoridae). *Journal of Morphology* **271**, 1260–1271.

Gribbins KM, Touzinsky KF, Siegel DS, Venable KJ, Hester GL, Elsey RM (2011) Ultrastructure of the Spermatozoon of the American Alligator, *Alligator mississippiensis* (Reptilia: Alligatoridae). *Journal of Morphology* **272**, 1281–1289.

Grigg GC (1987) Water relations of crocodilian eggs: management considerations. In *Wildlife Management: Crocodiles and Alligators*. (Eds GJW Webb, SC Manolis and PJ Whitehead) pp. 499–502. Surrey Beatty & Sons, Sydney.

Grigg GC, Beard LA (1985) Water loss and gain by eggs of *Crocodylus porosus*, related to incubation age and fertility. In *Biology of Australasian Frogs and Reptiles* (Eds GC Grigg, R Shine and H Ehmann) pp. 353–359. Surrey Beatty & Sons, Sydney.

Grigg GC, Wells RMG, Beard LA (1993) Allosteric control of oxygen binding by haemoglobin during embryonic development in the crocodile, *Crocodylus porosus*: the role of red cell organic phosphates and carbon dioxide. *Journal of Experimental Biology* **175**, 15–32.

Grigg GC, Thompson MJ, Beard LA, Harlow P (2010) Oxygen levels in mound nests of *Crocodylus porosus* and *Alligator mississippiensis* are high, and gas exchange occurs primarily by diffusion, not convection. *Australian Zoologist* **35(2)**, 235–244.

Guerrero SM, Calderón TM, de Pérez GR, Pinilla MPR (2004) Morphology of the male reproductive duct system of *Caiman crocodilis* (Crocodylia, Alligatoridae). *Annals of Anatomy* **186**, 235–245.

Guggisberg CAW (1972) *Crocodiles; Their Natural History, Folklore and Conservation*. Wren. Mount Eliza, Victoria.

Guillette LJ Jr, Milnes MR (2001) Recent observations on the reproductive physiology and toxicology of crocodilians. In *Crocodilian Biology and Evolution*. (Eds GC Grigg, F Seebacher and CE Franklin) pp. 199–213. Surrey Beatty & Sons, Sydney.

Guillette LJ Jr, Woodward AR, Qui Y-X, Cox MC, Matter JM, Gross TS (1995) Formation and regression of the corpus luteum of the American alligator (*Alligator mississippiensis*). *Journal of Morphology* **224**, 97–110.

Guillette LJ, Woodward AR, Crain DA, Masson GR, Palmer BD, Cox MC, *et al.* (1997) The reproductive cycle of the female American alligator (*Alligator mississippiensis*). *General and Comparative Endocrinology* **108**, 87–101.

Hamlin H, Lowers R, Guillette LJ Jr (2011) Seasonal and ontogenic variation in concentrations of plasma androgens in male alligators from a barrier island population – Merritt Island National Wildlife Refuge. *Biology of Reproduction* **85**, 1108–1113.

Harlow PS (2004) Temperature-dependent sex determination in lizards. In *Temperature-Dependent Sex Determination in Vertebrates.* (Eds N Valenzuela and VA Lance) pp. 42–52. Smithsonian Books, Washington DC.

Harvey KT, Hill GJ (2003) Mapping the nesting habitats of saltwater crocodiles (*Crocodylus porosus*) in Melacca swamp and Adelaide River wetlands, Northern Territory: an approach using remote sensing and GIS. *Wildlife Research* **30**, 365–375.

Heaphy L (1985) A study of the dorsal organs and cutaneous papillae of the Australian crocodiles. BSc (Hons) thesis, University of New South Wales, Sydney.

Hu Y, Wu XB (2010) Multiple paternity in Chinese Alligator (*Alligator sinensis*) clutches during a reproductive season at Xuanzhou Nature Reserve. *Amphibia-Reptilia* **31**, 419–424.

Huang Z, Watanabe ME (1986) Nest excavation and hatching behaviors of Chinese alligator and American alligator. *Acta Herpetologica Sinica* **5**, 5–9 [In Chinese with English summary]

Hunt RH (1975) Maternal behaviour in the Morelet's crocodile, *Crocodylus moreletii*. *Copeia* **1975**, 763–764.

Hunt RH (1977) Aggressive behavior by adult Morelet's crocodiles, *Crocodylus moreletii,* toward young. *Herpetologica* **33**, 195–201.

Hunt RH (1987) Nest excavation and neonate transport in wild *Alligator mississippiensis. Journal of Herpetology* **21**, 348–350.

Hunt RH, Watanabe ME (1982) Observations on the maternal behavior of the American alligator *Alligator mississippiensis. Journal of Herpetology* **16**(3), 235–239.

Hutton JM (1987) Incubation temperatures, sex ratios and sex determination in a population of Nile crocodiles (*Crocodylus niloticus*). *Journal of Zoology. London* **211**, 143–155.

Iungman J, Piña CI, Siroski P (2008) Embryological development of *Caiman latirostris* (Crocodylia: Alligatoridae). *Genesis (New York, N.Y.)* **46**, 401–417.

Jamieson BGM (1999) Spermatozoal phylogeny of the Vertebrata. In *The Male Gamete: From Basic Science to Clinical Applications.* (Ed. C Gagnon) pp. 303–331. Cache River Press, Vienna, Virginia.

Jamieson BGM (2007) Avian spermatozoa: structure and phylogeny. In *Reproductive Biology and Phylogeny of Birds. Volume 6 A and B.* (Ed. BGM Jamieson) pp. 349–511. Science Publishers, Enfield, New Hampshire.

Jamieson BGM, Scheltinga DM, Tucker AD (1997) The ultrastructure of spermatozoa of the Australian fresh water crocodile, *Crocodylus johnstoni* Krefft, 1873 (Crocodylidae, Reptilia). *Journal of Submicroscopic Cytology and Pathology* **29**, 265–274.

Jenkins NK (1975) Chemical composition of the eggs of the crocodile (*Crocodylus novaeguinea*) *Comparative Physiology and Biochemistry A* **51**, 891–895.

Joanen T (1969) Nesting ecology of alligators in Louisiana. *Proceedings of the 23rd Annual Conference of the Southeastern Association of Game and Fisheries Commissioners* **23**, 141–151.

Joanen T, McNease LL (1989) Ecology and physiology of nesting and early development of the American alligator. *American Zoologist* **29**, 987–998.

Joanen T, McNease L, Ferguson MWJ (1987) The effects of egg incubation temperature on post-hatching growth of American alligators. In *Wildlife management: Crocodiles and alligators,* (Eds GJW Webb, SC Manolis and PJ Whitehead), pp. 533–537. Surrey Beatty & Sons, Sydney.

Johnston S, Lever J, McLeod R, Oishi M, S Collins (2013) *Development of Breeding Techniques in the*

Crocodile Industry. Publication 11. Australian Government Rural Industries Research and Development Corporation, Canberra.

Johnston SD, Lever J, McLeod R, Qualischefski E, Oishi M, Omanga C, *et al* (2014a) Semen collection and seminal characteristics of the Australian saltwater crocodile (*Crocodylus porosus*). *Aquaculture* **422–423**, 25–35.

Johnston SD, Lever J, McLeod R, Qualischefski E, Brabazon S, Walton S, *et al.* (2014b) Extension, osmotic tolerance and cryopreservation of saltwater crocodile (*Crocodylus porosus*) spermatozoa. *Aquaculture* **426–427**, 213–221.

Kavaliers M, Ralph CL (1981) Encephalic photoreceptor involvement in the entrainment and control of circadian activity of young American alligators. *Physiology & Behavior* **26**, 413–418.

Kay WR (2004) Movements and home ranges of radio-tracked *Crocodylus porosus* in the Cambridge Gulf region of Western Australia. *Wildlife Research* **31**, 495–508.

Kelly DA (2013) Penile anatomy and hypotheses of erectile function in the American alligator (*Alligator mississippiensis*): muscular eversion and elastic retraction. *The Anatomical Record* **295**, 488–497.

Kern MD, Ferguson MW (1997) Gas permeability of American alligator eggs and its anatomical basis. *Physiological Zoology* **70**(5), 530–546.

King FW (2008) *Alligator Behavior: The Accuracy of William Bartram's Observations*. University of Florida, Gainesville, Florida, <http://web.uflib.ufl.edu/ufdc/?b=UF00088969>

Klein N, Scheyer T, Tütken T (2009) Skeletochronology and isotopic analysis of a captive individual of *Alligator mississippiensis* Daudin, 1802. *Fossil Record* **12**, 121–131.

Kofron CP (1990) The reproductive-cycle of the Nile crocodile (*Crocodylus niloticus*). *Journal of Zoology (London, England)* **221**, 477–488.

Kofron CP (1991) Courtship and mating of the Nile crocodile (*Crocodylus niloticus*). *Amphibia-Reptilia* **12**, 39–48.

Kushlan JA, Kushlan MS (1980) Function of nest attendance in the American alligator. *Herpetologica* **36**, 27–32.

Kushlan JA, Simon JC (1981) Egg manipulation by the American alligator. *Journal of Herpetology* **15**, 451–454.

Lance SL, Tuberville TD, Dueck L, Holz-Schietinger PL, Trosclair PL III, Elsey RM, Glenn TC (2009) Multi-year paternity and mate fidelity in the American Alligator, *Alligator mississippiensis*. *Molecular Ecology* **18**, 4508–4520.

Lance VA (1987). Hormonal control of reproduction in crocodilians. In *Wildlife Management: Crocodiles and Alligators*. (Eds GJW Webb, SC Manolis and PJ Whitehead). pp. 409–415. Surrey Beatty & Sons, Sydney.

Lance VA (1989) Reproductive-cycle of the American alligator. *American Zoologist* **29**, 999–1018.

Lance VA (2003) Alligator physiology and life history: the importance of temperature. *Experimental Gerontology* **38**, 801–805.

Lance VA (2009) Is regulation of aromatase expression in reptiles the key to understanding temperature-dependent sex determination? *Journal of Experimental Zoology* **311A**, 314–322.

Lance VA, Elsey R (2002) Sexual maturity in male American alligators: what can plasma testosterone tell us? In *Proceedings of the 16th Working Meeting of the Crocodile Specialist Group*. 7–10 October 2002, Gainesville, Florida. pp. 7-152. IUCN–The World Conservation Union, Gland, Switzerland and Cambridge, UK.

Lance VA, Vliet KA (1987) Effect of mammalian gonadotropins on testosterone secretion in male alligators. *Journal of Experimental Zoology* **241**, 91–94.

Lance VA, Vliet KA, Bolaffi JL (1985) Effect of mammalian luteinizing hormone-releasing hormone on plasma testosterone in male alligators with observations on the nature of alligator hypothalamic gonadotropic releasing hormone. *General and Comparative Endocrinology* **60**, 138–143.

Lance VA, Rostal DC, Elsey RM, Trosclair PL (2009) Ultrasonography of reproductive structures and hormonal correlates of follicular development in female American alligators, *Alligator mississippiensis*, in southwest Louisiana. *General and Comparative Endocrinology* **162**, 251–256.

Lang J (1987) Crocodilian behavior: implications for management. In *Wildlife Management: Crocodiles and Alligators* (Eds GJW Webb, SC Manolis and PJ Whitehead) pp. 273–294. Surrey Beatty & Sons, Sydney.

Lang JW, Andrews HV (1994) Temperature-dependent sex determination in crocodilians. *Journal of Experimental Zoology* **270**, 28–44.

Lang JW, Andrews H, Whitaker R (1989) Sex determination and sex ratios in *Crocodylus palustris*. *American Zoologist* **29**, 935–952.

Lewis JL, FitzSimmons NN, Jamerlan ML, Buchan JC, Grigg GC (2013) Mating systems and multiple paternity in the estuarine crocodile *(Crocodylus porosus)*. *Journal of Herpetology* **47**(1), 24–33.

Lillywhite HB, Ackerman RA (1984) Hydrostatic pressure, shell compliance and permeability to water vapor in flexible shelled eggs of the colubrid snake *Elaphe obsoleta*. In *Respiration and Metabolism of Embryonic Vertebrates*. (Ed. RS Seymour) pp. 121–135. Junk, Dordrecht, Netherlands.

Lindner G (2004) Crocodile management – Kakadu National Park. In *Proceedings of the 17th Working Meeting of the Crocodile Specialist Group*. 24–29 May 2004, Darwin. pp. 41–51. IUCN–The World Conservation Union, Gland, Switzerland and Cambridge UK.

Lutz PL, Dunbar-Cooper A (1984) The nest environment of the American crocodile *(Crocodylus acutus)*. *Copeia* **1984**, 153–161.

Maddock ST (2010) *Gavialis gangeticus* (Indian Gharial): behaviour. *Herpetological Bulletin* **113**, 39–40.

Magnusson WE (1979) Maintenance of temperature of crocodile nests (Reptilia, Crocodilidae). *Journal of Herpetology* **13**, 439–443.

Magnusson WE (1980a) Habitat required for nesting by *Crocodylus porosus* (Reptilia: Crocodilidae) in northern Australia. *Australian Wildlife Research* **7**, 149–156.

Magnusson WE (1980b) Hatching and crèche formation by *Crocodylus porosus*. *Copeia* **1980**(2), 359–362.

Magnusson WE, Sanaiotti TM (1995) Growth of *Caiman crocodilus crocodilus* in Central Amazonia, Brazil. *Copeia* **1995**, 498–501.

Magnusson WE, Taylor JA (1980) A description of developmental stages in *Crocodylus porosus*, for use in aging eggs in the field. *Australian Wildlife Research* **7**, 479–485.

Magnusson WE, Lima AP, Sampaio RA (1985) Sources of heat for nests from *Paleosuchus trigonatus* and a review of crocodilian nest temperatures. *Journal of Herpetology* **19**(2), 199–207.

Magnusson WE, Lima AP, Hero J, Sanaiotti TM, Yamakoshi M (1990) *Paleosuchus trigonatus* nests: sources of heat and sex ratios. *Journal of Herpetology* **24**(4), 397–400.

Matsumoto Y, Crews D (2012) Molecular mechanisms of temperature-dependent sex determination in the context of ecological developmental biology. *Molecular and Cellular Endocrinology* **354**, 103–110.

McVay JD, Rodriguez D, Rainwater TR, Dever JA, Platt SG, McMurry ST, *et al.* (2008) Evidence of multiple paternity in Morelet's crocodile *(Crocodylus moreletii)* in Belize, CA, inferred from microsatellite markers. *Journal of Experimental Zoology* **309A**, 643–648.

Meddle SL, Bentley GE, King VM (2002) Photoperiodism in birds and mammals. In *Biological Rhythms*. (Ed. V Kumar) pp. 192–206. Narosha Publishing House, New Delhi.)

Medem F (1981) *Los Crocodylia de sur America. Volume 1:Los Crocodylia de Colombia*. Colciencias, Bogotá, Colombia.

Mitchell NJ, Nelson NJ, Cree A, Pledger S, Keall SN, Daugherty CH (2006) Support for a unique pattern of temperature-dependent sex determination in archaic reptiles: evidence from two species of tuatara *(Sphenodon)*. *Frontiers in Zoology* **3**, 9.

Modha ML (1967) The ecology of the Nile crocodile (*Crocodylus niloticus* Laurenti) on Central Island, Lake Rudolf. *African Journal of Ecology* **5**, 74–95.

Moore B, Uribe MC, Boggs ASP, Guillette LJ Jr (2008) Developmental morphology of the neonatal alligator (*Alligator mississippiensis*) ovary. *Journal of Morphology* **269**, 302–312.

Moore BC, Mathavan K, Guillette LJ Jr (2012) Morphology and histochemistry of juvenile male American alligator (*Alligator mississippiensis*) phallus. *The Anatomical Record* **295**, 328–337.

Moulton TP, Magnusson WE, Queiroz Melo MT (1999) Growth of *Caiman latirostris* inhabiting a coastal environment in Brazil. *Journal of Herpetology* **33**(3), 479–484.

Mrosovsky N, Yntema CL (1980) Temperature dependence of sexual differentiation in sea turtles: implications for conservation practices. *Biological Conservation* **18**, 271–280.

Muniz FL, Da Silveira R, Campos Z, Magnusson WE, Hrbek T, Farias IP (2011) Multiple paternity in the Black Caiman (*Melanosuchus niger*) population in the Anavilhanas National Park, Brazilian Amazonia. *Amphibia-Reptilia* **32**(3), 428–434.

Neill WT (1971) *The Last of the Ruling Reptiles: Alligators, Crocodiles, and Their Kin.* Columbia University Press, New York.

Nelson NJ, Moore JA, Pillai S, Keall SN (2010) Thermosensitive period for sex determination in tuatara. *Herpetological Conservation and Biology* **5**(2), 324–329.

Ouboter PE, Nanhoe LMR (1987) Notes on nesting and parental care in *Caiman crocodilus crocodilus* in northern Suriname and an analysis of crocodilian nesting habitats. *Amphibia-Reptilia* **8**(4), 331–347.

Packard MJ, Packard GC (1989) Mobilization of calcium, phosphorus, and magnesium by embryonic Alligators *Alligator mississippiensis*. *The American Journal of Physiology* **257**, R1541–R1547.

Packard GC, Tracy CR, Roth JJ (1977) The physiological ecology of reptilian eggs and embryos, and the evolution of viviparity within the class Reptilia. *Biological Reviews of the Cambridge Philosophical Society* **52**, 71–105.

Packard GC, Taigen TL, Packard MJ, Schuman RD (1979) Water vapour conductance of testudinian and crocodilian eggs (Class Reptilia). *Respiration Physiology* **38**, 1–10.

Palmer BD, Guillette LJ (1992) Alligators provide evidence for an archosaurian mode of oviparity. *Biology of Reproduction* **46**, 39–47.

Passek KM, Gillingham JC (1999) Absence of kin discrimination in hatchling American alligators, *Alligator mississippiensis*. *Copeia* **1999**(3), 831–835.

Pen I, Uller T, Feldmeyer B, Harts A, While GM, Wapstra E (2010) Climate driven population divergence in sex determining systems. *Nature* **468**, 436–438.

Pieau C (1971) Sur la proportion sexuelle chex les embryons de deux Cheloniens (*Testudo graeca* L. et *Emys orbicularis* L.); issues d'oeufs incubes artificiellement. *Comptes Rendus de l'Académie des Sciences Paris* **272**(**D**), 3071–3074.

Platt SG, Rainwater TR, Thorbjarnarson JB, McMurry ST (2008) Reproductive dynamics of a tropical freshwater crocodilian: Morelet's crocodile in northern Belize. *Journal of Zoology (London, England)* **275**, 177–189.

Platt SG, Monyrath V, Sovannara H, Kheng L, Rainwater TR (2011) Nesting phenology and clutch characteristics of captive Siamese crocodiles (*Crocodylus siamensis*) in Cambodia. *Zoo Biology* **30**, 1–12.

Pooley AC (1962) The Nile Crocodile, *Crocodylus niloticus*. *The Lammergeyer* **2**, 1–55.

Pooley AC (1969) Preliminary studies on the breeding of the Nile crocodile *Crocodylus niloticus* in Zululand. *The Lammergeyer* **10**, 22–44.

Pooley AC (1974) Parental care in the Nile crocodile. *The Lammergeyer* **21**, 43–45.

Pooley AC (1982) *Discoveries of a Crocodile Man.* William Collins & Sons, London.

Pooley AC, Gans C (1976) The Nile crocodile. *Scientific American* **234**, 114–125.

Pooley AC (1977) Nest opening response of the Nile crocodile Crocodylus niloticus. *Journal of Zoology* **182**, 17–26.

Read MA, Grigg GC, Irwin SR, Shanahan D, Franklin CE (2007) Satellite tracking reveals long distance coastal travel and homing by translocated estuarine crocodiles, *Crocodylus porosus*. *PLoS ONE* **9**, 1–5

Reese AM (1915) *The Alligator and its Allies.* GP Putnam's Sons, New York.

Reisz RR, Scott D, Sues H-D, Evans DC, Raath MA (2005) Embryos of an early Jurassic prosauropod dinosaur and their evolutionary significance. *Science* **309**, 761–764.

Richardson KC, Park JY (2001) The histology of the dorsal integumentary glands in embryonic and young Estuarine crocodiles, *Crocodylus porosus* and Australian Freshwater crocodiles, *Crocodylus johnstoni*. In *Crocodilian Biology and Evolution* (Eds GC Grigg, F Seebacher and CE Franklin), pp. 180–187. Surrey Beatty & Sons, Sydney.

Riede T, Tokudas IT, Farmer CG (2011) Subglottal pressure and fundamental frequency control in contact calls of juvenile *Alligator mississippiensis*. *Journal of Experimental Biology* **214**, 3082–3095.

Riley J, Huchzermeyer FW (1999) African dwarf crocodiles in the Likouala swamp forests of the Congo Basin: habitat, density, and nesting. *Copeia* **1999**(2), 313–320.

Roth JJ, Gern WA, Roth EC, Ralph CL, Jacobson E (1980) Nonpineal melatonin in the alligator (*Alligator mississippiensis*). *Science* **210**, 548–550.

Saita A, Comazzi M, Perrotta E (1987) Electron microscope study of spermiogenesis in *Caiman crocodilus*. *Bollettino di Zoologia* **54**, 307–318.

Sarre SD, Georges A, Quinn A (2004) The ends of a continuum: genetic and temperature-dependent sex determination in reptiles. *BioEssays* **26**, 639–645.

Senter P (2008) Homology between and antiquity of stereotyped communicatory behaviors of crocodilians. *Journal of Herpetology* **42**(2), 354–360.

Seymour RS, Vleck D, Vleck CM (1986) Gas exchange in the incubation mounds of megapode birds. *Journal of Comparative Physiology. B, Biochemical, Systemic, and Environmental Physiology* **156**, 773–782.

Sharp PJ (2005) Photoperiodic regulation of seasonal breeding in birds. *Annals of the New York Academy of Sciences* **1040**, 189–199.

Shine R (1988) Parental care in reptiles. In *Biology of the Reptilia. Volume 16A.* (Eds C Gans and RB Huey) pp. 275–330. Alan R. Liss, New York.

Smith CA, McClive PJ, Western PS, Reed KJ, Sinclair AH (1999) Conservation of a sex-determining gene. *Nature* **402**, 601–602.

Somaweera R, Shine R (2012) Australian freshwater crocodiles (*Crocodylus johnstoni*) transport their hatchlings to the water. *Journal of Herpetology* **46**(3), 407–411.

Somaweera R, Webb J, Shine R (2011) It's a dog-eat-croc world: dingo predation on the nests of freshwater crocodiles in tropical Australia. *Ecological Research* **26**, 957–967.

Somaweera R, Brien M, Shine R (2013) The role of predation in shaping crocodilian natural history. *Herpetological Monograph* **27**(1), 23–51.

Staton MA, Dixon JR (1977) 'Breeding biology of the spectacled caiman, *Caiman crocodilus crocodilus*, in the Venezuelan Llanos'. Fish and Wildlife Service, Wildlife Research Report 5. United States Department of the Interior, Washington, DC.

Taplin LE (1984) Homeostasis of plasma electrolytes, sodium and water pools in the estuarine crocodile, *Crocodylus porosus*, from fresh, saline and hypersaline waters. *Oecologia* **63**, 63–70.

Teller C (2010) The evolutionary significance of temperature-dependent sex determination in reptiles. Rollins Undergraduate Research Journal 2(1), Article 5.

Thorbjarnarson JB (1996) Reproductive characteristics of the order Crocodylia. *Herpetologica* **52**(1), 8–24.

Thorbjarnarson JB, Hernández G (1993a) Reproductive ecology of the Orinoco crocodile (*Crocodylus intermedius*) in Venezuela. I. Nesting ecology and egg and clutch relationships. *Journal of Herpetology* **27**, 363–370.

Thorbjarnarson JB, Hernández G (1993b) Reproductive ecology of the Orinoco crocodile

(*Crocodylus intermedius*) in Venezuela. II. Reproductive and social behavior. *Journal of Herpetology* **27**, 371–379.

Thorbjarnarson JB, Wang X (2010) *The Chinese Alligator; Ecology, Behavior, Conservation, and Culture.* The Johns Hopkins University Press, Baltimore.

Thorbjarnarson JB, Wang XM, He LJ (2001) Reproductive ecology of the Chinese alligator (*Alligator sinensis*) and implications for conservation. *Journal of Herpetology* **35**, 553–558.

Todd NP (2007) Estimated source intensity and active space of the American alligator (Alligator mississippiensis) vocal display. *Journal of the Acoustical Society of America* **122**, 2906–2915.

Tucker AD, McCallum HI, Limpus CJ, McDonald KR (1998) Sex biased dispersal in a long-lived polygynous reptile (*Crocodylus johnstoni*). *Behavioral Ecology and Sociobiology* **44**, 85–90.

Tucker AD, Limpus CJ, McDonald KR, McCallum HI (2006) Growth dynamics of freshwater crocodiles (*Crocodylus johnstoni*) in the Lynd River, Queensland. *Australian Journal of Zoology* **54**, 409–415.

Uribe MCA, Guillette LJ Jr (2000) Oogenesis and ovarian histology of the American alligator *Alligator mississippiensis. Journal of Morphology* **245**, 225–240.

Varricchio DJ, Moore JR, Erickson GM, Norell MA, Jackson FD, Borkowski JJ (2008) Avian paternal care had dinosaur origin. *Science* **322**, 1826–1828.

Verdade LM (2001) Allometry of reproduction in broad-snouted caiman (*Caiman latirostris*). *Brazilian Journal of Biology* **61**(3), 431–435.

Vergne AL, Mathevon N (2008) Crocodile egg sounds signal hatching time. *Current Biology* **18**, R513–R514.

Vergne AL, Mathevon N, Pritz MB (2009) Acoustic communication in crocodilians: from behaviour to brain. *Biological Reviews of the Cambridge Philosophical Society* **84**, 391–411.

Vergne AL, Aubin T, Martin S, Mathevon N (2012) Acoustic communication in crocodilians: Information encoding and species specificity of juvenile calls. *Animal Cognition* **15**(6), 1095–1109.

Villamarín-Jurado F, Suárez E (2007) Nesting of the Black Caiman (Melanosuchus niger) in Northeastern Ecuador Journal of Herpetology. **41**(1), 164–167.

Vliet KA (1987) A quantitative analysis of the courtship behavior of the American alligator (*Alligator mississippiensis*). PhD thesis, University of Florida, Gainesville, Florida.

Vliet KA (1989) Social displays of the American alligator (*Alligator mississippiensis*). *American Zoologist* **29**, 1019–1031.

Vliet KA (2001) Courtship behaviour of American Alligators, *Alligator mississippiensis*. In *Crocodilian Biology and Evolution* (Eds GC Grigg, F Seebacher and CE Franklin), pp. 385–408. Surrey Beatty & Sons, Sydney.

Voeltzkow A (1899) Beiträge zur Entwicklungsgeschichte der Reptilien. I. Biologie und Entwicklung der äußeren Körperform von *Crocodilus madagascariensis. Abhandlungen der Senckenbergischen Naturforschenden Gesellschaft* **26**, 1–150.

Waitkuwait WE (1985) Investigations of the breeding biology of the West African slender-snouted crocodile *Crocodylus cataphractus* Cuvier, 1824. *Amphibia-Reptilia* **6**(4), 387–399.

Waitkuwait WE (1989) Present knowledge on the west African slender-snouted crocodile, *Crocodylus cataphractus* Cuvier 1824 and the west African dwarf crocodile *Osteolaemus tetraspis*, Cope 1861. In *Crocodiles: Their Ecology, Management, and Conservation.* (Eds PM Hall and RJ Bryant) pp. 260–275. A special publication of the IUCN–SSC Crocodile Specialist Group. IUCN, Gland, Switzerland.

Wang X, Wang D, Wu X, Wang C, Wang R, Xia T (2009a) Response specificity to advertisement vocalization in the Chinese Alligator (*Alligator sinensis*) *Ethology* **115**, 832–839.

Wang X, Wang D, Zhang S, Wang C, Wang R, Wu X (2009b) Why do Chinese alligators (Alligator sinensis) form bellowing choruses: a playback approach. Journal of the Acoustical Society of America. 126(4), 2082–2087.

Wang XY, Wang D, Wu XB, Wang RP, Wang CL (2006) Congregative effect of the Chinese alligator's bellowing chorus in mating season and its function in reproduction. *Acta Zoologica Sinica* **52**, 663–668.

Wang X, Wang D, Wu X, Wang R, Wang C (2007) Acoustic signals of the Chinese alligator (*Alligator sinensis*): social communication. *Journal of the Acoustical Society of America* **121**, 2984–2989.

Webb GJW, Manolis SC, Sack GC (1984) Cloacal sexing of hatchling crocodiles. *Australian Wildlife Research* **11**, 201–202.

Webb GJW,, Cooper-Preston H (1989) Effects of incubation temperature on crocodilians and the evolution of reptilian oviparity. *American Zoologist* **29**, 953–971.

Webb GJW, Smith AMA (1984) Sex ratio and survivorship in the Australian freshwater crocodile *Crocodylus johnstoni*. In *The Structure, Development and Evolution of Reptiles*. (Ed. MWJ Ferguson) pp. 319–355. Academic Press, London.

Webb GJW, Messel H, Magnusson W (1977) The nesting of *Crocodylus porosus* in Arnhem Land, northern Australia. *Copeia* **1977**, 238–249.

Webb GJW, Buckworth R, Manolis SC (1983a) *Crocodylus johnstoni* in the McKinlay River, N.T. VI. Nesting biology. *Australian Wildlife Research* **10**, 607–637.

Webb GJW, Buckworth R, Sack GC, Manolis SC (1983b) An interim method for estimating the age of *Crocodylus porosus* embryos. *Australian Wildlife Research* **10**, 563–570.

Webb GJW, Whitehead PJ, Manolis SC (1987a) Crocodile management in the Northern Territory of Australia. In *Wildlife Management: Crocodiles and Alligators*. (Eds GJW Webb, SC Manolis and PJ Whitehead) pp. 107–124. Surrey Beatty & Sons, Sydney.

Webb GJW, Manolis SC, Whitehead PJ (Eds) (1987b) *Wildlife Management: Crocodiles and Alligators*. Surrey Beatty & Sons, Sydney.

Webb GJW, Manolis SC, Dempsey KE, Whitehead PJ (1987c) Crocodilian eggs: a functional overview. In *Wildlife Management: Crocodiles and Alligators*. (Eds GJW Webb, SC Manolis and PJ Whitehead). pp. 417–422. Surrey Beatty & Sons, Sydney.

Webb GJW, Manolis SC, Whitehead PJ, Dempsey K (1987d) The possible relationship between embryo orientation, opaque banding and dehydration of the albumen in crocodile eggs. *Copeia* **1987**, 252–257.

Webb GJW, Beal AM, Manolis SC, Dempsey KE (1987e) The effects of incubation temperatures on sex determination and embryonic development rate in *Crocodylus johnstoni* and *C. porosus*. In *Wildlife Management: Crocodiles and Alligators*. (Eds GJW Webb, SC Manolis and PJ Whitehead) pp. 507–531. Surrey Beatty & Sons, Sydney.

Whitaker R, Andrews H (1986) Male parental care in mugger crocodiles. *National Geographic Research* **2**, 519–525.

Whitaker R (2007) The gharial: going extinct again (Gharial Multi-Task Force). *Iguana* **14**(1), 24–33.

Whitaker RM, Whitaker Z (1984) Reproductive biology of the mugger (*Crocodylus palustris*). *Journal of the Bombay Natural History Society* **81**, 297–312.

Whitehead P (1987) Respiration by *Crocodylus johnstoni* embryos. In *Wildlife Management, Crocodiles and Alligators*. (Eds GJW Webb, SC Manolis and PJ Whitehead) pp. 473–497. Surrey Beatty & Sons, Sydney.

Whitehead PJ, Webb GJW, Seymour RS (1990) Effect of incubation temperature on development of *Crocodylus johnstoni* embryos. *Physiological Zoology* **63**(5), 949–964.

Whitehead PJ, Seymour RS (1990) Patterns of metabolic rate in embryonic crocodilians *Crocodylus johnstoni* and *Crocodylus porosus*. *Physiological Zoology* **63**(2), 334–352.

Wibbels T (2003). Critical approaches to sex determination in sea turtles. In *The Biology of Sea Turtles. Volume 2*. (Eds PL Lutz, JA Musick and J Wyneken) pp. 103–134. CRC Press NY.

Wink CS, Elsey RM (1986) Changes in femoral morphology during egg-laying in *Alligator mississippiensis*. *Journal of Morphology* **189**, 183–188.

Wink CS, Elsey RM, Hill EM (1987) Changes in femoral robusticity and porosity during the reproductive cycle of the female alligator

(*Alligator mississippiensis*). *Journal of Morphology* **193**, 317–321.

Woodward A, Hines T, Abercrombie C, Hope C (1984) Spacing patterns in alligator nests. *Journal of Herpetology* **18**(1), 8–12.

Ziegler T, Olbort S (2007) Genital structures and sex identification in crocodiles. *Crocodile Specialist Group Newsletter* **26**, 16–17. <http://www.wmi. com.au/csgarticles/Genital_Structure_Sex_ Identification_L.pdf>

13

POPULATIONS AND POPULATION ECOLOGY

When Laurence Taplin came to work with us in Arnhem Land in 1979, he brought a technique he had learnt from Col Limpus in Queensland: catching 'small' crocodiles by jumping on them. This was a modification of the now familiar 'turtle rodeo' technique, in which marine turtles are captured by diving onto them out of a boat, steering them to the surface and hanging on until they can be brought on board. The first night out with us, Laurence showed us how to dive on crocs up to 1.5 m (about 5 feet). Protected (a bit) by a wetsuit, with the croc transfixed (hopefully) in the steady beam of a strong spotlight, he caught croc after croc, diving onto them from the boat, his arms outstretched in front, palms spread and thumbs locked, to descend onto and grasp the unsuspecting croc around its neck, then holding it aloft with its jaws snapping in mid air. We were very impressed! Before long, we were all employing this technique and I can clearly recall the rush of adrenaline before committing to the leap out of the boat.

Why mention this at the start of a chapter on population ecology? Because, on a recent trip to Australia's Northern Territory, David and I recalled some of those adventures and remarked that it would no longer be a wise way to catch small crocs in those waters ... the populations have recovered, or nearly so. C. porosus was hunted so hard after WWII that, by the early 1970s, adults were few and far between and, also, very shy ... they were the survivors of the hunting days. In the 1970s, we were catching their naïve offspring and the old, experienced

mommas and poppas were staying out of the way. With hunting stopped, numbers increased slowly in most river systems, young ones grew to adulthood and now they are not nearly as shy. Catching crocs by jumping on them was very effective, and loads of fun, but it is no longer to be recommended!

It is hard to know what is 'natural' in a crocodylian population and there are very few if any 'pristine' populations anywhere. In those populations recovering from hunting, the size and age structures are skewed by their history. Protection has ensured the recovery and security of some populations, but many have disappeared altogether as a result of human pressure (though no species yet, thankfully) and many are in decline. Studying the recovery has been very informative: two good examples being *C. porosus* (Webb *et al.* 2000, Fukuda *et al.* 2011) and *A. mississippiensis* (Elsey and Woodward 2010). The commercial value of their hides – the factor that was threatening their survival – has in the last 30 or so years become more of a stimulus for their conservation, for sustainable use and for increased levels of tolerance. These aspects will be discussed in Chapter 14. In this chapter, we will discuss populations and population ecology from a biological, rather than a management, perspective.

One thing well recognised about crocodylian populations is that their numbers can be reduced severely and quickly by substantial hunting of

---

American crocodiles, Crocodylus acutus, *in Costa Rica. Much like* C. porosus, *this species is associated with brackish or salt water and consequently has a wide distribution encompassing many islands and coastlines. (Photo Megan Higgie)*

adult animals. However, there can also be a strong recovery if hunting stops, unless prevented by changes in land use or other factors.

Whether aiming to improve the survival of a threatened species, set criteria for a safe, sustainable commercial harvest or manage risk to humans from a dangerous species, the most effective protocol will depend upon a good understanding of the forces that drive the dynamics of the population: that is, its population ecology. Even with that knowledge, however, the effect of a particular management strategy is not necessarily predictable because of the number of interacting variables. There are hundreds of studies that report the status of crocodylian populations, but substantially fewer that focus on their dynamics. This chapter will discuss the factors that shape the short- and long-term changes in population structure and numbers, and will present several case studies. I have deliberately included some details of the habitats, the study areas and methodologies, as well as results, in the hope of conveying a little of both the lifestyles of the animals and the nature of crocodylian field work.

The reader is referred to the excellent ecological reference books by Krebs (1994) and Caughley and Sinclair (1994).

POPULATIONS (AND SPECIES)

Biologically speaking, populations comprise members of the same species that occupy a particular geographic area and whose individuals can potentially interbreed (paraphrasing Krebs 1994). As with humans, crocodylian populations are often referred to by the political unit in which they live, rather than their actual distribution. Hence, we have reports about 'the status of alligators in Florida', or 'the population of C. porosus in Queensland', even though these species have much larger distributions. At one level, this seems inappropriate, but wildlife management is usually state- or nation-based, so it makes sense pragmatically. Similarly, and sensibly, populations are often defined by habitat entities, such as 'Caiman yacare in the Pantanal, South America', or 'C. niloticus in Lake Turkana, Kenya'.

Current species/population distributions are the product of long histories and change will continue. What we have now is just a snapshot in time. Successful dispersal can extend ranges. Changes in climate may fragment previously continuous habitat, segmenting a species/population into different geographic areas, between which movement of individuals is uncommon. Long periods of no or low interbreeding may give rise to genetic differences and sometimes populations become genetically distinct. With infrequent interbreeding or complete separation, perhaps as a result of changed environmental conditions forming a barrier, the genetic differences may become so large that separate species may result, which, if they were to re-join, would not interbreed. So, while talking about populations, it is important to realise that a particular species may be partially or completely segregated into genetically distinct entities – 'incipient-species' perhaps – whose existence may be important to recognise and conserve. The complex situation in the dwarf crocodiles, *Osteolaemus*, discussed in Chapter 1, provides an example. Molecular studies have recognised three species, two of which have genetically distinct populations within (Eaton *et al.* 2009; Shirley *et al.* 2014). Such populations are sometimes called 'evolutionarily significant units' (ESUs) *sensu* Moritz (1994): populations that can be considered sufficiently distinct to be worthy of conservation in their own right, regardless of the conservation status of the species as a whole. They may or may not evolve ultimately into different species and it is not straightforward to draw a line between an ESU and a new species.

Molecular techniques can be used also to shed light on affinities between species and aspects of past history such as hybridisations and natural or hunting-induced genetic 'bottlenecks' (when a population suffers a severe reduction in numbers and, thereby, genetic diversity).

DIVERSITY BETWEEN AND WITHIN CROCODYLIAN POPULATIONS

Two populations that are at least well on the way to evolving into separate species, and may have done so already, are the populations of the freshwater New Guinea crocodile, *Crocodylus novaeguineae* north and south of the New Guinea Highlands

(Fig. 13.1). The species was described from the Sepik River, north of the mountain range that divides the island longitudinally, and there is a question about whether the southern population is the same species or warrants separate status. Several authors have referred to consistent differences in morphology and reproductive biology between the two populations. Hall (1989) reviewed earlier material and examined several hundred specimens from both northern and southern populations, comparing palatal morphology and cranial and nuchal scalation. He noted also their allopatry, reproductive isolation and strikingly different nesting strategies and concluded that further investigation, including molecular studies, might 'yield more persuasive evidence for the partitioning of *C. novaeguineae* and lead to recognition of the southern form as a new taxon'. Jacob Gratten analysed mitochondrial DNA and microsatellite loci from ~40 individuals from each population. He found strong, but shallow,

population structure, suggesting that the Central Highlands have not been a long-term biogeographic barrier, but that 'contemporary gene flow is very limited' (Gratten 2004). He had one specimen from West Papua that was quite different and observed that although a case could be made for the southern population being a separate species, broader geographic sampling would be desirable and he preferred to regard the southern population as an ESU for management purposes. Likewise, analysis of mitochondrial DNA and microsatellite loci from the remnant population of *C. siamensis* in Borneo showed them to be strongly divergent from the remnants in Cambodia and Vietnam (Gratten 2004). At the very least, this highlights the importance of the Kalimantan population for conservation. Molecular studies do not always result in splitting. A freshwater crocodile was described in 1844 from a specimen collected in Brunei, northern Borneo and described as *Crocodylus raninus*. Ross (1990)

Fig. 13.1. *The New Guinea crocodile,* Crocodylus novaeguineae: *a freshwater species. Its populations north and south of the New Guinea highlands show genetic and morphological distinctness and little sign of interbreeding, so may be more appropriately recognised as separate species. (Photo Fritz Geller-Grimm via Wikimedia Commons)*

examined additional material and concluded that it was a valid species. However, questions remained. An editorial note accompanying Das and Charles (2002), a reprint of a 2000 paper, noted that *C. raninus* may not be different from *C. novaeguineae* and, on the basis of molecular data, Gratten (2004) came to that same conclusion.

Many crocodylian species have wide distributions and, although good at dispersing, adult individuals also show considerable site fidelity, so that genetic diversity between and within populations is common. Indeed, it seems that wherever molecular techniques are used to compare different populations, or apparently contiguous but geographically extensive populations, genetic differences are found.

The recent splitting of *Crocodylus niloticus* and *C. suchus* provides an interesting example, and one that sprang a few surprises. As discussed in Chapter 1, when Hekkala *et al.* (2010) analysed DNA microsatellites of samples of 'the Nile crocodile', '*C. niloticus*', collected in a series of watersheds

in Africa and Madagascar, they found significant genetic differentiation with quite large differences between river drainage systems (Fig. 13.2). Four microsatellite loci were sufficient to allow samples to be assigned to their population of origin. Put with other data, the populations fell into three genetic groups: West African, East African and Madagascan. With more sampling and further analysis, it turned out that the western populations are not only a cryptic species, but also less closely related to *C. niloticus* as might have been expected (Hekkala *et al.* 2011) (Fig. 2.1). The western species is part of an older lineage and, taking an earlier, available name, is now known as *C. suchus* (discussed in Chapter 1).

There could well be other surprises. Taxonomic judgements about species and phylogenies are undergoing revision for animals and plants worldwide by using molecular analytical techniques and mathematical analyses made possible by modern computing. Any currently recognised 'species' of crocodylian subject to such analysis is likely to be found to be genetically heterogeneous,

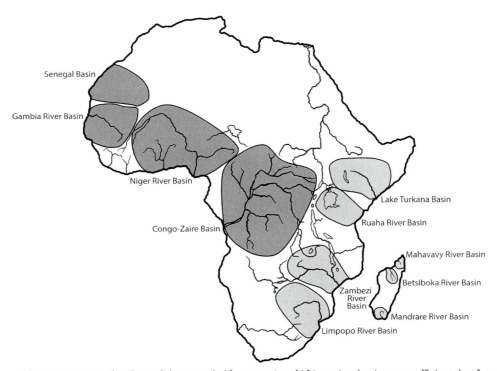

Fig. 13.2. *Genetic heterogeneity within Crocodylus sampled from a series of African river basins was sufficient that four microsatellite loci enabled samples to be assigned to their correct source (Hekkala et al. 2010). As discussed in Chapter 1, with further work it became clear that 'Nile crocodiles' – for a long time accepted as a single species – should be regarded as two: C. niloticus (light shading) and C. suchus (dark shading) (Hekkala et al. 2011). Refer also to Fig. 1.4 for a more comprehensive distribution map for these two species. (Figure adapted from Hekkala et al. 2010)*

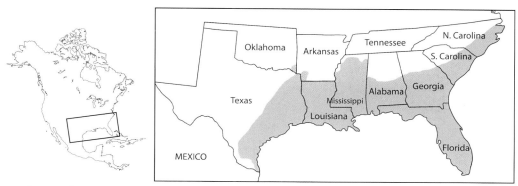

Fig. 13.3. *Distribution of* Alligator mississippiensis *in the USA. (Based on Elsey and Woodward 2010)*

and some may be sufficiently distinct to warrant separate species status.

The evolution of genetic heterogeneity does not depend necessarily on geographic isolation but can arise through 'isolation by distance' – clinal changes within a wide distribution. Thus, preliminary analyses have shown genetic differences between *C. porosus* sampled in Lakefield National Park north of Cairns and 700 km south on the Queensland coast near Townsville (FitzSimmons *et al.* 2001), even though they can easily travel long distances along coastlines (Read *et al.* 2007) and can live well in salt water (Chapter 11). *C. acutus* is also competent in salt water, yet the population on Cuba is genetically distinct from Costa Rica and Panama (Milián-García *et al.* 2011). *A. mississippiensis* too is genetically heterogeneous, with an apparent clinal change between Louisiana and Florida (Fig. 13.3), which is likely to fit an 'isolation by distance' model (Davis *et al.* 2001, 2002). In this case, however, the situation has been modified by hunting pressure, followed by the transport of thousands of individuals through conservation restocking (Chapter 14).

Some of this diversification within a continuous distribution arises because at least some crocodylians show strong site fidelity (Chapter 4, 12) and, despite skills at dispersal, genetic differentiation can occur over surprisingly short distances. Kay (2004a) sampled *C. porosus* in two adjacent rivers flowing into Cambridge Gulf in the Kimberley region of Western Australia (Fig. 13.4). Of 16 radio-tracked individuals, most moved extensively, but remained in the river where they were radio-tagged (Kay 2004b). Despite the proximity of the King and Ord Rivers, the genetic signatures of crocs in each river were different. Of 40 tissue samples from the Ord and 50 from the King, 72% and 80%, respectively, were assigned correctly to the river of origin (Kay 2004a). Tissue samples were included also from 33 crocodiles caught in the Glenelg River, 690 km to the west by the shortest sea route, and all but one were able to be assigned correctly. The 'misfit' was not assigned to any river and was judged to be a recent immigrant. Five of the 123 samples were judged to be from first-generation immigrants. Interestingly, genetic data in this study also suggested that the population had gone through a recent bottleneck, presumably associated with the heavy hunting pressure in the 30 years after the Second World War.

Among the three species of *Caiman* in South America (Figs 13.5, 13.6), *Caiman crocodilus* shows sufficient differentiation for there to be several sub-species recognised, albeit rather uncertainly, and whether any (additional to *Caiman latirostris* and *Caiman yacare*) should be elevated to full species status is under much discussion in the literature. However, Busack and Pandya (2001) analysed comparative morphologies and concluded that subdivision of *Caiman crocodilus*, even into sub-species, was not warranted, but they did support the recognition of *Caiman yacare* (formerly *Caiman crocodilus yacare*) as a valid species. There are already more than 35 publications dealing with distribution and taxonomy of the genus *Caiman* (Brazaitis, *pers. comm.*) and, with more research and more widespread sampling, it is likely that there will be more taxonomic revisions in the future.

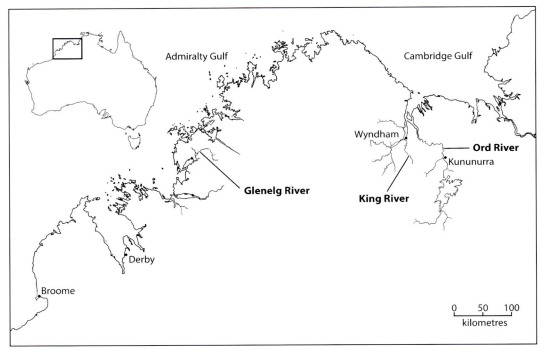

Fig. 13.4. *The Kimberley area of Western Australia. Radio-tracked* C. porosus *in the Ord and King Rivers mostly remained in those rivers and the populations in each river were found to be genetically distinct from each other and from crocs in the Glenelg River. One croc in the Glenelg was deemed to be a recent immigrant and five out of 123 crocs sampled across the three rivers were judged to be first-generation immigrants. (Adapted from Kay 2004a, with permission)*

The broad-snouted caiman, *Caiman latirostris*, provides another example of genetic differentiation over a short distance. One of the most widespread crocodylians (Figs 13.5, 13.6), it shows modest differentiation related to geographic distance. The most differentiated population is on Cardoso Island south of Sao Paulo, separated from the mainland by a marine strait only 1.5 km wide at the narrowest point (Villela *et al.* 2008). *Caiman latirostris* are tolerant of living in estuaries, as long as they can get periodic access to fresh water to drink (Grigg *et al.* 1998, Chapter 11) so this 'genetic gap' across a narrow channel, which does not seem to be much of a barrier, is interesting.

Field work to collect sufficient samples across the whole range of these species is extremely challenging, as is the subsequent analysis, but understanding genetic diversity and taxonomic questions can be very important for checking compliance with international trade requirements for crocodylian products because import regulations in the major importing nations vary between species, or sub-species, and between source countries (Chapter 14).

We can be sure that more crocodylian ESUs will be uncovered by further work, and perhaps more cryptic species too.

SPECIES IN SYMPATRY: INTERSPECIFIC COMPETITION

Adam Britton's website http//www.crocodilian.com provides a convenient reference to crocodylian distributions and a few minutes browsing shows that many species of crocodylian are sympatric, with overlapping or coincident geographic distributions. This raises questions about potential competition, and even hybridisation, between them.

Despite broad geographic overlap, differences in habitat, dietary preferences as a consequence of different body sizes, breeding seasons and other life history differences may result in minimal or no competition between apparently sympatric species. Therefore, geographic overlap does not necessarily mean sympatry at the level of the individual. For example, black caiman, *Melanosuchus niger*, and the two species of *Paleosuchus* have very similar geographic ranges, yet are found in different habitats.

Fig. 13.5. *The spectacled caiman,* Caiman crocodilus *and the broad-snouted caiman,* C. latirostris, *each show genetic differentiation within their enormous ranges, correlating partly with drainage basins, and modified by geographic distances. The genetic (and morphological) differentiation within* Caiman crocodilus *is such that several sub-species are recognised, somewhat uncertainly. One previous sub-species now recognised as a separate species is* Caiman yacare. *In* Caiman latirostris, *the most differentiated population is that found on Cardoso Island south of Sao Paulo, separated from the continental distribution by a 1.5 km sea water strait. (Maps modified from crocodilian.com with permission)*

P. trigonatus lives in forested uplands, away from the main water bodies. It is the only crocodylian to maintain self-sustaining populations in closed-canopy perennial streams. *Melanosuchus* occurs in the larger streams and *P. palpebrosus* occurs around lakes and flooded forest associated with large rivers. It is generally the only crocodylian resident in seasonal streams and streams running through savannas. Across and well beyond the broad geographic range of these three species, the most numerous species is probably *Caiman crocodilus*, with *P. trigonatus* perhaps the most numerous within its range (Magnusson *pers. comm.* 2011 and see 'Case histories' below). Despite these apparent habitat specialisations, the separations are only partial and there can be competition where they overlap. Da Silveira *et al.* (2013) found that *Melanosuchus niger* grew more slowly than *Caiman crocodilus* in an area with a high density of caiman and a higher productivity.

There are numerous examples of crocodylians in broad sympatry. In Australia, the ranges of

C. porosus and *C. johnstoni* overlap considerably, with the former aggressive to the latter (Fig. 13.7). Muggers (*C. palustris*), gharials (*G. gangeticus*) and *C. porosus* occur sympatrically in some northern Indian rivers. In West Africa, the geographic distributions of *Osteolaemus tetraspis, Mecistops* and *C. suchus* overlap, as do the distributions of *C. acutus* and *A. mississippiensis* in the United States; the last two can occur together in the same waterway. In Central America, the ranges of both *C. rhombifer* and *C. moreletii* are within the distribution of *C. acutus* (Fig. 13.8), with implications for their future through hybridisation (see below).

What are the implications of such apparent or actual co-occurrences? Diets tend to be similar, so there may be competition for food, basking sites or nesting sites if requirements are similar and these are in short supply. Competition between *Melanosuchus niger* and *Caiman crocodilus* was mentioned above, but interactions may be very much more direct, such as predation and/or territorial defence (Fig. 13.7), which may result in the victor feeding on the victim: a probable outcome in this example.

Most of these sympatries have eventuated naturally, and the interactions between the species and their effects are part of the normal ever-changing dynamics of ecosystems. Human interventions have led to opportunities to see examples of such changes. In Australia, the natural ranges of *C. johnstoni* and *C. porosus* overlap extensively in the upstream freshwaters, with *C. porosus* dominant downstream. The hunting of *C. porosus* in the Northern Territory (NT) in the 30 or so years before 1971 allowed *C. johnstoni* to extend further downstream into the tidal sections of rivers normally dominated by *C. porosus*. Webb *et al.* (1983a) suggested that competition from the recovering population of *C. porosus* in the Adelaide River, NT, (see below) was pushing the 'freshies' back upstream. Similar expansion upstream of *C. porosus* in the Daly, Roper and (probably) Victoria Rivers since their protection has been described also by Letnic and Connors (2006). Figure 13.7 shows an example of a *C. porosus* dealing with a 'freshie' in the Adelaide River where *C. porosus* now dominates, but where *C. johnstoni* were commonly sighted in the early 1980s (Webb *et al.* 1983a).

Fig. 13.6. *The four species of large caiman currently recognised: (top left) Caiman crocodilus Arquipelago Anavilhanas, Amazonas, Brazil; (top right) Caiman yacare, Pantanal, Brazil; (bottom left) Caiman latirostris, WCS Central Park Zoo; and (bottom right) Melanosuchus niger, Mamirauá Sustainable Development Reserve, Amazonas, Brazil. (Caiman photos DSK, Melanosuchus photo Cássia Camillo per Robinson Botero-Arias)*

Fig. 13.7. *A large estuarine crocodile, C. porosus, attacks and kills a freshwater crocodile, C. johnstoni in the Adelaide River, Northern Territory, Australia. This was photographed in an area now visited daily by tourists on the various 'jumping crocodiles' cruises. 'Freshies' were common this far downstream in the early 1980s after 'salties' had been depleted by hunting, but they are rarely seen there now, whereas 'salties' are very common. Whether this particular interaction subsequently became an act of predation is unknown. (Photos courtesy of Adam Bowman and Wallaroo Eco Tours http://www.litchfielddaytours.com.au)*

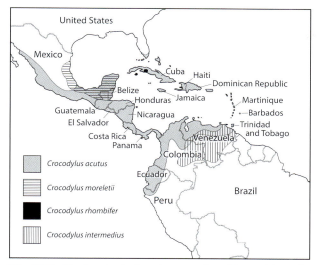

Fig. 13.8. *The four Latin American species of* Crocodylus. *The critically endangered Cuban crocodile* C. rhombifer *is at risk through hybridisation with the Cuban population of American crocodiles,* C. acutus, *which itself is different from* C. acutus *in Central America. In their range of overlap in Mexico and Belize, hybridisation between* C. acutus *and Morelet's crocodile,* C. moreletii, *is also of conservation interest. (Map modified from crocodilian.com, with permission)*

Fig. 13.9. *Morelet's crocodile,* C. moreletii, *is known to hybridise with* C. acutus *on the Yucatan peninsula in Belize. (Photo DSK)*

Crocodylians may face competition from non-crocodylian species as well. Luiselli *et al.* (1999) explored the possibility of competition between the dwarf crocodile, *Osteolaemus tetraspis,* and the Nile monitor lizard, *Varanus niloticus*, in their shared habitat: the freshwater ecosystems of the south-eastern Nigerian rainforest. Both species eat mostly crabs and a nearly 80% dietary overlap between these similar-sized species suggested a potential for competition for resources, even though crabs were common. The authors also observed predation by a *V. niloticus* on an *O. tetraspis*.

Another possible outcome of two or more crocodylians living in close sympatry is hybridisation.

HYBRIDISATION AND INTROGRESSION

Hybridisation and introgression (repeated backcrossing) may increase genetic variability and fitness, lead to the melding of two previously distinct lineages or, if the hybridisation is unidirectional, one lineage may cause the extinction of the other (Milián-García *et al.* 2011). Hybridisation between species in the wild is usually assumed to be undesirable, but it is worth remembering that it is part of natural processes and, even if it leads to extinctions, is not necessarily tragic. The yardstick for many people is that it should not be accelerated by or a consequence of human activities.

The ease with which some crocodylians hybridise in captivity obviously has conservation implications, particularly if restocking is mooted.

Hybridisation between captive members of *Crocodylus* is not uncommon. One famous hybrid is 'Yai', a 20 foot (6.1 m) cross between *C. porosus* and *C. siamensis* bred in the Samutprakan Crocodile Farm in Thailand (Chapter 1). Hybrids seem to be restricted to members of the same genus: no hybrids between alligatorids and crocodylids have yet been reported. Presumably the genetic differences are too large.

How common is hybridisation in the wild? Different habitat preferences and/or different breeding seasons, as well as behavioural or genetic impediments may make cross-species matings unlikely, but there are some well-documented examples. *C. acutus* and *C. moreletii* have been known to hybridise in captivity (Luis Sigler *pers. comm.* 2012) and specimens with intermediate scale patterns have often been reported from overlap zones (Fig. 13.8) in coastal Mexico and Belize, dating back to the 1920s (Ray *et al.* 2004). These authors analysed mitochondrial DNA and concluded that hybridisation between these two species was likely, particularly at one location in Belize. This was confirmed subsequently by Cedeño-Vásquez *et al.* (2008) and Rodriguez *et al.* (2008) using a variety of genetic approaches, with a focus on the Yucatan Peninsula. The implications of this are very interesting, because the hybridisation has presumably been occurring for a very long time and would seem to be posing a threat to the genetic integrity of both *C. moreletii* (Fig. 13.9) and *C. acutus* in their extensive area of overlap on the Yucatan Peninsula in particular (Fig. 13.8). Both sets of authors referred to this and identified a need for additional studies to resolve what is going on.

In Cuba, the critically endangered Cuban crocodile, *C. rhombifer*, is sympatric with the much more extensively distributed American crocodile, *C. acutus* (Fig. 13.8). These hybridise readily in captivity (Weaver *et al.* 2008). *C. rhombifer* – one of the world's most distinctive crocs (Fig. 13.10) with one of the smallest distributions (Fig. 13.8) – is nowadays restricted to one swamp in Cuba and a nearby island. Quaternary fossils from the Grand Cayman islands and Bahamas, as well as genetic studies, indicate a wider historic distribution (Milián-García *et al.* (2011). In their present distribution, they are sympatric with Cuban *C. acutus* at both sites and their breeding seasons overlap by a couple of months (Rodríguez-Soberón 2000). Milián-García *et al.* (2011) compared genetic markers from both species and putative hybrids in Cuba and on the coast of Panama and Costa Rica. They found evidence of hybridisation and introgression between Cuban *C. acutus* and *C. rhombifer* and suggested that this might be ancient. There was also evidence of recent hybridisation between Cuban *C. acutus* and *C. rhombifer*, which has implications for conservation of both 'evolutionarily significant units' (ESUs) (*sensu* Moritz 1994). Milián-García and colleagues also found consistent differences between the Cuban and Central American *C. acutus* (enough to wonder whether future research would lead to a taxonomic change) and, somewhat enigmatically, that *C. rhombifer* could be closer to Cuban *C. acutus* than the latter to Central American *C. acutus*.

Fig. 13.10. *A Cuban crocodile,* C. rhombifer. *This species is critically endangered, has a very small distribution on Cuba and hybridises with the sympatric local population of* C. acutus, *which is genetically distinct from* C. acutus *elsewhere. (Photo Charles Booher)*

Hybridisation between species of alligatorid seems to be much less common, even between captive individuals. Brazaitis *et al.* (1998) postulated hybridisation to explain the occurrence of individuals morphologically intermediate between *C. crocodilus* and *C. yacare* where they overlap in the Madeira River, Brazil (Fig. 13.5). A follow-up study by Hrbek *et al.* (2008) analysing mitochondrial DNA confirmed that conclusion while observing, however, that differentiation along a cline was also a possibility. Federico Medem speculated that hybridisation was likely between the two species of *Paleosuchus* (Magnusson *pers. comm.* 2011).

PRISTINE POPULATIONS?

It would be useful to know more about the demographic structure of 'pristine' populations of crocodylians, but few if any remain. Most studies have been of populations recovering from a severe reduction due to hunting or, unfortunately, have been documenting their decline. A recent study (Bourquin and Leslie 2011) looked at the demographics of an exploited population of *C. niloticus* in the Okavango Delta, Botswana, and employed some newer approaches (Chapter 14).

Information from one relatively pristine population was reported by Cott (1961) from Lake Mweru Wa Ntipa in what is now Zambia in the Great Rift Valley, where crocodiles had enjoyed complete protection. The numbers are relatively small, but, of 51 animals shot, 42 were adult (82%) and the adult sex ratio (22 males, 20 female) was essentially 1:1 (Fig. 13.31). I have not been able to find really good data on any pristine population and, indeed, I do not know of any. One possibility is *Paleosuchus trigonatus* in Brazil (see 'Case histories' below), but it is a somewhat atypical crocodylian. Ironically, the best guess about what a 'mature' population might look like comes from projections from nearly 40 years of monitoring the recovery of *C. porosus* in 12 tidal rivers in the NT (Fukuda *et al.* 2011). Ironic because, after intense hunting, (1945–1971) this population was reduced to a few percent of former adult abundance and was listed in Appendix 1 of the Convention of International Trade of Endangered Species (CITES) until 1985. There is more information on that recovering

population below. It has shown a striking reversal in the proportion of small to large animals during the recovery (Fukuda *et al.* 2011 and see 'Case histories' below). Soon after protection, 75–80% of the individuals in tidal rivers were smaller than 1.8 m. Now, crocs that size typically comprise only 20–40%. Some rivers, of course, have less breeding habitat and these show lower percentages of small individuals. Perhaps truly pristine populations have relatively few sub-adults, as in Lake Mweru Wa Ntipa: a consequence of competition and, often, cannibalism (see below).

It seems probable that 'pristine' crocodylian populations are skewed heavily towards adults, which produce hatchlings whose chance of survival to become part of the adult breeding population is very low, and that density-dependent factors, including resources (food, space, mates; Magnusson 1986) and cannibalism (see below), have a large influence on carrying capacity.

NEGATIVE GENETIC EFFECTS FROM OVEREXPLOITATION

Severe reduction of populations, by hunting for example, leads to a loss of genetic diversity and this can affect the capacity of the population to persist more than the lower numbers alone would imply. It has been argued, however, that in long-lived species this effect is minimised, although recovery can certainly take place from severely depleted populations. Bishop *et al.* (2009) reviewed this concept and asked whether the life history characteristics of Nile crocodiles, *C. niloticus*, in the Okanvango Delta, Botswana, can be expected to have buffered it against negative genetic effects arising from its history of overexploitation. Sampling genetic variation in the wild population, they estimated that the population's effective population size (the part of the population that is contributing to further generations) is one-fifth of pre-hunting levels and, further, is only 5% of the current population. Simulating the likely effects of genetic drift they found that, even if the current effective population size is maintained, the apparent retention of genetic variation as the numbers of juveniles increase is deceptive because it ignores the likelihood of continuing decline in allelic diversity and heterozygosity. This decline could be exacerbated by permitting further harvesting of adults, even if controlled.

The message is that, although conservation measures for *C. niloticus* may be increasing their numbers, the genetic diversity lost by overexploitation may continue to decline post-protection and could be exacerbated if adult harvesting continues. This has obvious implications for conservation of other crocodylians too and emphasises the value of genetic surveillance in parallel with conservation measures.

CROCODYLIAN LIFE HISTORIES IN GENERAL; NOT SIMPLY *R*-SELECTED OR *K*-SELECTED

Crocodylian life histories are diverse, but they do share many similarities. They produce reasonably large clutches of eggs (modal number in the 30s, range 12–48; Thorbjarnarson 1996), many of which do not hatch because of predation, disturbance, drying out, flooding, thermal stress or some other event during the typical 70–90 day (depending on species) incubation periods. There is often maternal care of the nest and usually a short period of maternal care after hatching, but survival of the hatchlings through to adulthood is comparatively low (mainly from predation and aggressive interactions) and, although there may be a respite during juvenile years, high mortality may continue right up to sub-adulthood, when males compete for a place in the breeding system. They take several years to reach sexual maturity, by which time they are quite large. Adult survivorship is high in a natural population and individuals are long-lived, with long reproductive lives (several decades). Dispersal is mostly by sub-adults (see definitions of life stages, Chapter 1), so these are the individuals that contribute to emigration and immigration. Adults of some species may range widely, but seem to show breeding site fidelity (more information is needed about this). Many habitats that are suitable for crocodylians and carry good populations do not have good nesting habitat, so we have the concept of 'source and sink' populations: the source populations in good breeding areas feeding the 'sink' areas by dispersal of sub-adults.

There is a generalisation in ecology that there is a dichotomy in lifestyles between animals that are large, long lived, and produce small numbers of young that they take good care of (usually called *K*-selected) and those that are smaller, shorter lived, and produce very large numbers of young, each of which grows rapidly but has only a low probability of surviving to maturity (*r*-selected species). The former, *K*-selected species are more typical of stable, predictable environments where populations often reach carrying capacity and, so, frequently encounter competition for dwindling resources. Humans, elephants and long-lived birds are examples of *K*-selected species. In contrast, *r*-selected species do well in unpredictable, unstable environments and can exploit opportunity, but rarely reach carrying capacity before suffering from environmental stress. Small rodents, many insects and weedy plant species provide examples. The concept is a useful one because it provides a good conceptual framework and also focusses attention on how interacting attributes of life history have big implications for the dynamics and survival of populations and species.

Bill Magnusson (1986) noted that crocodylian life histories do not easily fit this dichotomy, although they do have many of the characteristics of *K*-selected species. Crocodylians are large (adult females 10 –150+ kg) and long lived (many to 60 years and longer), with comparatively slow growth and long reproductive lives (*K*-selected traits). But most produce large clutches of eggs, have many clutches over a lifetime and neonate survival to adulthood is low (*r*-selected traits). Even though many show maternal care of the neonates, this does not typically continue past the first few weeks and data on its effectiveness are lacking. Cannibalism has been reported in several species and may be typical (see below). Some crocodylians, perhaps most, would be resource limited in a natural state: crocodylians often compete for space (basking sites, breeding pools, mates) and, because of the frequency of empty stomachs, some have been thought food limited (e.g. *C. niloticus*, Cott 1961; Graham 1968). Nevertheless, as we shall see, within the Crocodylia there is a range of lifestyles, a situation which is reminiscent of the *r–K* dichotomy in general, with faster growing 'weedy' species, such as *Caiman crocodilus*, at the more '*r*-selected' end and slower growing species, such as *Crocodylus johnstoni* and *C. porosus*, at the more '*K*-selected' end. For a more thorough and more thoughtful discussion of the difficulty of applying these terms to crocodylians, based on experiments, see Abercrombie *et al.* (2001).

Tony Tucker has pointed out that these life history characteristics can be seen as 'risk averse' or 'bet-hedging' (Tucker 2001), which serve them well in a stochastic environment. The chance of any particular hatchling reaching maturity is very low, but by having very long reproductive lives and large clutches the adults 'spread the risk' of reproductive failure. They are, however, very vulnerable to uncontrolled removal of the adults, about which more in Chapter 14. Tucker (2001) provided a useful review of stage-structured population models.

POPULATION DYNAMICS: CHANGES THROUGH TIME

A good understanding of an animal's population ecology requires knowledge of the population's size, age structure, survival rates at different ages/stages, reproductive rates and rates of emigration and immigration. Caughley and Sinclair (1994) provided a comprehensive general review of population ecology as it applies to wildlife management.

Writing about crocodylians in particular, Webb and Smith (1987) provided a thoughtful review based on considerable practical experience. They identified three levels of understanding of population dynamics. Is the population stable, decreasing or increasing? What is the age/size structure and survival rate at each stage and what is the reproductive rate as well as the extent of immigration and emigration (so that a simple model can be constructed)? Third, by what processes and how does the population respond to change, including an understanding of population regulatory processes such as competition for resources (food, space, mates) and cannibalism? Most crocodylian studies have focussed on the first level, some on the second and few on the third.

Effective techniques for monitoring change in both numbers and demographic structure are obviously fundamental and we will now turn our attention to that.

MONITORING TRENDS IN NUMBERS: DENSITY AND STRUCTURAL CHANGE

Whenever there is a crocodile attack in North Queensland, the newspapers are quick to report opinions that crocodile numbers are increasing and need to be controlled. This concern is understandable wherever large crocs and people co-occur. But are the numbers of crocodiles in North Queensland actually increasing? It is a point of local dispute and the data from insufficiently regular surveys are ambivalent. Reliable information about trends in numbers is an important responsibility for wildlife management agencies, not only for the large and dangerous crocodylians, but also for those which are less common and whose conservation status is uncertain. For most of them, effective programmes can be set up by which trends can be monitored to inform both public opinion and, if necessary, to prompt management actions.

In addition to trends in numbers and density distribution, biologists usually want to know trends in the size structure of the population so, on a survey, the size class of individuals sighted will usually be recorded also. We are lucky with crocodylians compared with, say, sea turtles, because size is a good indication of life stage, so reasonable estimates of the proportion of hatchlings, juveniles, sub-adults and adults can be made, even though the ease of sighting each life stage varies. Hatchlings are easy to see and approach – they have not yet learnt to be wary, whereas only the eyes of adults may be seen. The more information about the size structure the better, because all crocs do not contribute equally to the population.

The results are expressed as totals and density (number per unit of area or, in the case of riverine crocodylians, per km of shoreline), preferably broken down by size. In an interesting innovation, these data were translated into biomass, as kg/km in Fukuda *et al.* (2011), allowing an additional type of comparison between rivers and through time. This raises possibilities of exploring a more 'holistic' approach: how many tonnes of crocs can the system feed?

Crocodylians can be surveyed either by spotlighting eye shines at night from a boat, counting individuals from the air by day or counting the numbers of nests and/or anything else (indicators) that will give clues about the population trend. Even counting the number of tracks across mudbanks year after year can provide an index of change, including information about size structure because larger crocs leave a wider track. The raw counts provide an index of density and distribution. If surveys are conducted repeatedly, following exactly a standardised protocol, comparisons can be made between surveys and the population's trend determined. Converting those indices to actual numbers is far from easy and not always necessary. Bayliss (1987) provided a useful review of methodologies, particularly comparing boat and helicopter surveys. There is a large body of literature and surveys have to be modified to suit local situations.

Spotlight surveys

The most common method for surveying crocodylians is by spotlighting, usually from a boat (Fig. 13.11) and combined with approaching the animal to identify species and estimate its size. The method was instigated in Australia by Harry Messel and his colleagues (Messel *et al.* 1981), who also conducted substantial comparisons to assess its

Fig. 13.11. *A historic, but nevertheless typical, survey underway for* C. porosus *in a tidal river in the Northern Territory, Australia. The matt black boat makes it less visible. The team consisted of a spotter, a recorder and a driver, in this case Graeme Wells, George Vorlicek and Sandra Bourne, all members of the University of Sydney's legendary Crocodile Research Programme led by Harry Messel, who pioneered rigorous spotlight survey protocols in Australia. (Photo Bill Green)*

reliability. Eye shines can be seen from as far away as 400 m and Webb and Manolis (1992) provided a valuable review of the survey method. The method grew out of the way hunters looked for crocs from aluminium boats or canoes soon after the Second World War, using a battery operated spotlight. Normally cryptic both day and night, the croc's reflective tapetum lucidum gives them away, the eye bright and jewel like, the red and steady glow like a live coal in the distance (Chapter 5, Figs 5.7, 5.9). Of course, it is hard to tell the size of the croc from its eye shine, or the species (in areas where two or more species are present). Many a time a spotter approaches a croc, expecting from the deep, bright red reflection to see a big one and, upon getting close, suddenly discovers it is only a hatchling! More interesting is the reverse, when a hatchling is expected and suddenly, when almost on top of it, the tapetum's owner is revealed as a 4 m male!

Biologists have been looking for crocs by their eye shine for a long time. Hugh Cott wrote about it in Uganda in the 1950s and Alistair Graham (Graham 1968) wrote about wading in the shallows of Lake Rudolf (now Lake Turkana) with a flashlight, counting *C. niloticus* (!). In his paper, he advised that 'movement of the observer must be continuous, or the animals become disturbed' and that dark nights were essential to avoid being seen by the animals. Sound advice indeed!

Spotlight surveys are typically made from a boat, but can be adapted for swampy wetlands or wherever a boat or an airboat can travel. If the stream is too small, transects can be walked, as Bill Magnusson and Albertina Lima did in their study of *Paleosuchus trigonatus* (Magnusson and Lima 1991, see 'Case Histories' below) and Zilca Campos in her study of *Paleosuchus palpebrosus* (Campos *et al.* 2010). On tidal rivers, counting by spotlight is best done at low tide because the crocs are more likely to be visible when the water is lower, and in the cooler months because they are more likely to be in the water at night when the water is warmer than the air (Chapter 10).

The most unusual spotlighting I have done was with Tamir Ellis in Florida in 1980, driving along the busy freeway connecting the Florida Keys and sweeping the spotlight beam over the inshore ocean for a glimpse of *C. acutus*. Unfortunately, we didn't see any: they were then rare and critically endangered in Florida. They are now recovering and no longer on the endangered list (Chapter 14). David's first encounter with this species in the wild was along the same freeway 10 years later, and in even more unusual circumstances. He and a few local reptile enthusiasts, none of whom had seen *C. acutus* in the wild, drove from Miami to the keys in the hope of finding one. Spotting a 'likely looking' lagoon, they took a local biologist's advice, got down on their hands and knees near the water's edge and started barking like dogs! They felt so silly that they jumped to their feet and adopted a casual pose whenever a car drove past. But the trick worked! A croc came in close enough to give them a good look, before submerging and disappearing when they all stood up. Despite their success, David advises not to try this 'at home'!

As shown by Messel *et al.* (1981), it is crucial to standardise survey methods so that results can be compared across years. Anyone planning spotlight surveys should read Wood *et al.* (1985) who explored the practicality of using annual spotlight counts to monitor state wide trends in alligator numbers in Florida. They highlighted many of the difficulties and made cautionary recommendations about how to conduct surveys with enough precision to make comparisons between surveys and get useful data about population trends.

A rigorous analysis of factors that affect the 'sightability' of *C. porosus* in tidal riverine situations was undertaken by Harry Messel and his crew in Northern Australia during the 1970s. Their study involved repeated surveys under a diversity of conditions in the tidal reaches of the mangrove-lined Blyth River in Arnhem Land (Messel *et al.* 1981) and found that it is best to count with the maximum amount of mudbank exposed. Conveniently, the low tide develops progressively upwards from the mouth of tidal rivers, so an up-river survey can be undertaken over several hours of low tide exposure. Messel and crew were able to develop correction factors to translate observed numbers to total numbers. In that habitat, they concluded that the proportion of animals sighted varied with

crocodile size class: 63% of hatchlings, 67% of 2–6 feet (0.6–0.9 m) juveniles and averaging out at ~63% for all crocs (Messel *et al.* 1981). Some animals are simply missed, but some of course will be under the water or up among the riparian vegetation with their eyes hidden from the light.

One way to calibrate spotlight (or helicopter) surveys to obtain estimates of actual numbers is to apply mark–recapture methods. Animal populations are often assessed by mark–recapture techniques but capture makes crocs wary, and the larger they are the more wary they are (Webb and Messel 1979). Peter Bayliss modified the technique to 'mark–resight' by applying numbered tags to *C. porosus* in three sections of the Adelaide River, NT, either after hand capture or, for larger animals, using a skin tag harpooned into the crocodile (Bayliss *et al.* 1986; Bayliss 1987). Resighting on subsequent spotlight surveys was equated to recapture. Data were analysed using several different models and in the wider downstream section sighting factors of 61–67% were broadly similar to those of Messel *et al.* (1981), despite being devised in a different way. In narrower waterways, however, sighting proportions were lower and in the narrowest creeks ranged from 33–37% and 'sightability' was size-dependent: smaller crocs were more likely to be seen.

The essence of a wildlife monitoring programme, particularly of an exploited population, is that the precision of the monitoring must be sensitive enough to detect significant changes in time for management actions should that become necessary. Stirrat *et al.* (2001) assessed spotlight monitoring in tidal rivers in the NT and concluded that a decline of 10% could be detected by the method with 90% certainty within 4 years. As female *C. porosus* take 10–12 years to reach sexual maturity, so 4 years is about one-third of the generation time, so the established NT practice of spotlight monitoring is quite adequate for conservative management purposes.

Estimating size is an issue that takes practice, and results differ between observers (Magnusson 1983). Choquenot and Webb (1987) developed a photographic technique that aimed to put the estimates onto a more rigorous basis. It was used to calibrate observers and show how they are highly variable in their size estimates, but I am not aware that its use has become widespread. The emerging use of laser photogrammetry may be worth exploring for crocodylians. It has been used for measurement of whale sharks (Rohner *et al.* 2011) and also for ibex, elephants and manta rays.

Aerial surveys

In the 1960s, Alistair Graham counted *C. niloticus* in Lake Turkana from a Cessna 172 (Graham 1968). Light aircraft and helicopters were used by Ted Joanen in the 1970s in the Louisiana marshes, counting nests in particular. Rice *et al.* (2000) described and evaluated the application of helicopter surveys of alligator nests in Florida marshlands. Helicopters are the preferred choice for aerial survey when the budget permits. Aerial surveys for crocs are conducted during the day, so, of course, rely on seeing the animal, not its eye shine. A helicopter allows access to areas that cannot be reached by boat and can be used in both riverine and swamp situations (Fig. 13.12). Bayliss

et al. (1986) and Bayliss (1987) developed helicopter use for surveys in NT rivers. They flew at low level (20 m above ground) and low speed (50 knots = 93 km/h) or less) and compared the results with spotlight surveys and absolute numbers determined by a mark–resighting study. A smaller proportion of the population was counted, and hatchlings were not seen, but a much higher proportion of the adults were. Bayliss and colleagues pointed out that boat surveys are more time consuming, particularly in more remote areas, and calculated that helicopters are less expensive per unit survey area. However, in the same study in which they assessed the capacity of spotlight surveys to detect population change, Stirrat *et al.* (2001) questioned the effectiveness of helicopter survey. Nevertheless, the method is valuable for assessing trends.

The opportunity to locate and count riverside nests of *C. porosus* independently in aerial and ground surveys of the Liverpool and Tomkinson Rivers in the Northern Territory enabled Magnusson *et al.* (1978a) to develop a method for estimating the

Fig. 13.12. *A saltwater crocodile,* C. porosus, *at its nest site in a swamp, photographed from a helicopter. Nests such as this, surrounded completely by tall vegetation, could be found and counted only from the air. (Photo Grahame Webb)*

total number of nests. In practice, the sample sizes in this case were too small to allow a precise estimate, but the method is worth mentioning because the same approach may have a wider applicability. It may be applicable where two independent surveys of a population can be made contemporaneously, as long as it is possible to identify which entities were seen by both surveys. This is mandatory and it was possible in this case because all nests were mapped as they were located. Hence, it was possible to identify which of the total number seen in each of the surveys were seen by both. The method relies on binomial theory and is an adaptation of the Petersen estimate. A critical assumption is that the counts made in each survey are independent. It is less critical that the probability of seeing each entity is the same in each survey (or each survey method if the methods are different).

In the Brazilian Pantanal, where *Caiman yacare* concentrate around water bodies in the dry season, sometimes in enormous numbers (Fig. 4.2), Marcos Coutinho and Zilca Campos counted them successfully from an ultralight aircraft (Coutinho and Campos 1996). Flying with Marcos Coutinho in 1992 gave Lyn Beard and me the idea of using an ultra-light as a 'cheap' helicopter equivalent for counting kangaroos from the air (Grigg *et al.* 1997). This case of crocodylian studies spawning kangaroo studies has a twist, because it followed on from Guilherme Mourão, Marcos Coutinho and others using fixed-wing aircraft for aerial survey of caimans and other wildlife, following a visit to the Pantanal by Graeme Caughley, who had perfected the technique on kangaroos: a case of 'technology transfer' working in two directions, with kangaroo studies spawning crocodylian studies spawning kangaroo studies!

Another use of aerial survey is in locating nests either to get a useful indicator of population trends (as in Louisiana for alligator monitoring) or for broad-scale surveys of suitable nesting habitat (Figs 13.13, 13.14). In the wet seasons of 1975–76 and 1978–79, Bill Magnusson, Janet Taylor and I flew surveys of most of the rivers and wetlands east of Darwin across the top of Australia to Cape York, mapping suitable nesting habitat. We were

Fig. 13.13. *Nest of* C. porosus *on the Wildman River, Northern Territory. (Photo Buck Salau)*

Fig. 13.14. *A riverside nest site, C. porosus, Cadell River Northern Territory, February 1982. Although conspicuous from the air, nests such as this are quite likely to be missed on boat surveys. (Photo GCG)*

not really looking for nests, but for suitable habitat, with Bill making the judgment on the basis of his knowledge and experience (Magnusson *et al.* 1978b, 1980). Of course, the surveys had to be undertaken in the wet season when the potential habitat was looking its best. Accordingly, we spent day after day flying the sinuous rivers and sprawling wetlands in lazy left hand turns at ~100 feet above the ground to give Bill a good view, often under a low cloud base while rain or drizzle threatened. Bill spoke into a tape recorder, noting vegetation type and nests seen, and keying the descriptions to numbers, which Janet wrote onto the maps.

Counting sign

Sometimes other signs of crocodylian presence are scored as an indication of abundance. For example, tracks or belly slides across riverside sand or mud exposed at low tide are as good as a sighting to show a croc was there, and the size of the track reflects the size of the croc (Fig. 13.15). In the early 1970s in the

Liverpool River, the few large *C. porosus* that had survived the years of hunting were so wary that we saw their tracks but rarely saw them, except maybe a glimpsed eye shine. In the cool season, *C. porosus* frequently leave the water to bask (Chapter 7), so daytime surveys at low tide can be quite productive, even if the croc has slid into the water at the sound of the approaching boat.

Whichever survey method is chosen, not all the animals will be seen and apparent density will be less than the real density. Sometimes there is knowledge about the probability of sighting, from which a correction factor can be applied to convert apparent density into absolute density.

From knowledge of absolute density and the total habitat available, a total number of crocs in the population could, theoretically, be calculated. However, absolute density is almost never known and population ecologists are usually much more interested in the trends: up, down or stable. This is much cheaper to obtain and is likely to have

Fig. 13.15. *Tracks across (left) sand, being measured by Harry Messel, or (right) mud provide useful signs of crocodylian activity, in this case* C. porosus *in the Northern Territory, and the size of the animal can often be judged. (Photos Bill Green)*

smaller error terms. So a good indication about changes in populations can usually be gained from simply monitoring changes in relative density, determined by a standardised survey protocol from a boat, from the air, or even by counting some index of presence such as the number of tracks or the number of nests.

This brings us to the processes driving population dynamics.

POPULATION PROCESSES AND RELATED ISSUES

Before getting to some case histories, a short review of basic population ecology and some related issues (with a strong crocodylian flavour) will be useful. Any one of these topics could be the subject of a large review paper and none of what follows should be regarded as comprehensive: more like an annotated list.

At a simple level, the population dynamics of any organism are driven by natality (encompassing fecundity, 'birth rate' and recruitment), mortality (death rate), immigration and emigration. The relative magnitude of these determines whether the population is stable, decreasing or increasing and we can relate to that easily enough from our own experience of human populations.

How simple it sounds: but how challenging to quantify! I was involved for 25 years doing annual aerial surveys of kangaroos in a vast area (> 200 000 km²) of South Australia, and one thing we learned was that, even with 25 years' data, we would have liked more. Drought has a major impact on kangaroos and in all that time we had only one really significant drought in our study area. And kangaroos have a relatively short generation time, so imagine studying the population ecology of elephants!

Crocodylians are not easy to study either: typically they take a decade or more to reach maturity and live for several decades, so a detailed study following the passage of marked individuals would be a career-long commitment!

Life tables

In a perfect world, a crocodylian biologist would like information about the spatially explicit age structure of a population (numbers at each age and how they are distributed) and the survivorship at each age. This enables an age-based population model to be constructed. This is comparatively easy to do in human populations and is of great interest to life insurance companies. In crocodylians, age is much harder to know (but see below) and so stage-based models can be constructed instead, using rates of survival from egg to hatchling, to juvenile, to sub-adult and to adult, as best they can be determined, along with a quantitative understanding about how each of these may be affected by prevailing seasonal and other conditions.

Life tables such as these are very expensive, difficult to construct and require long-term monitoring of marked populations. It is important to recognise that the parameters of life tables are not constants and that populations are dynamic. In recovering populations, for example, increasing densities are likely to reduce rates of reproduction and survival but increase emigration. Conversely, Hines and Abercrombie (1987) reported that removing a substantial number of females from a population of alligators over several years in Orange Lake, Florida, had no effect on the number of nesting females: nesting was apparently density-dependent. Information about if and how populations compensate for losses is obviously very useful if harvesting is planned.

NATALITY, FECUNDITY, RECRUITMENT

In mammalian populations, it is appropriate to talk about birth rate, but this is inappropriate for animals that hatch from eggs (or plants that germinate from seeds) so the term natality is used instead, which has a broader, more encompassing meaning. Another word used to describe the rate of production of offspring is fecundity, which is an expression of 'potential' capacity for producing young. Thus, in crocodylians, egg production is an expression of fecundity, and the rate of successful hatching is the natality. Crocodylians are moderately fecund, with clutch sizes ranging from ~12 in *Paleosuchus* to 50 in some *Crocodylus* (Chapter 12). However, with such high mortality of both eggs and hatchlings, a more useful number to know for a population would be the rate at which individuals are recruited to the adult female component of the population, because not until they are sexually mature do crocs have

Fig. 13.16. *A grassy nest of* C. porosus *soon after construction and flooded. (Photos Bill Green)*

Fig. 13.17. *Peter Harlow contemplates a flooded nest on the Liverpool River, Northern Territory, while his companion tries his luck for a barramundi. (Photo GCG)*

the potential to contribute to population increase. It may sound trite but, in such long-lived animals, if predation or misadventure in the young stages is so high that there is no recruitment of new adults to the adult cohort, a population will die out no matter how many eggs are produced. The presence of large numbers of hatchlings does not necessarily imply that a population is healthy and on the increase.

Of course, the rates of all of these processes have implications for conservation, and for any considerations about harvesting (see Chapter 14).

Seasonal variation in egg production

Seasonal conditions can have a large influence on nesting. Variation in the timing of arrival and extent of rainfall associated with the monsoon in northern Australia means that egg production can vary greatly from year to year. Extreme weather events can reduce egg production. Elsey (2006) described reduced alligator nesting during a serious drought in Louisiana, exacerbated in that case by increased salinity in the marshes from storm surge associated with hurricanes Katrina and Rita, which brought saline water into the marshes.

Once hatched, recruitment to the adult population depends upon survival for many years in the face of significant mortality.

MORTALITY

Crocodylians die from different causes at different stages, becoming much less vulnerable once they become adult.

Fig. 13.18. *Fred Duncan's suggestion for crocodylian evolution. (Maningrida N.T. c. 1976; reproduced with the artist's permission)*

Mortality of eggs from flooding

Flooding of eggs is a major cause of loss in many species. In some habitats, mound nesters such as *Caiman crocodilus* and *C. porosus* build their mounds on mats of floating vegetation, which rise and fall with the changes in water level. In most situations, such as on river banks, many eggs simply drown when water levels rise following heavy rain (e.g. *C. porosus*, Figs 13.16, 13.17), often when exacerbated by high tides backing up tidal rivers. Mortality from flooding depends stochastically on the timing of flood rain events and may vary markedly between years. Death results from asphyxia, as the tiny pores in the shell become ineffective for diffusional exchange when under water. Even laying on a floating mat of vegetation may not provide sufficient protection from flooding. Campos (1993) noted mortality from flooding in the Brazilian Pantanal in nests of *Caiman yacare* on both floating mats and in marginal forest. Given the long time that crocs have

Fig. 13.19. *Hatchling crocodylians are vulnerable to predation by a wide range of animals. For the two Australian species of crocodile, these include (clockwise, from lower left): catfishes (*Arius *sp.); barramundi (*Lates calcarifer*); monitor lizards, such as Merten's monitor (*Varanus mertensi*) and the yellow-spotted monitor (*V.* panoptes*); raptors, such as the whistling kite (*Haliastur sphenurus*); the dingo (*Canis lupus dingo*); the black-necked stork (*Ephippiorhynchus asiaticus**); herons and egrets, such as the great egret (*Ardea alba*); snakes, such as the olive python (*Liasis olivaceus*) and turtles, such as the northern long-necked turtle (*Chelodina rugosa*). Source: Somaweera* et al. *(2013). (Photos DSK)*

been around, and with flooding such a source of loss, one might think that natural selection should by now have led to them nesting in trees like birds (Fig. 13.18).

Eggs die also from overheating, drying out, predation and disturbance by other animals and, in hole-nesting species (Chapter 12), nesting activity by other crocs.

Predation

Large crocodylians, once they have reached a good size, are minimally subject to predation and, in today's world (compared with the Mesozoic), they rarely suffer that fate unless hunted by humans. They may have gained their armoured protection in defence against predators that are now extinct. The smaller crocodylians may suffer predation even

as adults, from larger crocodylians (e.g. Fig. 13.7) and other predators: photographs of attacks by big cats and, very spectacularly, anacondas and similar constricting snakes frequently circulate on the internet. Regardless of adult size, however, all species are vulnerable when young. Eggs are an attractive food source for egg predators and even though there is nest guarding by some species (Chapter 12) this is not always effective.

Somaweera *et al.* (2013) scoured the literature and published an impressive review of predation on crocodylians at all life stages (including cannibalism). Varanid lizards and a range of mammals are the main predators of eggs. Marabou storks probe to locate the hole nests of *C. niloticus*, but most predation by birds occurs after a nest is opened. Fire ants have been reported colonising traps, found that dingoes were responsible for most of the 80 out of 111 nests that suffered predation.

Predation on hatchlings is high, and many of the habits and characteristics of crocodylians must have evolved for their defence, such as nest site selection and nest defence by the female, maternal care and crèche formation and the cryptic colouration of hatchlings. Somaweera (2011b) explored attributes of habitats chosen by hatchling *C. johnstoni* in Lake Argyle, Western Australia and concluded that availability of prey and protection from predation were important.

Cannibalism; a role in population regulation?

Cannibalism is a particular type of predation and its significance has been the subject of much discussion. Cott considered it very significant in *C. niloticus*, accounting for the low proportion of juveniles and that view was extended by Hutton (1989) in a review of reports to that time. Earlier on, Neill (1971) had been charmingly reluctant to attribute cannibalism to alligators. He commented that they were far less cannibalistic than man and that alligator parts in alligator stomachs were more likely consumed as carrion, 'after decomposition

Fig. 13.20. *This croc was found by the Messel crew, drowned in an illegally set barramundi net in the Wildman River, Northern Territory. It had three small crocodiles in its stomach (Messel et al. 1980). (Photo Bill Green)*

nests of *Caiman yacare* and *C. moreletii*, within the ants' native range and, with their invasion into the USA, *A. mississippiensis* and *C. acutus*.

Almost any predator given the opportunity will find hatchlings attractive (Fig. 13.19), and larger predators will take juveniles as well. The list includes crabs, fish, turtles, snakes, varanid lizards, many birds and more than 50 species of mammals. It includes crocodylians too, of a different or the same species (see below).

Not surprisingly, a species is likely to encounter different predators at different localities. To the east of Darwin, nests of the hole-nesting *C. johnstoni* are sought by varanid lizards and pigs (Webb *et al.* 1983b) whereas at Lake Argyle in Western Australia, home to more than 30 000 'freshies', Somaweera *et al.* (2011a), using movement-triggered camera

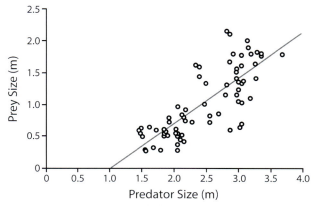

Fig. 13.21. *Studying cannibalism in an American alligator population in Louisiana, Rootes and Chabreck (1993) found a good relationship between the total lengths of cannibalistic predator and prey. Estimating that about half the mortality suffered by hatchlings and two-thirds of the mortality of alligators older than 11 months could be attributed to cannibalism, they concluded that the study population was significantly regulated by it. (From Rootes and Chabreck 1993 with permission, Allen Press Publishing Services)*

had destroyed its scent and appearance'. Bony alligator scutes were more likely to have been scavenged by living alligators as gastroliths, he wrote. However, Staton and Dixson (1975) reported cannibalism in *Caiman crocodilus* in the Venezuelan llanos and Messel *et al.* (1980) found three juvenile *C. porosus* in a large individual that was drowned in an illegally set fishing net (Fig. 13.20).

On the other hand, Magnusson (1986) reviewed dietary studies on a range of species and noted how few small crocs were found, which suggested that other crocs are not an important component of the diet. He concluded that cannibalism was unlikely to be important in regulating crocodylian populations. Many of those studies, however, were on juveniles and more evidence of cannibalism has accumulated since his review. Rootes and Chabreck (1993) examined more than 700 stomachs from commercially harvested alligators taken from a population in Louisiana with an extensive history of individuals being marked by clipping metal tags into foot webbing. They deduced cannibalism when tags were recovered in the stomach, which was quite common. There was a good relationship between predator size and prey size (Fig. 13.21) and it was estimated that 50% of hatchling mortality and 64% of all alligator mortality among those older than 11 months were the result of cannibalism. Nevertheless, cannibalism was only a small part of the alligator diet. Rootes and Chabreck considered that cannibalism was significant in regulating the population. Cannibalism has also been reported in the now recovering *C. acutus* (Richards and Wasilewski 2003).

Despite reluctance by some croc workers to acknowledge the reality of cannibalism, a review found it (including consumption of eggs) in more than 100 species of reptiles and amphibians (Polis and Myers 1985). Indeed, these authors noted a correlation between study duration and the likelihood of reporting cannibalism.

More recently, a broadly similar study by Delany *et al.* (2011) in Orange Lake, Florida, suggested that 6–7% of the juvenile population were being cannibalised annually and alligators up to 1.4 m were vulnerable. This study drew the same conclusion as Rootes and Chabreck (1993) that cannibalism may regulate the population. McNease and Joanen (1977) found that nine out of 75 adult female alligators from a sample of 314 large individuals in the southwest Louisiana marshes had alligator eggs, egg shells or 'well formed young' in their stomachs. The significance of this remains unknown.

Although Bill Magnusson in his 1986 paper discounted the significance of cannibalism, he recognised that severe competition directed by adult crocs to young adults and sub-adults, causing injury, would lead to high mortality, which, in turn, might lead to opportunistic, incidental cannibalism. How often cannibalism is significant enough to regulate populations remains unclear.

The exact motivation behind the event shown in Fig. 13.22 is uncertain. Is it cannibalism, a result of a large male alligator resenting the smaller one's presence, or a combination of the two? One possibility is that it started as a defence of space but morphed into a feeding opportunity. David noted that it came at the end of a long dry summer when food in the Everglades could have been short.

Mortality through competition

Competition can be either between or within species. Competition between *C. porosus* and *C. johnstoni* in zones of sympatry has been discussed above (and Fig. 13.7). The extent to which this is stimulated by interspecific competition or predation is not known, quite possibly either or both at different times, but, because *C. porosus* are cannibalistic, the ingestion of the 'prey' is the most likely outcome.

It is clear that pubescent individuals have to compete for space and a mate, and from the number of injuries that sub-adult crocodylians show it is clear that intraspecific competition can be intense, even fatal. A spectacular example is shown in Fig. 13.23 where a large male *C. porosus* attacked and killed a smaller (but still quite large!) individual, presumably as an act of aggression, and proceeded to eat much of it over the next few days (Garry Lindner and Buck Salau *pers. comm.* 2010). It is probably not uncommon that agonistic encounters such as this, which lead to a fatality, end

Fig. 13.22. *A large (~3.5 m) alligator tries to dismember a smaller (~1.6 m) alligator in Payne's Prairie State Park, Florida. After killing the smaller alligator, this large male rested with its victim's head in his mouth (first image) before garnering the energy to crack it like a whip, in an attempt to tear it apart for swallowing. The blur of the smaller alligator's tail, despite a shutter speed of 1/1000 s, gives an indication of the speed of this motion (centre image). Note also the alligator's bulging pseudotemporalis muscles and everted mandibular gland (Chapter 3), both of which spell 'effort'. The commotion attracted the attention of another, presumably hungry, alligator (bottom image), causing the larger animal to push its intended meal onto land as it turned to take it elsewhere. (Photos DSK)*

Fig. 13.23. *In the East Alligator River, Northern Territory, this 15 foot male (4.6 m)* C. porosus *killed the smaller 12 foot (3.6 m) male, presumably as an act of aggression and ate much of it over the next few days (G. Lindner and B. Salau pers. comm.). (Photo Tony de Groot)*

in predation. Valentine *et al.* (1972) found the front leg of an alligator in one of their stomach samples and attributed that to a fight. Fighting often leads to limb loss, a limb making an obvious 'handle' (Fig. 13.24).

Mortality from natural and human-related misadventure

Adults in unharvested populations have few predators, at least in today's world, but do suffer mortality from a range of misadventures such as overheating, hunting, drowning in fishing nets, cane toads and road kill. Figure 13.25 records an animal caught out on a flood plain by a retreat of

Fig. 13.24. *Fighting often leads to limb loss – a limb making an obvious 'handle', as attested by the missing right hindlimb of this American alligator, photographed in Everglades National Park. Note also the scarring along this animal's back. (Photo DSK)*

the tide and overheating to a point where it is no longer able to travel (Buck Salau and Garry Lindner *pers. comm.* 2010).

Direct mortality by hunting is a significant cause of death in crocodylians and must be the largest cause of deaths of adults in most places. Most of this is for food or hides (Chapter 14), but, although impossible to quantify, there is undoubtedly significant mortality from their direct destruction as 'pests' and also accidental drowning in fishing nets. Messel *et al.* (1981) reported being told by barramundi fisherman, while they were anchored in Port Musgrave in 1979, that they had taken 23 large drowned *C. porosus* from their nets in the previous 2 weeks. Undoubtedly, this is an exceptional example. Indirectly, alienation of habitat by changing land use is also a significant cause of population decline.

As this is being written, cane toads, *Rhinella marina*, which were introduced (successfully) into Australia in 1935 to control cane beetles (unsuccessfully) (see review by Shine 2010) are continuing to expand their range across the top of Australia from the east. They are now well established in northern Western Australia and still expanding westwards. Cane toads have poisonous secretions in parotid glands behind their eyes as a defensive mechanism, but predators of amphibians such as varanid lizards and native quolls are naïve to this and many have died after eating toads. They entered the Northern Territory in 1996, reached the Daly River west of

Fig. 13.25. *A* C. porosus *caught out by retreating water levels on the extensive flood plain of the East Alligator River, Northern Territory at the end of the 2009 wet season. As mud holes dried up, some crocs moved out, but some overheated and died on the baking mudflats before they reached the safety of the water (see also Fig. 10.23). (Photos Buck Salau)*

Darwin by 2003 and crossed into Western Australia in 2009. They are now sympatric with *C. johnstoni* throughout most of its range. *C. johnstoni* die after ingesting toads (Smith and Phillips 2006) and Letnic and Ward (2005) reported *C. johnstoni* preying on toads during their crocodile surveys on the Roper

Fig. 13.26. *An Australian freshwater crocodile,* C. johnstoni, *about to eat a cane toad,* Rhinella marina *(previously* Bufo marinus*), which has puffed itself up with air in self defence. Ingestion of the introduced toads kills 'freshies', but* C. porosus *are much more tolerant. Hatchling 'freshies' learn to avoid metamorphic toads, which apparently lack sufficient toxin to kill them and thereby may acquire behaviour that enables their survival when they grow large enough to ingest fully poisonous adult toads. This could account for the continuing existence of* C. johnstoni *in Queensland 'toad country' after 70 years of exposure. (Photo Mike Letnic)*

and Daly Rivers in 2005 (Fig. 13.26). As the toads spread westwards, there were numerous reports of dead *C. johnstoni* (Letnic *et al.* 2008) and these authors reported drops of up to 45% in sections of the Victoria River between 2005 and 2007, following the arrival of toads. The declines were highest in the drier, up-river sections and the authors suggested that this was because toads in these drier areas were obliged to use the river for rehydration, so exposing themselves to crocodiles. However, different effects on different populations are not always explained easily. Somaweera *et al.* (2012) were unable to find a satisfying explanation for the different effects of invasion on three geographically distant populations. Two populations of 'stunted' populations of *C. johnstoni* are known, and these have already been invaded by toads and affected significantly (see below).

Anywhere that 'freshies' live will be suitable habitat for toads too, so what will this mean for 'freshies' in the long term? Happily, there is room for optimism. Somaweera *et al.* (2011c) experimented with hatchling 'freshies' and found that they learned quite quickly to avoid eating metamorphic toads. These authors postulated that learning acquired by hatchlings on metamorphic toads, which are less toxic than adults, could stand them in good stead when they grow to a size large enough to predate the more toxic adult toads. It is worth noting that cane toads have been established for many years in north Queensland and coexist with freshies in many places. One such place is The Croc Hole in the Lynd River in north Queensland (Fig. 10.2). Even there, however, mortalities do occur: Mark Read and I removed a toad from the stomach of a dead 'freshie' there in 1998. Perhaps that one was a slow learner!

Interestingly, *C. porosus* seems to be less affected by the toxin produced by cane toads (Covacevich and Archer 1975; Smith and Phillips 2006). It has been suggested by Thomas Madsen (*pers. comm.*) that this is because the range of *C. porosus* includes South-East Asia where it co-evolved with other species of toxic Bufonidae. In contrast, *C. johnstoni* occurs in Australia only, where toads were absent until their introduction. Asian and African Varanidae have a gene that confers resistance to toad toxin, whereas this is absent in Australian varanids, making them susceptible to poisoning (Madsen *pers. comm.*; Ujvari *et al.* 2013). A similar gene is present in *C. porosus* but the situation in *C. johnstoni* is unknown. By analogy with the varanid study, it is likely that the gene conferring resistance is lacking in *C. johnstoni*.

Crocodylians show little road sense (Fig. 13.27) and are killed by road traffic, particularly in populated areas. Brien *et al.* (2008) reviewed deaths of *C. acutus* in Florida over 40 years from 1967 and concluded that ~70% of 143 recorded fatalities were likely to be from collisions with a motor vehicle.

SITE FIDELITY, DISPERSAL (EMIGRATION, IMMIGRATION)

The journeys made by salmon and eels to natal streams or patches of ocean are legendary. The reasons they make these spectacular journeys are less well understood, along with the navigational skills required.

Radio-telemetry and genetics data have shown that site fidelity is an important attribute of crocodylians as well. There are many anecdotal accounts implying site fidelity by crocodylians, including congregations by females at particular nesting areas (Dinets 2010). Data are accumulating that confirm this. As discussed in Chapter 4, satellite tracking of large male *C. porosus* has shown very convincingly that translocated individuals return home, even covering very large distances to do so (Read *et al.* 2007). Also in *C. porosus*, some telemeter-fitted individuals remained for months in the same small 'home range', whereas some left on coastal journeys and then returned. Some did not return. So far, the satellite tracking has focussed primarily on males, but studies now under way are remedying this deficiency. In a study using conventional VHF telemetry, Winston Kay found in the large Ord River in northern Australia (Fig. 13.4) that most females returned annually to the same nesting area, after spending much of the rest of the year elsewhere (Kay 2004a,b) (Chapter 12). The males too, in his study, spent most of their time in the same river, returning there if journeys were made elsewhere.

This strong site fidelity was confirmed by genetic data (Kay 2004a). Although the King and Ord rivers

Fig. 13.27. *A juvenile* C. porosus *taking its chance on the Arnhem Highway, Northern Territory, Australia. A nicely executed high walk (Chapter 4). (Photo Calvin Murakami)*

are not far distant, both flowing into Cambridge Gulf, and *C. porosus* are known to travel between them, the populations in each are sufficiently distinct to allow genetic assignment tests to discriminate between them in 74% of cases. Translocated juvenile American alligators too have been shown to return 'home' (Rodda 1984). It is clear that 'a sense of place' is very important to crocodylians, and probably particularly as adults (see Chapters 4 and 12). This makes a lot of sense because suitable places to find a mate and make a nest are valuable. Just like us, really!

Dispersal occurs by the movement of sub-adults and the driving force may be the commencement of puberty and the need to seek a suitable breeding location and mates. Dispersal is beneficial because it prevents inbreeding (see also, Social Structure below). Satellite and acoustic tracking data are adding to an understanding about how well crocodylians can move long distances (discussed in Chapter 4).

GROWTH

Growth curves and age determination

There is no simple way to age crocodylians in the wild and no reason to equate a size–age relationship of captive animals with what occurs naturally. Average growth curves against age can be constructed by measuring growth increments between successive captures from a marked population. Two such studies have yielded data: one from *Caiman crocodilus* where the Tapajos River enters the Amazon, near Alter do Chao in Brazil (Magnusson and Sanaiotti 1995), and another from an island population of the closely related *Caiman latirostris* on Ilha do Cardoso, south of Sao Paulo (Moulton *et al.* 1999). The resulting papers present useful discussions of the analysis of crocodylian growth data and the applicability of various mathematical growth models. See also Abercrombie (1992) and Tucker *et al.* (2006). The approach does not, however, yield an age for a particular individual because growth rates differ considerably with temperature history, feeding success and social position, which is much more typical of ectotherms than of endotherms (Chapter 10). More recently, Eaton and Link (2011) published an approach that considers variation between individuals, based on a 4 year capture–recapture study in Gabon. They combined incremental growth data with data from animals of known age and developed a model that allowed age to be predicted from a measurement of head length. They also provided a substantial review of the relevant literature.

It is worth noting here that body size is usually represented by snout–vent length (SVL) taken from the tip of the snout to the anterior end of the cloacal slit or, in some studies, to the posterior end. The latter is probably more logical because it represents more closely the end of the body cavity. Total length

is used less often because often animals lose some of their tail. Webb and Messel (1978) found that SVL can be predicted by any one of several linear dimensions in *C. porosus* to 2 m, such as head width and head length, if the sex is known. The larger the animal the less tight these relationships become and predicting total length from head length has become a controversial topic in relation to speculations about 'the largest crocs' (discussed in Chapter 1).

In addition to inferring age from measurements of growth between repeated recaptures of marked animals at known intervals, there have been attempts to measure ages directly by skeletochronology. This relies on growth being seasonally periodic, leading to the development of a laminar structure (growth rings) in bones. One of the most successful applications was that by Tony Tucker, working on

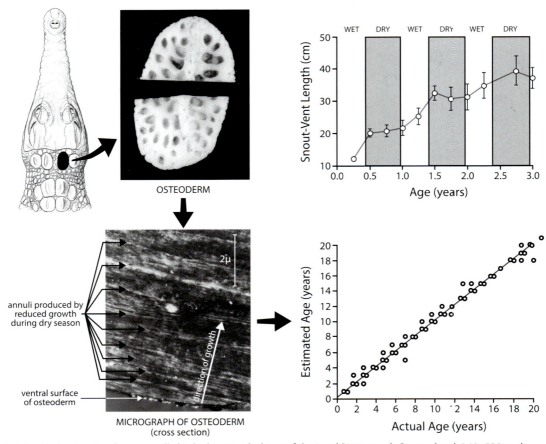

Fig. 13.28. *In* C. johnstoni *in the unusually high elevation habitat of the Lynd River, north Queensland, 360–520m above sea level, most of the annual growth occurs in the warmer wetter summers compared with the cool dry winters. Tucker (1997a,b) was able to age animals in his population study after validating apparent seasonal growth increments in the nuchal osteoderms against known-age animals. (Modified from Tucker 1997b with permission)*

C. johnstoni at an elevated site in north Queensland, where marked seasonal temperature cycles led to seasonal growth increments (Fig. 13.28) (Tucker 1997a,b).

Skeletochronology has been used or attempted in *C. niloticus* by Graham (1968) using layering in the dentary bone and also eye lens weight, and by Hutton (1986, 1987) using osteoderms or long bone growth laminae. In adult females, age estimates based on bone 'growth rings' have to be treated cautiously because remodelling of the bone, presumably associated with calcium mobilisation in eggshell production, can obscure the pattern. The use of skeletochronology in *A. mississippiensis* has produced somewhat uncertain results. Techniques that can be applied only to dead animals are much less useful. Appropriate validation against animals of known-age is obviously required in any application of the technique.

Age at maturity

The age and/or size at which crocodylians become sexually mature is of interest and is discussed in more detail in Chapter 12. Sexual maturity of females can be judged by observing nesting, seeing them with a pod of hatchlings or, as used by Tucker and Limpus (1997) and Lance *et al.* (2009), scanning the abdomen for eggs ultrasonically. After egg laying, a female's cloaca may remain flaccid for a few weeks and that may be a good indicator to an experienced observer (Webb *et al.* 1983c). In males, sexual maturity can be judged by sampling for sperm in the penile groove (Chapter 12 and Fig. 13.29).

As a generalisation, female crocodylians in the wild mature at ~12 years and males somewhat later and at a larger size. However, like most generalisations in biology, there are exceptions and large differences can occur not only between species, but also between and even within populations of the same species. For example, *Caiman crocodilus* mature earlier in the Llanos of Venezuela than in the cooler climate of Alter do Chao, near where the Tapajos River enters the Amazon in Brazil (Magnusson and Sanaiotti 1995). Females of the very closely related *Caiman latirostris* on Ilha do Cardoso, however, take much longer to mature: around 13 years, compared with 5–6 years at Alter do Chao and (Moulton

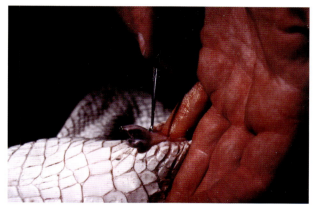

Fig. 13.29. *Colin Limpus runs a tiny scoop along the penile groove of a sexually mature* Crocodylus johnstoni *to collect a sample of fluid. The presence of sperm in a smear on a glass slide examined under the microscope would indicate reproductive status and preparedness for mating. (Photo GCG)*

et al. 1999) concluded that a slower maturation than *Caiman crocodilus* is a natural feature of the population, living at lower temperatures. Age at maturity does not necessarily correlate with adult body size (either within or between species). In *Paleosuchus trigonatus*, one of the world's smallest crocodylians, earliest maturation of females is 11 years and males 20 years (Magnusson and Lima 1991).

Food availability and stunted (dwarf) populations

Food shortage can stunt crocodylian growth and development. Graham (1968) considered that the population of *C. niloticus* in Lake Turkana was food limited, in comparison with data provided by Cott (1961) for the same species living elsewhere (see 'Case histories' below). Grahame Webb came across a population of what appeared to be stunted *C. johnstoni* in the upper reaches of the Liverpool River in northern Australia, above the waterfalls draining the sandstone escarpment and isolated from the downstream populations. They appeared emaciated, had slower growth rates and were sexually mature at sizes well below the size at which typical 'freshies' mature (Webb and Manolis 1989). Taken into captivity, they fed voraciously and put on weight but did not regain a normal growth trajectory. A second population of 'stunted' *C. johnstoni* is known from the Bullo

River, south-west of Darwin almost to the Western Australian border. Although the question of taxonomic distinctness has been raised, comparison of mitochondrial DNA haplotypes between dwarf and downstream populations has revealed that the dwarf populations do have some unique haplotypes, but the divergence is too small to imply taxonomic distinctness (Adam Britton *pers. comm.* 2013). The Bullo River population has been significantly affected by the arrival of cane toads (*Rhinella marina*) and its future is uncertain (Britton *et al.* 2013).

Social factors

Population processes can be very subtle, but also very significant. The influence of dominant males has long been recognised as important in shaping crocodylian behaviour (Lang 1987) and a good example comes from the Lynd River population of *C. johnstoni*. The presence of a dominant male in a lagoon appeared to be having an inhibitory effect on the growth of other males, keeping them cryptic and apparently delaying their maturation until there was 'a slot' (Tucker *et al.* 1998). Subsequently, in Frank Seebacher's study of the thermal relations of these upland animals (Chapter 10), it transpired that this was likely due at least partly to competition for suitable basking sites. Individuals of both sexes in the social hierarchy displaced subordinate animals from basking sites, compromising their

opportunities for reaching their preferred range of body temperature (Seebacher and Grigg 1997), which must compromise growth: their age at maturity was likely being postponed by social factors.

SEX RATIO

The sex ratio of the adult population will be the outcome of the sex ratio at hatching as modified by any differential mortality at each of the subsequent life stages. However, crocodylians are polygamous and females show multiple paternity (Chapter 12) so it is not easy to see a benefit from having any particular adult sex ratio.

The sex of crocodylians is determined by temperatures experienced by the embryo while the gonads are developing (i.e. temperature-dependent sex determination or TSD, Chapter 12) (review by Lang and Andrews 1994). Typically, low and high temperatures produce females, and intermediate temperatures produce males. However, the transition points have mostly been measured by experimental incubation at constant temperatures, which would rarely occur under natural conditions. Nevertheless, Webb and Cooper-Preston (1989) found good agreement between the sex ratios of hatchlings from eggs incubated in the laboratory at constant temperatures and those from eggs taken from the field after gender would have been determined (Fig. 13.30).

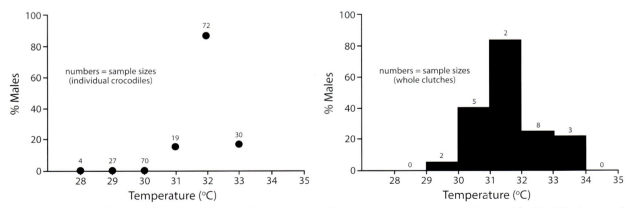

Fig. 13.30. *The sex of crocodylians depends upon the temperature they experience during incubation, specifically while the gonads are developing (Chapter 12). Webb and Cooper-Preston (1989) measured the proportions of male and female hatchlings from (left) C. porosus eggs incubated at a series of constant temperatures and (right) from eggs taken from field nests along with 'spot' temperatures taken '2–3 eggs deep' in the clutch after sex would have been determined and incubated to hatching. The results are in good agreement. (From Webb and Cooper-Preston 1989, with permission from Oxford University Press)*

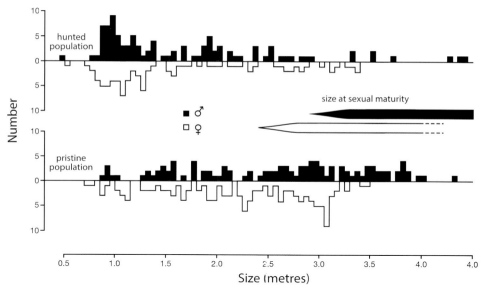

Fig. 13.31. *Length–frequency distributions of hunted and unhunted populations of* C. niloticus *in Zambia. (Upper panel) Hunted area, Luangwa Valley and Kafue Flats; n = 186, 14% adult, which were 'previously much shot over' and (lower panel) Upper Zambesi (Barotseland, 'where hunting had been discouraged') and Mweru Wa Ntipa ('where crocodiles have hitherto enjoyed complete protection', n = 235, 40% adult). At Mweru Wa Ntipa, the area with the most protection, 82% of the population comprised adults, with a 1:1 sex ratio. Cott noted that the hunters took all sizes and, elsewhere, that small individuals were very cryptic: a habit forced on them in his opinion by the cannibalism from larger animals. (Modified from Cott 1961, © 1961 The Zoological Society of London)*

One might expect hatchlings in the wild to be female-biased because males result from a narrower temperature range and, indeed, this may be the case. Natural nests produced female-skewed sex ratios in both South Carolina (males 21, 42, 25, 0, 24 and 39% in 1994–99, respectively) and Louisiana (males 11, 31, 26, 38 and 28% in 1995–99) (Rhodes and Lang 1995, 1996). In *C. johnstoni* Webb and Smith (1984) reported a significant female bias in hatchlings from 15 rivers in the NT: average 35% male, but ranging from 20 to 61% male. Despite the average female bias in this population, the extent of variability is very striking. Tucker *et al.* (1998) also found considerable variability in *C. johnstoni* in the Lynd River in north Queensland (see 'Case histories' below) but over the length of his study the hatchling sex ratio averaged 50:50.

The variability is at least partly due to seasonal temperature and the time within the season that the eggs were laid. In South Carolina, Rhodes and Lang (1996) reported more females from cooler seasons. Similarly, in *Caiman yacare* in the Pantanal, sex ratios were more variable from nests on floating grass mats than in adjacent forest

where temperatures were more stable (Campos 1993). Among Crocodylidae, Lang *et al.* (1989) reported variable, but female-biased (male 25%), hatchings of *C. palustris* from captive nesting in a large breeding enclosure at the famous Madras Crocodile Bank.

The sex ratio of juveniles, however, may or may not reflect a female bias, implying differential survival. Certainly incubation temperature can affect post-hatching fitness. Webb and Cooper-Preston (1989) showed that farmed *C. porosus* hatchlings incubated at 31–32°C survived better and grew faster than those incubated at higher or lower temperatures. They noted that temperature-dependent sex determination (TSD, Chapter 12) 'allocated' male gender at temperatures that convey these 'non-sexual' benefits and suggested this may confer a particular benefit of TSD over genetically determined sex. The benefits show up in the field as well. Lance *et al.* (2000) sexed nearly 3000 non-hatchling alligators from 11 locations in Louisiana over 6 years and (despite a few exceptions) found an overall male bias in the population (58%), overcoming the female bias displayed at hatching.

Lance and colleagues attributed this to differential mortality between males and females post-hatching and noted that this accords with one of Shine's (1999) differential fitness models, which might favour the evolution of TSD. An alternative explanation would be that it is incubation temperature *per se* which sets fitness quite independently of anything to do with TSD.

Further, not all juvenile crocodylian populations are male-biased. Webb *et al.* (1987) reported a 1:1 ratio in juvenile *C. porosus* captured in a tidal river in the NT. Likewise, in the 411 sexually immature (to 2.5 m total length) individuals from 651 *C. niloticus* measured and sexed by Cott (1961) from several East African sites, the sex ratio was 1:1 in all 25 cm size classes. Sampling is always an issue and we know little about how randomly Cott's samples were taken, except perhaps that, reading between the lines, he collected every sample he could, and it was from several populations that had not suffered too much reduction from hunting.

That makes Cott's data on adults of interest. He sexed 130 *C. niloticus* longer than 3 m and therefore sexually mature: 59% were males. This included some areas that had been subject to some hunting. At Lake Mweru Wa Ntipa, where crocodiles had been completely protected, 82% of the population were adult, with a sex ratio of 54% (*n* = 42, so essentially 1:1, but the number in the sample was small). Data on adults are often recorded from commercial hunts, which are biased towards larger individuals and, thus, towards males and of little value as a representative sample. However, Lance *et al.* (2000) identified four apparently unbiased studies on alligators in which the sex ratios of adults were 47% (Florida), 61% (South Carolina) and 68% and 72% male (Louisiana). On the other hand, a population of mature *C. johnstoni* was 17% male (Webb *et al.* 1987).

Can we find a generalisation about sex ratio? The pattern that emerges is that variability of sex ratio among crocodylians is routine and influenced strongly by environmental factors. Despite the variability, a female bias at hatching may be overcome by differential fitness between juvenile males and females, leading to a male-biased population of juveniles that survives into adulthood. However, there are exceptions and more data are needed, particularly on adult sex ratios, which are very difficult to get, particularly because males may be easier to sample than females in some habitats.

TEMPERATURE, GLOBAL WARMING

Temperature is discussed in more detail in Chapter 10. It can have direct effects on growth rate and indirect effects as well through influencing food availability. Thus, growth can be seasonal as a result of seasonal changes in temperature (as in Fig. 13.28). Webb *et al.* (1983c) found that growth of juvenile *C. johnstoni* in the McKinlay river was essentially negligible in the dry (winter) season, which they attributed to the paucity of insects and other food and there was probably also a direct effect of temperature on growth. Crocodylians with a wide latitudinal range grow more slowly in the cooler regions and grow faster and to larger sizes nearer the equator. Species such as *Caiman latirostris* and *A. mississippiensis* have ranges that extend a long way south or north into regions where winter temperatures are challenging (Chapter 10) and although they persist in these more challenging climates growth rate is slower.

As for the effects of global warming, no aquatic habitat on the planet with its current climate is too warm for crocodylians, but much of it is too cool, so global warming can be expected to allow expansion of the ranges of many species. Because their gender is determined by the temperature at which their embryos develop (TSD, Chapter 12) some authors have made alarming interpretations of the effects of the currently increasing world temperatures, implying that a shortage of females will jeopardise populations' survival. However, members of the Crocodylia have coped with changes of plus and minus at least 2–4°C since the Cretaceous and we can have confidence they will continue to choose suitable nest sites and accommodate to warmer conditions.

CASE HISTORIES

With the foregoing as background, we will now look at some case histories.

CROCODYLUS NILOTICUS IN ZAMBIA IN THE 1950S

Hugh Cott (1961), in his ground-breaking and wide ranging study on *Crocodilus niloticus*, published a comparison of length–frequency distribution of shot and (previously) unshot populations in the 1950s in what is now Zambia (Fig. 13.31). He gave little information to provide context, but the hunted populations are reminiscent of the situation I was familiar with when we started working in the Liverpool and Tomkinson Rivers in the NT in 1973, shortly after they were protected. Most of the crocodiles were hatchlings and juveniles. A 'mature', unharvested population in an area with nesting capacity can be expected to comprise more of the larger size classes, dominantly sexually mature adults and animals approaching what Cott saw in Namibia, in Barotseland on the Upper Zambesi and in the Mweru Wa Ntipa swamps in the Great Rift Valley (Fig. 13.31). The comparison showed that shooting for skins reduces adult numbers dramatically; the survival of hatchlings increases and the size distribution becomes skewed towards younger animals, as seen in the Kafue Flats and the Luangwa valley. It would be interesting to see comparable data from these locations now.

Hugh Cott (1961) noted at the time he was writing that *C. niloticus* were already being 'extirpated' from many areas and that the hunters often 'justify their activities on the grounds that they are performing a public service in eradicating a dangerous animal, or they meet criticism (from Cott presumably, who obviously had developed a great affection for crocs) with the contradictory assertion that 'the supply is inexhaustible'. Contradiction and controversy have been bedfellows of crocodylian politics for a very long time.

CROCODYLUS NILOTICUS IN LAKE RUDOLF (NOW TURKANA) IN THE 1960S

This study is particularly worthy of mention because it was one of the first studies actually aimed at crocodylian population ecology, on a population of *Crocodylus niloticus* in a closed system: the almost 300-km-long Lake Turkana (formerly Lake Rudolf)

in the Great Rift Valley in Kenya. Alistair Graham spent 14 months there in 1965–66, under contract to the Kenya Game Department, to assess the status of the *C. niloticus* population. At the time, this was the last relatively undisturbed population of *C. niloticus* in Kenya, and in fact all of Africa. The idea behind the study was to understand their ecology so that exploitation could be undertaken in a sustainable way. Apart from a report and a thesis, the work produced one of the most interesting combinations of biology, sociology, adventure and 'coffee table book' ever written: *Eyelids of Morning* (Graham and Beard 1973).

C. niloticus is similar in general appearance and habits to *C. porosus* and they get to be almost as large. Graham's project was funded solely by the sale of skins from a sample of 500 shot crocodiles, from which he obtained a lot of information about seasonal changes in reproductive organs and diet from dissection (as did Cott, who worked in with the shooters rather than doing it himself). Killing large numbers of animals to discover gonad state or stomach contents is rare these days (thankfully), except where opportunity is provided by a commercial harvest such as with *A. mississippiensis*. By aerial and spotlight survey, Graham estimated ~13 000 individuals over the 6870 km^2 of the lake, concentrated where the water was shallower and more sheltered from the prevailing strong winds. The crocs ate mostly fish and about half of the stomachs were empty. Females matured at 180 cm TL, males at 270 cm, and there was evidence of senescence in several very old females, and one male of 445 cm TL. As in *C. porosus*, they found that larger females produced larger clutches of eggs, but they were only about half the size of clutches reported elsewhere for *C. niloticus*. Graham was able to count growth rings in the dentary bones, finding a reasonable relationship with body size (different in males and females) and used that as an indication of age, although he advised caution with interpretation. The data suggested oldest ages of ~70 years.

As mentioned above, there was a high proportion of empty stomachs: much more than reported by Cott (1961) from 591 *C. niloticus* shot in Zambia. The surrounds of the lake are very barren, with little game, and the crocodiles depend on fish (not their favourite food, see Chapter 6). Growth rates were also lower than elsewhere, although the animals themselves were in good condition. Graham considered that the Lake Turkana population of *C. niloticus* was food limited.

In great contrast to the recovering population of *C. porosus* discussed below, the population of *C. niloticus* in Lake Turkana in the 1960s was apparently a 'mature', essentially stable population, at carrying capacity and comprising mainly immature animals. Mostly, nests were successful and there were few egg predators, apart from humans at just a couple of locations. Even though there was a preponderance of immature animals in this population of 13 000, with an annual production of ~20 000 hatchlings, there must have been very high hatchling mortality (Alistair Graham *pers. comm.* 2011). Indeed, on a trip to North Island where Graham knew of 18 nests that had hatched in the previous season, equating to ~400 hatchlings, only one crocodile less than 90 cm in length could be found.

CROCODYLUS POROSUS IN AUSTRALIA

The story of the Australian *C. porosus* population is a fascinating one, not only because of its longevity and its scientific results, but because of the people involved. Two names dominate – Harry Messel and Grahame Webb – both strong-minded individuals

who have each been likened at times to bull crocodiles. Harry came on the scene in 1970 with an interest in developing radio-tracking technology and sufficient research funds to make it happen. He reasoned that if equipment could be made that would work on large crocodiles in the challenging environment of Australia's tropical rivers and billabongs, it would work anywhere. Grahame was appointed as a biologist to the programme in 1973 and, very shortly, work on population ecology began in parallel with the technological development. The initial focus was on the Liverpool and Tomkinson Rivers, at whose mouth the Crocodile Research Programme's headquarters were based, in Maningrida. Harry saw the need for surveys over a wider area and, never one to be daunted, recognised that a large boat would enable the survey area to be extended right across the top of Australia, from the Kimberley to Cape York. A purpose-built, shallow draught craft was designed and built in Sydney, 21 m in length and 120 tonnes (Fig. 13.32), with bunks for eight, a small laboratory and, very significantly, a hydraulic crane, which enabled its two specially designed, aluminium 5 m rounded-bow, black painted crocodile survey boats (Fig. 13.11) to be lifted in and out of the water. The ship was inevitably nick-named 'The Messel vessel'. Indeed, it was formally named the *RV Harry Messel* by the University of Sydney's Science Foundation, which had been created in 1954 by Harry Messel (Fig. 13.33) and which continues today, actively

Fig. 13.32. *Two views of the shallow draught* RV Harry Messel *and its two work/survey boats in the Liverpool River, Northern Territory, Australia. The ship was extremely well equipped as a mobile research base, with accommodation for eight researchers and crew, a laboratory opening onto the rear deck where crocs could be 'processed' and a crane mounted aft, which enabled the workboats to be brought on board or deployed. (Photos GCG, Bill Green)*

Fig. 13.33. *Professor Harry Messel at work on board the* RV Harry Messel. *He was Professor of Physics at the University of Sydney (1952–1987) and instigator and Director of the University's Crocodile Research Programme from 1971 to 1987. (Photo Bill Green)*

promoting science with a series of Summer Schools for High School students. Using this ship as a floating, mobile base, standardised spotlight surveys were conducted in all of the accessible waterways, from the Kimberley in Western Australia to the tip of Cape York in Queensland, over a period of 17 years. This was an astonishing achievement! The results are written up in a series of 19 monographs, listed individually in Messel *et al.* (1981). They take up 26 cm on my bookshelves and weigh 13 kg! Beautifully illustrated by Bill Green's photographs and containing numerous detailed tables of results, they record much more than crocodile survey data and include even detailed maps and navigational, tidal, salinity and mangrove vegetation information about most of the waterways across the top of Australia (Messel *et al.* 1982).

The timing of the first survey work was pretty well perfect: it recorded a comprehensive baseline for the riverine component of *C. porosus* populations soon after the species became protected, in 1970 in Western Australia, 1971 in NT and 1974 in Queensland. Protection followed the recognition that *C. porosus* had become severely depleted by unregulated hunting for its skin. Interestingly, their precarious future at that stage was drawn to the attention of the authorities by some of the very same people who had been responsible: the crocodile shooters themselves. Nobody knows how numerous crocs were before the hunting began but,

from anecdotal accounts combined with records of hides sold, *C. porosus* were clearly very common. Soldiers stationed in and around Darwin in the Second World War provided some of the stories. I have been given two very similar accounts, quite independently, from airmen who vividly recall flying low along Arnhem Land beaches and seeing numerous large crocs running down the beach ahead of the aircraft and splashing into the water. Flying low along the same coastline in the late 1970s and 1980s, I saw nothing like that: a croc on a beach was then an unusual sighting.

Grahame Webb made a comprehensive attempt to estimate probable pre-hunting populations of *C. porosus* in the NT, reckoning something like 80 000–100 000 as an upper limit for the population in 1945 (Webb *et al.* 1984). By the late 1960s, hunting had depleted the populations and Webb and colleagues estimated that the population in the NT at protection in 1971 was in the vicinity of 3000–5000, with perhaps as few as 500 adults.

Hunting was stimulated after the Second World War, particularly in the NT and coincident with the availability of 4WD vehicles capable of travelling the new bush tracks, aluminium dinghies and battery-powered spotlights. Ion Idriess' book *In Crocodile Land* published in 1946 provides a very readable account of those early days; indeed its opening words are prophetic – 'How about coming crocodile shooting?' suggested the Skipper. 'There's money in skins at present'. 'That tempts me', I answered. Other books worth reading about the early days in the NT, when Europeans were few and far between, are Syd Kyle-Little's *Whispering Wind – Adventures in Arnhem land* published in 1957 and the legendary Donald Thompson's *Donald Thomson in Arnhem Land* published in 2005 (and compiled by Nicholas Peterson). Fishermen operating around the coast were serious hunters of crocodiles in tidal rivers accessible from the sea. The *modus operandi* of hunting from a small boat or canoe was straightforward: crocs were picked up in a spotlight beam, shot and an Aboriginal assistant threw a spear or used a gaff to secure the croc before it sank and was lost. Skinning was done the next day and the skins salted quickly for preservation: a skin would not last long in the warm, humid climate and many were lost before they could be transported the long distances over rough tracks to Darwin. Salted skins were exported mainly to Singapore (salties were called 'Singapore small scale' in the international trade) and from there to France and European tanneries, ending up as expensive handbags and shoes. Depletion is thought to have been most rapid from 1945 to 1951, with a slower decline from then until 1971 when it was stopped. About 300 000 skins were harvested in Australia and sold in that 27-year period (Webb *et al.* 2000).

Starting soon after protection, major parts of the NT population have been surveyed on a regular basis since 1974, the stimulus being conservation concerns and, latterly, because of interest in commercialisation. Grahame Webb has been a major advocate for scientifically based commercialisation as a tool to encourage crocodile conservation. The idea is that if crocodiles are commercially valuable, more effort will be put into conserving them. Indeed, as the NT populations recovered, that philosophy has permeated the policies of the IUCN's Crocodile Specialist Group (Chapter 14), with both Grahame Webb and Harry Messel as driving forces. In the NT, the University of Sydney study was followed by annual surveys undertaken and/or funded by the NT government (the agency now known as the Department of Land Resource Management) and the Federal Government (the agency now known as the Department of Environment), as well as Grahame's company Wildlife Management International (WMI). Grahame's involvement extended from 1973 to 1999 and data collection continues today by the relevant government agency. Fewer data are available from Queensland or Western Australia, but the data from the NT rivers tell the interesting story of a spectacular recovery (Webb *et al.* 2000; Fukuda *et al.* 2011) and also shed considerable light on the population ecology of this species in particular and, by analogy, on crocodylian populations in general.

By 1971, when protection of crocodiles was brought in, adults were both rare and wary. Rivers close to Darwin such as the Adelaide River and the Mary River had very low densities of all size classes and nesting habitat was degraded from overgrazing by cattle and feral water buffalo. Recovery in these rivers would depend initially on recruitment by

dispersal from other river systems. In more remote rivers, such as the Liverpool and Tomkinson, there were more animals in all size classes, although still very low, and nesting habitat was in better condition, so population increases could have been expected from young produced locally.

Messel and his team started systematic spotlight surveys of many NT rivers in 1974–75, with estimates of size (as a surrogate for age/life stage) made on each animal sighted, offering the opportunity to follow changes in the population's size (and life stage) structure through time. Adults in the early surveys were rarely able to be approached to get a size estimate and there is good reason to assume that many of the 'eyes only' counts represented adult animals (Webb *et al.* 1989). Hatchlings (< 2 feet, i.e. < 60 cm) were (and are) naïve enough to be approached, and even picked up, and their presence gave a good indication of the whereabouts of a nest/s.

Despite the known high mortality of eggs, estimated to be ~75% in the Liverpool and Tomkinson Rivers (Webb and Manolis 1989), spotlight surveys showed that there was a dramatic increase in crocodile numbers within the first 4–5 years after protection, mostly in animals '4 foot' (1.2 m) or less, as might be expected. Adult numbers, judged mostly from 'eyes only' sightings, remained low. This was to be expected: with *C. porosus* taking 12–16 years to reach maturity, eggs hatched after 1971 could not be expected to become part of the breeding population until the mid 1980s.

It is very lucky that such extensive data were collected by Messel and the University of Sydney programme so shortly after *C. porosus* in Australia were afforded protection from hunting (Messel *et al.* 1984) and continued for such a long period since (40 years, so far!). Monitoring the recovery since the depletion by hunting has provided a wealth of information about the processes which drive the dynamics of crocodylian populations. A striking finding was the role of density-dependent factors. For example, survivorship of hatchlings in those early years was quite high. Interestingly, Webb and Manolis (1992) reported higher survival rates, up to 80%, if overall hatchling numbers were low, and lower survival rates, perhaps only 20%, if

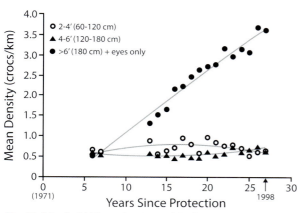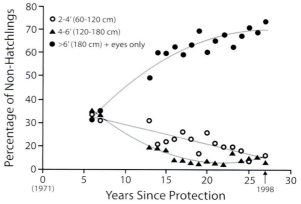

Fig. 13.34. *(Left) Mean density and (right) percentage contribution of different size classes (hatchlings excluded) of* C. porosus *seen on spotlight surveys in 11 Northern Territory rivers in the years since protection in 1971 until 1998. As the recovery proceeded, the proportion of large animals increased markedly. (From Webb* et al. *2000, with permission)*

density was high. Webb *et al.* (2000) speculated that this could be explained by lower predation by non-crocodylian predators such as birds, fish, snakes and goannas. Perhaps such predators are more attracted to hatchlings as a source of prey when there are more hatchlings about, through development of a search image. Survival in their second year was lower if a river had higher numbers of crocs > 2 m and perhaps cannibalism was a factor (see Fig. 13.20). The rate of increase in the number of juveniles at and above 3–4 years of age (~3–4'; 0.9–1.2m) was slowed by the presence of higher numbers of non-hatchlings in the river, and the interpretation was made that 'space' becomes limiting, stimulating emigration from the nesting (source) rivers. There is

no doubt that dispersal from good nesting areas is a significant phenomenon, as realised early by Messel *et al.* (1984). The continuing arrival of medium-sized *C. porosus* into Darwin Harbour (see Chapter 14) provides an interesting example.

Webb *et al.* (2000) reported a detailed analysis of changes in the density and size structure of populations in 11 rivers between 1978 and 1998, and this has been extended subsequently by Fukuda *et al.* (2011) to accommodate data up to 2009. The results are fascinating. The increase in density began to slow after 20 or so years (Fig. 13.34) (Webb *et al.* 2000) in a manner reminiscent of the theoretical graphs seen in ecology textbooks and that has continued (Fig. 13.35) (Fukuda *et al.* 2011). Some of

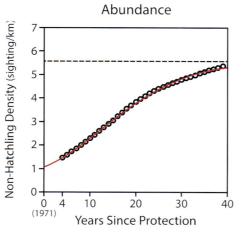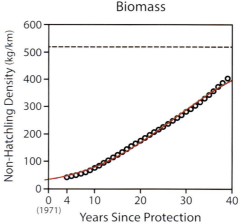

Fig. 13.35. *Summary graphs to show estimated recovery in (left) abundance and (right) biomass of* C. porosus *in Northern Territory rivers since 1971 when they received protection from hunting. The graphs come from extensive surveys of nearly 700 km of 12 river systems, most surveyed more than 20 times (range 11–29) between the mid 1970s and 2008–2009. The broken lines are predicted asymptotes. (Adapted from Fukuda* et al. *2011, © The Wildlife Society, 2011)*

the rivers, at least, are approaching what might be thought of as a 'carrying capacity'. Not only that, but the size structure has changed as well: larger crocodiles have gradually become a larger and larger proportion of the total population (Figs 13.34, 13.36). Although the number of adults has increased, producing more nests and more hatchlings, the survivorship of the younger stages is affected by the density-dependent factors mentioned above: predation, including cannibalism, and competition for space must be prompting dispersal. The figures presented here represent summaries of data from all of the rivers: the same general patterns show up in most rivers when they are considered individually.

What is the future of populations in the tidal rivers? Are they really approaching 'carrying capacity'? The numbers appear to be approaching a plateau, but biomass is not (Fig. 13.35). Fukuda *et al.* (2011) identified a plateau in biomass of ~520 kg/km of river, based on actual sightings. This might translate to three–four adult crocs per kilometre, which seems low in comparison with anecdotal accounts by pilots who flew the northern Australian beaches during the Second World War and also the descriptions by Cott (1961) of the abundance of similarly sized adult *C. niloticus* in parts of the Victoria Nile in Uganda. The current projections could be a significant underestimate. On the other hand, those anecdotal reports may have concentrated on the more spectacular concentrations, which have since come to be regarded as typical of the times.

The number of successful nests may vary considerably from season to season. In rivers where there were still sufficient adults after the hunting

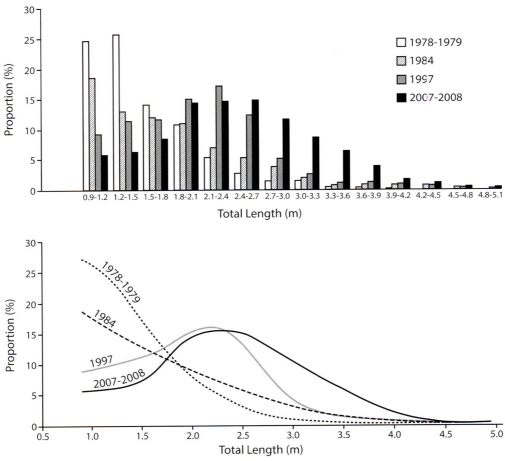

Fig. 13.36. *Changes in proportion of sizes of* C. porosus *pooled from spotlight surveys in 6–12 tidal rivers in the Northern Territory, Australia from 1978–79 to 2008–9 (modified from Fukuda* et al. *2011). The population structure has changed dramatically since hunting was stopped in 1971, when it was much more strongly juvenile dominated. (Upper panel adapted from Fukuda* et al. *2011, © The Wildlife Society, 2011)*

ended, such as the Liverpool and Tomkinson, there were many hatchlings in some years, very few in others. Nest success was related to the timing of the rains and, most importantly, the extent of flooding. Flooding often resulted when flood rains coincided with king tides. Another significant 'learning' from the monitoring of *C. porosus* is that density-dependent factors are significantly at play (Messel *et al.* 1981; Messel and Vorlicek 1986).

Most of the research on *C. porosus* in the NT has focussed in the tidal rivers. What of the future for *C. porosus* in other habitats? Interesting studies by Letnic and Connors (2006) and Letnic *et al.* (2011) analysed changes in *C. porosus* populations since the 1980s in those areas normally considered peripheral to the tidal rivers and wetlands where most nesting occurs and that are usually regarded as the core areas. They focussed on historical and current survey data from the upstream, freshwater sections of the Victoria, Daly and Roper River systems. In all three, populations of *C. porosus* have increased in the last 20 years and the increase is continuing. What is more, the crocs sighted in these areas were of a good size: mostly larger than 2.1 m (7 feet). There is apparently also a strong male bias: of 53 crocs removed from the Katherine River between 1994 and 2004 (40% 1.2–2.4m, 52% 2.1–3.4m), all but one was male. The upstream limits seem to be set by physical barriers such as escarpment waterfalls, but obviously food availability, as well as the presence of sufficient water and other habitat attributes, will have a strong influence. Letnic *et al.* (2011) discussed the particular implications of *C. porosus* becoming more common in those parts of the Katherine River, which have been a strong focus for recreational activity, including canoeing and swimming. The possible effect on public safety is significant because areas thought previously to be *C. porosus*-free and safe for swimming are increasingly being found unsafe. This will be discussed further in Chapter 14.

These observations are consistent with the idea of dispersal by immature males being stimulated by interactions with adult males in the breeding habitat to expand the range into other suitable habitat. We saw a similar spatial segregation of the population in *Caiman latirostris* in Brazil, where the sub-adults were out in the estuary where there was plenty of food, but away from the breeding pools, which were inhabited by the (presumably) dominant males (Grigg *et al.* 1998) (see Chapter 11).

There has been much less survey effort for *C. porosus* in Queensland, but what there is shows that recovery has been slower than in the NT. Fukuda *et al.* (2007) explored whether and to what extent habitat influences or human activity might have been a factor affecting this. After extensive biogeographic analysis, relating crocodile abundance data to environmental attributes in 55 different river systems across northern Australia, from Cambridge Gulf in Western Australia to Proserpine in Queensland, they came to the conclusion that it is habitat availability rather than human disturbance that is the significant factor. Strongly seasonal summer rainfall, warm winter temperatures and the proportion of favourable wetland vegetation (paperbarks, grasses and, sedges) are all important attributes. The first two are common to all three states. Proportions of suitable vegetation and wetland habitat, however, are well recognised to be lower in Queensland. Certainly we noticed a conspicuous difference between the coastal wetlands and rivers on the west coast of Cape York and those in the NT, from Coburg Peninsular eastwards, when we did our airborne surveys in 1977 and 1979 (Magnusson *et al.* 1978b, 1980). Queensland simply has less suitable habitat. Does this mean that historical populations of *C. porosus* were lower in Western Australia and Queensland? Letnic and Connors (2006) have shown that non-breeding sized *C. porosus* are expanding their distribution upstream in many rivers and other freshwater areas in the NT, and we know from their continued removal by the authorities in Darwin that there is a steady stream of sub-adults and young adults moving coastally. Presumably these long-lived and mobile animals will distribute themselves wherever there are rivers and wetland areas with suitable warmth and seasonal rainfall. The carrying capacity of an area will not be determined by proximity to nesting habitat, but, more likely, by food availability. We must conclude that there were previously much higher populations in Queensland and in Western Australia and, even if recovery is slower because it will rely primarily on immigration, that we can anticipate a return to higher numbers in these areas as well in due course unless inhibited by human actions. The east coast of Queensland will be an exception. Populations of *C. porosus* are likely to remain remnant from approximately Port Douglas south because of a combination of nesting habitat alienation and active removal (Chapter 14) in this area with its higher and continually expanding densities of human habitation and land use.

The NT data show that unharvested populations are likely to have a high proportion of adult individuals, with a small survival of hatchlings and juveniles. It also shows that, although unharvested populations are very vulnerable to removal of large numbers of adults (which is comparatively simple to do!), the populations do have a high capacity

Fig. 13.37. *Igarapé Acará in Reserva Ducke, Amazonas, Brazil. This tannin-stained rainforest stream may seem an unlikely habitat for a crocodylian, yet 10–15 kg adult* Paleosuchus trigonatus *can be found in caves and holes in its undercut banks. (Photo Lyn Beard)*

for recovery if harvesting ceases, in the absence of other threats such as habitat alienation.

These results and speculations about the likely future of the *C. porosus* populations in the tidal rivers of the NT prompt questions about the demographics of unshot/unharvested populations of crocodylians. There are not many studies of 'pristine' populations.

One species which is likely to have many pristine populations is the smooth-fronted caiman, *Paleosuchus trigonatus*: a very cryptic denison of Amazonian rainforest streams. In fact, it may have more pristine populations than any extant crocodylian because the area of the Amazon forest is huge (about the size of Australia), and there are very low densities of humans away from major rivers and highways (Magnusson *pers. comm.* 2011). For someone who learned about crocs in the wide, sinuous, turbid and tidal rivers of northern Australia, a swiftly flowing, tannin-stained and shaded mountain stream (Fig. 13.37) seems an unlikely habitat for a crocodylian, but Lyn Beard and I went to Reserva Ducke on the outskirts of Manaus with Bill Magnusson in 1992 and he soon showed us some very different crocs …

PALEOSUCHUS TRIGONATUS IN THE BRAZILIAN AMAZON (FIG. 13.38)

This is a type of crocodylian that most people have never heard of, which lives cryptically in rainforest streams, even some very small ones, and under a closed canopy. It scored a mention in Chapter 10 because of its minimal thermoregulatory behaviour and because it does not rely on basking, and again in Chapter 12 because of the female's preference for building nests up against termite mounds where she can use the warmth from the mound to accelerate the development of her eggs.

Bill Magnusson's and Albertina Lima's 8-year study of *Paleosuchus trigonatus* in Reserva Florestal Adolfo Ducke, 25 km east of Manaus, Brazil, by direct observation, spotlighting, mark–recapture and radio-telemetry, is truly inspirational. *P. trigonatus* is a very cryptic species. It has little or no commercial value and, because the indigenous inhabitants were (sadly) exterminated

Fig. 13.38. *The smooth-fronted caiman,* Paleosuchus trigonatus. *Both species of* Paleosuchus *are primarily forest dwellers and have a very large area of geographic overlap. (Photo William E Quatman)*

from Reserva Ducke long ago, there has been no hunting for the last few hundred years; it therefore provides an example of a fairly 'pristine' population. It is very difficult to believe that these clear, vegetation-stained streams, in many places only a few centimetres deep and surrounded by forest with little vegetation at ground level, could support 'high densities of 10–15 kg caimans with very bright eye shines'. Such an animal was very difficult to study. Hatchlings and sub-adults were located by walking along streams at night using a head light to detect eye shines (Figs 13.39). Adults

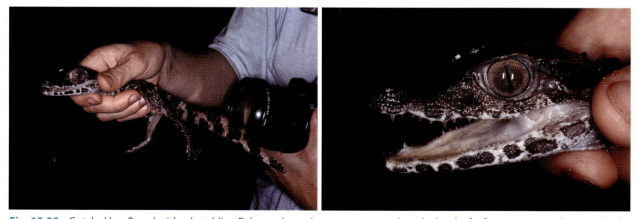

Fig. 13.39. *Gotcha! Lyn Beard with a hatchling* Paleosuchus trigonatus *captured on the bank of a forest stream in Reserva Ducke, Amazonas, Brazil, after it was located from a distance by its bright eye shine. (Photos GCG)*

Fig. 13.40. *Bill Magnusson with mask and snorkel, searching undercut banks with a waterproof flashlight for adult* Paleosuchus trigonatus. *(Photo GCG)*

were located 'by diving the entire lengths of the streams during the day and exploring caves under the banks with a waterproof flashlight' (Figs 13.40). They were captured by setting snares in front of underwater lairs. Their mound nests were located by intense searches along the banks in the nesting season, particularly in the vicinity of hatchlings, when observed. The study showed that growth is slow and *P. trigonatus* are very slow to mature: at least 11 years for females and 20 years for males. Adults are long-lived and sedentary, males are

territorial and females have small home ranges within the territory of at least one male (Fig. 13.41). Females are more tolerant of each other than males are and they nest about every third year, laying 15 eggs into a mound nest. Egg mortality is thought to be low, but hatchling mortality is high; the female does not remain for long with the young, who disperse from the nest vicinity within 1–2 weeks, typically moving upstream.

P. trigonatus provides a nice example of the way ectotherms can have a higher biomass than endotherms. Magnusson and Lima calculated that in their patch of forest in Reserva Ducke, the caimans had a biomass of 35–60 kg/km^2, 6–10 times higher than a much better known predator of about the same size that occurs throughout their range: the dwarf leopard or ocelot, *Leopardus pardalis*. It is much greater too than the combined biomasses of the larger (> 4 kg) rainforest mammalian carnivores in Panama and south-west Gabon. In pointing this out, Magnusson and Lima wrote 'Despite the current interest in the functioning of the Amazonian ecosystem, this large predator with the largest biomass has been almost completely overlooked'.

Another very informative study of a reasonably pristine population, also remarkable for both its

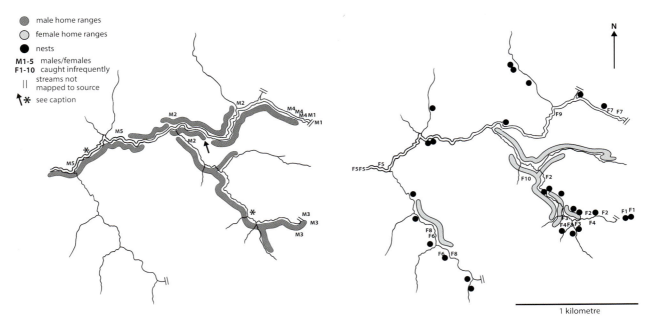

Fig. 13.41. *Distribution of (left) adult males and (right) females and nests in a section of the Igarapé Acará in Reserva Ducke, Amazonas, Brazil. Shaded stripes indicate home ranges of individuals captured three or more times. The male marked by an arrow moved as far as shown by the asterisks. (Adapted from Magnusson and Lima 1991, copyrighted material included courtesy of SSAR)*

Fig. 13.42. *A 'freshie', C. johnstoni, relaxes in the morning sunshine beside Yellow Water billabong in Kakadu National Park, Northern Territory. (Photo DSK)*

longevity and its scope, was undertaken by Col Limpus and Anton Tucker on *Crocodylus johnstoni*, another freshwater crocodylian, Australia's 'freshie'.

CROCODYLUS JOHNSTONI IN THE LYND R, NORTH QUEENSLAND

This is Australia's 'other' crocodile – the Johnston River croc or freshie (Figs 13.42, 13.43) – also referred to sometimes (but only by people who study it) as 'the thinking man's crocodile'. It is widely distributed in freshwater sections of rivers and in billabongs across northern Australia, sometimes occurring in brackish and even hyperosmotic tidal waters (Chapter 11).

An outstanding 20 (plus) year mark–recapture study of *Crocodylus johnstoni* was undertaken in ~60 km of the Lynd River and Fossilbrook Creek in North Queensland. The study was initiated in 1976 by Col Limpus from Queensland Parks and Wildlife Service and then extended and brought to completion in 1995 by Anton (Tony) Tucker in a PhD study through the University of Queensland (Tucker 1997a and a series of subsequent publications). The Lynd River flows westwards from the highlands at the base of Cape York and joins the Mitchell River, which then flows into the Gulf of Carpentaria ~500 km distant. The study area is rocky and elevated, at 360–520 m above sea level. The river is spring fed, subject to substantial additional and swift inflow during the wet season (late summer monsoon) and which retreats to permanent pools connected by shallow rapids and short cascades in the winter (dry) season. Field trips were usually made twice a year, in the nesting (spring) and hatching (summer) seasons (Figs 13.44, 13.45). Longitudinal data are available for an astonishing 970 known-age males and 1025 females, marked originally as hatchlings. Intervals between marking and recapture ranged from 6 months to 18 years.

Fig. 13.43. *A well-camouflaged mature* C. johnstoni *photographed in 2013 in the South Alligator River east of Darwin, well above the tidal influence. This is in an area that was invaded by cane toads several years earlier. (Photo Buck Salau)*

Seeking a way to age them, Tony Tucker found that annuli visible in nuchal osteoderms under the electron microscope (Fig. 13.28) were able to be validated as reliable indicators of annual growth increments in this very seasonal climate (Tucker 1997a,b). This enabled him to determine with confidence the ages of a further 198 males and 191 females. Such seasonal growth increments are not necessarily typical of crocodylians in general: the marked seasonality of the climate in the Lynd River study area undoubtedly had a big influence and being able to age the animals facilitated the study considerably.

This population has an unknowable hunting history. Crocodiles were not protected in Queensland until 1974, and, although *C. johnstoni* were never as much favoured by shooters as *C. porosus* because their hides are less valuable, they were hunted widely across Queensland, probably including the Lynd. This was an extensive mark–recapture study

Fig. 13.44. *Catching* Crocodylus johnstoni *in 'The Croc Hole', Lynd River, north Queensland. (Left) Nets are deployed across the waterhole and (right)* C. johnstoni *driven into the net by thumping and prodding the rocky bottom with sticks. (Photos GCG)*

Fig. 13.45. *Tony Pople reads the clock-face balance. Length–weight relationships reveal the body condition of an individual (Chapter 1). Length in crocodylians is usually measured as snout–vent length (SVL), from the tip of the snout to the anterior end of the cloacal slit. Many practitioners choose to measure to the posterior end of the slit, which probably makes more sense because it better marks the end of the body cavity. (Photo GCG)*

and individual animals were followed over many years as they dispersed (Tucker *et al.* 1998) and grew (Tucker *et al.* 2006) choosing different habitat types within the stream as they grew and as their status changed within the population (Tucker *et al.* 1997). Crocodiles were captured by any and all available techniques, standard measurements were made, sex was determined manually and reproductive status determined by diagnostic ultrasonography of females (allowing ovarian follicle size and/or egg number to be scored) (Tucker and Limpus 1997 and see Lance *et al.* 2009) after its validation by laparoscopy and, in males, microscopic examination for sperm of a smear taken from the penile groove (Fig. 13.29). Diet was assessed by stomach flushing (Chapter 6) (Tucker *et al.* 1996). All of this added up to an astonishing dataset on a population in an interesting habitat and which, over the 20 years, was judged to be essentially stable, if not increasing slightly.

The Limpus–Tucker study provided a vast amount of information and understanding about *C. johnstoni* and included some interesting and provocative results, particularly about just how subtle some influential population processes can be.

Adults were found mainly in pools and that, too, is where nesting was focussed. Nests were dug into sandbanks in spring, towards the end of

Fig. 13.46. *Col Limpus and others explore a large burrow system used by* C. johnstoni *at 'The Croc Hole', Lynd River north Queensland. (Photo GCG)*

the dry season when the river levels are low. Adult males also spent a lot of time in substantial burrows dug into the banks (Fig. 13.46). Small crocodiles were found mainly in rapids and stream margins, males dispersing rather more than females, perhaps driven by competition from adult males. A consequence, of course, is that the chance of inbreeding is reduced. Males become larger than females, through continuing to grow for longer. This is interesting because growth in reptiles is often regarded as indeterminate – that is, continuing until death – yet adult males and females were found in this study which grew not at all over long periods. From that and a careful analysis of size and age data, Tucker concluded that growth is determinate, although both the rate of growth and the final size will depend upon feeding opportunities and other environmental factors, plus biotic factors such as the presence of other individuals. This was one of

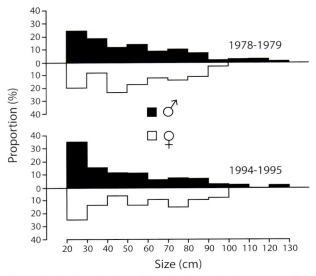

Fig. 13.47. *Population size structure of the population (excluding hatchlings) after the 1978–79 and 1994–95 nesting seasons, as sampled in a long-term mark–recapture study of C. johnstoni in 62 km of the Lynd River and Fossilbrook Creek in northern Queensland. (Modified from Tucker 1997a, with permission)*

the provocative results. The same observation has been made in another very long-lived reptile: New Zealand's Tuatara.

Population size structure was broadly very similar throughout the 20+ years of the study. Representative examples are shown in Fig. 13.47. Concerning the life history 'statistics': on average 85% of females produced clutches each year, averaging 12.8 eggs per clutch. About 10% of nests did not produce hatchlings because of non-viable eggs, flooding or goanna predation. Sex ratios were quite variable between years, but averaged 50:50 in the long term: more males (18%) were produced in El Nino years. Females matured at 12–22 years (average 19) Males matured at 12–24 years (average 21). Combining skeletochronology and growth rate analyses, maximum longevity is likely to be 55 years for females and 65 for males. The range difference of a decade in age at maturity is striking. Because of the territoriality of adult males, growth and development of smaller individuals can be inhibited. In a study of the thermal biology of *C. johnstoni* in one particular waterhole of the Lynd River (Chapter 10), Frank Seebacher noted that there was competition for basking sites, with

the more aggressive males and females chasing subordinate animals away, preventing them from basking and, in effect, condemning them to a lower temperature existence and thus a slower rate of development (Seebacher and Grigg 1997). That same waterhole was controlled by a large male for several years, with younger mature males' aspirations and their growth and maturity being suppressed as a consequence. Social factors are quite hard to study and almost certain to be overlooked, but are likely to exert subtle, but very significant, influences over crocodylian population ecology.

Tucker reported that hatching and juvenile survival was quite low except in very favourable conditions. Survivorship in their second decade, however, as sub-adults mostly, was very high: ~95%. Male survivorship fell in the ages 18–30 years after sexual maturity, presumably because of injury associated with fighting higher status males while seeking breeding rights to females. In good conditions, population increase was ~8.0% per annum. In less favourable conditions, adults matured more slowly and population increase was reduced to zero. These findings became the basis for a stage-based population model for the species (Tucker 2001). The keys to persistence of the population – sometimes referred to as key survival 'strategies' (a term I dislike, with its implication of design or forethought) – are the high adult survivorship, delayed maturity and great longevity of individuals. This is interpreted as a 'bet-hedging strategy', which allows populations to survive periods of harsh conditions, with significant recruitment of hatchlings occurring only sporadically, when conditions are good, yet still resulting in adequate recruitment to maintain the breeding population.

Crocodylians are flexible in their behaviours and different habitats offer different predation threats and different feeding opportunities, so life history statistics are always likely to vary between habitats. In an extensive study of *C. johnstoni* in a very different habitat – the McKinlay River in the NT – Grahame Webb and co-workers found, for example, higher nest mortality, earlier onset of maturity and a higher intrinsic rate of increase (Webb and Smith 1984) than in the population studied by Limpus and Tucker.

ALLIGATOR MISSISSIPPIENSIS IN THE USA

The American alligator provides a striking example of a species once endangered but now thriving. Throughout its recovery, the commercial value of alligator hides has been one of the driving forces behind the implementation of measures to ensure its conservation. It provides a striking example of how sustainable commercial use and conservation of a wildlife species can co-exist.

Alligators were hunted extensively throughout their range (Fig. 13.3) from the 1800s, and by the 1960s were considered endangered. They were listed as such in 1967. The species was then listed, in 1975, on Appendix I of the Convention on International Trade in Endangered Species of Wild Fauna and Flora (CITES), meaning that alligator products (i.e. hides) could enter international trade only if bred in captivity. Protection followed in all the range states, but commercial closed cycle farming was encouraged and continues in Louisiana, Florida, Texas and Georgia. With hunting made illegal, wildlife management agencies established monitoring programmes and were able to document recovering populations. Their management is under the control of each of the range states separately, and active research programmes in Louisiana (Joanen and McNease 1987) and Florida (Hines and Abercrombie 1987) in particular developed ways to manage commercial use effectively and sustainably, and set the scene for the management systems in place today (reviewed by Elsey and Woodward 2010). Early on, in Florida particularly, alligators causing potential threat to humans ('nuisance alligators') were captured or shot by licenced hunters and the skins allowed to be sold. As the recovery proceeded, alligators were downlisted to Appendix II of the CITES list, meaning that their products can be exported as long as the management authority provides a certification that the export will not be detrimental to the survival of the species. The alligator population in the wild was recently estimated to be 2–3 million (Elsey and Woodward 2010). Tens of thousands of alligators are now hunted annually, under permit, and hundreds of thousands are held in numerous farms in the US southern states. Some

of the farming is 'ranching': that is, rearing eggs or hatchlings taken from the wild, some of it is 'closed cycle' and some a mixture of both. More will be said about this in the next chapter.

What needs to be said here, however, is that experimental manipulations of alligator populations have revealed the difficulty of developing population models that take into account perturbations such as harvesting. In the late 1970s in Florida, in response to growing numbers of 'nuisance alligator' reports once populations started to recover (sometimes 5000 reports per year), the Fish and Wildlife Conservation Commission embarked on a study of alligator population dynamics (Hines and Abercrombie 1987). They had in mind potential harvesting and wanted to put its management for sustainability onto a scientific basis. Accordingly, they monitored nest numbers in several Florida lakes, harvesting in some and keeping others as controls, starting in 1981. The study across all the lakes was continued for 10 years, until June 1991. The harvest was 70% male and averaged 13% annually of the alligators large enough to be of commercial interest (Woodward *et al.* 1992). Over time, the harvest proportion of small alligators increased, nest production either remained stable or increased and the animals became more wary, but the conclusion was that there were no measurable negative effects on the population.

In a related study, removal of 50% of the annual production of eggs and hatchlings was attempted in three other central Florida lakes from 1981 to 1986, with a fourth lake identified as a control (Jennings *et al.* 1988). Nest numbers in each lake were monitored by aerial survey and the populations by spotlighting, and no change could be detected in the demographic structure of any of the populations that could be attributed to the removal of eggs and hatchlings. The 50% annual removals were continued on two of the lakes until 1991, affording much more time for any impact to show up in the population size structure (Rice *et al.* 1999). To give an idea of the scope of the study, nearly 46 000 eggs or hatchlings were removed from these two lakes (Griffin and Jesup): the produce of about a thousand nests. Nest numbers increased in both the control lake (Woodruff) and in Lake Jesup, with a hint of the same in Lake Griffin. The total populations increased in all lakes. Juvenile densities were stable in the harvested lakes, but increased in the control. Despite this, densities of sub-adults increased in the harvested lakes. The authors concluded that a 50% harvest of eggs/hatchlings did not affect recruitment into the sub-adult or adult size classes. They speculated about density-dependent factors operating, suggesting that the experimental harvests may have made more food available to the remaining animals, resulting in faster growth rates and increased survival.

These results show the resilience of a population of crocodylians to egg harvest, at least in a productive habitat, backing the experience gained already by crocodile farms in many places that 'ranch' eggs brought in from the wild.

The sustainable commercial use of crocodylians and its potential as a conservation tool will be discussed in Chapter 14.

REFERENCES

Abercrombie CL (1992) Fitting curves to crocodilian age-size data: some hesitant recommendations. In *Proceedings of the 11th Working Meeting, Crocodile Specialists Group.* 2–7 August 1992, Victoria Falls, Zambia. pp. 5–21. IUCN, Gland, Switzerland.

Abercrombie CL, Rice KG, Hope CA (2001) The great alligator-caiman debate: meditations on crocodilian life-history strategies. In *Crocodilian Biology and Evolution.* (Eds GC Grigg, F Seebacher and CE Franklin), pp. 409–418. Surrey Beatty & Sons, Sydney.

Bayliss P (1987) Survey methods and monitoring within crocodile management programs. In *Wildlife Management: Crocodiles and Alligators.* (Eds GJW Webb, SC Manolis and PJ Whitehead) pp. 157–175. Surrey Beatty & Sons, Sydney.

Bayliss P, Webb GJW, Whitehead PJ, Dempsey KE, Smith AMA (1986) Estimating the abundance of saltwater crocodile, *Crocodylus porosus* Schneider, in tidal wetlands of the N.T.: A mark-recapture experiment to correct spotlight counts to absolute numbers and the calibration of helicopter and spotlight counts. *Australian Wildlife Research* **13**, 309–320.

Bishop JM, Leslie AJ, Bourquin S, Badenhorst L, O'Ryan C (2009) Overexploitation and the declining effective population size of a top predator. *Biological Conservation* **142**(10), 2335–2341.

Bourquin SL, Leslie AJ (2011) Estimating demographics of the Nile crocodile (*Crocodylus niloticus* Laurenti) in the panhandle region of the Okavango Delta, Botswana. *African Journal of Ecology.*

Brazaitis P, Rebêlo GH, Yamashita C (1998) The distribution of *Caiman crocodilus crocodilus* and *Caiman yacare* populations in Brazil. *Amphibia-Reptilia* **19**, 193–201.

Brien ML, Cherkiss MS, Mazotti FJ (2008) American crocodile, *Crocodylus acutus*, mortalities in Florida. *Florida Field Naturalist* **36**(3), 55–59.

Britton ARC, Britton EK, McMahon CR (2013) Impact of a toxic invasive species on freshwater crocodile (*Crocodylus johnstoni*) populations in upstream escarpments. *Wildlife Research* **40**, 312–317.

Busack SD, Pandya S (2001) Geographic variation in *Caiman crocodilus* and *Caiman yacare* (Crocodylia: Alligatoridae): systematics and legal implications. *Herpetologica* **57**, 294–312.

Campos Z (1993) Effect of habitat on survival of eggs and sex ratio of hatchling Caiman crocodilus yacare in the Pantanal, Brazil. *Journal of Herpetology* **27**, 127–132.

Campos Z, Sanaiotti T, Magnusson WE (2010) Maximum size of dwarf caiman, *Paleosuchus palpebrosus* (Cuvier, 1807) in the Amazon and habitats surrounding the Pantanal, Brazil. *Amphibia - Reptilia* **31**, 439–442.

Caughley GJ, Sinclair ARE (1994) *Wildlife Ecology and Management*. Blackwell Scientific, Boston, Massachusetts.

Cedeño-Vázquez JR, Rodriguez D, Calmé S, Ross JP, Densmore LD, III, Thorbjarnarson JB (2008) Hybridization between *Crocodylus acutus* and *Crocodylus moreletii* in the Yucatan Peninsula: I. evidence from mitochondrial DNA and morphology. *The Journal of Experimental Zoology* **309A**, 661–673.

Choquenot D, Webb GJW (1987) A photographic technique for estimating the size of crocodiles seen in spotlight surveys and for quantifying observer bias. In *Wildlife Management: Crocodiles and Alligators.* (Eds GJW Webb, SC Manolis and PJ Whitehead) pp. 217–224. Surrey Beatty & Sons, Sydney.

Cott HB (1961) Scientific results of an enquiry into the ecology and economic status of the Nile Crocodile (*Crocodylus niloticus*) in Uganda and Northern Rhodesia. *Transactions of the Zoological Society of London* **29**, 211–356.

Coutinho M, Campos Z (1996) Effect of habitat and seasonality on the densities of caiman in southern Pantanal, Brazil. *Journal of Tropical Ecology* **12**, 741–747.

Covacevich J, Archer M (1975) The distribution of the cane toad, *Bufo marinus*, in Australia and its effects on indigenous vertebrates. *Memoirs of the Queensland Museum* **17**, 305–310.

Das I, Charles J (2002) New record of a freshwater crocodile from Brunei. IUCN/SSC Crocodile Specialist Group Newsletter 21, 10–11.

Da Silveira R, Campos Z, Thorbjarnarson J, Magnusson WE (2013) Growth rates of black caiman (*Melanosuchus niger*) and spectacled caiman (*Caiman crocodilus*) from two different Amazonian flooded habitats. *Amphibia-Reptilia* **34**, 437–449.

Davis LM, Glenn TC, Elsey RM, Brisbin IL, Rhodes WE, Dessauer HC, *et al.* (2001) Genetic structure of six populations of American alligators: a microsatellite analysis. In *Crocodilian Biology and Evolution.* (Eds GC Grigg, F Seebacher and CE Franklin). pp 38–50. Surrey Beatty & Sons, Sydney.

Davis LM, Glenn TC, Strickland DC, Guillette LJ, Elsey RM, Rhodes WE, *et al.* (2002) Microsatellite DNA analyses support on East-West phylogeographic split of American alligator populations. *Journal of Experimental Zoology (Molecular and Developmental Evolution)* **294**, 352–372.

Delany MF, Woodward AR, Kiltiei RA, Moore CT (2011) Mortality of American alligators attributed to cannibalism. *Herpetologica* **67**(2), 174–185.

Dinets V (2010) Nocturnal behavior of the American alligator (*Alligator mississippiensis*) in the wild during the mating season. *Herpetological Bulletin* **111**, 4–11.

Eaton MJ, Link WA (2011) Estimating age from recapture data: integrating incremental growth measures with ancillary data to infer age-at-length. *Ecological Applications* **21**, 2487–2497.

Eaton MJ, Martin AP, Thorbjarnarson J, Amato G (2009) Species-level diversification of African dwarf crocodiles (Genus *Osteolaemus*): a geographic and phylogenetic perspective. *Molecular Phylogenetics and Evolution* **50**, 496–506.

Elsey R (2006) Alligator nesting decreased by lingering hurricane effects and drought. *Crocodile Specialist Group Newsletter* **25**(4), 19–22.

Elsey RM, Woodward AR (2010) American Alligator, *Alligator mississippiensis.* In *Crocodiles. Status Survey and Conservation Action Plan.* 3rd edn (Eds SC Manolis and C Stevenson) pp. 1–4. Crocodile Specialist Group, Darwin.

FitzSimmons NN, Tanksley S, Forstner MRJ, Louis EE, Daglish R, Gratten J, *et al.* (2001) Microsatellite markers for *Crocodylus*: new genetic tools for population genetics, mating system studies and forensics. In *Crocodilian Biology and Evolution.* (Eds GC Grigg, F Seebacher and CE Franklin) pp. 51–57. Surrey Beatty and Sons, Sydney.

Fukuda Y, Whitehead P, Boggs G (2007) Broad scale environmental influences on the abundance of saltwater crocodiles, *Crocodylus porosus Australian Wildlife Research* **34**, 167–176.

Fukuda Y, Whitehead P, Letnic M, Manolis C, Delaney R, Lindner G, *et al.* (2011) Recovery of saltwater crocodiles, *Crocodylus porosus*, following the cessation of hunting in tidal rivers of the Northern Territory, Australia. *Journal of Wildlife Management* **75**, 1253–1266.

Graham A (1968) 'The Lake Rudolf crocodile (*Crocodylus niloticus* Laurenti) population'. Report to the Kenya Game Department by

Wildlife Services Limited. Nairobi, Kenya, <http://ufdcimages.uflib.ufl.edu/aa/00/00/75/90/00001/lakerudolfcrocodilesalistair graham.pdf>

Graham A, Beard P (1973). *Eyelids of Morning: The Mingled Destinies of Crocodiles and Men*. New York Graphic Society, Greenwich, Connecticut.

Gratten J (2004) The molecular systematics, phylogeography and population genetics of Indo-Pacific *Crocodylus*. Unpublished PhD Thesis, The University of Queensland, Brisbane.

Grigg GC, Pople AR, Beard LA (1997) Application of an ultra-light aircraft to aerial surveys of kangaroos on grazing properties. *Wildlife Research* **24**, 359–372.

Grigg GC, Beard LA, Moulton T, Queirol Melo MT, Taplin LE (1998) Osmoregulation by the broad-snouted caiman, *Caiman latirostris*, in estuarine habitat in southern Brazil. *Journal of Comparative Physiology. B, Biochemical, Systemic, and Environmental Physiology* **168**, 445–452.

Hall PM (1989) Variation in geographic isolates of the New Guinea crocodile (*Crocodylus novaeguineae* Schmidt) compared with the similar, allopatric, Philippine crocodile (*C. mindorensis* Schmidt). *Copeia* **1989**, 71–80.

Hekkala ER, Amato G, Desalle R, Blum MJ (2010) Molecular assessment of population differentiation and individual assignment potential of Nile crocodile (*Crocodylus niloticus*) populations. *Conservation Genetics* **11**(4), 1435–1443.

Hekkala ER, Shirley MH, Amato G, Austin JD, Charter S, Thorbjarnason J, *et al.* (2011) An ancient icon reveals new mysteries: mummy DNA resurrects a cryptic species within the Nile crocodile. *Molecular Ecology* **20**, 4199–4215.

Hines TC, Abercrombie CL (1987) The management of alligators in Florida, USA. In *Wildlife Management: Crocodiles and Alligators.* (Eds GJW Webb, SC Manolis and PJ Whitehead) pp. 43–47. Surrey Beatty & Sons, Sydney.

Hrbek T, Vasconcelos WR, Rebelo G, Farias IP (2008) Phylogenetic relationships of South American alligatorids and the caiman of Madeira River. *Journal of Experimental Zoology* **309A**, 588–599.

Hutton JM (1986) Age determination of living Nile crocodiles from the cortical stratification of bone. *Copeia* **1986**, 332–341.

Hutton JM (1987) Techniques for aging wild crocodilians. In *Wildlife Management: Crocodiles and Alligators.* (Eds GJW Webb, SC Manolis and PJ Whitehead) pp. 211–216. Surrey Beatty & Sons, Sydney.

Hutton JM (1989) Movements, home range, dispersal and the separation of size classes in Nile Crocodiles. *American Zoologist* **29**(3), 1033–1049.

Idriess IL (1946) *In Crocodile Land; Wandering in Northern Australia.* Angus and Robertson, Sydney.

Jennings ML, Percival HF, Woodward AR (1988) Evaluation of alligator hatching and egg removal from three Florida lakes. *Proceedings of the Annual Conference of Southeast Fish and Wildlife Agencies* **42**, 283–294.

Joanen T, Mc Nease L (1987) The management of alligators in Louisiana, USA. In *Wildlife Management: Crocodiles and Alligators.* (Eds GJW Webb, SC Manolis and PJ Whitehead) pp. 33–42. Surrey Beatty & Sons, Sydney.

Kay WR (2004a) Population ecology of *Crocodylus porosus* (Schneider 1801) in the Kimberley region of Western Australia. PhD thesis, University of Queensland, Brisbane.

Kay WR (2004b) Movements and home ranges of radio-tracked *Crocodylus porosus* in the Cambridge Gulf region of Western Australia. *Wildlife Research* **31**, 495–508.

Krebs CJ (1994) *Ecology: The Experimental Analysis of Distribution and Abundance.* Harper Collins College Publishers, New York.

Kyle-Little S (1957) *Whispering Wind – Adventures in Arnhem Land.* Hutchinson, London.

Lance VA, Elsey RM, Lang J (2000) Sex ratios of American alligators (Crocodylidae): male or female biased? *Journal of Zoology (London, England)* **252**, 71–78.

Lance VA, Rostal DC, Elsey RM, Trosclair PL III (2009) Ultrasonography of reproductive structures and hormonal correlates of follicular development in female American Alligators,

Alligator mississippiensis, in southwest Louisiana. *General and Comparative Endocrinology* **162**, 251–256.

Lang JW (1987) Crocodilian behavior: implications for management. In *Wildlife Management: Crocodiles and Alligators*. (Eds GJW Webb, SC Manolis and P J Whitehead.) pp. 273–294. Surrey Beatty & Sons, Sydney.

Lang JW, Andrews HV (1994) Temperature-dependent sex determination in crocodilians. Journal of Experimental Zoology **270**, 28–44.

Lang JW, Andrews H, Whitaker R (1989) Sex determination and sex ratios in *Crocodylus palustris*. American Zoologist **29**, 935–952.

Letnic M, Connors G (2006) Changes in the distribution and abundance of saltwater crocodiles (*Crocodylus porosus*) in the upstream, freshwater reaches of rivers in the Northern Territory, Australia. *Wildlife Research* **33**, 529–538.

Letnic M, Ward S (2005) Observations of freshwater crocodiles (*Crocodylus johnstoni*) preying upon cane toads (*Bufo marinus*) in the Northern Territory. *Herpetofauna* **35**, 98–100.

Letnic M, Webb JK, Shine R (2008) Invasive cane toads (*Bufo marinus*) cause mass mortality of freshwater crocodiles (*Crocodylus johnstoni*) in tropical Australia. *Biological Conservation* **141**, 1773–1782.

Letnic M, Carmody P, Burke J (2011) Problem crocodiles (*Crocodylus porosus*) in the freshwater, Katherine River, Northern Territory, Australia. *Australian Zoologist* **35**(3), 858–863.

Luiselli L, Akani GC, Capizzi D (1999) Is there any interspecific competition between dwarf crocodiles (*Osteolaemus tetraspis*) and Nile monitors (Varanus niloticus ornatus) in the swamps of central Africa? A study from south-eastern Nigeria. *Journal of Zoology* **247**, 127–131.

Magnusson WE (1983) Size estimates of crocodilians. *Journal of Herpetology* **17**, 86–88.

Magnusson WE (1986) The peculiarities of crocodilian population dynamics and their possible importance for management strategies. In *Proceedings of the 7th Working Meeting of the IUCN–SSC Crocodile Specialist Group*. 21–28

October 1984m Caracas, Venezuela. pp. 434–442. IUCN–World Conservation Union, Gland, Switzerland and Cambridge, UK.

Magnusson WE, Lima AP (1991) The ecology of a cryptic predator, *Paleosuchus trigonatus*, in a tropical rainforest. *Journal of Herpetology* **25**, 41–48.

Magnusson WE, Sanaiotti TM (1995) Growth of *Caiman crocodilus crocodilus* in central Amazonia, Brazil. *Copeia* **1995**, 498–501.

Magnusson W, Caughley GJ, Grigg GC (1978a) A double-survey estimate of population size from incomplete counts. *Journal of Wildlife Management* **42**, 174–176.

Magnusson W, Grigg GC, Taylor JA (1978b) An aerial survey of potential nesting areas of the Saltwater Crocodile (*Crocodylus porosus* Schneider) on the north coast of Arnhem Land, northern Australia. *Australian Wildlife Research* **5**, 401–415.

Magnusson W, Grigg GC, Taylor JA (1980) An aerial survey of potential nesting areas of *Crocodylus porosus* on the west coast of Cape York Peninsula. *Australian Wildlife Research* **7**, 465–478.

McNease L, Joanen T (1977) Alligator diet in relation to marsh salinity. *Proceedings of the Annual Conference of Southeastern Association of Fish and Wildlife Agencies* **31**, 36–40.

Messel H, Vorlicek GC (1986) Population dynamics and status of *Crocodylus porosus* in the tidal waterways of northern Australia. *Australian Wildlife Research* **13**, 71–111.

MesseI H, Vorlicek GC, Wells AG, Green WJ (1980) *Monograph 14. Surveys of Tidal River Systems in the Northern Territory of Australia and their Crocodile populations*. Pergamon Press, Sydney, Australia.

MesseI H, Vorlicek GC, Green WJ, Onley IC (1984) *Monograph 18. Surveys of Tidal River Systems in the Northern Territory of Australia. Population Dynamics of* Crocodylus porosus *and Status, Management and Recovery. Update 1979–1983*. Pergamon Press, Sydney.

Messel H, Vorlicek GC, Wells GA, Green WJ (1981) *Monograph 1. Surveys of the Tidal Systems in the Northern Territory of Australia and their Crocodile*

Populations. The Blyth-Cadell River Systems Study and the Status of Crocodylus porosus Populations in the Tidal Waterways of Northern Australia. Pergamon Press, Sydney.

Messel H, Green WJ, Vorlicek GC, Wells AG (1982) *Monograph 15. Surveys of Tidal River Systems in the Northern Territory of Australia and their Crocodile populations. Work Maps of Tidal Waterways in Northern Australia.* Pergamon Press, Sydney.

Milián-García Y, Venegas-Anaya M, Frias-Soler R, Crawford AJ, Ramos-Targarona R, Rodríguez-Soberón R, *et al.* (2011) Evolutionary history of Cuban crocodiles *Crocodylus rhombifer* and *Crocodylus acutus* inferred from multilocus markers. *Journal of Experimental Zoology A Ecological Genetics and Physiology* **315**, 358–375.

Moritz CM (1994) Defining "evolutionary significant units" for conservation. *Trends in Ecology & Evolution* **9**, 373–375.

Moulton TP, Magnusson WE, Melo MTQ (1999) Growth of *Caiman latirostris* inhabiting a coastal environment in Brazil. *Journal of Herpetology* **33**, 479–484.

Neill WT (1971) *The Last of the Ruling Reptiles.* Columbia University Press, New York.

Polis GA, Myers CA (1985) A survey of intraspecific predation among reptiles and amphibians. *Journal of Herpetology* **19**(1), 99–107.

Ray DA, Dever JA, Platt SG, Rainwater TR, Finger AG, McMurry ST, *et al.* (2004) Low levels of nucleotide diversity in *Crocodylus moreletii* and evidence of hybridization with *C. acutus. Conservation Genetics* **5**, 449–462.

Read MA, Grigg GC, Irwin SR, Shanahan D, Franklin CE (2007) Satellite tracking reveals long distance coastal travel and homing by translocated Estuarine Crocodiles, *Crocodylus porosus. PLoS ONE* **9**, 1–5.

Rhodes WE, Lang JW (1995) Sex ratios of naturally incubated alligator hatchlings: field techniques and initial results. *Proceedings of the Annual Conference of Southeast Association of Fish and Wildlife Agencies* **49**, 640–646.

Rhodes WE, Lang JW (1996) Alligator nest temperatures and hatchling sex ratios in coastal South Carolina. *Proceedings of the Annual Conference of Southeast Asociation of Fish and Wildlife Agencies* **50**, 521–531.

Rice KG, Percival FH, Woodward AR, Jennings ML (1999) Effects of egg and hatchling harvest on American alligators in Florida. *Journal of Wildlife Management* **63**(4), 1193–1200.

Rice KG, Percival HF, Woodward AR (2000) Estimating sighting proportions of American alligator nests during helicopter survey. *Proceedings of the Annual Conference of Southeast Asociation of Fish and Wildlife Agencies* **54**, 314–321.

Richards PM, Wasilewski J (2003) *Crocodylus acutus* (American Crocodile) cannibalism. *Herpetological Review* **34**, 371.

Rodda GH (1984) Homeward paths of displaced juvenile alligators as determined by radiotelemetry. *Behavioral Ecology and Sociobiology* **14**, 241–246.

Rodriguez D, Cedeño-Vázquez JR, Forstner MRJ, Densmore LD III (2008) Hybridization between *Crocodylus acutus* and *Crocodylus moreletii* in the Yucatan Peninsula: II. Evidence from microsatellites. *Journal of Experimental Zoology* **309A**, 661–673.

Rodríguez-Soberón R (2000) Situación actual de *Crocodylus acutus* en Cuba. In *Proceedings of the 15th Working Meeting of the Crocodile Specialist Group.* 17–20 January 2000, Varadero, Cuba. pp. 17–32. IUCN–The World Conservation Union, Gland, Switzerland and Cambridge, UK.

Rohner CA, Richardson A, Marshall AD, Weeks SJ, Pierce SJ (2011) How large is the world's largest fish? Measuring whale sharks *Rhincodon typus* with laser photogrammetry. *Journal of Fish Biology* **78**(1), 378–385.

Rootes WL, Chabreck RH (1993) Cannibalism in the American alligator. *Herpetologica* **49**, 99–107.

Ross CA (1990) *Crocodylus raninus* S. Muller and Schlegel, A valid species of crocodile (Reptilia: Crocodylidae) from Borneo. *Proceedings of the Biological Society of Washington* **103**, 955–961.

Seebacher F, Grigg GC (1997) Patterns of body temperature in wild freshwater crocodiles, *Crocodylus johnstoni*: thermoregulation vs. conformity, seasonal acclimatisation and the effect of social interactions. *Copeia* **1997**(3), 549–557.

Shine R (1999) Why is sex determined by nest temperature in many reptiles? *Trends in Ecology and Evolution* 14, 186–189.

Shine R (2010) The ecological impact of invasive cane toads (*Bufo marinus*) in Australia. *The Quarterly Review of Biology* **85**, 253–291.

Shirley MH, Villanova V, Vliet KA, Austin JD (2014) Genetic barcoding facilitates captive and wild management of three cryptic African crocodile species complexes. *Animal Conservation*, (In press).

Smith JG, Phillips BL (2006) Toxic tucker: the potential impact of cane toads on Australian reptiles. *Pacific Conservation Biology* **12**, 40–49.

Somaweera R, Webb JK, Shine R (2011a) It's a dog-eat-croc world: dingo predation on the nests of freshwater crocodiles in tropical Australia. *Ecological Research* **26**, 957–967.

Somaweera R, Webb JK, Shine R (2011b) Determinants of habitat selection by hatchling Australian freshwater crocodiles. *PLoS ONE* **6**(12), e28533.

Somaweera R, Webb J, Brown G, Shine R (2011c) Hatchling Australian freshwater crocodiles rapidly learn to avoid toxic invasive cane toads *Behaviour* **148**(4), 501–517.

Somaweera R, Shine R, Webb J, Dempster T, Letnic M (2012) Why does vulnerability to toxic invasive cane toads vary among populations of Australian freshwater crocodiles? *Animal Conservation* **16**, 86–96.

Somaweera R, Brien M, Shine R (2013) The role of predation in shaping crocodilian natural history. *Herpetological Monograph* **27**(1), 23–51.

Staton MA, Dixon JR (1975) Studies on the dry season biology of *Caiman crocodilus crocodilus* from the Venezuelan llanos. *Memorias de la Sociedad de Ciencias Naturales "La Salle"* **101**, 237–265.

Stirrat SC, Lawson D, Freeland WJ, Morton R (2001) Monitoring *Crocodylus porosus* populations in the Northern Territory of Australia: a retrospective power analysis. *Wildlife Research* **28**, 547–554.

Thomson DF (2005) *Donald Thomson in Arnhem Land*. Compiled by Nicholas Peterson. Miegunyah Press, Melbourne.

Thorbjarnarson JB (1996) Reproductive characteristics of the order Crocodylia. *Herpetologica* **52**, 8–24.

Thorbjarnarson J, Wang X (2010) *The Chinese Alligator; Ecology, Behavior, Conservation, and Culture.* The Johns Hopkins University press, Baltimore, Maryland.

Tucker AD (1997a) Ecology and demography of freshwater crocodiles (*Crocodylus johnstoni*) in the Lynd River of north Queensland. PhD thesis, University of Queensland, Brisbane.

Tucker AD (1997b) Skeletochronology of post-occipital osteoderms for age validation of Australian freshwater crocodiles (*Crocodylus johnstoni*). *Marine and Freshwater Research* **48**, 343–351.

Tucker AD (2001) Sensitivity analysis of stage-structured demographic models for freshwater crocodiles. In *Crocodilian Biology and Evolution.* (Eds GC Grigg, F Seebacher and CE Franklin). pp. 349–363. Surrey Beatty & Sons, Sydney.

Tucker A, Limpus C (1997) Assessment of reproductive status in Australian freshwater crocodiles (*Crocodylus johnstoni*) by ultrasound imaging. *Copeia* **1997**(4), 851–857.

Tucker AD, Limpus CJ, McCallum HI, McDonald KR (1996) Ontogenetic dietary partitioning by *Crocodylus johnstoni* during the dry season. *Copeia* **1996**, 978–988.

Tucker AD, McCallum HI, Limpus CJ (1997) Habitat use by *Crocodylus johnstoni* in the Lynd River, Queensland. *Journal of Herpetology* **31**, 114–121.

Tucker AD, McCallum HI, Limpus CJ, McDonald KR (1998) Sex biased dispersal in a long-lived polygynous reptile (*Crocodylus johnstoni*). *Behavioral Ecology and Sociobiology* **44**, 85–90.

Tucker AD, Limpus CJ, McDonald KR, McCallum HI (2006) Growth dynamics of freshwater crocodiles (*Crocodylus johnstoni*) in the Lynd River, Queensland. *Australian Journal of Zoology* **54**, 409–415.

Ujvari B, Mun HC, Conigrave AD, Bray A, Osterkamp J, Halling P, Madsen T (2013) Isolation breeds naivety: island living robs Australian varanid lizards of toad-toxin immunity via four-base-pair mutation. *Evolution* **67**(1), 289–294.

Valentine JM, Walther JR, McCartney KM, Ivy LM (1972) Alligator diets on the Sabine National Wildlife Refuge, Louisiana. *Journal of Wildlife Management* **36**, 809–815.

Villela PMS, Coutinho LL, Piña CI, Verdade LM (2008) Macrogeographic genetic variation in broad-snouted caiman (*Caiman latirostris*). *Journal of Experimental Zoology A Ecological Genetics and Physiology* **309**(10), 628–636.

Weaver JP, Rodriguez D, Venegas-Anaya M, Cedeño-Vázquez JR, Forstner MRJ, Densmore LD III (2008) Genetic characterization of captive Cuban crocodiles (*Crocodylus rhombifer*) and evidence of hybridization with the American crocodile (*Crocodylus acutus*). *Journal of Experimental Zoology A Ecological Genetics and Physiology* **309**, 649–660.

Webb GJW, Cooper-Preston H (1989) Effects of incubation temperature on crocodiles and the evolution of reptilian oviparity. *American Zoologist* **29**, 953–971.

Webb GJW, Manolis SC (1989) *Crocodiles of Australia*. Reed Books, Sydney.

Webb GJW, Manolis SC (1992) Monitoring saltwater crocodiles (*Crocodylus porosus*) in the Northern Territory of Australia. In *Wildlife 2001: Populations*. (Eds DR McCullough and RH Barrett) pp. 404–418. Elsevier Applied Science, New York.

Webb GJW, Messel H (1979) Wariness in *Crocodylus porosus*. *Australian Wildlife Research* **6**, 227–234.

Webb GJW, Messel H (1978) Morphometric analysis of *Crocodylus porosus* from the north coast of Arnhem Land, Northern Australia. *Australian Journal of Zoology* **26**, 1–27.

Webb GJW, Smith AMA (1984) Sex ratio and survivorship in the Australian freshwater crocodile, *Crocodylus johnstoni*. *Symposia of the Zoological Society of London* **52**, 319–355.

Webb GJW, Smith AMA (1987) Life history parameters, population dynamics and the management of crocodilians. In *Wildlife Management: Crocodiles and Alligators*.(Eds GJW Webb, SC Manolis and PJ Whitehead) pp. 199–210. Surrey Beatty & Sons, Sydney.

Webb GJW, Manolis SC, Sack GC (1983a) *Crocodylus johnstoni* and *C. porosus* coexisting in a tidal river. *Australian Wildlife Research* **10**, 639–650.

Webb GJW, Buckworth R, Manolis SC (1983b) *Crocodylus johnstoni* in the McKinlay River. N.T. VI. Nesting Biology. *Australian Wildlife Research* **10**, 607–637.

Webb GJW, Buckworth R, Manolis SC (1983c) *Crocodylus johnstoni* in the McKinlay River area, N.T. III. Growth, movement and the population age structure. *Australian Wildlife Research* **10**, 383–401.

Webb GJW, Manolis SC, Whitehead PJ, Letts GA (1984) 'A proposal for the transfer of the Australian population of *Crocodylus porosus* Schneider (1801), from Appendix I to Appendix II of C.I.T.E.S'. Conservation Commission of the Northern Territory Technical Report. Northern Territory Government, Darwin.

Webb GJW, Whitehead PJ, Manolis SC (1987) Crocodile management in the Northern Territory of Australia. In *Wildlife Management: Crocodile and Alligators*. (Eds GJW Webb, SC Manolis and PJ Whitehead) pp. 107–124. Surrey Beatty & Sons, Sydney.

Webb GJW, Bayliss PG, Manolis SC (1989) Population research on crocodiles in the Northern Territory, 1984–86. In *Crocodiles: Proceedings of the 8th Working Meeting of the IUCN–SSC Crocodile Specialist Group*. 13–18 October 1986, Quito, Ecuador. IUCN, Gland, Switzerland.

Webb GJW, Britton ARC, Manolis SC, Ottley B, Stirrat S (2000) The recovery of *Crocodylus porosus* in the Northern Territory of Australia: 1971–1998. In *Crocodiles. Proceedings of the 15th Working Meeting of the IUCN–SSC Crocodile Specialist Group*. 17–20 January 2000.

Varadero, Cuba. pp. 195–234. IUCN, Gland, Switzerland.

Wood JM, Woodward AR, Humphrey SR, Hines TC (1985) Night counts as an index of American alligator population trends. *Wildlife Society Bulletin* **13**, 262–273.

Woodward A, Moore CT, Delany MF (1992) 'Experimental alligator harvest'. Final report to the Florida Game and Freshwater Fish Commission. Florida Game and Freshwater Fish Commission, Gainesville, Florida.

CONSERVATION, COMMERCIALISATION AND CONFLICT

I was telephoned from Darwin in August 2012 by a friend to tell me about a scary event he had experienced a few days earlier over on Western Australia's Kimberley coast. He and his wife were a quarter of the way along their 51 day 4100 km voyage in a 20-foot aluminium open boat following the spectacular and remote coastline from Darwin to Derby. They were camped 50–60 m inland from the high water mark on a sandy beach in Parry Harbour, south of Cape Bougainville. At 2 a.m. he clambered out of the tent with a small flashlight (as one does) and found himself staring into the eyes of a 'huge' croc just 2 m (6 feet) away. He yelled and the crocodile did an immediate U-turn and took off at full speed towards the water. Tension in his bladder had probably saved his life. In the torchlight, there was another big croc at the low tide waters edge and a straight line of footprints led about 90 m from the water, directly into the south-east breeze and directly toward their tent. He and his wife were very unnerved and remained vigilant, sitting by the fire for the rest of the night. In the morning, they examined the tracks more closely (Fig. 14.1) and, with the tide higher, saw the two crocs patrolling around their boat. They estimated the size at 14 feet (4.3 m). They got away eventually in the afternoon, continued their journey and, once back in Darwin, phoned me in Brisbane. Even in the re-telling, some weeks later, my friend sounded anxious. Were they, he wanted to know, being stalked? No doubt they'd had a very lucky escape:

the animal had probably picked up their scent and followed it to the tent. A person had been dragged out of a tent in North Queensland in 2004. But I was not surprised that the croc had bolted: they are generally wary of humans. In hindsight, of course, my friends feel they should have set camp further from the water. But what distance is safe?

This 'crocodile–human interaction' had a happy ending. But many do not and, worldwide, a few hundred people die every year from 'croc attack' (Sideleau and Britton 2014). Of the 27 or so species of crocodylian, fewer than half have caused fatalities and only two are regarded as routine 'man eaters', yet this reputation colours most people's perception of them all. But crocodylians are also wonderful, fascinating, soul stirring and, in some eyes, beautiful, so they prompt, demand even, a desire to be conserved and protected. They are also valuable as a drawcard for tourists, for their skins and for their meat. Their danger and their commercial value have each contributed to their decline in today's world, augmented by ever-increasing loss of habitat with the expansion of agriculture and the human population explosion.

So, crocodylians are simultaneously a pest, a resource, and worthy of our admiration and care. This chapter discusses how we humans try to deal with their threat, lament and resist their diminution yet exploit them commercially and, often, try to use their commercial value as a force for their conservation.

The Orinoco crocodile, Crocodylus intermedius, *is the rarest of the New World crocodylians and is listed on the IUCN Red List as Critically Endangered. (Photo DSK taken at Zoo Miami)*

Fig. 14.1. *The left track suggests the slow and deliberate emergence of the croc from the water. The right, widely spaced track was made by the crocodile, hastily scampering back to the water at high speed after being disturbed outside the tent in the middle of the night. (Photo Joan Fitzsimmons)*

INTRODUCTION

This chapter is about the impacts that crocodylians and humans have on each other and how we deal with them. Some of the effects are very direct: they kill us and we kill them. Sometimes we kill them directly, through hunting or revenge. Sometimes we kill them indirectly, through habitat alienation or accidentally, such as in fishing nets. Crocodylians apparently perceive us either as something to be wary of or as a territorial invader and/or prey. Historically, humans have perceived crocodylians as a source of food and leather and, the large ones, as a threat. And it was that way for many millennia, except that some human cultures perceived crocodylians also in a spiritual sense and revered them.

Then, from the mid 1800s, and increasingly since, crocodylian hides have come to be perceived as a marketable commodity, particularly serving the fashion industry. Consequently, there was a long period of unregulated hunting and many

crocodylian populations suffered very severe reductions, some to a point of scarcity of supply and some to near extinction. In the nick of time, it seems, the second half of the 20th century saw the growth of a 'conservation ethic' in which all plant and animal species came to be seen as worth conserving, even if they eat you. This changed perception came alongside a better understanding of ecological processes and crocodylians are today recognised as being intrinsically, as well as instrumentally, valuable. Although many species are still at significant risk, conservation of crocodylians is now a very significant factor driving current attitudes and practices.

Because of this, government wildlife management agencies are now in the interesting position of being required to have three potentially contradictory responsibilities: to conserve them and promote respect for them; to regulate their commercial use; and, also, to keep us safe from them. How, and how well, these separate tasks are being managed are the particular subjects of this chapter.

When I became interested in crocodylians in the early 1970s, the prospect of survival for most species was considered dim. With hindsight, the decade of the 1970s was the turning point. Until then, most populations were hunted without constraint, but by the end of that decade many range states were taking active steps towards conserving their crocodylians and, as I write this, the situation is much better than it was, but still with some major worries.

Two of the three major topics in this chapter are the conservation and commercialisation of crocs and, perhaps surprisingly, an improved situation for some species has gone hand in hand with regulated commercial use, rather than with full protection. This is in tune with the concept of achieving conservation of wildlife species by making them commercially valuable: that is, by harvesting them at sustainable rates ('sustainable use of wildlife', SUW). This is a concept that developed in the 1970s and grew strength in the 1980s. The idea is sometimes encapsulated in the somewhat inadequate phrase 'use it or lose it', but it is firmly underpinned by sound ecological theory. Harvesting theory rests upon populations being regulated by some combination of density-dependent reproduction and mortality. Thus, the harvest is compensated for to some extent by lowered natural mortality rates and increased fecundity. The concept has been well described and tested for large mammals (Fowler 1987) and density-dependent regulation is known to apply in *C. porosus* (Chapter 13) and, very likely, crocodylians in general. A sustainable harvest of crocodylians can, therefore, be expected to be supported at some level (at least), by the intrinsic rate of increase. The harvest yield will be largest from a population at carrying capacity, but there are good examples of recovering crocodylian populations being harvested without the recovery being compromised (Chapter 13).

Following conservation and commercialisation, the third topic of discussion will be conflict between crocodylians and humans: that is, human safety. This has always been an issue, but it is intensifying in some places because conservation successes have led to higher numbers of some of the large and potentially most dangerous species, including some in urban areas. Thus, in northern Australia, we progressed from the 'uncontrolled hunting' phase to the 'crocodiles are sacred' phase but, as croc populations recover and as human populations within crocodylian habitats grow, the likelihood of an attack can only strengthen. Increasingly, conservation ethics are colliding with human safety considerations and it poses significant political and wildlife management challenges. Danger from the larger crocodylians is a stark reality in many countries. How can humans be protected from them? This book is in praise of crocodylians: they are magnificent animals, worthy of much admiration and I hope I have stimulated that recognition in preceding chapters. But I certainly prefer not to be eaten by one!

HISTORICAL PERSPECTIVE

PRE-1970: THE YEARS OF UNCONTROLLED HARVEST

As humans arose in Africa, we did so alongside several species of crocodile, including some now extinct. Marks on bones record crocodile predation on early humans (Njau and Blumenschine 2012). Those early humans surely used them as a source of food, recognised the danger posed by the large ones and, not surprisingly, crocodylians appeared in myths and rituals. Just as they are iconic for us, so they were for some human cultures. As humans expanded their range and invaded areas inhabited by different crocodylians, they either brought with them or re-discovered knowledge about how to live with them and make use of them, and how to accept the inevitable fatalities. Early humans were hunter gatherers, had small populations and probably made less impact on crocodiles than crocodiles made on them. This was probably the situation for many millennia.

The agricultural and industrial revolutions led to human populations increasing gradually and then more rapidly, and these developing cultures began to explore and expand. From among the European countries, Portugal, Spain, France and England all colonised 'new' lands in Africa and the New World tropics, many of them having crocs. Entrepreneurs soon saw opportunities for attractive leather and, after some false starts, crocodylian leather became fashionable before the end of the 1800s. Thorbjarnarson (1999) has provided a useful and thoughtful overview of the commercialisation of crocodylians and its prospects and limitations.

Significant commercial use of the skins for leather began in the 1800s, with the first major use being harvests of American alligators for durable boots and saddles in the American Civil War. Later in the same century, because of its attractive patterning, crocodylian leather became a high fashion item – used for boots, belts, wallets, handbags, and so on – and, as demand grew in Europe, uncontrolled harvesting led to severe reduction of most wild crocodylian populations, including American alligators, and other sources were sought. By the 1930s, harvesting was expanding into northern South America and diversifying to different species. The extent of the reduction of the Australian *C. porosus* populations was described in Chapter 13. Although most of this occurred in the years following the Second World War, there was significant hunting in the 1930s as well (Fig. 14.2). After the War, hunting technologies (spotlights, outboard motors and 4WD vehicles) and ease of access improved. Significant harvesting occurred wherever crocodylians were accessible in sufficient numbers to be economic and hide hunting expanded throughout South-East Asia, Africa, South America and Australia. Air cargo facilities also became available to ship raw skins away from tropical climates, thus minimising spoilage.

Hugh Cott (1961) wrote about the extent of the depredations in Africa by the 1950s and Smith (1980) reported that 7.5 million caiman skins were taken from Amazonas State in Brazil between 1950 and 1965. The discussion about crocodylian population ecology in Chapter 13 showed how vulnerable crocodylian populations are to a harvest of adults, and spotlighting so effectively broke through a crocodylian's main defence – its crypsis – that all sizes were able to be targeted very effectively.

Typically, skins were salted and exported in their raw state. Different species presented different decorative opportunities, but also different challenges, for the tanneries, so prices paid by the tanneries,

Fig. 14.2. *Crocodile hunting in the 1930s in the Daly River, south-west of Darwin, Northern Territory, Australia. (Photo TG Crooks)*

mostly in France and Germany, varied by species. Species such as the two *Paleosuchus* (Figs 3.14, 9.1) *Caiman crocodilus, Caiman yacare* and *Caiman latirostris* (Fig. 3.13) in South America had such bony belly scale osteoderms (Chapter 3) that they were initially of little value for leather. However, *Caiman yacare* populations suffered considerable depletion because the animal had flank regions that were somewhat less bony than that of the common caiman. The tanners found that its skin could be processed to an aesthetically pleasing leather by dissolving and thinning the belly osteoderms, with the short wide body and larger size providing more useable leather per cost of processing. The black caiman, *Melanosuchus niger*, was nearly exterminated from the wild. A large inhabitant of the Amazon basin, often reaching 4–5 m in length, it too had belly osteoderms, but they are easily removed in processing and the resulting skin could be finished to substitute satisfactorily, to an unwary buyer, for more costly and better quality alligator or crocodile leather. The effect on it has been that this has become one of the most critically endangered species (Table 14.1). *C. johnstoni,* which also has a bony hide (Fig. 3.15), was ignored early on until tanneries worked out a way to process the skins well. With *C. porosus* readily available in the same areas and more valuable, due to its greatly reduced dermal ossification (Fig. 3.17), there is much less interest in harvesting or farming freshies (Table 14.2): the salties are saving their skins! The most valuable skins are those of *C. porosus* which, with *A. mississippiensis* and *C. niloticus,* comprise the most sought after, so called 'classic' crocodylian skins (Table. 14.2).

By the mid 1960s, serious concern was being raised about crocodylian futures, sometimes by the hunters themselves. Many heavily harvested populations had declined to the point where hunting was no longer profitable and, often contemporaneously, nation states implemented legislation to protect them, so the 'classic' skins were suddenly in seriously short supply. As these species became depleted, some tanneries closed down, but demand continued, particularly for the classic skins, making them more expensive. This led to slightly less attractive and less valuable species being substituted in trade and these too became heavily exploited (*C. moreletii, C. siamensis, M. niger* and *C. intermedius*). Because it was, and is, so numerous, *Caiman crocodilus* came to comprise the largest proportion of the market, by volume (Table 14.2), even though caiman skins are worth much less than the classic skins.

Caiman crocodilus remains numerous in the wild in many regions, although suffering locally from habitat loss and human incursion and consumption. Highly prolific and resistant to exploitation and adverse environmental conditions, it breeds well in captivity and provides the industry with the major source of cheap skins, primarily from Colombia. Caiman skin comprises more than a million skins in trade annually: more than the skins of all other species of crocodylians combined, so this species is the mainstay of the exotic leather industry.

HALTING THE DECLINES

Right from the earliest days of trying to implement crocodylian conservation, there was interest in looking for ways to have both crocodylians and the industries supported by them. Most of the declines had come from harvesting, so regulatory efforts targeted that. A typical pattern was for governments to bring in local protection, carry out population assessments and, with varying degrees of success, develop some sort of a regulatory system that identified allowable harvests estimated to be sustainable. Some species not targeted for trade were also in serious decline because of habitat alienation and increased use for subsistence as human populations increased: these will be discussed separately.

The Crocodile Specialist Group (CSG; see Box 14.1) was formed in 1971 and engaged with industry players from the outset. It grew in numbers and diversity of membership, but all wanted the same thing: more crocodylians, whether the primary driver for that was for species conservation or to sustain the crocodile leather industry. Individual species management programmes were developed, which were designed to sustain both the industry and the crocs, and there was encouragement for crocodylian research across a wide front. Support for crocodylian industries was not without controversy because some members were of a more protectionist frame of mind but, in the long run, the more conservative players were won over to a more pragmatic view and sustainable commercial harvesting has been supported strongly by the CSG since the late 1980s. With most of the tanneries and high fashion industry being in Europe, but the skins coming from elsewhere, the trade in hides was mostly international and this created a particular opportunity for regulation. The formation of the Convention on International Trade in Endangered Species of Wild Fauna and Flora (CITES) in 1975 and its (then) component TRAFFIC provided a significant step forward. CITES established a protocol for regulating international trade in wildlife and TRAFFIC monitored its magnitude,

BOX 14.1   FOUR RELEVANT NGOS: IUCN, CSG, CITES AND TRAFFIC

Four international non-government organisations are of particular relevance to the conservation and management of crocodylians: the IUCN (International Union for the Conservation of Nature); CITES (the Convention on International Trade in Endangered Species of Wild Fauna and Flora); the CSG (Crocodile Specialist Group) and TRAFFIC, which, among other things, monitors international trade in wildlife. IUCN was formed in 1948 and has grown into a global conservation network with 1200 member organisations (both government and non-government) operating in more than 160 countries (www.iucn.org). Conserving biodiversity is central to its mission. One of its six commissions is the Species Survival Commission (SSC), which advises IUCN on 'technical aspects of species conservation and mobilises action for those species that are threatened with extinction'. The SSC is the umbrella organisation for ~120 specialist groups made up of people with expertise in particular taxa, the one most relevant to crocodylians being the Crocodile Specialist group (CSG). TRAFFIC (www.traffic.org) is an international non-governmental network that began in 1976 as a specialist group within the SSC and is now a partner organisation to both IUCN and the World Wildlife Fund (WWF). According to its website, 'TRAFFIC, the wildlife trade monitoring network, works to ensure that trade in wild plants and animals is not a threat to the conservation of nature'. It works closely with IUCN and its specialist groups and with CITES. CITES and the CSG are the two organisations most immediately relevant to crocodylian use and conservation.

The IUCN created and maintains the Red List of Threatened Species (http://www.iucnredlist.org/), which categorises species according to threat as assessed by particular criteria (http://www.iucnredlist.org/documents/RedListGuidelines.pdf). The categories that are relevant to crocodylians are:

Critically Endangered, Endangered and **Vulnerable** (collectively '**Threatened**'): terms to some extent self explanatory, defined by reference to quantitative criteria that are designed to reflect varying degrees of threat of extinction (see Guidelines).

Least Concern: taxa that do not qualify as Threatened or Near Threatened (does not imply that these species are of no conservation concern).

Data Deficient: self explanatory.

The current Red List does not yet take into account the taxonomic changes identified in Chapter 1, but, of the 23 species it does list, 11 are Threatened (eight Endangered or Critically Endangered), 11 are of Least Concern (sometimes, and in my opinion, better known as Lower Risk) and one Data Deficient (Table 14.1). This is a healthier situation than 30 years ago, the result of a combination of better information for some species and good management of others.

Crocodile Specialist Group (CSG) (www.iucncsg.org)

The CSG was formed in 1971. There were 15 members at its first meeting in New York. Hugh Cott and Tony Pooley, who featured in Chapter 1, were both there, as well as several other luminaries whose crocodylian research has graced the pages of this book. Significantly, the crocodile industry was represented and that involvement has continued and is one of its strengths. There are now ~450 members. Meetings are held every couple of years, many in developing countries and typically they attract a couple of hundred attendees. Formation of the CSG coincided with growing concern about the effect of hide hunting on crocodylian populations and, also, a time when conservation 'movements' were emerging in developed countries. Early on, the CSG focus was on 'protection', but almost from the outset the CSG engaged with people involved in the leather and fashion industries, recognising that because they had a lot to lose if the supply of skins dried up they had a vested interest in croc populations surviving. Clearly a 'no more harvest' strategy was not going to be popular, even though some tanneries were facing shortages of product (Thorbjarnarson 1999) and the

BOX 14.1   CONTINUED

CSG gradually came to adopt a pragmatic approach to harvesting. It embraced the emerging concept of achieving conservation goals through the regulation of harvesting to be within sustainable limits. Of course the CSG had no regulatory authority and had to rely on the force of good arguments, based on survey data and common sense, to persuade government agencies and practitioners to the idea that there was benefit in harvesting at sustainable rates. More about this below. The CSG has also initiated many crocodylian surveys – usually in out-of-the-way places – and many conservation projects. It publishes quarterly newsletters and the proceedings of its (approximately) bi-annual meetings, conservation Action Plans and many special reports, all of which are freely available on the CSG website. Anyone with a new interest in crocodylians should spend at least a couple of hours familiarising themselves with what it holds. Anyone with a serious interest will be visiting it often.

Most of the leather processors and fashion goods manufacturers were, and are, in Europe, Japan and Singapore, but none of the crocodylians are, so the trade in crocodylian hides was inevitably international, and that is where CITES comes in.

CITES (http://www.cites.org/)

The Convention on International Trade in Endangered Species of Wild Fauna and Flora (CITES) is based in Washington DC and is 'an agreement between signatory governments to ensure that international trade in specimens of wild animals and plants does not threaten their survival'. At the time of writing, 179 nations are signatories.

Particularly, CITES has categorised species into three lists, the Appendices, according to the survival risk they face and the extent to which their status may be threatened by trade. Those that are threatened with extinction and are, or may be, threatened by trade are listed in Appendix I. Trade in wild captured Appendix I species is illegal, except in special circumstances. There are ~1200 species on Appendix I, including 16 of the 23 listed crocodylian species. The list will be revised as the new species are evaluated. Of the Appendix I species, seven have populations in certain countries that have been judged sufficiently secure to have those populations listed on Appendix II. This illustrates an important point: there can be 'split-listing'. For example, *C. porosus* is split listed: the Australian and PNG populations are listed on Appendix II whereas populations elsewhere are listed in Appendix I. Species listed on Appendix II have been judged to be not necessarily threatened with extinction, but which could decline sufficiently to meet the IUCN criteria for Threatened unless trade is strictly regulated. Trade in Appendix II species is legal only under strict regulations designed not to compromise their continued survival in the wild. Decisions about which list is appropriate for a particular species are made at triennial conferences of the signatories to the convention and any listing or case for listing or a change to the listing requires a substantial case to be made, backed up by a scientific data. The Australian population of *C. porosus*, for example, was downlisted to Appendix II in 1985 as a result of extensive research and monitoring that had been conducted, and knowledge about its extensive recovery (Webb *et al.* 1984) (Chapter 13). Table 14.1 shows the CITES listing of all crocodylian species: unregulated trade is not permitted in any of them and most are on Appendix I for which no trade of animals taken in the wild is allowed.

It should be noted that Carpenter (2011) has drawn attention to a range of CITES' limitations, citing particularly the implied transfer of ownership of wildlife away from local communities and the weak enforcement of wildlife regulations in many countries. His paper criticises CITES for a reluctance to embrace community-based conservation and concludes by encouraging the parties to urge the adoption of community-based conservation in addition to its customary approaches. The paper does not refer to any crocodylian examples, but some of the criticisms are relevant and as we shall see community-based conservation will be crucial for some species, particularly those in which there is no commercial trade.

TABLE 14.1 CONSERVATION STATUS OF EXTANT CROCODYLIAN SPECIES IN 1982 AND AS SHOWN CURRENTLY IN THE IUCN RED LIST

See Box 14.1 for an explanation of the Red List threat categories. Several species need re-assessment (marked with an asterisk). Note: In 1982 the Red List had no 'Critically Endangered' category. All six species now listed as Critically Endangered would almost certainly have been listed in that category in 1982, had it existed.

	Situation in 1982 (Groombridge 1982)		Current situation IUCN Red List (and notes) (see also CSG Action Plans for all species at http://www.iucncsg.org/pages/Publications.html for the reviews of conservation status and actions)	
Family Gavialidae (gharial and false gharial)				
Gavialis gangeticus Gharial	Endangered (there was no Critically Endangered category in 1982)	Appendix I	Critically Endangered. May be fewer than 200 mature adults remaining.	Appendix I
Tomistoma schlegelii False or Malayan gharial	Endangered	Appendix I	*Endangered. Maybe fewer than 2500 mature adults in wild.	Appendix I
Family Alligatoridae (alligators and caimans)				
USA and China				
Alligator mississippiensis American alligator	Not Threatened	Appendix II	*Lower Risk/Least Concern	Appendix II
Alligator sinensis Chinese alligator	Endangered (there was no Critically Endangered category in 1982	Appendix I	*Critically Endangered. < 200 adults in wild, large numbers in farms	I (Captive bred treated as Appendix 2)
Tropical America				
Caiman crocodilus (several sub-species) Spectacled caiman	Vulnerable	Appendix II	*Lower Risk/Least Concern	II (sub-species *apaporiensis* Appendix I)
Caiman yacare Yacare or Paraguayan caiman	Indeterminate	Appendix II	*Lower Risk/Least Concern	Appendix II
Caiman latirostris Broad-snouted caiman	Endangered	Appendix I	*Lower Risk/Least Concern	I (Argentina population Appendix II)
Melanosuchus niger Black caiman	Endangered	Appendix I	Lower Risk, conservation dependent	I (Except population in Brazil is Appendix II)
Paleosuchus palpebrosus Cuvier's dwarf caiman	Not Threatened	Appendix II	*Lower Risk/Least Concern	Appendix II
Paleosuchus trigonatus Schneider's dwarf, or smooth-fronted, caiman	Not Threatened	Appendix II	*Lower Risk Least Concern	Appendix II
Family Crocodylidae ('true' crocodiles)				
Tropical America				
Crocodylus acutus American crocodile	Endangered	Appendix I	Vulnerable. Population increasing in Florida	I (Except Cuban population in Appendix II)
Crocodylus intermedius Orinoco crocodile	Endangered (there was no Critically Endangered category in 1982	Appendix I	*Critically Endangered	Appendix I

TABLE 14.1 CONTINUED				
Crocodylus moreletii Morelet's crocodile	Endangered	Appendix I	Lower Risk/Least Concern. Population trend: stable	I(Belize and Mexico populations Appendix II with zero quota)
Crocodylus rhombifer Cuban crocodile	Endangered (there was no Critically Endangered category in 1982	Appendix I	*Critically Endangered	Appendix I
South-East Asia				
Crocodylus palustris Mugger or marsh crocodile	Vulnerable	Appendix I	*Vulnerable.	Appendix I
Crocodylus siamensis Siamese crocodile	Endangered (there was no Critically Endangered category in 1982	Appendix I	*Critically Endangered. Population decreasing	Appendix I
Crocodylus mindorensis Philippine freshwater crocodile	Endangered (there was no Critically Endangered category in 1982	Appendix I	*Critically Endangered	Appendix I
Papua-New Guinea and Australia				
Crocodylus novaeguineae New Guinea freshwater crocodile	Vulnerable	Appendix II	*Lower Risk/Least Concern	Appendix II
Crocodylus johnstoni Freshwater or Johnston's crocodile	Vulnerable	Appendix II	*Lower Risk/Least Concern	Appendix II
Crocodylus porosus Indo-Pacific, estuarine or saltwater crocodile	Endangered	Appendix I and II	*Lower Risk/Least Concern	I (Australian, Indonesian, PNG populations Appendix II)
Africa				
Crocodylus niloticus (including 'C. suchus') Nile crocodile		Appendix I and II	*Lower Risk/Least Concern (but assessment needed of each species)	I (+ various populations are Appendix II, see note below)
Mecistops spp. (proposed to comprise two discrete taxa, Chapter 1). African slender-snouted crocodile	Indeterminate	Appendix I	*Data Deficient (assessment needed of each species)	Appendix I
Osteolaemus (likely three species; Chapter 1) Dwarf crocodile	Indeterminate	Appendix I	*Vulnerable (assessment needed of each species)	Appendix I

*Needs updating

Note: Crocodylus niloticus is on Appendix I, except the populations of Botswana, Egypt (subject to a zero quota for wild specimens traded for commercial purposes), Ethiopia, Kenya, Madagascar, Malawi, Mozambique, Namibia, South Africa, Uganda, the United Republic of Tanzania (subject to an annual export quota of no more than 1600 wild specimens including hunting trophies, in addition to ranched specimens), Zambia and Zimbabwe, which are included in Appendix II.

among other things (see Box 14.1). The CSG, as a specialist group within the International Union for the Conservation of Nature (IUCN) could work in with CITES and range states, providing a respected scientific advisory service and liaison between the international agencies and relevant governments. According to their conservation status, crocodylians were listed on either Appendix I or Appendix II of

TABLE 14.2 WORLD TRADE IN CROCODYLIAN SKINS, 2001–2010, AS COLLATED BY THE UNITED NATIONS ENVIRONMENT PROGRAMME'S WORLD CONSERVATION MONITORING CENTRE (UNEP-WCMC), CAMBRIDGE, UK, IN COLLABORATION WITH THE LOUISIANA ALLIGATOR ADVISORY COUNCIL

Their source of data is the CITES Trade Database, from annual reports submitted by the member Parties to CITES. The figures are underestimates because of under-reporting, but they do emphasise the size of the trade. It is noteworthy that six of the 11 species for which hide traffic data are presented are listed on Appendix I. Four of these, however, have split listings, with certain populations downlisted to Appendix II. (From Caldwell 2012 with permission).

Species	2001	2002	2003	2004	2005	2006	2007	2008	2009	2010
A. mississippiensis	343 116	237 840	341 734	368 409	356 393	421 220	262 133	230 464	297 187	369 731
C. acutus	100	630	830	227	204	120	404	1371	1460	200
C. johnstoni	0	2	0	0	65	0	0	0	0	0
C. moreletii	2430	1591	997	549	855	158	11	724	485	0
C. niloticus	150 757	159 970	148 553	140 497	151 491	166 307	161 185	168 764	149 082	161 840
C. novaeguineae	30 634	30 749	27 308	39 796	32 002	38 645	28 663	27 543	26 095	25 055
C. porosus	28 223	24 278	26 564	30 728	37 441	34 152	45 215	53 918	47 575	53 470
C. siamensis	4422	3580	10 982	20 930	31 517	47 972	54 331	63 471	34 373	33 094
Subtotal of 'classic' skins	**559 682**	**458 640**	**556 968**	**601 136**	**578 451**	**708 574**	**551 942**	**546 255**	**556 257**	**643 390**
Caiman crocodilus crocodilus	*25 510	22 709	34 636	70 722	65 078	69 574	44 894	36 989	43 638	24 643
Caiman crocodilus fuscus	710 113	552 077	572 059	621 691	603 223	972 941	670 828	533 549	408 754	651 121
Caiman latirostris	88	90	165	215	2752	1669	1125	809	304	1933
Caiman yacare	32 128	78 811	60 288	41 882	53 241	52 998	65 452	61 297	47 208	24 546
Caiman total	**767 839**	**653 687**	**667 148**	**734 510**	**724 924**	**1 097 182**	**782 299**	**632 644**	**499 994**	**702 243**
Grand total	**1 327 521**	**1 112 327**	**1 224 116**	**1 335 646**	**1 303 375**	**1 805 756**	**1 334 241**	**1 178 899**	**1 056 251**	**1 345 633**

CITES. No international trade was (and is) allowed in the most 'at risk', Appendix I species, whereas trade is allowed in Appendix II species, subject to certain conditions being met. Typically, the conditions require a satisfactory management plan that covers sustainability and includes ongoing population monitoring. Examples of such plans are Leach *et al.* (2009) and Delaney *et al.* (2010) for *C. porosus* and *C. johnstoni* in the Northern Territory, respectively. An important aspect of the system is that species can be 'split-listed' so, for example, the Australian population of *C. porosus*, originally on Appendix I, was downlisted to Appendix II, whereas in other jurisdictions, where it is more vulnerable, it remains on Appendix I.

Listing on Appendix I does not necessarily mean that there can be no commercial exploitation of that species, even of populations not identified specifically for Appendix II status. Under CITES guidelines, trade is permitted in second generation (F2), captive-bred Appendix I species. Removal of animals from the wild to start a captive breeding enterprise was (and is) generally not permitted, unless that could be done without detriment to the wild population. However, there were already many crocodylians in captivity by the 1970s and these could be made the nucleus of a commercial operation. The Edward River Crocodile Farm at Pormpuraaw on the west coast of Cape York Peninsula is one example. It had started before CITES and it had to re-gear to sell only second generation produce after *C. porosus* was listed on Appendix I in 1979. With crocodylians taking about a decade to mature, producing second generation skins for sale is a long process and identification of individual stock and proof of their F2 status provides additional burden, so this was not entirely satisfactory. For species on Appendix II, however, there are many more options. Appendix II species can be taken directly from the wild – that is, a wild harvest – subject to the harvest being regulated and usually accompanied by a monitoring programme to ensure that the harvesting is not putting the population in jeopardy. A 'downlisting' from Appendix I to II requires a good case to be put before a triennial meeting of representatives of CITES signatory nations.

As the various crocodylian-based industries adjusted to the new, regulated regime in the 1970s, there was an early emphasis on industry activities tailor-made to particular situations, often involving 'ranching', in which eggs or hatchlings could be taken from the wild, grown up in captivity and sold. Signatories to CITES are obliged to disallow entry of hides unless they can be certified as having been taken legally. This includes entry for trans-shipment to third countries. Presumably because they could see it was in their long-term interest, the big tanneries agreed to accept only legally available skins and, quite often, provided direct support for crocodile conservation and management programmes in developing countries. For Nile crocodiles, despite their Appendix I status, ranching programmes were approved in many range countries, sometimes with a conditional requirement to return a proportion of animals to the wild after they had been raised in captivity from wild-collected eggs. Thus, the higher success rate from artificially incubated eggs would offset any loss of recruitment as a result of removing eggs from wild nests. In due course, as populations in the wild recovered or were better documented, Nile crocodiles in some African countries were downlisted to Appendix II (see Table 14.1) and regulated trade became permitted.

CURRENT STATUS

The current status of the world's crocodylians is shown in Table 14.1, as well as their status in 1982, as then understood. The situation has not changed a great deal, but most of the changes have been in a positive direction. All species, without reference to the situation in particular range states, have the same Appendix listing as they did 30 years ago, but populations in some countries have been downlisted from Appendix I to Appendix II. All of these changes have been the result of positive conservation action by the relevant governments. Happily, no species has been 'uplisted' from 2 to 1 and that can be said to be a consequence of positive actions. Less happily, seven species are listed as Endangered or Critically Endangered on the IUCN Red List and the list has not yet accommodated or assessed the new African species. The splitting of *Mecistops* into two species and *Osteolaemus* into three (probably), could well lead to the addition of more Critically Endangered species to be concerned about. Likewise, the status of *C. suchus* is of concern. Hekkala *et al.* (2010, 2011) noted that until its separate status is recognised it will continue to be treated under a regulatory regime designed for *C. niloticus*, which is not appropriate. The same general point applies to the *Osteolaemus* and *Mecistops* species (Shirley *et al.* 2014): they too need to be considered independently. One of the problems is the lag between discovery of new species and their subsequent legal recognition and incorporation into amended regulations.

Although Table 14.1 suggests that not a lot has changed since 1982, the situation is actually much better than it was. The real comparison would be with what it would have looked like if the pre-1970 practices had persisted. Commercially exploited crocodylians are now managed routinely by many nations in ways that combine conservation and commercial use and many populations are now much more secure. Additionally, the status of some of the Critically Endangered species is less dire because of conservation actions. The Chinese alligator now has official recognition within China as a species of concern, so it is protected and the subject of research and a re-introduction programme. Likewise gharials and Philippine (see below), Cuban and Siamese crocodiles are the focus of active conservation efforts. Recent reviews of conservation status are available in the CSG Action Plans for each species, accessible at http://www.iucncsg.org/pages/Publications.html.

This is not to say that everything is rosy. Clearly the CITES provisions and aspirations have many good features but, as with the experience of prohibiting sales of alcohol, prohibition of trade in a species is not always successful. It may have unanticipated and unwanted side effects: for example, protection of the 'classic' species may put additional pressure on other species. And, as with managing alcohol prohibition, it depends upon continuing and adequate compliance checking. In the crocodylian skin trade, this depends upon each skin having a non-removable tag attached to it at source. This establishes the origin of the skin and the tag has to remain secured through the tanning process and beyond, until the skin is manufactured into a handbag, boots, belt, watchband or whatever. Compliance also depends upon inspection at the point of entry to the country where the tanning will be undertaken, or for trans-shipment. This is more complicated than it sounds. In species whose distribution spans several countries, split listings open the possibility for skins to be border-hopped from a country where the population has Appendix I status to a country whose population has an Appendix II listing. Because there may be slight, but consistent, differences in scalation between populations in different countries, a skilled inspector may be able to detect such a ruse, perhaps assisted by the CITES identification guide (CITES 1995). This is a lot more difficult than it sounds and, in future, identification by DNA analysis is desirable. Peter Brazaitis, now retired from the Wildlife Conservation Society, NY, is an expert in hide identification. In addition to his past positions as Superintendent of Reptiles at the Bronx Zoo and Curator of the New York Central Park Zoo in New York, he has for many years acted as a consultant to the US Fish and Wildlife Service and other wildlife law enforcement agencies as a forensic herpetologist to identify bundles of suspicious consignments of crocodylian and other reptilian skins. Some of his stories about this are collected in his very engaging autobiography *You Belong in a Zoo!* (Brazaitis 2003): a 'must read' for every herpetologist.

COMMERCIAL USE OF CROCODYLIANS

There are many ways that money can be made from crocodylians and every species can have value. If not bred and raised, or harvested, for hides and meat, they can be targeted for local sale as food (bush meat, see below). These are consumptive uses. In addition, there is wildlife viewing: a non-consumptive commercial use and a lucrative adjunct to tourism, one which has a distinct positive conservation feedback. An offshoot of wildlife tourism is trophy hunting, which is definitely consumptive, and a common element of more widely targeted safari hunting in Africa and, to a much less extent, Australia. Crocodylians provide many things to sell.

SKINS, MEAT, CURIOS, *MATERIA MEDICA*, PET TRADE

The main saleable product from crocodylians is their skin, with meat and curios as by-products. The skins makes excellent leather with distinctive markings and a high gloss and it is used to make handbags, boots and shoes, wallets, belts, hat bands and more – even jackets and full length coats. This is high fashion and prices are premium, especially from a recognised house. Handbags typically cost several thousand dollars, and a full length coat US$100 000 and upwards. The main by-product is meat, which is white and, to my perception, a bit like chicken but even more dependent upon the nature of the accompanying spices or sauce. Of course, appreciation of the meat may depend upon one's motivation: the meat is claimed by some to have aphrodisiac properties. Another by-product that has come up for discussion, and experimentation, is the potential use of alligator fat as a source of biodiesel. Teeth, skulls, pickled heads, and various other 'bits' can be sold as curios. Hatchings are subjected to taxidermy and sold, sometimes in grotesque poses. However, the Florida Fish and Wildlife Conservation Commission has a proposal under consideration to prohibit the sale of stuffed baby alligators that depict them in 'an unnatural body or body part positioning'. The aim of this is to deter the harvesting of hatchlings for that purpose. Other by-products include those reckoned to be useful medicinally, with blood, fat, oil and collagen variously used in traditional and holistic medicine to treat a diversity of ills: rheumatism, skin infections, skin repair, stroke, epilepsy, eczema, hypertension, cancer and incontinence. In some countries, crocodylians are also sold as pets.

HARVESTING, RANCHING AND FARMING

Some of the crocodylians used commercially come from wild harvest at a skinnable size, some from animals taken from the wild as eggs, hatchlings or juveniles and 'grown out' in captivity (a practice known as ranching) and some come from closed-cycle farming. It is not a small trade. Table 14.2 shows the extent of that proportion of the international trade in hides that is reported, by species, as collated by Caldwell (2012). Because of significant under-reporting, the trade is thought to be much larger than the table suggests.

Many of the skins produced nowadays come from ranching Appendix II species collected from the wild as eggs. This takes advantage of the crocodylians' population ecology, by which a large number of eggs are produced, yet each with only a small chance of surviving to sub-adulthood or beyond (Chapter 13). Farms can therefore be stocked by collecting eggs from the wild – with little or no measurable detriment to the wild population – incubating them and then rearing them to a size suitable for skinning. Skins can be salted and stored for shipping (Fig. 14.3). This is deceptively simple sounding, as though it can be accomplished with a comparatively modest investment and without significant technology, as long as the essential requirements of suitable temperatures, plenty of water and access to a food source are available (farms are often located in association with chicken farms, but additional supplementation is required for best results). The reality is somewhat different. As in all livestock production, crocodile and alligator farmers need to cope with providing adequate nutrition and disease management and, because crocs have not (yet) been bred to humbly accept domestication, manage aggressive behavioural traits as well (Brien *et al*. 2013). These factors frequently inhibit satisfactory growth rates and lead to production of blemished, less valuable skins. A lot of research has gone into determining the best form of housing and nutritional requirements in captivity (Chapter 6) and also the veterinary management of diseases to which crocodylians are subject when kept at high densities (Huchzermeyer 2003). Farmers also need to have a good appreciation of the cost of production in relation to market value, with both of these subject to uncertainties. Not surprisingly, the larger, better resourced farms are more successful and better able to ride market fluctuations. The farm run by Mainland Holdings in Lae in Papua New Guinea provides a good example (Fig. 5.9). This is the world's largest *C. porosus* farm, with more than 40 000 animals, and the same company provides ~65% of PNG's market for fresh and frozen chicken.

Fig. 14.3. *Farmed or harvested skins can be salted and stored for shipment. (Photo GCG)*

Most crocodylians are well able not only to be reared in captivity but also to breed, so ranching operations typically grow into and include captive breeding as well. In some cases, only closed-cycle farming is practised. The role, or not, of crocodile and alligator farms in conservation will be discussed below.

Wild harvest is another source of skins traded on the world market and Webb and Manolis (2003) provided a comprehensive review of many of the practical aspects of managing crocodylian harvests. They reviewed the diversity of approaches taken by various agencies to ensure that harvesting is not to a population's detriment and discussed many of the current programmes. Management agency websites are also useful and convenient sources of information. This topic is also taken up in more detail below.

Only three examples of ongoing and sustainable commercial skin production will be mentioned

here. As discussed in Chapter 13, the messages from American alligators and estuarine crocodiles are that, if crocodylians have places to feed and breed and are left to themselves, they can rebuild their populations from a low base quite successfully. What's more, they can do this with a very substantial harvest of their eggs (*C. porosus*) and even with a regulated harvest of adults (*A. mississippiensis*). The important word in that sentence is 'regulated', and 'population monitoring' should be included as well, because, without that, the effectiveness of the regulation cannot be assessed and, if necessary, modified.

Alligator mississippiensis

American alligators occur in southern USA (Fig. 13.3) and they provide a conservation success story. Most of their recovery has occurred alongside controlled commercial use, regulated by government agencies in each of the individual states and, for export, under CITES permits. Thought to be at risk as early as the 1940s, American alligators can now be regarded as recovered and they are the basis of a thriving industry with many facets, as well as accommodating recreational hunting. They remain protected at the edges of their range, but where they are numerous (Louisiana, Florida, Georgia, Texas and South Carolina), state-based management programmes have led to spectacular recovery. In Louisiana, concern about low numbers led to a complete closure to harvesting in 1962 and full protection was maintained until 1972. Pioneers in developing management programmes for American alligators were Ted Joanen, Robert Chabreck and Larry McNease in Louisiana, whose studies during the 1960s indicated that the population should be able to sustain a controlled harvest. They thought that the best way to conserve alligators would be to permit harvesting, giving it an economic and also a social value. They estimated a population of ~170 000 alligators in 1970, mostly on privately owned marshes. The first harvest was approved in 1972 and by 1979 numbers had increased sufficiently that alligators were downlisted to Appendix II. By 1993 it was estimated that there were nearly a million alligators in the wild in Louisiana. The current estimate is ~1.5 million. Under Louisiana's management programme, survey data form the basis for annual harvest quotas and the harvest is controlled by allocating tags. The Louisiana Department of Wildlife and Fisheries publishes reports (e.g. Louisiana Department of Wildlife and Fisheries 2010) that detail the management programme's operations and complexities. About 33 000 skins are harvested from the wild annually. Apart from this, Louisiana encourages ranching of eggs taken from the wild, as well as captive breeding, and there are numerous alligator farms that sell meat and hides. In the coastal lowland marshlands, which hold most of the alligators, nests are easily visible from the air so population trends can be monitored by aerial surveys. Compared with fewer than 10 000 nests in the marshlands in the early 1970s, the average number in the first 10 years of this century was 34 620. Hurricanes lower nest numbers: for example, after Hurricane Rita in September 2005, only 20 000 nests were estimated in 2006, and fewer than 25 000 in 2009 following Hurricane Ike in September 2008. In the period 1999–2008, an average of ~393 000 eggs were taken annually for ranching (range 236 000–460 000), the equivalent of nearly 10 000 nests a year. Because this is such a high proportion of the available nests, ranchers are required to return a proportion of these eggs as young alligators within 2 years: ~12% if they return small ones and fewer if large animals are returned, acknowledging an increase in survivorship with body size. This is somewhat controversial because of uncertainty about their survival once released and an extensive study is underway to assess this. Far more skins are produced in farms than are taken from the wild: for example, 287 000 in 2008 compared with a wild harvest of 35 627. Louisiana also maintains an active programme to remove 'nuisance' alligators (those 'at least four feet in length and … believed to pose a threat to people, pets or property', see below). Licenced hunters remove the problem animals and are allowed to sell the hide and meat. On average, 5000 complaints are dealt with each year in Louisiana, with half of these leading to a slaughter. These figures show the scale of commercial and non-commercial alligator usage in just one state.

Florida also has farming and hunting, plus, because of much higher human populations in alligator habitat, it has bigger issues related to the control of 'nuisance alligators', as discussed below. Population studies in Florida have shown that the American alligator population is likely to remain stable if there is a annual male-focussed harvest of ~13% of individuals greater than 4 feet long (1.2 m) and if eggs are taken for ranching from 50% of all located nests. From wild harvest and ranching, ~30 000–40 000 skins are produced annually. Webb and Manolis (2003) provided a substantial review of alligator management in Louisiana and Florida.

Commercial use in Georgia is more recent, with farming and wild harvest (one alligator longer than 4 feet long per hunter; hunters chosen by ballot). As elsewhere, the manner of take is strictly controlled and successful hunters can sell the hide. Many, however, opt to make a trophy mounted by a taxidermist. Texas also allows hunting, egg collection and farming, controlled by a permit system, with population monitoring by aerial survey of nests.

In short, ecological studies, combined with annual monitoring and a complex licensing system to ensure control over the harvest, have ensured that the recovery of alligator populations has occurred contemporaneously with commercial use. Ross (1998) considered that the only remaining threat to alligators is really habitat loss to expanding agriculture and residential development, pollution and changing water regimes. He concluded that 'alligator populations under these management programmes are certainly stable or even increasing' and that 'the sustainable use of alligators in the USA generates more than $60M annually (i.e. in 1998) providing a substantial incentive to retain habitat and tolerate alligators'.

Caiman crocodilus

About half of all crocodylian skins entering international trade come from *Caiman crocodilus* (Table 14.2). This is despite its belly skin having bony osteoderms that complicate the tanning process and for a long time caused the focus of use to be just on the strips of skin on the flanks, which are generally used for smaller fashion products such as belts, watch straps and wallets. More recently, improved tanning and processing allow products from caiman skins to be sold for almost as much as genuine crocodile. *Caiman crocodilus* ranges through 17 countries from southern Mexico to Peru and Brazil and differentiates geographically to the extent that four or five sub-species (or species) can be recognised. The species is listed on Appendix II, with the exception of one of the sub-species, *C.c. apaporiensis,* which is of conservation concern and is listed on Appendix I. *Caiman yacare*, previously regarded as a sub-species, is now generally regarded as a full species, but that debate may not be settled. Indeed, there are opinions that taxonomic categorisation of the whole genus needs review. Wild harvest, farming and ranching all contribute to the enormous annual production of *Caiman crocodilus* hides exported from South American countries, each of which has different regulations. More than half come from Colombia, which has reported an average of 600 000 skins per year over the decade to 2010, with nearly a million in 2006 (Caldwell 2012). Captive breeding is a significant source: the CITES national export quotas (http://www.cites.org/eng/resources/quotas/) indicate production of *Caiman crocodilus* skins from Colombia for 2013. (It should be noted that the quotas are not set by CITES but by the parties themselves, if satisfied that the proposed export will not be detrimental to the survival of the species.)

Venezuela was the first South American country to aim to put caiman harvesting on a controlled, legal basis. This was implemented in 1983, with a conservation benefit through SUW in mind, and a large commercial harvest developed on privately owned cattle ranches in the enormous wetland of the llanos. This has been compromised by recent political instability and by competition with captive propagation in Colombia and the production trend has been downwards in the last decade. Papers by Thorbjarnarson and Velasco (1999), Thorbjarnarson (1999), Velasco *et al.* (2003) and Webb and Manolis (2003) provide a good summary of the situation up to those dates. The aim was to create an economic incentive to landholders to maintain healthy populations of caiman and also to maintain their habitat. The harvest is restricted

to adult males, 1.8 m and longer and quotas set at 20% (2003 figure) of the estimated number of males in that size class. The reasoning behind that harvest strategy was that adult males were considered to be constraining the development of sub-adult males, so their removal was anticipated to lead to a compensatory increase in the population. There is some evidence that this has occurred: Velasco *et al.* (2003) reported higher densities in harvested than non-harvested areas and, although harvesting may have focussed on already richer areas, it was encouraging that adult males were still in good supply. The population is monitored by spotlight counts in sample areas and by tracking skin size of the harvest. There seems to be little doubt that the Venezuelan harvesting is sustainable, showing a robustness attributable in large measure to the life history attributes of *Caiman crocodilus* (Chapter 13). Whether the economic benefit from caiman harvests encouraged landholders to maintain populations or suitable caiman habitat is doubtful (Thorbjarnarson 1999): the economic return will be high in relation to cost, but small in comparison with the return from cattle. Interestingly, caiman have apparently benefitted from the additional ponds created for use by cattle.

Crocodylus porosus

The skin of this species has relatively little dermal ossification and is consequently the most desirable, so it commands a premium price. Production from Australia and Papua New Guinea, (but mostly the former), dominate supply. The Australian population of *C. porosus* was listed originally on Appendix II, along with the population in Papua New Guinea, but was transferred to Appendix I in 1979. With the extent of the recovery following protection in 1970 becoming obvious, and with so much research work on the status of the population, it was downlisted again in 1985: the result of a case spearheaded by Grahame Webb and colleagues (Webb *et al.* 1984). This opened the way for ranching operations, as well as regulated wild harvests. The decline of the *C. porosus* population and the research and management actions that led to its subsequent recovery in the Northern Territory were reviewed in Chapter 13. In Australia, export of CITES-listed species requires a management plan to be developed and approved, and the current plan (2009–2014) for the Northern Territory population is available on the web (Leach *et al*. 2009). The plan allows for commercial activity through ranching (mostly via egg collection), limited wild harvest and tourism activities and these generate significant employment opportunities and economic benefit. A proposal for federal approval of trophy hunting has so far not been successful (see below). The philosophy driving the Northern Territory programme is that of achieving conservation alongside regulated commercial harvest. Most of the harvest has been of eggs, and the recovery of the population in the Northern Territory has occurred so convincingly that early requirements to return a proportion of hatchlings to the wild were not implemented. With the benefit of hindsight, this is not surprising. Rice *et al*. (1999) harvested about half of the *A. mississippinesis* egg and hatchling production from two Florida lakes over 11 years and found that this did not adversely affect recruitment to either sub-adult or adult size classes.

About 50 000 eggs are taken annually: more than are generated by captive breeding in Northern Territory crocodile farms. Payment per egg to landholders is reckoned to provide an incentive for Indigenous and other landholders to conserve crocodile nesting habitat. The quota for 2013–2014 is 70 000 eggs, distributed across nine regional catchments. There are also quotas for 500 hatchlings, 400 juveniles and 500 adults. The previous management plan allowed wild harvest of 600 adults, but fewer than 100 adults were taken each year, so the quota has been reduced. Six farms were together exporting 10 000–20 000 skins each year between 2001 and 2006, but there has been an upward trend since, with ~30 000 exported in 2010 (Caldwell 2012).

WILD HARVESTING FOR FOOD: BUSH MEAT

Crocodylians have been part of human diet for many hundreds of thousands of years at least. Indirect evidence for that comes by extrapolation from the reverse: evidence of crocodylian predation on early humans. Njau and Blumenschine (2012) described tooth marks and damage indicative of

both a crocodylian and a leopard-like carnivore on fossilised skeletal remains of two early humans in Plio-Pleistocene beds in Olduvai Gorge, Tanzania. The species of early human in each case is uncertain, but both *Australopithecus boisei* and *Homo habilis* are candidates. The timeframe for the predation or scavenging is ~1.8 Mya. Aware of the findings by Njau and Blumenschine, Brochu *et al*. (2010) named a newly discovered species of fossil crocodylid from the Olduvai Gorge *Crocodylus anthropophagus* in honour of that component of its diet. A couple of million years ago, the habitat in what is now the Olduvai Gorge is thought to have been river basins and marshlands with plenty of water and edible plants, suitable for both early humans and crocodiles. The extrapolation is that if there was enough human–crocodile interaction for the crocs to prey on the humans, we can be pretty sure that the humans ate crocodile eggs and crocs of a catchable size as well, just as many of us do in similar shared habitats today.

Crocodylians continue to be captured for food in both South America and Africa. Reviewing wildlife utilisation in Latin America, Ojasti (1996) reported indigenous subsistence hunting for *Caiman crocodilus* and both species of *Paleosuchus*, noting that this type of use was the most common use of *Paleosuchus* because their heavily ossified skins are unsuitable for leather. Hunting for food is also undertaken commercially. Mendonça *et al*. (2010) described the hunting and trade of dried-salted meat from *Caiman crocodilus* and *Melanosuchus niger* in the Purus River, a tributary of the Amazon. Following a survey conducted in 2008 in the Piagaçu-Purus Sustainable Development Reserve, they concluded that the trade is one of the largest illegal exploitations of wildlife in the world. The dried-salted meat sold then for US$0.60–0.75 per kg and is a year-round activity that is economically important for local people.

Both subsistence and unregulated commercial hunting are also widespread in Central Africa. John Thorbjarnarson and Mitchell Eaton undertook a preliminary assessment of the effect of the bush meat trade on crocodylians in coastal Gabon and northern Congo. The account of their 5 week journey makes very interesting reading (Thorbjarnarson and Eaton

2004). In the swamp forests of both countries, they found a significant harvest of *Osteolaemus tetraspis* for local consumption and for sale in the population centres. This is a small and not particularly aggressive species and, because they can be kept alive for long periods while tied up, they are much easier to transport than mammals and birds (Fig. 14.4). In Gabon, with ice and outboard motors more available, they are more likely to be killed first. *C. suchus* and *M. cataphractus* also occur in these countries. Thorbjarnarson and Eaton reported the former to be rare, with *M. cataphractus* much less common in the bush meat markets than *O. tetraspis*.

The following methods of capturing *O. tetraspis* were recorded:

- Hooking out of burrows with a hook attached to a pole or liana (mainly used in the dry season).

- Baited hook (principal technique in the wet season). A likely site is identified, usually from tracks or other signs, or by banging a paddle on the side of the canoe and hearing a crocodile respond. A large fish hook baited with decomposing fish or a frog is then left at the edge of the water and tied to a nearby tree. The crocodile swallows the hook and is captured when the hunter returns. (The same technique is commonly used to catch American alligators.)

- Hunter attracts crocodiles by imitating adult/juvenile distress calls, or by making a bubbling noise by blowing into a bamboo tube with the tip underwater. Crocodiles that approach are captured with a fish spear, a forked stick, a net, or grabbed with the bare hands.

- Hunting at night from a canoe with a flashlight and a multipronged fish spear.

- Finding incidentally in forest pools and grabbing with bare hands.

Fig. 14.4. *A 9.1 kg female* Osteolaemus osborni *captured at night using a flashlight by a hunter from the village of Mokengui in the Lac Tele Community Reserve, Republic of Congo. (Photo Mitchell Eaton)*

- Blocking up burrows with leaves/mud and pulling out dead crocodiles after they have drowned.

- A cloth soaked with gasoline is pushed into a dry burrow with a stick, causing the crocodile to abandon the burrow.

- Using a spear to stab and kill the animal while it is in the burrow.

TROPHY HUNTING

Hunting wildlife for trophies is distasteful to many people, but in some circumstances good arguments can be made in its support. For example, in many African countries wildlife habitat is being turned over to cattle, causing land degradation and loss of the wildlife. In some of these countries, this process is being reversed as landowners find it is economically more rewarding to keep or replace the wildlife and open it up for commercial safari hunting. One of the earliest entrants in this was Zimbabwe's CAMPFIRE (Communal Areas Management Program for Indigenous Resources). This was initiated in 1982, with an agreement drawn up between Indigenous land owners and the Department of National Parks. A 2009 review tracks its history and diversification since then, combining hunting, wildlife viewing, wildlife photographic safaris and tourist lodges, providing income for the community (http://www.africahunting. com/hunting-africa/1780-zimbabwe-campfire-programs.html). This is outside our present scope, but readers wishing to explore this concept further could consult Cousins *et al.* (2008) who conducted an evaluation of the effectiveness of private wildlife ranching as a conservation tool, finding both pros and cons. Concerning trophy hunting in particular, Lindsey *et al.* (2006) reported that at least 1.4 million km² of 23 countries in sub-Saharan Africa are under the custodianship of trophy hunting operators serving ~20 000 clients per annum. Many people would think this appalling, but it is appropriate to be appalled also by the steady replacement of wildlife by cattle grazing. There seems to be little doubt that conservation benefits are accruing, despite many shortcomings (Lindsey *et al.* 2006, 2007; McGranahan 2011) and pragmatic

conservationists who might otherwise deplore trophy hunting seem to be willing to overcome their personal distaste.

With skilful taxidermy, large crocodylians make desirable trophies, either as life-like mounts or as skulls, and in African countries Nile crocodiles feature prominently in brochures and advertisements for safaris. One of them offers a 10-day combined hippo and crocodile safari in Mozambique for US$9300, including licence fees (US$1000 for the hippo, US$300 for the croc). Estuarine crocodiles too, as the world's largest, are sought after and Australia is a logical potential source. Export of wildlife or wildlife products from Australia requires the approval of the Federal Government and that depends upon there being an approved management plan. Trophy hunting *C. porosus* would attract clients from Europe and North America more than locally, and the Northern Territory Government has several times applied to include a provision for trophy hunting in the management plan under which crocodile skins are exported (Leach *et al.* 2009). That approval has been denied so far, but trophy hunting operates in a limited way by declaring the trophies as within the approved harvest and the relevant website advises that the animals are first captured by trapping (http://www.australiawidesafaris.com.au/). Alligator hunting is common in the USA, regulated by tags issued by state agencies, and practised mostly by individuals on private land. There are also commercial guides. Hunting rules differ between states and a great diversity of techniques are approved, depending on the jurisdiction. The range includes baited hooks on set lines (as in Central and West Africa), bow and arrow, gig (a barbed trident), snare, harpoon with float line, snatch-hook/snagging (use of a treble hook, usually with a casting rod), bang stick with power head (pole with a shotgun cartridge mounted, used to dispatch an alligator caught on a baited set line) and, in some states, specified types of firearm.

The issues raised by trophy hunting are animal welfare concerns, conservation issues and potential conflict between hunting and other land uses, particularly crocodile tourism. Conflict between land uses can be overcome by identifying suitable areas where it could be permitted and avoiding those where tourism ventures operate. Animal welfare issues are a big concern and some of the methods currently sanctioned for wild harvest would not meet with approval by animal welfare agencies. Although many people are unconcerned about any need for a crocodylian to be killed humanely, there is little doubt that many others rate that very highly as a requirement for any wild harvest, no matter what its motivation, sport or commerce. (The concern quite properly extends to slaughter of farmed crocodylians too: on veterinary advice the CSG recommends direct shot to the brain or cervical dislocation and immediate pithing).

Much trophy hunting is by shooting, but it is not completely straightforward to dispatch a large crocodylian humanely with a single shot and have it remain accessible for trophy retrieval, particularly if there is a desire to have the head undamaged or minimally damaged as a souvenir. In the proposal to allow trophy hunting in the Northern Territory, it is envisioned that a trophy harvest will always be small and therefore not a conservation concern. Indeed, in the Northern Territory an argument can be made that there is now a superabundance of crocodiles, to the extent that in some areas the number of large animals needs to be reduced in response to concern about public safety (see below). In recent years, however, an additional, conservation-related or, perhaps better put, an ecology-related concern about trophy hunting has arisen. This is the growing realisation about 'boss crocs' and their role in crocodile social life (discussed in Chapter 12) and it is, of course, these large males that will always be targeted specifically as trophies. Some opinion has it that removal of the large males will de-stabilise the social hierarchy, leading to a release of aggressive behaviour by younger males, with uncertain effects. On this point, a survey in the Northern Territory by Tisdell and Nantha (2007) reported that most pastoralists support trophy hunting (six out of nine interviewed) because it could add value to their crocodiles, but two of them said that the big (trophy-sized) crocodiles are 'not the problem' and that the smaller animals should be hunted too. One said that the big crocodiles keep the smaller ones at bay and so reduce the number of cattle taken. There is certainly room for carefully

conducted behavioural research on crocodylian social systems and the role of the very large males.

TOURISM: WILDLIFE VIEWING

Think of Kruger, the Serengeti, Royal Chitwan, the Pantanal, Galapagos, Yellowstone, Great Barrier Reef and the Seychelles. All are famous for their wildlife. We are all familiar with wildlife tourism and well aware that in today's world many countries rely on wildlife tourism dollars. Crocodylians are high on the list of 'must see' species. The caiman are a draw card to the Pantanal, Nile crocs to African parks and big 'salties' to Australia's Northern Territory and Queensland, particularly the Northern Territory, which has somehow captured the crocodile mojo. The Darwin daily newspaper the *NT News* has a well deserved reputation for crocodile stories. The 2 November 2012 issue managed to have five crocodile stories on its front page. Among the most well-known tourist attractions are those on the Adelaide River just east of Darwin where crocs have learnt to jump for meat dangled from the end of a long pole (Fig. 14.5).

There are tourism ventures to see saltwater crocodiles in North Queensland, the Northern Territory and in the Kimberley region of Western Australia. Of course, David and I have no need to wonder why people want to see crocodiles, nor that they want to see salties more than freshies. How could it be otherwise? But Chris Ryan from the University of Waikato in New Zealand did wonder and took on a research project to determine exactly why it is (Ryan 1998). His project was to determine which of the relative attributes of their potential threat, danger, power, links with the prehistoric and survivorship were important attractants, enabling him to draw a 'perceptual map'. In other words, he wanted to analyse the nature of their attraction. He noted that 'it might be argued that dolphins attract through their intelligence, gorillas and monkeys through distant genetic relation, furry animals through cuddly connotation, and lions through their feline grace. Saltwater crocodiles do not possess any of these attributes'. Ryan interviewed 50 visitors to the Northern Territory Wildlife Park which, quite appropriately for his study, does not have a crocodile focus. He took a

Fig. 14.5. *One of the 'jumping crocodile' tourist attractions on the Adelaide River, Northern Territory. See also Figs 4.19, 4.20. (Photos GCG)*

'phenomenographic approach' which he described as 'a post-positivist mode of research as its ontology and epistemology assumes a consensus external reality which can be discovered by the researcher'. Okay. Analysis of the results clustered 'dangerous', 'large' and 'scared' together, with another cluster comprising 'alone' and 'prehistoric' linked to 'awesome', 'amazing' and 'ferocious'. Among the points drawn attention to in the discussion of results were that crocodiles attract fascination because of their potential threat to humans and that many people feel respect for them because of their living link to the prehistoric. Another finding was that visitors were uneasy about commercial attractions that were more like sideshows and would rather see crocs in natural situations. On the other hand, attractions with an educative slant met with more approval, and Crocodylus Park was singled out as an example.

In a follow up study by mail to 7000 people who had visited Kakadu National Park in the preceding 3 years, 2334 responded (Ryan and Harvey 2000). They fell into three categories: crocodile 'likers', crocodile 'indifferents' and crocodile 'dislikers'. These categories were not distributed randomly among socioeconomic groups. There was a trend

for 'dislikers' to be female, and 'likers' were over-represented among the over 50s. 'Likers' did not visit crocodile attractions more than 'dislikers'. It was cheering to note that the people in the survey did like to see crocodiles. I'll vote for that.

There are others, of course, but the best place that I know to see saltwater crocodiles is on one of the Yellow Water Cruises at Cooinda in Kakadu National Park: a tourism venture owned by the Indigenous Gagadju people. Their boats take visitors on a large billabong in Jim Jim Creek, upstream of where it joins the South Alligator River. The crocodile population is high, as is the boat traffic, and the crocs have become habituated. To give you an idea, David leaned over the side of the boat to snap the photo in Fig. 14.6. Also, I shot some very nice video, almost at water level, behind a large croc swimming at the surface and keeping pace with the boat, its tail scutes breaking the water surface as the tail swept slowly back and forth in a wide sinuous arc (http://crocodilian.com/books/grigg-kirshner). On another occasion I photographed a croc with a freshly captured fish in its mouth, defending it from two determined jabiru storks, while the boat hove to a few metres away so that 50 or so tourists could get a good look.

Fig. 14.6. *The crocodiles are so habituated that you don't always need a telephoto lens on the Yellow Waters Cruise in Kakadu National Park. (Photo DSK)*

SUSTAINABLE USE (SUW) AS A TOOL FOR CONSERVATION

Concept

The CAMPFIRE programme in Zimbabwe, discussed above in relation to trophy hunting, was one of the earliest to explore using wildlife as a source of commercial gain and employing that as a stimulus for conserving it. In that example, the concept was that land that would otherwise have been turned over for cattle production, alienating the wildlife and causing land degradation, was used instead for trophy hunting and for photographic and wildlife viewing safaris. Thus the wildlife, instead of cattle, became the economic base. Such practices are often referred to as 'use it or lose it', and as Sustainable Use of Wildlife, SUW (Grigg *et al.* 1995). A cynic once put it as 'killing animals for their own good' (!) and there is a minority of people who do not condone any harvest of wild animals, no matter what benefit may accrue. There are two quite

different aspects to SUW: one benefits the exploited species, the other is any collateral benefit. Thus, there can be incentive to harvest a wildlife species at sustainable rates in order to ensure that the harvest can continue indefinitely: for example, the Australian kangaroo harvest, with annual harvest quotas based on aerial surveys of the populations (Pople and Grigg 1999; Grigg and Pople 2001). Wild harvests of American alligators, discussed above, provide another example. The other aspect is that a wildlife manager may look towards a secondary, collateral benefit from the harvest. In the case of kangaroos, I have argued for many years that if wool growers could come to see kangaroos as a resource instead of a pest, which would probably happen if kangaroo meat came to be valued more highly, they would be able to carry fewer sheep, reducing the total grazing pressure and, thus, reducing the land degradation caused by overgrazing (Grigg 1987, 2002). A crocodylian example of a secondary benefit is that harvesting eggs from the wild as a source of animals to be 'ranched', with a payment per egg to land owners, may put a value on retaining the wetland habitat so more harvestable eggs can be produced. This is the current practice in the Northern Territory, but whether it will provide sufficient incentive to retain the 'natural' state of the land if and when an alternate, incompatible use for it arises, is unknown. I suppose it will depend on the comparative economics.

To put the concept of 'SUW for conservation' another way, the principle is simple: ascribing economic value to wildlife can lead to its being taken care of. If a wildlife product can be harvested directly and sustainably from a particular habitat, then the commercial use of that product puts a value not only on the product itself but also on the habitat that supports the wildlife generating the product. This may lead to the habitat being conserved, conserving the exploited wildlife and, in passing, conserving all of the other components of the fauna and flora that comprise that habitat. This is in sharp contrast to what is all too often the case, where natural habitat is cleared in preparation for planting a traditional crop or introducing grazing livestock (Pople and Grigg 1999).

In short, whereas much wildlife conservation effort was aimed early on at 'preservation' and 'protection', usually by enacting regulations against hunting and setting aside nature reserves, the last 25 years have seen the growth of the idea and application of using SUW as a conservation tool. To deal with significant internal controversy about this topic, the Australasian Wildlife Management Society developed a position paper on this topic and because it includes many elements useful generically it is reproduced here (with modifications) in Box 14.2.

SUW and crocodylian conservation

As discussed earlier, the thrust behind crocodylian conservation since the early days of the CSG has been to strive for recovery of populations alongside regulated commercial use, and this has been the case in the two most spectacular recoveries from low level: American alligators and estuarine crocodiles ('salties') in Australia. These recoveries and the contemporaneous commercial harvests

have been discussed above (alligators) and in Chapter 13 (estuarine crocodiles). Ted Joanen, then working for the Louisiana Department of Wildlife and Fisheries in Louisiana and based in 'alligator country' at the Rockefeller Wildlife Refuge was a prime mover in the former and Grahame Webb the latter. Ted, now retired, was a major pioneer in the science and politics behind the development of alligator management for commercial use. Grahame Webb, of Wildlife Management International in Darwin and current Chairman of the CSG, has been a long-term campaigner in support of SUW (Webb and Manolis 1993; Webb et al. 2004). He was the driving force in the push to have the Australian population of C. porosus downlisted from Appendix I to Appendix II (Webb et al. 1984) and the main architect of commercial harvesting of crocodiles in the Northern Territory as it now operates (Webb et al. 1987).

A major collateral benefit from these alligator and crocodile harvesting programmes has probably been in the greatly increased public tolerance for these large and dangerous animals. In Australia, the attitude to crocodiles in the Northern Territory, which provide substantial local economic benefit from crocodile tourism and skin exports, is very much more positive than it is in Queensland where the multi-faceted 'crocodile industry' is much less developed so far, much less publicised and much less appreciated. If the competition for tags is any indication, with many more applications than the number of tags available and specialist retails shops selling appropriate hunting tackle, alligator hunting in the USA appears to be much appreciated.

A note of caution needs to be sounded. In contradiction to hoping for a conservation benefit to flow from a regulated, sustainable wild harvest, the trend in commercial production of crocodylian skins is to closed-cycle farming. This is partly driven by economics and partly because farm-reared skins are more likely to be blemish free. The trend is less pronounced where there is an abundant source of wild eggs that can be harvested for ranching, but their collection costs money, sites are often distant, there is an element of danger at the nest and, more and more, breeding-sized wild adults are being removed to farms because they

BOX 14.2 AWMS' POSITION PAPER ON THE SUSTAINABLE COMMERCIAL USE OF WILDLIFE (REPRODUCED WITH PERMISSION)

Wildlife is used to encompass undomesticated native animals and uncultivated native plants.

Sustainability is taken to mean the capacity for long-term commercial use without reducing the species' geographic range, changing existing patterns of genetic variability, or radically altering community structure and function).

STATEMENT

THE AUSTRALASIAN WILDLIFE MANAGEMENT SOCIETY:

RECOGNISES that modern human communities place a high value on the conservation of native plants and animals and that decisions about wildlife use are always the consequence of an amalgam of facts and values;

Is CONCERNED that, despite this interest, biodiversity continues to be lost, due largely to land use priorities which favour exotic species at the expense of native ones;

AGREES that, in developing a policy in relation to any particular wildlife management issue, society must place particular emphasis upon the application of scientific information and methodology but, in doing so, should not ignore values, and should strive to find a consensus view reflecting the values held by a majority of its members;

SUPPORTS the concept of achieving habitat and species conservation goals through the sustainable use of wildlife, whether consumptive or non-consumptive, as spelled out in the resolution adopted at the December 1990 General Assembly of the International Union for the Conservation of Nature (IUCN), which recognised, *inter alia*, that '... ethical, wise and sustainable use of some wildlife can provide an alternative or supplementary means of productive land use, and can be consistent with and encourage conservation, where such use is in accordance with adequate safeguards...';

RECOGNISES the need to develop suitable guidelines to ensure that the commercial use of a particular species or habitat is sustainable;

ACCEPTS that landowners are more likely to expend resources conserving wildlife that is economically valuable to them, than wildlife with a neutral or negative economic value;

ACKNOWLEDGES that the commercial use of wildlife provides special opportunities for the sustainable economic development of rural people, especially in remote areas;

Is AWARE that it is now technically and scientifically possible to sustain uses of wildlife for commercial purposes without endangering species or their supporting ecosystems;

RECOGNISES that any species which has the capacity to support commercially profitable, ecologically sustainable harvests is a potential candidate for commercial activity;

RECOGNISES that there are some species for which cultural considerations prohibit any consumptive use and, further, that there is a need to consider each case individually, on its merits and, further;

CONSIDERS that the sanction of a harvest or other use should take into account potential benefits to the conservation of the species' habitat and other species in its community.

ACCORDINGLY, IT IS RECOMMENDED THAT:

1. Commercial use of wildlife be restricted to those species with a capacity to sustain a commercially viable use and, where conservation goals can be addressed, these should be maximised. The goal should be the conservation of ecosystems rather than single species, and the maintenance of existing biodiversity, or its increase.

BOX 14.2 CONTINUED

2. The acceptability to the public of specific commercial consumptive uses of wildlife species should be determined through a process of consultation.

3. A management plan always be drawn up to define and demonstrate the criteria raised in point 1 and those that follow. The plan should be subject to periodic public review.

4. The management plan should be developed in consultation with relevant interest groups and be available for public scrutiny. It is highly desirable that programmes be structured in such a way as to return profits to the local area and, wherever possible, should return benefits directly to the landholder.

5. The management plan should be adaptive, based upon firm scientific principles and the best available knowledge, and include ongoing monitoring, reporting and research to ensure that the use is economically ecologically and culturally sustainable.

6. There be guidelines in place and mechanisms available to enable a discontinuation of the commercial use of wildlife if conservation objectives are compromised.

7. The government department directly responsible for conservation be the regulatory body and be entirely independent of the industry.

8. The marketplace should be allowed to operate as freely as possible within a clearly identified regulatory framework set by the relevant government agency to ensure that conservation goals are not compromised.

9. There be minimal waste in consumptive use. In the case of non- consumptive use, disturbance to the exploited species and its habitat should be minimised.

10. There be guidelines to ensures humane practices in the use of animals.

11. Care be taken to avoid impact upon 'look-alike' species which may be taken mistakenly or deliberately.

are judged to be 'problem' or 'nuisance' animals. Additionally, genetic studies are under way to identify characteristics that produce desirable traits in the skins, such as colour and scale pattern. Humans have a long history of modifying the characteristics of domesticated animals through selective breeding and we seem to be on the edge of that with at least some species of crocodylian. Using wildlife sustainably only works as a conservation tool while it puts pressure on maintaining animals in the wild. From a conservation point of view, skin production by closed-cycle farming is of less benefit for conservation than a wild harvest, other than as a source of animals to reintroduce to the wild in a desperate situation. Farming can also create a potential hazard, because hybridisation has led to many farms having stock that are no longer suitable for re-introduction.

A sceptic about the value of SUW might say that its introduction – putting crocodylian harvesting onto a scientific basis – coincided with the growth of the conservation ethic and the populations would have recovered anyway, so that the SUW approach does not deserve the credit it gets. But this negative view is hard to sustain. It seems clear enough that the commercial contribution from alligators in the USA has been a strong force behind their current level of abundance, but also the extent to which they are tolerated – a tolerance that has led to their celebration. Southern USA is proud of its alligators. The recovery of *C. porosus* in northern Australia is more a consequence of their protection than their commercial use, but their protection has been tolerated better because of it. Here, the industry has grown by taking advantage of the enormous surplus egg production, while crocodiles in the rivers have been left alone to do whatever comes naturally. Territorians are proud of their crocodiles and much of the benefit from SUW in the NT has been collateral rather than direct. And the sceptics should be reminded that these positive attitudes to crocodiles and alligators are a very big turn around from the negative attitudes that most people had towards them 50 years ago.

John Thorbjarnarson and Alvaro Velasco attempted a more objective appraisal of the SUW approach as applied to the Venezuelan programme for regulating the harvesting of spectacled caiman, *Caiman crocodilus* (Thorbjarnarson and Velasco 1999). The aim of this at its introduction in 1983 was to create an economic incentive for landholders to maintain healthy populations of caiman and also to maintain their habitat. They noted that, by 1995, a million caiman had been harvested, with an export value more than US$115 million. Caiman provided a high return in relation to little or no investment, but it was small in relation to that from cattle. The only evidence of landholders managing caiman populations positively was a preference to sell skins taken illegally from elsewhere instead of their own, avoiding any risk to future quota allocations. Thorbjarnarson (1999) took a wider view of the benefits and limitations of a SUW approach to crocodylian management and conservation. He noted that by, the end of the 1980s, ~40 countries had crocodylian management programmes based on regulated commercial use, but that their success depended heavily on the price they could get for the skins. He noted the risk in having conservation programmes so dependent on the luxury fashion market, so that when prices crashed, so did some of the programmes. An example with a more positive outlook was presented at the 21st meeting of the CSG in Manila in 2012. Larriera *et al.* (2012) reported that the eggs of broad-snouted caiman, *Caiman latirostris*, at the southern end of their range in Argentina, were being collected and sold to ranching operations by gauchos on local beef cattle properties. The result was cash for them and a positive attitude to the caiman: they had become re-categorised from 'pest' to 'resource'. The concept is undoubtedly a valid one. Sometimes it will be a conservation success story, and sometimes not. The trend to vertical integration from production to retail, however, will have outcomes that are not easy to predict and may not be positive for crocodylian conservation.

Every case must be judged on its merits, and no two are likely to be exactly the same. But, despite the difficulty of making objective assessments about the efficacy of the SUW approach as a conservation tool, there can be no denying that if crocodylians are going to be harvested, it should be done sustainably.

CONSERVATION OF THE NON-COMMERCIAL SPECIES

SUW and its potential conservation benefit have little or no applicability to the species whose wild populations are the most endangered. Although a commercial benefit may provide incentive to harvest abundant species sustainably, rather than catastrophically, that leaves many species out: Chinese alligators, gharials and Philippine crocodiles are stand out examples, and we can add *Tomistoma* and the Orinoco, Cuban and Siamese crocodiles. These are all Endangered or Critically Endangered and already too thin on the ground for anyone to contemplate a regulated harvest. The situation with the three *Osteolaemus*, both of the *Mecistops* and *C. suchus* is less clear, but of serious concern: opinion is that all of these are probably at risk. A full review of conservation efforts that do not embody the SUW philosophy is beyond the present scope but a couple of examples will highlight some of the issues.

There is abundant evidence that crocodylians can recover well if there is suitable habitat and if they are left to themselves, so simply affording full protection can do wonders. This has worked well for the Florida population of *C. acutus*. It was put on the US Endangered Species list in 1975 with just 200–300 adults, but under this protection the population has grown to a couple of thousand (Mazotti 2013). Apart from being given legal protection, the main conservation activity was provision of some artificial nesting areas to offset significant losses as a result of the extensive human development in Southern Florida. The crocs also found suitable nesting habitat adjoining the warmer waters in the cooling ponds for the Florida Power and Light Company's power plant at Turkey Point. Like the healthy breeding population of *C. palustris* in the Vishwamitri River in north-western India where it flows through a city of 2 million people (Chapter 12), *C. acutus* has accommodated well to canals and other Florida waterways, even in some heavily populated areas.

American crocodiles may be benefitting from Florida's famous tolerance of alligators. The conservation message here is that crocodylians can recover if they have the resources they need and are left alone, even in such an unlikely habitat as heavily built up Southern Florida. But few people in Florida would be targeting *C. acutus* for subsistence, and there is a very active wildlife management agency that ensures a high level of compliance. Enacting protective legislation is important, but it can be effective by itself only in certain cultures.

Determined conservation efforts above and beyond legal protection are ongoing for some of the Critically Endangered crocodylians. Unfortunately, two of the most unusual crocs, *Tomistoma* and *Gavialis,* are also among the most endangered, both particularly from loss or alteration of habitat. Gharials in the Chambal River in northern India, their last 'stronghold', number only a couple of hundred breeding adults and have to contend with riverbank sand mining, agriculture, grazing and other disturbances, and, in the water, entanglement in fishing nets, baited hooks set for turtles and pollution, plus destruction at the hands of unsympathetic fishermen. The Gharial Conservation Alliance (http://www.gharialconservationalliance.org/) – a project of the Madras Crocodile Bank Trust – is an international organisation 'dedicated to saving gharials from extinction and ensuring the establishment of sustainable wild populations'. Gharial conservation has been underway since the 1970s, with a focus on captive breeding and also 'head-starting' individuals for release into the wild. The efficacy of these practices has come under serious question in a study by Nair *et al.* (2012) who conducted boat surveys over 75 km of the Chambal River. In the process of developing improvements to surveys methods, including the addition of photo-identification as a way to improve mark–recapture estimates of absolute densities, they came to the conclusion that the population size and structure are such that artificial supplementation is unnecessary. They considered that these actions were diverting resources from other approaches, and the GCA's website now

notes a shift towards habitat protection, as identified at the top of the CSG's most recent Action Plan (Stevenson and Whitaker 2012) and to enforcement of protected areas, education and cooperation with local people.

Tomistoma was once widespread in lowland freshwater swamps in Borneo, eastern Sumatra and Peninsular Malaysia, but is now severely restricted, with only a few scattered populations, most in Sumatra and Kalimantan (Bezuijen *et al.* 2010). The Tomistoma Task Force was set up in 2003 within the IUCN–SSC Crocodile Specialist Group. Its aim is to quantify the status of the Malay gharial, *Tomistoma schlegelii,* in the wild, identify the threats to which they are exposed and promote conservation actions (http://tomistoma.org/pa/). *Tomistoma* is legally protected throughout its range, but the TTF website documents severe ongoing threats from illegal logging, forest fires which destroy nesting habitat, and incidental capture by fishermen. TTF is constrained by shortages of funds. Although there are many *Tomistoma* in captivity, indeed in crocodile farms, it is hard to be optimistic about their long-term future in the wild unless something changes dramatically.

The Philippine crocodile (Fig. 14.7) is the focus of ongoing conservation programmes and one is particularly noteworthy because of the way it has harnessed support from the local population (Van der Ploeg *et al.* 2011a,b). Philippine crocodiles are Critically Endangered, with only a few small populations holding on in remote, but still populated, country on the northern island of Luzon and the southern island of Mindanao (Van Weerd 2012). Unregulated hunting had reduced the population to ~500 adults by 1981 and, despite legal protection, without a sympathetic attitude by landowners the next estimate showed a decline to ~100 by 1996 (Ross 1998). With conservation in mind, the Philippine Government established a Crocodile Farming Institute (CFI) and a captive breeding programme was initiated in 1987 for both *C. porosus* and *C. mindorensis* and sought to establish a leather industry. The history of this initiative has been summarised by Van der Ploeg *et al.* (2011a). Very briefly, the idea was to loan or sell juvenile crocodiles to numerous private commercial

Fig. 14.7. *An adult Philippine crocodile,* Crocodylus mindorensis, *in the Disulap River, San Mariano on the island of Luzon. Note the radio-transmitter attached on this animal's tail; if a reserve for crocodiles is to be created, it is a good idea to know how far the adults are likely to travel around. This species is Critically Endangered and since 2003 there has been an active conservation and rehabilitation programme run by the Mabuwaya Foundation with support from Isabela State University and Leiden University of the Netherlands (http://www.mabuwaya.org/). Employing a diversity of approaches within communities in north-east Luzon the programme focusses on the intrinsic values of these animals, highlighting the theme of 'the Philippine crocodile something to be proud of!' (Photo Merlijn Van Weerd)*

crocodile ranchers who would raise them and then share the profit. CFI would also regulate hunting, engage rural communities in the collection of eggs, and improve the processing and marketing of skins. The earnings would partly be used by CFI to pay for crocodile conservation programmes in the wild and the benefits earned with egg harvesting by communities would provide an incentive for protection of wild crocodiles. An optimistic report to the CSG by Anon. (2000) noted that 'six highly qualified companies had been selected … to breed crocodiles … the government approved the distribution of 110 crocodiles from CFI to the private breeders and the program is expected to generate huge revenues for the government and provide employment to many rural farmers'. In practice, it did not work out the way it was

planned, but instead produced several large closed-cycle crocodile farms, mainly producing skins of *C. porosus*. As mentioned earlier, closed-cycle crocodile farming does not provide a direct incentive to protect wild crocodiles as compared with ranching or egg-harvesting programmes. However, on the conservation positive side, in 2000, the six farms formed an NGO, the CPPI (Crocodylus porosus Philippines Inc.), an association that led to captive-bred Philippine crocodiles being available for release to the wild and in 2006 CPPI formed the Philippine Crocodile Research and Conservation Program and commenced a programme of introductions. Three wild-acclimated adults were released in 2009 to a swamp on Mindanao, as reported by Van Weerd *et al.* (2011) in a paper evaluating the practice of re-introductions as a

Fig. 14.8. *This public notice marks the Philippine crocodile sanctuary of Disulap River in the Municipality of San Mariano, Luzon, Philippines. (Photo Merlijn Van Weerd)*

conservation tool. In March 2013, and as part of the same project, 36 juvenile Philippine crocodiles were introduced to Paghongawan Marsh, in a protected area on Siargao Island off the coast of the Southern Philippines (Mercado *et al*. 2013) and several months later appeared to be doing well. Of particular interest, both of these were 'soft releases', with the animals having been held elsewhere for a couple of years in semi-wild conditions without supplementary feeding before release.

Meanwhile, Jan Van der Ploeg and Merlijn Van Weerd had taken an entirely different approach, with a community-based programme aimed at conserving a remnant population of Philippine crocodiles in the Northern Sierra Madre mountains in Luzon. The programme began in 1999, initiated with support from Isabela State University and Leiden University of the Netherlands, and the Mabuwaya Foundation was established in 2003. It promotes the aims of the organisation through an education programme using posters, calendars, comic books, newsletters, billboards, community theatre shows and interactive communication means such as school lectures and community consultations. Its efforts have led to three crocodile sanctuaries, a head-start programme that raises wild-born juveniles for release and there is also now a much more positive attitude to crocodiles within the community (Figs 14.8–14.12). Such programmes can be called CEPA programmes (communication, education and public awareness: the term having been coined by the IUCN Commission on Education and Communication), essentially tapping into and promoting community pride in having a rare and iconic species living in their midst. Given the potential danger that crocodiles represent and the extent to which they are regarded by many people as threats or vermin, achieving a positive attitude towards crocodiles is a significant feat. The story of the Mabuwaya Foundation has been written up in a substantial and very attractive and informative book (Van Weerd and Van der Ploeg 2012).

Fig. 14.9. *Mabuwaya Foundation members and residents place a notice in Lamang, San Mariano to inform the community about Philippine crocodiles and environmental laws. (Photo Willem Van de Ven)*

Is their approach leading to an increase in crocodiles? Van der Ploeg *et al.* (2011a) noted that the number of non-hatchling crocs in north-east Luzon increased from 12 in 2000 to 64 in 2009 and they are no longer killed deliberately. Some die accidentally of course, in traps and nets. The main threats now are the clearing of forest for agriculture and conversion of freshwater habitat for growing rice. The authors noted that the programme will need to be continued for a long time if it is to recover the population of Philippine crocodiles, and that funding it will remain a major challenge.

It is quite possible that an even larger contribution from the Mabuwaya Foundation's efforts may be seen in outcomes achieved elsewhere. What the people spearheading this campaign in Luzon have done, and are still doing, is being noticed. The Foundation won the Castillo Award for Crocodile Conservation from the IUCN Crocodile Specialist Group in 2006 and similar programmes are being implemented in Cambodia (*C. siamensis*), India

(*Gavialis*) and in the planning in Indonesia for *Tomistoma* conservation.

It is greatly to their credit also that they have written extensively about their project, including a very interesting paper dealing with the question 'Why conserve crocodiles anyway?' The paper was prompted by a question raised at a village beside a creek that happened to have the largest reproducing population of wild *C. mindorensis* anywhere. During the presentation of a proposal to declare the creek a protected area, someone asked 'Why?' I think the presenters may have got a bit of a shock to realise that they had no immediate and good answer and it set them on an analysis of the validity of reasons others had given for nature conservation (Van der Ploeg *et al.* 2011b). They noted that answers given by conservationists often lack scientific basis or are irrelevant from a local perspective. They noted a tendency to rely on economic benefits: that is, instrumental or utilitarian values, while dismissing aesthetic and moral considerations, assuming

Fig. 14.10. *Elementary school children visit the Municipal Philippine crocodile rearing station in San Mariano. Engagement with school children is an important part of the Mabuwaya Foundation's activities and has been shown to be effective in promoting the crocodile conservation message. (Photo Willem Van de Ven)*

wrongly that such values are little appreciated in impoverished communities in developing countries. The thrust of their paper is that some of the utilitarian arguments are over simplified and sometimes inaccurate, such as the potentially flawed notion that the presence of crocodylians increases fish numbers or improves the balance between desirable and undesirable species, or plays an important role in fertilising rivers. The paper argues that intrinsic and cultural values, such as pride, love and curiosity, may offer a more realistic and honest foundation on which to encourage preservation of a species and it is a paper that anybody involved in conservation practices is well advised to read.

The contrast between SUW and CEPA is an interesting one. SUW gives an answer based on utilitarian (instrumental) value, whereas CEPA's answer highlights crocodylians' intrinsic values. Depending upon circumstances, either approach to crocodylian conservation may be the better choice,

or both together: the intrinsic value can underlie them both.

CONFLICT BETWEEN HUMANS AND CROCODYLIANS

The positive attitude towards crocodylians that pervades this book would be perplexing to many people. In many rural communities in croc country, routine day-to-day activities such as collecting water, bathing and fishing put people at risk. Prevailing experiences of crocodiles in rural Africa are that they are dangerous to humans and cause hardship because of attacks on livestock, competition for fish and damage to fish nets (Lamarque *et al.* 2009). Perusal of Table 14.3, based on a study in Mozambique (Anderson and Pariela 2005), makes it easy to see why crocs are perceived so negatively by so many of the people who have the most dealings with them. The table could

Fig. 14.11. *Victorino Montanes and Romy Aggabao releasing a marked, head-started crocodile in Dunoy Lake, San Mariano. Survival rates over their first year have been satisfactory. (Photo Jordy Groffen)*

represent the situation in other some countries too. It puts into perspective the question considered by Van der Ploeg *et al.* (2011b), discussed in the last section: why conserve crocodiles?

Alongside these hardships and the negative attitudes borne in experience are beliefs about supernatural connections between humans and crocodiles (Fergusson 2008). A widely held belief is that crocodile attacks are the work of witches: either directing the crocodiles or turning into crocodiles, to deliver punishment. In another, more positive, set of beliefs, some people see crocodiles as the spirits of past generations, make them their totem and will not kill or eat them. Both sets of beliefs promote, and may explain, the fatalism that many Africans display and makes it difficult to persuade people to take steps to protect themselves.

Crocodylians, of course, are not the only wildlife that cause conflicts with people. Human–wildlife conflict (HWC) is a global problem and it may be at its worst in Africa, with a range of problems caused by many other large animals, in addition to crocodiles, such as elephants, lions, hippopotamus and buffalo that may also kill people or, along with a wide array of smaller animals, may make life difficult for farmers growing crops.

THE SCALE AND NATURE OF ATTACKS ON HUMANS

Concern about conflict between humans and crocodylians inevitably focusses on fatalities because crocs are such skilful and opportunistic predators: the large ones can and do see humans as prey and fatal attacks are more newsworthy. But the non-fatal attacks should be borne in mind too, and they include injuries inflicted by croc species too small to see humans as routine prey. That is not to say that small crocodylians cannot kill humans: a 2 year old child in Florida was killed by a 1.8 m alligator and, of course, crocs do like dogs. Wherever crocodylians and humans co-exist, there are likely to be negative consequences. Although conflict

Fig. 14.12. *Students from Isabela State University perform a Philippine crocodile dance show during the barangay fiesta of Cadsalan. Such events are important for community engagement and apparently the show is much enjoyed, particularly the part about reproductive behaviour. (Photo Jan Van der Ploeg)*

with crocs extends to those causing economic loss such as damaged fishing nets, this section will focus on attacks by crocodylians on humans.

Attacks are ongoing. Sideleau and Britton (2014) collated all reports they could find of attacks reported between January 2008 and mid 2013. Their sources were online media reports, local wildlife officials, crocodylian experts and relevant publications. They found 1237 attacks, of which 674 were fatal, by 15 species (Table 14.4). There is no doubt that these numbers are underestimates, particularly the number of non-fatal attacks, but, despite this, more than half of the crocodylians made an appearance on the list and none with a bad reputation failed to make it. Sideleau and Britton have also launched a website: the Worldwide Crocodilian Attack Database http://www.crocodile-attack.info.

Most of the species listed in Table 14.4 do not normally prey on humans: *C. porosus* and *C. niloticus* are the two species with the particular

reputation as 'man-eaters'. The largest alligatorid species, the black caiman, *Melanosuchus niger*, was also a threat in times past when they were more abundant, whereas the equally large, or larger, Orinoco crocodile, *C. intermedius*, seems not to have had a reputation for attacks on humans. Gharials are also reported to have attacked humans (Bustard and Singh 1982), but the chance of that in today's world is very low (would it be unfair to add 'unfortunately'?). The attacks by the mugger, *C. palustris*, have been mostly in Gujarat state where a large population has accommodated to living in the river flowing through a city of about 2 million people (Bayani *et al.* 2011). The Malay gharial, *Tomistoma schlegelii*, is not reputed as a predator on humans, but one certain fatal attack in 2008 was reported by Rachmawan and Brend (2009). According to the report, the victim was pushing a raft of small logs towards a waiting boat on a river in Central Kalimantan when the croc attacked, after

TABLE 14.3 THE COST–BENEFITS OF CROCODILES IN COMMUNAL FARMING AREAS IN MOZAMBIQUE (ANDERSON AND PARIELA 2005)

Social costs	Social benefits
Severe trauma in losing a family member to a crocodile	None
Loss of working member causes severe hardship to family	
If a mother is killed, children have to take her household duties and this can have an impact on their education	
Environmental costs	**Environmental benefits**
Not quantified	Unknown and complicated by the impacts of other factors: dams, erosion, alien species, pollution
Economic costs	**Economic benefits**
Damage to fishing nets	Crocodile farming
Loss of livestock severe to poor households	
Loss of labour when family member killed	
Political costs	**Political benefits**
Affected communities blame governments for not reacting to their problems	None

which he screamed and disappeared. Subsequently a large (~4 m) female (!) *Tomistoma* was captured and killed and the man's remains were found inside, along with a proboscis monkey (*Nasalis larvatus*) and a long-tailed macaque (*Macaca fascicularis*). Apart from predation, attacks may be motivated by territorial or nest defence, the best evidence being increased reports of attack in breeding seasons. Attacks by smaller species are often in response to capture attempts or being stood on. Australia's Johnston River or freshwater crocodile, *C. johnstoni*, is normally regarded in that category and can cause nasty injuries. Several attacks have been recorded, probably an underestimate, and have been reviewed by Hines and Skroblin (2010) following an attack on one of them in Western Australia's Kimberley region in 2008 after she had bumped into one while swimming. She was gripped firmly on the knee and the animal attempted to roll. Her companion gouged its eye and it released her. On the way to the bank of the waterhole, a second animal lacerated her arm. These authors tabulated several previous attacks and considered that the danger presented by 'freshies' was underestimated

by their reputation for being harmless. An incident in Lake Argyle in 2009, also in the Kimberley, was reported by Somaweera (2011) and judged to have been unprovoked. The man was swimming with friends in deep water near their boat and was grabbed as he did a backward somersault to change direction for his return to the boat. He took injuries to his chest and arm and cracked ribs from the force of the attack. Little can be said about the crocodile's motivation, but both events emphasise the need for detailed records of attacks to be kept and the need for treating freshies with appropriate respect.

One report of *C. porosus* attack demands special mention. There have been persistent reports of the deaths of almost 1000 Japanese soldiers on Ramree Island off the western coast of Burma (Myanmar) in 1945, towards the end of the Second World War. The story goes that, as part of an operation to recapture Rangoon (Yangon), a force of British and Indian troops made an amphibious assault on the tip of the island, forcing the 1000 or Japanese defenders to retreat into mangrove swamps. They hoped to find their way from there to the mainland and join the main Japanese force, but this was blocked by

TABLE 14.4 REPORTED ATTACKS BY CROCODYLIANS BETWEEN JANUARY 2008 AND JULY 2013 (SIDELEAU AND BRITTON 2014)

How representative these numbers are is not known, but they are undoubtedly underestimates. A study in Zambia found that only one-third of the fatal attacks there had been reported to authorities (Wallace *et al.* 2011) and the proportion of non-fatal attacks that get reported is certain to be much lower. See also Worldwide Crocodilian Attack Database http://www.crocodile-attack.info

	Attacks	Fatal	Locations (number of attacks, followed by fatalities)
C. porosus	494	285 (58%)	Indonesia 121, 65 (East Timor, Sumatra and East Kalimantan); Malaysia 57, 32 (most in Sarawak); India 54, 31 (Orissa); Papua-New Guinea 50, 40; Sri Lanka 21, 12; Australia, 29, 8.
C. niloticus	428	309 (72%)	African countries
C. palustris	98	50 (51%)	India 83, 43 (particularly within Gujarat state); Sri Lanka 11, 4; Nepal 4, 3.
C. acutus	69	13 (19%)	Mexico 37, 2; Costa Rica 14, 5; Panama 8, 3.
A. mississippiensis	47	0	
M. niger	36	9 (25%)	Brazil 28,9 (most in Amazonas); Peru 3, 0; Ecuador 3, 0; Guyana 1, 0.
C. moreletii	16	2	Mexico 11, 2 (Tamaulipas); Guatemala 3, 0; Belize 2, 0.
T. schlegelii	8	4	Central Kalimantan 4, 2; East Kalimantan 1,1; Sumatra 3, 1.
C. johnstoni	6	0	Western Australia 4, 0; Northern Territory 2, 0.
C. siamensis	2	0	Vietnam 1, 0; East Kalimantan 1, 0.
C. mindorensis	2	0	Philippines 2, 0 (Luzon)
C. intermedius	1	0	Venezuela 1, 0 (tributary of the Orinoco River)
Caiman yacare	2	0	Brazil 3, 0; Argentina 2, 0.
Caiman latirostris	5	0	Brazil 2, 0 (both provoked)
Caiman crocodilus	15	0	Brazil, 8, 0; Colombia 5, 0; Suriname 1, 0; Trinidad 1, 0. (some may have been provoked)
Species undetermined	8	2	
TOTALS	1237	674 (54%)	

a Royal Navy flotilla. Only ~20 survived, the rest having supposedly been attacked and eaten by crocodiles. There is a colourful account of a night pierced by screams as soldiers were crushed, and vultures arriving at dawn to clean up any remains. The story was investigated by Platt *et al.* (2001), interviewing residents on Ramree Island who were there at the time and evaluating historical records. They came to the conclusion that, although many Japanese soldiers died on Ramree Island, it was over a period of time and from a diversity of causes, mainly disease, combat, starvation and even shark attack. Apparently 10–15 were attacked by crocodiles during the retreat: not insignificant, but a far cry from the 1000 mentioned in the report.

An extensive review of human–wildlife conflict in Africa concluded that crocodiles are responsible for more deaths than any other large animal, even more than the hippopotamus (Lamarque *et al.* 2009). Backing that up, official records in Mozambique for the 27 months from July 2006 to September 2008 show that of 204 deaths caused by wildlife where the perpetrator was known, 134 were caused by *C. niloticus* (Dunham *et al.* 2010). Lamarque and colleagues did not report an estimate of the annual Africa-wide crocodile death toll from *C. niloticus*, but reports from individual countries or regions imply that it must be in the hundreds. We can be sure that official reports are underestimates in developing countries: a consequence of administrative issues and crocodile attacks being routine to the point of being almost unremarkable. Some quantitative idea of this comes from interviews with villagers compared with examination of Zambia Wildlife Authority records in a study to estimate the scale of human–crocodile conflict (HCC) in the western Chiawa Game Management Area in Zambia. The area of the GMA is 2344 km^2 and had a human population of ~20 000, mostly in 15 villages along less than 50 km of the Zambezi River. Over 10 years (2000–2009) 98 attacks were reported, of which 62% were fatal. However, only 32% of the fatal attacks had been reported to the relevant government authority (Wallace *et al.* 2011).

It is worth noting that reports of African fatalities are primarily from within the range of *C. niloticus*, not from West Africa where the large crocodylian is *C. suchus* (Fig. 1.6). This species is less aggressive and is afforded a much higher regard, to the extent that many villages have sacred crocodile pools. The village of Bazoulé in Burkina Faso has become a tourist destination on the strength of crocodiles tolerating close contact with people. Crocodylians have a special place in the myths, legends, rituals and taboos of many other cultures too. They feature heavily in masks and carvings in Papua New Guinea and in carvings and on bark paintings by Australian Aborigines, for some of whom they have strong totemic significance. In some cultures, there are beliefs about connections between crocodiles and family ancestors and, for this or other reasons, tribal laws prohibit killing crocs, or include rules about how and when they can be killed such as in revenge for a fatal attack. These beliefs commonly dictate behaviour in crocodile habitat and may explain some of the fatalism that some cultures seem to show towards exposure to risk.

Caldicott *et al.* (2005) compiled a detailed review of attacks by *C. porosus* on humans in northern Aus-

tralia and, in the process, reviewed aspects of crocodylian attacks in general. A book by Edwards (1998) provides narrative accounts of some of the more newsworthy of these attacks and related events. Caldicott and colleagues noted 62 definite unprovoked attacks in the wild by *C. porosus* between 1971 and 2004 (excluding those in response to egg collection or other artificial provocation). Seventeen of these were fatal. Sixty-three per cent were in the Northern territory, 24% in Queensland and 13% in Western Australia. They noted a trend towards attacks increasing with time, as the annual rate went up 7-fold from ~0.5 in the 1970s to 3.8 in the early half of the 2000s, while population increased by only 50%. That trend has continued, with the attack rate over the period reported by Sideleau and Britton (2014) 6.4 per annum. During the last 40 years, the numbers of both *C. porosus* and humans in crocodile country in Australia have increased, so an increasing rate of conflict since 1971 is not surprising. Also, with its recovering population, *C. porosus* in the Northern Territory has been extending its range further up rivers to areas long thought to be free of saltwater crocodiles (Letnic and Connors 2006; Letnic *et al.* 2011) but now requiring regular surveillance.

For all species, most attacks are recorded during daylight hours, which is probably just a reflection of when humans choose to be in or near the water. They occur when people are swimming, bathing, crossing a river with cattle, collecting water, drinking or fishing at water's edge. Thomas and Leslie (2006) analysed 125 attacks on humans in the Okavango Delta of Botswana and reported that they were more common where there were more people. There is a trend for more attacks in the warmer months of the year when crocs are more active and coincident with breeding activity when territorial defence is more likely to be in play. In northern Australia, breeding by *C. porosus* occurs in the warm summer months, but the main tourist season is in the cooler dry season and so there is not a strong seasonal correlation in Australia.

C. porosus density in much of the rest of its current range is much lower than in Australia, but attacks are still common, likely to be many more than the 100 or so per annum reported in Table 14.4. The larger number of attacks is probably a consequence of much higher human populations in crocodile habitat and intense and more extensive land use. Many attacks are recorded from Sarawak and one that caused much public anxiety was recorded by Ritchie and Jong (1993) in a book *Bujang Senang, Terror of Batang Lupar,* which reflects the flavour of crocodile sentiment in that country.

Caldicott and colleagues noted that attacks by *C. porosus* are rare on people in boats and this applies to other species too. One exception is the famous Sweetheart 'saga' (Stringer 1986): a 5.1 m croc in the Finnis River to the west of Darwin that attacked outboard motor propellers, perhaps confusing them with other males roaring. He overturned several boats without, however, ever attacking the people thrown into the water. Sweetheart was captured – the idea being to relocate him to a croc farm – but unfortunately he died as he was being transported while drugged. Getting the dosage right and managing transport of very large crocs is a tricky business (see Chapter 5). Sweetheart suffered taxidermy and is on display in the Northern Territory Museum and Art Gallery in Darwin.

There is another example where a motor seemed to have provoked the attack, rather than the people. Richardson and Livingstone (1962) featured in an attack by a Nile crocodile on a lake within the Mweru Wa Ntipa marshes in what is now Zambia. These two scientists had been depth sounding the lake in a 2 m inflatable dinghy powered by a small outboard motor. On their way to shore, the boat was attacked repeatedly by a large crocodile. The animal 'rose up and threshed around … the whole boat was alive with its activity … raked the senior author across one buttock with two teeth … the junior author attempted to kick the boat off its back through the canvas floor'. With the boat deflating around them they engaged in a desperate struggle with the crocodile, 'the senior author aimed a blow with a hollow aluminium oar at the eye' and generally beat it about the head, to no avail; it kept attacking the boat. 'We took to the water and began swimming for shore, rapidly and quietly at first, then more slowly when it became apparent that we were not going to be eaten immediately'. They reached shore after 2 h swimming. It must have been a desperate situation but, scientists that they were, they wrote it

up properly and collected other accounts in which attacks seemed to have been stimulated by the noise of the motor and in which the focus of the attack was the boat itself and not the passengers.

Three other scientists recorded being the target of an attack by a *C. porosus* in northern Australia (Webb *et al*. 1978), although it would not qualify as an unprovoked attack. Grahame Webb, Michael Yerbury and Vic Onions were trying to locate a 3.71 m saltie that had been fitted with a radio-transmitter 4 months earlier in late 1975 in the Tomkinson River. They wanted to have a look at the croc and the transmitter to check all was okay. Having acquired the radio signal, they climbed a tree out over the mangrove-lined water's edge in the hope of spotting the animal. They soon saw it 30 m away and splashed in the water surface with a stick to attract it a little closer so they could get a better look. They succeeded. The croc turned towards them, dived, and then surfaced immediately below them in their tree, oriented towards them. It paused for a few seconds, opened its mouth slightly (according to one author) and launched itself at them. It came out of the water as far as its hind legs, tail presumably thrashing. But its jaws snagged in a fork in the tree and clamped briefly on the trunk a few centimetres from the foot of one of the authors and it splashed back into the water. The authors put the attack down to territorial defence.

Although it seems likely that you will be fairly safe in your boat (just as well given the popularity of barramundi fishing in Northern Territory waters!), you should also be aware that a fisherman was taken out of his boat on the South Alligator River in the NT in early 2014, apparently while leaning over its stern. Attacks on canoes or small pirogues (canoe-like boats such as are commonly used by duck hunters in the marshes of Louisiana) are recorded for *A. mississippiensis*, *C. niloticus* and *C. porosus*. It has been suggested that, from below, a canoe may look like another crocodile. That may be, but I see no reason to assume that a hungry crocodile cannot judge that attacking a person in a canoe is a more feasible proposition than a person in a boat.

Most attacks, of course, are not fatal. If you manage to get bitten but not killed, expect serious infection. Swabs from *C. porosus* and *C. johnstoni* mouths identified *Aeromonas hydrophila*, *Pseudomonas aeruginosa*, *Proteus vulgaris* and *P. mirabilis*, *Salmonella* spp., and several more (Caldicott *et al.* 2005). It must be assumed that any crocodylian bite will be infected and it should be treated accordingly. Traumatic injury needs immediate attention of course, with emphasis on the usual immediate actions to maintain air flow and manage bleeding. The paper's senior author, David Caldicott, was for many years in the Trauma Unit of Adelaide Hospital's Department of Emergency Medicine and there is a substantial section providing advice about rendering first aid to a rescued victim, as well as hospital and surgical management. It would be nice to think that the paper is well read in hospitals throughout the tropics …

American alligators are very numerous and widely distributed in the southern states of the USA, having recovered from a very low base in the 1960s (Chapter 13). With high human population densities in many places and the many and diverse uses of waterways by humans for commercial and leisure activities, conflict with alligators is common and increasing. Langley (2005) reviewed the trend, as well as the nature, of the conflicts between humans and alligators. He noted 376 attacks causing injury and 15 deaths between 1948 and August 2004 and there have been a few more since then. The nature of conflicts with alligators are rather different from those reported for Nile crocodiles: a reflection of the different ways in which waterways are used. An analysis of 305 attacks by alligators in Florida revealed that the highest proportion were of swimmers or snorkelers (21%), followed by responses to attempts to capture them (17.4%), fishing (9.9%), retrieving golf balls (9.5%) and wading (5.3%), whereas pulling weeds/planting, walking/standing/sitting on the bank, working on or falling out of a boat, water-skiing and canoeing each comprised less than 4%.

MITIGATION OF ATTACKS ON HUMANS

Protection at the water's edge

Authors writing about conflicts with wildlife in Africa usually mention the fear and dislike people show towards Nile crocodiles and, often with surprise, that few steps are taken to avoid risk and to ensure public safety. Most fatal attacks in African rural communities are on people carrying out their normal daily activities: getting water, bathing, fishing or swimming. Anderson and Pariela (2005) discussed possible mitigating actions, identifying education, barriers at the water's edge and provision of wells or water-lifting devices to provide a water source that avoids a need to wade in with a bucket. They also advocated strategic capture and removal of crocodiles to farms. The same considerations apply elsewhere as well.

Barriers at the water's edge could be applied to manage risk wherever lakes or rivers are used for dipping water, bathing and washing. There was a focus on Crocodile Exclusion Enclosures (CEEs) at the 22nd CSG Working meeting in Sri Lanka in May 2013, with a report on the CSG website. It noted that different types of barricades and palisades are, or have been, used in Venezuela, Sri Lanka, Africa and Papua New Guinea, over a range of sizes and made of a variety of materials. In North Queensland, swimming enclosures on beaches were once commonplace, mainly as protection from sharks, and apparently in 1929 a man was bitten by a *C. porosus* that had managed to get inside the enclosure. The drawbacks to CEEs are the difficulty of accommodating changing water levels, damage during flooding, the expense of installation and maintenance, as well as their comparatively infrequent use.

Some good advice about safe behaviour in crocodile country was provided by Lindner (2004). Garry was writing about estuarine crocodiles, but the advice is widely applicable. 'Some activities dangerously attract the interest of crocodiles. Such activities include: cleaning fish; leaving offal or food scraps near the water's edge; camping close to the water's edge; fishing while standing in the water; getting captured fish by hand from the water; holding fish in the water for some time before releasing them; going very close to crocodiles in boats; feeding crocodiles; annoying crocodiles; boating in unsafe or small craft; adults or children wading and splashing at the water's edge; and, swimming in areas where crocodiles may be'.

In countries with a more westernised lifestyle, exposure to crocodile or alligator attack is more likely to come from leisure activities or when workmen or police need to work in the water during flooding or when retrieving a corpse. For working in water where there are crocodiles, occupation health and safety guidelines are available on the web, but complete safety can rarely be assured.

Australia has a lot of experience with *C. porosus* and Florida and the other US alligator States have much experience with American alligators.

Crocodylus porosus *in Australia*

Of 62 recorded attacks, 81% were in or at the edge of the water (wading, swimming, snorkelling, fishing, spear-fishing and similar), 10% were associated with boating or canoeing and 19% were on land, asleep in a tent, on the beach or near water, including one near a crocodile nest (Caldicott *et al.* 2005). All 15 of the fatal attacks were made on a victim in the water

and they were distributed nearly evenly between day time and night time.

To try to minimise attacks, the main thrusts in northern Australia are by active education and by crocodile removal programmes, with the city of Darwin in the Northern Territory and the most populous parts of the North Queensland coast the main areas of concern. Both are major tourism destinations, so education needs to be continuous to accommodate the continual arrival of new faces. Interestingly, more than 80% of the fatal attacks in the Northern Territory have been on locals: the tourists apparently are getting the message, but the locals may have become complacent through familiarity. In both Queensland and the Northern Territory, the public awareness campaign takes the form of signage at waterways (Fig. 14.13), activities in schools and publicity in a range of media about safe behaviour. Both jurisdictions have a 'Crocwise' programme (easily found via any web browser), which

Fig. 14.13. *Not all signs are as large as this one, but there are hundreds of signs in northern Australia warning about danger from crocodiles. (Photo Garry Lindner)*

informs about safe and unsafe behaviour and encourages people to report suspicious crocodile activity.

Both jurisdictions remove 'problem crocodiles' (Northern Territory) (Nichols and Letnic 2008) and 'crocodiles of concern' (Queensland). These each have specific definitions but, essentially, they are crocs whose behaviour, size or location makes them a threat or potential threat. Both jurisdictions have declared specified areas close to high densities of human habitation from which removals are made. Aboriginal people too used to remove estuarine crocodiles from places where they swam (Lindner 2004). The Northern Territory programme has been running longer and, as the crocodile population has recovered, the number of problem animals removed annually has increased steadily from 116 in 1999 to 318 in 2012. It takes a lot of effort and resources. Most animals are caught in floating traps (e.g. Fig. 1.27) of which a large number are set at fixed locations and which have to be baited and checked regularly. When a situation arises that needs an immediate response an animal may be taken by harpooning (Chapter 1). In Queensland, most crocodiles are removed to a farm or for display, whereas in the Northern Territory they are killed and the skin and meat can be sold. Translocation is no longer practised, because their capacity for homing is too well known (discussed in Chapter 4). The internationally recognised tourist destination of Kakadu National Park is in the Northern Territory, but it is managed federally and is outside the jurisdiction of the Northern Territory Government. An active education programme runs in Kakadu too, detailed in Lindner (2004), and park staff carry out regular active surveillance of waterways and remove problem crocodiles there as necessary.

It must be noted that, although crocodile removal programmes can lower the chance of an attack, they do not provide any guarantee of safety. As our satellite tracking data show (Chapter 4), a big male a hundred or more kilometres away may decide, for reasons so far unknown, to undertake a journey that will have him cruising past a popular beach resort a few days later. This does not make removal practices futile, but serves as a reminder that public awareness of crocodiles must be maintained so that users of waterways are not lulled into a false sense of security.

Despite the active public awareness campaigns, misbehaviour is too common, sometimes alcohol related, and over-enthusiastic fishermen are also prone to choosing unwise places to try their luck (Fig. 14.14). I have heard of swimmers hanging their clothes on a crocodile warning sign before taking to the cool waters of a shady billabong. Fatalism about crocodiles is not, it seems restricted to 'developing' countries.

American alligators

Florida residents are quite used to seeing alligators in local waterways, including urban ponds, and there is considerable tolerance for and an element of pride in their alligators. I recall in 1985 spending time on the University of Florida campus in Gainesville and alligators were commonly seen on campus, and they still are. Nevertheless, although American alligators are nowhere near as likely to be aggressive as *C. porosus* or *C. niloticus*, injuries and occasional deaths do occur and the Florida Fish and Wildlife Conservation Commission (FWC) maintains a State-wide Nuisance Alligator Programme (SNAP). The programme has been described by Woodward and Cook (2000). From the FWC website, an alligator is defined as a nuisance 'if it is at least four feet in length and is believed to pose a threat to people, pets or property'. There is a toll-free telephone hotline and the problem will be dealt with by a 'nuisance alligator trapper', licenced by the FWC. The trappers capture the animal and remove it, rather than kill it in public view. They are compensated by selling the meat and hides, making the programme essentially self funding. The numbers handled are very large, and have increased in each decade since the 1970s: ~28 000 in the 1980s, 45 000 in the 1990s and nearly 80 000 from 2000 to the end of 2009. Other alligator States too have 'nuisance alligator' programmes.

American crocodiles

The increasing population of *C. acutus* in Florida is wonderful from a conservation point of view, but there is a safety issue that will have to be managed. Much to their credit, Florida residents are quite tolerant of alligators, but crocodiles are

Fig. 14.14. *Fishermen do not always choose sensible places to fish. These photos were taken at the (top left) Magela Creek crossing or at Cahills Crossing on the East Alligator River, all in Kakadu National Park. Crocodiles are common, with densities in this part of the East Alligator River in September – December (pre-wet season) of 40 to 50 non-hatchling animals per kilometre, mostly 2 m to 3.5 m, some > 4 m. (Top right) This causeway carries the main road eastwards into Arnhem Land. (Photos: upper Garry Lindner; lower Tony De Groot and Sandra Jaeschke)*

a bit different and there is a growing need for people to learn about that. Although *C. acutus* have a reputation for being less aggressive than either *C. porosus* or *C. niloticus*, they can grow almost as large and in the last 5 years 13 fatal attacks have been recorded from Mexico, Costa Rica (Fig. 14.15) and Panama (Table 14.4). Reports of crocodile sightings are becoming more common, some in quite urbanised areas. In the Florida Keys, some crocs have even learned to use swimming pools as occasional sources of fresh water, sometimes with the tolerance of the owners. The crocs turn up in the pool every couple weeks, take a drink, and then leave (Perran Ross *pers. comm.* 2013).

At the time of writing, nuisance crocodiles too can be reported to the toll-free nuisance alligator hotline for referral for appropriate action, one difference being that they are not killed. As the population grows, and as individuals in the recovering population grow larger, the need for such a programme is only likely to increase.

THE FUTURE FOR CROCODYLIANS?

There can be no general answer, nor any easy answer to this question. It will be different for different species and it is tied up with the unknowable futures of the many human populations living within crocodylian ranges, as well as the futures of human civilisations, currently looking unsustainable. With the Crocodylia stretching back to the Mesozoic and the extant species for millions of years (Fig. 2.1), a long-term perspective would be appropriate, but also nigh on impossible. A best guess might be that many of the current species will become extinct in competition with the expansion of the human population, but that some will survive.

Fig. 14.15. *American crocodiles relaxing on a riverbank in Costa Rica.* C. acutus *has been responsible for at least 14 attacks (five fatal) in Costa Rica since 2008, out of 69 (13 fatal) from within its range. (Photo Megan Higgie)*

Of course, crocodylians have already survived the collapse of numerous human civilisations, but that is no guide to their future because the current human population level and its magnitude and rate of resource consumption are unprecedented. The best guide is what happens at a local level, and we know already that crocodylians become locally extinct in response to high-density human population and intensive land use. The current populations of the Chinese alligator and Siamese crocodile, as well as the local extinctions of *C. porosus* throughout much of its former range, are good examples, but there are many more. The implication is that crocodylians will survive the present human *status quo* (unsustainable population increase and resource use) only in remote wetland habitats in which humans can see no benefit.

The conservation status and philosophy about commercial use of the world's crocodylians has undergone dramatic changes in the last 40 years. In 1976, at the third meeting of the IUCN SSC Crocodile Specialist Group, held in Maningrida in the Northern Territory, most of the news for crocodylians was bad news. Some crocodylians that might have become extinct already have not yet done so because of the growth of the conservation ethic during the second half of the last century. But how long can a 'conservation ethic' survive in the face of the expansion of requirements for land and resources that accompany the increase in human populations? If some crocodylian populations are surviving now because of the commercial value of their hides, and if that is dependent upon its use in high fashion, that is a fragile underpinning. Can we imagine that national parks and reserves will survive the onslaught of human population growth and its implications, with the possibility of ensuring the survival of a representative collection of crocodylians?

On a more positive note for crocodylians, but not for us, rising temperatures and rising sea levels from anthropogenic climate change will likely create more habitat that will be suitable for them. None of the current wetland habitats seems to be too hot for them, so they are much less likely than we are to suffer negatively from the current climatic trend.

So much for the longer term. What about the next few decades? Pilots who flew low along Australia's northern beaches in the early 1940s during the Second World War tell of seeing scores of large crocodiles dashing into the water ahead of them. In the mid 1970s, when I first heard such a story, I was flying those same beaches myself and I was disinclined to believe them. In those days a crocodile on a beach was a rarity: an event that would trigger a gut wrenching turn and a wingover for a closer look. But I heard similar stories about abundant crocs on beaches from several more Second World War pilots, and gradually came to believe them: as an exaggeration, of course, but as a realistic recollection of what they had seen at some times and in some places. The stories recall the way it must have been before the easy availability of aluminium dinghies, battery-powered spotlights and outboard motors led to significant hunting of crocodiles for skins and the removal of most of the big animals. Those stories may also provide a glimpse of a possible future, in Australia at least, where numbers are again on the increase (see Chapter 13) as a result of their protection. Are we looking at a future in which Australia's *C. porosus* populations will again be numerous enough to prompt such stories? The data seem to be pointing in that direction. In recovering populations in Northern Territory rivers, the proportion of larger animals is gradually increasing, along with the increase in total numbers (Chapter 13) and, if media reports are any guide, the same is happening in northern Western Australia.

Will Australians tolerate such large numbers of crocodiles? How well will Florida residents adjust to American crocodiles being common? People seem to have difficulty adjusting their lives to accommodate crocodiles, as if they feel it is a natural right to swim, surf, snorkel or dive without concern or restriction, even in prime crocodile habitat. So the crocodile removal programmes around human habitation are likely to increase and, as the costs of

these increase, it is easy to imagine that problem animals will be shot, not captured and relocated. Australian *C. porosus*, however, is one of the luckier species: human populations across Australia's north are unlikely to increase in the next few decades to the point where there will be nowhere left where a crocodile can be a crocodile.

This brings us back to the first paragraph. Although it is hard to see all of the current species having a significantly long-term future, some of them (including *C. porosus* in Australia) probably will, and some may even prosper. For the time being, we can celebrate the ones we have and the human endeavour being dedicated to their conservation and security.

REFERENCES

Anderson J, Pariela F (2005) 'Strategies to mitigate human-wildlife conflict in Mozambique. FAO Wildlife Management Working Paper Number 8'. Report for the National Directorate of Forests and Wildlife, Mozambique. < http://www.fao.org/docrep/010/ai547e/ai547e00.HTM>

Anon. (2000) Increased production of crocodiles. *Crocodile Specialist Group Newsletter* **19**(4), 14–15.

Bayani AS, Trivedi JN, Suresh B (2011) Nesting behavior of *Crocodylus palustris* (Lesson) and probable survival benefits due to varied nest structures. *Electronic Journal of Environmental Sciences* **4**, 85–90.

Bezuijen MR, Shwedick BM, Sommerlad R, Stevenson C, Steubing RB (2010) Tomistoma *Tomistoma schlegelii*. In *Crocodiles. Status Survey and Conservation Action Plan*. 3rd edn (Eds SC Manolis and C Stevenson) pp. 133–138. Crocodile Specialist Group, Darwin.

Brazaitis P (2003) *You Belong in a Zoo*. Villard Books, New York.

Brien ML, Webb GJ, Lang JW, McGuinness KA, Christian KA (2013) Born to be bad: agonistic behaviour in hatchling saltwater crocodiles (*Crocodylus porosus*). *Behaviour* **150**, 737–762.

Brochu CA, Njau J, Blumenschine RJ, Densmore LD (2010) A new horned crocodile from the Plio-

Pleistocene hominid sites at Olduvai Gorge, Tanzania. *PLoS ONE* **5**(2), e9333.

Bustard HR, Singh LAK (1982) Gharial attacks on man. *Journal of the Bombay Natural History Society* **78**, 610–611.

Caldicott DGE, Croser D, Manolis C, Webb G, Britton A (2005) Crocodile attack in Australia, an analysis of its incidence, and review of the pathology and management of crocodilian attacks in general. *Wilderness & Environmental Medicine* **16**, 143–159.

Caldwell J (2012) *World Trade in Crocodilian Skins 2008–2010*. UNEP–WCMC, Cambridge, UK.

Carpenter S (2011) The devolution of conservation: why CITES must embrace community-based resource management.*The Arizona Journal of Law and Environmental Policy***2**(1), 1–51.

CITES (1995) *CITES Identification Guide – Crocodilians*. Environment Canada, Ottawa, Canada. <http://www.ec.gc.ca/Publications/default.asp?lang=En&xml=839625F8-BF8A-4169-BFB0-399233745A49>

Cott HB (1961) Scientific results of an inquiry into the ecology and economic status of the Nile crocodile (*Crocodilus niloticus*) in Uganda and Northern Rhodesia. *Transactions of the Zoological Society of London* **29**, 211–356.

Cousins JA, Sadler JP, Evans J (2008) Exploring the role of private wildlife ranching as a conservation tool in South Africa: stakeholder perspectives. *Ecology and Society* **13**(2), 43.

Delaney R, Neave H, Fukuda Y, Saalfeld WK (2010) *Management Program for the Freshwater Crocodile (Crocodylus Johnstoni) in the Northern Territory of Australia, 2010–2015*. Northern Territory Department of Natural Resources, Environment, the Arts and Sport, Darwin, <http://www.environment.gov.au/biodiversity/wildlife-trade/sources/management-plans/nt-freshwater-crocodile.html>

Dunham KM, Ghiurghi A, Cumbi R, Urbano F (2010) Human–wildlife conflict in Mozambique: a national perspective, with emphasis on wildlife attacks on humans. *Oryx* **44**, 185–193.

Edwards H (1998) *Crocodile Attack in Australia*. J.B. Books, Adelaide.

Fergusson RA (2008) *Techniques for Mitigation of Crocodile Attacks on Rural Communities in Africa*. FAO, Rome, Italy.

Fowler CW (1987) A review of density dependence in populations of large mammals. In *Current Mammalogy. Volume 1*. (Ed. HH Genoways) pp. 401–41. Plenum Press, New York.

Grigg GC (1987) Kangaroos – a better economic base for our marginal grazing lands? *Australian Zoologist* **24**(1), 73–80.

Grigg GC, Hale PT, Lunney D (1995) *Conservation Through the Sustainable Use of Wildlife*. Centre for Conservation Biology. The University of Queensland, Brisbane.

Grigg GC (2002) Conservation benefit from harvesting kangaroos: status report at the start of a new millennium. A paper to stimulate discussion and research. In *A Zoological Revolution: Using Native Fauna to Assist in its Own Survival*. (Eds D Lunney and CR Dickman) pp. 52–76.Royal Zoological Society of New South Wales, Sydney.

Grigg GC, Pople AR (2001) Sustainable use and pest control: kangaroos, a case study. In *Conservation of Exploited Species* (Eds RD Reynolds, G Mace and KH Redford) pp. 403–423. Cambridge University Press, Cambridge, UK.

Groombridge B (1982) *The IUCN Amphibia-Reptilia Red Data Book*. IUCN, Gland, Switzerland.

Hekkala ER, Amato G, DeSalle R, Blum MJ (2010) Molecular assessment of population differentiation and individual assignment potential of Nile crocodile (*Crocodylus niloticus*) populations. *Conservation Genetics* **11**, 1435–1443.

Hekkala ER, Shirley MH, Amato G, Austin JD, Charter S, Thorbjarnason J, *et al*. (2011) An ancient icon reveals new mysteries: mummy DNA resurrects a cryptic species within the Nile crocodile. *Molecular Ecology* **20**, 4199–4215.

Hines KN, Skroblin A (2010) Australian freshwater crocodile (*Crocodylus johnstoni*) attacks on humans. *Herpetological Review* **41**(4), 430–433.

Huchzermeyer FW (2003) *Crocodiles: Biology, Husbandry and Diseases.* CABI Publishing, Oxford, UK.

Lamarque F, Anderson J, Fergusson R, Lagrange M, Osei-Owusu Y, Bakker L (2009) Human-wildlife conflict in Africa: causes, consequences and management strategies. FAO Forestry Paper No. 157. FAO, Rome, <http://www.fao.org/docrep/012/i1048e/i1048e00.pdf>

Langley RL (2005) Alligator attacks on humans in the United States. *Wilderness & Environmental Medicine* **16**(3), 119–124.

Larriera A, Siroski P, Piña C, Imhof A (2012) Ranching of *Caiman latirostris* and *Caiman yacare* in Argentina: Where a Problem Becomes a Livelihood. *Proceedings of the IUCN SSC Crocodile Specialist Group* **21**, 89–93.

Leach GJ, Delaney R, Fukuda Y (2009) *Management Program for the Saltwater Crocodile in the Northern Territory of Australia, 2009-2014.* Northern Territory Department of Natural Resources, Environment, The Arts and Sport, Darwin, <http://www.environment.gov.au/biodiversity/wildlife-trade/sources/management-plans/nt-crocodile-plan.html>

Letnic M, Connors G (2006) Changes in the distribution and abundance of saltwater crocodiles (*Crocodylus porosus*) in the upstream, freshwater reaches of rivers in the Northern Territory, Australia. *Wildlife Research* **33**, 529–538.

Letnic M, Carmody P, Burke J (2011) Problem crocodiles (*Crocodylus porosus*) in the freshwater, Katherine River, Northern Territory *Australian Zoologist* **35**(3), 858–863.

Lindner G (2004) Crocodile management – Kakadu National Park. In *Proceedings of the 17th Working Meeting of the Crocodile Specialist Group*, 24–29 May 2004, Darwin. pp. 41–51. IUCN–The World Conservation Union, Gland, Switzerland and Cambridge UK.

Lindsey PA, Frank LG, Alexander R, Mathieson A, Romañach SS (2006) Trophy hunting and conservation in Africa: problems and one potential solution. *Conservation Biology* **21**(3), 880–883.

Lindsey PA, Roulet PA, Romanach SS (2007) Economic and conservation significance of the trophy hunting industry in sub-Saharan Africa. *Biological Conservation* **134**, 455–469.

Louisiana Department of Wildlife and Fisheries (2010) *Louisiana's Alligator Management Program 2009–2010 Annual Report.* <http://www.wlf.louisiana.gov/wildlife/alligator-program-annual-reports>

Mazotti FJ (2013) American Crocodiles (*Crocodylus acutus*) in Florida. University of Florida, Gainesville, Florida, <http://www.edis.ifas.ufl.edu/uw157 >

McGranahan DA (2011) Identifying ecological sustainability assessment factors for ecotourism and trophy hunting operations on private rangeland in Namibia *Journal of Sustainable Tourism* **19**(1), 115–131.

Mendonça WCS, Silveira RD, Marioni B, Magnusson WE, Thorbjarnarson J (2010) Effects of water level, distance from community and hunting effort on caiman meat production. In *Crocodiles. Proceedings of the 20th Working Meeting of the Crocodile Specialist Group*. P. 48. IUCN, Gland, Switzerland.

Mercado V, Alcala A, Belo W, Manalo R, Diesmos A, De Leon J (2013) Soft release introduction of the Philippine Crocodile (*Crocodylus mindorensis*, Schmidt 1935) in Paghongawan Marsh, Siargao Island Protected Landscape and Seascape, Southern Philippines. *Crocodile Specialist Group Newsletter* **32**(2), 13–15.

Nair T, Thorbjarnarson JB, Aust P, Krishnaswamy J (2012) Rigorous gharial population estimation in the Chambal: implications for conservation and management of a globally threatened crocodilian. *Journal of Applied Ecology* **46**, 1046–1054.

Nichols T, Letnic M (2008) Problem crocodiles: reducing the risk of attacks by *Crocodylus porosus* in Darwin Harbour, Northern Territory, Australia. In *Urban Herpetology. Herpetological Conservation Volume 3.* (Eds JC Mitchell, RE Jung Brown and B Bartholomew) pp. 503–511. Society for the Study of Amphibians and Reptiles, Salt Lake City, Utah.

Njau JK, Blumenschine RJ (2012) 2012 Crocodylian and mammalian carnivore feeding traces on hominid fossils from FLK 22 and FLK NN 3,Plio-Pleistocene, Olduvai Gorge, Tanzania. *Journal of Human Evolution* **63**, 408–417.

Ojasti J (1996) Wildlife Utilization in Latin America: current situation and prospects for sustainable management. FAO Conservation Guide 25. FAO, Rome.

Platt SG, Ko WK, Kalyar Myo M, Khaing LL, Rainwater T (2001) Man eating by estuarine crocodiles: the Ramree Island massacre revisited. *Herpetological Bulletin* **75**, 15–18.

Pople AR, Grigg GC (1999) *Commercial Harvesting of Kangaroos in Australia.* Department of the Environment, Canberra, <http://www.environment.gov.au/biodiversity/wildlife-trade/publications/kangaroo/harvesting/index.html>

Rachmawan D, Brend S (2009) Human-*Tomistoma* interactions in Central Kalimantan, Indonesian Borneo. *Crocodile Specialist Group Newsletter* **28**(1), 9–11.

Rice KG, Percival FH, Woodward AR, Jennings ML (1999) Effects of egg and hatchling harvest on American alligators in Florida *Journal of Wildlife Management* **63**(4), 1193–1200.

Richardson J, Livingstone D (1962) An attack by a Nile crocodile on a small boat. *Copeia* **1962**(1), 201–204.

Ritchie J, Jong J (1993) *Bujang Senang, Terror of Batang Lupar.* Samasa Press, Kuching, Indonesia.

Ross PJ (1998) *Crocodiles: Status Survey and Conservation Action Plan.* 2nd edn. IUCN–SSC Crocodile Specialist Group, Gland, Switzerland.

Ryan C (1998) Saltwater crocodiles as tourist attractions. *Journal of Sustainable Tourism* **6**, 314–327.

Ryan C, Harvey K (2000) Who likes saltwater crocodiles? Analysing socio-demographics of those viewing tourist wildlife attractions based on saltwater crocodiles *Journal of Sustainable Tourism* **8**(5), 426–433.

Shirley MH, Vliet KA, Carr AN, Austin JD (2014) Rigorous approaches to species delimitation have significant implications for African crocodilian systematics and conservation. *Proceedings Biological Sciences* **281**, 20132483.

Sideleau BM, Britton ARC (2014) An analysis of crocodilian attacks worldwide for the period of 2008–July 2013. In *Crocodiles. Proceedings of the 22nd Working Meeting of the IUCN–SSC Crocodile Specialist Group.* Negombo, Sri Lanka, 21–23 May, 2013. pp. 110–113. IUCN, Gland, Switzerland.

Smith NJH (1980) Caimans, capybaras, otters, manatees, and man in Amazonia. *Biological Conservation* **19**, 177–187.

Somaweera R (2011) Unprovoked attacks by Australian freshwater crocodiles with a probable new case report from Lake Argyle in Western Australia. *Australian Zoologist* **35**, 973–976.

Stevenson C, Whitaker R (2012) *Gharial* Gavialis gangeticus *Action Plan.* IUCN–SSC Crocodile Specialist Group, Gland, Switzerland.

Stringer C (1986) *The Saga of Sweetheart.* Adventure Publications, Darwin.

Thomas GD, Leslie AJ (2006) Human-Crocodile Conflict (Nile Crocodile: *Crocodylus niloticus*) in the Okavango Delta, Botswana. *Proceedings of the 18th Working Meeting of the Crocodile Specialist Group.* 19–23 June 2006, Montélimar, France. P. 84. IUCN, Gland, Switzerland.

Thorbjarnarson J (1999) Crocodile tears and skins: international trade, economic constraints, and limits to the sustainable use of crocodilians. *Conservation Biology* **13**(3), 465–470.

Thorbjarnarson JB, Eaton MJ (2004) Preliminary examination of crocodile bushmeat issues in the Republic of Congo and Gabon. In *Crocodiles: Proceedings of the 17th Working Meeting of the IUCN–SSC Crocodile Specialist Group.* 24–29 May 2004, Darwin. pp. 236–247. IUCN, Gland, Switzerland.

Thorbjarnarson J, Velasco A (1999) Economic incentives for management of Venezuelan caiman. *Conservation Biology* **13**, 397–406.

Tisdell CA, Nantha HS (2007) Management, conservation and farming of saltwater crocodiles: an Australian case study of sustainable commercial use. In *Perspectives in Animal Ecology and Reproduction. Volume 4.* (Eds VK Gupta and AK Verma) pp. 233–264. Daya Publishing House, Delhi.

Van der Ploeg J, Cauilan-Cureg M, Van Weerd M, Persoon G (2011a) Why must we protect crocodiles? Explaining the value of the Philippine Crocodile to rural communities. *Journal of Integrative Environmental Sciences* **8**, 287–298.

Van der Ploeg J, Araño RR, Van Weerd M (2011b) What local people think about crocodiles: challenging environmental policy narratives in the Philippines. *Journal of Environment & Development* **20**(3), 218–244.

Van Weerd M (2012) *Philippine Crocodile* Crocodylus mindorensis *Action Plan.* IUCN–Crocodile Specialist Group. Gland, Switzerland.

Van Weerd M, Van der Ploeg J (2012) *The Philippine Crocodile: Ecology, Culture and Conservation.* Mabuwaya Foundation, Philippines.

Van Weerd M, Balbas M, Telan S, Rodriguez D, Guerrero J, Van de Ven W (2011) Philippine Crocodile Reintroduction Workshop. Crocodile Specialist Group Newsletter **30**(2), 10–12.

Velasco A, Colomina G, De Sola R, Villarroel G (2003) Effects of harvests on wild populations of *Caiman crocodilus crocodilus* in Venezuela Interciencia **28**, 544–548.

Wallace KM, Leslie AJ, Coulson T (2011) Living with predators: a focus on the issues of human-crocodile conflict within the lower Zambezi valley. *Wildlife Research* **38**(8), 747–755.

Webb GJW, Manolis SC (1993) Conserving Australia's crocodiles through commercial incentives. In *Herpetology in Australia. A Diverse Discipline.* (Eds D Lunney and D Ayers) pp. 250–256. Surrey Beatty & Sons, Sydney.

Webb GJW, Manolis SC (2003) *Guidelines on the Harvesting and Management of Wild Crocodilian Populations and the Determination of "Detriment" within the Context of CITES.* CITES, Gland, Switzerland.

Webb GJW, Yerbury M, Onions V (1978) A record of a *Crocodylus porosus* (Reptilia, Crocodylidae) attack. *Journal of Herpetology* **12**, 267–268.

Webb GJW, Manolis SC, Whitehead PJ, Letts GA (1984) 'A proposal for the transfer of the Australian population of *Crocodylus porosus* Schneider (1801), from Appendix I to Appendix II of C.I.T.E.S'. Technical Report No. 21. Conservation Commission of the Northern Territory, Darwin.

Webb GJW, Whitehead PJ, Manolis SC (1987) Crocodile management in the Northern Territory of Australia. In *Wildlife Management: Crocodiles and Alligators.* (Eds GJW Webb, SC Manolis and PJ Whitehead) pp. 107–124. Surrey Beatty & Sons, Sydney.

Webb GJW, Brook B, Whitehead P, Manolis SC (2004) Wildlife management principles and practices in crocodile conservation and sustainable use. In *Crocodiles. Proceedings of the 17th Working Meeting of the IUCN–SSC Crocodile Specialist Group.* 24–29 May 2004, Darwin. pp. 84–91. IUCN, Gland, Switzerland.

Woodward AR, Cook BL (2000) Nuisance-alligator (*Alligator mississippiensis*) control in Florida, USA. In *Crocodiles. Proceeding of the 15th Working Meeting of the IUCN–SSC Crocodile Specialist Group.* 17–20 January 2000, Varadero, Cuba. pp. 446–455 IUCN, Gland, Switzerland.

INDEX

Italics refer to a figure or a table, bolded type indicates a main treatment